Table of Powers and Roots

n	n^2	$\sqrt{n}$	n^3	$\sqrt[3]{n}$	n	n^2	$\sqrt{n}$	n^3	$\sqrt[3]{n}$
1	1	1.000	1	1.000	51	2,601	7.141	132,651	3.708
2	4	1.414	8	1.260	52	2,704	7.211	140,608	3.733
3	9	1.732	27	1.442	53	2,809	7.280	148,877	3.756
4	16	2.000	64	1.587	54	2,916	7.348	157,464	3.780
5	25	2.236	125	1.710	55	3,025	7.416	166,375	3.803
6	36	2.449	216	1.817	56	3,136	7.483	175,616	3.826
7	49	2.646	343	1.913	57	3,249	7.550	185,193	3.849
8	64	2.828	512	2.000	58	3,364	7.616	195,112	3.871
9	81	3.000	729	2.080	59	3,481	7.681	205,379	3.893
10	100	3.162	1,090	2.154	60	3,600	7.746	216,000	3.915
11	121	3.317	1,331	2.224	61	3,721	7.810	226,981	3.936
12	144	3.464	1,728	2.289	62	3,844	7.874	238,328	3.958
13	169	3.606	2,197	2.351	63	3,969	7.937	250,047	3.979
14	196	3.742	2,744	2.410	64	4,096	8.000	262,144	4.000
15	225	3.873	3,375	2.466	65	4,225	8.062	274,625	4.021
16	256	4.000	4,096	2.520	66	4,356	8.124	287,496	4.041
17	289	4.123	4,913	2.571	67	4,489	8.185	300,763	4.062
18	324	4.243	5,832	2.621	68	4,624	8.246	314,432	4.082
19	361	4.359	6,859	2.668	69	4,761	8.307	328,509	4.102
20	400	4.472	8,000	2.714	70	4,900	8.367	343,000	4.121
21	441	4.583	9,261	2.759	71	5,041	8.426	337,911	4.141
22	484	4.690	10,648	2.802	72	5,184	8.485	373,248	4.160
23	529	4.796	12,167	2.844	73	5,329	8.544	389,017	4.179
24	576	4.899	13,824	2.884	74	5,476	8.602	405,224	4.198
25	625	5.000	15,625	2.924	75	5,625	8.660	421,875	4.217
26	676	5.099	17,576	2.962	76	5,776	8.718	438,976	4.236
27	729	5.196	19,683	3.000	77	5,929	8.775	456,533	4.254
28	784	5.292	21,952	3.037	78	6,084	8.832	474,552	4.273
29	841	5.385	24,389	3.072	79	6,241	8.888	493,039	4.291
30	900	5.477	27,000	3.107	80	6,400	8.944	512,000	4.309
31	961	5.568	29,791	3.141	81	6,561	9.000	531,441	4.327
32	1,024	5.657	32,768	3.175	82	6,724	9.055	551,368	4.344
33	1,089	5.745	35,937	3.208	83	6,889	9.100	571,787	4.362
34	1,156	5.831	39,304	3.240	84	7,056	9.165	592,704	4.380
35	1,225	5.916	42,875	3.271	85	7,225	9.220	614,125	4.397
36	1,296	6.000	46,656	3.302	86	7,396	9.274	636,056	4.414
37	1,369	6.083	50,653	3.332	87	7,569	9.327	658,503	4.431
38	1,444	6.164	54,872	3.362	88	7,744	9.381	681,472	4.448
39	1,521	6.245	59,319	3.391	89	7,921	9.434	704,969	4.465
40	1,600	6.325	64,000	3.420	90	8,100	9.487	729,000	4.481
41	1,681	6.403	68,921	3.448	91	8,281	9.539	753,571	4.498
42	1,764	6.481	74,088	3.476	92	8,464	9.592	778,688	4.514
43	1,849	6.557	79,507	3.503	93	8,649	9.644	804,357	4.531
44	1,936	6.633	85,184	3.530	94	8,836	9.695	830,584	4.547
45	2,025	6.708	91,125	3.557	95	9,025	9.747	857,375	4.563
46	2,116	6.782	97,336	3.583	96	9,216	9.798	884,736	4.579
47	2,209	6.856	103,823	3.609	97	9,409	9.849	912,673	4.595
48	2,304	6.928	110,592	3.634	98	9,604	9.899	941,192	4.610
49	2,401	7.000	117,649	3.659	99	9,801	9.950	970,299	4.626
50	2,500	7.071	125,000	3.684	100	10,000	10.000	1,000,000	4.642

Algebra for
College Students

Algebra for College Students

Second Edition

Jerome E. Kaufmann
Western Illinois University

PWS-KENT Publishing Company
Boston

PWS–KENT
Publishing Company

20 Park Plaza
Boston, Massachusetts 02116

PWS Publishers is a division of Wadsworth, Inc.

Portions of this book also appear in *Intermediate Algebra for College Students, Second Edition* by Jerome E. Kaufmann, copyright © 1986 by PWS Publishers.

Library of Congress Cataloging-in-Publication Data

Kaufmann, Jerome E.
   Algebra for college students.

   Includes index.
   1. Algebra.   I. Title.
QA154.2.K36   1987   512.9   86-22639
ISBN 0-87150-017-5

Printed in the United States of America.
    88 89 90 91—10 9 8 7 6 5 4

Sponsoring Editor: *Chuck Glaser*
Production Coordinator: *Susan Graham*
Interior Design: *Elise Kaiser*
Production: *Lifland et al.*
Interior Illustration: *Julia Gecha and Deborah Schneck*
Cover Design: *Julia Gecha*
Composition: *Polyglot Pte. Ltd.*
Cover Printer: *Phoenix Color Corporation*
Text Printer/Binder: *R. R. Donnelley*

Cover photo © 1984 Bill Gallery/Stock, Boston, Inc.

Preface

This text was written for those college students who need an algebra course to bridge the gap between elementary algebra and the more advanced courses in precalculus mathematics. The first six chapters contain intermediate algebra topics, and the last eight chapters contain a blend of intermediate and college algebra topics. All of the material is written at an intermediate level.

The basic concepts of algebra are presented in a simple and straightforward manner. The concepts are motivated by examples and are continuously reinforced by additional examples. Algebraic ideas are developed in a logical sequence and in an easy-to-read manner without excessive technical vocabulary and formalism.

As in the first edition, there is a common thread throughout the book: namely, "learn a skill," then "use the skill to solve equations and inequalities," and then "use equations and inequalities to solve word problems." This thread influenced the organization of the text as follows:

1. Word problems are scattered throughout the text. Every effort has been made to start with easy ones, in an effort to build students' confidence in solving word problems.

2. Numerous problem-solving suggestions are offered throughout, and there are special discussions in several sections. The key issue is the various problem-solving techniques used, not the types of word problems presented.

3. Newly acquired skills are used to solve equations and inequalities as soon as possible; the concept of solving equations and inequalities is introduced early and developed throughout the text. The concepts of factoring, solving equations, and solving word problems are tied together in Chapter 3.

Numerous examples demonstrate a large variety of situations; others are left for the students to think about in the problem sets. I have tried in the examples to guide students in organizing their work, and also to help them decide when a shortcut might be used.

The problem sets were carefully constructed on an odd/even basis; that is, all variations of skill development exercises are contained in both the even- and the odd-numbered problems. Thus, either the "evens" or the "odds" constitutes a meaningful assignment. A double dosage can be provided by assigning all of them. Also taken into consideration was the fact that some skills are more difficult to develop and therefore need more drill than others.

I tried to assign the calculator its rightful place in the study of mathematics—that of a tool, useful at times, unnecessary at other times. No special problems were created just so that students could use the calculator. Instead, some of the usual algebra problems, which lend themselves to the use of the calculator, are labeled as calculator problems.

Specific Comments About Some of the Chapters

1. Chapter 1 was written so that it can be covered quickly—and on an individual basis, if desired—by those needing only a brief review of some basic algebraic concepts.

2. In Chapter 5, exponents and radicals are developed separately and then merged to unify rational exponents and roots. The general concept of the "nth root" is discussed, but the problems concentrate on square root and cube root.

3. Three points should be made about Chapter 6, Quadratic Equations and Inequalities. First, the process of completing the square is introduced as a viable equation-solving process for certain types of quadratic equations. The emphasis here on completing the square pays additional dividends in Chapter 7, where it is used in graphing parabolas and circles. Second, the often-overlooked relationships involving the sum and product of roots of a quadratic equation are used to provide an effective checking procedure. Third, complex numbers are introduced in the last three sections of Chapter 6, so that instructors may include or not include this material as they see fit. Later uses of the complexes—such as in Chapter 11 in solving systems involving nonlinear equations—are restricted to the problem sets.

4. Chapter 7 was written on the premise that students at this level need more work with coordinate geometry concepts—specifically graphing techniques—*before* being introduced to the idea of a function. My personal experiences indicate a need for students in this course to become proficient at graphing straight lines, parabolas, and circles. Furthermore, at least a little work with graphing ellipses and hyperbolas seems appropriate.

5. Chapter 8 presents a straightforward approach to the function concept. The entire chapter is devoted to functions; the issue is not clouded by jumping back and forth between functions and relations that are not functions.

6. Chapter 9 was written with two primary objectives in mind: to develop some additional techniques for solving equations and to expand graphing capabilities.

7. Chapter 10 presents a modern-day version of exponents and logarithms. The emphasis is on making the concepts and their applications understood; the calculator is used as a tool to help with the computational aspects.

8. Chapters 11 and 12 have been written so as to allow the instructor some flexibility as to choice of topics. Chapter 11 contains the standard material on solving systems of equations by the methods of elimination by addition and substitution. Section 11.5 provides another look at problem solving via some elementary linear programming problems. The use of matrices and determinants for solving systems of linear equations is the focus of Chapter 12.

Significant Changes Made in the Second Edition

1. Many of the problem sets have been increased in size, in part through a significant increase in the number of word problems.

2. The Chapter Tests have been changed to Review Problem Sets and increased significantly in size. This provides a more comprehensive vehicle for students to use to pull together the concepts of a chapter.

3. A small group of problems called "Miscellaneous Problems" has been added to some of the problem sets. Basically, these problems are designed to broaden the students' background with respect to a particular topic. For example, in Problem Set 7.3 the coefficient relationships for parallel and perpendicular lines are used to write equations of lines. These problems can all be omitted without destroying the continuity of the text.

4. As recommended by the American Mathematical Association of Two-Year Colleges, some basic geometric concepts that are used in subsequent courses have been integrated in a problem-solving setting as follows:

   Section 2.2: Complementary and supplementary angles; sum of angles of triangle equals $180°$

   Section 2.4: Area and volume formulas

   Section 3.4: More on area, volume, perimeter, and circumference formulas

   Section 3.7: Pythagorean Theorem

   Section 6.1: More on Pythagorean Theorem, including work with isosceles right triangles and $30°–60°$ right triangles

5. More problem-solving suggestions have been offered throughout the text, in the belief that looking at the same problem in different ways can sometimes enhance one's understanding of the problem.

6. Sections 3.6 and 3.7 have been rewritten to include a more systematic technique for factoring trinomials of the form $ax^2 + bx + c$. More emphasis has been placed on using factoring techniques to solve quadratic equations.

7. Chapter 6 has been reorganized so as to give students an opportunity to learn, on a day-to-day basis, different techniques for solving quadratic equations. Then in Section 6.4 suggestions are offered as to when it is appropriate to use a particular technique. Also in Section 6.4 are

discussions of many applications of quadratic equations. In Section 6.5 the approach to solving quadratic inequalities combines the use of "test numbers" with observation of the sign behavior of the factors.

8. Chapter 7 has been reorganized to put all of the work with straight lines in the first three sections. Then parabolas, circles, ellipses, and hyperbolas are graphed. A final section contains the graphing of "new" curves—that is, some curves that might not be recognizable from the equation.

9. Chapter 9 has been rewritten to pull together the two sections on solving polynomial equations. A section (Section 9.5) on graphing some fairly simple rational functions has been added.

10. The first two sections of Chapter 12 have been rewritten to be a little more systematic in the use of matrices to solve systems of linear equations. The triangular form and the reduced echelon form play a more important role than before.

I would like to take this opportunity to thank all of the people who served as reviewers for this manuscript and to extend my special thanks to the reviewers of the first edition of *Algebra for College Students*: Donald Bellairs (Grossmont College), Thomas Hale (California Polytechnic State University), Ray Knodel (Bemidji State University), Norman Mittman (Northeastern Illinois University), W. Richard Slinkman (Bemidji State University), and Kay Weiss (Kansas City Community College).

I would also like to thank the following reviewers of the current edition: Charles Miles (Hillsborough Community College), Sabbah Al-Hadad (California Polytechnic State University, San Luis Obispo), Robert Horvath (El Camino College), Kathy Wahl (University of Illinois), Jan Vandever (South Dakota State University), and Thomas Hale (California Polytechnic State University, San Luis Obispo).

I am very grateful to Chuck Glaser, my editor, and to the staff of Prindle, Weber, and Schmidt for their continuous cooperation and assistance throughout this project. In particular, I would like to thank Susan Graham for her complete cooperation in her role as production editor. A special word of thanks is due Sally Lifland and her staff at Lifland et al., Bookmakers.

Special thanks again are due to my wife, Arlene, who continues to spend numerous hours proofreading and typing the manuscript along with the answer book, solutions manual, and accompanying sets of tests.

Contents

Algebra for
College Students

1 Basic Concepts and Properties

Algebra is often described as a **generalized arithmetic**. That description may not tell the whole story, but it does indicate an important idea: a good understanding of arithmetic provides a sound basis for the study of algebra. Be sure that you thoroughly understand the basic concepts reviewed in this first chapter.

1.1 Sets, Real Numbers, and Numerical Expressions

In arithmetic, symbols such as $6, \frac{2}{3}, 0.27,$ and π are used to represent numbers. The basic operations of addition, subtraction, multiplication, and division are commonly indicated by the symbols $+$, $-$, $\cdot$, and $\div$, respectively. Thus, specific **numerical expressions** can be formed. For example, the indicated sum of six and eight is written as $6 + 8$.

In algebra, the concept of a variable provides the basis for generalizing arithmetic ideas. For example, by using x and y to represent *any* numbers, the expression $x + y$ can be used to represent the indicated sum of *any* two numbers. The x and y in such an expression are called **variables** and the phrase $x + y$ is called an **algebraic expression**.

Many of the notational agreements made in arithmetic are extended to algebra, with a few modifications. The following chart summarizes these notational agreements pertaining to the four basic operations.

Note the different ways of indicating a product, including the use of parentheses. The *ab form* is the simplest and probably most widely used form. Expressions such as *abc*, *6xy*, and *14xyz* all indicate multiplication. We also call your attention to the various forms used to indicate division. In algebra, the fractional form $\frac{x}{y}$ is usually used, although the other forms do serve a purpose at times.

Operation	Arithmetic	Algebra	Vocabulary
addition	$4 + 6$	$x + y$	The **sum** of x and y
subtraction	$14 - 10$	$a - b$	The **difference** of a minus b
multiplication	$7 \cdot 5$ or 7×5	$a \cdot b$ or $a(b)$ or $(a)b$ or $(a)(b)$ or ab	The **product** of a and b
division	$8 \div 4$ or $\frac{8}{4}$ or $4\overline{)8}$	$x \div y$ or $\frac{x}{y}$ or $y\overline{)x}$	The **quotient** of x divided by y

Use of Sets

Some of the basic vocabulary and symbolism associated with the concept of sets can be used effectively in the study of algebra. A **set** is a collection of objects and the objects are called **elements** or **members** of the set. In arithmetic and algebra the elements of a set are usually numbers.

The use of set braces, { }, to enclose the elements or a description of the elements, and the use of capital letters to name sets provide a convenient way to communicate about sets. For example, a set A, consisting of the vowels of the alphabet can be represented as

$A = \{$vowels of the alphabet$\}$ word description

or $A = \{a, e, i, o, u\}$ list or roster description

or $A = \{x | x$ is a vowel$\}$ set builder notation

The listing approach can be modified if the number of elements is quite large. For example, all of the letters of the alphabet can be listed as

$$\{a, b, c, \ldots, z\}.$$

We simply begin by writing enough elements to establish a pattern; then the three dots indicate that the set continues in that pattern. The final entry indicates the last element of the pattern. If we write

$$\{1, 2, 3, \ldots\}$$

the set begins with the counting numbers 1, 2, and 3. The three dots indicate that it continues in a like manner forever; there is no last element. A set consisting of no elements is called the **null set** (written $\varnothing$).

The **set builder notation** combines the use of braces and the concept of a variable. For example, $\{x | x$ is a vowel$\}$ is read "the set of all x such that x is a vowel." Note that the vertical line is read "such that." The set $\{1, 2, 3, \ldots\}$ can be described using set builder notation as $\{x \mid x > 0 \text{ and } x \text{ is a whole number}\}$.

The symbol $\in$ is used to denote set membership. Thus, if $A = \{a, e, i, o, u\}$, we can write $e \in A$ which we read as "e is an element of A." The slash symbol, $/$, is commonly used in mathematics as a negation symbol. Therefore, $m \notin A$ is read as "m is *not* an element of A."

Two sets are said to be *equal* if they contain exactly the same elements. For example,

$$\{1, 2, 3\} = \{2, 1, 3\},$$

because both sets contain the same elements; the order in which the elements are written doesn't matter. The slash mark through the equality symbol denotes "not equal to." Thus, if $A = \{1, 2, 3\}$ and $B = \{1, 2, 3, 4\}$, we can write $A \neq B$, which is read as "set A is not equal to set B."

Real Numbers

Most of the algebra that we will study in this text is referred to as the *algebra of real numbers*. This simply means that the variables represent real numbers. Therefore, it is necessary for us to be familiar with the various terminology used to classify different types of real numbers.

$\{1, 2, 3, 4, \ldots\}$	natural numbers; counting numbers; positive integers
$\{0, 1, 2, 3, \ldots\}$	whole numbers; nonnegative integers
$\{\ldots -3, -2, -1\}$	negative integers
$\{\ldots -3, -2, -1, 0\}$	nonpositive integers
$\{\ldots -3, -2, -1, 0, 1, 2, 3, \ldots\}$	integers

A **rational number** is defined as any number that can be expressed in the form $\dfrac{a}{b}$, where a and b are integers and b is not zero. The following are examples of rational numbers.

$$\frac{2}{3}; \quad -\frac{3}{4}; \qquad\qquad 6 \quad \text{because } 6 = \frac{6}{1};$$

$$-4 \quad \text{because } -4 = \frac{-4}{1} = \frac{4}{-1}; \qquad 0 \quad \text{because } 0 = \frac{0}{1} = \frac{0}{2} = \frac{0}{3} \cdots;$$

$$0.3 \quad \text{because } 0.3 = \frac{3}{10}; \qquad 6\frac{1}{2} \quad \text{because } 6\frac{1}{2} = \frac{13}{2}.$$

A rational number can also be defined in terms of a decimal representation. Before doing so, let's briefly review the different possibilities for decimal representations.

Decimals can be classified as **terminating**, **repeating**, or **nonrepeating**. Some examples of each are as follows.

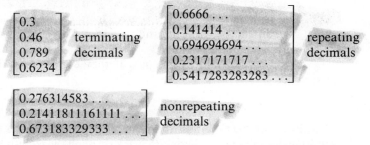

$$\begin{bmatrix} 0.3 \\ 0.46 \\ 0.789 \\ 0.6234 \end{bmatrix} \begin{array}{l} \text{terminating} \\ \text{decimals} \end{array} \qquad \begin{bmatrix} 0.6666\ldots \\ 0.141414\ldots \\ 0.694694694\ldots \\ 0.2317171717\ldots \\ 0.5417283283283\ldots \end{bmatrix} \begin{array}{l} \text{repeating} \\ \text{decimals} \end{array}$$

$$\begin{bmatrix} 0.276314583\ldots \\ 0.21411811161111\ldots \\ 0.673183329333\ldots \end{bmatrix} \begin{array}{l} \text{nonrepeating} \\ \text{decimals} \end{array}$$

A repeating decimal has a block of digits that repeats indefinitely. This repeating block of digits may be of any number of digits and may or may not begin immediately after the decimal point. A small horizontal bar is commonly used to indicate the repeat block. Thus, $0.6666\ldots$ is written as $0.\overline{6}$ and $0.2317171717\ldots$ as $0.23\overline{17}$.

In terms of decimals, a **rational number** is defined as a number that has either a terminating or repeating decimal representation. The following examples illustrate some rational numbers written in $\dfrac{a}{b}$ form and in decimal form.

$$\frac{3}{4} = 0.75; \qquad \frac{3}{11} = 0.\overline{27}; \qquad \frac{1}{8} = 0.125; \qquad \frac{1}{7} = 0.\overline{142857}; \qquad \frac{1}{3} = 0.\overline{3}.$$

An **irrational number** is defined as a number that cannot be expressed in $\dfrac{a}{b}$ form, where a and b are integers and b is not zero. Furthermore, an irrational number has a nonrepeating decimal representation. Some examples of irrational numbers and a partial decimal representation for each are as follows.

$$\sqrt{2} = 1.414213562373095\ldots; \qquad \sqrt{3} = 1.73205080756887\ldots;$$
$$\pi = 3.14159265358979\ldots.$$

The entire set of *real numbers* is composed of the rational numbers along with the irrationals. The following tree-diagram can be used to summarize the various classifications of the real number system.

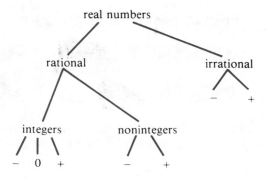

Any real number can be traced down through the diagram as follows.

7 is real, rational, an integer, and positive;

$-\dfrac{2}{3}$ is real, rational, noninteger, and negative;

$\sqrt{7}$ is real, irrational, and positive;

0.38 is real, rational, noninteger, and positive.

The concept of subset is convenient to use at this time. A set A is a **subset** of a set B if and only if every element of A is also an element of B. This is written as $A \subseteq B$ and is read "A is a subset of B." For example, if $A = \{1, 2, 3\}$ and $B = \{1, 2, 3, 5, 9\}$, then $A \subseteq B$ because every element of A is also an element of B. The slash mark again denotes negation, so if $A = \{1, 2, 5\}$ and $B = \{2, 4, 7\}$, we can say that "A is not a subset of B" by writing $A \nsubseteq B$.

The following kinds of statements can be made about the real number system using the subset vocabulary.

1. The set of rational numbers is a subset of the set of real numbers.
2. The set of irrational numbers is a subset of the set of real numbers.
3. The set of integers is a subset of the set of rational numbers. The set of integers is also a subset of the set of real numbers.

Equality

The relation **equality** plays an important role in many facets of mathematics—especially when manipulating with real numbers and algebraic expressions representing real numbers. An equality is a statement that two symbols, or groups of symbols, are names for the same number. The symbol $=$ is used to express an equality. Thus, we can write

$$6 + 1 = 7; \qquad 18 - 2 = 16; \qquad 36 \div 4 = 9.$$

(The symbol $\neq$ means *is not equal to*.) The following four basic properties of equality are rather self-evident, but they do need to be kept in mind. (This list will be extended in Chapter 2 when we work with solutions of equations.)

Properties of Equality

(1) Reflexive Property

For any real number a,

$$a = a.$$

Examples $14 = 14$, $x = x$, $a + b = a + b$.

(2) Symmetric Property

For any real numbers a and b,

$$\text{if } a = b, \quad \text{then } b = a.$$

Examples   If $13 + 1 = 14$,   then $14 = 13 + 1$.
If $3 = x + 2$,   then $x + 2 = 3$.

(3) Transitive Property

For any real numbers a, b, and c,

$$\text{if } a = b \quad \text{and} \quad b = c, \quad \text{then } a = c.$$

Examples   If $3 + 4 = 7$ and $7 = 5 + 2$,   then $3 + 4 = 5 + 2$.
If $x + 1 = y$ and $y = 5$,   then $x + 1 = 5$.

(4) Substitution Property

For any real numbers a and b, if $a = b$ then a may be replaced by b, or b may be replaced by a, in any statement without changing the meaning of the statement.

Examples   If $x + y = 4$ and $x = 2$,   then $2 + y = 4$.
If $a - b = 9$ and $b = 4$,   then $a - 4 = 9$.

Numerical Expressions

Let's conclude this section by **simplifying some numerical expressions** involving whole numbers. The following examples illustrate several important ideas pertaining to this process. Study them carefully and be sure that you agree with each result.

Example 1   Simplify $18 + 16 - 9 + 14 - 12 - 10$.

Solution   The additions and subtractions are to be done from left to right in the order in which they appear. Thus, $18 + 16 - 9 + 14 - 12 - 10$ simplifies to 17.   ●

Example 2   Simplify $7 \cdot 4 \div 2 \cdot 3 \cdot 2 \div 4$.

Solution   The multiplications and divisions are to be done from left to right in the order in which they appear. Thus, $7 \cdot 4 \div 2 \cdot 3 \cdot 2 \div 4$ simplifies to 21.   ●

Example 3   Simplify $5 \cdot 3 + 4 \div 2 - 2 \cdot 6 - 28 \div 7$.

Solution   *First*, we do the multiplications and divisions in the order in which they appear. Then we do the additions and subtractions in the order in which they appear. Our work may take on the following format.

$$5 \cdot 3 + 4 \div 2 - 2 \cdot 6 - 28 \div 7 = 15 + 2 - 12 - 4$$
$$= 1.$$

●

Example 4   Simplify $(4 + 6)(7 + 8)$.

Solution   The parentheses are used to indicate the *product* of the quantities $4 + 6$ and $7 + 8$. Perform the additions inside the parentheses first and then multiply.

$$(4 + 6)(7 + 8) = (10)(15) = 150.$$

●

Example 5   Simplify $(3 \cdot 2 + 4 \cdot 5)(6 \cdot 8 - 5 \cdot 7)$.

Solution   First, we do the multiplications inside the parentheses.

$$(3 \cdot 2 + 4 \cdot 5)(6 \cdot 8 - 5 \cdot 7) = (6 + 20)(48 - 35).$$

Then, we do the addition and subtraction inside the parentheses.

$$(6 + 20)(48 - 35) = (26)(13).$$

Then, we find the final product.

$$(26)(13) = 338.$$

●

Example 6   Simplify $6 + 7[3(4 + 6)]$.

Solution   Brackets are used for the same purposes as parentheses. In such a problem we need to simplify *from the inside out*; perform the operations in the innermost parentheses first. We would thus obtain

$$6 + 7[3(4 + 6)] = 6 + 7[3(10)]$$
$$= 6 + 7[30]$$
$$= 6 + 210 = 216.$$

●

Example 7   Simplify $\dfrac{6 \cdot 8 \div 4 - 2}{5 \cdot 4 - 9 \cdot 2}$.

Solution   First, we perform the operations above and below the fraction bar.

$$\frac{6 \cdot 8 \div 4 - 2}{5 \cdot 4 - 9 \cdot 2} = \frac{12 - 2}{20 - 18} = \frac{10}{2}.$$

Then, we find the final quotient.

$$\frac{10}{2} = 5.$$

●

Remark   Using parentheses, we could also write the problem in Example 7 as $(6 \cdot 8 \div 4 - 2) \div (5 \cdot 4 - 9 \cdot 2)$.

Here is a summary of the ideas presented in the previous examples pertaining to simplifying numerical expressions. When evaluating a numerical expression the operations should be performed in the following order.

1. Perform the operations inside the symbols of inclusion (parentheses, brackets, and braces) and above and below each fraction bar. Start with the innermost inclusion symbol.

2. Perform all multiplications and divisions in the order in which they appear from left to right.

3. Perform all additions and subtractions in the order in which they appear from left to right.

Problem Set 1.1

Identify each of the following as true or false.

1. Every rational number is a real number.
2. Every irrational number is a real number.
3. Every real number is a rational number.
4. If a number is real, then it is irrational.
5. Some irrational numbers are also rational numbers.
6. All integers are rational numbers.
7. The number zero is a rational number.
8. Zero is a positive integer.
9. Zero is a negative integer.
10. All whole numbers are integers.

From the list $0, \sqrt{5}, -\sqrt{2}, \frac{7}{8}, -\frac{10}{13}, 7\frac{1}{8}, 0.279, 0.4\overline{67}, -\pi, -14, 46,$ and $6.75,$ identify each of the following.

11. The natural numbers
12. The whole numbers
13. The integers
14. The rational numbers
15. The irrational numbers
16. The nonnegative integers
17. The nonpositive integers
18. The real numbers

From problems 19–32, use the following set designations.

$$N = \{x|x \text{ is a natural number}\}.$$
$$W = \{x|x \text{ is a whole number}\}.$$
$$I = \{x|x \text{ is an integer}\}.$$
$$Q = \{x|x \text{ is a rational number}\}.$$
$$H = \{x|x \text{ is an irrational number}\}.$$
$$R = \{x|x \text{ is a real number}\}.$$

Place ⊆ or ⊄ in each blank to make a true statement.

19. N _____ R

20. R _____ N

21. N _____ I

22. I _____ Q

23. H _____ Q

24. Q _____ H

25. W _____ I

26. N _____ W

27. I _____ W

28. I _____ N

29. $\{0, 2, 4, \ldots\}$ _____ W

30. $\{1, 3, 5, 7, \ldots\}$ _____ I

31. $\{-2, -1, 0, 1, 2\}$ _____ W

32. $\{0, 3, 6, 9, \ldots\}$ _____ N

List the elements of each set. For example, the elements of $\{x|x$ is a natural number less than 4$\}$ can be listed as $\{1, 2, 3\}$.

33. $\{x|x$ is a natural number less than 2$\}$

34. $\{x|x$ is a natural number greater than 5$\}$

35. $\{n|n$ is a whole number less than 4$\}$

36. $\{y|y$ is an integer greater than $-3\}$

37. $\{y|y$ is an integer less than 2$\}$

38. $\{n|n$ is a positive integer greater than $-4\}$

39. $\{x|x$ is a whole number less than 0$\}$

40. $\{x|x$ is a negative integer greater than $-5\}$

41. $\{n|n$ is a nonnegative integer less than 3$\}$

42. $\{n|n$ is a nonpositive integer greater than 1$\}$

Replace each question mark to make the given statement an application of the indicated property of equality. For example, $16 = ?$ becomes $16 = 16$ because of the reflexive property of equality.

43. If $5 = n + 3$, then $n + 3 = ?$; symmetric property of equality

44. $5x = ?$; reflexive property of equality

45. If $t = 4$ and $s + t = 9$, then $s + ? = 9$; substitution property of equality

46. If $4 = 3x + 1$ then $? = 4$; symmetric property of equality

47. If $y = x$ and $x = z + 2$, then $y = ?$; transitive property of equality

48. If $n = 2$ and $3n + 4 = 7$, then $3(?) + 4 = 7$; substitution property of equality

49. $5x + 7 = ?$; reflexive property of equality

50. If $y = x$ and $x = -6$, then $y = ?$; transitive property of equality

Simplify each of the following numerical expressions.

51. $15 + 8 - 6 + 9 - 12$

52. $19 + 14 - 12 - 2 + 10 - 7 + 4$

53. $21 \div 7 \cdot 4 \div 3 \cdot 8$

54. $6 \cdot 8 \div 3 \div 4 \cdot 5$

55. $6 + 8 \cdot 7$

56. $14 - 3 \cdot 2$

57. $5 \cdot 4 + 3 \cdot 2 - 2 \cdot 9$

58. $9 \cdot 7 - 4 \cdot 5 - 3 \cdot 2 + 4 \cdot 7$

59. $(7 + 9)(8 - 3)$

60. $(14 - 11)(13 - 6)$

61. $(5 + 6)(4 - 2)(7 + 3)$

62. $14 + (7 - 1)(8 - 2)$

63. $(3 \cdot 4 + 2 \cdot 1)(5 \cdot 2 + 6 \cdot 7)$

64. $(5 \cdot 9 - 3 \cdot 4)(6 \cdot 9 - 2 \cdot 7)$

65. $13 + 5[7(8 - 2)]$

66. $6[4(7 - 3)] - 50$

67. $3[4(6 + 7)] + 2[3(4 - 2)]$

68. $[3 + 2(4 \cdot 1 - 2)][18 - (2 \cdot 4 - 7 \cdot 1)]$

69. $[27 - (4 \cdot 2 + 5 \cdot 2)][(5 \cdot 6 - 4) - 20]$

70. $17 + 3\left(\dfrac{8 - 2}{12 - 9}\right) - 2\left(\dfrac{9 - 1}{19 - 15}\right)$

71. $4\left(\dfrac{8 + 12}{12 - 2}\right) + 5\left(\dfrac{8 + 7}{8 - 7}\right) - 47$

72. $[7 + 2 \cdot 3 \cdot 5 - 5] \div 8$

73. $\dfrac{4 \cdot 9 - 3 \cdot 5 - 3}{18 - 12}$

74. $\dfrac{3 \cdot 8 - 4 \cdot 3}{5 \cdot 7 - 34} + 19$

75. You should be able to do calculations like those in Problems 51–74 *with* and *without* a calculator. Different types of calculators handle the *priority of operations* issue in different ways. Be sure you can do Problems 51–74 with *your* calculator.

1.2 Addition and Subtraction of Integers

Before reviewing the four basic operations with integers let's briefly discuss some concepts and terminology commonly used with this material.

The symbol, -1, can be read as *negative one, opposite of one,* or *additive inverse of one.* The *opposite of* and *additive inverse of* terminology is especially meaningful when working with variables. For example, the symbol, $-x$, read as "opposite of x" or "additive inverse of x" emphasizes an important issue. Since x can be any real number, $-x$ (opposite of x) can be zero, positive, or negative. If x is a positive number, then $-x$ is negative. If x is a negative number, then $-x$ is positive. If x is zero, then $-x$ is zero. These facts are illustrated as follows.

$-(6) = -6$ The opposite of six (additive inverse of six) is negative six.

$-(-4) = 4$ The opposite of negative four is four.

$-(0) = 0$ The opposite of zero is zero.

In general, it can be stated that *the opposite of the opposite of any real number is the real number itself.* This is symbolically expressed as $-(-a) = a$ and is sometimes referred to as the *double negative* property.

Absolute Value

The concept of **absolute value** can be used to precisely describe how to operate with positive and negative numbers. Geometrically, the absolute value of any number is the distance between the number and zero on the number line. For example, the absolute value of 2 is 2. The absolute value of -3 is 3. The absolute value of 0 is 0 (see Figure 1.1).

Figure 1.1

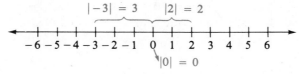

Symbolically, absolute value is denoted with vertical bars. Thus, we write

$$|2| = 2; \qquad\qquad |-3| = 3; \qquad\qquad |0| = 0.$$

More formally, the concept of absolute value can be defined as follows.

Definition 1.1	For all real numbers a, **1.** If $a \geq 0$, then $|a| = a$. **2.** If $a < 0$, then $|a| = -a$.

According to Definition 1.1 we obtain

$$|6| = 6 \qquad \text{by applying part \textbf{1.} of Definition 1.1;}$$
$$|0| = 0 \qquad \text{by applying part \textbf{1.} of Definition 1.1;}$$
$$|-7| = -(-7) = 7 \qquad \text{by applying part \textbf{2.} of Definition 1.1.}$$

Notice that the absolute value of a positive number is the number itself, but the absolute value of a negative number is its opposite. Thus, the absolute value of any number, except 0, is positive and the absolute value of 0 is 0.

Adding Integers

Addition of integers can be described using various physical models. For example, profits and losses pertaining to investments provide a meaningful model. A loss of $25 (written as -25) on one investment along with a profit of $60 (written as 60) on a second investment produces an overall profit of $35. This could be written as $(-25) + 60 = 35$. Think in terms of profits and losses for each of the following examples.

$$50 + 75 = 125; \qquad\qquad 20 + (-30) = -10;$$
$$(-10) + (-40) = -50; \qquad\qquad (-50) + 75 = 25;$$
$$100 + (-50) = 50; \qquad\qquad (-50) + (-50) = -100.$$

Although all problems involving addition of integers could be done using the *profit and loss* interpretation, it is sometimes convenient to be able to give a more precise description of the addition process. For this purpose we can use the concept of absolute value. Suppose that we want to describe the process of adding -50 and 75. We could say "subtract the absolute value of -50 from the absolute value of 75." Thus,

$$(-50) + 75 = |75| - |-50|$$
$$= 75 - 50$$
$$= 25.$$

The addition problem $20 + (-30)$ could be described as: subtract the absolute value of 20 from the absolute value of -30 and then take the opposite. Thus,

$$20 + (-30) = -(|-30| - |20|)$$
$$= -(30 - 20)$$
$$= -(10)$$
$$= -10.$$

The addition problem $(-10) + (-40)$ could be described as: add the absolute value of -10 and the absolute value of -40 and then take the opposite. We would write

$$(-10) + (-40) = -(|-10| + |-40|)$$
$$= -(10 + 40)$$
$$= -(50)$$
$$= -50.$$

In general, **addition of integers** can be described as follows.

Two positive integers   The sum of two positive integers is the sum of their absolute values.

Two negative integers   The sum of two negative integers is the opposite of the sum of their absolute values.

One positive and one negative integer   The sum of a positive integer and a negative integer can be found by subtracting the smaller absolute value from the larger absolute value and giving the result the sign of the original number having the larger absolute value. If the integers have the same absolute value, then their sum is 0.

Zero and another integer   The sum of 0 and any integer is the integer itself.

Now look over the following examples and think in terms of the previous description for adding integers.

$$(-6) + (-8) = -(|-6| + |-8|) = -(6 + 8) = -14;$$
$$(-19) + (-11) = -(|-19| + |-11|) = -(19 + 11) = -30;$$
$$18 + (-14) = (|18| - |-14|) = 18 - 14 = 4;$$
$$(-17) + 25 = (|25| - |-17|) = 25 - 17 = 8;$$
$$14 + (-21) = -(|-21| - |14|) = -(21 - 14) = -7;$$
$$(-32) + 17 = -(|-32| - |17|) = -(32 - 17) = -15;$$
$$19 + (-19) = 0; \qquad -49 + 0 = -49;$$
$$-21 + 21 = 0; \qquad 0 + (-72) = -72.$$

It is true that this *absolute value approach* precisely describes the process of adding integers, but don't forget about the profit and loss interpretation. The next problem set also includes other physical models describing the addition of integers. Some people find such models very helpful.

Subtracting Integers

Subtraction of integers is described in terms of addition as follows.

Subtraction of integers If a and b are integers, then $a - b = a + (-b)$.

It may be helpful for you to read $a - b = a + (-b)$ as "*a* minus *b* is equal to *a* plus the opposite of *b*." In other words, every subtraction problem can be changed to an equivalent addition problem. Consider the following examples.

$$7 - 9 = 7 + (-9) = -2; \qquad -5 - (-13) = -5 + 13 = 8;$$
$$8 - (-12) = 8 + 12 = 20; \qquad -16 - (-11) = -16 + 11 = -5;$$
$$-9 - 6 = -9 + (-6) = -15.$$

It should be apparent that addition of integers is a key operation. Being able to add integers effectively is a necessary skill for future algebraic work.

Simplifying numerical expressions involving addition and subtraction of integers can be done by first changing all subtractions to additions and then performing the additions.

Example 1   Simplify $7 - 9 - 14 + 12 - 6 + 4$.

Solution
$$7 - 9 - 14 + 12 - 6 + 4 = 7 + (-9) + (-14) + 12 + (-6) + 4$$
$$= -6.$$

●

Example 2   Simplify $-12 + 17 - (-10) + 9 - 3$.

Solution
$$-12 + 17 - (-10) + 9 - 3 = -12 + 17 + 10 + 9 + (-3)$$
$$= 21.$$

●

It is helpful if you can *mentally* convert subtractions to additions. In the next two examples the work shown in the dashed boxes might be done mentally.

Example 3   Simplify $4 - 9 - 18 + 13 - 10$.

Solution
$$4 - 9 - 18 + 13 - 10 = \boxed{4 + (-9) + (-18) + 13 + (-10)}$$
$$= -20.$$

●

Example 4   Simplify $(3 - 7) - (4 - 9)$.

Solution
$$(3 - 7) - (4 - 9) = [3 + (-7)] - [4 + (-9)]$$
$$= (-4) - (-5)$$
$$= -4 + 5$$
$$= 1.$$

Problem Set 1.2

Perform the following additions and subtractions.

1. $(-7) + (-8)$	**2.** $(-9) + (-5)$	**3.** $7 + (-13)$
4. $5 + (-12)$	**5.** $9 + (-6)$	**6.** $14 + (-7)$
7. $-11 + 14$	**8.** $-14 + 21$	**9.** $-7 - 4$
10. $-8 - 9$	**11.** $6 - 14$	**12.** $8 - 19$
13. $12 - (-11)$	**14.** $6 - (-5)$	**15.** $-9 - (-4)$
16. $-8 - (-11)$	**17.** $23 + (-32)$	**18.** $17 + (-28)$
19. $-31 + (-24)$	**20.** $-24 + (-33)$	**21.** $14 - 32$
22. $17 - (-18)$	**23.** $-48 - 17$	**24.** $-36 - 21$
25. $56 + (-29)$	**26.** $63 + (-36)$	**27.** $-37 - (-19)$
28. $-43 - 72$	**29.** $-49 - (-68)$	**30.** $-52 - (-96)$

Simplify each of the following.

31. $|-24| + |36|$	**32.** $|-48| - |29|$	**33.** $|-36| - |-48|$
34. $|-14| - |-25|$	**35.** $|-14| + |-12| + |-2|$	**36.** $|-4| + |-7| + |-21|$
37. $|4 + (-6)|$	**38.** $|7 - 4 - 9|$ **39.** $-|-4|$	**40.** $-|7|$

Simplify each of the following numerical expressions.

41. $6 - 8 + 9 - 14 + 12$	**42.** $8 - 9 - 12 + 10 - 14 - 1$
43. $-4 - 6 - 7 + 8 + 9 - 1$	**44.** $-7 - 6 - 5 + 9 + 4 - 2$
45. $8 + (-7) - (-6) + 9 - (-1)$	**46.** $12 + (-14) + 16 - (-11) + 9$
47. $-12 - (-12) + 11 - 14 + (-9)$	**48.** $-21 + (-14) - 16 + 14 - (-10)$
49. $-4 - 7 - 8 - 9 - 10$	**50.** $-7 - 6 - 5 - 4 + 3 - 9$
51. $5 - (7 - 9)$	**52.** $-6 - (8 - 12)$
53. $(1 - 2) - (6 + 3)$	**54.** $(5 - 4 - 6) - (-2 - 1 + 4)$
55. $(17 - 12 - 18) - (19 - 22 + 27)$	**56.** $14 - 18 - 21 - (27 - 32 - 34)$
57. $17 - [14 - (21 - 31)]$	**58.** $16 - 18 + 19 - [14 - 22 - (31 - 41)]$
59. $(-12) + (-14) + [-21 - (14 - 17)]$	**60.** $[14 - (16 - 18)] - [32 - (8 - 9)]$

61. A game like football can also be used to interpret addition of integers. A *gain* of 7 yards on one play followed by a *loss* of 3 yards on the next play places the ball 4 yards *ahead* of the original line of scrimmage. This can be expressed as $7 + (-3) = 4$. Use this "football interpretation" to find the following sums.

(a) $6 + (-4)$ **(b)** $4 + (-3)$ **(c)** $5 + (-8)$

(d) $3 + (-9)$ **(e)** $(-1) + (-5)$ **(f)** $(-6) + (-4)$

(g) $4 + (-12)$ **(h)** $5 + (-17)$ **(i)** $-14 + 23$

62. The number line is a convenient pictorial aid for interpreting addition of integers. Consider the following examples.

$3 + (-2)$ 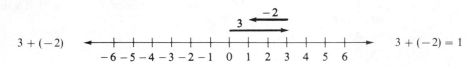 $3 + (-2) = 1$

$-2 + 5$ 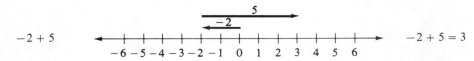 $-2 + 5 = 3$

$-1 + (-2)$ 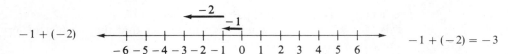 $-1 + (-2) = -3$

Once a feeling for movement on the number line is acquired, then merely forming a mental image of the movement is sufficient. Use the "number line interpretation" to find the following sums.

(a) $5 + (-9)$ **(b)** $6 + (-7)$ **(c)** $-8 + 5$

(d) $-7 + 1$ **(e)** $(-5) + (-6)$ **(f)** $(-12) + (-4)$

(g) $14 + (-23)$ **(h)** $-18 + 27$ **(i)** $(-14) + (-19)$

63. Be sure that you can handle addition and subtraction of integers on your calculator by doing Problems 41–60 with your calculator.

1.3 Multiplication and Division of Integers

Multiplication of whole numbers may be interpreted as repeated addition. For example, $3 \cdot 2$ means three 2's; thus, $3 \cdot 2 = 2 + 2 + 2 = 6$. This same "repeated addition" interpretation of multiplication can be used to find the product of a positive integer and a negative integer as illustrated by the following examples

$$2(-3) = -3 + (-3) = -6; \qquad 3(-2) = -2 + (-2) + (-2) = -6;$$
$$4(-5) = -5 + (-5) + (-5) + (-5) = -20.$$

Note the use of parentheses to indicate multiplication. Sometimes both numbers are enclosed in parentheses, as with $(3)(-4)$.

When multiplying whole numbers we realize that the order in which we multiply two factors does not change the product. For example, $2(3) = 6$ and $3(2) = 6$. Using this idea, we can handle a negative integer times a positive integer as follows.

$$(-2)(3) = (3)(-2) = (-2) + (-2) + (-2) = -6;$$
$$(-3)(4) = (4)(-3) = (-3) + (-3) + (-3) + (-3) = -12;$$
$$(-4)(3) = (3)(-4) = (-4) + (-4) + (-4) = -12.$$

Finally, let's consider the product of two negative integers. The following pattern helps with the reasoning for this situation.

$$4(-2) = -8;$$
$$3(-2) = -6;$$
$$2(-2) = -4;$$
$$1(-2) = -2;$$
$$0(-2) = \quad 0; \quad \text{(The product of zero and any number is zero.)}$$
$$(-1)(-2) = \quad ?$$

Certainly, to continue this pattern the product of -1 and -2 has to be 2. In general, this type of reasoning would help us realize that the product of any two negative integers is a positive integer.

Using the concept of absolute value it is quite easy to precisely describe **multiplication of integers**.

1. The product of two positive integers or two negative integers is the product of their absolute values.

2. The product of a positive and a negative integer (either order) is the opposite of the product of their absolute values.

3. The product of zero and any integer is zero.

The following examples illustrate this description of multiplication.

$$(-6)(-7) = |-6| \cdot |-7| = 6 \cdot 7 = 42;$$
$$(8)(-9) = -(|8| \cdot |-9|) = -(8 \cdot 9) = -72;$$
$$(-5)(12) = -(|-5| \cdot |12|) = -(5 \cdot 12) = -60;$$
$$(-19)(0) = 0; \quad (0)(-31) = 0.$$

The previous examples illustrated a step-by-step process for multiplying integers. In practice, however, the key issue is to remember whether the product is positive or negative. In other words, we need to remember that *the product of two positive or two negative integers is positive* and *the product of a positive and a negative integer (either order) is negative*.

Dividing Integers

The relationship between multiplication and division provides the basis for dividing integers. For example, we know that $8 \div 2 = 4$ because $2 \cdot 4 = 8$. In other words, the quotient of two numbers can be found by looking at a related multiplication problem.

In the following examples we have used this same type of reasoning to determine some quotients involving integers.

$$\frac{6}{-2} = -3 \quad \text{because } (-2)(-3) = 6;$$

$$\frac{-12}{3} = -4 \quad \text{because } (3)(-4) = -12;$$

$$\frac{-18}{-2} = 9 \quad \text{because } (-2)(9) = -18;$$

$$\frac{0}{-5} = 0 \quad \text{because } (-5)(0) = 0;$$

$$\frac{-8}{0} \text{ is undefined.} \quad \text{(Remember that division by zero is undefined!)}$$

A precise description for **division of integers** follows.

1. The quotient of two positive or two negative integers is the quotient of their absolute values.

2. The quotient of a positive integer and a negative integer or a negative integer and a positive integer is the opposite of the quotient of their absolute values.

3. The quotient of zero and any nonzero integer (zero divided by any nonzero integer) is zero.

The following examples illustrate this description of division.

$$\frac{-16}{-4} = \frac{|-16|}{|-4|} = \frac{16}{4} = 4; \qquad \frac{28}{-7} = -\left(\frac{|28|}{|-7|}\right) = -\left(\frac{28}{7}\right) = -4;$$

$$\frac{-21}{3} = -\left(\frac{|-21|}{|3|}\right) = -\left(\frac{21}{3}\right) = -7; \qquad \frac{0}{-9} = 0.$$

Again, for practical purposes the key idea is to remember whether the quotient is positive or negative. Let's consider simplifying some numerical expressions involving integers.

Example 1   Simplify $-6 + 8(-3) - (-4)(-3)$.

Solution
$$-6 + 8(-3) - (-4)(-3)$$
$$= -6 + (-24) - 12 \qquad \text{Do the multiplications first.}$$
$$= -42.$$

Example 2   Simplify $-24 \div 4 + 8(-5) - (-5)(3)$.

Solution
$$-24 \div 4 + 8(-5) - (-5)(3)$$
$$= -6 + (-40) - (-15) \qquad \text{Do the multiplications and}$$
$$\qquad\qquad\qquad\qquad\qquad\qquad \text{divisions first.}$$
$$= -31.$$

Example 3   Simplify $-7 - 5[-3(6 - 8)]$.

Solution
$$-7 - 5[-3(6 - 8)]$$
$$= -7 - 5[-3(-2)] \quad \text{Start with the innermost parentheses.}$$
$$= -7 - 5[6]$$
$$= -7 - 30$$
$$= -37.$$

Example 4   Simplify $[3(-7) - 2(9)][5(-7) + 3(9)]$.

Solution
$$[3(-7) - 2(9)][5(-7) + 3(9)] = [-21 - 18][-35 + 27]$$
$$= [-39][-8]$$
$$= 312.$$

The last two sections reviewed the basic operations with integers; however, we should realize that the descriptions given for manipulating with integers apply to real numbers in general. For example, the sum of two negative real numbers is the opposite of the sum of their absolute values.

$$\left(-\frac{1}{5}\right) + \left(-\frac{2}{5}\right) = -\left(\left|-\frac{1}{5}\right| + \left|-\frac{2}{5}\right|\right)$$
$$= -\left(\frac{1}{5} + \frac{2}{5}\right)$$
$$= -\frac{3}{5}.$$

Likewise, the product of two negative real numbers is the product of their absolute values.

$$(-0.2)(-0.8) = |-0.2| \cdot |-0.8|$$
$$= (0.2)(0.8)$$
$$= 0.16.$$

Problem Set 1.3

Perform the following multiplications and divisions.

1. $(-7)(-4)$	**2.** $(-6)(-9)$	**3.** $8(-4)$
4. $6(-9)$	**5.** $(-7)(12)$	**6.** $(-5)(13)$
7. $(-15) \div (-5)$	**8.** $(-42) \div (-6)$	**9.** $(-72) \div 6$
10. $(-104) \div 13$	**11.** $\dfrac{38}{-19}$	**12.** $\dfrac{51}{-17}$

13. $(-2)(-3)(5)$

14. $(-4)(3)(-6)$

15. $(-1)(-5)(-9)$

16. $(-4)(-2)(-11)$

17. $(8)(-2)(5)$

18. $(9)(-3)(6)$

19. $0 \div (-9)$

20. $(-21) \div 0$

21. $\dfrac{-17}{0}$

22. $0 \div (-14)$

23. $\dfrac{-144}{-12}$

24. $\dfrac{156}{-13}$

25. $(-8) \div (-4) \div (-2)$

26. $(-24) \div (2) \div (-3)$

27. $(36) \div (-3) \div 4$

28. $56 \div (-4) \div 7$

29. $-\left(\dfrac{-8}{4}\right)$

30. $-\left(\dfrac{-56}{-7}\right)$

31. $-\left(\dfrac{-72}{-9}\right)$

32. $-\left(\dfrac{21}{-7}\right)$

Simplify each of the following numerical expressions.

33. $7 - (3)(-4)$

34. $14 - (-2)(9)$

35. $-5 + (-1)(6) - (-2)(4)$

36. $-8 - (3)(-4) + (-8)(5)$

37. $4(-7) - (-3)(9)$

38. $(-6)(8) + (-9)(9)$

39. $(-7)(-5) + (-8)(12)$

40. $(-6)(-6) - (-7)(13)$

41. $2(4 - 7) - (4)(-1)$

42. $6(7 - 12) + (-3)(8)$

43. $(6 - 12)(3 - 8)$

44. $(5 - 9)(-3 - 4)$

45. $-5(-3 - 8)$

46. $-7(-4 - 10)$

47. $(-56) \div (8) - (-4) \div (-1)$

48. $(60) \div (-12) - (-18)(-1) + (-24) \div (4)$

49. $3[6 - (-4)] - 2(-8 - 3)$

50. $-2(-8 + 11) + 5(-9 - 3)$

51. $\dfrac{6 + (-15)}{-3} + \dfrac{-5}{-4 - 1}$

52. $\dfrac{-8 + 12}{-2} + \dfrac{-6 - 8}{7}$

53. $[4(-5) - (6)(-9)][7(-2) + 8(-1)]$

54. $[(-3)(4) - (-2)(-1)][(-4)(7) - (-9)(0)]$

55. $(7 - 9)[4 - (3 - 6)]$

56. $(-6 - 5)[-3 - (2 - 11)]$

57. $(-9 - 4)(8 - 13) - (6 - 10)(-4 - 3) - 14$

58. Do Problems 47–57 with your calculator.

1.4 Properties of Real Numbers and Use of Exponents

This section begins by listing and briefly discussing some of the basic properties of real numbers. Be sure that you understand these properties, for not only do they facilitate manipulations with real numbers, but they also serve as the basis for many algebraic computations.

Closure Property for Addition

If a and b are real numbers, then $a + b$ is a real number.

Closure Property for Multiplication

If a and b are real numbers, then ab is a real number.

The set of real numbers is said to be *closed* with respect to addition and also with respect to multiplication. That is, the sum of two real numbers is a real number and the product of two real numbers is a real number.

Commutative Property of Addition

If a and b are real numbers, then

$$a + b = b + a.$$

Commutative Property of Multiplication

If a and b are real numbers, then

$$ab = ba.$$

Addition and multiplication are said to be commutative operations. This means that the order in which you add or multiply two numbers does not affect the result. For example, $6 + (-8) = (-8) + 6$ and $(-4)(-3) = (-3)(-4)$. It is also important to realize that subtraction and division *are not* commutative operations; order does make a difference. For example, $3 - 4 = -1$ but $4 - 3 = 1$. Likewise, $2 \div 1 = 2$ but $1 \div 2 = \frac{1}{2}$.

Associative Property of Addition

If a, b, and c are real numbers, then

$$(a + b) + c = a + (b + c).$$

Associative Property of Multiplication

If a, b, and c are real numbers, then

$$(ab)c = a(bc).$$

Addition and multiplication are associative operations. The associative properties are grouping properties. For example, $(-8 + 9) + 6 = -8 + (9 + 6)$; changing the grouping of the numbers does not affect the final sum. This is also true for multiplication, which is illustrated by $[(-4)(-3)](2) = (-4)[(-3)(2)]$. Subtraction and division *are not* associative operations. For example, $(8 - 6) - 10 = -8$ but $8 - (6 - 10) = 12$. An example showing that division is not associative is $(8 \div 4) \div 2 = 1$, but $8 \div (4 \div 2) = 4$.

Identity Property of Addition

If a is any real number, then

$$a + 0 = 0 + a = a.$$

Zero is called the identity element for addition. This merely means that the sum of any real number and zero is identically the same real number. For example, $-87 + 0 = 0 + (-87) = -87$.

Identity Property of Multiplication

If a is any real number, then

$$a(1) = 1(a) = a.$$

One is called the identity element for multiplication. The product of any real number and one is identically the same real number. For example, $(-119)(1) = (1)(-119) = -119$.

Additive Inverse Property

For every real number a, there exists a real number $-a$ such that

$$a + (-a) = -a + a = 0.$$

The real number, $-a$, is called the *additive inverse* of a or the *opposite* of a. For example, 16 and -16 are additive inverses and their sum is 0. The additive inverse of 0 is 0.

Multiplication Property of Zero

If a is any real number, then

$$(a)(0) = (0)(a) = 0.$$

The product of any real number and zero is zero. For example, $(-17)(0) = 0(-17) = 0$.

Multiplication Property of Negative One

If a is any real number, then

$$(a)(-1) = (-1)(a) = -a.$$

The product of any real number and -1 is the opposite of the real number. For example, $(-1)(52) = (52)(-1) = -52$.

Multiplicative Inverse Property

For every nonzero real number a, there exists a real number $\dfrac{1}{a}$, such that

$$a\left(\frac{1}{a}\right) = \frac{1}{a}(a) = 1.$$

The number $\dfrac{1}{a}$ is called the *multiplicative inverse* or the *reciprocal* of a. For example, the reciprocal of 2 is $\dfrac{1}{2}$ and $2\left(\dfrac{1}{2}\right) = \dfrac{1}{2}(2) = 1$. Likewise, the reciprocal of $\dfrac{1}{2}$ is $\dfrac{1}{\frac{1}{2}} = 2$. Therefore, 2 and $\dfrac{1}{2}$ are said to be reciprocals (or multiplicative inverses) of each other. Since division by zero is undefined, zero does not have a reciprocal.

Distributive Property

If a, b, and c are real numbers, then

$$a(b + c) = ab + ac.$$

The distributive property ties together the operations of addition and multiplication. We say that *multiplication distributes over addition*. For example, $7(3 + 8) = 7(3) + 7(8)$. Since $b - c = b + (-c)$, it follows that *multiplication also distributes over subtraction*. This can be symbolically expressed as $a(b - c) = ab - ac$. For example, $6(8 - 10) = 6(8) - 6(10)$.

Now let's consider some examples that illustrate the use of the properties of real numbers to facilitate certain types of manipulations.

Example 1

Simplify $(74 + (-36)) + 36$.

Solution   In such a problem it is much more advantageous to group -36 and 36.

$$(74 + (-36)) + 36 = 74 + ((-36) + 36) \qquad \text{(By using the associative property}$$
$$= 74 + 0 = 74. \qquad\qquad\qquad \text{for addition)}$$

Example 2   Simplify $[(-19)(25)](-4)$.

Solution   It is much easier to group 25 and -4. Thus,

$$[(-19)(25)](-4) = (-19)[(25)(-4)] \qquad \text{(By using the associative}$$
$$= (-19)(-100) \qquad\qquad \text{property for multiplication)}$$
$$= 1900.$$

Example 3   Simplify $17 + (-14) + (-18) + 13 + (-21) + 15 + (-33)$.

Solution   One could add in the order in which the numbers appear. However, since addition is commutative and associative we could change the order and group in any convenient way. For example, we could add all of the positive integers and add all of the negative integers and then find the sum of these two results. It might be convenient to use the vertical format for this as follows.

$$
\begin{array}{rrr}
 & -14 & \\
17 & -18 & \\
13 & -21 & -86 \\
15 & -33 & 45 \\
\underline{} & \underline{-86} & \overline{-41.} \\
45 & -86 &
\end{array}
$$

Example 4   Simplify $-25(-2 + 100)$.

Solution   For this problem it might be easiest to first apply the distributive property and then to simplify as follows.

$$-25(-2 + 100) = (-25)(-2) + (-25)(100)$$
$$= 50 + (-2500)$$
$$= -2450.$$

Example 5   Simplify $(-87)(-26 + 25)$.

Solution   For this problem we are better off not applying the distributive property, but merely adding the numbers inside the parentheses first and then finding the indicated product.

$$(-87)(-26 + 25) = (-87)(-1)$$
$$= 87.$$

Example 6   Simplify $37(104) + 37(-4)$.

Solution   Remember that the distributive property allows us to change from the form $a(b + c)$ to $ab + ac$ or from $ab + ac$ to $a(b + c)$. In this problem we want to use the latter change. Thus,

$$37(104) + 37(-4) = 37(104 + (-4))$$
$$= 37(100)$$
$$= 3700.$$

Examples 4, 5, and 6 illustrate an important issue. Sometimes the form $a(b + c)$ is more convenient, but at other times the form $ab + ac$ is better. In these cases, as well as in the cases of other properties, you should *think first* and decide whether or not the properties can be used to make the manipulations easier.

Exponents

Exponents are used to indicate repeated multiplication. For example, $4 \cdot 4 \cdot 4$ can be written as 4^3 where the "raised 3" indicates that 4 is to be used as a factor 3 times. The following general definition is helpful.

Definition 1.2	If n is a positive integer and b is any real number, then

$$b^n = \underbrace{bbb \cdots b.}_{n \text{ factors of } b}$$

The b is referred to as the **base** and n as the **exponent**. The expression b^n can be read as "b to the nth power." The terms *squared* and *cubed* are commonly associated with exponents of 2 and 3, respectively. For example, b^2 is read "b squared" and b^3 as "b cubed." An exponent of 1 is usually not written, so b^1 is written as b.

The following examples illustrate Definition 1.2.

$$2^3 = 2 \cdot 2 \cdot 2 = 8; \qquad \left(\frac{1}{2}\right)^5 = \frac{1}{2} \cdot \frac{1}{2} \cdot \frac{1}{2} \cdot \frac{1}{2} \cdot \frac{1}{2} = \frac{1}{32};$$

$$3^4 = 3 \cdot 3 \cdot 3 \cdot 3 = 81; \qquad (0.7)^2 = (0.7)(0.7) = 0.49;$$

$$-5^2 = -(5 \cdot 5) = -25; \qquad (-5)^2 = (-5)(-5) = 25.$$

Please take special note of the last two examples. Note that $(-5)^2$ means -5 is the base and is to be used as a factor twice. However, -5^2 means that 5 is the base and after it is squared then we take the opposite of that result.

Simplifying numerical expressions containing exponents creates no trouble if we merely keep in mind that exponents are used to indicate repeated multiplication. Let's consider some examples.

Example 7   Simplify $3(-4)^2 + 5(-3)^2$.

Solution $\begin{aligned} 3(-4)^2 + 5(-3)^2 &= 3(16) + 5(9) \qquad \text{Find the powers.} \\ &= 48 + 45 \\ &= 93. \end{aligned}$

Example 8   Simplify $[3(-1) - 2(1)]^3$.

Solution $\begin{aligned} [3(-1) - 2(1)]^3 &= [-3 - 2]^3 \\ &= [-5]^3 \\ &= -125. \end{aligned}$

Example 9   Simplify $2(2)^2(-3) - 3(2)(-3)^2 - (-3)^3$.

Solution $\begin{aligned} 2(2)^2(-3) - 3(2)(-3)^2 - (-3)^3 &= 2(4)(-3) - 3(2)(9) - (-27) \\ &= -24 - 54 + 27 \\ &= -51. \end{aligned}$

Problem Set 1.4

In Exercises 1–14, state the property that justifies each of the statements. For example, $3 + (-4) = (-4) + 3$ because of the commutative property of addition.

1. $x(2) = 2(x)$
2. $(7 + 4) + 6 = 7 + (4 + 6)$
3. $1(x) = x$
4. $43 + (-18) = (-18) + 43$
5. $(-1)(93) = -93$
6. $109 + (-109) = 0$
7. $5(4 + 7) = 5(4) + 5(7)$
8. $-1(x + y) = -(x + y)$
9. $[(-8)(4)](25) = (-8)[(4)(25)]$
10. $7yx = 7xy$
11. $(x + 2) + (-2) = x + (2 + (-2))$
12. $6(4) + 7(4) = (6 + 7)(4)$
13. $\left(\dfrac{2}{3}\right)\left(\dfrac{3}{2}\right) = 1$
14. $[(17)(8)](25) = (17)[(8)(25)]$

Simplify each of the following numerical expressions. Don't forget to take advantage of the properties whenever they can be used to make the computation easier.

15. $42 + (-14) + (-16) + 36 + (-21) + 15$
16. $-35 + 43 + 41 + 35 + (-43) - 42$
17. $(72 + (-46)) + 47$
18. $34 + (68 - 35)$
19. $(25)(-17)(4)$
20. $(18)(-4)(16)(25)$
21. $13(96) + 13(4)$
22. $-75(48 + (-46))$

23. $16 - 14 - 13 - 18 + 19 + 14 - 17 + 21$

24. $14 - 12 - 21 - 14 + 17 - 18 + 19 - 32$

25. $(50)(17)(-2) - (4)(13)(-25)$ **26.** $(-7)(2)(5) - (5)(19)(-20)$

Simplify each of the following numerical expressions.

27. $3^2 - 2^3$ **28.** $2^3 - 3^2$ **29.** $-4^2 + 5^2$

30. $-6^2 - 2^5$ **31.** $(-3)^3 - 3^3$ **32.** $(-2)^4 + 2^4$

33. $2(-1)^3 - 3(2)^4$ **34.** $5(-2)^3 + 4(-1)^3$

35. $5(3)^2 + 4(-2)^3$ **36.** $3(-1)^3 - 2(3)^2$

37. $-2(-3)^3 + 3(-1)^4$ **38.** $-3(-2)^3 - (-3)^2$

39. $(-4)^2 - 2(-4)(3) + 3^2$ **40.** $(-1)^2 - 2(-1)(-4) - (-4)^2$

41. $1^3 + 3(1)^2(-1) - 3(1)(-1)^2$ **42.** $2(-3)^3 - 3(2)^4 - 4(-1)^5$

43. $[3(-4) + 4(2)]^3$ **44.** $[-2(-3) - 3(1)]^4$

45. $[2(-1)^2 - 3(2)^2]^2$ **46.** $[4(-1)^3 + 2(-2)^3]^2$

47. $(-2)^3 + 2(-2)^2 - 3(-2) - 1$ **48.** $2(-1)^3 - 3(-1)^2 + 4(-1) - 5$

49. $3(-3)^3 + 4(-3)^2 - 5(-3) + 7$ **50.** $2^4 - 2(2)^3 - 3(2)^2 + 7(2) - 10$

51. Be sure you can handle the use of exponents on your calculator by doing Problems 41–50 with your calculator.

52. Use your calculator to evaluate each of the following.

(a) 2^{10} **(b)** 3^7 **(c)** $(-2)^8$

(d) $(-2)^{11}$ **(e)** -4^9 **(f)** -5^6

(g) $(3.14)^3$ **(h)** $(1.41)^4$ **(i)** $(1.73)^5$

1.5 Algebraic Expressions

Algebraic expressions such as

$$2x, \qquad 8xy, \qquad 3xy^2, \qquad -4a^2b^3c, \qquad \text{and} \qquad z$$

are called **terms**. A term is an indicated product and may have any number of factors. The variables involved in a term are called **literal factors** and the numerical factor is called the **numerical coefficient**. Thus, in $8xy$, the x and y are literal factors and 8 is the numerical coefficient. The numerical coefficient of the term $-4a^2bc$ is -4. Since $1(z) = z$, the numerical coefficient of the term z is understood to be 1. Terms having the same literal factors are called **similar terms** or **like terms**. Some examples of similar terms are

$$3x \quad \text{and} \quad 14x; \qquad\qquad 5x^2 \quad \text{and} \quad 18x^2;$$
$$7xy \quad \text{and} \quad -9xy; \qquad\qquad 9x^2y \quad \text{and} \quad -14x^2y;$$
$$2x^3y^2, \quad 3x^3y^2, \quad \text{and} \quad -7x^3y^2.$$

By the symmetric property of equality the distributive property can be written as
$$ab + ac = a(b + c).$$
Then the commutative property of multiplication can be applied to change the form to
$$ba + ca = (b + c)a.$$
This latter form provides the basis for simplifying algebraic expressions by **combining similar terms**. Consider the following examples.

$$3x + 5x = (3 + 5)x \qquad\qquad -6xy + 4xy = (-6 + 4)xy$$
$$= 8x. \qquad\qquad\qquad\qquad = -2xy.$$

$$5x^2 + 7x^2 + 9x^2 = (5 + 7 + 9)x^2 \qquad 4x - x = 4x - 1x$$
$$= 21x^2. \qquad\qquad\qquad = (4 - 1)x = 3x.$$

More complicated expressions might first require some rearranging of the terms by applying the commutative property for addition.

$$7x + 2y + 9x + 6y = 7x + 9x + 2y + 6y$$
$$= (7 + 9)x + (2 + 6)y$$
$$= 16x + 8y.$$

$$6a - 5 - 11a + 9 = 6a + (-5) + (-11a) + 9$$
$$= 6a + (-11a) + (-5) + 9$$
$$= (6 + (-11))a + 4$$
$$= -5a + 4.$$

As soon as you feel that you understand the various simplifying steps you may want to do the steps mentally. Then you could go directly from the given expression to the simplified form as follows.

$$14x + 13y - 9x + 2y = 5x + 15y;$$
$$3x^2y - 2y + 5x^2y + 8y = 8x^2y + 6y;$$
$$-4x^2 + 5y^2 - x^2 - 7y^2 = -5x^2 - 2y^2.$$

Sometimes an algebraic expression can be simplified by applying the distributive property to remove parentheses and then to combine similar terms as the next examples illustrate.

$$4(x + 2) + 3(x + 6) = 4(x) + 4(2) + 3(x) + 3(6)$$
$$= 4x + 8 + 3x + 18$$
$$= 4x + 3x + 8 + 18$$
$$= (4 + 3)x + 26$$
$$= 7x + 26.$$

$$-5(y + 3) - 2(y - 8) = -5(y) - 5(3) - 2(y) - 2(-8)$$
$$= -5y - 15 - 2y + 16$$
$$= -5y - 2y - 15 + 16$$
$$= -7y + 1.$$

$$5(x - y) - (x + y) = 5(x - y) - 1(x + y) \quad \text{(Remember } -a = -1(a).)$$
$$= 5(x) - 5(y) - 1(x) - 1(y)$$
$$= 5x - 5y - 1x - 1y$$
$$= 5x - 1x - 5y - 1y$$
$$= 4x - 6y.$$

When multiplying two terms such as 3 and $2x$ the associative property for multiplication provides the basis for simplifying the product.

$$3(2x) = (3 \cdot 2)x = 6x.$$

This idea can be put to use in the following example.

$$3(2x + 5y) + 4(3x + 2y) = 3(2x) + 3(5y) + 4(3x) + 4(2y)$$
$$= 6x + 15y + 12x + 8y$$
$$= 6x + 12x + 15y + 8y$$
$$= 18x + 23y.$$

After you are sure of each step, a more simplified format may be used, as the following examples illustrate.

Be careful with this sign.

$$5(a + 4) - 7(a + 3) = 5a + 20 - 7a - 21$$
$$= -2a - 1.$$

$$3(x^2 + 2) + 4(x^2 - 6) = 3x^2 + 6 + 4x^2 - 24$$
$$= 7x^2 - 18.$$

$$2(3x - 4y) - 5(2x + 6y) = 6x - 8y - 10x - 30y$$
$$= -4x - 38y.$$

Evaluating Algebraic Expressions

An algebraic expression takes on a numerical value whenever each variable in the expression is replaced by a real number. For example, if x is replaced by 5 and y by 9, the algebraic expression $x + y$ becomes the numerical expression $5 + 9$, which simplifies to 14. We say that $x + y$ *has a value* of 14 when x equals 5 and y equals 9. If $x = -3$ and $y = 7$ then $x + y$ has a value of $-3 + 7 = 4$.

The following examples illustrate the process of finding a value of an algebraic expression. The process is commonly referred to as **evaluating algebraic expressions**.

Example 1   Find the value of $3x - 4y$ when $x = 2$ and $y = -3$.

Solution
$$3x - 4y = 3(2) - 4(-3) \quad \text{when } x = 2 \text{ and } y = -3$$
$$= 6 + 12$$
$$= 18.$$

Example 2   If $a = -1$ and $b = 4$, find the value of $2a^2 - 5b^2$.

Solution $2a^2 - 5b^2 = 2(-1)^2 - 5(4)^2$ when $a = -1$ and $b = 4$
$$= 2(1) - 5(16)$$
$$= 2 - 80$$
$$= -78.$$

Example 3   Evaluate $x^2 - 2xy + y^2$ for $x = -2$ and $y = -5$.

Solution $x^2 - 2xy + y^2 = (-2)^2 - 2(-2)(-5) + (-5)^2$ when $x = -2$ and $y = -5$
$$= 4 - 20 + 25$$
$$= 9.$$

Example 4   Evaluate $(3x + 2y)(2x - y)$ for $x = 4$ and $y = -1$.

Solution $(3x + 2y)(2x - y) = (3(4) + 2(-1))(2(4) - (-1))$ when $x = 4$ and $y = -1$
$$= (12 - 2)(8 + 1)$$
$$= (10)(9)$$
$$= 90.$$

Simplifying by combining similar terms will help you evaluate some algebraic expressions. The following examples illustrate this idea.

Example 5   Evaluate $7x - 2y + 4x - 3y$ for $x = -3$ and $y = 6$.

Solution   Let's first simplify the given expression.

$7x - 2y + 4x - 3y = 11x - 5y.$

Now we can evaluate for $x = -3$ and $y = 6$.

$11x - 5y = 11(-3) - 5(6)$ when $x = -3$ and $y = 6$
$$= -33 - 30$$
$$= -63.$$

Example 6   Evaluate $2(3x + 1) - 3(4x - 3)$ for $x = -6$.

Solution $2(3x + 1) - 3(4x - 3) = 6x + 2 - 12x + 9$
$$= -6x + 11.$$

Substituting $x = -6$, we obtain

$-6x + 11 = -6(-6) + 11$
$$= 36 + 11$$
$$= 47.$$

Example 7   Evaluate $2(a^2 + 1) - 3(a^2 + 5) + 4(a^2 - 1)$ for $a = 10$.

Solution

$$2(a^2 + 1) - 3(a^2 + 5) + 4(a^2 - 1) = 2a^2 + 2 - 3a^2 - 15 + 4a^2 - 4$$
$$= 3a^2 - 17.$$

Substituting $a = 10$, we obtain

$$3a^2 - 17 = 3(10)^2 - 17$$
$$= 3(100) - 17$$
$$= 300 - 17$$
$$= 283.$$

Translating from English to Algebra

To be able to use the tools of algebra for solving problems, we must be able to translate from English to algebra. Part of this translation process requires that we recognize key phrases in the English language that translate into algebraic expressions involving the operations of addition, subtraction, multiplication, and division. Some of these key phrases and their algebraic counterparts are listed in the following table. The variable n represents the number being referred to in each phrase.

English phrase	Algebraic expression
Addition	
the sum of a number and 4	$n + 4$
7 more than a number	$n + 7$
a number plus 10	$n + 10$
a number increased by 6	$n + 6$
8 added to a number	$n + 8$
Subtraction	
14 minus a number	$14 - n$
12 less than a number	$n - 12$
a number decreased by 10	$n - 10$
the difference between a number and 2	$n - 2$
a number subtracted from 13	$13 - n$
Multiplication	
14 times a number	$14n$
the product of 4 and a number	$4n$
$\frac{3}{4}$ of a number	$\frac{3}{4}n$
twice a number	$2n$
multiply a number by 12	$12n$

Division

the quotient of 6 and a number	$\dfrac{6}{n}$
the quotient of a number and 6	$\dfrac{n}{6}$
a number divided by 9	$\dfrac{n}{9}$
the ratio of a number and 4	$\dfrac{n}{4}$

Mixture of Operations

4 more than three times a number	$3n + 4$
5 less than twice a number	$2n - 5$
3 times the sum of a number and 2	$3(n + 2)$
2 more than the quotient of a number and 12	$\dfrac{n}{12} + 2$
7 times the difference of 6 and a number	$7(6 - n)$

Often an English statement may not contain any of the key words such as sum, difference, product, or quotient. Instead, the statement may describe a physical situation and from this description we must deduce the operations involved. Some suggestions for handling such situations are made in the following examples.

Example 8   Sonya can type 65 words per minute. How many words will she type in m minutes?

Solution   The total number of words typed equals the product of the rate per minute and the number of minutes. Therefore, Sonya should be able to type $65m$ words in m minutes. ●

Example 9   Russ has n nickels and d dimes. Express this amount of money in cents.

Solution   Each nickel is worth 5 cents and each dime is worth 10 cents. Thus, the amount, in cents, is represented by $5n + 10d$. ●

Example 10   The cost of a 50-pound sack of fertilizer is d dollars. How much is the cost per pound for the fertilizer?

Solution   The price per pound is calculated by dividing the total cost by the number of pounds. Thus, the price per pound is represented by $\dfrac{d}{50}$. ●

The English statement to be translated to algebra may contain some geometric ideas. Tables 1.1 and 1.2 contain some of the basic relationships pertaining to linear measurement in the English and metric systems, respectively.

Table 1.1   **English system**

12 inches	= 1 foot
3 feet	= 1 yard
1760 yards	= 1 mile

Table 1.2   **Metric system**

1 kilometer	=	1000 meters
1 hectometer	=	100 meters
1 dekameter	=	10 meters
1 decimeter	=	0.1 meter
1 centimeter	=	0.01 meter
1 millimeter	=	0.001 meter

Example 11   The distance between two cities is k kilometers. Express this distance in meters.

Solution   Since 1 kilometer equals 1000 meters, the distance in meters is represented by $1000k$.   ●

Example 12   The length of a rope is y yards and f feet. Express this length in inches.

Solution   Since 1 foot equals 12 inches and 1 yard equals 36 inches, the length of the rope, in inches, can be represented by $36y + 12f$.   ●

Example 13   The length of a rectangle is l centimeters and the width is w centimeters. Express the perimeter of the rectangle in meters.

Solution   A sketch of the rectangle with the given information indicated may be helpful.

l centimeters

w centimeters

The perimeter of a rectangle is the sum of the lengths of the four sides. Thus, the perimeter, in centimeters, is $l + w + l + w$, which simplifies to $2l + 2w$. Now, since 1 centimeter equals 0.01 of a meter, the perimeter, expressed in meters, is $0.01(2l + 2w)$. This could also be written as $\dfrac{2l + 2w}{100}$.   ●

Problem Set 1.5

Simplify each of the following algebraic expressions by combining similar terms.

1. $-5x + 8x$

2. $7x - 9x + 4x$

3. $3a^2 - 4a^2$

4. $11b^3 - 14b^3$

5. $3n - 5n - 7n$

6. $8n + 10n - 6n$

7. $6x - 7x + 2y$

8. $14x - 16y - 13y + x$

9. $-4a^2 + 5b^2 - 3a^2 - 6b^2$

10. $xy - z + 9xy + 5z$

11. $7x - 4 + 6x - 9$

12. $3x - 2 + 4x - 3 - 6x - 8$

13. $3a^2b - 2ab^2 + 5a^2b$

14. $7xy^2 - 9x^2y - 3x^2y + xy^2$

Simplify each of the following algebraic expressions by removing parentheses and combining similar terms.

15. $2(x + 1) + 3(x + 2)$

16. $5(x - 3) + 6(x + 4)$

17. $-4(a - 3) - 2(a + 1)$

18. $-5(a + 2) - 6(a - 7)$

19. $5(n^2 + 3) - 7(n^2 - 2)$

20. $3(n^2 + 4) + (n^2 - 2)$

21. $-3(x^2 - 1) - (x^2 - 4)$

22. $4(x + y) - 3(x - y)$

23. $3(4x - 3) + 2(5x - 6)$

24. $4(2x - 1) + 7(3x + 4)$

25. $5(3x - 1) - 4(2x - 3)$

26. $2(2x - 3) - 3(4x - 1)$

27. $-2(n^2 - 1) - 3(2n^2 + 5)$

28. $-3(2n^2 - 5) - 4(n^2 + 8)$

29. $6(2x + 3y) - 5(3x - 4y)$

30. $-5(4x - 3y) + 3(7x - y)$

Evaluate each of the following algebraic expressions for the given values of the variables.

31. $5x + 3y$; $x = -2$ and $y = -4$

32. $7x - 4y$; $x = -1$ and $y = 6$

33. $3x^2 - 2y^2$; $x = 2$ and $y = -3$

34. $4a^2 + 3b^2$; $a = 3$ and $b = 4$

35. $a^2 + 2ab + b^2$; $a = -2$ and $b = -1$

36. $x^2 - 2xy + y^2$; $x = 4$ and $y = -2$

37. $3x^2 - xy - 2y^2$; $x = 1$ and $y = -1$

38. $5x^2 + xy - y^2$; $x = 2$ and $y = -2$

39. $2xy - x^2y^2 + y^2$; $x = -1$ and $y = 5$

40. $x^2y^3 - xy - x^2y^2$; $x = 1$ and $y = -2$

41. $5a + 7b - 9a - 6b$; $a = -7$ and $b = 8$

42. $-5x + 8y + 7y + 8x$; $x = 5$ and $y = -6$

43. $-5a^2 + 6 + 7a^2 - 5$; $a = -3$

44. $5x + 4y - 9y - 2y$; $x = 2$ and $y = -8$

45. $-3a + 6b - 7a + 4b$; $a = -4$ and $b = 9$

46. $5(x - 1) + 7(x + 4)$; $x = 3$

47. $2(3x + 4) - 3(2x - 1)$; $x = -2$

48. $-4(2x - 1) - 5(3x + 7)$; $x = -1$

49. $3(x^2 - 1) - 2(2x^2 + 3) - (x^2 + 1)$; $x = -2$

50. $2(n^2 + 3) - 3(n^2 - 1) - 4(n^2 + 5)$; $n = 3$

For Problems 51–56, use your calculator and evaluate each of the algebraic expressions for the indicated values. Express the final answers to the nearest tenth.

51. πr^2; $\pi = 3.14$ and $r = 2.1$

52. πr^2; $\pi = 3.14$ and $r = 8.4$

53. $\pi r^2 h$; $\pi = 3.14, r = 1.6$, and $h = 11.2$

54. $\pi r^2 h$; $\pi = 3.14, r = 4.8$, and $h = 15.1$

55. $2\pi r^2 + 2\pi rh$; $\pi = 3.14, r = 3.9$, and $h = 17.6$

56. $2\pi r^2 + 2\pi rh$; $\pi = 3.14, r = 7.8$, and $h = 21.2$

For Problems 57–71, translate each English phrase into an algebraic expression using n to represent the unknown number.

57. The sum of a number and 2

58. A number increased by 10

59. A number decreased by 15

60. Four less than a number

61. A number subtracted from 50

62. The product of a number and 25

63. One-half of a number

64. Two less than three-fourths of a number

65. Five more than twice a number

66. The quotient of a number and 10

67. The quotient of 100 and a number

68. Three less than five times a number

69. Eight more than one-half of a number

70. Nine times the difference of a number and 17

71. Ten times the sum of a number and 5

For Problems 72–92, answer the question with an algebraic expression.

72. Brian is y years old. How old will he be in 10 years?

73. Becky is y years old. How old was she 15 years ago?

74. Pam is *t* years old and her mother is 7 less than twice as old as Pam. What is the age of Pam's mother?

75. The sum of two numbers is 65 and one of the numbers is *x*. What is the other number?

76. The difference of two numbers is 47 and the smaller number is *n*. What is the other number?

77. The product of two numbers is 98 and one of the numbers is *n*. What is the other number?

78. The quotient of two numbers is 8 and the smaller number is *y*. What is the other number?

79. The perimeter of a square is *c* centimeters. How long is each side of the square?

80. The perimeter of a square is *m* meters. How long, in centimeters, is each side of the square?

81. Eric has *n* nickels, *d* dimes, and *q* quarters in his bank. How much money, in cents, does he have in his bank?

82. Tina has *c* cents which is all in quarters. How many quarters does she have?

83. If *n* represents a whole number, what represents the next larger whole number?

84. If *n* represents an odd integer, what represents the next larger odd integer?

85. If *n* represents an even integer, what represents the next larger even integer?

86. The cost of a 5-pound box of candy is *c* cents. What is the price per pound?

87. Larry's annual salary is *d* dollars. What is his monthly salary?

88. Kim's monthly salary is *d* dollars. What is her annual salary?

89. The perimeter of a square is *i* inches. What is the perimeter expressed in feet?

90. The perimeter of a rectangle is *y* yards and *f* feet. What is the perimeter expressed in feet?

91. The length of a line segment is *d* decimeters. How long is the line segment expressed in meters?

92. The distance between two cities is *m* miles. How far is this, expressed in feet?

Chapter 1 Summary

(1.1) A **set** is a collection of objects; the objects are called **elements** or **members** of the set. Set *A* is a subset of set *B* if and only if every member of *A* is also a member of *B*. The sets of **natural numbers**, **whole numbers**, **integers**, **rational numbers**, and **irrational numbers** are all subsets of the set of **real numbers**.

Numerical expressions are evaluated by performing the operations in the following order.

1. Perform the operations inside the parentheses and above and below fraction bars.

2. Find all powers or convert them to indicated multiplication.

3. Perform all multiplications and divisions in the order in which they appear from left to right.

4. Perform all additions and subtractions in the order that they appear from left to right.

(1.2)   The **absolute value** of a real number a is defined by (1) if $a \geq 0$, then $|a| = a$, and (2) if $a < 0$, then $|a| = -a$.

The sum of two positive integers is the sum of their absolute values. The sum of two negative integers is the opposite of the sum of their absolute values. The sum of a positive and a negative integer is given by: (a) If the positive integer has the larger absolute value, then their sum is the difference of their absolute values found by subtracting the smaller absolute value from the larger. (b) If the negative integer has the larger absolute value, then their sum is the opposite of the difference of their absolute values where this difference is found by subtracting the smaller absolute value from the larger.

The statement $a - b = a + (-b)$ changes every subtraction problem to an equivalent addition problem.

(1.3)   The **product** of two positive or two negative integers is the product of their absolute values. The **product** of one positive and one negative integer is the opposite of the product of their absolute values.

The **quotient** of two positive or two negative integers is the quotient of their absolute values. The quotient of one positive and one negative integer is the opposite of the quotient of their absolute values.

(1.4)   The following basic properties of real numbers help with numerical manipulations and serve as a basis for algebraic computations:

Closure Properties $a + b$ is a real number

ab is a real number

Commutative Properties $a + b = b + a$

$ab = ba$

Associative Properties $(a + b) + c = a + (b + c)$

$(ab)c = a(bc)$

Identity Properties $a + 0 = 0 + a = a$

$a(1) = 1(a) = a$

Additive Inverse Property $a + (-a) = (-a) + a = 0$

Multiplication Property of Zero $a(0) = 0(a) = 0$

Multiplication Property of Negative One $-1(a) = a(-1) = -a$

Multiplicative Inverse Property $a\left(\dfrac{1}{a}\right) = \left(\dfrac{1}{a}\right)a = 1.$

Distributive Properties $a(b + c) = ab + ac$

$a(b - c) = ab - ac$

(1.5)   Algebraic expressions such as $2x$, $8xy$, $3xy^2$, $-4a^2b^3c$, and z are called **terms**. A term is an indicated product and may have any number of factors. The variables involved in a term are called **literal factors** and the numerical factor is called the **numerical coefficient**. Terms having the same literal factors are called **similar** or **like terms**.

The distributive property in the form $ba + ca = (b + c)a$ serves as the basis for **combining similar terms**. For example,

$$3x^2y + 7x^2y = (3 + 7)x^2y = 10x^2y.$$

To translate English phrases into algebraic expressions, we must be familiar with the standard vocabulary of *sum*, *difference*, *product*, and *quotient*, as well as other terms used to express the same ideas.

Chapter 1 Review Problem Set

1. From the list

$$0, \quad \sqrt{2}, \quad \frac{3}{4}, \quad -\frac{5}{6}, \quad 8\frac{1}{3}, \quad -\sqrt{3}, \quad -8, \quad 0.34, \quad 0.2\overline{3}, \quad 67, \quad \text{and} \quad \frac{9}{7}$$

identify each of the following:

   (a) The natural numbers
   (b) The integers
   (c) The nonnegative integers
   (d) The rational numbers
   (e) The irrational numbers

For Problems 2–10, state the property of equality or the property of real numbers that justifies each of the statements. For example, $6(-7) = -7(6)$ because of the *commutative property for multiplication* and "if $2 = x + 3$, then $x + 3 = 2$" is true because of the *symmetric property of equality*.

2. $7 + (3 + (-8)) = (7 + 3) + (-8)$
3. If $x = 2$ and $x + y = 9$, then $2 + y = 9$.
4. $-1(x + 2) = -(x + 2)$
5. $3(x + 4) = 3(x) + 3(4)$
6. $[(17)(4)](25) = (17)[(4)(25)]$
7. $x + 3 = 3 + x$
8. $3(98) + 3(2) = 3(98 + 2)$
9. $\left(\dfrac{3}{4}\right)\left(\dfrac{4}{3}\right) = 1$
10. If $4 = 3x - 1$, then $3x - 1 = 4$.

For Problems 11–22, simplify each of the numerical expressions.

11. $-8 + (-4) - (-6)$ **12.** $9 - 12 + (-4) - (-1)$
13. $-8 - 7 + 4 + 6 - 10 + 8 - 3$ **14.** $4(-3) - 12 \div (-4) + (-2)(-1) - 8$
15. $-3(2 - 4) - 4(7 - 9) + 6$ **16.** $(48 + (-73)) + 74$

17. $[5(-2) - 3(-1)][-2(-1) + 3(2)]$

18. $-4^2 - 2^3$

19. $(-2)^4 + (-1)^3 - 3^2$

20. $2(-1)^2 - 3(-1)(2) - 2^2$

21. $[4(-1) - 2(3)]^3$

22. $3 - [-2(3 - 4)] + 7$

For Problems 23–32, simplify each of the algebraic expressions by combining similar terms.

23. $3a^2 - 2b^2 - 7a^2 - 3b^2$

24. $4x - 6 - 2x - 8 + x + 12$

25. $ab^2 - 3a^2b + 4ab^2 + 6a^2b$

26. $3(x + 1) - 2(x - 4) + 5(x + 4)$

27. $3(2n^2 + 1) + 4(n^2 - 5)$

28. $-2(3a - 1) + 4(2a + 3) - 5(3a + 2)$

29. $-(n - 1) - (n + 2) + 3$

30. $3(2x - 3y) - 4(3x + 5y) - x$

31. $4(a - 6) - (3a - 1) - 2(4a - 7)$

32. $-5(x^2 - 4) - 2(3x^2 + 6) + (2x^2 - 1)$

For Problems 33–42, evaluate each of the algebraic expressions for the given values of the variables.

33. $-5x + 4y$ for $x = 3$ and $y = -4$

34. $3x^2 - 2y^2$ for $x = -1$ and $y = -2$

35. $-5(2x - 3y)$ for $x = 1$ and $y = -3$

36. $(3a - 2b)^2$ for $a = -2$ and $b = 3$

37. $a^2 + 3ab - 2b^2$ for $a = 2$ and $b = -2$

38. $3n^2 - 4 - 4n^2 + 9$ for $n = 7$

39. $3(2x - 1) + 2(3x + 4)$ for $x = 4$

40. $-4(3x - 1) - 5(2x - 1)$ for $x = -2$

41. $2(n^2 + 3) - 3(n^2 + 1) + 4(n^2 - 6)$ for $n = 3$

42. $5(3n - 1) - 7(-2n + 1) + 4(3n - 1)$ for $n = -5$

For Problems 43–50, translate each English phrase into an algebraic expression using n to represent the unknown number.

43. Four increased by twice a number

44. Fifty subtracted from three times a number.

45. Six less than two-thirds of a number

46. Ten times the difference of a number and 14

47. Eight more than five times a number

48. The quotient of a number and three less than the number

49. Three less than five times the sum of a number and 2

50. Three-fourths of the sum of a number and 12

For Problems 51–60, answer the question with an algebraic expression.

51. The sum of two numbers is 37 and one of the numbers is n. What is the other number?

52. Tina can type w words in an hour. What is her typing rate per minute?

53. Harry is y years old. His brother is 7 years less than twice as old as Harry. How old is Harry's brother?

54. If n represents a multiple of three, what represents the next largest multiple of three?

55. Celia has p pennies, n nickels, and q quarters. How much, in cents, does Celia have?

56. The perimeter of a square is i inches. How long, in feet, is each side of the square?

57. The length of a rectangle is y yards and the width is f feet. What is the perimeter of the rectangle expressed in inches?

58. The length of a piece of wire is d decimeters. What is the length expressed in centimeters?

59. Joan is f feet and i inches tall. How tall is she, expressed in inches?

60. The perimeter of a rectangle is 50 centimeters. If the rectangle is c centimeters long, how wide is it?

2 Equations and Inequalities

Throughout this text we are using a common thread of developing algebraic skills, using the skills to help solve equations and inequalities, and then using equations and inequalities to solve applied problems. This chapter will review and extend a variety of concepts pertaining to this thread.

2.1 Solving First-Degree Equations

Section 1.1 stated that *an equality (equation) is a statement that two symbols, or groups of symbols, are names for the same number.* It should be further stated that an equation may be true or false. For example, the equation $3 + (-8) = -5$ is true, but the equation $-7 + 4 = 2$ is false.

Algebraic equations contain one or more variables. The following are examples of algebraic equations.

$$3x + 5 = 8; \qquad 4y - 6 = -7y + 9; \qquad x^2 - 5x - 8 = 0;$$
$$3x + 5y = 4; \qquad x^3 + 6x^2 - 7x - 2 = 0.$$

An algebraic equation such as $3x + 5 = 8$ is neither true nor false as it stands and is often referred to as an *open sentence.* Each time that a number is substituted for x, the algebraic equation $3x + 5 = 8$ becomes a numerical statement which is true or false. For example, if $x = 0$, then $3x + 5 = 8$ becomes $3(0) + 5 = 8$, which is a false statement. If $x = 1$, then $3x + 5 = 8$ becomes $3(1) + 5 = 8$, which is a true statement. **Solving an equation** refers to the process of finding the number (or numbers) that makes an algebraic equation a true numerical statement. Such numbers are called the **solutions** or **roots** of the equation and are said to **satisfy** *the equation.* The set of all solutions of an equation is called its **solution set**. Thus, $\{1\}$ is the solution set of $3x + 5 = 8$.

In this chapter we shall consider techniques for solving **first degree equations in one variable**. This means that the equations contain only one variable and that this variable has an exponent of one. The following are examples of first degree equations in one variable.

$$3x + 5 = 8; \qquad \frac{2}{3}y + 7 = 9;$$

$$7a - 6 = 3a + 4; \qquad \frac{x - 2}{4} = \frac{x - 3}{5}.$$

Equivalent equations are equations having the same solution. For example,

1. $3x + 5 = 8$
2. $3x = 3$
3. $x = 1$

are all equivalent equations since $\{1\}$ is the solution set of each.

The general procedure for solving an equation is to continue replacing the given equation with equivalent, *but simpler*, equations until an equation of the form *variable = constant* or *constant = variable* is obtained. Thus, in the example above, $3x + 5 = 8$ has been simplified to $3x = 3$ which has been further simplified to $x = 1$, from which the solution set $\{1\}$ is obvious.

Techniques for solving equations are centered around the use of the various properties of equality. In addition to the *reflexive, symmetric, transitive*, and *substitution* properties listed in Section 1.1, the following properties of equality play an important role.

Addition Property of Equality

For all real numbers a, b, and c,

$$\text{if } a = b, \quad \text{then } a + c = b + c.$$

Multiplication Property of Equality

For all real numbers a, b, and c, where $c \neq 0$,

$$\text{if } a = b, \quad \text{then } ac = bc.$$

The Addition Property of Equality states that *any number can be added to both sides of an equation and an equivalent equation is produced.* The Multiplication Property of Equality states that *an equivalent equation is obtained whenever both sides of an equation are multiplied by the same **nonzero** real number.* The following examples illustrate the use of these properties to solve equations.

Example 1   Solve $3x + 5 = 8$.

Solution
$$3x + 5 = 8$$
$$3x + 5 + (-5) = 8 + (-5) \qquad \text{Add } -5 \text{ to both sides.}$$
$$3x = 3$$
$$\frac{1}{3}(3x) = \frac{1}{3}(3) \qquad \text{Multiply both sides by } \frac{1}{3}.$$
$$x = 1.$$

The solution set is $\{1\}$.   ●

To *check* an apparent solution we can substitute it in the original equation and see if a true numerical statement is obtained.

Check $3x + 5 = 8$
$$3(1) + 5 \overset{?}{=} 8$$
$$3 + 5 \overset{?}{=} 8$$
$$8 = 8.$$

Now we know that $\{1\}$ is the solution set of $3x + 5 = 8$. We will not show our checks for every example in this text, but do remember that checking is a way of detecting arithmetic errors.

Example 2   Solve $-7 = -5a + 9$.

Solution
$$-7 = -5a + 9$$
$$-7 + (-9) = -5a + 9 + (-9) \qquad \text{Add } -9 \text{ to both sides.}$$
$$-16 = -5a$$
$$-\frac{1}{5}(-16) = -\frac{1}{5}(-5a) \qquad \text{Multiply both sides by } -\frac{1}{5}.$$
$$\frac{16}{5} = a.$$

The solution set is $\left\{\frac{16}{5}\right\}$.   ●

Notice that in Example 2 the final equation is $\frac{16}{5} = a$ instead of $a = \frac{16}{5}$. Technically, the symmetric property of equality (if $a = b$, then $b = a$) would permit us to change from $\frac{16}{5} = a$ to $a = \frac{16}{5}$, but such a change is not necessary to determine that the solution is $\frac{16}{5}$. It should also be recognized that the symmetric property could be used at the very beginning to change $-7 = -5a + 9$ to $-5a + 9 = -7$. Some people prefer working with the variable on the left side of the equation.

Another point of clarification should be made. We stated the properties of equality in terms of only two operations, addition and multiplication. The operations of subtraction and division could also be included in the statements of the properties. That is to say, we could think in terms of "subtracting the same number from both sides of an equation" and also in terms of "dividing both sides of an equation by the same nonzero number." For example, in the solution of Example 2, we could subtract 9 from both sides rather than add -9 to both sides. Likewise, we could divide both sides by -5 instead of multiplying both sides by $-\dfrac{1}{5}$.

Example 3

Solve $7x - 3 = 5x + 9$.

Solution

$$7x - 3 = 5x + 9$$
$$7x - 3 + (-5x) = 5x + 9 + (-5x) \qquad \text{Add } -5x \text{ to both sides.}$$
$$2x - 3 = 9$$
$$2x - 3 + 3 = 9 + 3 \qquad \text{Add 3 to both sides.}$$
$$2x = 12$$
$$\frac{1}{2}(2x) = \frac{1}{2}(12) \qquad \text{Multiply both sides by } \frac{1}{2}.$$
$$x = 6.$$

The solution set is $\{6\}$.

Example 4

Solve $4(y - 1) + 5(y + 2) = 3(y - 8)$.

Solution

$$4(y - 1) + 5(y + 2) = 3(y - 8)$$
$$4y - 4 + 5y + 10 = 3y - 24 \qquad \text{Remove parentheses by applying distributive property.}$$
$$9y + 6 = 3y - 24 \qquad \text{Simplify left side by combining similar terms.}$$
$$9y + 6 + (-3y) = 3y - 24 + (-3y) \qquad \text{Add } -3y \text{ to both sides.}$$
$$6y + 6 = -24$$
$$6y + 6 + (-6) = -24 + (-6) \qquad \text{Add } -6 \text{ to both sides.}$$
$$6y = -30$$
$$\frac{1}{6}(6y) = \frac{1}{6}(-30) \qquad \text{Multiply both sides by } \frac{1}{6}.$$
$$y = -5.$$

The solution set is $\{-5\}$.

The process of solving first degree equations in one variable can be summarized as follows.

Step 1   Simplify both sides of the equation as much as possible.

Step 2   Use the addition property of equality to isolate a term containing the variable on one side and a constant on the other side of the equation.

Step 3   Use the multiplication property of equality to make the coefficient of the variable 1. That is, multiply both sides of the equation by the reciprocal of the numerical coefficient of the variable. The solution set should now be obvious.

Step 4   Check each solution by substituting it in the original equation and verifying that the resulting numerical statement is true.

The ability to use the tools of algebra to solve problems makes it necessary to be able to translate back and forth between the English language and the language of algebra. More specifically, at this time we need to translate *English sentences* into *algebraic equations*. Such translations allow us to use our knowledge of *equation solving* to solve word problems. Let's consider an example.

Problem 1

If 27 is subtracted from three times a certain number, the result is 18. Find the number.

Solution

Let n represent the number to be found. The sentence "if 27 is subtracted from three times a certain number, the result is 18" translates into the equation $3n - 27 = 18$. Solving this equation, we obtain

$$3n - 27 = 18$$
$$3n = 45 \qquad \text{Add 27 to both sides.}$$
$$n = 15. \qquad \text{Multiply both sides by } \frac{1}{3}.$$

The number to be found is 15.   ●

The statement "let n represent the number to be found" is often referred to as **declaring the variable**. We need to choose a letter to use as a variable and indicate what it represents for a specific problem. This may seem like an insignificant idea, but as the problems become more complex the process of declaring the variable becomes even more important. Furthermore, it is true that you could probably solve a problem such as Problem 1 without setting up an algebraic equation. However, as problems increase in difficulty the translation from English to algebra becomes a key issue. Therefore, even with these relatively easy problems we suggest that you concentrate on the translation process.

The next example involves the use of integers. Remember that the set of integers consists of $\{\ldots -2, -1, 0, 1, 2, \ldots\}$. Furthermore, the integers can be classified as *even*, $\{\ldots -4, -2, 0, 2, 4, \ldots\}$ or *odd*, $\{\ldots -3, -1, 1, 3, \ldots\}$. .

Problem 2   The sum of three consecutive integers is 13 greater than twice the smallest of the three integers. Find the integers.

Solution   Since consecutive integers differ by 1, we will represent them as follows: let n represent the smallest of the three consecutive integers; then $n + 1$ represents the second largest, and $n + 2$ represents the largest.

the sum of the three
consecutive integers   13 greater than twice the smallest

$$n + (n + 1) + (n + 2) = 2n + 13$$
$$3n + 3 = 2n + 13$$
$$n = 10.$$

The three consecutive integers are 10, 11, and 12.   ●

To *check* our answers for Problem 2 we must determine whether or not they satisfy the conditions stated in the original problem. Since 10, 11, and 12 are consecutive integers whose sum is 33 and since twice the smallest plus 13 is also 33 ($2(10) + 13 = 33$), we know that our answers are correct. (Remember, when checking a result for a word problem it is *not* sufficient to check the result in the equation used to solve the problem because the equation itself may be in error!)

In the two previous problems, the equation formed was almost a direct translation of a sentence in the statement of the problem. Now let's consider a situation where we need to think in terms of a *guideline* that may not be explicitly stated in the problem.

Problem 3   Danny received a car repair bill for $106. This included $23 for parts, $22 per hour for each hour of labor, and $6 for taxes. Find the number of hours of labor.

Solution   Let h represent the number of hours of labor. Then $22h$ represents the total charge for labor. A guideline of "charge for parts plus charge for labor plus tax equals the total bill" can be used to set up the following equation.

parts labor tax   total bill
↓   ↓   ↓   ↓
23 + 22h + 6   =   106.

Solving this equation we obtain

$$22h + 29 = 106$$
$$22h = 77$$
$$h = 3\frac{1}{2}.$$

Danny was charged for $3\frac{1}{2}$ hours of labor.   ●

Problem Set 2.1

Solve each of the following equations.

1. $4x + 3 = 11$ 2. $5x + 4 = 19$ 3. $3x + 4 = -8$

4. $6x + 5 = -13$ 5. $-x - 1 = 13$ 6. $7 - x = -1$

7. $5y - 2 = 28$ 8. $7y - 3 = 32$ 9. $2x - 9 = 12$

10. $3x + 7 = 15$ 11. $-5 = 3a + 7$ 12. $-13 = 6a + 17$

13. $-4y - 7 = 3$ 14. $-6y - 5 = 10$ 15. $5x - 1 = 3x + 9$

16. $7x - 3 = 4x + 21$ 17. $6y + 4 = 2y + 3$ 18. $8y + 1 = 5y - 6$

19. $2x + 7 = 6x - 1$ 20. $3x + 2 = 7x - 2$ 21. $-3a - 5 = -4a + 6$

22. $-7a + 6 = -8a + 14$ 23. $2x - 1 - x = 3x - 5$

24. $5x + 3 - 2x = x - 15$ 25. $5y + 5 + 3y = 4y + 1$

26. $6y + 18 + y = 2y + 3$ 27. $3x + 4 + 2x = 7x - 1 - 3x$

28. $4x - 3 + 2x = 8x - 3 - x$ 29. $10n - 3 - 2n = 3n - 13 + 2n$

30. $6n - 4 - 3n = 3n + 10 + 4n$ 31. $3(x + 1) = -12$

32. $4(x - 3) = -20$ 33. $-2(x - 4) = 7$

34. $-3(x - 2) = 11$ 35. $4(3x - 2) = 3(3x - 1)$

36. $5(2x + 1) = 4(3x - 7)$ 37. $3x - 2(x - 1) = 13$

38. $5x - 4(x - 6) = -11$

39. $-3(2x - 5) = 2(4x + 7)$

40. $-2(3x - 1) - 3 = -4$

41. $4(x - 2) - 3(x - 1) = 2(x + 6)$

42. $3(2a - 1) - 2(5a + 1) = 4(3a + 4)$

43. $-(a - 1) - (3a - 2) = 6 + 2(a - 1)$

44. $-2(n - 4) - (3n - 1) = -2 + (2n - 1)$

Solve each of the following problems by setting up and solving an algebraic equation.

45. If 13 is subtracted from four times a certain number the result is 11. Find the number.

46. If 2 is added to five times a certain number the result is the same as if 16 is subtracted from twice the number. Find the number.

47. Find three consecutive integers such that twice the sum of the first two is 11 more than three times the largest.

48. Find four consecutive integers whose sum is -86.

49. Find three consecutive even integers such that four times the first minus the third is 6 more than twice the second.

50. Find three consecutive odd integers such that three times the second minus the third is 11 more than the first.

51. The sum of two numbers is 109. The larger number is one less than four times the smaller number. Find the two numbers.

52. The difference of two numbers is 33. The larger number is one more than three times the smaller number. Find the numbers.

53. A plumbing repair bill, not including tax, was $95. This included $17 for parts and an amount for 3 hours of labor. Find the hourly rate that was charged for labor.

54. Doug is paid double time for each hour worked over 40 hours in a week. Last week he worked 46 hours and earned $468. What is his normal hourly rate?

55. Mona has 41 coins consisting of nickels, dimes, and quarters. The number of dimes is one more than twice the number of nickels and the number of quarters is two less than three times the number of nickels. How many coins of each kind does she have?

56. Joe has a collection of pennies, nickels, and dimes totaling 230 coins. The number of nickels is 10 more than the number of pennies and the number of dimes is twice the number of nickels. How many coins of each kind does he have?

57. In a class of 62 students, the number of females is one less than twice the number of males. How many females and how many males are there in the class?

58. The selling price of a ring is $750. This represents $150 less than three times the cost of the ring. Find the cost of the ring.

59. Barry sells bicycles on a salary-plus-commission basis. He receives a monthly salary of $300 and a commission of $15 for each bicycle that he sells. How many bicycles must he sell in a month to have a total monthly income of $750?

60. An apartment complex contains 230 apartments each having one, two, or three bedrooms. The number of two-bedroom apartments is 10 more than twice the number of three-bedroom apartments. The number of one-bedroom apartments is twice the number of two-bedroom apartments. How many apartments of each kind are in the complex?

Miscellaneous Problems

61. Verify that for any three consecutive integers, the sum of the smallest and largest is equal to twice the middle integer. (Hint: Use n, $n + 1$, and $n + 2$ to represent the three consecutive integers.)

62. Verify that no four consecutive integers can be found such that the product of the smallest and largest is equal to the product of the other two integers.

2.2 Equations Involving Fractional Forms

To solve equations involving fractions it is usually easiest to begin by *clearing the equation of all fractions*. This can be accomplished by multiplying both sides of the equation by the least common multiple of all the denominators in the equation. Remember that the least common multiple of a set of whole numbers is the smallest nonzero whole number that is divisible by each of the numbers. For example, the least common multiple of 2, 3, and 6 is 12. When working with fractions, the least common multiple of a set of denominators is referred to as the **least common denominator** (LCD). Let's consider some equations involving fractions.

Example 1   Solve $\frac{1}{2}x + \frac{2}{3} = \frac{3}{4}$.

Solution

$$\frac{1}{2}x + \frac{2}{3} = \frac{3}{4}$$

$$12\left(\frac{1}{2}x + \frac{2}{3}\right) = 12\left(\frac{3}{4}\right) \qquad \text{Multiply both sides by 12, which is the LCD of 2, 3, and 4.}$$

$$12\left(\frac{1}{2}x\right) + 12\left(\frac{2}{3}\right) = 12\left(\frac{3}{4}\right) \qquad \text{Apply distributive property on left side.}$$

$$6x + 8 = 9$$

$$6x = 1$$

$$x = \frac{1}{6}.$$

The solution set is $\left\{\frac{1}{6}\right\}$.

Check

$$\frac{1}{2}x + \frac{2}{3} = \frac{3}{4}$$

$$\frac{1}{2}\left(\frac{1}{6}\right) + \frac{2}{3} \overset{?}{=} \frac{3}{4}$$

$$\frac{1}{12} + \frac{2}{3} \overset{?}{=} \frac{3}{4}$$

$$\frac{1}{12} + \frac{8}{12} \overset{?}{=} \frac{3}{4}$$

$$\frac{9}{12} \overset{?}{=} \frac{3}{4}$$

$$\frac{3}{4} = \frac{3}{4}.$$

Example 2   Solve $\frac{x}{2} + \frac{x}{3} = 10$.

Solution

$$\frac{x}{2} + \frac{x}{3} = 10 \qquad \text{Recall that } \frac{x}{2} = \frac{1}{2}x.$$

$$6\left(\frac{x}{2} + \frac{x}{3}\right) = 6(10) \qquad \text{Multiply both sides by the LCD.}$$

$$6\left(\frac{x}{2}\right) + 6\left(\frac{x}{3}\right) = 6(10) \qquad \text{Apply distributive property on left side.}$$

$$3x + 2x = 60$$

$$5x = 60$$

$$x = 12.$$

The solution set is $\{12\}$.

As you study the examples of this section, pay special attention to the steps being shown in the solutions. Certainly, there are no rules as to which steps should be performed mentally; this is an individual decision. When you solve problems show enough steps to allow the flow of the process to be understood and to minimize the chances of making careless computational errors.

Example 3 Solve $\dfrac{x-2}{3} + \dfrac{x+1}{8} = \dfrac{5}{6}$.

Solution

$$\frac{x-2}{3} + \frac{x+1}{8} = \frac{5}{6}$$

$$24\left(\frac{x-2}{3} + \frac{x+1}{8}\right) = 24\left(\frac{5}{6}\right) \qquad \text{Multiply both sides by the LCD.}$$

$$24\left(\frac{x-2}{3}\right) + 24\left(\frac{x+1}{8}\right) = 24\left(\frac{5}{6}\right) \qquad \text{Apply distributive property to left side.}$$

$$8(x-2) + 3(x+1) = 20$$

$$8x - 16 + 3x + 3 = 20$$

$$11x - 13 = 20$$

$$11x = 33$$

$$x = 3.$$

The solution set is $\{3\}$.

●

Example 4 Solve $\dfrac{3t-1}{5} - \dfrac{t-4}{3} = 1$.

Solution

$$\frac{3t-1}{5} - \frac{t-4}{3} = 1$$

$$15\left(\frac{3t-1}{5} - \frac{t-4}{3}\right) = 15(1) \qquad \text{Multiply both sides by LCD.}$$

$$15\left(\frac{3t-1}{5}\right) - 15\left(\frac{t-4}{3}\right) = 15(1) \qquad \text{Apply distributive property to left side.}$$

$$3(3t-1) - 5(t-4) = 15$$

$$9t - 3 - 5t + 20 = 15 \qquad \text{Be careful with this sign!}$$

$$4t + 17 = 15$$

$$4t = -2$$

$$t = -\frac{2}{4} = -\frac{1}{2}. \qquad \text{Reduce!}$$

The solution set is $\left\{-\dfrac{1}{2}\right\}$.

Problem Solving

As we expand our skills for solving equations, we also expand our capabilities for solving verbal problems. There is no one definite procedure that will ensure success at solving word problems, but the following suggestions can be helpful.

Suggestions for Solving Word Problems

1. Read the problem carefully, making certain that you understand the meanings of all of the words. Be especially alert for any technical terms used in the statement of the problem.

2. Read the problem a second time (perhaps even a third time) to get an overview of the situation being described. Determine the known facts as well as what is to be found.

3. Sketch any figure, diagram, or chart that might be helpful in analyzing the problem.

4. Choose a meaningful variable to represent an unknown quantity in the problem (perhaps t, if time is an unknown quantity) and represent any other unknowns in terms of that variable.

5. Look for a *guideline* that can be used to set up an equation. A guideline might be a formula such as "distance equals rate times time," or a statement of a relationship such as "the sum of the two numbers is 28."

6. Form an equation containing the variable that translates the conditions of the guideline from English to algebra.

7. Solve the equation and use the solution to determine all facts requested in the problem.

8. *Check* all answers back into the *original statement of the problem.*

Keep these suggestions in mind as we continue to solve problems. Some of these suggestions will be elaborated on at different times throughout the text. Now let's consider some problems.

Problem 1   Find a number such that three-eighths of the number minus one-half of it is 14 less than three-fourths of the number.

Solution   Let n represent the number to be found.

$$\frac{3}{8}n - \frac{1}{2}n = \frac{3}{4}n - 14$$

$$8\left(\frac{3}{8}n - \frac{1}{2}n\right) = 8\left(\frac{3}{4}n - 14\right)$$

$$8\left(\frac{3}{8}n\right) - 8\left(\frac{1}{2}n\right) = 8\left(\frac{3}{4}n\right) - 8(14)$$

$$3n - 4n = 6n - 112$$
$$-n = 6n - 112$$
$$-7n = -112$$
$$n = 16.$$

The number is 16. (Check it!)   ●

Problem 2

The width of a rectangular lot is 8 feet less than three-fifths of the length. The perimeter of the lot is 400 feet. Find the length and width of the lot.

Solution

Let l represent the length of the lot. Then $\frac{3}{5}l - 8$ represents the width.

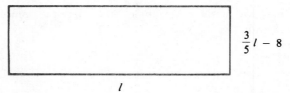

A *guideline* for this problem is the formula *perimeter of a rectangle equals twice the length plus twice the width* ($P = 2l + 2w$). Using this formula we can form the following equation.

$$\begin{array}{ccc} P & = 2l + & 2w \\ \downarrow & \downarrow & \downarrow \end{array}$$
$$400 = 2l + 2\left(\frac{3}{5}l - 8\right).$$

Solving this equation we obtain

$$400 = 2l + \frac{6l}{5} - 16$$
$$5(400) = 5\left(2l + \frac{6l}{5} - 16\right)$$
$$2000 = 10l + 6l - 80$$
$$2000 = 16l - 80$$
$$2080 = 16l$$
$$130 = l.$$

The length of the lot is 130 feet and the width is $\frac{3}{5}(130) - 8 = 70$ feet.   ●

In Problems 1 and 2 notice the use of different letters as variables. It is helpful to choose a variable that has significance for the problem you are working on. For example, in Problem 2 the choice of l to represent the length seems natural and meaningful. Certainly this is another matter of personal preference, but you might try it.

In Problem 2 a geometric relationship ($P = 2l + 2w$) serves as a guideline for setting up the equation. The following geometric relationships pertaining to angle measure may also serve as guidelines.

1. *Two* angles whose sum of measures is $90°$ are called *complementary angles*.
2. *Two* angles whose sum of measures is $180°$ are called *supplementary angles*.
3. The sum of the measures of the three angles of a triangle is $180°$.

Problem 3

One of two complementary angles is $6°$ larger than one-half of the other angle. Find the measure of each of the angles.

Solution

Let a represent the measure of one of the angles. Then $\frac{1}{2}a + 6$ represents the measure of the other angle. Since they are complementary angles, the sum of their measures is $90°$.

$$a + \left(\frac{1}{2}a + 6\right) = 90$$
$$2a + a + 12 = 180$$
$$3a + 12 = 180$$
$$3a = 168$$
$$a = 56.$$

If $a = 56$, then $\frac{1}{2}a + 6$ becomes $\frac{1}{2}(56) + 6 = 34$. The angles have measures of $34°$ and $56°$.   ●

Problem 4

Find the measures of the three angles of a triangle if the smallest angle is one-sixth of the largest and the other angle is twice the smallest angle.

Solution

Let a represent the measure of the largest angle. Then $\frac{1}{6}a$ represents the measure of the smallest angle and $2\left(\frac{1}{6}a\right)$, which simplifies to $\frac{1}{3}a$, represents the other angle. Since the sum of the measures of the angles of a triangle is $180°$, we can set up and solve the following equation.

$$a + \frac{1}{6}a + \frac{1}{3}a = 180$$
$$6a + a + 2a = 1080$$
$$9a = 1080$$
$$a = 120.$$

If $a = 120$, then $\frac{1}{6}a$ becomes $\frac{1}{6}(120) = 20$ and $\frac{1}{3}a$ becomes $\frac{1}{3}(120) = 40$. The angles have measures of $120°$, $20°$, and $40°$.   ●

Problem Set 2.2

Solve each of the following equations.

1. $\frac{2}{3}x = 6$

2. $\frac{3}{5}x = 9$

3. $\frac{-3x}{4} = \frac{3}{2}$

4. $\frac{-6x}{7} = 12$

5. $\frac{n}{2} - \frac{1}{4} = \frac{5}{6}$

6. $\frac{n}{3} + \frac{3}{8} = \frac{5}{12}$

7. $\frac{3n}{4} - \frac{n}{12} = 6$

8. $\frac{2n}{3} - \frac{n}{5} = 7$

9. $\frac{a}{5} - 2 = \frac{a}{2} + 1$

10. $\frac{4a}{5} - 7 = \frac{a}{10}$

11. $\frac{h}{3} + \frac{h}{4} = 1$

12. $\frac{h}{5} + \frac{h}{6} = 1$

13. $\frac{h}{2} + \frac{h}{3} - \frac{h}{4} = 1$

14. $\frac{2h}{3} + \frac{3h}{7} = 1$

15. $\frac{x-1}{3} + \frac{x+2}{9} = \frac{1}{3}$

16. $\frac{x+5}{7} + \frac{x-3}{4} = \frac{5}{14}$

17. $\frac{x+3}{5} - \frac{x-2}{6} = 0$

18. $\frac{x+7}{8} - \frac{x-3}{12} = -2$

19. $\frac{n+3}{6} - \frac{n-5}{4} = \frac{3}{8}$

20. $\frac{3n-1}{9} - \frac{n-2}{8} = \frac{1}{2}$

21. $\frac{2y-3}{3} + \frac{y+1}{2} = 3$

22. $\frac{y}{3} + \frac{y-5}{10} = \frac{4y+3}{5}$

23. $\frac{2x+3}{6} - \frac{x-9}{4} = 5$

24. $\frac{4x-1}{10} - \frac{5x+2}{4} = -3$

25. $\frac{2x+7}{9} - 4 = \frac{x-7}{12}$

26. $\frac{3x-1}{8} - 2 = \frac{2x+5}{7}$

27. $\frac{5a+2}{5} + \frac{3a-4}{4} + \frac{2a-1}{10} = 3$

28. $\frac{2a-3}{6} + \frac{3a-2}{4} + \frac{5a+6}{12} = 4$

29. $\frac{2x+7}{8} + x - 2 = \frac{x-1}{2}$

30. $x + \frac{3x-1}{9} - 4 = \frac{3x+1}{3}$

31. $\frac{2a+1}{14} - \frac{3a+4}{7} = \frac{a-1}{2}$

32. $\frac{x+3}{2} + \frac{x+4}{5} = \frac{3}{10}$

33. $\frac{x-1}{4} - \frac{x-2}{6} = \frac{2}{3}$

34. $n + \frac{2n-3}{9} - 2 = \frac{2n+1}{3}$

35. $\frac{2}{3}(n+3) - \frac{3}{7}(n+4) = 1$

36. $\frac{3}{4}(t-2) - \frac{2}{5}(2t-3) = \frac{1}{5}$

37. $\frac{2}{3}(x-1) + \frac{3}{4}(x-2) = 1$

38. $\frac{1}{2}(2x-1) - \frac{1}{3}(5x+2) = 3$

39. $\frac{5}{6}(x+1) - \frac{2}{5}(x-1) = \frac{1}{2}$

40. $(3x-1) + \frac{2}{7}(7x-2) = -\frac{11}{7}$

Solve each of the following problems by setting up and solving an algebraic equation.

41. Find a number such that five-sixths of the number is 4 more than two-thirds of the number.

42. Three-fourths of a number plus two-fifths of the number is 13 less than one-half of the number. Find the number.

43. The width of a rectangle is 2 centimeters less than its length. One-half of the length plus two-thirds of the width equals 8 centimeters. Find the length and width of the rectangle.

44. The width of a rectangle is 2 meters more than three-fourths of the length. The perimeter of the rectangle is 60 meters. Find the length and width of the rectangle.

45. Barry is paid $1\frac{1}{2}$ times his normal hourly rate for each hour worked over 40 hours in a week. Last week he worked 45 hours and earned $380. What is his normal hourly rate?

46. Find three consecutive integers such that the sum of the first plus one-third of the second plus three-eighths of the third is 25.

47. Jody has a collection of 116 coins consisting of dimes, quarters, and silver dollars. The number of quarters is 5 less than three-fourths of the number of dimes. The number of silver dollars is 7 more than five-eighths of the number of dimes. How many coins of each kind are in her collection?

48. A board 20 feet long is cut into two pieces such that the length of one piece is two-thirds of the length of the other piece. Find the length of the shortest piece of board.

49. The average of the salaries of Tim, Ann, and Eric is $24,000 per year. If Ann earns $10,000 more than Tim and Eric's salary is $2000 more than twice Tim's salary, find the salary of each person.

50. The sum of the present ages of Angie and her mother is 64 years. In eight years Angie will be three-fifths as old as her mother at that time. Find the present ages of Angie and her mother.

51. Donna's present age is two-thirds of Jesse's present age. In 12 years the sum of their ages will be 54 years. Find their present ages.

52. One of two supplementary angles is 4° more than one-third of the other angle. Find the measure of each of the angles.

53. If one-half of the complement of an angle plus three-fourths of the supplement of the angle equals 110°, find the measure of the angle.

54. If the complement of an angle is 5° less than one-sixth of its supplement, find the measure of the angle.

55. In $\triangle ABC$, angle B is 8° less than one-half of angle A and angle C is 28° larger than angle A. Find the measures of the three angles of the triangle.

56. Aura took three biology exams and had an average score of 88. Her second exam was 10 points better than her first and her third exam was 4 points better than her second exam. What were her three exam scores?

2.3 Equations Involving Decimals

Certainly the equation $x - 0.17 = 0.18$ can be solved by adding 0.17 to both sides. However, as equations containing decimals become more complex, it is often easier to first *clear the equation of all decimals* by multiplying both sides by an appropriate power of 10. Let's consider some examples.

Example 1   Solve $0.6x = 1.2$.

Solution

$$0.6x = 1.2$$
$$10(0.6x) = 10(1.2) \qquad \text{Multiply both sides by 10.}$$
$$6x = 12$$
$$x = 2.$$

Check $0.6x = 1.2$

$$0.6(2) \overset{?}{=} 1.2$$

$$1.2 = 1.2.$$

The solution set is $\{2\}$.

Example 2

Solve $0.07x + 0.11x = 3.6$.

Solution

$$0.07x + 0.11x = 3.6$$
$$100(0.07x + 0.11x) = 100(3.6) \qquad \text{Multiply both sides by 100.}$$
$$7x + 11x = 360$$
$$18x = 360$$
$$x = 20.$$

Check $0.07x + 0.11x = 3.6$

$$0.07(20) + 0.11(20) \overset{?}{=} 3.6$$
$$1.4 + 2.2 \overset{?}{=} 3.6$$
$$3.6 = 3.6.$$

The solution set is $\{20\}$.

Example 3

Solve $s = 1.95 + 0.35s$.

Solution

$$s = 1.95 + 0.35s$$
$$100(s) = 100(1.95 + 0.35s) \qquad \text{Multiply both sides by 100.}$$
$$100s = 195 + 35s$$
$$65s = 195$$
$$s = 3.$$

The solution set is $\{3\}$. (Check it!)

Example 4

Solve $0.12x + 0.11(7000 - x) = 790$.

Solution

$$0.12x + 0.11(7000 - x) = 790$$
$$100[0.12x + 0.11(7000 - x)] = 100(790) \qquad \text{Multiply both sides by 100.}$$
$$12x + 11(7000 - x) = 79000$$
$$12x + 77000 - 11x = 79000$$
$$x + 77000 = 79000$$
$$x = 2000.$$

The solution set is $\{2000\}$.

Many consumer problems can be solved by using an algebraic approach. For example, let's consider some discount sale problems involving the relationship *original selling price minus discount equals discount sale price*.

Problem 1   Karyl bought a dress at a 35% discount sale for $32.50. What was the original price of the dress?

Solution   Let p represent the original price of the dress. Using the discount sale relationship as a guideline, the problem translates into an equation as follows.

original selling price	minus	discount	equals	discount sale price
↓		↓		↓
$(100\%)(p)$	$-$	$(35\%)(p)$	$=$	$\$32.50$.

Switching this equation to decimal form and solving, we obtain

$$(100\%)(p) - (35\%)(p) = 32.50$$
$$(65\%)(p) = 32.50$$
$$0.65p = 32.50$$
$$65p = 3250$$
$$p = 50.$$

The original price of the dress was $50. ●

Problem 2   A pair of jogging shoes that was originally priced at $25 is on sale for 20% off. Find the discount sale price of the shoes.

Solution   Let s represent the discount sale price.

original price	minus	discount	equals	sale price
↓		↓		↓
$\$25$	$-$	$(20\%)(\$25)$	$=$	s .

Solving this equation we obtain
$$25 - (20\%)(25) = s$$
$$25 - (0.2)(25) = s$$
$$25 - 5 = s$$
$$20 = s.$$

The shoes are on sale for $20. ●

Another basic relationship pertaining to consumer problems is *selling price equals cost plus profit*. Profit, also called markup, markon, and margin of profit, may be stated in different ways. It may be stated as a percent of the selling price, a percent of the cost, or simply in terms of dollars and cents. We shall consider some problems for which the profit is calculated either as a percent of the cost or as a percent of the selling price.

Problem 3   A retailer has some shirts that cost $20 each. She wants to sell them at a profit of 15% of the cost. What selling price should be marked on the shirts?

Solution   Let s represent the selling price. The *selling price equals cost plus profit* relationship can be used as a guideline.

selling price   equals   cost   plus   profit
   ↓   ↓   ↓
   s $=$ $20 $+$ $(15\%)(\$20).$

Solving this equation yields

$$s = 20 + (15\%)(20)$$
$$s = 20 + (0.15)(20)$$
$$s = 20 + 3$$
$$s = 23.$$

The selling price should be $23.   ●

Problem 4   A retailer of sporting goods bought a putter for $25. He wants to price the putter to make a profit of 20% of the selling price. What price should he mark on the putter?

Solution   Let s represent the selling price.

selling price   equals   cost   plus   profit
   ↓   ↓   ↓
   s $=$ $25 $+$ $(20\%)(s).$

Solving this equation yields

$$s = 25 + (20\%)(s)$$
$$s = 25 + 0.2s$$
$$10s = 250 + 2s \qquad \text{Multiply both sides by 10.}$$
$$8s = 250$$
$$s = 31.25.$$

The selling price should be $31.25.   ●

Certain types of investment problems can be solved by using an algebraic approach. Consider the following examples.

Problem 5   A man invests $8000: part of it at 11% and the remainder at 12%. His total yearly interest from the two investments is $930. How much does he have invested at each rate?

Solution   Let x represent the amount invested at 11%. Then $8000 - x$ represents the amount invested at 12%. The following guideline can be used.

interest earned   interest earned   total amount
  from 11% $+$ from 12% $=$ of interest
 investment   investment   earned
      ↓   ↓   ↓
 $(11\%)(x)$ $+ (12\%)(8000 - x) =$ $930.$

Solving this equation yields

$$(11\%)(x) + (12\%)(8000 - x) = 930$$
$$0.11x + 0.12(8000 - x) = 930$$
$$11x + 12(8000 - x) = 93000 \qquad \text{Multiply both sides by 100.}$$
$$11x + 96000 - 12x = 93000$$
$$-x + 96000 = 93000$$
$$-x = -3000$$
$$x = 3000.$$

Therefore, $8000 - x = 5000$.

$3000 was invested at 11% and $5000 at 12%.   ●

Problem 6

A certain amount of money is invested at 8% and $1500 more than that amount is invested at 9%. The annual interest from the 9% investment exceeds the annual interest from the 8% investment by $160. How much is invested at each rate?

Solution

Let x represent the amount invested at 8%. Then $x + 1500$ represents the amount invested at 9%. The following guideline can be used.

$$\underset{\text{from } 9\% \text{ investment}}{\text{interest earned}} = \underset{\text{from } 8\% \text{ investment}}{\text{interest earned}} + \$160$$
$$\downarrow \qquad\qquad \downarrow \qquad\qquad \downarrow$$
$$(9\%)(x + 1500) = (8\%)(x) + \$160.$$

Solving this equation, we obtain

$$(9\%)(x + 1500) = (8\%)(x) + 160$$
$$0.09(x + 1500) = 0.08(x) + 160$$
$$9(x + 1500) = 8x + 16000 \qquad \text{Multiply both sides by 100.}$$
$$9x + 13500 = 8x + 16000$$
$$x = 2500.$$

Therefore, $x + 1500 = 4000$.

$2500 was invested at 8% and $4000 at 9%.   ●

Don't forget to check word problems; we must determine whether the answers satisfy the conditions stated in the original problem. A check for Problem 6 is as follows.

Check   We claim that $2500 was invested at 8% and $4000 at 9% and this satisfies the condition "$1500 more was invested at 9% than at 8%." The $2500 at 8% produces $200 of interest and the $4000 at 9% produces $360. Therefore, the interest from the 9% investment exceeds the interest from the 8% investment by $160. The conditions of the problem are satisfied and our answers are correct.

Problem Set 2.3

Solve each of the following equations.

1. $0.12x = 2.4$
2. $0.11x = 5.5$
3. $0.09y = 36$
4. $0.08y = 56$
5. $0.3t = 0.7t - 2$
6. $n + 0.4n = 56$
7. $s = 1.2 + 0.2s$
8. $s = 9 + 0.25s$
9. $s = 1.95 + 0.35s$
10. $s = 3.3 + 0.45s$
11. $0.09x + 0.1(500 - x) = 48$
12. $0.11x + 0.12(900 - x) = 104$
13. $0.09(x + 100) = 0.08x + 11$
14. $0.08(x + 200) = 0.07x + 20$
15. $0.8(t - 2) = 0.5(9t + 10)$
16. $0.12t - 2.1 = 0.07t - 0.2$
17. $0.3(2n - 5) = 11 - 0.65n$
18. $0.92 + 0.9(x - 0.3) = 2x - 5.95$
19. $0.8x + 0.9(850 - x) = 715$
20. $0.1d + 0.11(d + 1500) = 795$
21. $0.10t + 0.12(t + 1000) = 560$
22. $0.12x + 0.1(5000 - x) = 560$
23. $0.09x = 1650 - 0.12(x + 5000)$
24. $0.08(x + 200) = 0.07x + 20$
25. $0.5(3t + 0.7) = 20.6$
26. $0.3(2t + 0.1) = 8.43$
27. $0.2(x + 0.2) + 0.5(x - 0.4) = 5.44$
28. $0.1(x - 0.1) - 0.4(x + 2) = -5.31$

Solve each of the following problems by setting up and solving an algebraic equation.

29. Jim bought a pair of slacks at a 30% discount sale for $28. What was the original price of the slacks?
30. Judy bought a coat at a 15% discount sale for $68. What was the original price of the coat?
31. Find the discount sale price of a $49 item that is on sale for 20% off.
32. Mark bought a $45 pair of shoes on sale for 35% off. How much did he pay for the shoes?
33. The owner of a pizza parlor wants to make a profit of 30% of the cost for each pizza sold. If it costs $3 to make a pizza, at what price should it be sold?
34. A retailer has some skirts that cost $30 each. She wants to sell them at a profit of 15% of the cost. What price should she charge for the skirts?
35. If a head of lettuce costs a retailer $0.45, at what price should it be sold to make a profit of 40% on the selling price?
36. If a ring costs a jeweler $300, at what price should it be sold to make a profit of 50% on the selling price?
37. A retailer has some skirts that cost her $24 each. If she sells them for $31.20 per skirt, find her rate of profit based on the cost.
38. If the cost for a pair of shoes for a retailer is $32 and he sells them for $44.80, what is his rate of profit based on the cost?
39. Don bought a car with 6% tax included for $13,674. What was the price of the car without the tax?
40. Robin's salary for next year is to be $29,700. This represents an 8% increase over this year's salary. Find Robin's present salary.
41. A total of $4000 was invested, part of it at 8% interest and the remainder at 9%. If the total yearly interest amounted to $350, how much was invested at each rate?
42. Cindy invested a certain amount of money at 10% interest and $1500 more than that amount at 11%. Her total yearly interest was $795. How much did she invest at each rate?

43. A sum of $2000 is split between two investments, one paying 7% interest and the other 8%. If the return on the 8% investment exceeds that on the 7% investment by $40 per year, how much is invested at each rate?

44. If $500 is invested at 6% interest, how much additional money must be invested at 9% so that the total return for both investments averages 8%?

45. Sarah has a collection of nickels, dimes, and quarters worth $15.75. She has 10 more dimes than nickels and twice as many quarters as dimes. How many coins of each kind does she have?

46. Javier has a handful of coins consisting of pennies, nickels, and dimes worth $2.63. The number of nickels is one less than twice the number of pennies and the number of dimes is 3 more than the number of nickels. How many coins of each kind does he have?

47. Abby has 37 coins, consisting only of dimes and quarters, worth $7.45. How many dimes and how many quarters does she have?

48. A collection of 70 coins consisting of dimes, quarters, and half-dollars has a value of $17.75. There are three times as many quarters as dimes. Find the number of each kind of coin.

Solve each of the following equations expressing solutions in decimal form. Check all of your solutions. Use your calculator whenever it seems helpful.

49. $1.2x + 3.4 = 5.2$

50. $0.12x - 0.24 = 0.66$

51. $0.12x + 0.14(550 - x) = 72.5$

52. $0.14t + 0.13(890 - t) = 67.95$

53. $0.7n + 1.4 = 3.92$

54. $0.14n - 0.26 = 0.958$

55. $0.3(d + 1.8) = 4.86$

56. $0.6(d - 4.8) = 7.38$

57. $0.8(2x - 1.4) = 19.52$

58. $0.5(3x + 0.7) = 20.6$

2.4 Formulas

To find the distance traveled in 4 hours at a rate of 55 miles per hour we multiply the rate times the time; thus, the distance is $55(4) = 220$ miles. The rule *distance equals rate times time* is commonly stated as a formula: $d = rt$. *Formulas* are rules stated in symbolic form and usually expressed as equations.

The techniques we have considered for solving equations can be used to solve a formula for a specified variable if we are given numerical values for the other variables in the formula. Let's consider some examples.

Example 1   If P dollars is invested at r percent for t years, the amount of simple interest i is given by the formula $i = Prt$. Find the amount of interest earned by $500 at 7% for 2 years.

Solution   By substituting $500 for P, 7% for r, and 2 for t, we obtain

$i = Prt$

$i = (500)(7\%)(2)$

$i = (500)(0.07)(2)$

$i = 70.$

Thus, $70 is the interest earned.   ●

Example 2   If P dollars is invested at a simple rate of r percent, then the amount A accumulated after t years is given by the formula $A = P + Prt$. If $500 is invested at 8%, how many years would it take to accumulate $600?

Solution   Substituting $500 for P, 8% for r, and $600 for A, we obtain

$A = P + Prt$

$600 = 500 + 500(8\%)(t).$

Solving this equation for t yields

$600 = 500 + 500(0.08)(t)$

$600 = 500 + 40t$

$100 = 40t$

$2\dfrac{1}{2} = t.$

It will take $2\dfrac{1}{2}$ years to accumulate $600.   ●

When using a formula it is sometimes convenient to first change its form. For example, suppose we are to use the *perimeter* formula for a rectangle ($P = 2l + 2w$) to complete the following chart.

perimeter (P)	32	24	36	18	56	80	
length (l)	10	7	14	5	15	22	(all in centimeters)
width (w)	?	?	?	?	?	?	

Since w is the unknown quantity, it would simplify the computational work if we first solved the formula for w in terms of the other variables as follows.

$P = 2l + 2w$

$P - 2l = 2w$ Add $-2l$ to both sides.

$\dfrac{P - 2l}{2} = w$ Multiply both sides by $\dfrac{1}{2}$.

$w = \dfrac{P - 2l}{2}.$ Apply the symmetric property of equality.

Now for each value for P and l, the corresponding value for w can be easily determined. Be sure you agree with the following values for w: 6, 5, 4, 4, 13, and 18.

The formula $P = 2l + 2w$ can also be solved for l in terms of P and w as follows.

$$P = 2l + 2w$$

$P - 2w = 2l$ Add $-2w$ to both sides.

$\dfrac{P - 2w}{2} = l$ Multiply both sides by $\dfrac{1}{2}$.

$l = \dfrac{P - 2w}{2}.$ Apply symmetric property of equality.

Let's consider some other often-used formulas and see how the properties of equality can be used to alter their forms. Throughout this section, we will identify formulas when we first use them.

Example 3

Solve $A = \dfrac{1}{2}bh$ for h. (Area of a triangle)

Solution

$A = \dfrac{1}{2}bh$

$2A = bh$ Multiply both sides by 2.

$\dfrac{2A}{b} = h$ Multiply both sides by $\dfrac{1}{b}$.

$h = \dfrac{2A}{b}.$ Apply symmetric property of equality.   ●

Example 4

Solve $A = P + Prt$ for t.

Solution

$A = P + Prt$

$A - P = Prt$ Add $-P$ to both sides.

$\dfrac{A - P}{Pr} = t$ Multiply both sides by $\dfrac{1}{Pr}$.

$t = \dfrac{A - P}{Pr}.$ Apply symmetric property of equality.   ●

Example 5

Solve $A = P + Prt$ for P.

Solution

$A = P + Prt$

$A = P(1 + rt)$ Apply distributive property to right side.

$\dfrac{A}{1 + rt} = P$ Multiply both sides by $\dfrac{1}{1 + rt}$.

$P = \dfrac{A}{1 + rt}.$ Apply symmetric property of equality.   ●

Example 6   Solve $A = \dfrac{1}{2}h(b_1 + b_2)$ for b_1. (Area of a trapezoid)

Solution

$$A = \frac{1}{2}h(b_1 + b_2)$$

$2A = h(b_1 + b_2)$ Multiply both sides by 2.

$2A = hb_1 + hb_2$ Apply distributive property to right side.

$2A - hb_2 = hb_1$ Add $-hb_2$ to both sides.

$\dfrac{2A - hb_2}{h} = b_1$ Multiply both sides by $\dfrac{1}{h}$.

$b_1 = \dfrac{2A - hb_2}{h}.$ Apply symmetric property of equality.   ●

In Example 5, notice that the distributive property was used to change from a form of $P + Prt$ to $P(1 + rt)$. However, in Example 6 the distributive property was used to change $h(b_1 + b_2)$ to $hb_1 + hb_2$. In both problems the key issue is to *isolate the term* containing the variable being solved for so that an appropriate application of the multiplication property of equality will produce the desired result. Also note the use of *subscripts* to identify the two bases of a trapezoid. Subscripts allow us to use the same letter b to identify the bases, but b_1 represents one base and b_2 the other.

Sometimes we are faced with equations such as $ax + b = c$, where x is the variable and a, b, and c are referred to as *arbitrary constants*. Again the properties of equality can be used to solve the equation for x as follows.

$$ax + b = c$$

$ax = c - b$ Add $-b$ to both sides.

$x = \dfrac{c - b}{a}$ Multiply both sides by $\dfrac{1}{a}$.

In Chapter 7 we will be working with equations such as $2x - 5y = 7$, which are called equations of *two* variables in x and y. Often we need to change the form of such equations by "solving for one variable in terms of the other variable." The properties of equality provide the basis for doing this.

Example 7   Solve $2x - 5y = 7$ for y in terms of x.

Solution

$$2x - 5y = 7$$

$-5y = 7 - 2x$ Add $-2x$ to both sides.

$y = \dfrac{7 - 2x}{-5}$ Multiply both sides by $-\dfrac{1}{5}$.

$y = \dfrac{2x - 7}{5}.$ Multiply numerator and denominator of the fraction on the right by -1. (The final step would not be absolutely necessary, but usually we prefer to have a positive number as a denominator.)   ●

Equations of two variables may also contain arbitrary constants. For example, the equation $\dfrac{x}{a} + \dfrac{y}{b} = 1$ contains the variables x and y, and the arbitrary constants a and b.

Example 8

Solve the equation $\dfrac{x}{a} + \dfrac{y}{b} = 1$ for x.

Solution

$$\frac{x}{a} + \frac{y}{b} = 1$$

$$ab\left(\frac{x}{a} + \frac{y}{b}\right) = ab(1) \qquad \text{Multiply both sides by } ab.$$

$$bx + ay = ab$$

$$bx = ab - ay \qquad \text{Add } -ay \text{ to both sides.}$$

$$x = \frac{ab - ay}{b}. \qquad \text{Multiply both sides by } \frac{1}{b}.$$ ●

Formulas and Problem Solving

Formulas are often used as *guidelines* for setting up an appropriate algebraic equation when solving a word problem. Let's consider an example to illustrate this point.

Problem 1

How long will it take $500 to double itself if it is invested at 8% simple interest?

Solution

Let t represent the number of years it will take $500 to earn $500 in interest (double itself). The formula $i = Prt$ can be used as a guideline.

$$i = Prt$$

$$500 = 500(8\%)(t).$$

Solving this equation we obtain

$$500 = 500(0.08)(t)$$

$$1 = 0.08t$$

$$100 = 8t$$

$$12\frac{1}{2} = t.$$

It will take $12\frac{1}{2}$ years. ●

Sometimes formulas are used within the analysis of a problem but do not serve as the main guideline for setting up the equation. For example, uniform motion problems involve the formula $d = rt$, but the main guideline for setting up an equation for such problems is usually a statement about either *times*, *rates*, or *distances*. Let's consider an example to illustrate this idea.

Problem 2   Lori starts jogging at 5 miles per hour. One-half hour later, Karen starts jogging on the same route at 7 miles per hour. How long will it take Karen to catch Lori?

Solution   First, let's sketch a diagram and record some information.

Lori at 5 mph.

Karen at 7 mph, but starts $\frac{1}{2}$ hour later.

If we let t represent Karen's time, then $t + \frac{1}{2}$ represents Lori's time. The statement "Karen's distance equals Lori's distance" can be used as a guideline.

Karen's distance   Lori's distance
$$7t = 5\left(t + \frac{1}{2}\right).$$

Solving this equation we obtain

$$7t = 5t + \frac{5}{2}$$

$$2t = \frac{5}{2}$$

$$t = \frac{5}{4}.$$

Karen should catch Lori in $1\frac{1}{4}$ hours.   ●

Some people find it helpful to use a chart to organize the known and unknown facts in a uniform motion problem. Let's illustrate this approach with an example.

Problem 3   Two trains leave a city at the same time, one traveling east and the other traveling west. At the end of $9\frac{1}{2}$ hours, they are 1292 miles apart. If the rate of the train traveling east is 8 miles per hour faster than the other train, find their rates.

Solution   If we let r represent the rate of the westbound train, then $r + 8$ represents the rate of the eastbound train. Now we can record the *times* and *rates* in a chart and then use the distance formula ($d = rt$) to represent the *distances*.

	rate	time	distance ($d = rt$)
westbound train	r	$9\frac{1}{2}$	$\frac{19}{2}r$
eastbound train	$r + 8$	$9\frac{1}{2}$	$\frac{19}{2}(r + 8)$

Since the distance that the westbound train travels plus the distance that the eastbound train travels equals 1292 miles, we can set up and solve the following equation.

$$\frac{19r}{2} + \frac{19(r + 8)}{2} = 1292$$

$$19r + 19(r + 8) = 2584$$

$$19r + 19r + 152 = 2584$$

$$38r = 2432$$

$$r = 64.$$

The westbound train travels at a rate of 64 miles per hour and the eastbound train at a rate of $64 + 8 = 72$ miles per hour. ●

Now let's consider a problem that is often referred to as a *mixture-type* problem. There is no basic formula that applies to all of these problems, but the suggestion that you "think in terms of a pure substance" is often helpful in setting up a guideline. Also keep in mind that a statement such as "a 40% solution of some substance" means that the solution contains 40% of that particular substance and 60% of something else that is mixed with it. For example, a 40% salt solution contains 40% salt and the other 60% is something else, probably water. Now let's illustrate what is meant by the suggestion "think in terms of a pure substance."

Problem 4   How many liters of pure alcohol must be added to 20 liters of a 40% solution to obtain a 60% solution?

Solution   The key idea to solving such a problem is to recognize the following guideline.

$$\begin{pmatrix} \text{amount of pure} \\ \text{alcohol in the} \\ \text{original solution} \end{pmatrix} + \begin{pmatrix} \text{amount of pure} \\ \text{alcohol to be} \\ \text{added} \end{pmatrix} = \begin{pmatrix} \text{amount of pure} \\ \text{alcohol in the} \\ \text{final solution} \end{pmatrix}$$

Letting l represent the number of liters of pure alcohol to be added, the guideline translates into the following equation.

$$(40\%)(20) + l = 60\%(20 + l).$$

Solving this equation yields

$$0.4(20) + l = 0.6(20 + l)$$
$$8 + l = 12 + 0.6l$$
$$0.4l = 4$$
$$l = 10.$$

We need to add 10 liters of pure alcohol. (Perhaps you should check this answer back ino the original statement of the problem!) ●

Problem Set 2.4

1. Solve $i = Prt$ for i, given that $P = \$200$, $r = 9\%$, and $t = 4$ years.

2. Solve $i = Prt$ for i, given that $P = \$350$, $r = 12\%$, and $t = 2\frac{1}{2}$ years.

3. Solve $i = Prt$ for t, given that $i = \$54$, $P = \$200$, and $r = 9\%$.

4. Solve $i = Prt$ for t, given that $i = \$200$, $P = \$400$, and $r = 10\%$.

5. Solve $i = Prt$ for r, given that $i = \$60$, $P = \$250$, and $t = 3$ years. Express r as a percent.

6. Solve $i = Prt$ for r, given that $i = \$275$, $P = \$1000$, and $t = 2\frac{1}{2}$ years. Express r as a percent.

7. Solve $A = P + Prt$ for A, given that $P = \$750$, $r = 9\%$, and $t = 10$ years.

8. Solve $A = P + Prt$ for A, given that $P = \$2000$, $r = 8\frac{1}{2}\%$, and $t = 5$ years.

9. Solve $A = P + Prt$ for t, given that $A = \$1204$, $P = \$700$, and $r = 9\%$.

10. Solve $A = P + Prt$ for t, given that $A = \$213$, $P = \$150$, and $r = 7\%$.

11. Use the formula $P = 2l + 2w$ and complete the following chart. (You may want to change the form of the formula.)

perimeter (P)	28	18	12	34	68	(centimeters)
width (w)	6	3	2	7	14	(centimeters)
length (l)	?	?	?	?	?	(centimeters)

12. Use the formula $A = \frac{1}{2}h(b_1 + b_2)$ and complete the following chart.

Area (A)	98	104	49	162	$16\frac{1}{2}$	$38\frac{1}{2}$	(square feet)
height (h)	14	8	7	9	3	11	(feet)
one base (b_1)	8	12	4	16	4	5	(feet)
other base (b_2)	?	?	?	?	?	?	(feet)

Solve each of the following formulas for the indicated variable.

13. $A = lw$ for l (Area of a rectangle)

14. $V = Bh$ for h (Volume of a prism)

15. $V = \frac{1}{3}Bh$ for B (Volume of a pyramid)

16. $V = \pi r^2 h$ for h (Volume of a circular cylinder)

17. $A = 2\pi r^2 + 2\pi rh$ for h (Surface area of a circular cylinder)

18. $C = 2\pi r$ for r (Circumference of a circle)

19. $A = \frac{1}{2}h(b_1 + b_2)$ for h (Area of a trapezoid)

20. $I = \frac{100M}{C}$ for C (Intelligence Quotient)

21. $C = \frac{5}{9}(F - 32)$ for F (Fahrenheit to Celsius)

22. $F = \frac{9}{5}C + 32$ for C (Celsius to Fahrenheit)

For Problems 23–32, solve each of the equations for x.

23. $ax - b = c$

24. $mx + b = e$

25. $ax + bx = c$

26. $l - kx = mx$

27. $\frac{x}{a} + 1 = b$

28. $\frac{x - a}{b} = c$

29. $a(x + b) = c$

30. $a(x + b) = b(x - c)$

31. $x(a - b) = m(x - c)$

32. $\frac{1}{2}x - a = \frac{1}{3}b$

For Problems 33–42, solve each of the equations for the indicated variable.

33. $3x + 5y = 9$ for x

34. $2x + 7y = 9$ for y

35. $7x - 2y = -1$ for y

36. $-2x + 5y = 4$ for x

37. $y = mx + b$ for x

38. $x = \frac{y - b}{m}$ for y

39. $\frac{x - a}{b} = \frac{y - a}{c}$ for x

40. $\frac{y - a}{b} = \frac{x + b}{c}$ for x

41. $(y - 2)(a + 1) = x$ for y

42. $(y + 1)(a - 3) = x - 2$ for y

Solve each of the following problems by setting up and solving an appropriate algebraic equation.

43. The length of a rectangle is 2 meters more than three times the width. The perimeter of the rectangle is 44 meters. Find the length and width of the rectangle.

44. The perimeter of a triangle is 32 inches. The second side of the triangle is 1 inch less than three times the first side. The third side is 7 inches more than twice the first side. Find the lengths of the three sides of the triangle.

45. How long will it take $1000 to double itself at 9% simple interest?

46. How long will it take P dollars to double itself at 8% simple interest?

47. How long will it take $500 to triple itself at 10% simple interest?

48. How long will it take $200 to be worth $500 if it is invested at 7% simple interest?

49. Dave leaves city A on a moped traveling toward city B at 18 miles per hour. At the same time, Tina leaves city B on a bicycle traveling toward city A at 14 miles per hour. The distance between the two cities is 112 miles. How long will it take before Dave and Tina meet?

50. Two airplanes leave St. Louis at the same time and fly in opposite directions. If one travels at 500 kilometers per hour and the other at 600 kilometers per hour, how long will it take for them to be 2475 kilometers apart?

51. A car leaves a town at 60 kilometers per hour. How long will it take a second car traveling at 75 kilometers per hour to catch the first car if it leaves 1 hour later?

52. Dennis starts walking at 4 miles per hour. An hour and a half later Cathy starts jogging along the same route at 6 miles per hour. How long will it take Cathy to catch up with Dennis?

53. Bret starts on a 70 mile bicycle ride at 20 miles per hour. After a time he becomes a little tired and slows down to 12 miles per hour for the rest of the trip. The entire trip of 70 miles took $4\frac{1}{2}$ hours. How far had Bret ridden when he reduced his speed to 12 miles per hour?

54. How many milliliters of pure acid must be added to 150 milliliters of a 30% solution of acid to obtain a 40% solution?

55. How many cups of grapefruit juice must be added to 40 cups of punch containing 5% grapefruit juice to obtain a punch that is 10% grapefruit juice?

56. Suppose that you have a supply of a 30% solution of alcohol and a 70% solution of alcohol. How many quarts of each should be mixed to produce 20 quarts that is 40% alcohol?

57. How many gallons of a 12% salt solution must be mixed with 6 gallons of a 20% salt solution to obtain a 15% salt solution?

58. A 16-quart radiator contains a 50% solution of antifreeze. How much needs to be drained out and replaced with pure antifreeze to obtain a 60% antifreeze solution?

For Problems 59–66, use your calculator to help solve each formula for the indicated variable.

59. Solve $i = Prt$ for i, given that $P = \$875$, $r = 12\frac{1}{2}\%$, and $t = 4$ years.

60. Solve $i = Prt$ for i, given that $P = \$1125$, $r = 13\frac{1}{4}\%$, and $t = 4$ years.

61. Solve $i = Prt$ for t, given that $i = \$453.25$, $P = \$925$, and $r = 14\%$.

62. Solve $i = Prt$ for t, given that $i = \$243.75$, $P = \$1250$, and $r = 13\%$.

63. Solve $i = Prt$ for r, given that $i = \$356.50$, $P = \$1550$, and $t = 2$ years. Express r as a percent.

64. Solve $i = Prt$ for r, given that $i = \$159.50$, $P = \$2200$, and $t = 0.5$ of a year. Express r as a percent.

65. Solve $A = P + Prt$ for P, given that $A = \$1423.50$, $r = 9\frac{1}{2}\%$, and $t = 1$ year.

66. Solve $A = P + Prt$ for P, given that $A = \$2173.75$, $r = 8\frac{3}{4}\%$, and $t = 2$ years.

2.5 Inequalities

Just as we use the symbol $=$ to represent *is equal to*, we also use the symbols $<$ and $>$ to represent *is less than* and *is greater than*, respectively. Thus, various **statements of inequality** can be made as follows.

$a < b$ means a is less than b;

$a \leq b$ means a is less than or equal to b;

$a > b$ means a is greater than b;

$a \geq b$ means a is greater than or equal to b.

The following are examples of **numerical statements of inequality**.

$$7 + 8 > 10; \qquad\qquad -4 + (-6) \geq -10;$$
$$-4 > -6; \qquad\qquad 7 - 9 \leq -2;$$
$$7 - 1 < 20; \qquad\qquad 3 + 4 > 12;$$
$$8(-3) < 5(-3); \qquad\qquad 7 - 1 < 0.$$

Notice that only $3 + 4 > 12$ and $7 - 1 < 0$ are *false*; the other six are *true* numerical statements.

Algebraic inequalities contain one or more variables. The following are examples of algebraic inequalities.

$$x + 4 > 8; \qquad\qquad 3x + 2y \leq 4;$$
$$3x - 1 < 15; \qquad\qquad x^2 + y^2 + z^2 \geq 7;$$
$$y^2 + 2y - 4 \geq 0.$$

An algebraic inequality such as $x + 4 > 8$ is neither true nor false as it stands and is called an **open sentence**. For each numerical value substituted for x, the algebraic inequality $x + 4 > 8$ becomes a numerical statement of inequality which is true or false. For example, if $x = -3$ then $x + 4 > 8$ becomes $-3 + 4 > 8$, which is false. If $x = 5$, then $x + 4 > 8$ becomes $5 + 4 > 8$, which is true. **Solving an inequality** refers to the process of finding the numbers that make an algebraic inequality a true numerical statement. Such numbers are called the *solutions* of the inequality and are said to *satisfy* it.

The general process for solving inequalities closely parallels that for solving equations. We continue to replace the given inequality with *equivalent, but simpler,* inequalities. For example,

$$3x + 4 > 10 \tag{1}$$
$$3x > 6 \tag{2}$$
$$x > 2 \tag{3}$$

are all equivalent inequalities; that is, they all have the same solutions. By inspection we see that the solutions for (3) are "all numbers greater than 2." Thus, (1) has the same solutions.

The exact procedure for simplifying inequalities so that the solutions can be determined is primarily based on two properties. The first of these is the **addition property of inequality**.

Addition Property of Inequality

For all real numbers a, b, and c,

$$\text{if } a > b, \quad \text{then } a + c > b + c.$$

The addition property of inequality states *any number can be added to both sides of an inequality and an equivalent inequality is produced*. The property has been stated in terms of $>$, but analogous properties exist for $<$, $\geq$, and $\leq$.

Before stating the multiplication property of inequality let's look at some numerical examples.

$2 < 5$	multiply by 4 $\longrightarrow$	$4(2) < 4(5)$;
$-3 > -7$	multiply by 2 $\longrightarrow$	$2(-3) > 2(-7)$;
$-4 < 6$	multiply by 10 $\longrightarrow$	$10(-4) < 10(6)$;
$4 < 8$	multiply by -3 $\longrightarrow$	$-3(4) > -3(8)$;
$3 > -2$	multiply by -4 $\longrightarrow$	$-4(3) < -4(-2)$;
$-4 < -1$	multiply by -2 $\longrightarrow$	$-2(-4) > -2(-1)$.

Notice that in the first three examples, multiplying both sides of an inequality by a *positive number* produces an inequality of the *same sense*. That means that if the original inequality is *less than*, then the new inequality is *less than*; and if the original inequality is *greater than*, then the new inequality is *greater than*. The last three examples illustrate that multiplying both sides of an inequality by a *negative number* produces an inequality of the *opposite sense*.

The multiplication property of inequality can be stated as follows.

Multiplication Property of Inequality

For all real numbers a, b, and c,

1. if $a > b$ and $c > 0$, then $ac > bc$
2. if $a > b$ and $c < 0$, then $ac < bc$.

Similar properties hold if each inequality is reversed or if $>$ is replaced with $\geq$ and $<$ with $\leq$. For example, if $a \leq b$ and $c < 0$, then $ac \geq bc$.

Now let's use the addition and multiplication properties of inequality to help solve some inequalities.

Example 1

Solve $3x - 4 > 8$.

Solution

$$3x - 4 > 8$$
$$3x - 4 + 4 > 8 + 4 \qquad \text{Add 4 to both sides.}$$
$$3x > 12$$
$$\frac{1}{3}(3x) > \frac{1}{3}(12) \qquad \text{Multiply both sides by } \frac{1}{3}.$$
$$x > 4.$$

The solution set is $\{x | x > 4\}$. (Remember that the set $\{x | x > 4\}$ is read as "the set of all x such that x is greater than 4.")

In Example 1, once we obtained the simple inequality $x > 4$, the solution set $\{x | x > 4\}$ became obvious. Solution sets for inequalities may also be expressed on a number line graph. The graph of the solution set for Example 1 is given in Figure 2.1.

Figure 2.1

The unshaded circle around 4 indicates that 4 is not a solution and the shaded portion indicates that all numbers greater than 4 are solutions.

Example 2

Solve $-2x + 1 > 5$ and graph the solutions.

Solution

$$-2x + 1 > 5$$
$$-2x + 1 + (-1) > 5 + (-1) \qquad \text{Add } -1 \text{ to both sides.}$$
$$-2x > 4$$
$$-\frac{1}{2}(-2x) < -\frac{1}{2}(4) \qquad \text{Multiply both sides by } -\frac{1}{2}.$$
$$\qquad\qquad\qquad\qquad \text{Notice that the sense}$$
$$x < -2. \qquad\qquad \text{of the inequality}$$
$$\qquad\qquad\qquad\qquad \text{has been reversed.}$$

The solution set is $\{x | x < -2\}$ which can be illustrated on a number line as in Figure 2.2.

Figure 2.2

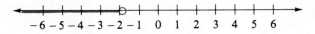

Many of the same techniques used to solve equations, such as removing parentheses and combining similar terms, may be used to solve inequalities. However, we must be extremely careful when using the multiplication property of inequality. Study each of the following examples very carefully. The format used highlights the major steps of a solution.

Example 3   Solve $-3a + 5a - 2 \geq 8a - 7 - 9a$.

Solution

$$-3a + 5a - 2 \geq 8a - 7 - 9a$$

$$2a - 2 \geq -a - 7 \qquad \text{Combine similar terms on both sides.}$$

$$3a - 2 \geq -7 \qquad \text{Add } a \text{ to both sides.}$$

$$3a \geq -5 \qquad \text{Add 2 to both sides.}$$

$$\frac{1}{3}(3a) \geq \frac{1}{3}(-5) \qquad \text{Multiply both sides by } \frac{1}{3}.$$

$$a \geq -\frac{5}{3}.$$

The solution set is $\left\{ a \mid a \geq -\dfrac{5}{3} \right\}$.

Example 4   Solve $-5(x - 1) \leq 10$ and graph the solutions.

Solution

$$-5(x - 1) \leq 10$$

$$-5x + 5 \leq 10 \qquad \text{Apply distributive property on left side.}$$

$$-5x \leq 5 \qquad \text{Add } -5 \text{ to both sides.}$$

$$-\frac{1}{5}(-5x) \geq -\frac{1}{5}(5) \qquad \text{Multiply both sides by } -\frac{1}{5}, \text{ which}$$

$$\text{reverses the inequality.}$$

$$x \geq -1.$$

The solution set $\{x \mid x \geq -1\}$ is illustrated in Figure 2.3.

Figure 2.3

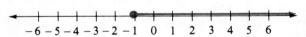

$$-6 \ -5 \ -4 \ -3 \ -2 \ -1 \ \ 0 \ \ 1 \ \ 2 \ \ 3 \ \ 4 \ \ 5 \ \ 6$$

Example 5   Solve $4(n - 3) > 9(n + 1)$.

Solution

$$4(n - 3) > 9(n + 1)$$

$$4n - 12 > 9n + 9 \qquad \text{Apply distributive property on both sides.}$$

$$-5n - 12 > 9 \qquad \text{Add } -9n \text{ to both sides.}$$

$$-5n > 21 \qquad \text{Add 12 to both sides.}$$

$$-\frac{1}{5}(-5n) < -\frac{1}{5}(21) \qquad \text{Multiply both sides by } -\frac{1}{5}.$$

$$n < -\frac{21}{5}.$$

The solution set is $\left\{ n \mid n < -\dfrac{21}{5} \right\}$.

The next example will solve the inequality without indicating the justification for each step. Be sure that you can supply the reasons for the steps.

Example 6   Solve $3(2x + 1) - 2(2x + 5) < 5(3x - 2)$.

Solution

$$3(2x + 1) - 2(2x + 5) < 5(3x - 2)$$
$$6x + 3 - 4x - 10 < 15x - 10$$
$$2x - 7 < 15x - 10$$
$$-13x - 7 < -10$$
$$-13x < -3$$
$$-\frac{1}{13}(-13x) > -\frac{1}{13}(-3)$$
$$x > \frac{3}{13}.$$

The solution set is $\left\{ x \mid x > \dfrac{3}{13} \right\}$.   ●

Checking solutions for an inequality presents a problem. Obviously, we cannot check all of the infinitely many solutions for a particular inequality. However, by checking at least one solution, especially when the multiplication property has been used, we might catch a mistake of forgetting to change the *sense* of an inequality. In Example 6 we are claiming that *all numbers greater than* $\dfrac{3}{13}$ will satisfy the original inequality. Let's check one such number, say 1.

$$3(2x + 1) - 2(2x + 5) < 5(3x - 2)$$
$$3(2(1) + 1) - 2(2(1) + 5) \overset{?}{<} 5(3(1) - 2)$$
$$3(3) - 2(7) \overset{?}{<} 5(1)$$
$$9 - 14 \overset{?}{<} 5$$
$$-5 < 5.$$

Thus, 1 satisfies the original inequality. Had we forgotten to switch the sense of the inequality when both sides were multiplied by $-\dfrac{1}{13}$, then our answer would have been $x < \dfrac{3}{13}$, and such an error would be detected by the check.

Problem Set 2.5

Solve each of the following inequalities and graph the solutions.

1. $x - 1 > -2$ **2.** $x + 3 < 5$ **3.** $-3x \geq 6$

4. $-2x \leq -6$ **5.** $4x \leq -4$ **6.** $5x \geq -5$

7. $3x + 1 < 7$

10. $3x - 4 < -4$

13. $3 + 4x > -1$

16. $6 - 3x < 12$

19. $3x - 4 \geq 2x - 3$

8. $4x + 2 > 10$

11. $-5x - 1 \leq 4$

14. $1 + 5x > -9$

17. $4(x - 1) > 8$

20. $5x + 1 \leq 4x - 2$

9. $2x - 3 > -3$

12. $-2x - 3 \geq 5$

15. $7 - 2x < 11$

18. $3(x + 2) < -6$

Solve each of the following inequalities.

21. $3x - 2 > 5$

23. $-7x + 1 < -15$

25. $2(3x + 1) \geq 10$

27. $-3(2x - 5) \leq 9$

29. $5a - 3 > 3a - 7$

31. $4n + 5 < 7n - 4$

33. $3(n - 1) \geq -(n + 4)$

35. $-2(n - 4) < 5(n - 1)$

37. $-3(2x + 1) > -2(x + 4)$

39. $4(2x - 1) - 3(3x + 4) \geq 0$

41. $3(y + 2) + 4 < -2y + 14 + y$

43. $5(x - 6) - 6(x + 2) < 0$

45. $3(x - 1) - (x - 2) > -2(x + 4)$

47. $-4(2t - 1) - 3(t + 2) \geq 0$

49. $3(x + 2) - 4(x - 1) < 6$

22. $4x - 1 < 10$

24. $3 - 8x > -8$

26. $3(2x - 1) \leq 12$

28. $-2(4x - 3) \geq 22$

30. $7a + 3 < 4a - 9$

32. $3n - 5 > 8n + 5$

34. $4(n - 3) \leq -2(n + 1)$

36. $-3(n + 2) > 2(n - 6)$

38. $-5(3x + 4) < -2(7x - 1)$

40. $3(x - 2) - 5(2x - 1) \geq 0$

42. $-5(y - 1) + 3 > 3y - 4 - 4y$

44. $7(x + 1) - 8(x - 2) < 0$

46. $-(x - 3) + 2(x - 1) < 3(x + 4)$

48. $-3(3t + 2) - 2(4t + 1) \geq 0$

50. $5(x - 4) - 6(x + 2) < 4$

2.6 More on Inequalities

To solve equations involving fractions we found that *clearing the equation of all fractions* by multiplying both sides by the least common denominator of all of the denominators in the equation was frequently an effective technique. This same basic approach also works very well with inequalities involving fractions, as the next examples illustrate.

Example 1   Solve $\dfrac{2}{3}x - \dfrac{1}{2}x > \dfrac{3}{4}$.

Solution

$$\frac{2}{3}x - \frac{1}{2}x > \frac{3}{4}$$

$$12\left(\frac{2}{3}x - \frac{1}{2}x\right) > 12\left(\frac{3}{4}\right) \qquad \text{Multiply both sides by 12, which is the LCD of 3, 2, and 4.}$$

$$12\left(\frac{2}{3}x\right) - 12\left(\frac{1}{2}x\right) > 12\left(\frac{3}{4}\right) \qquad \text{Apply distributive property.}$$

$$8x - 6x > 9$$

$$2x > 9$$

$$x > \frac{9}{2}.$$

The solution set is $\left\{x \mid x > \dfrac{9}{2}\right\}$.

Example 2

Solve $\dfrac{x+2}{4} + \dfrac{x-3}{8} < 1$.

Solution

$$\frac{x+2}{4} + \frac{x-3}{8} < 1$$

$$8\left(\frac{x+2}{4} + \frac{x-3}{8}\right) < 8(1) \qquad \text{Multiply both sides by 8, which is the LCD of}$$
$$\text{4 and 8.}$$

$$8\left(\frac{x+2}{4}\right) + 8\left(\frac{x-3}{8}\right) < 8(1)$$

$$2(x+2) + (x-3) < 8$$

$$2x + 4 + x - 3 < 8$$

$$3x + 1 < 8$$

$$3x < 7$$

$$x < \frac{7}{3}.$$

The solution set is $\left\{x \mid x < \dfrac{7}{3}\right\}$.

Example 3

Solve $\dfrac{n}{2} - \dfrac{n-1}{5} \geq \dfrac{n+2}{10} - 4$.

Solution

$$\frac{n}{2} - \frac{n-1}{5} \geq \frac{n+2}{10} - 4$$

$$10\left(\frac{n}{2} - \frac{n-1}{5}\right) \geq 10\left(\frac{n+2}{10} - 4\right)$$

$$10\left(\frac{n}{2}\right) - 10\left(\frac{n-1}{5}\right) \geq 10\left(\frac{n+2}{10}\right) - 10(4)$$

$$5n - 2(n-1) \geq n + 2 - 40$$

$$5n - 2n + 2 \geq n - 38$$
$$3n + 2 \geq n - 38$$
$$2n + 2 \geq -38$$
$$2n \geq -40$$
$$n \geq -20.$$

The solution set is $\{n|n \geq -20\}$.   ●

The idea of *clearing of all decimals* also works with inequalities in much the same way as it does with equations. We can multiply both sides of an inequality by an appropriate power of ten and then proceed to solve in the usual way. The next two examples illustrate this procedure.

Example 4   Solve $s \geq 1.6 + 0.2s$.

Solution

$$s \geq 1.6 + 0.2s$$
$$10(s) \geq 10(1.6 + 0.2s) \qquad \text{Multiply both sides by 10.}$$
$$10s \geq 16 + 2s$$
$$8s \geq 16$$
$$s \geq 2.$$

The solution set is $\{s|s \geq 2\}$.   ●

Example 5   Solve $0.08x + 0.09(x + 100) \geq 43$.

Solution

$$0.08x + 0.09(x + 100) \geq 43$$
$$100(0.08x + 0.09(x + 100)) \geq 100(43) \qquad \text{Multiply both sides by 100.}$$
$$8x + 9(x + 100) \geq 4300$$
$$8x + 9x + 900 \geq 4300$$
$$17x + 900 \geq 4300$$
$$17x \geq 3400$$
$$x \geq 200.$$

The solution set is $\{x|x \geq 200\}$.   ●

Compound Statements

The words *and* and *or* are also used in mathematics to form **compound statements**. The following are examples of some compound numerical statements using *and*. Such statements are called **conjunctions**. We agree to call a conjunction true only if all of its component parts are true. Statements **1.** and **2.** below are true, but **3.**, **4.**, and **5.** are false.

1. $3 + 4 = 7$ and $-4 < -3$.   (True)
2. $-3 < -2$ and $-6 > -10$.   (True)
3. $6 > 5$ and $-4 < -8$.   (False)
4. $4 < 2$ and $0 < 10$.   (False)
5. $-3 + 2 = 1$ and $5 + 4 = 8$.   (False)

Compound statements using *or* are called **disjunctions**. The following are some examples of disjunctions involving numerical statements.

6. $0.14 > 0.13$ or $0.235 < 0.237$.   (True)

7. $\dfrac{3}{4} > \dfrac{1}{2}$ or $-4 + (-3) = 10$.   (True)

8. $-\dfrac{2}{3} > \dfrac{1}{3}$ or $(0.4)(0.3) = 0.12$.   (True)

9. $\dfrac{2}{5} < -\dfrac{2}{5}$ or $7 + (-9) = 16$.   (False)

A disjunction is true if at least one of its component parts is true. In other words, disjunctions are false only if all of the component parts are false. In the statements above **6.**, **7.**, and **8.** are true, but **9.** is false.

Now let's consider finding solutions for some compound statements involving algebraic inequalities. Keep in mind that our previous agreements for labeling conjunctions and disjunctions true or false form the basis for our reasoning.

Example 6   Solve the conjunction $x > -1$ *and* $x < 3$ and graph its solution set on a number line.

Solution   The key word is *and*, so both inequalities need to be satisfied. Thus, all numbers between -1 and 3 are solutions and the solution set is expressed as $\{x \mid -1 < x < 3\}$. The statement $-1 < x < 3$ is read *negative one is less than x and x is less than three.* In other words, x is clamped between -1 and 3. The graph of the solution set is given in Figure 2.4.

Figure 2.4

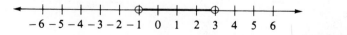

Example 6 illustrates another concept pertaining to sets. The set of all elements common to two sets is called the **intersection** of the two sets. Thus, in Example 6 we found the intersection of the two sets $\{x \mid x > -1\}$ and $\{x \mid x < 3\}$ to be the set $\{x \mid -1 < x < 3\}$. In general, the intersection of two sets is defined as follows:

Definition 2.1	The *intersection* of two sets A and B (written $A \cap B$) is the set of all elements that are both in A and in B. Using set builder notation we can write $$A \cap B = \{x \mid x \in A \ and \ x \in B\}.$$

Example 7   Solve the conjunction $3x + 1 > -5 \ and \ 2x + 5 > 7$ and graph its solution set on a number line.

Solution   First, let's simplify both inequalities.

$$3x + 1 > -5 \quad and \quad 2x + 5 > 7$$
$$3x > -6 \quad and \quad 2x > 2$$
$$x > -2 \quad and \quad x > 1.$$

Since it is a conjunction, we must satisfy both inequalities. Thus, all numbers greater than 1 are solutions and the solution set can be expressed as $\{x \mid x > 1\}$. The graph of the solution set is indicated in Figure 2.5.

Figure 2.5

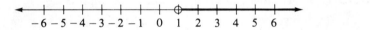

A conjunction such as "$3x + 1 > -3$ and $3x + 1 < 7$," in which the same algebraic expression is contained in both inequalities, can be conveniently solved using the compact form as follows.

$$3x + 1 > -3 \quad and \quad 3x + 1 < 7 \text{ becomes}$$
$$-3 < 3x + 1 < 7$$
$$-4 < 3x < 6 \qquad \text{Add } -1 \text{ to the left side, middle, and right side.}$$
$$-\frac{4}{3} < x < 2. \qquad \text{Multiply through by } \frac{1}{3}.$$

The solution set is $\left\{x \mid -\dfrac{4}{3} < x < 2\right\}$.

The word *and* is the key idea that ties the concept of a conjunction to the set concept of intersection. In a like manner, the word *or* links the idea of a disjunction to the set concept of **union**. The union of two sets is defined as follows.

Definition 2.2	The union of two sets A and B (written $A \cup B$) is the set of all elements that are in A or in B, or in both. Using set builder notation we can write $$A \cup B = \{x \mid x \in A \ or \ x \in B\}.$$

Example 8   Solve the disjunction $x < -1$ *or* $x > 2$ and graph its solution set on a number line.

Solution   The key word is *or*, so any numbers satisfying either inequality (or both) are solutions. Thus, all numbers less than -1, along with all numbers greater than 2, are solutions. The solution set can be expressed as $\{x|x < -1 \text{ or } x > 2\}$ and its graph is indicated in Figure 2.6.

Figure 2.6

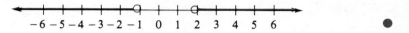

Example 8 illustrates that in terms of set vocabulary, the solution set of a disjunction is the union of the solution sets of the component parts of the disjunction. It should also be noted that there is *no compact form* for writing $x < -1$ or $x > 2$ *or for any disjunction.*

Example 9   Solve the disjunction $2x - 5 < -11$ *or* $5x + 1 \geq 6$ and graph its solution set on a number line.

Solution   First, let's simplify both inequalities.

$$2x - 5 < -11 \quad \text{or} \quad 5x + 1 \geq 6$$
$$2x < \ -6 \ \text{ or} \qquad 5x \geq 5$$
$$x < \ -3 \ \text{ or} \qquad x \geq 1.$$

Since it is a disjunction, all numbers less than -3, along with all numbers greater than or equal to 1, will satisfy it. Thus, the solution set is $\{x|x < -3 \text{ or } x \geq 1\}$ and its graph is indicated in Figure 2.7.

Figure 2.7

In summary, to solve a compound sentence involving inequality, proceed as follows:

1. Solve separately each inequality in the compound sentence.
2. If it is a *conjunction*, the solution set is the *intersection* of the solution sets of each inequality.
3. If it is a *disjunction*, the solution set is the *union* of the solution sets of each inequality.

Let's conclude this section by considering some word problems that contain inequality statements.

Problem 1   Rhonda had scores of 94, 84, 86, and 88 on her first four exams of the semester. What score must she obtain on the 5th exam to have an average of 90 or better for the five exams?

Solution   Let s represent the score needed on the 5th exam. Since the average is computed by adding all scores and dividing by the number of scores, we have the following inequality to solve.

$$\frac{94 + 84 + 86 + 88 + s}{5} \geq 90.$$

Solving this inequality we obtain

$$\frac{352 + s}{5} \geq 90$$

$$5\left(\frac{352 + s}{5}\right) \geq 5(90) \qquad \text{Multiply both sides by 5.}$$

$$352 + s \geq 450$$

$$s \geq 98.$$

She must receive a score of 98 or better.   ●

Problem 2   An investor has $1000 to invest. Suppose she invests $500 at 8% interest. At what rate must she invest the other $500 so that the two investments together yield more than $100 of yearly interest?

Solution   Let r represent the unknown rate of interest. The following guideline can be used to set up an inequality.

$$\underset{\downarrow}{\begin{array}{c}\text{interest from}\\8\%\text{ investment}\end{array}} + \underset{\downarrow}{\begin{array}{c}\text{interest from } r\\\text{percent investment}\end{array}} \underset{\downarrow}{>} \$100$$

$$(8\%)(\$500) \quad + \quad r(\$500) \qquad > \$100.$$

Solving this inequality yields

$$40 + 500r > 100$$

$$500r > 60$$

$$r > \frac{60}{500}$$

$$r > 0.12. \qquad \text{Change to a decimal.}$$

The other $500 must be invested at a rate greater than 12%.   ●

Problem 3   If the temperature for a 24-hour period ranged between 41°F and 59°F, inclusive, (that is, $41 \leq F \leq 59$) what was the range in Celsius degrees?

Solution   Using the formula $F = \frac{9}{5}C + 32$, we have the following compound inequality to solve.

$$41 \leq \frac{9}{5}C + 32 \leq 59.$$

Solving this yields

$$9 \leq \frac{9}{5}C \leq 27 \qquad \text{Add } -32.$$

$$\frac{5}{9}(9) \leq \frac{5}{9}\left(\frac{9}{5}C\right) \leq \frac{5}{9}(27) \qquad \text{Multiply by } \frac{5}{9}.$$

$$5 \leq C \leq 15.$$

The range is between $5°C$ and $15°C$, inclusive.

Problem Set 2.6

Solve each of the following inequalities.

1. $\frac{3}{5}x + \frac{2}{3}x > \frac{19}{5}$

2. $\frac{3}{4}x - \frac{2}{3}x < 2$

3. $x - \frac{3}{7} < \frac{x}{3} + 1$

4. $x + \frac{2}{9} > \frac{x}{3} - 4$

5. $\frac{n-1}{3} + \frac{n+2}{8} \geq 1$

6. $\frac{n+3}{4} + \frac{n-5}{5} \leq 2$

7. $\frac{a-4}{6} - \frac{a-2}{9} \leq \frac{5}{18}$

8. $\frac{a+3}{8} - \frac{a+5}{5} \geq \frac{3}{10}$

9. $\frac{3x+2}{9} - \frac{2x+1}{3} > -1$

10. $\frac{4x-3}{6} - \frac{2x-1}{12} < -2$

11. $s \geq 2.1 + 0.3s$

12. $s \geq 3.4 + 0.15s$

13. $0.07x + 0.08(x + 100) > 38$

14. $0.09x + 0.1(x + 200) > 77$

15. $0.08x + 0.09(2x) \geq 130$

16. $0.06x + 0.08(250 - x) \geq 19$

Graph the solutions for each of the following compound inequalities.

17. $x > -2 \text{ and } x < 1$

18. $x > 2 \text{ and } x < 5$

19. $x \leq 3 \text{ and } x > -2$

20. $x \leq 4 \text{ and } x \geq 0$

21. $x > 3 \text{ or } x < -2$

22. $x > 2 \text{ or } x < -1$

23. $x \leq 2 \text{ or } x > 5$

24. $x < -3 \text{ or } x \geq 0$

25. $x > 2 \text{ and } x > 0$

26. $x > -1 \text{ and } x > 3$

27. $x < 0 \text{ and } x > 2$

28. $x < 3 \text{ and } x \leq -1$

29. $x > -1 \text{ or } x > 3$

30. $x > 2 \text{ or } x > -2$

31. $x > -2 \text{ or } x < 1$

32. $x > 0 \text{ or } x < 4$

Solve and graph the solutions for each of the following compound inequalities.

33. $x - 3 > -2 \text{ and } x - 3 < 2$

34. $x + 4 > -1 \text{ and } x + 4 < 1$

35. $x + 1 > 4 \text{ or } x + 1 < -4$

36. $x - 2 > 3 \text{ or } x - 2 < -3$

37. $3x - 2 \geq 4 \text{ and } x > 0$

38. $4x + 1 > 13 \text{ and } x \geq 0$

39. $3x - 2 < 0 \text{ and } 2x - 1 > 0$

40. $4x - 3 < 0 \text{ and } x + 2 > 0$

41. $5x + 2 < -1 \text{ or } 5x + 2 > 1$

42. $7x - 3 < -2 \text{ or } 7x - 3 > 2$

Solve the following compound inequalities using the compact form.

43. $-3 < 2x - 1 < 7$

44. $1 < 3x - 2 < 7$

45. $-20 \le 4t + 2 \le -10$ **46.** $0 < 2t + 5 \le 15$

47. $-4 < 3x + 1 < 4$ **48.** $-7 \le 2x + 3 \le 7$

49. $-6 < \dfrac{x - 2}{4} < 6$ **50.** $-2 < \dfrac{x + 1}{3} < 2$

51. $-3 < -x + 4 < 3$ **52.** $-6 < -2x - 1 < 6$

Solve each of the following by setting up and solving an appropriate inequality.

53. Ron has scores of 52, 84, 65, and 74 on his first four math exams. What score must he make on the 5th exam to have an average of 70 or better for the five exams?

54. The average height of the two forwards and the center of a basketball team is 6 feet and 8 inches. What must the average height of the two guards be so that the team average is at least 6 feet and 4 inches?

55. Mona invests \$100 at 8% yearly interest. How much does she have to invest at 9% so that the total yearly interest from the two investments exceeds \$26?

56. Lance has \$500 to invest. If he invests \$300 at 9% interest, at what rate must he invest the remaining \$200 so that the two investments yield more than \$47 in yearly interest?

57. The temperature for a 24-hour period ranged between $-4°F$ and $23°F$, inclusive. What was the range in Celsius degrees? $\left(\text{Use } F = \dfrac{9}{5}C + 32.\right)$

58. Oven temperatures for baking various foods usually range between $325°F$ and $425°F$, inclusive. Express this range in Celsius degrees. (Round answers to the nearest degree.)

59. A person's intelligence quotient (IQ) is found by dividing mental age (M), as indicated by standard tests, by the chronological age (C), and then multiplying this ratio by 100. The formula $IQ = \dfrac{100M}{C}$ can be used. If the IQ range of a group of 11-year-olds is given by $80 \le IQ \le 140$, find the mental age range of this group.

60. Repeat Problem 59 for an IQ range of 70 to 125, inclusive, for a group of 9-year-olds.

61. Debbie had scores of 95, 82, 93, and 84 on her first four exams of the semester. What score must she obtain on the fifth exam to have an average of 90 or better for the five exams?

62. Marsha bowled 142 and 170 in her first two games. What must she bowl in the third game to have an average of at least 160 for the three games?

63. Scott shot rounds of 82, 84, 78, and 79 on the first four days of a golf tournament. What must he shoot on the fifth day of the tournament to average 80 or less for the five days?

2.7 Equations and Inequalities Involving Absolute Value

In Section 1.2 the absolute value of a real number was defined by

$$|a| = \begin{cases} a, & \text{if } a \ge 0 \\ -a, & \text{if } a < 0 \end{cases}.$$

We also interpreted the absolute value of any real number to be the distance between the number and zero on a number line. For example, $|6| = 6$ interprets as 6 units between 6 and 0. Likewise, $|-8| = 8$ interprets as 8 units between -8 and 0.

Interpreting absolute value as distance on a number line provides a straight-forward approach to solving a variety of equations and inequalities involving absolute value. First, let's consider some equations.

Example 1   Solve $|x| = 2$.

Solution   Thinking in terms of *distance between the number and zero*, we see that x must be 2 or -2. That is, the equation

$|x| = 2$ is equivalent to

$x = -2$ or $x = 2$.

The solution set is $\{-2, 2\}$.   ●

Example 2   Solve $|x + 2| = 5$.

Solution   The number, $x + 2$, must be -5 or 5. Thus,

$|x + 2| = 5$ is equivalent to

$x + 2 = -5$ or $x + 2 = 5$.

Solving each equation of the disjunction yields

$x + 2 = -5$ or $x + 2 = 5$

$x = -7$ or $x = 3$.

The solution set is $\{-7, 3\}$.

Check $\quad |x + 2| = 5 \qquad |x + 2| = 5$

$|-7 + 2| \overset{?}{=} 5 \qquad |3 + 2| \overset{?}{=} 5$

$|-5| \overset{?}{=} 5 \qquad\quad |5| \overset{?}{=} 5$

$5 = 5. \qquad\qquad 5 = 5.$ ●

The following general property should seem reasonable from the distance interpretation of absolute value.

Property 2.1	$|ax + b| = k$ is equivalent to $ax + b = -k$ or $ax + b = k$, where k is a positive number.

Therefore, our format for solving equations of the form $|ax + b| = k$ can be illustrated in Example 3.

Example 3   Solve $|5x + 3| = 7$.

Solution

$$|5x + 3| = 7$$
$$5x + 3 = -7 \quad \text{or} \quad 5x + 3 = 7$$
$$5x = -10 \quad \text{or} \quad 5x = 4$$
$$x = -2 \quad \text{or} \quad x = \frac{4}{5}.$$

The solution set is $\left\{-2, \dfrac{4}{5}\right\}$.   (Check these solutions!)

The *distance interpretation* for absolute value also provides a good basis for solving some inequalities involving absolute value. Consider the following examples.

Example 4   Solve $|x| < 2$ and graph the solution set.

Solution   The number, x, must be *less than two units away from zero*. Thus,

$|x| < 2$ is equivalent to

$x > -2 \quad \text{and} \quad x < 2.$

The solution set is $\{x | -2 < x < 2\}$ and its graph is illustrated in Figure 2.8.

Figure 2.8

Example 5   Solve and graph the solutions for $|x + 3| < 1$.

Solution   Let's continue to think in terms of "distance" on a number line. The number, $x + 3$, must be *less than one unit away from zero*. Thus,

$|x + 3| < 1$ is equivalent to

$x + 3 > -1 \quad \text{and} \quad x + 3 < 1.$

Solving this conjunction yields

$x + 3 > -1 \quad \text{and} \quad x + 3 < 1$
$x > -4 \quad \text{and} \quad\quad x < -2.$

The solution set is $\{x | -4 < x < -2\}$ and its graph is illustrated in Figure 2.9.

Figure 2.9

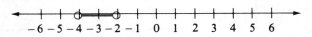

Take another look at Examples 4 and 5. The following general property should seem reasonable.

Property 2.2	$|ax + b| < k$ is equivalent to $ax + b > -k$ and $ax + b < k$, where k is a positive number.

Remember that a conjunction such as $ax + b > -k$ and $ax + b < k$ can be written in the compact form $-k < ax + b < k$. This compact form provides a very convenient format for solving inequalities such as $|3x - 1| < 8$, as Example 6 illustrates.

Example 6   Solve and graph the solutions for $|3x - 1| < 8$.

Solution

$$|3x - 1| < 8$$
$$-8 < 3x - 1 < 8$$
$$-7 < 3x < 9 \qquad \text{Add 1 to left side, middle, and right side.}$$
$$\frac{1}{3}(-7) < \frac{1}{3}(3x) < \frac{1}{3}(9) \qquad \text{Multiply through by } \frac{1}{3}.$$
$$-\frac{7}{3} < x < 3.$$

The solution set is $\left\{x \mid -\dfrac{7}{3} < x < 3\right\}$ and its graph is indicated in Figure 2.10.

Figure 2.10

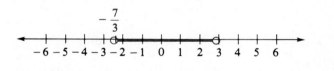

The distance interpretation also helps make clear a property pertaining to *greater than* situations involving absolute value. Consider the following examples.

Example 7   Solve and graph the solutions for $|x| > 1$.

Solution   The number, x, must be *more than one unit away from zero*. Thus,

$|x| > 1$ is equivalent to

$x < -1 \quad \text{or} \quad x > 1.$

The solution set is $\{x \mid x < -1 \text{ or } x > 1\}$ and its graph is indicated in Figure 2.11.

Figure 2.11

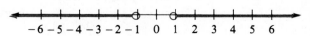

Example 8   Solve and graph the solutions for $|x - 1| > 3$.

Solution   The number, $x - 1$, must be *more than three units away from zero*. Thus, $|x - 1| > 3$ is equivalent to

$$x - 1 < -3 \quad \text{or} \quad x - 1 > 3.$$

Solving this disjunction yields

$$x - 1 < -3 \quad \text{or} \quad x - 1 > 3$$
$$x < -2 \quad \text{or} \quad x > 4.$$

The solution set is $\{x \mid x < -2 \text{ or } x > 4\}$ and its graph is given in Figure 2.12.

Figure 2.12

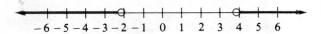

Examples 7 and 8 illustrate the following general property.

Property 2.3	$|ax + b| > k$ is equivalent to $ax + b < -k$ or $ax + b > k$, where k is a positive number.

Therefore, solving inequalities of the form $|ax + b| > k$ can take on the format illustrated in Example 9.

Example 9   Solve and graph the solutions for $|3x - 1| > 2$.

Solution
$$|3x - 1| > 2$$
$$3x - 1 < -2 \quad \text{or} \quad 3x - 1 > 2$$
$$3x < -1 \quad \text{or} \quad 3x > 3$$
$$x < -\frac{1}{3} \quad \text{or} \quad x > 1.$$

The solution set is $\left\{x \mid x < -\frac{1}{3} \text{ or } x > 1\right\}$ and its graph is given in Figure 2.13.

Figure 2.13

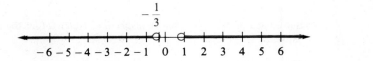

Therefore, as you solve equations and inequalities involving absolute value, you will find the following properties very helpful.

$|ax + b| = k$ is equivalent to $ax + b = -k$ or $ax + b = k$.

$|ax + b| < k$ is equivalent to $-k < ax + b < k$.

$|ax + b| > k$ is equivalent to $ax + b < -k$ or $ax + b > k$.

(k is a positive number.)

However, if at any time you become doubtful as to what property applies, don't forget the distance interpretation for absolute value. In fact, it is the distance interpretation that allows us to *solve by inspection* such equations and inequalities as $|2x - 5| = -3$, $|x - 7| < -4$, and $|2x - 1| > -1$. (Notice that in each of these examples, k is a negative number; therefore, our previous properties do not apply.)

$|2x - 5| = -3$ has *no solutions* because the absolute value (distance) cannot be negative. (The solution set is $\emptyset$, the null set.)

$|x - 7| < -4$ has *no solutions* because we cannot obtain an absolute value less than -4. (The solution set is $\emptyset$.)

$|2x - 1| > -1$ is *satisfied by all real numbers* because the absolute value of $(2x - 1)$, regardless of what number is substituted for x, will always be greater than -1. (The solution set is $\{x|x \text{ is a real number}\}$.)

Problem Set 2.7

Solve and graph the solutions for each of the following.

1. $|x| < 3$ **2.** $|x| < 5$ **3.** $|x| \leq 1$ **4.** $|x| \leq 2$

5. $|x| > 3$ **6.** $|x| > 2$ **7.** $|x| \geq 2$ **8.** $|x| \geq 1$

9. $|x + 1| < 2$ **10.** $|x + 2| < 4$ **11.** $|x - 2| < 4$ **12.** $|x - 1| < 3$

13. $|x + 2| \geq 3$ **14.** $|x + 3| \geq 1$ **15.** $|x - 4| > 1$ **16.** $|x - 5| > 2$

Solve each of the following.

17. $|x| = 4$ **18.** $|x| = 10$ **19.** $|x - 1| = 3$ **20.** $|x - 2| = 1$

21. $|x + 3| = 8$ **22.** $|x + 4| = 10$ **23.** $|2x - 1| = 11$ **24.** $|3x - 2| = 13$

25. $|4x + 3| = 7$ **26.** $|5x + 3| = 11$ **27.** $|2x - 1| < 9$ **28.** $|3x - 2| < 10$

29. $|2x + 5| > 9$ **30.** $|3x + 2| > 5$ **31.** $|4x + 1| \leq 3$ **32.** $|3x + 5| \leq 6$

33. $|3x - 2| \geq 5$ **34.** $|4x - 3| \geq 17$ **35.** $|3 - x| < 2$ **36.** $|5 - x| > 4$

37. $\left|x - \dfrac{2}{3}\right| = \dfrac{3}{4}$ **38.** $\left|x + \dfrac{1}{4}\right| = \dfrac{2}{5}$ **39.** $|6x + 7| > 14$ **40.** $|5x - 9| < 14$

Solve each of the following *by inspection.* **Don't forget the distance interpretation for absolute value.**

41. $|x + 4| = -7$ **42.** $|3x - 2| = -1$ **43.** $|x - 2| < -5$

44. $|x + 4| < -2$ **45.** $|x + 1| > -4$ **46.** $|2x - 3| > -6$

47. $|x| = 0$ **48.** $|x - 2| = 0$ **49.** $|x + 8| < 0$

50. $|x + 4| > 0$

Miscellaneous Problems

Solve each of the following equations.

51. $|x - 2| = |x + 6|$ **52.** $|x + 1| = |x - 4|$ **53.** $|x + 1| = |x - 1|$

54. $|2x - 1| = |x - 3|$ **55.** $|3x + 1| = |2x + 3|$ **56.** $|-2x - 3| = |x + 1|$

Chapter 2 Summary

(2.1) **Solving an algebraic equation** refers to the process of finding the number (or numbers) that makes the algebraic equation a true numerical statement. Such numbers are called the **solutions** or **roots** of the equation and are said to **satisfy** the equation. The set of all solutions of an equation is called the **solution set**.

The general procedure for solving an equation is to continue replacing the given equation with **equivalent, but simpler,** equations until we arrive at one that can be solved by inspection. Two properties of equality play an important role in the process of solving equations.

Addition Property of Equality If $a = b$, then $a + c = b + c$.

Multiplication Property of Equality If $a = b$ and $c \neq 0$, then $ac = bc$.

(2.2) To solve an equation involving fractions, first **clear the equation of all fractions**. It is usually easiest to begin by multiplying both sides of the equation by the least common multiple of all of the denominators in the equation (by the *least common denominator* or LCD).

Keep the following suggestions in mind as you solve word problems:

1. Read the problem carefully.
2. Sketch any figure, diagram, or chart that might be helpful.
3. Choose a meaningful variable.
4. Look for a guideline.
5. Form an equation or inequality.
6. Solve the equation or inequality.
7. Check your answers.

(2.3) To solve equations containing decimals, you can **clear the equation of all decimals** by multiplying both sides by an appropriate power of ten.

(2.4) **Formulas** are rules frequently stated in symbolic form using equations. A formula such as $P = 2l + 2w$ can be solved for l $\left(l = \dfrac{P - 2w}{2} \right)$ or for

$w\left(w = \dfrac{P - 2l}{2}\right)$ by applying the addition and multiplication properties of equality.

Formulas are often used as **guidelines** for solving word problems.

(2.5)   **Solving an algebraic inequality** refers to the process of finding the numbers that make the algebraic inequality a true numerical statement. Such numbers are called the **solutions** and the set of all solutions is called the **solution set**.

The general procedure for solving an inequality is to continue replacing the given inequality with **equivalent, but simpler**, inequalities until we arrive at one that can be solved by inspection. The following properties form the basis for solving algebraic inequalities:

1. If $a > b$, then $a + c > b + c$.   (Addition property)
2. (a) If $a > b$ and $c > 0$, then $ac > bc$. $\left(\begin{array}{c}\text{Multiplication} \\ \text{properties}\end{array}\right)$
   (b) If $a > b$ and $c < 0$, then $ac < bc$.

(2.6)   To solve compound sentences involving inequalities, we proceed as follows:

1. Solve separately each inequality in the compound sentence.
2. If it is a **conjunction**, the solution set is the **intersection** of the solution sets of each inequality.
3. If it is a **disjunction**, the solution set is the **union** of the solution sets of each inequality.

The intersection and union of two sets are defined as follows:

Intersection $A \cap B = \{x | x \in A \text{ and } x \in B\}$.
Union $A \cup B = \{x | x \in A \text{ or } x \in B\}$.

(2.7)   The **absolute value** of a number can be interpreted on the number line as the distance between that number and zero. The following properties form the basis for solving equations and inequalities involving absolute value:

1. $|ax + b| = k$ is equivalent to $ax + b = -k$ or $ax + b = k$.
2. $|ax + b| < k$ is equivalent to $-k < ax + b < k$.
3. $|ax + b| > k$ is equivalent to $ax + b < -k$ or $ax + b > k$.

(k is a positive number.)

Chapter 2 Review Problem Set

For Problems 1–15, solve each of the equations.

1. $5(x - 6) = 3(x + 2)$
2. $2(2x + 1) - (x - 4) = 4(x + 5)$
3. $-(2n - 1) + 3(n + 2) = 7$
4. $2(3n - 4) + 3(2n - 3) = -2(n + 5)$

5. $\dfrac{3t - 2}{4} = \dfrac{2t + 1}{3}$

6. $\dfrac{x + 6}{5} + \dfrac{x - 1}{4} = 2$

7. $1 - \dfrac{2x - 1}{6} = \dfrac{3x}{8}$

8. $\dfrac{2x + 1}{3} + \dfrac{3x - 1}{5} = \dfrac{1}{10}$

9. $\dfrac{3n - 1}{2} - \dfrac{2n + 3}{7} = 1$

10. $|3x - 1| = 11$

11. $0.06x + 0.08(x + 100) = 15$

12. $0.4(t - 6) = 0.3(2t + 5)$

13. $0.1(n + 300) = 0.09n + 32$

14. $0.2(x - 0.5) - 0.3(x + 1) = 0.4$

15. $|2n + 3| = 4$

For Problems 16–20, solve each equation for x.

16. $ax - b = b + 2$

17. $ax = bx + c$

18. $m(x + a) = p(x + b)$

19. $5x - 7y = 11$

20. $\dfrac{x - a}{b} = \dfrac{y + 1}{c}$

For Problems 21–24, solve each of the formulas for the indicated variable.

21. $A = \pi r^2 + \pi r s$ for s

22. $A = \dfrac{1}{2}h(b_1 + b_2)$ for b_2

23. $R = \dfrac{R_1 R_2}{R_1 + R_2}$ for R_1

24. $\dfrac{1}{R} = \dfrac{1}{R_1} + \dfrac{1}{R_2}$ for R

For Problems 25–36, solve each of the inequalities.

25. $5x - 2 \geq 4x - 7$

26. $3 - 2x < -5$

27. $2(3x - 1) - 3(x - 3) > 0$

28. $3(x + 4) \leq 5(x - 1)$

29. $\dfrac{5}{6}n - \dfrac{1}{3}n < \dfrac{1}{6}$

30. $\dfrac{n + 4}{5} + \dfrac{n - 3}{6} > \dfrac{7}{15}$

31. $s \geq 4.5 + 0.25s$

32. $0.07x + 0.09(500 - x) \geq 43$

33. $|2x - 1| < 11$

34. $|3x + 1| > 10$

35. $-3(2t - 1) - (t + 2) > -6(t - 3)$

36. $\dfrac{2}{3}(x - 1) + \dfrac{1}{4}(2x + 1) < \dfrac{5}{6}(x - 2)$

For Problems 37–43, graph the solutions of each compound inequality.

37. $x > -1$ and $x < 1$

38. $x > 2$ or $x \leq -3$

39. $x > 2$ and $x > 3$

40. $x < 2$ or $x > -1$

41. $2x + 1 > 3$ or $2x + 1 < -3$

42. $2 \leq x + 4 \leq 5$

43. $-1 < 4x - 3 \leq 9$

Solve each of the following problems by setting up and solving an appropriate equation or inequality.

44. The width of a rectangle is 2 meters more than one-third of the length. The perimeter of the rectangle is 44 meters. Find the length and width of the rectangle.

45. A total of $500 was invested, part of it at 7% and the remainder at 8% interest. If the total yearly interest from both investments amounted to $38, how much was invested at each rate?

46. Susan's average score for her first three psychology exams is 84. What must she get on the 4th exam so that her average for the four exams is 85 or better?

47. Find three consecutive integers such that the sum of one-half of the smallest and one-third of the largest is one less than the other integer.

48. Pat is paid time-and-a-half for each hour worked over 36 hours in a week. Last week he worked 42 hours and earned a total of $472.50. What is his normal hourly rate?

49. Louise has a collection of nickels, dimes, and quarters worth $24.75. The number of dimes is 10 more than twice the number of nickels and the number of quarters is 25 more than the number of dimes. How many coins of each kind does she have?

50. If the complement of an angle is one-tenth of the supplement of the angle, find the measure of the angle.

51. A retailer has some sweaters that cost her $38 each. She wants to sell them at a profit of 20% of her cost. What price should she charge for the sweaters?

52. Nora scored 16, 22, 18, and 14 points for each of the first four basketball games. How many points does she need to score in the 5th game so that her average for the first five games is at least 20 points per game?

53. Gladys leaves a town driving at a rate of 40 miles per hour. Two hours later, Heidi leaves from the same place traveling the same route and catches Gladys in 5 hours and 20 minutes. How fast was Heidi traveling?

54. In $1\frac{1}{4}$ hours more time, Rita, riding her bicycle at 12 miles per hour, rode 2 miles further than Sonya, who was riding her bicycle at 16 miles per hour. How long did each girl ride?

55. How many cups of orange juice must be added to 50 cups of a punch that is 10% orange juice to obtain a punch that is 20% orange juice?

3 Polynomials

The main thrust of this text is to develop algebraic skills, to use skills to solve equations and inequalities, and to use equations and inequalities to solve word problems. This should be readily apparent throughout this chapter. The work will center around a class of algebraic expressions called **polynomials**.

3.1 Polynomials: Sums and Differences

Recall that algebraic expressions such as $5x$, $-6y^2$, $7xy$, $14a^2b$, and $-17ab^2c^3$ are called terms. A **term** is an indicated product and may contain any number of factors. The variables involved in a term are called **literal factors** and the numerical factor is called the **numerical coefficient**. Thus, in $7xy$ the x and y are literal factors, 7 is the numerical coefficient, and the term is *in two variables* (x and y).

Terms containing variables with only nonnegative integers as exponents are called **monomials**. The previously listed terms, $5x$, $-6y^2$, $7xy$, $14a^2b$, and $-17ab^2c^3$, are all monomials. (We shall work with some algebraic expressions later such as $7x^{-1}y^{-1}$ and $6a^{-2}b^{-3}$ which are not monomials.)

The **degree** of a monomial is the sum of the exponents of the literal factors.

$7xy$ is of degree 2; $5x$ is of degree 1;

$14a^2b$ is of degree 3; $-6y^2$ is of degree 2;

$-17ab^2c^3$ is of degree 6.

If the monomial contains only one variable, then the exponent of the variable is the degree of the monomial. The two examples on the right illustrate this point. Any nonzero constant term is said to be of degree zero.

A **polynomial** is a monomial or a finite sum (or difference) of monomials. Thus,

$$4x^2, \qquad 3x^2 - 2x - 4, \qquad 7x^4 - 6x^3 + 4x^2 + x - 1,$$

$$3x^2y - 2xy^2, \qquad \frac{1}{5}a^2 - \frac{2}{3}b^2, \qquad \text{and } 14$$

are examples of polynomials. In addition to calling a polynomial with one term a **monomial**, we also classify polynomials with two terms as **binomials** and those with three terms as **trinomials**.

The *degree of a polynomial* is the degree of the term with the highest degree in the polynomial. The following examples help illustrate some of this terminology.

The polynomial $4x^3y^4$ is a monomial in two variables of degree 7.

The polynomial $4x^2y - 2xy$ is a binomial in two variables of degree 3.

The polynomial $9x^2 - 7x + 1$ is a trinomial in one variable of degree 2.

Combining Similar Terms

Remember that *similar* or *like terms* are terms having the same literal factors. In the preceding chapters we have frequently used the idea of simplifying algebraic expressions by combining similar terms, as the next examples illustrate.

$$2x + 3y + 7x + 8y = \boxed{\begin{aligned} 2x + 7x + 3y + 8y \\ (2 + 7)x + (3 + 8)y \end{aligned}}$$
$$= 9x + 11y.$$

The steps in the "dashed boxes" are usually done mentally.

$$4a - 7 - 9a + 10 = \boxed{\begin{aligned} 4a + (-7) + (-9a) + 10 \\ 4a + (-9a) + (-7) + 10 \\ (4 + (-9))a + (-7) + 10 \end{aligned}}$$
$$= -5a + 3.$$

Both adding and subtracting polynomials rely on basically the same ideas. The commutative, associative, and distributive properties provide the basis for rearranging, regrouping, and combining similar terms. Let's consider some examples.

Example 1   Add $4x^2 + 5x + 1$ and $7x^2 - 9x + 4$.

Solution   The horizontal format is most commonly used for such work. Thus,

$$(4x^2 + 5x + 1) + (7x^2 - 9x + 4) = (4x^2 + 7x^2) + (5x - 9x) + (1 + 4)$$
$$= 11x^2 - 4x + 5.$$

Example 2   Add $5x - 3$, $3x + 2$, and $8x + 6$.

Solution $$(5x - 3) + (3x + 2) + (8x + 6) = (5x + 3x + 8x) + (-3 + 2 + 6)$$
$$= 16x + 5.$$

Example 3 Find the indicated sum: $(-4x^2y + xy^2) + (7x^2y - 9xy^2) + (5x^2y - 4xy^2)$.

Solution

$$(-4x^2y + xy^2) + (7x^2y - 9xy^2) + (5x^2y - 4xy^2)$$
$$= (-4x^2y + 7x^2y + 5x^2y) + (xy^2 - 9xy^2 - 4xy^2)$$
$$= 8x^2y - 12xy^2.$$

●

The idea of subtraction as *adding the opposite* ($a - b = a + (-b)$) extends to polynomials in general. The opposite of a polynomial can be formed by taking the opposite of each term. For example, the opposite of $3x^2 - 7x + 1$ is $-3x^2 + 7x - 1$. Symbolically this is expressed as

$$-(3x^2 - 7x + 1) = -3x^2 + 7x - 1.$$

Now consider the following subtraction problems.

Example 4 Subtract $3x^2 + 7x - 1$ from $7x^2 - 2x - 4$.

Solution

Using the horizontal format, we obtain

$$(7x^2 - 2x - 4) - (3x^2 + 7x - 1)$$
$$= (7x^2 - 2x - 4) + (-3x^2 - 7x + 1)$$
$$= (7x^2 - 3x^2) + (-2x - 7x) + (-4 + 1)$$
$$= 4x^2 - 9x - 3.$$

●

Example 5 Subtract $-3y^2 + y - 2$ from $4y^2 + 7$.

Solution

$$(4y^2 + 7) - (-3y^2 + y - 2) = (4y^2 + 7) + (3y^2 - y + 2)$$
$$= (4y^2 + 3y^2) + (-y) + (7 + 2)$$
$$= 7y^2 - y + 9.$$

●

The work can be shown using a vertical format as the next example illustrates.

Example 6 Subtract $4x^2 - 7xy + 5y^2$ from $3x^2 - 2xy + y^2$.

Solution

$$\begin{array}{l} 3x^2 - 2xy + y^2 \\ \underline{4x^2 - 7xy + 5y^2} \end{array}$$
  Notice which polynomial goes on the bottom and how the similar terms are aligned.

Now we can *mentally form the opposite of the bottom polynomial* and add.

$$\begin{array}{l} 3x^2 - 2xy + y^2 \\ \underline{4x^2 - 7xy + 5y^2} \\ -x^2 + 5xy - 4y^2 \end{array}$$
  The opposite of $4x^2 - 7xy + 5y^2$ is $-4x^2 + 7xy - 5y^2$.

●

The distributive property along with the properties "$a = 1(a)$" and "$-a = -1(a)$" can also be used when adding and subtracting polynomials. The next examples illustrate this approach.

Example 7   Perform the indicated operations: $(5x - 2) + (2x - 1) - (3x + 4)$.

Solution
$$(5x - 2) + (2x - 1) - (3x + 4) = 1(5x - 2) + 1(2x - 1) - 1(3x + 4)$$
$$= 1(5x) - 1(2) + 1(2x) - 1(1) - 1(3x) - 1(4)$$
$$= 5x - 2 + 2x - 1 - 3x - 4$$
$$= 5x + 2x - 3x - 2 - 1 - 4$$
$$= 4x - 7. \qquad \bullet$$

Certainly some of the steps can be done mentally and our format may be simplified as illustrated by the next two examples.

Example 8   Perform the indicated operations: $(5a^2 - 2b) - (2a^2 + 4) + (-7b - 3)$.

Solution
$$(5a^2 - 2b) - (2a^2 + 4) + (-7b - 3) = 5a^2 - 2b - 2a^2 - 4 - 7b - 3$$
$$= 3a^2 - 9b - 7. \qquad \bullet$$

Example 9   Simplify $(4t^2 - 7t - 1) - (t^2 + 2t - 6)$.

Solution
$$(4t^2 - 7t - 1) - (t^2 + 2t - 6) = 4t^2 - 7t - 1 - t^2 - 2t + 6$$
$$= 3t^2 - 9t + 5. \qquad \bullet$$

Remember that a polynomial in parentheses preceded by a negative sign can be written without parentheses by replacing each term in the parentheses with its opposite.

Problem Set 3.1

1. Determine the degree of each of the following polynomials.

(a) $5xy^2 + 7x$

(b) $-6x^2y - 2xy$

(c) $5a^2b^2 + 7a^2b + 6ab$

(d) $8x^2y^2 - 9xy^2 + 10x^2y$

(e) $4x^3 - 3x^2 + 2x + 1$

(f) $-2x^4 - 6x - 2$

(g) $6x^7 - 2$

(h) $4y^6 - 3y^2 - 2$

(i) -14

(j) xy

In Problems 2–11, add the polynomials.

2. $5x - 2$ and $3x + 5$

3. $-4x + 7$ and $8x - 3$

4. $-3t - 5$ and $-9t + 9$

5. $5x^2 - 7x - 1$ and $4x^2 + 6x + 9$

6. $-2x^2 + 3x - 2$ and $6x^2 - 7x - 1$

7. $8x^2 y - 7xy$ and $-14x^2 y + xy$

8. $-7a^2 b^2 + 5ab^2$ and $4a^2 b^2 - 7ab^2$

9. $10x^2 - 11x + 5$ and $-4x^2 - 7$

10. $3x - 7, 5x - 4$, and $-2x - 1$

11. $-2x^2 + 6x - 1, 7x + 4$, and $-3x^2 - 2$

In Problems 12–21, subtract the polynomials using the horizontal format.

12. $7x - 1$ from $10x + 3$

13. $5x + 4$ from $3x - 2$

14. $-3a - 2$ from $4a + 1$

15. $4a + 1$ from $-3a - 2$

16. $2x^2 - 4x - 3$ from $5x^2 + 8x + 6$

17. $4x^2 - 3x - 2$ from $2x^2 - 7x - 1$

18. $5a^2 - 7a - 9$ from $-8a^2 + 7a + 12$

19. $-a^2 - a + 1$ from $a^2 + a + 4$

20. $3x^3 - 2x^2 - x + 4$ from $4x^3 + 5x^2 + 3$

21. $7x^3 - 2$ from $4x^3 + 3x^2 - 5x - 1$

In Problems 22–27, subtract the polynomials using the vertical format.

22. $3x - 1$ from $8x + 7$

23. $-7x - 2$ from $12x - 13$

24. $9x^2 - 7x - 1$ from $10x^2 + 7x - 2$

25. $-3a^2 - 7a - 4$ from $-5a^2 + 6a + 1$

26. $8x^2 + 5x - 7$ from $2x + 3$

27. $-x^2 - 2x - 4$ from $4x^4 - 3x^3 - 2x - 6$

In Problems 28–35, perform the indicated operations.

28. $(5x - 1) + (7x - 2) - (6x - 3)$

29. $(3x - 4) - (6x + 3) + (9x - 4)$

30. $(7a - 2) - (8a - 1) - (10a - 2)$

31. $(8x^2 - 6x - 2) + (x^2 - x - 1) - (3x^2 - 2x + 4)$

32. $(12x^2 + 7x - 2) - (3x^2 + 4x + 5) + (-4x^2 - 7x - 2)$

33. $(4x^3 - 2x - 1) - (7x^2 + 2x - 1) - (3x^3 - x - 1)$

34. $(5a^2 - 7a + 10) - (a^3 - a^2 - 4) - (7a^2 + 3)$

35. $(3t^2 - 3t + 1) - (-4t + 5) - (2t^3 - t - 1)$

36. Subtract $4x^2 - 7x - 1$ from the sum of $3x^2 - 2x - 1$ and $5x^2 + 8x - 1$.

37. Subtract $-3x^2 - 9x + 11$ from the sum of $-4x^2 - 7x + 2$ and $2x^2 + x + 1$.

38. Subtract $5t^2 - t - 1$ from the sum of $3t - 2$ and $2t^2 + 8$.

39. Subtract the sum of $3n^2 - n - 2$ and $5n^2 + 2n + 8$ from $2n^2 - n - 9$.

40. Subtract the sum of $5x^2 - 2x - 11$ and $-6x^2 + 3x + 9$ from $-x^2 - x + 4$.

In Problems 41–50 simplify by removing the inner parentheses first and working outward.

41. $x - [3x - (x - 2)]$

42. $x^2 - [4x^2 - (x^2 + 5)]$

43. $(5x + 2) - [3x - (2x + 4) - 6]$

44. $(7x - 4) - [2x + (3x - 1) - 2]$

45. $(2x^2 - (3x - 1)] - [5x^2 - (4x + 3)]$

46. $[4t^2 - (2t + 1) + 3] - [3t^2 + (2t - 1) - 5]$

47. $-(3n^2 - 2n + 4) - [2n^2 - (n^2 + n + 3)]$

48. $[2n^2 - (2n^2 - n + 5)] + [3n^2 + (n^2 - 2n - 7)]$

49. $3x^2 - [4x^2 - 2x - (x^2 - 2x + 6)]$

50. $[7xy - (2x - 3xy + y)] - [3x - (x - 10xy - y)]$

3.2 Products and Quotients of Monomials

Suppose that we want to find the product of two monomials such as $3x^2y$ and $4x^3y^2$. Using the properties of real numbers and the idea that exponents indicate repeated multiplication we can proceed as follows.

$$(3x^2y)(4x^3y^2) = (3 \cdot x \cdot x \cdot y)(4 \cdot x \cdot x \cdot x \cdot y \cdot y)$$
$$= 3 \cdot 4 \cdot x \cdot x \cdot x \cdot x \cdot x \cdot y \cdot y \cdot y$$
$$= 12x^5y^3.$$

Such an approach could be used to find the product of any two monomials. However, there are some basic properties of exponents that make the process of multiplying monomials a much easier task. Let's consider each of these properties and illustrate its use when multiplying monomials.

The following examples lead into the first property.

$$x^2 \cdot x^3 = (x \cdot x)(x \cdot x \cdot x) = x^5;$$
$$a^4 \cdot a^2 = (a \cdot a \cdot a \cdot a)(a \cdot a) = a^6;$$
$$b^3 \cdot b^4 = (b \cdot b \cdot b)(b \cdot b \cdot b \cdot b) = b^7.$$

In general,

$$b^n \cdot b^m = \underbrace{(b \cdot b \cdot b \cdots b)}_{\substack{n \text{ factors} \\ \text{of } b}} \underbrace{(b \cdot b \cdot b \cdots b)}_{\substack{m \text{ factors} \\ \text{of } b}}$$

$$= \underbrace{b \cdot b \cdot b \cdots b}_{(n+m) \text{ factors of } b}$$

$$= b^{n+m}.$$

The first property can be stated as follows.

Property 3.1	If b is any real number and n and m are positive integers, then $$b^n \cdot b^m = b^{n+m}.$$

Property 3.1 states "to find the product of two positive integral powers of the same base, add the exponents and use this sum as the exponent of the common base."

$$x^7 \cdot x^8 = x^{7+8} = x^{15}; \qquad\qquad y^6 \cdot y^4 = y^{6+4} = y^{10};$$
$$2^3 \cdot 2^8 = 2^{3+8} = 2^{11}; \qquad\qquad (-3)^4 \cdot (-3)^5 = (-3)^{4+5} = (-3)^9;$$
$$\left(\frac{2}{3}\right)^7 \cdot \left(\frac{2}{3}\right)^5 = \left(\frac{2}{3}\right)^{7+5} = \left(\frac{2}{3}\right)^{12}.$$

The following examples illustrate the use of Property 3.1 along with the commutative and associative properties of multiplication to form the basis for multiplying monomials. The steps enclosed in the dashed boxes might be performed mentally.

Example 1
$$(3x^2y)(4x^3y^2) = \boxed{\; 3 \cdot 4 \cdot x^2 \cdot x^3 \cdot y \cdot y^2 \;}$$
$$= \boxed{\; 12x^{2+3}y^{1+2} \;}$$
$$= 12x^5y^3.$$

Example 2
$$(-5a^3b^4)(7a^2b^5) = \boxed{\; -5 \cdot 7 \cdot a^3 \cdot a^2 \cdot b^4 \cdot b^5 \;}$$
$$= \boxed{\; -35a^{3+2}b^{4+5} \;}$$
$$= -35a^5b^9.$$

Example 3
$$\left(\frac{3}{4}xy\right)\left(\frac{1}{2}x^5y^6\right) = \boxed{\; \frac{3}{4} \cdot \frac{1}{2} \cdot x \cdot x^5 \cdot y \cdot y^6 \;}$$
$$= \boxed{\; \frac{3}{8}x^{1+5}y^{1+6} \;}$$
$$= \frac{3}{8}x^6y^7.$$

Example 4
$$(-ab^2)(-5a^2b) = \boxed{\; (-1)(-5)(a)(a^2)(b^2)(b) \;}$$
$$= \boxed{\; 5a^{1+2}b^{2+1} \;}$$
$$= 5a^3b^3.$$

Example 5
$$(2x^2y^2)(3x^2y)(4y^3) = \boxed{\; 2 \cdot 3 \cdot 4 \cdot x^2 \cdot x^2 \cdot y^2 \cdot y \cdot y^3 \;}$$
$$= \boxed{\; 24x^{2+2}y^{2+1+3} \;}$$
$$= 24x^4y^6.$$

Another useful property of exponents is illustrated by the following examples.

$$(x^2)^3 = x^2 \cdot x^2 \cdot x^2 = x^{2+2+2} = x^6;$$
$$(a^3)^2 = a^3 \cdot a^3 = a^{3+3} = a^6;$$
$$(b^4)^3 = b^4 \cdot b^4 \cdot b^4 = b^{4+4+4} = b^{12}.$$

In general,

$$(b^n)^m = \underbrace{b^n \cdot b^n \cdot b^n \cdots b^n}_{m \text{ factors of } b^n}$$

$$= b^{\overbrace{n+n+n+ \cdots +n}^{m \text{ of these}}}$$

$$= b^{mn}.$$

The property can be stated as follows.

Property 3.2   If b is any real number and m and n are positive integers, then

$$(b^n)^m = b^{mn}.$$

Property 3.2 can be used to find "the power of a power" as follows.

$$(x^4)^5 = x^{5(4)} = x^{20}; \qquad\qquad (y^6)^3 = y^{3(6)} = y^{18};$$
$$(2^3)^7 = 2^{7(3)} = 2^{21}.$$

A third property of exponents pertains to raising a monomial to a power. Consider the following examples, which are used to introduce the property.

$$(3x)^2 = (3x)(3x) = 3 \cdot 3 \cdot x \cdot x = 3^2 \cdot x^2;$$
$$(4y^2)^3 = (4y^2)(4y^2)(4y^2) = 4 \cdot 4 \cdot 4 \cdot y^2 \cdot y^2 \cdot y^2 = (4)^3(y^2)^3;$$
$$(-2a^3b^4)^2 = (-2a^3b^4)(-2a^3b^4) = (-2)(-2)(a^3)(a^3)(b^4)(b^4)$$
$$= (-2)^2(a^3)^2(b^4)^2.$$

In general,

$$(ab)^n = \underbrace{(ab)(ab)(ab) \cdots (ab)}_{n \text{ factors of } ab}$$

$$= \underbrace{(a \cdot a \cdot a \cdot a \cdots a)}_{\substack{n \text{ factors} \\ \text{of } a}} \underbrace{(b \cdot b \cdot b \cdots b)}_{\substack{n \text{ factors} \\ \text{of } b}}$$

$$= a^n b^n.$$

The property can be formally stated as follows.

Property 3.3	If a and b are real numbers and n is a positive integer, then
	$$(ab)^n = a^n b^n.$$

Property 3.3, along with Property 3.2, forms the basis for raising a monomial to a power, as the next examples illustrate.

Example 6

$(x^2 y^3)^4 = (x^2)^4 (y^3)^4$ Use $(ab)^n = a^n b^n$.

$\qquad\qquad = x^8 y^{12}$.   Use $(b^n)^m = b^{mn}$.   ●

Example 7

$(3a^5)^3 = (3)^3 (a^5)^3$

$\qquad\quad = 27 a^{15}$.   ●

Example 8

$(-2xy^4)^5 = (-2)^5 (x)^5 (y^4)^5$

$\qquad\qquad = -32 x^5 y^{20}$.   ●

Dividing Monomials

Developing an effective process for dividing by a monomial relies on yet another property of exponents. This property is also a direct consequence of the definition of an exponent and is illustrated by the following examples.

$$\frac{x^4}{x^3} = \frac{x \cdot x \cdot x \cdot x}{x \cdot x \cdot x} = x; \qquad \frac{x^3}{x^3} = \frac{x \cdot x \cdot x}{x \cdot x \cdot x} = 1;$$

$$\frac{a^5}{a^2} = \frac{a \cdot a \cdot a \cdot a \cdot a}{a \cdot a} = a^3; \qquad \frac{y^5}{y^5} = \frac{y \cdot y \cdot y \cdot y \cdot y}{y \cdot y \cdot y \cdot y \cdot y} = 1;$$

$$\frac{y^8}{y^4} = \frac{y \cdot y \cdot y \cdot y \cdot y \cdot y \cdot y \cdot y}{y \cdot y \cdot y \cdot y} = y^4.$$

The general property can be stated as follows.

Property 3.4	If b is any nonzero real number and m and n are positive integers, then
	1. $\dfrac{b^n}{b^m} = b^{n-m}$ when $n > m$
	2. $\dfrac{b^n}{b^m} = 1$ when $n = m$.

Applying Property 3.4 to the previous examples yields

$$\frac{x^4}{x^3} = x^{4-3} = x^1 = x; \qquad\qquad \frac{x^3}{x^3} = 1;$$

$$\frac{a^5}{a^2} = a^{5-2} = a^3; \qquad\qquad \frac{y^5}{y^5} = 1;$$

$$\frac{y^8}{y^4} = y^{8-4} = y^4.$$

(It should be noted that the situation when $n < m$ will be discussed in a later chapter.)

Property 3.4, along with our knowledge of dividing integers, provides the basis for dividing monomials. The following examples illustrate the process.

$$\frac{24x^5}{3x^2} = 8x^{5-2} = 8x^3; \qquad\qquad \frac{-36a^{13}}{-12a^5} = 3a^{13-5} = 3a^8;$$

$$\frac{-56x^9}{7x^4} = -8x^{9-4} = -8x^5; \qquad\qquad \frac{72b^5}{8b^5} = 9 \qquad \left(\frac{b^5}{b^5} = 1\right);$$

$$\frac{48y^7}{-12y} = -4y^{7-1} = -4y^6; \qquad\qquad \frac{12x^4y^7}{2x^2y^4} = 6x^{4-2}y^{7-4} = 6x^2y^3.$$

Problem Set 3.2

Find each of the following products.

1. $(2x^2)(5x^3)$ **2.** $(3x^4)(4x^5)$

3. $(-4x^3)(6x)$ **4.** $(2xy^2)(-5x^2y)$

5. $(-6a^2b^3)(-7ab^2)$ **6.** $(-5a^3b^5)(-7a^4b)$

7. $(x^2yz^3)(-2xy^2z^3)$ **8.** $(-xy^2z)(-x^2y^3z^2)$

9. $(3xy)(-x^4y^3)$ **10.** $(x^2y)(5xy^2)$

11. $(4ab^3)(-3a^2b^2)$ **12.** $(-6a^2b^3)(-7a^3b^4)$

13. $(mn^2)(-m^2n^2)$ **14.** $(7m^2n)(-8m^3n)$

15. $\left(\frac{2}{3}x^4y\right)\left(\frac{4}{5}x^2y^2\right)$ **16.** $\left(\frac{3}{5}xy^6\right)\left(\frac{1}{2}x^5y^2\right)$

17. $\left(-\frac{3}{4}a\right)\left(\frac{2}{3}a^2b^3\right)$ **18.** $\left(\frac{5}{6}ab^2\right)\left(-\frac{4}{5}a^4\right)$

19. $\left(\frac{1}{2}xy\right)\left(-\frac{1}{3}x^2y\right)$ **20.** $\left(\frac{2}{3}x^2y^3\right)(xy^2)$

21. $(2x)(-3x^2)(5x^3)$ **22.** $(4x)(7x^2)(-2x^4)$

23. $(-2n^2)(5n^3)(-8n^4)$ **24.** $-a^2(a^2b)(2a)(-3b)$

25. $(x^2y)(-3xy^2)(x^3y^4)$ **26.** $(-r^3)(-r^2t^2)(-rt^4)$

27. $(-2y^3)(-y^2)(-3y^2)$ **28.** $(y^2z)(-5yz^2)(-y^2z)$

29. $(3ab)(-4a^2b)(5a)$ **30.** $(4b)(-2a^2b^2)(-6a)$

31. $(-a^2b^2)(-3ab^2)(-2a^2b^3)$ **32.** $(-5a^2b^3)(ab)(7a)$

33. $\left(\frac{1}{2}xy\right)(2x^2y)(-3x^2y^3)$ **34.** $(-4xy^2)\left(\frac{1}{4}xy^2\right)(-2x)$

Raise each of the following monomials to the indicated power.

35. $(-2xy)^5$ **36.** $(-3x^2y^2)^4$ **37.** $(2ab^2)^5$ **38.** $(a^2b^3)^6$

39. $(-2x^2y)^3$ **40.** $(-4x^3y^4)^3$ **41.** $(a^2b^2c)^5$ **42.** $(-xyz^3)^4$

43. $(8x^2y^3)^2$ **44.** $(9x^3y^5)^2$ **45.** $(2xy^2)^4$ **46.** $(-x^2y)^3$

47. $(-a^2b^3)^3$ **48.** $(-2ab^3)^4$ **49.** $(-5m^2n^3)^2$ **50.** $(-6m^4n^5)^2$

51. $-(x^2y^3)^2$ **52.** $-(2xy^3)^4$ **53.** $(3x^3yz^2)^3$ **54.** $(2a^2bc^3)^3$

Find each of the following quotients.

55. $\dfrac{4x^3y^2}{2xy}$ **56.** $\dfrac{6x^2y^5}{2xy^2}$ **57.** $\dfrac{18x^4y^3}{-6x^3y}$ **58.** $\dfrac{35x^5y^4}{-7x^2y^2}$

59. $\dfrac{-28ab^2c^3}{-4bc^2}$ **60.** $\dfrac{-42a^2bc^5}{-6ac^3}$ **61.** $\dfrac{-15x^2y^2z^3}{xy^2z}$ **62.** $\dfrac{-21x^4y^3z^5}{x^2y^3z^2}$

63. $\dfrac{r^3s^2t^4}{-rst}$ **64.** $\dfrac{a^3b^4c}{-a^2bc}$ **65.** $\dfrac{-63xy^2}{-9xy^2}$ **66.** $\dfrac{-56x^2y^2}{8x^2y}$

67. $\dfrac{12ab^2}{-12ab^2}$ **68.** $\dfrac{25a^5b^6}{-5a^4}$ **69.** $\dfrac{-24x^4y^3}{-y^3}$ **70.** $\dfrac{28x^3y^7}{-xy}$

Find each of the following products. Assume that the variables in the exponents represent positive integers.

$$\text{Example: } (x^{2n})(x^{3n}) = x^{2n+3n} = x^{5n}.$$

71. $(x^n)(x^{4n})$ **72.** $(x^{2n})(x^{3n+1})$ **73.** $(a^{3n-1})(a^{4n+3})$

74. $(a^{2n-1})(a^{2n+1})$ **75.** $(x^{3n+2})(x^{2n-3})$ **76.** $(x^{n+2})(x^{n+3})$

77. $(a^{6n-1})(a^2)$ **78.** $(a^3)(a^{3n+4})$ **79.** $(2x^n)(-3x^n)$

80. $(2x^{n+1})(3x^{2n-1})$ **81.** $(-2a^2)(-3a^{3n+1})$ **82.** $(x^n)(x^{2n+1})(x^{3n})$

83. $(2x^{2n})(3x^n)(4x^{n-1})$ **84.** $(x^{3n-2})(x^{2n+3})$ **85.** $(x^{5n+1})(2x^{3n-4})$

3.3 Multiplying Polynomials

The distributive property is usually stated as "$a(b + c) = ab + ac$"; however, it can be extended as follows.

$$a(b + c + d) = ab + ac + ad;$$
$$a(b + c + d + e) = ab + ac + ad + ae; \quad \text{etc.}$$

The commutative and associative properties, the properties of exponents, and the distributive property work together to form a basis for finding the product of a monomial and a polynomial. The following examples illustrate this idea.

Example 1
$$3x^2(2x^2 + 5x + 3) = 3x^2(2x^2) + 3x^2(5x) + 3x^2(3)$$
$$= 6x^4 + 15x^3 + 9x^2.$$

●

Example 2
$$-2xy(3x^3 - 4x^2y - 5xy^2 + y^3) = -2xy(3x^3) - (-2xy)(4x^2y) - (-2xy)(5xy^2)$$
$$+ (-2xy)(y^3)$$
$$= -6x^4y + 8x^3y^2 + 10x^2y^3 - 2xy^4.$$

●

Extending from finding the product of a monomial and a polynomial to finding the product of two polynomials is again based on the distributive property. Consider the following examples.

Example 3
$$(x + 2)(y + 5) = x(y + 5) + 2(y + 5)$$
$$= x(y) + x(5) + 2(y) + 2(5)$$
$$= xy + 5x + 2y + 10.$$

●

Notice that each term of the first polynomial is multiplied times each term of the second polynomial.

Example 4
$$(x - 3)(y + z + 3) = x(y + z + 3) - 3(y + z + 3)$$
$$= xy + xz + 3x - 3y - 3z - 9.$$

●

Frequently, multiplying polynomials will produce similar terms that can be combined to simplify the resulting polynomial.

Example 5
$$(x + 5)(x + 7) = x(x + 7) + 5(x + 7)$$
$$= x^2 + 7x + 5x + 35$$
$$= x^2 + 12x + 35.$$

●

Example 6
$$(x - 2)(x^2 - 3x + 4) = x(x^2 - 3x + 4) - 2(x^2 - 3x + 4)$$
$$= x^3 - 3x^2 + 4x - 2x^2 + 6x - 8$$
$$= x^3 - 5x^2 + 10x - 8.$$

●

Example 7
$$(3x - 2y)(x^2 + xy - y^2) = 3x(x^2 + xy - y^2) - 2y(x^2 + xy - y^2)$$
$$= 3x^3 + 3x^2y - 3xy^2 - 2x^2y - 2xy^2 + 2y^3$$
$$= 3x^3 + x^2y - 5xy^2 + 2y^3.$$

●

It is helpful to be able to find the product of two binomials without showing all of the intermediate steps. This is quite easy to do by developing a *three-step shortcut pattern* as demonstrated by the following examples.

Example 8

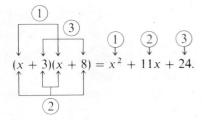

$$(x + 3)(x + 8) = x^2 + 11x + 24.$$

Step ① Multiply $x \cdot x$.
Step ② Multiply $3 \cdot x$ and $8 \cdot x$ and combine.
Step ③ Multiply $3 \cdot 8$.

Example 9

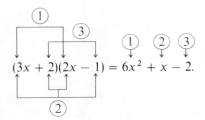

$$(3x + 2)(2x - 1) = 6x^2 + x - 2.$$

Now see if you can use the pattern to find the following products.

$$(x + 2)(x + 6) = ?$$
$$(x - 3)(x + 5) = ?$$
$$(2x + 5)(3x + 7) = ?$$
$$(3x - 1)(4x - 3) = ?$$

Your answers should be $x^2 + 8x + 12$, $x^2 + 2x - 15$, $6x^2 + 29x + 35$, and $12x^2 - 13x + 3$.

Keep in mind that this shortcut pattern applies only to finding the product of two binomials. For other situations, such as finding the product of a binomial and a trinomial, we suggest showing some intermediate steps as in the next example.

Example 10

$$
\begin{aligned}
(2x - 1)(x^2 - 4x + 6) &= 2x(x^2 - 4x + 6) - 1(x^2 - 4x + 6) \\
&= 2x^3 - 8x^2 + 12x - x^2 + 4x - 6 \\
&= 2x^3 - 9x^2 + 16x - 6.
\end{aligned}
$$

Exponents are used to indicate repeated multiplication of polynomials. For example, $(x + 3)^2$ means $(x + 3)(x + 3)$ and $(x - 4)^3$ means $(x - 4)(x - 4)(x - 4)$. To square a binomial we can simply write it as the product of two equal binomials and apply the shortcut pattern. Thus,

$$(x + 3)^2 = (x + 3)(x + 3) = x^2 + 6x + 9;$$
$$(x - 6)^2 = (x - 6)(x - 6) = x^2 - 12x + 36;$$
$$(3x - 4)^2 = (3x - 4)(3x - 4) = 9x^2 - 24x + 16.$$

When squaring binomials, be careful not to forget the middle term. That is to say, $(x + 3)^2 \neq x^2 + 3^2$; instead, $(x + 3)^2 = x^2 + 6x + 9$.

The following examples suggest a format to use when cubing a binomial or raising a binomial to a power greater than three.

Example 11

$$\begin{aligned} (x + 4)^3 &= (x + 4)(x + 4)(x + 4) \\ &= (x + 4)(x^2 + 8x + 16) \\ &= x(x^2 + 8x + 16) + 4(x^2 + 8x + 16) \\ &= x^3 + 8x^2 + 16x + 4x^2 + 32x + 64 \\ &= x^3 + 12x^2 + 48x + 64. \end{aligned}$$

●

Example 12

$$\begin{aligned} (n - 2)^4 &= (n - 2)(n - 2)(n - 2)(n - 2) \\ &= (n^2 - 4n + 4)(n^2 - 4n + 4) \\ &= n^2(n^2 - 4n + 4) - 4n(n^2 - 4n + 4) + 4(n^2 - 4n + 4) \\ &= n^4 - 4n^3 + 4n^2 - 4n^3 + 16n^2 - 16n + 4n^2 - 16n + 16 \\ &= n^4 - 8n^3 + 24n^2 - 32n + 16. \end{aligned}$$

●

When multiplying binomials, there are some special patterns that you should recognize. These patterns can be used to find products and will also be used later when factoring polynomials. Each of the patterns will be stated in general terms along with some examples to illustrate the use of each pattern.

Pattern

$$(a + b)^2 = (a + b)(a + b) = a^2 + \qquad 2ab \qquad + \qquad b^2.$$

square of first term of binomial + twice the product of the two terms of binomial + square of second term of binomial

Examples

$$(x + 4)^2 = x^2 + 8x + 16;$$
$$(2x + 3y)^2 = 4x^2 + 12xy + 9y^2;$$
$$(5a + 7b)^2 = 25a^2 + 70ab + 49b^2.$$

Pattern $(a - b)^2 = (a - b)(a - b) = a^2 - \quad 2ab \quad + \quad b^2.$

square of
first term $-$ twice the
product of $+$ square of
second term
of binomial   the two terms   of binomial
of binomial

Examples $(x - 8)^2 = x^2 - 16x + 64;$
$(3x - 4y)^2 = 9x^2 - 24xy + 16y^2;$
$(4a - 9b)^2 = 16a^2 - 72ab + 81b^2.$

Pattern $(a + b)(a - b) = a^2 - \quad b^2.$

square of
first term $-$ square of
second term
of binomials   of binomials

Examples $(x + 7)(x - 7) = x^2 - 49;$
$(2x + y)(2x - y) = 4x^2 - y^2;$
$(3a - 2b)(3a + 2b) = 9a^2 - 4b^2.$

Problem Set 3.3

Find the indicated products. Remember the shortcut for multiplying binomials and the other special patterns discussed in this section.

1. $3xy(4x^2y + 5xy^2)$
2. $4y(5xy - 6y^2)$
3. $-2ab^2(3a^2b - 4ab^3)$
4. $-5a^2b(7ab - 8a^2b^2)$
5. $6a^3b^2(5ab - 4a^2b + 3ab^2)$
6. $7ab^4(3a^2b - 2b - 4a^2)$
7. $-xy^4(5x^2y - 4xy^2 + 3x^2y^2)$
8. $-ab(4b^2 - 3a^2b - 2a^2b^2)$
9. $(a + b)(c + d)$
10. $(s - t)(x + y)$
11. $(s + t)(s - t)$
12. $(s - t)(s - t)$
13. $(x + 8)(x + 12)$
14. $(x + 1)(x + 11)$
15. $(y - 4)(y + 9)$
16. $(y - 6)(y + 8)$
17. $(n + 3)(n - 8)$
18. $(n + 2)(n - 7)$
19. $(x - 5)(x + 5)$
20. $(t - 10)(t + 10)$
21. $(x - 9)^2$
22. $(x - 1)^2$
23. $(x - 5)(x - 6)$
24. $(x - 4)(x - 10)$
25. $(x - 1)(x + 2)(x - 3)$
26. $(x + 1)(x - 3)(x + 5)$
27. $(x - 3)(x + 3)(x - 6)$
28. $(x - 1)(x + 1)(x - 9)$
29. $(t + 6)^2$
30. $(t + 9)^2$
31. $(y - 5)^2$
32. $(y - 1)^2$
33. $(3x + 4)(4x + 5)$
34. $(5x + 2)(2x + 7)$
35. $(2y + 1)(2y - 1)$
36. $(4y + 7)(4y - 7)$
37. $(5x - 1)(x + 3)$
38. $(4x - 3)(7x + 9)$
39. $(t + 1)(2t - 5)$
40. $(4t + 1)(3t - 7)$
41. $(3t + 5)^2$
42. $(4t + 3)^2$
43. $(4x - 3)(4x + 3)$
44. $(5n - 1)(5n + 1)$

45. $(6x - 5)^2$ **46.** $(5x - 3)^2$ **47.** $(7x - 4)(3x - 5)$

48. $(6x - 5)(5x - 6)$ **49.** $(3t + 2s)(2t - 3s)$ **50.** $(4t - 5s)(t + 2s)$

51. $(3x - a)(3x + a)$ **52.** $(4x - 3y)(4x + 3y)$ **53.** $(t + 1)(t^2 - 2t - 4)$

54. $(t - 1)(t^2 + t + 1)$ **55.** $(x - 2)(x^2 + 6x - 3)$ **56.** $(x + 4)(2x^2 - 3x - 1)$

57. $(2x - 1)(x^2 + 4x + 3)$ **58.** $(3x + 2)(2x^2 - x - 1)$

59. $(3x - 2)(2x^2 + 3x + 4)$ **60.** $(4x - 3)(5x^2 + 3x - 4)$

61. $(x^2 + 2x + 1)(x^2 + 3x + 4)$ **62.** $(x^2 - x + 4)(x^2 - 7x - 3)$

63. $(2x^2 + x - 2)(x^2 + 4x - 5)$ **64.** $(3x^2 - 2x - 4)(2x^2 + 3x - 7)$

The following two special patterns result from cubing a binomial.

$$(a + b)^3 = a^3 + 3a^2b + 3ab^2 + b^3;$$
$$(a - b)^3 = a^3 - 3a^2b + 3ab^2 - b^3.$$

Use these patterns to cube each of the following.

65. $(x + 1)^3$ **66.** $(x + 2)^3$ **67.** $(x - 2)^3$ **68.** $(x - 3)^3$

69. $(2x + 1)^3$ **70.** $(3x + 2)^3$ **71.** $(3x - 1)^3$ **72.** $(2x - 5)^3$

73. $(4x + 3)^3$ **74.** $(3x - 4)^3$

Miscellaneous Problems

75. We have used the following two multiplication patterns:

$$(a + b)^2 = a^2 + 2ab + b^2;$$
$$(a + b)^3 = a^3 + 3a^2b + 3ab^2 + b^3.$$

By multiplying, these patterns can be extended as follows:

$$(a + b)^4 = a^4 + 4a^3b + 6a^2b^2 + 4ab^3 + b^4;$$
$$(a + b)^5 = a^5 + 5a^4b + 10a^3b^2 + 10a^2b^3 + 5ab^4 + b^5.$$

Based on these results, see if you can determine a pattern that will allow you to complete each of the following without using the long multiplication process.

(a) $(a + b)^6$ (b) $(a + b)^7$

(c) $(a - b)^8$ (Remember that $a - b = a + (-b)$.)

(d) $(2x + y)^4$ (e) $(3x + 2y)^5$

(f) $(x - 2)^6$ (g) $(2x - 1)^7$

For Problems 76–85, find the indicated products. Assume all variables appearing as exponents represent positive integers.

76. $(x^a - 2)(x^a + 2)$ **77.** $(x^{2a} - 3)(x^{2a} + 3)$ **78.** $(x^a + 3)(x^a - 5)$

79. $(x^a + 7)(x^a - 2)$ **80.** $(2x^a - 3)(3x^a + 2)$ **81.** $(3x^a + 5)(4x^a + 1)$

82. $(3x^m - y)(4x^m + y)$ **83.** $(x^m + 3)(x^{2m} - 1)$ **84.** $(2x^m + 5)^2$

85. $(3x^m - 2)^2$

3.4 Factoring: Use of the Distributive Property

Recall that 2 and 3 are said to be *factors* of 6 because the product of 2 and 3 is 6. Likewise, in an indicated product such as $7ab$, the 7, a, and b are called factors of the product. If a positive integer greater than 1 has no factors that are positive integers other than itself and 1, then it is called a **prime number**. Thus, the prime numbers less than 20 are 2, 3, 5, 7, 11, 13, 17, and 19. A positive integer greater than 1 that is not a prime number is called a **composite number**. The composite numbers less than 20 are 4, 6, 8, 9, 10, 12, 14, 15, 16, and 18.

Every composite number can be expressed as the product of prime numbers. Consider the following examples.

$$4 = 2 \cdot 2;\qquad\qquad\qquad 63 = 3 \cdot 3 \cdot 7;$$
$$12 = 2 \cdot 2 \cdot 3;\qquad\qquad 121 = 11 \cdot 11;$$
$$35 = 5 \cdot 7.$$

The indicated product form containing only prime factors is called the **prime factorization form** of a number. Thus, the prime factorization form of 63 is $3 \cdot 3 \cdot 7$. We also say that the number has been **completely factored** when it is expressed in the prime factorization form.

In general, factoring is the reverse of multiplication. Previously, we have used the distributive property to find the product of a monomial and a polynomial as the next examples illustrate.

$$3(x + 2) = 3(x) + 3(2) = 3x + 6;$$
$$5(2x - 1) = 5(2x) - 5(1) = 10x - 5;$$
$$x(x^2 + 6x - 4) = x(x^2) + x(6x) - x(4) = x^3 + 6x^2 - 4x.$$

Now we shall also use the distributive property (in the form "$ab + ac = a(b + c)$") to reverse the process, that is, to factor a given polynomial. Consider the following examples. (The steps indicated in the dashed boxes can be done mentally.)

$$3x + 6 = \boxed{3(x) + 3(2)} = 3(x + 2);$$
$$10x - 5 = \boxed{5(2x) - 5(1)} = 5(2x - 1);$$
$$x^3 + 6x^2 - 4x = \boxed{x(x^2) + x(6x) - x(4)} = x(x^2 + 6x - 4).$$

Note that in each example a given polynomial has been factored into the product of a monomial and a polynomial.

Obviously, polynomials could be factored in a variety of ways. Consider some factorizations of $3x^2 + 12x$.

$$3x^2 + 12x = 3x(x + 4); \quad \text{or} \quad 3x^2 + 12x = 3(x^2 + 4x); \quad \text{or}$$
$$3x^2 + 12x = x(3x + 12); \quad \text{or} \quad 3x^2 + 12x = \frac{1}{2}(6x^2 + 24x).$$

We are, however, primarily interested in the first of the above factorization forms and shall refer to it as the **completely factored form**. A polynomial with integral coefficients is in completely factored form if:

1. It is expressed as a product of polynomials with *integral coefficients*.
2. No polynomial, other than a monomial, within the factored form can be further factored into polynomials with integral coefficients.

Do you see why only the first of the above factored forms of $3x^2 + 12x$ is said to be in completely factored form? In each of the other three forms the polynomial inside the parentheses can be further factored. Furthermore, in the last form, $\frac{1}{2}(6x^2 + 24x)$, the condition of using only integral coefficients is violated.

The factoring process being discussed in this section ($ab + ac = a(b + c)$) is often referred to as **factoring out the highest common monomial factor**.

The key idea in this process is to recognize the monomial factor that is common to all terms. For example, observe that each term of the polynomial $2x^3 + 4x^2 + 6x$ has a factor of $2x$. Thus, we can write

$$2x^3 + 4x^2 + 6x = 2x(\qquad\qquad)$$

and insert within the parentheses the appropriate polynomial factor. The terms of this polynomial factor are determined by dividing each term of the original polynomial by the factor of $2x$. The final completely factored form is

$$2x^3 + 4x^2 + 6x = 2x(x^2 + 2x + 3).$$

The following examples further illustrate this process of factoring out the highest common monomial factor.

$$12x^3 + 16x^2 = 4x^2(3x + 4); \qquad 30x^3 + 42x^4 - 24x^5 = 6x^3(5 + 7x - 4x^2);$$
$$8ab - 18b = 2b(4a - 9); \qquad 8y^3 + 4y^2 = 4y^2(2y + 1);$$
$$6x^2y^3 + 27xy^4 = 3xy^3(2x + 9y).$$

Note that in each example the common monomial factor itself is not expressed in a completely factored form. For example, $4x^2(3x + 4)$ is not written as $2 \cdot 2 \cdot x \cdot x \cdot (3x + 4)$.

Sometimes there may be a common binomial factor rather than a common monomial factor. For example, each of the two terms of the expression $x(y + 2) + z(y + 2)$ has a binomial factor of $(y + 2)$. Thus, we can factor $(y + 2)$ from each term and our result is as follows:

$$x(y + 2) + z(y + 2) = (y + 2)(x + z).$$

Consider a few more examples involving a common binomial factor.

$$a^2(b + 1) + 2(b + 1) = (b + 1)(a^2 + 2);$$
$$x(2y - 1) - y(2y - 1) = (2y - 1)(x - y);$$
$$x(x + 2) + 3(x + 2) = (x + 2)(x + 3).$$

It may be that the original polynomial exhibits no apparent common monomial or binomial factor as is the case with $ab + 3a + bc + 3c$. However, by factoring a from the first two terms and c from the last two terms it can be expressed as

$$ab + 3a + bc + 3c = a(b + 3) + c(b + 3).$$

Now a common binomial factor of $(b + 3)$ is obvious and we can proceed as before.

$$a(b + 3) + c(b + 3) = (b + 3)(a + c).$$

This factoring process is referred to as **factoring by grouping**. Let's consider a few more examples of this type.

$ab^2 - 4b^2 + 3a - 12 = b^2(a - 4) + 3(a - 4)$ Factor b^2 from first two terms and 3 from last two terms.

$\qquad\qquad\qquad = (a - 4)(b^2 + 3).$ Factor common binomial from both terms.

$x^2 - x + 5x - 5 = x(x - 1) + 5(x - 1)$ Factor x from first two terms and 5 from last two terms.

$\qquad\qquad\qquad = (x - 1)(x + 5).$ Factor common binomial from both terms.

$x^2 + 2x - 3x - 6 = x(x + 2) - 3(x + 2)$ Factor x from first two terms and -3 from last two terms.

$\qquad\qquad\qquad = (x + 2)(x - 3).$ Factor common binomial factor from both terms.

It may be necessary to rearrange some terms first before applying the distributive property. Terms containing common factors need to be grouped together and this may be done in more than one way. The next example illustrates this idea.

$$4a^2 - bc^2 - a^2b + 4c^2 = 4a^2 - a^2b + 4c^2 - bc^2$$
$$= a^2(4 - b) + c^2(4 - b)$$
$$= (4 - b)(a^2 + c^2);$$

or $\qquad 4a^2 - bc^2 - a^2b + 4c^2 = 4a^2 + 4c^2 - bc^2 - a^2b$
$$= 4(a^2 + c^2) - b(c^2 + a^2)$$
$$= 4(a^2 + c^2) - b(a^2 + c^2)$$
$$= (a^2 + c^2)(4 - b).$$

Equations and Problem Solving

One reason factoring is an important algebraic skill is that it provides the basis for extending our techniques for solving equations. Each time that we examine a factoring technique we will then use it to help solve certain types of equations.

Before considering some equations for which the *highest common factor* technique is useful, we need another property of equality. Suppose that we are told that the product of two numbers is zero. Do you agree that we can conclude that at least one of the numbers must be zero? Let's state a property that formalizes this idea.

Property 3.5

Let a and b be real numbers.

$$\text{If } ab = 0, \quad \text{then } a = 0 \quad \text{or} \quad b = 0.$$

Property 3.5, along with the *highest common factor* pattern, provides us with another technique for solving equations.

Example 1   Solve $x^2 + 6x = 0$.

Solution

$$x^2 + 6x = 0$$
$$x(x + 6) = 0 \qquad \text{Factor left side.}$$
$$x = 0 \quad \text{or} \quad x + 6 = 0 \qquad \text{Use Property 3.5.}$$
$$x = 0 \quad \text{or} \qquad x = -6.$$

Thus, both 0 and -6 will satisfy the original equation and the solution set is $\{-6, 0\}$. ●

Example 2   Solve $a^2 = 11a$.

Solution

$$a^2 = 11a$$
$$a^2 - 11a = 0 \qquad \text{Add } -11a \text{ to both sides.}$$
$$a(a - 11) = 0 \qquad \text{Factor left side.}$$
$$a = 0 \quad \text{or} \quad a - 11 = 0 \qquad \text{Use Property 3.5.}$$
$$a = 0 \quad \text{or} \qquad a = 11.$$

The solution set is $\{0, 11\}$. ●

Example 3   Solve $3n^2 - 5n = 0$.

Solution

$$3n^2 - 5n = 0$$
$$n(3n - 5) = 0$$
$$n = 0 \quad \text{or} \quad 3n - 5 = 0$$
$$n = 0 \quad \text{or} \qquad 3n = 5$$
$$n = 0 \quad \text{or} \qquad n = \frac{5}{3}.$$

The solution set is $\left\{0, \frac{5}{3}\right\}$. ●

Example 4

Solve $3ax^2 + bx = 0$ for x.

Solution

$$3ax^2 + bx = 0$$
$$x(3ax + b) = 0$$
$$x = 0 \quad \text{or} \quad 3ax + b = 0$$
$$x = 0 \quad \text{or} \quad 3ax = -b$$
$$x = 0 \quad \text{or} \quad x = -\frac{b}{3a}.$$

The solution set is $\left\{ 0, -\dfrac{b}{3a} \right\}$. ●

Many of the problems that we are going to solve in the next few sections have a geometric setting. So let's briefly review a few formulas from geometry that we can put to use in a problem-solving situation.

Rectangle

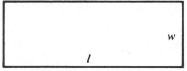

$$A = lw \qquad P = 2l + 2w$$

A area
P perimeter
l length
w width

Square

$$A = s^2 \qquad P = 4s$$

A area
P perimeter
s length of a side

Triangle

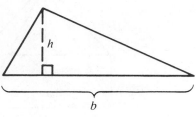

$$A = \frac{1}{2} bh$$

A area
b base
h altitude (height)

Parallelogram

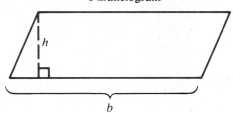

$$A = bh$$

A area
b base
h altitude (height)

Trapezoid

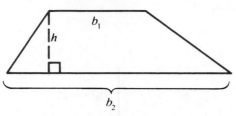

$$A = \frac{1}{2} h (b_1 + b_2)$$

A area
b_1, b_2 bases
h altitude

Circle

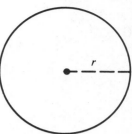

$$A = \pi r^2 \qquad c = 2\pi r$$

A area
c circumference
r radius

Sphere

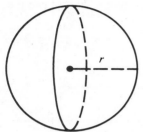

$$S = 4\pi r^2 \qquad V = \frac{4}{3} \pi r^3$$

S surface area
V volume
r radius

Right circular cylinder

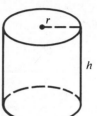
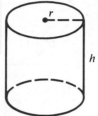

$$V = \pi r^2 h \qquad S = 2\pi r^2 + 2\pi rh$$

V volume
S total surface area
r radius
h altitude (height)

Right circular cone

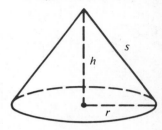

$$V = \frac{1}{3} \pi r^2 h \qquad S = \pi r^2 + \pi rs$$

V volume
S total surface area
r radius
h altitude (height)
s slant height

Prism

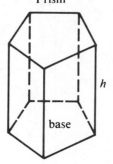

$$V = Bh$$

V volume
B area of base
h altitude (height)

Pyramid

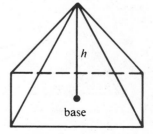

$$V = \frac{1}{3} Bh$$

V volume
B area of base
h altitude (height)

Finally, let's consider a word problem that translates into an equation that can be solved using the techniques of this section.

Problem 1   The area of a square is three times its perimeter. Find the length of a side of the square.

Solution   Let s represent the length of a side of the square.

The area is represented by s^2 and the perimeter by $4s$. Thus,

$$s^2 = 3(4s) \qquad \text{(The area is to be three times the perimeter.)}$$

$$s^2 = 12s$$
$$s^2 - 12s = 0$$
$$s(s - 12) = 0$$
$$s = 0 \quad \text{or} \quad s = 12.$$

Since 0 is not a "reasonable" solution, it must be a 12-by-12 square. (Be sure to check this answer in the *original statement of the problem!*) ●

Problem Set 3.4

1. Classify each of the following numbers as *prime* or *composite*.

(a) 29   (b) 39   (c) 51

(d) 47   (e) 77   (f) 71

(g) 83   (h) 87   (i) 91

2. Factor each of the following composite numbers into the product of prime numbers. For example, $30 = 2 \cdot 3 \cdot 5$.

(a) 28   (b) 39   (c) 44

(d) 49   (e) 56   (f) 64

(g) 72   (h) 84   (i) 87

(j) 91   (k) 100   (l) 144

Factor completely each of the following.

3. $4x + 6y$ **4.** $5x + 15y$ **5.** $6x^2 + 9x$

6. $4x^2 + 12x$ **7.** $5y^2 - 25y$ **8.** $7y^2 - 28y$

9. $14xy - 21x$ **10.** $12xy - 20y$ **11.** $6x^3 + 8x^2$

12. $10x^4 + 35x^2$ **13.** $18a^2b + 27ab^2$ **14.** $24a^3b^2 + 36a^2b$

15. $12x^3y^4 - 39x^4y^3$ **16.** $15x^4y^2 - 45x^5y^4$ **17.** $8x^4 + 12x^3 - 24x^2$

18. $6x^5 - 18x^3 + 24x$ **19.** $5x + 7x^2 + 9x^4$ **20.** $9x^2 - 17x^4 + 21x^5$

21. $15x^2y^3 + 20xy^2 + 35x^3y^4$ **22.** $8x^5y^3 - 6x^4y^5 + 12x^2y^3$

23. $x(z + 3) + y(z + 3)$ **24.** $a(b + c) + 3(b + c)$

25. $2a(x - y) + 3b(x - y)$ **26.** $5(x + y) + a(x + y)$

27. $3x(2a + b) - 2y(2a + b)$ **28.** $4a(2x + 3y) - 5b(2x + 3y)$

Factor each of the following by grouping.

29. $3x + 3y + ax + ay$ **30.** $bx + 2b + cx + 2c$

31. $2ax - ay + 2bx - by$ **32.** $3ax - bx + 3ay - by$

33. $2xy + 2y + 3ax + 3a$ **34.** $2ax + 2a + 3bx + 3b$

35. $x + y + ax + ay$ **36.** $ac + bc + a + b$

37. $ax - ay - bx + by$ **38.** $x^2 - ax - bx + ab$

39. $ax^2 - 2x^2 + 3a - 6$ **40.** $5x + 20 + a^2x + 4a^2$

41. $2bx + yc + xc + 2by$ **42.** $2ac + 3bd + 2bc + 3ad$

43. $2a^2 - 3bc - 2ab + 3ac$ **44.** $ax - by + bx - ay$

45. $x^2 + 4x + 2x + 8$ **46.** $x^2 + x + 5x + 5$

47. $x^2 - 3x - 2x + 6$ **48.** $x^2 - 4x - 5x + 20$

Solve each of the following equations.

49. $x^2 + 3x = 0$ **50.** $x^2 + 8x = 0$ **51.** $x^2 - 20x = 0$

52. $x^2 - x = 0$ **53.** $a^2 = 2a$ **54.** $b^2 = 5b$

55. $3y = y^2$ **56.** $-4x = x^2$ **57.** $4x^2 = 8x$

58. $6x^2 = -24x$ **59.** $7x = -x^2$ **60.** $9a = -a^2$

61. $3x^2 = 7x$ **62.** $5x^2 + 9x = 0$ **63.** $-2x = 7x^2$

64. $-3x = 11x^2$

Solve each of the following for the indicated variable.

65. $ax^2 + bx = 0$ for x **66.** $5bx^2 - 3ax = 0$ for x

67. $3ay^2 = by$ for y **68.** $2by^2 = -3ay$ for y

69. $x^2 + ax + bx + ab = 0$ for x **70.** $y^2 - ay + 2by - 2ab = 0$ for y

Set up an equation and solve each of the following problems.

71. The area of a square is four times its perimeter. Find the length of a side of the square.

72. The square of a number equals five times that number. Find the number.

73. Three times the square of a number equals four times the number. Find the number.

74. The area of a circular region is three times the circumference of the region. Find the length of a radius of the region.

75. Four times the square of a number equals -16 times the number. Find the number.

76. Suppose that the area of a circle is numerically equal to twice the perimeter of a square and that the length of a radius of the circle is equal to the length of a side of the square. Find the length of a side of the square. Express your answer in terms of π.

77. Find the length of a radius of a circle such that the circumference of the circle is numerically equal to the area of the circle.

78. Find the length of a radius of a sphere such that the surface area of the sphere is numerically equal to the volume of the sphere.

79. The area of a square lot is twice the area of an adjoining rectangular plot of ground. If the rectangular plot is 50 feet wide and its length is the same as the length of a side of the square lot, find the dimensions of both the square and the rectangle.

80. The area of a square is one-fourth as large as the area of a triangle. One side of the triangle is 16 inches long and the altitude to that side is the same length as a side of the square. Find the length of a side of the square.

Miscellaneous Problems

For Problems 81–86, factor each expression assuming that all variables appearing as exponents represent positive integers.

81. $2x^{2a} - 3x^a$

82. $6x^{2a} + 8x^a$

83. $y^{3m} + 5y^{2m}$

84. $3y^{5m} - y^{4m} - y^{3m}$

85. $2x^{6a} - 3x^{5a} + 7x^{4a}$

86. $6x^{3a} - 10x^{2a}$

3.5 The Difference of Two Squares

In Section 3.3 we examined some special multiplication patterns. One of these patterns was the following:

$$(a + b)(a - b) = a^2 - b^2.$$

This same pattern, but viewed as a factoring pattern, is referred to as:

> **Difference of Two Squares**
>
> $$a^2 - b^2 = (a + b)(a - b).$$

Applying the pattern is a fairly simple process as these next examples illustrate. Again the steps included in dashed boxes are usually performed mentally.

$$x^2 - 16 = \boxed{(x)^2 - (4)^2} = (x + 4)(x - 4);$$

$$4x^2 - 25 = \boxed{(2x)^2 - (5)^2} = (2x + 5)(2x - 5);$$

$$16x^2 - 9y^2 = \boxed{(4x)^2 - (3y)^2} = (4x + 3y)(4x - 3y);$$

$$1 - a^2 = \boxed{(1)^2 - (a)^2} = (1 + a)(1 - a).$$

Since multiplication is commutative, the order of writing the factors is not important. For example, $(x + 4)(x - 4)$ can also be written as $(x - 4)(x + 4)$.

You must be careful not to assume an analogous factoring pattern for the *sum* of two squares; *it does not exist*. For example, $x^2 + 4 \neq (x + 2)(x + 2)$ because $(x + 2)(x + 2) = x^2 + 4x + 4$. We say that a polynomial such as $x^2 + 4$ is a **prime polynomial** or that it is *not factorable using integers*.

Sometimes the *difference of two squares* pattern can be applied more than once as the next examples illustrate.

$$x^4 - y^4 = (x^2 + y^2)(x^2 - y^2) = (x^2 + y^2)(x + y)(x - y);$$
$$16x^4 - 81y^4 = (4x^2 + 9y^2)(4x^2 - 9y^2) = (4x^2 + 9y^2)(2x + 3y)(2x - 3y).$$

It may also be that the squares are other than simple monomial squares as these next three examples illustrate.

$$(x + 3)^2 - y^2 = ((x + 3) + y)((x + 3) - y) = (x + 3 + y)(x + 3 - y);$$
$$4x^2 - (2y + 1)^2 = (2x + (2y + 1))(2x - (2y + 1))$$
$$= (2x + 2y + 1)(2x - 2y - 1);$$
$$(x - 1)^2 - (x + 4)^2 = ((x - 1) + (x + 4))((x - 1) - (x + 4))$$
$$= (x - 1 + x + 4)(x - 1 - x - 4)$$
$$= (2x + 3)(-5).$$

It is possible that both the technique of *factoring out a common monomial factor* and the pattern of the *difference of two squares* can be applied to the same problem. *In general, it is best to first look for a common monomial factor.* Consider the following examples.

$$2x^2 - 50 = 2(x^2 - 25)$$
$$= 2(x + 5)(x - 5);$$
$$48y^3 - 27y = 3y(16y^2 - 9)$$
$$= 3y(4y + 3)(4y - 3);$$
$$9x^2 - 36 = 9(x^2 - 4)$$
$$= 9(x + 2)(x - 2).$$

Word of Caution   The polynomial $9x^2 - 36$ can be factored as follows:

$$9x^2 - 36 = (3x + 6)(3x - 6)$$
$$= 3(x + 2)(3)(x - 2)$$
$$= 9(x + 2)(x - 2).$$

However, when taking this approach there seems to be a tendency to stop at the step $(3x + 6)(3x - 6)$. Therefore, remember the suggestion to *first look for a common monomial factor*.

The following examples should help you summarize all of our factoring techniques thus far.

$7x^2 + 28 = 7(x^2 + 4);$

$4x^2y - 14xy^2 = 2xy(2x - 7y);$

$x^2 - 4 = (x + 2)(x - 2);$

$18 - 2x^2 = 2(9 - x^2) = 2(3 + x)(3 - x);$

$y^2 + 9$ is not factorable using integers;

$5x + 13y$ is not factorable using integers;

$x^4 - 16 = (x^2 + 4)(x^2 - 4) = (x^2 + 4)(x + 2)(x - 2).$

The Sum and Difference of Two Cubes

As pointed out before, no *sum of squares* pattern exists analogous to the *difference of squares* factoring pattern. That is to say, a polynomial such as $x^2 + 9$ is not factorable using integers. However, there do exist patterns for both the *sum and difference of two cubes*. These patterns are as follows:

Sum and Difference of Two Cubes

$$a^3 + b^3 = (a + b)(a^2 - ab + b^2);$$
$$a^3 - b^3 = (a - b)(a^2 + ab + b^2).$$

Note how these patterns are applied in the next three examples.

$x^3 + 27 = (x)^3 + (3)^3 = (x + 3)(x^2 - 3x + 9);$

$x^3 - 8y^3 = (x)^3 - (2y)^3 = (x - 2y)(x^2 + 2xy + 4y^2);$

$8a^3 + 125b^3 = (2a)^3 + (5b)^3 = (2a + 5b)(4a^2 - 10ab + 25b^2).$

Equations and Problem Solving

Remember that each time we pick up a new factoring technique we also develop more power for solving equations. Let's consider how the *difference of two squares* factoring pattern can be used to help solve certain types of equations.

Example 1   Solve $x^2 = 16$.

Solution

$$x^2 = 16$$
$$x^2 - 16 = 0$$
$$(x + 4)(x - 4) = 0$$
$$x + 4 = 0 \quad \text{or} \quad x - 4 = 0$$
$$x = -4 \quad \text{or} \quad x = 4.$$

The solution set is $\{-4, 4\}$. (Be sure to check these solutions in the original equation!)

Example 2   Solve $9x^2 = 64$.

Solution

$$9x^2 = 64$$
$$9x^2 - 64 = 0$$
$$(3x + 8)(3x - 8) = 0$$
$$3x + 8 = 0 \quad \text{or} \quad 3x - 8 = 0$$
$$3x = -8 \quad \text{or} \quad 3x = 8$$
$$x = -\frac{8}{3} \quad \text{or} \quad x = \frac{8}{3}.$$

The solution set is $\left\{-\frac{8}{3}, \frac{8}{3}\right\}$.

Example 3   Solve $7x^2 - 7 = 0$.

Solution

$$7x^2 - 7 = 0$$
$$7(x^2 - 1) = 0$$
$$x^2 - 1 = 0 \qquad \text{Multiply both sides by } \frac{1}{7}.$$
$$(x + 1)(x - 1) = 0$$
$$x + 1 = 0 \quad \text{or} \quad x - 1 = 0$$
$$x = -1 \quad \text{or} \quad x = 1.$$

The solution set is $\{-1, 1\}$.

In the previous examples we have been using the property "if $ab = 0$, then $a = 0$ or $b = 0$." This property can be extended to any number of factors whose product is zero. Thus, for three factors the property could be stated "if $abc = 0$, then $a = 0$ or $b = 0$ or $c = 0$." The next two examples illustrate this idea.

Example 4   Solve $x^4 - 16 = 0$.

Solution

$$x^4 - 16 = 0$$
$$(x^2 + 4)(x^2 - 4) = 0$$
$$(x^2 + 4)(x + 2)(x - 2) = 0$$
$$x^2 + 4 = 0 \quad \text{or} \quad x + 2 = 0 \quad \text{or} \quad x - 2 = 0$$
$$x^2 = -4 \quad \text{or} \quad x = -2 \quad \text{or} \quad x = 2.$$

The solution set is $\{-2, 2\}$. (Since no real numbers when squared will produce -4, the equation $x^2 = -4$ yields no additional real solutions.)

Example 5

Solve $x^3 - 49x = 0$.

Solution

$$x^3 - 49x = 0$$
$$x(x^2 - 49) = 0$$
$$x(x + 7)(x - 7) = 0$$
$$x = 0 \quad \text{or} \quad x + 7 = 0 \quad \text{or} \quad x - 7 = 0$$
$$x = 0 \quad \text{or} \quad x = -7 \quad \text{or} \quad x = 7.$$

The solution set is $\{-7, 0, 7\}$. ●

The more that we know about solving equations the better off we are for solving word problems.

Problem 1

The combined area of two squares is 40 square centimeters. Each side of one square is three times as long as a side of the other square. Find the dimensions of each of the squares.

Solution

Let s represent the length of a side of the smaller square. Then $3s$ represents the length of a side of the larger square.

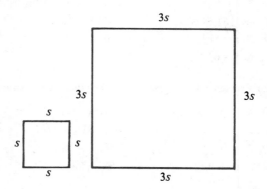

$$s^2 + (3s)^2 = 40$$
$$s^2 + 9s^2 = 40$$
$$10s^2 = 40$$
$$s^2 = 4$$
$$s^2 - 4 = 0$$
$$(s + 2)(s - 2) = 0$$
$$s + 2 = 0 \quad \text{or} \quad s - 2 = 0$$
$$s = -2 \quad \text{or} \quad s = 2.$$

Since s represents the length of a side of a square, the solution -2 has to be disregarded. Thus, the length of a side of the small square is 2 centimeters and the large square has sides of $3(2) = 6$ centimeters. ●

Problem Set 3.5

Use the *difference of squares* pattern to factor each of the following.

1. $x^2 - 4$

2. $x^2 - 49$

3. $9x^2 - 25$

4. $4x^2 - 49$

5. $4x^2 - 25y^2$

6. $36n^2 - 25$

7. $9x^2y^2 - 64$

8. $a^2b^2 - c^2d^4$

9. $x^2 - y^4$

10. $x^6 - y^2$

11. $1 - 81n^2$

12. $9 - 64n^2$

13. $(x + 4)^2 - y^2$

14. $(2x - 1)^2 - y^2$

15. $x^2 - (y - 1)^2$

16. $x^2 - (y + 2)^2$

17. $4a^2 - (3b + 1)^2$

18. $9s^2 - (2t - 1)^2$

19. $(x + 3)^2 - (x + 6)^2$

20. $(x - 2)^2 - (x - 5)^2$

Factor completely. Indicate any that are not factorable using integers. Don't forget to first look for a common monomial factor.

21. $5x^2 - 25$

22. $3x^2 - 27$

23. $2x^2 + 18$

24. $5x^2 + 45$

25. $24y^2 - 54$

26. $18y^2 - 50$

27. $a^3b - 16ab$

28. $x^3y - x^2y$

29. $4x^2 + 9$

30. $a^4 - 1$

31. $x^4 - 81$

32. $9a^2 + 25$

33. $6x^3 + 8x^2$

34. $12x^2y - 18xy^3$

35. $12x^3 - 27xy^2$

36. $4x^3y - 64xy^3$

37. $1 - 16x^4$

38. $6x - 6x^3$

39. $20x - 5x^3$

40. $1 - x^4y^4$

41. $9x^2 - 36y^2$

42. $4x^2 - 64y^2$

For Problems 43–54, use the *sum or difference of cubes* patterns to factor each of the following.

43. $x^3 + 8$

44. $x^3 + 1$

45. $a^3 - 27$

46. $a^3 - 64$

47. $8x^3 + 27y^3$

48. $27x^3 + 64y^3$

49. $1 - 8x^3$

50. $1 - 27a^3$

51. $125x^3 + 27y^3$

52. $x^3y^3 - 1$

53. $x^6 + y^6$

54. $x^6 - y^6$

For Problems 55–68, solve each of the equations.

55. $x^2 - 49 = 0$

56. $x^2 - 81 = 0$

57. $4x^2 - 25 = 0$

58. $9y^2 = 16$

59. $5x^2 - 20 = 0$

60. $4x^2 - 36 = 0$

61. $x^2 = 7x$

62. $x^2 = -9x$

63. $x^3 - 4x = 0$

64. $a^3 - 16a = 0$

65. $y^4 - 1 = 0$

66. $x^4 - 81 = 0$

67. $3x^3 = 75x$

68. $2x^3 = 72x$

Set up an equation and solve each of the following problems.

69. The cube of a number equals the square of the same number. Find the number.

70. The cube of a number equals four times the same number. Find the number.

71. The combined area of two squares is 26 square meters. The sides of the larger square are five times as long as the sides of the smaller square. Find the dimensions of each of the squares.

72. The combined area of two circles is 80π square centimeters. The length of a radius of one circle is twice the length of a radius of the other circle. Find the length of the radius of each circle.

73. The length of a rectangle is one and one-third times as long as its width. The area of the rectangle is 48 square centimeters. Find the length and width of the rectangle.

74. A rectangle is twice as long as it is wide and its area is 50 square meters. Find the length and width of the rectangle.

75. The total surface area of a right circular cone is 72π square feet. If the slant height of the cone is equal in length to a radius of the base, find the length of a radius.

76. The total surface area of a right circular cylinder is 54π square inches. If the altitude of the cylinder is twice the length of a radius, find the altitude of the cylinder.

77. The length of an altitude of a triangle is one-third of the length of the side to which it is drawn. If the area of the triangle is 6 square centimeters, find the length of that altitude.

78. The sum of the areas of a circle and a square is $(16\pi + 64)$ square yards. If a side of the square is twice the length of a radius of the circle, find the length of a side of the square.

3.6 Factoring Trinomials

Expressing a trinomial as the product of two binomials is one of the most common types of factoring used in algebra. Like before, to develop a factoring technique we first look at some multiplication ideas. Let's consider the product $(x + a)(x + b)$ using the distributive property to show how each term of the resulting trinomial is formed.

$$(x + a)(x + b) = x(x + b) + a(x + b)$$
$$= x(x) + x(b) + a(x) + a(b)$$
$$= x^2 + (a + b)x + ab.$$

Notice that the coefficient of the middle term is the *sum* of a and b and the last term is the *product* of a and b. These two relationships can be used to factor trinomials. Let's consider some examples.

Example 1   Factor $x^2 + 8x + 12$.

Solution   We need to complete the following with two integers whose sum is 8 and whose product is 12.

$$x^2 + 8x + 12 = (x + \underline{\quad})(x + \underline{\quad}).$$

The possible pairs of factors of 12 are 1(12), 2(6), and 3(4). Since $6 + 2 = 8$, we can complete the factoring as follows:

$$x^2 + 8x + 12 = (x + 6)(x + 2).$$

To *check* our answer we find the product of $(x + 6)$ and $(x + 2)$.   ●

Example 2   Factor $x^2 - 10x + 24$.

Solution   We need two integers whose product is 24 and whose sum is -10. Let's use a small table to organize our thinking.

Product	Sum
$(-1)(-24) = 24$	$-1 + (-24) = -25$
$(-2)(-12) = 24$	$-2 + (-12) = -14$
$(-3)(-8) = 24$	$-3 + (-8) = -11$
$(-4)(-6) = 24$	$-4 + (-6) = -10$

The bottom line contains the numbers that we need. Thus,

$$x^2 - 10x + 24 = (x - 4)(x - 6).$$

Example 3   Factor $x^2 + 7x - 30$.

Solution   We need two integers whose product is -30 and whose sum is 7.

Product	Sum
$(-1)(30) = -30$	$-1 + 30 = 29$
$1(-30) = -30$	$1 + (-30) = -29$
$2(-15) = -30$	$2 + (-15) = -13$
$-2(15) = -30$	$-2 + 15 = 13$
$-3(10) = -30$	$-3 + 10 = 7$

(No need to search any further.)

The numbers that we need are -3 and 10 and we can complete the factoring.

$$x^2 + 7x - 30 = (x + 10)(x - 3).$$

Example 4   Factor $x^2 + 7x + 16$.

Solution   We need two integers whose product is 16 and whose sum is 7.

Product	Sum
$1(16) = 16$	$1 + 16 = 17$
$2(8) = 16$	$2 + 8 = 10$
$4(4) = 16$	$4 + 4 = 8$

Since we have exhausted all possible pairs of factors of 16 and no two factors have a sum of 7, we conclude that $x^2 + 7x + 16$ is *not factorable using integers*.   ●

The tables in Examples 2, 3, and 4 have been used to illustrate one way of organizing your thoughts for such problems. Normally you will probably do such factoring problems mentally without taking the time to formulate a table. Notice, however, that in Example 4 the table helped us to be absolutely sure that we tried all the possibilities. Whether or not you use the table, keep in mind that the key ideas are the product and sum relationships.

Example 5   Factor $n^2 - n - 72$.

Solution   Notice that the coefficient of the middle term is -1. So, we are looking for two integers whose product is -72, and since their sum is -1, the absolute value of the negative number must be one larger than the positive number. The numbers are -9 and 8, and we can complete the factoring.

$$n^2 - n - 72 = (n - 9)(n + 8)$$ ●

Example 6   Factor $t^2 + 2t - 168$.

Solution   We need two integers whose product is -168 and whose sum is 2. Since the absolute value of the constant term is rather large, it might help to look at it in prime factored form.

$$168 = 2 \cdot 2 \cdot 2 \cdot 3 \cdot 7.$$

Now we can mentally form two numbers by using all of these factors in different combinations. Using two 2's and a 3 in one number, and the other 2 and the 7 in the other number, produces $2 \cdot 2 \cdot 3 = 12$ and $2 \cdot 7 = 14$. Since the coefficient of the middle term of the trinomial is 2, we know that we must use 14 and -12. Thus, we obtain

$$t^2 + 2t - 168 = (t + 14)(t - 12).$$ ●

Trinomials of the Form $ax^2 + bx + c$

We have been factoring trinomials of the form $x^2 + bx + c$. When factoring trinomials for which the coefficient of the squared term does not equal one ($ax^2 + bx + c$), one approach is an extension of the ideas used in the previous examples. To see the basis of this technique, let's look at the following product.

$$(px + r)(qx + s) = px(qx) + px(s) + r(qx) + r(s)$$
$$= (pq)x^2 + (ps + rq)x + rs.$$

Notice that the product of the coefficient of the x^2 term and the constant term is $pqrs$. Likewise, the product of the two coefficients of x, ps and rq, is also $pqrs$. Therefore, when factoring the trinomial $(pq)x^2 + (ps + rq)x + rs$, the two coefficients of x must have a sum of $(ps) + (rq)$ and a product of $pqrs$. Now let's see how this works in some specific examples.

Example 7   Factor $2x^2 + 11x + 5$.

Solution $2x^2 + 11x + 5$ sum of 11

product of $2 \cdot 5 = 10$

We need to find two integers whose product is 10 and whose sum is 11. Obviously, 1 and 10 satisfy these conditions. Thus, the middle term, $11x$, can be expressed as $x + 10x$ and we can proceed as follows.

$$2x^2 + 11x + 5 = 2x^2 + x + 10x + 5$$
$$= x(2x + 1) + 5(2x + 1)$$
$$= (2x + 1)(x + 5).$$

●

Example 8   Factor $10t^2 - 17t + 3$.

Solution $10t^2 - 17t + 3$ sum of -17

product of $10 \cdot 3 = 30$

We need two integers whose product is 30 and whose sum is -17. These conditions are satisfied by -2 and -15, and we can proceed as follows.

$$10t^2 - 17t + 3 = 10t^2 - 2t - 15t + 3$$
$$= 2t(5t - 1) - 3(5t - 1)$$
$$= (5t - 1)(2t - 3).$$

●

Example 9   Factor $4x^2 - 4x - 15$.

Solution $4x^2 - 4x - 15$ sum of -4

product of $4(-15) = -60$

We need two integers whose product is -60 and whose sum is -4. The integers -10 and 6 satisfy these conditions and we can factor as follows.

$$4x^2 - 4x - 15 = 4x^2 - 10x + 6x - 15$$
$$= 2x(2x - 5) + 3(2x - 5)$$
$$= (2x - 5)(2x + 3).$$

●

Example 10   Factor $4x^2 + 6x + 9$.

Solution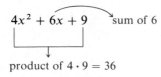

$4x^2 + 6x + 9$ sum of 6

product of $4 \cdot 9 = 36$

We need two integers whose product is 36 and whose sum is 6. The only possible pairs of factors of 36 are $1 \cdot 36$, $2 \cdot 18$, $3 \cdot 12$, $4 \cdot 9$, and $6 \cdot 6$. A sum of 6 cannot be produced by any of these pairs and therefore the trinomial $4x^2 + 6x + 9$ is *not factorable using integers*.   ●

Example 11   Factor $20x^2 + 39x + 18$.

Solution $20x^2 + 39x + 18$ sum of 39

product of $20 \cdot 18 = 360$

We need two integers whose product is 360 and whose sum is 39. To help find these integers, let's prime factor 360.

$360 = 2 \cdot 2 \cdot 2 \cdot 3 \cdot 3 \cdot 5$

Now by grouping these factors in various ways we find that $2 \cdot 2 \cdot 2 \cdot 3 = 24$ and $3 \cdot 5 = 15$, and $24 + 15 = 39$. So the numbers are 15 and 24 and we can factor as follows.

$$20x^2 + 39x + 18 = 20x^2 + 15x + 24x + 18$$
$$= 5x(4x + 3) + 6(4x + 3)$$
$$= (4x + 3)(5x + 6).$$ ●

One final point should be made relative to factoring trinomials. We have been using a rather systematic approach. You may find that as you become more proficient at factoring, a less systematic, trial and error process may be convenient to use at times. For example, suppose that we want to factor $5n^2 + 21n + 4$. By looking at the first term, $5n^2$, and the two positive signs, we know that the binomials are of the form

$$(n + \underline{\quad})(5n + \underline{\quad}).$$

Since the last term, 4, can be factored as $1 \cdot 4$, $4 \cdot 1$, or $2 \cdot 2$, we have only three possibilities to try.

$$(n + 1)(5n + 4) \quad \text{or} \quad (n + 4)(5n + 1) \quad \text{or} \quad (n + 2)(5n + 2)$$

By mentally determining the middle term formed in each product, we find that the second possibility yields the correct middle term of $21n$. Thus, we obtain

$$5n^2 + 21n + 4 = (n + 4)(5n + 1).$$

Summary of Factoring Techniques

Before summarizing our work with factoring techniques let's look at two more special factoring patterns. In Section 3.3 we used the following two patterns to square binomials.

$$(a + b)^2 = a^2 + 2ab + b^2; \qquad (a - b)^2 = a^2 - 2ab + b^2.$$

These patterns can also be used for factoring purposes.

$$a^2 + 2ab + b^2 = (a + b)^2; \qquad a^2 - 2ab + b^2 = (a - b)^2.$$

The trinomials on the left sides are called **perfect square trinomials**; they are the result of squaring a binomial. Perfect square trinomials can always be factored using the usual techniques for factoring trinomials. However, they are easily recognized by the nature of their terms. For example, $4x^2 + 12x + 9$ is a perfect square trinomial because

1. the first term is a perfect square, $\qquad\qquad (2x)^2$
2. the last term is a perfect square, $\qquad\qquad (3)^2$
3. the middle term is *plus* twice the product of the quantities being squared in the first and last terms. $\qquad 2(2x)(3)$

Likewise, $9x^2 - 30x + 25$ is a perfect square trinomial because

1. the first term is a perfect square, $\qquad\qquad (3x)^2$
2. the last term is a perfect square, $\qquad\qquad (5)^2$
3. the middle term is *minus* twice the product of the quantities being squared in the first and last terms. $\qquad -2(3x)(5)$

Once we know that we have a perfect square trinomial, then the factors follow immediately from the two basic patterns. Thus,

$$4x^2 + 12x + 9 = (2x + 3)^2; \qquad 9x^2 - 30x + 25 = (3x - 5)^2.$$

Here are some additional examples of perfect square trinomials and their factored forms.

$$
\begin{aligned}
x^2 + 14x + 49 &= (x)^2 + 2(x)(7) + (7)^2 = (x + 7)^2; \\
n^2 - 16n + 64 &= (n)^2 - 2(n)(8) + (8)^2 = (n - 8)^2; \\
36a^2 + 60ab + 25b^2 &= (6a)^2 + 2(6a)(5b) + (5b)^2 = (6a + 5b)^2; \\
16x^2 - 8xy + y^2 &= (4x)^2 - 2(4x)(y) + (y)^2 = (4x - y)^2.
\end{aligned}
$$

Perhaps you will want to do this step mentally after you feel comfortable with the process.

As indicated previously, factoring is an important algebraic skill. We have learned some basic factoring techniques one at a time, but you must be able to apply them as the situation presents itself. So let's review the techniques and consider a variety of examples illustrating their use.

In this chapter we have discussed factoring

1. by using the distributive property to factor out a common monomial (or binomial) factor
2. by applying the *difference of two squares* pattern
3. by applying the *sum or difference of two cubes* patterns
4. of trinomials into the product of two binomials (the *perfect square trinomial* pattern is a special case of this technique).

As a general guideline, *always look for a common monomial factor first and then proceed with the other techniques.* Study the following examples carefully and be sure that you agree with the indicated factors.

$$2x^2 + 20x + 48 = 2(x^2 + 10x + 24)$$
$$= 2(x + 4)(x + 6).$$
$$3x^3y^3 + 27xy = 3xy(x^2y^2 + 9).$$
$$2x^3 - 16 = 2(x^3 - 8)$$
$$= 2(x - 2)(x^2 + 2x + 4).$$

$$16a^2 - 64 = 16(a^2 - 4)$$
$$= 16(a + 2)(a - 2).$$
$$30n^2 - 31n + 5 = (5n - 1)(6n - 5).$$
$$t^4 + 3t^2 + 2 = (t^2 + 2)(t^2 + 1).$$

$x^2 + 3x - 21$ is not factorable using integers.

Problem Set 3.6

Factor completely each of the following. Indicate any that are not factorable using integers.

1. $x^2 + 10x + 24$
2. $x^2 + 14x + 48$
3. $x^2 - 6x + 5$
4. $x^2 - 9x + 8$
5. $a^2 + 5a - 14$
6. $a^2 + 5a - 24$
7. $y^2 + 19y + 90$
8. $y^2 + 20y + 96$
9. $x^2 - 5x - 14$
10. $x^2 - 4x - 32$
11. $x^2 + 6x + 20$
12. $40 - 6x - x^2$
13. $15 - 2x - x^2$
14. $x^2 + 7x - 36$
15. $x^2 + 5xy + 6y^2$
16. $x^2 - 4xy - 5y^2$
17. $a^2 - ab - 42b^2$
18. $a^2 + ab - 72b^2$
19. $2x^2 + 7x + 6$
20. $4x^2 + 17x + 15$
21. $5x^2 - 7x + 2$
22. $3x^2 - 11x + 10$
23. $6a^2 + 7a - 3$
24. $6a^2 + 7a - 5$
25. $3x^2 + 8x + 2$
26. $3x^2 - 11x - 4$
27. $2x^2 - 7x - 30$
28. $6x^2 + 9x + 5$
29. $20n^2 + 9n - 18$
30. $18n^2 - 27n - 5$
31. $10x^2 - 33x - 7$
32. $12x^2 + x - 1$
33. $30x^2 - x - 1$
34. $35x^2 - 38x + 8$
35. $8y^2 + 22y - 21$
36. $20y^2 + 31y - 9$
37. $12n^2 - n - 35$
38. $24n^2 - 2n - 5$
39. $7n^2 + 31n + 12$
40. $5n^2 + 33n + 18$
41. $x^2 + 21x + 108$
42. $x^2 + 25x + 150$
43. $n^2 - 26n + 168$
44. $n^2 - 36n + 320$
45. $t^2 - 2t - 143$

46. $t^2 + 3t - 180$ **47.** $t^4 + 10t^2 + 24$ **48.** $t^4 - 5t^2 + 6$

49. $3x^4 + 7x^2 - 6$ **50.** $10x^4 + 3x^2 - 4$ **51.** $x^4 - x^2 - 12$

52. $x^4 - 9x^2 + 8$ **53.** $4n^4 + 3n^2 - 27$ **54.** $18n^4 + 25n^2 - 3$

55. $x^4 - 13x^2 + 36$ **56.** $x^4 - 17x^2 + 16$

The following problems should help you pull together all of the factoring techniques of this chapter. Factor completely each of the following, indicating any that are not factorable using integers.

57. $x^2 + 3x - 54$ **58.** $x^2 - 7x - 8$ **59.** $4x^2 + 16$

60. $n^3 - 49n$ **61.** $x^3 - 9x$ **62.** $12n^2 + 59n + 72$

63. $9a^2 - 42a + 49$ **64.** $1 - 16x^4$ **65.** $2n^3 + 6n^2 + 10n$

66. $x^2 - (y - 7)^2$ **67.** $10x^2 + 39x - 27$ **68.** $3x^2 + x - 5$

69. $x^5 - x$ **70.** $6x^2 + 54$ **71.** $2n^2 - n - 5$

72. $36a^2 - 12a + 1$ **73.** $(x + 5)^2 - y^2$ **74.** $n^2 - 17n + 60$

75. $n^2 + 18n + 77$ **76.** $18n^3 + 39n^2 - 15n$ **77.** $8x^2 + 2xy - y^2$

78. $12x^2 + 7xy - 10y^2$ **79.** $14w^2 - 29w - 15$ **80.** $2t^2 - 8$

81. $25t^2 - 100$ **82.** $2n^3 + 14n^2 - 20n$ **83.** $4x^2 - 37x + 40$

84. $25n^2 + 64$ **85.** $10x^3 + 15x^2 + 20x$ **86.** $18w^2 + 9w - 35$

87. $6x^4 - 5x^2 - 21$ **88.** $x^4 - 5x^2 - 36$ **89.** $18x^2 - 12x + 2$

90. $x^4 + 6x^2 + 9$ **91.** $x^3 + 125$ **92.** $3x^4 - 81x$

Miscellaneous Problems

For Problems 93–98, factor each trinomial assuming that all variables appearing as exponents represent positive integers.

93. $x^{2a} + 2x^a - 24$ **94.** $x^{2a} + 10x^a + 21$ **95.** $6x^{2a} - 7x^a + 2$

96. $4x^{2a} + 20x^a + 25$ **97.** $12x^{2n} + 7x^n - 12$ **98.** $20x^{2n} + 21x^n - 5$

Consider the following approach to factoring $(x - 2)^2 + 3(x - 2) - 10$:

$$(x - 2)^2 + 3(x - 2) - 10 = y^2 + 3y - 10 \qquad \text{Replace } x - 2 \text{ with } y.$$
$$= (y + 5)(y - 2) \qquad \text{Factor.}$$
$$= (x - 2 + 5)(x - 2 - 2) \qquad \text{Replace } y \text{ with } x - 2.$$
$$= (x + 3)(x - 4).$$

Use this approach to factor Problems 99–104.

99. $(x - 3)^2 + 10(x - 3) + 24$ **100.** $(x + 1)^2 - 8(x + 1) + 15$

101. $(2x + 1)^2 + 3(2x + 1) - 28$ **102.** $(3x - 2)^2 - 5(3x - 2) - 36$

103. $6(x - 4)^2 + 7(x - 4) - 3$ **104.** $15(x + 2)^2 - 13(x + 2) + 2$

3.7 Equations and Problem Solving

The techniques presented in the previous section for factoring trinomials provide us with more power to solve equations. That is to say, the property "if $ab = 0$, then $a = 0$ or $b = 0$" continues to play an important role as we solve equations involving trinomials that can be factored. Let's consider some examples.

Example 1   Solve $x^2 - 11x - 12 = 0$.

Solution

$$x^2 - 11x - 12 = 0$$
$$(x - 12)(x + 1) = 0$$
$$x - 12 = 0 \quad \text{or} \quad x + 1 = 0$$
$$x = 12 \quad \text{or} \quad x = -1.$$

The solution set is $\{-1, 12\}$. (Check these solutions in the original equation.)   ●

Example 2   Solve $20x^2 + 7x - 3 = 0$.

Solution

$$20x^2 + 7x - 3 = 0$$
$$(4x - 1)(5x + 3) = 0$$
$$4x - 1 = 0 \quad \text{or} \quad 5x + 3 = 0$$
$$4x = 1 \quad \text{or} \quad 5x = -3$$
$$x = \frac{1}{4} \quad \text{or} \quad x = -\frac{3}{5}.$$

The solution set is $\left\{-\dfrac{3}{5}, \dfrac{1}{4}\right\}$.   ●

Example 3   Solve $-2n^2 - 10n + 12 = 0$.

Solution

$$-2n^2 - 10n + 12 = 0$$
$$-2(n^2 + 5n - 6) = 0$$
$$n^2 + 5n - 6 = 0 \qquad \text{Multiply both sides by } -\frac{1}{2}.$$
$$(n + 6)(n - 1) = 0$$
$$n + 6 = 0 \quad \text{or} \quad n - 1 = 0$$
$$n = -6 \quad \text{or} \quad n = 1.$$

The solution set is $\{-6, 1\}$.   ●

Example 4   Solve $16x^2 - 56x + 49 = 0$.

Solution

$$16x^2 - 56x + 49 = 0$$
$$(4x - 7)^2 = 0$$
$$(4x - 7)(4x - 7) = 0$$
$$4x - 7 = 0 \quad \text{or} \quad 4x - 7 = 0$$
$$4x = 7 \quad \text{or} \quad 4x = 7$$
$$x = \frac{7}{4} \quad \text{or} \quad x = \frac{7}{4}.$$

The only solution is $\frac{7}{4}$; thus, the solution set is $\left\{\frac{7}{4}\right\}$.   ●

Sometimes the original form of an equation needs to be changed before a factoring technique becomes apparent. The next two examples illustrate situations of this type.

Example 5   Solve $9a(a + 1) = 4$.

Solution

$$9a(a + 1) = 4$$
$$9a^2 + 9a = 4$$
$$9a^2 + 9a - 4 = 0$$
$$(3a + 4)(3a - 1) = 0$$
$$3a + 4 = 0 \quad \text{or} \quad 3a - 1 = 0$$
$$3a = -4 \quad \text{or} \quad 3a = 1$$
$$a = -\frac{4}{3} \quad \text{or} \quad a = \frac{1}{3}.$$

The solution set is $\left\{-\frac{4}{3}, \frac{1}{3}\right\}$. (Perhaps you should check these solutions in the original equation.)   ●

Example 6   Solve $(x - 1)(x + 9) = 11$.

Solution

$$(x - 1)(x + 9) = 11$$
$$x^2 + 8x - 9 = 11$$
$$x^2 + 8x - 20 = 0$$
$$(x + 10)(x - 2) = 0$$
$$x + 10 = 0 \quad \text{or} \quad x - 2 = 0$$
$$x = -10 \quad \text{or} \quad x = 2.$$

The solution set is $\{-10, 2\}$.   ●

Problem Solving

As we might expect, increasing our power to solve equations broadens our base for solving problems. Now we are ready to tackle some problems that can be solved by using equations of the types presented in the previous examples of this section.

Problem 1   A room contains 78 chairs. The number of chairs per row is 1 more than twice the number of rows. Find the number of rows and the number of chairs per row.

Solution   Let r represent the number of rows. Then $2r + 1$ represents the number of chairs per row.

$$r(2r + 1) = 78 \qquad \text{The number of rows times the number of chairs}$$
$$\text{per row yields the total number of chairs.}$$

$$2r^2 + r = 78$$
$$2r^2 + r - 78 = 0$$
$$(2r + 13)(r - 6) = 0$$
$$2r + 13 = 0 \quad \text{or} \quad r - 6 = 0$$
$$2r = -13 \quad \text{or} \quad r = 6$$
$$r = -\frac{13}{2} \quad \text{or} \quad r = 6.$$

The solution $-\dfrac{13}{2}$ must be disregarded, so there are 6 rows and $2r + 1$ or $2(6) + 1 = 13$ chairs per row.   ●

Problem 2   A strip of uniform width is to be cut off of both sides and both ends of a sheet of paper that is 8 inches by 11 inches in order to reduce the size of the paper to an area of 40 square inches. Find the width of the strip.

Solution   Let x represent the width of the strip, as indicated in the following diagram.

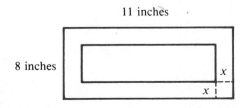

The length of the paper after the strips of width x are cut from both ends will be $11 - 2x$

and the width of the newly formed rectangle will be $8 - 2x$. Since the area $(A = lw)$ is to be 40 square inches, we can set up and solve the following equation.

$$(11 - 2x)(8 - 2x) = 40$$
$$88 - 38x + 4x^2 = 40$$
$$4x^2 - 38x + 48 = 0$$
$$2x^2 - 19x + 24 = 0$$
$$(2x - 3)(x - 8) = 0$$
$$2x - 3 = 0 \quad \text{or} \quad x - 8 = 0$$
$$2x = 3 \quad \text{or} \quad x = 8.$$
$$x = \frac{3}{2} \quad \text{or} \quad x = 8.$$

The solution of 8 must be discarded since the width of the original sheet is only 8 inches. Therefore, the strip to be cut off of all four sides must be $1\frac{1}{2}$ inches wide. (Check this answer!)   ●

The Pythagorean Theorem, an important theorem pertaining to right triangles, can sometimes serve as a guideline for solving problems dealing with right triangles. The Pythagorean Theorem states that *in any right triangle, the square of the longest side* (called the hypotenuse) *is equal to the sum of the squares of the other two sides* (called legs). Let's use this relationship to help solve a problem.

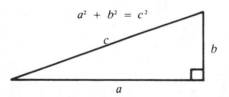

Problem 3   One leg of a right triangle is 2 centimeters more than twice the length of the other leg. The hypotenuse is 1 centimeter longer than the longer of the two legs. Find the lengths of the three sides of the right triangle.

Solution   Let l represent the length of the shortest leg. Then $2l + 2$ represents the length of the other leg and $2l + 3$ represents the length of the hypotenuse. Using the Pythagorean Theorem as a guideline, we can set up and solve the following equation.

$$l^2 + (2l + 2)^2 = (2l + 3)^2$$
$$l^2 + 4l^2 + 8l + 4 = 4l^2 + 12l + 9$$
$$l^2 - 4l - 5 = 0$$
$$(l - 5)(l + 1) = 0$$
$$l - 5 = 0 \quad \text{or} \quad l + 1 = 0$$
$$l = 5 \quad \text{or} \quad l = -1.$$

The negative solution must be discarded, so the length of one leg is 5 centimeters; the other leg is $2(5) + 2 = 12$ centimeters long and the hypotenuse is $2(5) + 3 = 13$ centimeters long.   ●

Problem Set 3.7

Solve each of the following equations. You will need to use factoring techniques discussed throughout this chapter.

1. $x^2 + 3x + 2 = 0$

2. $x^2 + 5x + 6 = 0$

3. $n^2 + 11n + 28 = 0$

4. $n^2 + 11n + 30 = 0$

5. $n^2 - 10n + 16 = 0$

6. $n^2 - 11n + 24 = 0$

7. $x^2 + 4x - 21 = 0$

8. $x^2 + 5x - 36 = 0$

9. $w^2 - 5w - 50 = 0$

10. $s^2 - 6s - 72 = 0$

11. $x^2 + 20x + 96 = 0$

12. $x^2 + 20x + 75 = 0$

13. $2t^2 + 19t + 35 = 0$

14. $4t^2 + 19t + 12 = 0$

15. $6x^2 + 17x - 14 = 0$

16. $6x^2 - 13x - 28 = 0$

17. $8n^2 - 6n - 5 = 0$

18. $35n^2 - 18n - 8 = 0$

19. $24n^2 - 38n + 15 = 0$

20. $28n^2 - 47n + 15 = 0$

21. $7x^2 + 62x - 9 = 0$

22. $8n^2 - 47n - 6 = 0$

23. $20x^2 + 41x + 20 = 0$

24. $15x^2 + 34x + 15 = 0$

25. $x^2 + x - 2 = 0$

26. $3x^2 = 75$

27. $2x^3 = 50x$

28. $n^2 + 7n - 44 = 0$

29. $16t^2 - 72t + 81 = 0$

30. $16 - x^4 = 0$

31. $3w^3 - 24w^2 + 36w = 0$

32. $(x - 5)(x + 3) = 9$

33. $a(a - 1) = 2$

34. $2n^3 = 72n$

35. $(x + 1)^2 - 4 = 0$

36. $-6n^2 + 13n - 2 = 0$

37. $1 - x^4 = 0$

38. $3t(t - 4) = 0$

39. $-18n^2 - 15n + 7 = 0$

40. $4a(a + 1) = 3$

41. $(x + 8)(x - 6) = -24$

42. $(3x - 1)^2 - 16 = 0$

43. $5x(3x - 2) = 0$

44. $12x^3 + 46x^2 + 40x = 0$

45. $x^4 + 5x^2 - 36 = 0$

46. $2x^4 + x^2 - 3 = 0$

47. $x^4 - 9x^2 = 0$

48. $3x^4 - 46x^2 - 32 = 0$

49. $4x^4 - 13x^2 + 9 = 0$

50. $9x^4 - 37x^2 + 4 = 0$

51. $n(n + 1) = 182$

52. $n(n + 2) = 360$

53. $5x^2 = 43x - 24$

54. $-3x^2 - 19x + 14 = 0$

Set up an appropriate equation and solve each of the following problems.

55. Find two consecutive integers whose product is 56.

56. Find two consecutive even whole numbers whose product is 48.

57. Find two integers whose product is 91 such that one of the integers is one less than twice the other integer.

58. Find two integers whose product is 150 such that one of the integers is one more than four times the other integer.

59. The length of a rectangle is one centimeter more than four times its width. If the area of the rectangle is 39 square centimeters, find its length and width.

60. The perimeter of a rectangle is 30 meters and the area is 56 square meters. Find the length and width of the rectangle.

61. An apple orchard contains 90 trees. The number of trees in each row is 3 more than twice the number of rows. Find the number of rows and the number of trees per row.

62. The lengths of the three sides of a right triangle are represented by consecutive whole numbers. Find the lengths of the three sides.

63. The area of the floor of a rectangular room is 175 square feet. The length of the room is $1\frac{1}{2}$ feet longer than the width. Find the length of the room.

64. The length of one leg of a right triangle is 3 inches more than the length of the other leg. If the length of the hypotenuse is 15 centimeters, find the lengths of the two legs.

65. The lengths of the three sides of a right triangle are represented by consecutive even whole numbers. Find the lengths of the three sides.

66. The area of a triangular sheet of paper is 28 square inches. One side of the triangle is two inches more than three times the length of the altitude to that side. Find the length of that side and the altitude to the side.

67. A strip of uniform width is shaded along both sides and both ends of a rectangular poster that is 12 inches by 16 inches. How wide is the shaded strip if one-half of the poster is shaded?

68. The sum of the squares of two consecutive integers is 85. Find the integers.

69. The sum of the areas of two circles is 65π square feet. The length of a radius of the larger circle is one foot less than twice the length of a radius of the smaller circle. Find the length of a radius of each circle.

70. The combined area of a square and a rectangle is 64 square centimeters. The width of the rectangle is 2 centimeters more than the length of a side of the square and the length of the rectangle is 2 centimeters more than its width. Find the dimensions of the square and the rectangle.

Chapter 3 Summary

(3.1) A **term** is an indicated product and may contain any number of factors. The variables involved in a term are called **literal factors** and the numerical factor is called the **numerical coefficient**. Terms containing variables with only non-negative integers as exponents are called **monomials**. The **degree** of a monomial is the sum of the exponents of the literal factors.

A **polynomial** is a monomial or a finite sum (or difference) of monomials. We classify polynomials as follows:

polynomial with one term $\longrightarrow$ monomial

polynomial with two terms $\longrightarrow$ binomial

polynomial with three terms $\longrightarrow$ trinomial

Similar or **like terms** have the same literal factors. The commutative, associative, and distributive properties provide the basis for rearranging, regrouping, and combining similar terms.

(3.2)   The following properties provide the basis for multiplying and dividing monomials.

1. $b^n \cdot b^m = b^{n+m}$

2. $(b^n)^m = b^{mn}$

3. $(ab)^n = a^n b^n$

4. (a) $\dfrac{b^n}{b^m} = b^{n-m}$ if $n > m$

   (b) $\dfrac{b^n}{b^m} = 1$ if $n = m$

(3.3)   The commutative and associative properties, the properties of exponents, and the distributive property work together to form a basis for multiplying polynomials.

The following can be used as multiplication and factoring patterns.

$$(a + b)^2 = a^2 + 2ab + b^2;$$
$$(a - b)^2 = a^2 - 2ab + b^2;$$
$$(a + b)(a - b) = a^2 - b^2.$$

(3.4)   If a positive integer greater than 1 has no factors that are positive integers other than itself and 1, then it is called a **prime number**. A positive integer greater than 1 that is not a prime number is called a **composite number**.

The indicated product form containing only prime factors is called the **prime factorization form** of a number.

An expression such as $ax + bx + ay + by$ can be factored as follows:

$$ax + bx + ay + by = x(a + b) + y(a + b)$$
$$= (a + b)(x + y)$$

This is called *factoring by grouping*.

The distributive property in the form "$ab + ac = a(b + c)$" forms the basis for **factoring out the highest common monomial factor**.

Expressing polynomials in factored form and then applying the property "if $ab = 0$, then $a = 0$ or $b = 0$," provides us with another technique for solving equations.

(3.5)   The factoring pattern

$$a^2 - b^2 = (a + b)(a - b)$$

is called the **difference of two squares**.

The *difference of two squares* factoring pattern, along with the property "if $ab = 0$, then $a = 0$ or $b = 0$," provides us with another technique for solving equations.

The factoring patterns

$$a^3 + b^3 = (a + b)(a^2 - ab + b^2), \text{ and}$$
$$a^3 - b^3 = (a - b)(a^2 + ab + b^2)$$

are called the **sum and difference of two cubes**.

(3.6)   Expressing a trinomial, for which the coefficient of the squared term is one, as a product of two binomials is based on the following relationship.

$$(x + a)(x + b) = x^2 + (a + b)x + ab.$$

The coefficient of the middle term is the *sum* of a and b, and the last term is the *product* of a and b.

If the coefficient of the squared term of a trinomial does not equal one, then the following relationship holds.

$$(px + r)(qx + s) = (pq)x^2 + (ps + rq)x + rs.$$

The two coefficients of x, ps and rq, must have a sum of $(ps) + (rq)$ and a product of $pqrs$. Thus, to factor something like $6x^2 + 7x - 3$, we need to find two integers whose product is $6(-3) = -18$ and whose sum is 7. The integers are 9 and -2 and we can factor as follows.

$$\begin{aligned} 6x^2 + 7x - 3 &= 6x^2 + 9x - 2x - 3 \\ &= 3x(2x + 3) - 1(2x + 3) \\ &= (2x + 3)(3x - 1). \end{aligned}$$

A *perfect square trinomial* is the result of squaring a binomial. There are two basic perfect square trinomial factoring patterns as follows:

$$a^2 + 2ab + b^2 = (a + b)^2$$
$$a^2 - 2ab + b^2 = (a - b)^2$$

(3.7)   The factoring techniques discussed in this chapter along with the property "if $ab = 0$, then $a = 0$ or $b = 0$" provide the basis for expanding our equation-solving processes.

Being able to solve more types of equations increases our capabilities for problem solving.

Chapter 3 Review Problem Set

Perform the indicated operations and simplify each of the following.

1. $(3x - 2) + (4x - 6) + (-2x + 5)$ 2. $(8x^2 + 9x - 3) - (5x^2 - 3x - 1)$

3. $(6x^2 - 2x - 1) + (4x^2 + 2x + 5) - (-2x^2 + x - 1)$

4. $(-5x^2y^3)(4x^3y^4)$ 5. $(-2a^2)(3ab^2)(a^2b^3)$

6. $5a^2(3a^2 - 2a - 1)$ 7. $(4x - 3y)(6x + 5y)$

8. $(x + 4)(3x^2 - 5x - 1)$ 9. $(4x^2y^3)^4$

10. $(3x - 2y)^2$ 11. $(-2x^2y^3z)^3$

12. $\dfrac{-39x^3y^4}{3xy^3}$ 13. $[3x - (2x - 3y + 1)] - [2y - (x - 1)]$

14. $(x^2 - 2x - 5)(x^2 + 3x - 7)$ 15. $(7 - 3x)(3 + 5x)$

16. $-(3ab)(2a^2b^3)^2$ 17. $\left(\dfrac{1}{2}ab\right)(8a^3b^2)(-2a^3)$

18. $(7x - 9)(x + 4)$ 19. $(3x + 2)(2x^2 - 5x + 1)$

20. $(3x^{n+1})(2x^{3n-1})$

Factor completely each of the following. Indicate any that are not factorable using integers.

21. $x^2 + 3x - 28$ 22. $2t^2 - 18$

23. $4n^2 + 9$ 24. $12n^2 - 7n + 1$

25. $x^6 - x^2$ 26. $x^3 - 6x^2 - 72x$

27. $6a^3b + 4a^2b^2 - 2a^2bc$ 28. $x^2 - (y - 1)^2$

29. $8x^2 + 12$ 30. $12x^2 + x - 35$

31. $16n^2 - 40n + 25$ 32. $4n^2 - 8n$

33. $3w^3 + 18w^2 - 24w$ 34. $20x^2 + 3xy - 2y^2$

35. $16a^2 - 64a$ 36. $3x^3 - 15x^2 - 18x$

37. $n^2 - 8n - 128$ 38. $t^4 - 22t^2 - 75$

39. $35x^2 - 11x - 6$ 40. $15 - 14x + 3x^2$

41. $64n^3 - 27$ 42. $16x^3 + 250$

Solve each of the following equations.

43. $4x^2 - 36 = 0$ 44. $x^2 + 5x - 6 = 0$

45. $49n^2 - 28n + 4 = 0$ 46. $(3x - 1)(5x + 2) = 0$

47. $(3x - 4)^2 - 25 = 0$ 48. $6a^3 = 54a$

49. $x^5 = x$ 50. $-n^2 + 2n + 63 = 0$

51. $7n(7n + 2) = 8$ 52. $30w^2 - w - 20 = 0$

53. $5x^4 - 19x^2 - 4 = 0$ 54. $9n^2 - 30n + 25 = 0$

55. $n(2n + 4) = 96$ 56. $7x^2 + 33x - 10 = 0$

57. $(x + 1)(x + 2) = 42$ 58. $x^2 + 12x - x - 12 = 0$

59. $2x^4 + 9x^2 + 4 = 0$ 60. $30 - 19x - 5x^2 = 0$

61. $3t^3 - 27t^2 + 24t = 0$ 62. $-4n^2 - 39n + 10 = 0$

Set up an equation and solve each of the following problems.

63. Find three consecutive integers such that the product of the smallest and largest is 1 less than 9 times the middle integer.

64. Find two integers whose sum is 2 and whose product is −48.

65. Find two consecutive odd whole numbers whose product is 195.

66. Two cars leave an intersection at the same time, one traveling north and the other traveling east. Some time later, they are 20 miles apart and the car going east had traveled 4 miles further than the other car. How far had each car traveled?

67. The perimeter of a rectangle is 32 meters and its area is 48 square meters. Find the length and width of the rectangle.

68. A room contains 144 chairs. The number of chairs per row is 2 less than twice the number of rows. Find the number of rows and the number of chairs per row.

69. The area of a triangle is 39 square feet. The length of one side is 1 foot more than twice the altitude to that side. Find the length of that side and the altitude to the side.

70. A rectangular-shaped pool 20 feet by 30 feet has a sidewalk of uniform width around the pool. The area of the sidewalk is 336 square feet. Find the width of the sidewalk.

71. The sum of the areas of two squares is 89 square centimeters. The length of a side of the larger square is 3 centimeters more than the length of a side of the smaller square. Find the dimensions of each square.

72. The total surface area of a right circular cylinder is 32π square inches. If the altitude of the cylinder is three times the length of a radius, find the altitude of the cylinder.

4 Rational Expressions

Rational expressions are to algebra what rational numbers are to arithmetic. Most of the work that we will do with rational expressions in this chapter parallels the work you have previously done with arithmetic fractions. The same basic properties used to explain reducing, adding, subtracting, multiplying, and dividing arithmetic fractions will serve as a basis for our work with rational expressions. The techniques of factoring studied in the previous chapter will also play an important role in this chapter. We will conclude the chapter by working with some fractional equations containing rational expressions.

4.1 Simplifying Rational Expressions

Any number that can be written in the form $\frac{a}{b}$, where a and b are integers and b is not zero, is called a **rational** number. The following are examples of rational numbers.

$$\frac{1}{2}, \qquad \frac{3}{4}, \qquad \frac{15}{7}, \qquad \frac{-5}{6}, \qquad \frac{7}{-8}, \qquad \frac{-12}{-17}.$$

Numbers such as 6, -4, 0, $4\frac{1}{2}$, 0.7 and 0.21 are also rational because they can be expressed as the indicated quotient of two integers. For example,

$$6 = \frac{6}{1} = \frac{12}{2} = \frac{18}{3} \qquad \text{and so on;} \qquad 4\frac{1}{2} = \frac{9}{2};$$

$$-4 = \frac{4}{-1} = \frac{-4}{1} = \frac{8}{-2} \qquad \text{and so on;} \qquad 0.7 = \frac{7}{10};$$

$$0 = \frac{0}{1} = \frac{0}{2} = \frac{0}{3} \qquad \text{and so on;} \qquad 0.21 = \frac{21}{100}.$$

Our work with division of integers helps with the next examples.

$$\frac{8}{-2} = \frac{-8}{2} = -\frac{8}{2} = -4; \qquad\qquad \frac{12}{3} = \frac{-12}{-3} = 4.$$

The following general properties should be observed.

Property 4.1	**1.** $\dfrac{-a}{b} = \dfrac{a}{-b} = -\dfrac{a}{b}$
	2. $\dfrac{-a}{-b} = \dfrac{a}{b}$

Therefore, a rational number such as $\dfrac{-2}{5}$ can also be written as $\dfrac{2}{-5}$ or $-\dfrac{2}{5}$.

The following property, often referred to as the **fundamental principle of fractions**, is used to reduce fractions or express fractions in simplest or reduced form.

Property 4.2	If b and k are nonzero integers and a is any integer, then $$\frac{a \cdot k}{b \cdot k} = \frac{a}{b}.$$

Let's consider some examples that illustrate various forms that are used when applying Properties 4.1 and 4.2.

Example 1   Reduce $\dfrac{18}{24}$ to lowest terms.

Solution $\dfrac{18}{24} = \dfrac{3 \cdot 6}{4 \cdot 6} = \dfrac{3}{4}.$ ●

Example 2   Change $\dfrac{40}{48}$ to simplest form.

Solution $\dfrac{\overset{5}{\cancel{40}}}{\underset{6}{\cancel{48}}} = \dfrac{5}{6}.$ (A common factor of 8 has been divided out of both numerator and denominator.)   ●

Example 3   Express $\dfrac{-36}{63}$ in reduced form.

Solution $\dfrac{-36}{63} = -\dfrac{36}{63} = -\dfrac{4 \cdot 9}{7 \cdot 9} = -\dfrac{4}{7}.$

Example 4   Reduce $\dfrac{72}{-90}.$

Solution $\dfrac{72}{-90} = -\dfrac{72}{90} = -\dfrac{2 \cdot 2 \cdot 2 \cdot 3 \cdot 3}{2 \cdot 3 \cdot 3 \cdot 5} = -\dfrac{4}{5}.$

In Example 4, notice the use of the prime factored forms to help recognize the common factors of the numerator and denominator.

Rational Expressions

A **rational expression** is defined as the indicated quotient of two polynomials. The following are examples of rational expressions.

$$\frac{3x^2}{5}, \qquad \frac{x-2}{x+3}, \qquad \frac{x^2+5x-1}{x^2-9}, \qquad \frac{xy^2+x^2y}{xy}, \qquad \frac{a^3-3a^2-5a-1}{a^4+a^3+6}.$$

Because division by zero must be avoided, no values that create a denominator of zero can be assigned to variables. Thus, the rational expression $\dfrac{x-2}{x+3}$ is meaningful for all values of x except for $x = -3$. Rather than making restrictions for each individual expression, we will merely assume that all denominators represent nonzero real numbers.

Property 4.2 $\left(\dfrac{a \cdot k}{b \cdot k} = \dfrac{a}{b}\right)$ also serves as the basis for simplifying rational expressions, as the next examples illustrate.

Example 5   Simplify $\dfrac{15xy}{25y}.$

Solution $\dfrac{15xy}{25y} = \dfrac{3 \cdot 5 \cdot x \cdot y}{5 \cdot 5 \cdot y} = \dfrac{3x}{5}.$

Example 6   Simplify $\dfrac{-9}{18x^2y}.$

Solution $\dfrac{-9}{18x^2y} = -\dfrac{\overset{1}{9}}{\underset{2}{18}x^2y} = -\dfrac{1}{2x^2y}.$ (A common factor of 9 was divided out of numerator and denominator.)

Example 7   Simplify $\dfrac{-28a^2b^2}{-63a^2b^3}$.

Solution $\dfrac{-28a^2b^2}{-63a^2b^3} = \dfrac{4 \cdot \not7 \cdot \not a^2 \cdot \not b^2}{9 \cdot \not7 \cdot \not a^2 \cdot \underset{b}{\not b^3}} = \dfrac{4}{9b}$.

The factoring techniques from Chapter 3 can be used to factor numerators and/or denominators so that the property $\dfrac{a \cdot k}{b \cdot k} = \dfrac{a}{b}$ can be applied. Several examples should clarify this process.

Example 8   Simplify $\dfrac{x^2 + 4x}{x^2 - 16}$.

Solution $\dfrac{x^2 + 4x}{x^2 - 16} = \dfrac{x(\not{x + 4})}{(x - 4)(\not{x + 4})} = \dfrac{x}{x - 4}$.

Example 9   Simplify $\dfrac{4a^2 + 12a + 9}{2a + 3}$.

Solution $\dfrac{4a^2 + 12a + 9}{2a + 3} = \dfrac{(\not{2a + 3})(2a + 3)}{1(\not{2a + 3})} = \dfrac{2a + 3}{1} = 2a + 3$.

Example 10   Simplify $\dfrac{5n^2 + 6n - 8}{10n^2 - 3n - 4}$.

Solution $\dfrac{5n^2 + 6n - 8}{10n^2 - 3n - 4} = \dfrac{(\not{5n - 4})(n + 2)}{(\not{5n - 4})(2n + 1)} = \dfrac{n + 2}{2n + 1}$.

Example 11   Simplify $\dfrac{6x^3y - 6xy}{x^3 + 5x^2 + 4x}$.

Solution $\dfrac{6x^3y - 6xy}{x^3 + 5x^2 + 4x} = \dfrac{6xy(x^2 - 1)}{x(x^2 + 5x + 4)} = \dfrac{6xy(\not{x + 1})(x - 1)}{x(\not{x + 1})(x + 4)} = \dfrac{6y(x - 1)}{x + 4}$.

Note that in Example 11 we left the numerator of the final fraction in factored form. This is often done if expressions other than monomials are involved. Either $\dfrac{6y(x - 1)}{x + 4}$ or $\dfrac{6xy - 6y}{x + 4}$ is an acceptable answer.

Remember that the quotient of any nonzero real number and its opposite is -1. For example, $\dfrac{6}{-6} = -1$ and $\dfrac{-8}{8} = -1$. Likewise, the indicated quotient of any polynomial and its opposite is equal to -1. For example,

$$\frac{a}{-a} = -1 \qquad \text{because } a \text{ and } -a \text{ are opposites;}$$

$$\frac{a - b}{b - a} = -1 \qquad \text{because } a - b \text{ and } b - a \text{ are opposites;}$$

$$\frac{x^2 - 4}{4 - x^2} = -1 \quad \text{because } x^2 - 4 \text{ and } 4 - x^2 \text{ are opposites.}$$

The final example of this section illustrates the use of this idea when simplifying rational expressions.

Example 12 Simplify $\dfrac{6a^2 - 7a + 2}{10a - 15a^2}$.

Solution $\dfrac{6a^2 - 7a + 2}{10a - 15a^2} = \dfrac{(2a - 1)(3a - 2)}{5a\,(2 - 3a)} \qquad \left(\dfrac{3a - 2}{2 - 3a} = -1\right)$

$$= (-1)\left(\frac{2a - 1}{5a}\right)$$

$$= -\frac{2a - 1}{5a}. \qquad\qquad •$$

Problem Set 4.1

Express each of the following rational numbers in reduced form.

1. $\dfrac{12}{18}$ 2. $\dfrac{27}{54}$ 3. $\dfrac{48}{72}$ 4. $\dfrac{-24}{56}$

5. $\dfrac{32}{-84}$ 6. $\dfrac{39}{-65}$ 7. $\dfrac{-28}{-77}$ 8. $\dfrac{-80}{-96}$

Simplify each of the following.

9. $\dfrac{8x}{18xy}$ 10. $\dfrac{14x^2y}{21xy}$ 11. $\dfrac{9a^2b^2}{39ab^3}$ 12. $\dfrac{15a^3b^2}{45a^2b}$

13. $\dfrac{-18x^3y}{54x^4}$ 14. $\dfrac{-26xy^2}{65y}$ 15. $\dfrac{66ab^2c}{-72ab^3}$ 16. $\dfrac{44x^2yz^3}{-54x^3y^2z}$

17. $\dfrac{-34x^2y^2}{-51xy^5}$ 18. $\dfrac{-63xy^4}{-81x^2y}$ 19. $\dfrac{x^2 - 2x}{x^2 - 4}$ 20. $\dfrac{x^2 - y^2}{x^2 + xy}$

21. $\dfrac{10x + 15}{30x + 10}$ 22. $\dfrac{6x - 18}{8x + 32}$ 23. $\dfrac{a^2 + 7a + 12}{a^2 - 6a - 27}$

24. $\dfrac{a^2 + 7a - 8}{2a^2 + a - 3}$

25. $\dfrac{n^2 + 3n - 40}{3n^2 + 23n - 8}$

26. $\dfrac{n^2 - 12n + 35}{n^2 - 3n - 10}$

27. $\dfrac{x^2}{x^2 + 4}$

28. $\dfrac{6x^2 + x - 15}{8x^2 - 10x - 3}$

29. $\dfrac{12x^2 + 11x - 15}{20x^2 - 23x + 6}$

30. $\dfrac{5x^2 + 45}{21x}$

31. $\dfrac{n^2 - 8n + 16}{n^2 - 16}$

32. $\dfrac{n^2 - 1}{n^2 + 2n + 1}$

33. $\dfrac{3x^3 - 6x^2 - 24x}{9x^2 + 27x + 18}$

34. $\dfrac{2x^3 + 3x^2 - 14x}{x^2 y + 7xy - 18y}$

35. $\dfrac{9y^2 - 1}{3y^2 + 11y - 4}$

36. $\dfrac{5y^2 + 22y + 8}{25y^2 - 4}$

37. $\dfrac{3x^3 + 12x}{9x^2 + 18x}$

38. $\dfrac{15x^3 - 15x^2}{5x^3 + 5x}$

39. $\dfrac{16x^3 y + 24x^2 y^2 - 16xy^3}{24x^2 y + 12xy^2 - 12y^3}$

40. $\dfrac{4x^2 y + 8xy^2 - 12y^3}{18x^3 y - 12x^2 y^2 - 6xy^3}$

41. $\dfrac{5n^2 + 18n - 8}{3n^2 + 13n + 4}$

42. $\dfrac{3n^2 + 14n - 24}{7n^2 + 44n + 12}$

43. $\dfrac{3 + x - 2x^2}{2 + x - x^2}$

44. $\dfrac{8 + 18x - 5x^2}{10 + 31x + 15x^2}$

45. $\dfrac{x^4 - 2x^2 - 15}{2x^4 + 9x^2 + 9}$

46. $\dfrac{6x^4 - 11x^2 + 4}{2x^4 + 17x^2 - 9}$

47. $\dfrac{n^2 - 2n - 80}{n^2 + 24n + 128}$

48. $\dfrac{n^2 + 24n + 140}{n^2 - 6n - 160}$

49. $\dfrac{-6x^3 - 21x^2 + 12x}{-18x^3 - 42x^2 + 120x}$

50. $\dfrac{-40x^3 + 24x^2 + 16x}{20x^3 + 28x^2 + 8x}$

Simplify each of the following. You will need to use *factoring by grouping*.

51. $\dfrac{xy + 2y + 3x + 6}{xy + 2y + 4x + 8}$

52. $\dfrac{xy + ay + bx + ab}{xy + ay + cx + ac}$

53. $\dfrac{x^2 - 2x + ax - 2a}{x^2 - 2x + 3ax - 6a}$

54. $\dfrac{ax - 3x + 2ay - 6y}{2ax - 6x + ay - 3y}$

55. $\dfrac{x^2 + 3x + 4x + 12}{2x^2 + 6x - x - 3}$

56. $\dfrac{5x^2 + 5x + 3x + 3}{5x^2 + 3x - 30x - 18}$

57. $\dfrac{nr - 6 - 3n + 2r}{nr + 10 + 2r + 5n}$

58. $\dfrac{2st - 30 - 12s + 5t}{3st - 6 - 18s + t}$

Simplify each of the following. You may want to refer to Example 12 in this section.

59. $\dfrac{3x - 2}{2 - 3x}$

60. $\dfrac{5a - 7}{7 - 5a}$

61. $\dfrac{n^2 - 25}{5 - n}$

62. $\dfrac{7 - y}{y^2 - 49}$

63. $\dfrac{3x - x^2}{x^2 - 9}$

64. $\dfrac{2y - 2xy}{x^2 y - y}$

65. $\dfrac{x^2 + 2x - 24}{20 - x - x^2}$

66. $\dfrac{n^2 - 5n - 24}{40 + 3n - n^2}$

67. $\dfrac{x^2 - (y - 1)^2}{(y - 1)^2 - x^2}$

68. $\dfrac{2x^3 - 8x}{4x - x^3}$

4.2 Multiplying and Dividing Rational Expressions

Multiplication of rational numbers in common fraction form is defined as follows.

Definition 4.1	If a, b, c, and d are integers with b and d not equal to zero, then $$\frac{a}{b} \cdot \frac{c}{d} = \frac{a \cdot c}{b \cdot d} = \frac{ac}{bd}.$$

To multiply rational numbers in common fraction form we merely *multiply numerators and multiply denominators* as the next examples illustrate. (The steps in the dashed boxes are usually done mentally.)

$$\frac{2}{3} \cdot \frac{4}{5} = \boxed{\frac{2 \cdot 4}{3 \cdot 5}} = \frac{8}{15};$$

$$\frac{-3}{4} \cdot \frac{5}{7} = \boxed{\frac{-3 \cdot 5}{4 \cdot 7} = \frac{-15}{28}} = -\frac{15}{28};$$

$$-\frac{5}{6} \cdot \frac{13}{3} = \boxed{\frac{-5}{6} \cdot \frac{13}{3} = \frac{-5 \cdot 13}{6 \cdot 3} = \frac{-65}{18}} = -\frac{65}{18}.$$

We also agree, when multiplying rational numbers, to express the final product in reduced form. The following examples illustrate some different formats used to *multiply and simplify* rational numbers.

$$\frac{3}{4} \cdot \frac{4}{7} = \frac{3 \cdot \cancel{4}}{\cancel{4} \cdot 7} = \frac{3}{7};$$

$$\overset{1}{\underset{1}{\frac{\cancel{8}}{\cancel{9}}}} \cdot \overset{3}{\underset{4}{\frac{\cancel{27}}{\cancel{32}}}} = \frac{3}{4};$$ A common factor of 9 has been divided out of 9 and 27, and a common factor of 8 has been divided out of 8 and 32.

$$\left(-\frac{28}{25}\right)\left(-\frac{65}{78}\right) = \frac{2 \cdot 2 \cdot 7 \cdot \cancel{5} \cdot \cancel{13}}{\cancel{5} \cdot 5 \cdot \cancel{2} \cdot 3 \cdot \cancel{13}} = \frac{14}{15}.$$ We should recognize that a *negative times a negative is positive*. Also, notice the use of prime factors to help recognize common factors.

Multiplication of rational expressions follows the same basic pattern as multiplication of rational numbers in common fraction form. That is to say, *we multiply numerators and multiply denominators and express the final product in simplified or reduced form.* Let's consider some examples.

$$\frac{3x}{4y} \cdot \frac{8y^2}{9x} = \frac{\cancel{3} \cdot \overset{2}{\cancel{8}} \cdot \cancel{x} \cdot \overset{y}{\cancel{y^2}}}{\underset{3}{\cancel{4} \cdot \cancel{9}} \cdot \cancel{x} \cdot \cancel{y}} = \frac{2y}{3};$$ Notice the use of the commutative property of multiplication to rearrange factors in a more convenient form for recognizing common factors of the numerator and denominator.

$$\frac{-4a}{6a^2b^2} \cdot \frac{9ab}{12a^2} = -\frac{\overset{3}{\cancel{4}} \cdot \cancel{9} \cdot \cancel{a^2} \cdot \cancel{b}}{\underset{2 \quad 3 \quad a^2 \quad b}{\cancel{6} \cdot \cancel{12} \cdot \cancel{a^4} \cdot \cancel{b^2}}} = -\frac{1}{2a^2b};$$

$$\frac{12x^2y}{-18xy} \cdot \frac{-24xy^2}{56y^3} = \frac{\overset{2}{\cancel{12}} \cdot \overset{3}{\cancel{24}} \cdot \overset{x^2}{\cancel{x^3}} \cdot \cancel{y^3}}{\underset{3 \quad 7 \quad y}{\cancel{18} \cdot \cancel{56} \cdot \cancel{x} \cdot \cancel{y^4}}} = \frac{2x^2}{7y}.$$

You should recognize that the first fraction is equivalent to $-\dfrac{12x^2y}{18xy}$ and the second to $-\dfrac{24xy^2}{56y^3}$; thus, the product is positive.

If the rational expressions contain polynomials (other than monomials) that are factorable, then our work may take on the following format.

Example 1   Multiply and simplify $\dfrac{y}{x^2 - 4} \cdot \dfrac{x + 2}{y^2}$.

Solution $$\frac{y}{x^2 - 4} \cdot \frac{x + 2}{y^2} = \frac{\cancel{y}(\cancel{x + 2})}{\underset{y}{\cancel{y^2}}(\cancel{x + 2})(x - 2)} = \frac{1}{y(x - 2)}.$$

In Example 1, notice that the steps of multiplying numerators and denominators and factoring the polynomials were combined. Also, notice that the final answer was left in factored form. Either $\dfrac{1}{y(x - 2)}$ or $\dfrac{1}{xy - 2y}$ would be acceptable answers.

Example 2   Multiply and simplify $\dfrac{x^2 - x}{x + 5} \cdot \dfrac{x^2 + 5x + 4}{x^4 - x^2}$.

Solution $$\frac{x^2 - x}{x + 5} \cdot \frac{x^2 + 5x + 4}{x^4 - x^2} = \frac{x(\cancel{x - 1})(\cancel{x + 1})(x + 4)}{(x + 5)(\underset{x}{\cancel{x^2}})(\cancel{x - 1})(\cancel{x + 1})} = \frac{x + 4}{x(x + 5)}.$$

Example 3   Multiply and simplify $\dfrac{6n^2 + 7n - 5}{n^2 + 2n - 24} \cdot \dfrac{4n^2 + 21n - 18}{12n^2 + 11n - 15}$.

Solution $$\frac{6n^2 + 7n - 5}{n^2 + 2n - 24} \cdot \frac{4n^2 + 21n - 18}{12n^2 + 11n - 15}$$

$$= \frac{(\cancel{3n + 5})(2n - 1)(\cancel{4n - 3})(\cancel{n + 6})}{(\cancel{n + 6})(n - 4)(\cancel{3n + 5})(\cancel{4n - 3})} = \frac{2n - 1}{n - 4}.$$

Dividing Rational Expressions

Division of rational numbers in common fraction form is defined as follows.

Definition 4.2

If a, b, c, and d are integers with b, c, and d not equal to zero, then

$$\frac{a}{b} \div \frac{c}{d} = \frac{a}{b} \cdot \frac{d}{c} = \frac{ad}{bc}.$$

Definition 4.2 states that to divide two rational numbers in fractional form we *invert the divisor and multiply*. The numbers $\dfrac{c}{d}$ and $\dfrac{d}{c}$ are called *reciprocals* or *multiplicative inverses* of each other because their product is one. Thus, division can also be described as *to divide by a fraction, multiply by its reciprocal*. The following examples illustrate the use of Definition 4.2.

$$\frac{7}{8} \div \frac{5}{6} = \frac{7}{\underset{4}{\cancel{8}}} \cdot \frac{\overset{3}{\cancel{6}}}{5} = \frac{21}{20}; \qquad \frac{-5}{9} \div \frac{15}{18} = -\frac{\cancel{5}}{\cancel{9}} \cdot \frac{\overset{2}{\cancel{18}}}{\cancel{15}} = -\frac{2}{3};$$

$$\frac{14}{-19} \div \frac{21}{-38} = \left(-\frac{14}{19}\right) \div \left(-\frac{21}{38}\right) = \left(-\frac{\cancel{14}}{\cancel{19}}\right)\left(-\frac{\overset{2}{\cancel{38}}}{\underset{3}{\cancel{21}}}\right) = \frac{4}{3}.$$

Division of algebraic rational expressions is defined in the same way as division of rational numbers. That is, the quotient of two rational expressions is the product of the first expression times the reciprocal of the second. Consider the following examples.

Example 4

Divide and simplify $\dfrac{16x^2y}{24xy^3} \div \dfrac{9xy}{8x^2y^2}.$

Solution

$$\frac{16x^2y}{24xy^3} \div \frac{9xy}{8x^2y^2} = \frac{16x^2y}{24xy^3} \cdot \frac{8x^2y^2}{9xy} = \frac{16 \cdot 8 \cdot \overset{x^2}{\cancel{x^4}} \cdot \cancel{y^3}}{\underset{3}{\cancel{24}} \cdot 9 \cdot \cancel{x^2} \cdot \underset{y}{\cancel{y^4}}} = \frac{16x^2}{27y}.$$

Example 5

Divide and simplify $\dfrac{3a^2 + 12}{3a^2 - 15a} \div \dfrac{a^4 - 16}{a^2 - 3a - 10}.$

Solution

$$\frac{3a^2 + 12}{3a^2 - 15a} \div \frac{a^4 - 16}{a^2 - 3a - 10} = \frac{3a^2 + 12}{3a^2 - 15a} \cdot \frac{a^2 - 3a - 10}{a^4 - 16}$$

$$= \frac{\cancel{3}(a^2 + 4)(a - 5)(a + 2)}{\cancel{3}a(a - 5)(a^2 + 4)(a + 2)(a - 2)}$$

$$= \frac{1}{a(a - 2)}.$$

Example 6   Divide and simplify $\dfrac{28t^3 - 51t^2 - 27t}{49t^2 + 42t + 9} \div (4t - 9)$.

Solution

$$\dfrac{28t^3 - 51t^2 - 27t}{49t^2 + 42t + 9} \div \dfrac{4t - 9}{1} = \dfrac{28t^3 - 51t^2 - 27t}{49t^2 + 42t + 9} \cdot \dfrac{1}{4t - 9}$$

$$= \dfrac{t(7t + 3)(4t - 9)}{(7t + 3)(7t + 3)(4t - 9)}$$

$$= \dfrac{t}{7t + 3}.$$

●

In a problem such as Example 6, it may be helpful to write the divisor with a denominator of 1. Thus, $4t - 9$ can be written as $\dfrac{4t - 9}{1}$; then its reciprocal is obviously $\dfrac{1}{4t - 9}$.

Let's consider one final example involving both multiplication and division.

Example 7   Perform the indicated operations and simplify

$$\dfrac{x^2 + 5x}{3x^2 - 4x - 20} \cdot \dfrac{x^2 y + y}{2x^2 + 11x + 5} \div \dfrac{xy^2}{6x^2 - 17x - 10}.$$

Solution

$$\dfrac{x^2 + 5x}{3x^2 - 4x - 20} \cdot \dfrac{x^2 y + y}{2x^2 + 11x + 5} \div \dfrac{xy^2}{6x^2 - 17x - 10}$$

$$= \dfrac{x^2 + 5x}{3x^2 - 4x - 20} \cdot \dfrac{x^2 y + y}{2x^2 + 11x + 5} \cdot \dfrac{6x^2 - 17x - 10}{xy^2}$$

$$= \dfrac{x(x + 5)(y)(x^2 + 1)(2x + 1)(3x - 10)}{(3x - 10)(x + 2)(2x + 1)(x + 5)(x)(y^2)}$$

$$= \dfrac{x^2 + 1}{y(x + 2)}.$$

●

Problem Set 4.2

Perform the following indicated operations involving rational numbers. Express final answers in reduced form.

1. $\dfrac{6}{15} \cdot \dfrac{12}{30}$ 2. $\dfrac{7}{8} \cdot \dfrac{12}{14}$ 3. $\dfrac{-5}{6} \cdot \dfrac{8}{25}$ 4. $\dfrac{-9}{24} \cdot \dfrac{36}{54}$

5. $\dfrac{4}{-6} \cdot \dfrac{-8}{18}$ 6. $\dfrac{-10}{12} \cdot \dfrac{16}{-15}$ 7. $\dfrac{3}{7} \div \dfrac{9}{7}$ 8. $\dfrac{5}{8} \div \dfrac{11}{3}$

9. $\dfrac{-7}{4} \div \dfrac{21}{8}$ 10. $\dfrac{1}{6} \div \dfrac{9}{-8}$ 11. $\dfrac{2}{3} \cdot \dfrac{6}{7} \div \dfrac{8}{3}$ 12. $\dfrac{4}{9} \cdot \dfrac{6}{11} \div \dfrac{4}{15}$

Perform the following indicated operations involving rational expressions. Express final answers in simplest form.

13. $\dfrac{4x^2}{5y^2} \cdot \dfrac{15xy}{24x^2y^2}$

14. $\dfrac{5xy}{8y^2} \cdot \dfrac{18x^2y}{15}$

15. $\dfrac{10a^2}{5b^2} \cdot \dfrac{15b^3}{2a^4}$

16. $\dfrac{5a^2b^2}{11ab} \cdot \dfrac{22a^3}{15ab^2}$

17. $\dfrac{-14xy^4}{18y^2} \cdot \dfrac{24x^2y^3}{35y^2}$

18. $\dfrac{6xy}{9y^4} \cdot \dfrac{30x^3y}{-48x}$

19. $\dfrac{3ab^3}{4c} \div \dfrac{21ac}{12bc^3}$

20. $\dfrac{9a^2c}{12bc^2} \div \dfrac{21ab}{14c^3}$

21. $\dfrac{7x^2y}{9xy^3} \div \dfrac{3x^4}{2x^2y^2}$

22. $\dfrac{5x^4}{12x^2y^3} \div \dfrac{9}{5xy}$

23. $\dfrac{5xy}{7a} \cdot \dfrac{14a^2}{15x} \cdot \dfrac{3a}{8y}$

24. $\dfrac{9x^2y^3}{14x} \cdot \dfrac{21y}{15xy^2} \cdot \dfrac{10x}{12y^3}$

25. $\dfrac{5xy}{x+6} \cdot \dfrac{x^2-36}{x^2-6x}$

26. $\dfrac{3x+6}{5y} \cdot \dfrac{x^2+4}{x^2+10x+16}$

27. $\dfrac{2a^2+6}{a^2-a} \cdot \dfrac{a^3-a^2}{8a-4}$

28. $\dfrac{5a^2+20a}{a^3-2a^2} \cdot \dfrac{a^2-a-12}{a^2-16}$

29. $\dfrac{10n^2+21n-10}{5n^2+33n-14} \cdot \dfrac{2n^2+6n-56}{2n^2-3n-20}$

30. $\dfrac{3n^2+15n-18}{3n^2+10n-48} \cdot \dfrac{12n^2-17n-40}{8n^2+2n-10}$

31. $\dfrac{7xy}{x^2-4x+4} \div \dfrac{14y}{x^2-4}$

32. $\dfrac{9y^2}{x^2+12x+36} \div \dfrac{12y}{x^2+6x}$

33. $\dfrac{x^2+5xy-6y^2}{xy^2-y^3} \cdot \dfrac{2x^2+15xy+18y^2}{xy+4y^2}$

34. $\dfrac{x^2-4xy+4y^2}{7xy^2} \div \dfrac{4x^2-3xy-10y^2}{20x^2y+25xy^2}$

35. $\dfrac{t^4-81}{t^2-6t+9} \cdot \dfrac{6t^2-11t-21}{5t^2+8t-21}$

36. $\dfrac{10t^3+25t}{20t+10} \cdot \dfrac{2t^2-t-1}{t^5-t}$

37. $\dfrac{21t^2+22t-8}{5t^2-43t-18} \div \dfrac{12t^2+7t-12}{20t^2-7t-6}$

38. $\dfrac{6x^2-35x+25}{4x^2-11x-45} \div \dfrac{18x^2+9x-20}{24x^2+74x+45}$

39. $\dfrac{2x^4+x^2-3}{2x^4+5x^2+2} \cdot \dfrac{3x^4+10x^2+8}{3x^4+x^2-4}$

40. $\dfrac{3x^4+2x^2-1}{3x^4+14x^2-5} \cdot \dfrac{x^4-2x^2-35}{x^4-17x^2+70}$

41. $\dfrac{6-n-2n^2}{12-11n+2n^2} \cdot \dfrac{24-26n+5n^2}{2+3n+n^2}$

42. $\dfrac{5-14n-3n^2}{1-2n-3n^2} \cdot \dfrac{9+7n-2n^2}{27-15n+2n^2}$

43. $\dfrac{xy+xc+ay+ac}{xy-2xc+ay-2ac} \cdot \dfrac{2x^3-8x}{12x^3+20x^2-8x}$

44. $\dfrac{nr+3n+2r+6}{nr+3n-3r-9} \cdot \dfrac{n^2-9}{n^3-4n}$

45. $\dfrac{9n^2-12n+4}{n^2-4n-32} \cdot \dfrac{n^2+4n}{3n^3-2n^2}$

46. $\dfrac{4t^2+t-5}{t^3-t^2} \cdot \dfrac{t^4+6t^3}{16t^2+40t+25}$

47. $\dfrac{4xy^2}{7x} \cdot \dfrac{14x^3y}{12y} \div \dfrac{7y}{9x^3}$

48. $\dfrac{x^2-x}{4y} \cdot \dfrac{10xy^2}{2x-2} \div \dfrac{3x^2+3x}{15x^2y^2}$

49. $\dfrac{2x^2+3x}{2x^3-10x^2} \cdot \dfrac{x^2-8x+15}{3x^3-27x} \div \dfrac{14x+21}{x^2-6x-27}$

50. $\dfrac{a^2-4ab+4b^2}{6a^2-4ab} \cdot \dfrac{3a^2+5ab-2b^2}{6a^2+ab-b^2} \div \dfrac{a^2-4b^2}{8a+4b}$

4.3 Adding and Subtracting Rational Expressions

Addition and subtraction of rational numbers can be defined as follows.

Definition 4.3	If a, b, and c are integers and b is not zero, then

$$\frac{a}{b} + \frac{c}{b} = \frac{a+c}{b}; \qquad \text{(Addition)}$$

$$\frac{a}{b} - \frac{c}{b} = \frac{a-c}{b}. \qquad \text{(Subtraction)}$$

We say that *rational numbers with a common denominator can be added or subtracted by adding or subtracting the numerators and placing the result over the common denominator.* The following examples illustrate Definition 4.3.

$$\frac{2}{9} + \frac{3}{9} = \frac{2+3}{9} = \frac{5}{9};$$

$$\frac{7}{8} - \frac{3}{8} = \frac{7-3}{8} = \frac{4}{8} = \frac{1}{2}; \qquad \text{(Don't forget to reduce!)}$$

$$\frac{4}{6} + \frac{-5}{6} = \frac{4+(-5)}{6} = \frac{-1}{6} = -\frac{1}{6};$$

$$\frac{7}{10} + \frac{4}{-10} = \frac{7}{10} + \frac{-4}{10} = \frac{7+(-4)}{10} = \frac{3}{10}.$$

This same *common denominator* approach is used when adding or subtracting rational expressions, as these next examples illustrate.

$$\frac{3}{x} + \frac{9}{x} = \frac{3+9}{x} = \frac{12}{x};$$

$$\frac{8}{x-2} - \frac{3}{x-2} = \frac{8-3}{x-2} = \frac{5}{x-2};$$

$$\frac{9}{4y} + \frac{5}{4y} = \frac{9+5}{4y} = \frac{14}{4y} = \frac{7}{2y}; \qquad \begin{array}{l}\text{(Don't forget to simplify}\\ \text{the final answer!)}\end{array}$$

$$\frac{n^2}{n-1} - \frac{1}{n-1} = \frac{n^2-1}{n-1} = \frac{(n+1)(n-1)}{n-1} = n+1;$$

$$\frac{6a^2}{2a+1} + \frac{13a+5}{2a+1} = \frac{6a^2+13a+5}{2a+1} = \frac{(2a+1)(3a+5)}{2a+1} = 3a+5.$$

Technically, in each of the previous examples involving rational expressions, we should restrict the variables so as to exclude division by zero. For example, $\dfrac{3}{x} + \dfrac{9}{x} = \dfrac{12}{x}$ is true for all real number values for x, *except* $x = 0$. Likewise, $\dfrac{8}{x-2} - \dfrac{3}{x-2} = \dfrac{5}{x-2}$ as long as x does not equal 2. Rather than taking the time and space to write down restrictions for each problem we will merely assume that such restrictions exist.

If rational numbers are to be added or subtracted that do not have a common denominator, then we apply the fundamental principle of fractions $\left(\dfrac{a}{b} = \dfrac{ak}{bk}\right)$ to obtain equivalent fractions with a common denominator. Equivalent fractions are fractions such as $\dfrac{1}{2}$ and $\dfrac{2}{4}$ that name the same number. Consider the following example.

$$\frac{1}{2} + \frac{1}{3} = \frac{3}{6} + \frac{2}{6} = \frac{3+2}{6} = \frac{5}{6}.$$

$$\left(\begin{array}{c}\dfrac{1}{2} \text{ and } \dfrac{3}{6} \\ \text{are equivalent} \\ \text{fractions.}\end{array}\right)\left(\begin{array}{c}\dfrac{1}{3} \text{ and } \dfrac{2}{6} \\ \text{are equivalent} \\ \text{fractions.}\end{array}\right)$$

Notice that we chose 6 as our common denominator and 6 is the **least common multiple** of the original denominators 2 and 3. (The least common multiple of a set of whole numbers is the smallest nonzero whole number divisible by each of the numbers.) In general, we use the least common multiple of the denominators of the fractions to be added or subtracted as a *least common denominator* (LCD).

A least common denominator may be found by inspection or by using the prime factored forms of the numbers. Let's consider some examples using each of these techniques.

Example 1   Subtract $\dfrac{5}{6} - \dfrac{3}{8}$.

Solution   By inspection we can see that the LCD is 24. Thus, both fractions can be changed to equivalent fractions, each having a denominator of 24.

$$\frac{5}{6} - \frac{3}{8} = \left(\frac{5}{6}\right)\left(\frac{4}{4}\right) - \left(\frac{3}{8}\right)\left(\frac{3}{3}\right) = \frac{20}{24} - \frac{9}{24} = \frac{11}{24}. \qquad\bullet$$

$$\underset{\text{form of 1}}{\uparrow} \qquad \underset{\text{form of 1}}{\uparrow}$$

In Example 1, notice that we used the fact that the fundamental principle of fractions, "$\dfrac{a}{b} = \dfrac{a \cdot k}{b \cdot k}$," can be written as "$\dfrac{a}{b} = \left(\dfrac{a}{b}\right)\left(\dfrac{k}{k}\right)$." This latter form emphasizes the fact that one is the multiplication identity element.

Example 2   Perform the indicated operations $\dfrac{3}{5} + \dfrac{1}{6} - \dfrac{13}{15}$.

Solution   Again by inspection we can determine that the LCD is 30. Thus, we can proceed as follows.

$$\frac{3}{5} + \frac{1}{6} - \frac{13}{15} = \left(\frac{3}{5}\right)\left(\frac{6}{6}\right) + \left(\frac{1}{6}\right)\left(\frac{5}{5}\right) - \left(\frac{13}{15}\right)\left(\frac{2}{2}\right)$$

$$= \frac{18}{30} + \frac{5}{30} - \frac{26}{30}$$

$$= \frac{18 + 5 - 26}{30}$$

$$= \frac{-3}{30} = -\frac{1}{10}. \qquad \text{(Don't forget to reduce!)}$$

Example 3   Add $\dfrac{7}{18} + \dfrac{11}{24}$.

Solution   Let's use the prime factored forms of the denominators to help find the LCD.

$$18 = 2 \cdot 3 \cdot 3 \qquad 24 = 2 \cdot 2 \cdot 2 \cdot 3.$$

The LCD must contain three factors of 2 since 24 contains three 2's. The LCD must also contain two factors of 3 since 18 has two 3's. Thus, the

$$\text{LCD} = 2 \cdot 2 \cdot 2 \cdot 3 \cdot 3 = 72.$$

Now we can proceed as usual.

$$\frac{7}{18} + \frac{11}{24} = \left(\frac{7}{18}\right)\left(\frac{4}{4}\right) + \left(\frac{11}{24}\right)\left(\frac{3}{3}\right) = \frac{28}{72} + \frac{33}{72} = \frac{61}{72}.$$

Adding and subtracting rational expressions with different denominators follows the same basic routine as adding or subtracting rational numbers with different denominators. Study the following examples carefully and notice the similarity to our previous work with rational numbers.

Example 4   Add $\dfrac{x + 2}{4} + \dfrac{3x + 1}{3}$.

Solution   By inspection we see that the LCD is 12.

$$\frac{x + 2}{4} + \frac{3x + 1}{3} = \left(\frac{x + 2}{4}\right)\left(\frac{3}{3}\right) + \left(\frac{3x + 1}{3}\right)\left(\frac{4}{4}\right)$$

$$= \frac{3(x + 2)}{12} + \frac{4(3x + 1)}{12}$$

$$= \frac{3x + 6 + 12x + 4}{12}$$

$$= \frac{15x + 10}{12}.$$

Notice the final result in Example 4. The numerator, $15x + 10$, could be factored as $5(3x + 2)$. However, since this produces no common factors with the denominator, the fraction cannot be simplified. Thus, the final answer can be left as $\dfrac{15x + 10}{12}$, or it would also be acceptable to express it as $\dfrac{5(3x + 2)}{12}$.

Example 5   Subtract $\dfrac{a - 2}{2} - \dfrac{a - 6}{6}$.

Solution   By inspection we see that the LCD is 6.

$$\frac{a - 2}{2} - \frac{a - 6}{6} = \left(\frac{a - 2}{2}\right)\left(\frac{3}{3}\right) - \frac{a - 6}{6}$$

$$= \frac{3(a - 2) - (a - 6)}{6} \qquad \text{(Be careful with this sign as you move to the next step!)}$$

$$= \frac{3a - 6 - a + 6}{6}$$

$$= \frac{2a}{6} = \frac{a}{3}. \qquad \text{(Don't forget to simplify.)}$$

Example 6   Perform the indicated operations $\dfrac{x + 3}{10} + \dfrac{2x + 1}{15} - \dfrac{x - 2}{18}$.

Solution   If you cannot determine the LCD by inspection, then use the prime factored forms of the denominators.

$$10 = 2 \cdot 5; \qquad 15 = 3 \cdot 5; \qquad 18 = 2 \cdot 3 \cdot 3.$$

The LCD must contain one factor of 2, two factors of 3, and one factor of 5. Thus, the LCD is $2 \cdot 3 \cdot 3 \cdot 5 = 90$.

$$\frac{x + 3}{10} + \frac{2x + 1}{15} - \frac{x - 2}{18} = \left(\frac{x + 3}{10}\right)\left(\frac{9}{9}\right) + \left(\frac{2x + 1}{15}\right)\left(\frac{6}{6}\right) - \left(\frac{x - 2}{18}\right)\left(\frac{5}{5}\right)$$

$$= \frac{9(x + 3) + 6(2x + 1) - 5(x - 2)}{90}$$

$$= \frac{9x + 27 + 12x + 6 - 5x + 10}{90}$$

$$= \frac{16x + 43}{90}.$$

Having the denominators contain variables does not create any serious difficulties; our approach remains basically the same.

Example 7   Add $\dfrac{3}{2x} + \dfrac{5}{3y}$.

Solution   Using an LCD of $6xy$ we can proceed as follows.

$$\frac{3}{2x} + \frac{5}{3y} = \left(\frac{3}{2x}\right)\left(\frac{3y}{3y}\right) + \left(\frac{5}{3y}\right)\left(\frac{2x}{2x}\right)$$

$$= \frac{9y}{6xy} + \frac{10x}{6xy}$$

$$= \frac{9y + 10x}{6xy}.$$

Example 8   Subtract $\dfrac{7}{12ab} - \dfrac{11}{15a^2}$.

Solution   We can prime factor the numerical coefficients of the denominators to help find the LCD.

$$\left.\begin{array}{l} 12ab = 2 \cdot 2 \cdot 3 \cdot a \cdot b \\ 15a^2 = 3 \cdot 5 \cdot a^2 \end{array}\right) \quad \begin{array}{l} \text{LCD} = 2 \cdot 2 \cdot 3 \cdot 5 \cdot a^2 \cdot b \\ = 60a^2 b \end{array}$$

$$\frac{7}{12ab} - \frac{11}{15a^2} = \left(\frac{7}{12ab}\right)\left(\frac{5a}{5a}\right) - \left(\frac{11}{15a^2}\right)\left(\frac{4b}{4b}\right)$$

$$= \frac{35a}{60a^2 b} - \frac{44b}{60a^2 b}$$

$$= \frac{35a - 44b}{60a^2 b}.$$

Example 9   Add $\dfrac{x}{x-3} + \dfrac{4}{x}$.

Solution   By inspection the LCD is $x(x-3)$.

$$\frac{x}{x-3} + \frac{4}{x} = \left(\frac{x}{x-3}\right)\left(\frac{x}{x}\right) + \left(\frac{4}{x}\right)\left(\frac{x-3}{x-3}\right)$$

$$= \frac{x^2}{x(x-3)} + \frac{4(x-3)}{x(x-3)}$$

$$= \frac{x^2 + 4x - 12}{x(x-3)} \quad \text{or} \quad \frac{(x+6)(x-2)}{x(x-3)}.$$

Problem Set 4.3

Perform the following indicated operations involving rational numbers. Be sure to express answers in reduced form.

1. $\dfrac{3}{4} + \dfrac{1}{6}$

2. $\dfrac{4}{5} + \dfrac{3}{7}$

3. $\dfrac{7}{8} - \dfrac{2}{5}$

4. $\dfrac{11}{12} - \dfrac{1}{6}$

5. $\dfrac{5}{6} + \dfrac{1}{-4}$

6. $\dfrac{4}{-9} + \dfrac{3}{5}$

7. $\dfrac{7}{15} + \dfrac{9}{25}$

8. $\dfrac{3}{14} - \dfrac{11}{21}$

9. $\dfrac{2}{3} - \dfrac{7}{8} + \dfrac{1}{4}$

10. $\dfrac{1}{5} + \dfrac{5}{6} - \dfrac{7}{15}$

11. $\dfrac{5}{6} - \dfrac{7}{9} - \dfrac{3}{10}$

12. $\dfrac{1}{3} - \dfrac{1}{4} - \dfrac{3}{14}$

Add or subtract the following rational expressions as indicated. Be sure to express answers in simplest form.

13. $\dfrac{5x}{x - 2} + \dfrac{3}{x - 2}$

14. $\dfrac{2x}{x + 1} - \dfrac{1}{x + 1}$

15. $\dfrac{3a}{a + 4} + \dfrac{12}{a + 4}$

16. $\dfrac{5a}{a - 1} - \dfrac{5}{a - 1}$

17. $\dfrac{7}{5y} - \dfrac{22}{5y}$

18. $\dfrac{2(3x - 4)}{7x^2} - \dfrac{7x - 8}{7x^2}$

19. $\dfrac{x - 2}{4} + \dfrac{x - 3}{3}$

20. $\dfrac{x + 4}{6} + \dfrac{2x - 1}{4}$

21. $\dfrac{a + 4}{5} + \dfrac{2a - 4}{6}$

22. $\dfrac{a - 3}{6} + \dfrac{2a - 4}{8}$

23. $\dfrac{n + 5}{6} - \dfrac{n - 1}{9}$

24. $\dfrac{3n - 1}{9} - \dfrac{n + 2}{12}$

25. $\dfrac{4x + 3}{10} - \dfrac{3x - 2}{6}$

26. $\dfrac{2x - 5}{6} - \dfrac{3x - 1}{14}$

27. $\dfrac{x + 1}{4} + \dfrac{x - 3}{6} - \dfrac{x - 2}{8}$

28. $\dfrac{x - 2}{5} - \dfrac{x + 3}{6} + \dfrac{x + 1}{15}$

29. $\dfrac{5}{6x} + \dfrac{1}{8x}$

30. $\dfrac{7}{9x} - \dfrac{5}{6x}$

31. $\dfrac{8}{9x} - \dfrac{5}{4y}$

32. $\dfrac{3}{10x} - \dfrac{5}{4y}$

33. $\dfrac{3}{4x} + \dfrac{2}{3y} - 1$

34. $\dfrac{5}{6x} - \dfrac{3}{4y} + 2$

35. $\dfrac{3}{5x^2} + \dfrac{5}{6x}$

36. $\dfrac{7}{6a^2} - \dfrac{5}{9a}$

37. $\dfrac{6}{8n^2} - \dfrac{3}{5n}$

38. $\dfrac{10}{7n} - \dfrac{12}{4n^2}$

39. $\dfrac{1}{n^2} + \dfrac{3}{4n} - \dfrac{5}{6}$

40. $\dfrac{3}{n^2} - \dfrac{2}{5n} + \dfrac{4}{3}$

41. $\dfrac{1}{x} - \dfrac{3}{2x^2} - \dfrac{4}{5x}$

42. $\dfrac{5}{3x^2} - \dfrac{4}{5x} - \dfrac{2}{x}$

43. $\dfrac{5}{7t} + \dfrac{3}{4t^2} + \dfrac{1}{14t}$

44. $\dfrac{6}{5t^2} - \dfrac{4}{7t^3} + \dfrac{9}{5t^3}$

45. $\dfrac{9}{14x^2y} - \dfrac{4x}{7y^2}$

46. $\dfrac{7}{9xy^3} - \dfrac{4}{3x} + \dfrac{5}{2y^2}$

47. $\dfrac{7}{16a^2b} + \dfrac{3a}{20b^2}$

48. $\dfrac{5b}{24a^2} - \dfrac{11a}{32b}$

49. $\dfrac{3x}{x + 1} + \dfrac{2}{x}$

50. $\dfrac{2x}{x-5} - \dfrac{4}{x}$

51. $\dfrac{a+1}{a} - \dfrac{2}{a+1}$

52. $\dfrac{a-2}{a} - \dfrac{3}{a+4}$

53. $\dfrac{3}{2x+1} + \dfrac{2}{3x+4}$

54. $\dfrac{5}{3x-2} + \dfrac{6}{4x+5}$

55. $\dfrac{5}{x-1} - \dfrac{3}{2x-3}$

56. $\dfrac{7}{3x-5} - \dfrac{5}{2x+7}$

57. $\dfrac{-3}{4x+3} + \dfrac{5}{2x-5}$

58. $\dfrac{-1}{x+4} + \dfrac{4}{7x-1}$

59. $\dfrac{-2}{n-6} - \dfrac{6}{2n+3}$

60. $\dfrac{-3}{4n+5} - \dfrac{8}{3n+5}$

61. Recall that the indicated quotient of a polynomial and its opposite is -1. For example, $\dfrac{x-2}{2-x}$ simplifies to -1. Keep this idea in mind as you add or subtract the following rational expressions.

(a) $\dfrac{1}{x-1} - \dfrac{x}{x-1}$

(b) $\dfrac{3}{2x-3} - \dfrac{2x}{2x-3}$

(c) $\dfrac{4}{x-4} - \dfrac{x}{x-4} + 1$

(d) $-1 + \dfrac{2}{x-2} - \dfrac{x}{x-2}$

62. Consider the addition problem $\dfrac{8}{x-2} + \dfrac{5}{2-x}$. Notice that the denominators are opposites of each other. If the property $\dfrac{a}{-b} = -\dfrac{a}{b}$ is applied to the second fraction, we obtain $\dfrac{5}{2-x} = -\dfrac{5}{x-2}$. Thus, we proceed as follows.

$$\frac{8}{x-2} + \frac{5}{2-x} = \frac{8}{x-2} - \frac{5}{x-2} = \frac{8-5}{x-2} = \frac{3}{x-2}$$

Use this approach to do the following problems.

(a) $\dfrac{7}{x-1} + \dfrac{2}{1-x}$

(b) $\dfrac{5}{2x-1} + \dfrac{8}{1-2x}$

(c) $\dfrac{4}{a-3} - \dfrac{1}{3-a}$

(d) $\dfrac{10}{a-9} - \dfrac{5}{9-a}$

(e) $\dfrac{x^2}{x-1} - \dfrac{2x-3}{1-x}$

(f) $\dfrac{x^2}{x-4} - \dfrac{3x-28}{4-x}$

4.4 More on Rational Expressions and Complex Fractions

In this section we shall expand our work with adding and subtracting rational expressions and also discuss the process of simplifying complex fractions. Before we begin, however, this seems like an appropriate time to offer a bit of advice regarding your study of algebra. Success in algebra depends upon having a good understanding of the concepts as well as being able to perform the various computations. Regarding the computational work, you should adopt a carefully organized format showing as many steps as *you need* in order to minimize the chances of making careless errors. Don't be over-anxious to find short-cuts for certain computations before you have a thorough understanding of the steps involved in the process. This is especially appropriate advice at the beginning of this section.

Study the following examples very carefully. Notice the same basic procedure for each problem of: (1) finding the LCD, (2) changing each fraction to an equivalent fraction having the LCD as its denominator, (3) adding or subtracting numerators and placing this result over the LCD, and (4) looking for possibilities to simplify the resulting fraction.

Example 1   Add $\dfrac{8}{x^2 - 4x} + \dfrac{2}{x}$.

Solution

$$\left.\begin{array}{l} x^2 - 4x = x(x - 4) \\[4pt] \quad\quad x = x \end{array}\right) \quad \longrightarrow \text{LCD is } x(x - 4)$$

$$\frac{8}{x(x - 4)} + \frac{2}{x} = \frac{8}{x(x - 4)} + \left(\frac{2}{x}\right)\left(\frac{x - 4}{x - 4}\right)$$

$$= \frac{8}{x(x - 4)} + \frac{2(x - 4)}{x(x - 4)}$$

$$= \frac{8 + 2x - 8}{x(x - 4)}$$

$$= \frac{2x}{x(x - 4)}$$

$$= \frac{2}{x - 4}.$$

Example 2   Subtract $\dfrac{a}{a^2 - 4} - \dfrac{3}{a + 2}$.

Solution

$$\left.\begin{array}{l} a^2 - 4 = (a + 2)(a - 2) \\[4pt] a + 2 = a + 2 \end{array}\right) \quad \longrightarrow \text{LCD is } (a + 2)(a - 2)$$

$$\frac{a}{(a + 2)(a - 2)} - \frac{3}{a + 2} = \frac{a}{(a + 2)(a - 2)} - \left(\frac{3}{a + 2}\right)\left(\frac{a - 2}{a - 2}\right)$$

$$= \frac{a}{(a + 2)(a - 2)} - \frac{3(a - 2)}{(a + 2)(a - 2)}$$

$$= \frac{a - 3a + 6}{(a + 2)(a - 2)}$$

$$= \frac{-2a + 6}{(a + 2)(a - 2)} \quad \text{or} \quad \frac{-2(a - 3)}{(a + 2)(a - 2)}.$$

Example 3   Add $\dfrac{3n}{n^2 + 6n + 5} + \dfrac{4}{n^2 - 7n - 8}$.

Solution

$$\left.\begin{array}{l} n^2 + 6n + 5 = (n + 5)(n + 1) \\ n^2 - 7n - 8 = (n - 8)(n + 1) \end{array}\right) \longrightarrow \text{LCD is } (n + 1)(n + 5)(n - 8)$$

$$\dfrac{3n}{(n + 5)(n + 1)} + \dfrac{4}{(n - 8)(n + 1)}$$

$$= \left(\dfrac{3n}{(n + 5)(n + 1)}\right)\left(\dfrac{n - 8}{n - 8}\right) + \left(\dfrac{4}{(n - 8)(n + 1)}\right)\left(\dfrac{n + 5}{n + 5}\right)$$

$$= \dfrac{3n(n - 8)}{(n + 5)(n + 1)(n - 8)} + \dfrac{4(n + 5)}{(n + 5)(n + 1)(n - 8)}$$

$$= \dfrac{3n^2 - 24n + 4n + 20}{(n + 5)(n + 1)(n - 8)}$$

$$= \dfrac{3n^2 - 20n + 20}{(n + 5)(n + 1)(n - 8)}.$$

Example 4   Perform the indicated operations.

$$\dfrac{2x^2}{x^4 - 1} + \dfrac{x}{x^2 - 1} - \dfrac{1}{x - 1}.$$

Solution

$$\left.\begin{array}{l} x^4 - 1 = (x^2 + 1)(x + 1)(x - 1) \\ x^2 - 1 = (x + 1)(x - 1) \\ x - 1 = x - 1 \end{array}\right) \longrightarrow \text{LCD is } (x^2 + 1)(x + 1)(x - 1)$$

$$\dfrac{2x^2}{(x^2 + 1)(x + 1)(x - 1)} + \dfrac{x}{(x + 1)(x - 1)} - \dfrac{1}{x - 1}$$

$$= \dfrac{2x^2}{(x^2 + 1)(x + 1)(x - 1)} + \left(\dfrac{x}{(x + 1)(x - 1)}\right)\left(\dfrac{x^2 + 1}{x^2 + 1}\right)$$

$$\qquad - \left(\dfrac{1}{x - 1}\right)\left(\dfrac{(x^2 + 1)(x + 1)}{(x^2 + 1)(x + 1)}\right)$$

$$= \dfrac{2x^2}{(x^2 + 1)(x + 1)(x - 1)} + \dfrac{x(x^2 + 1)}{(x^2 + 1)(x + 1)(x - 1)}$$

$$\qquad - \dfrac{(x^2 + 1)(x + 1)}{(x^2 + 1)(x + 1)(x - 1)}$$

$$= \dfrac{2x^2 + x^3 + x - x^3 - x^2 - x - 1}{(x^2 + 1)(x + 1)(x - 1)}$$

$$= \dfrac{x^2 - 1}{(x^2 + 1)(x + 1)(x - 1)}$$

$$= \dfrac{\cancel{(x + 1)}\cancel{(x - 1)}}{(x^2 + 1)\cancel{(x + 1)}\cancel{(x - 1)}} = \dfrac{1}{x^2 + 1}.$$

Complex Fractions

Fractional forms that contain rational numbers or rational expressions in the numerators and/or denominators are called **complex fractions**. The following are examples of complex fractions.

$$\frac{\dfrac{3}{5}}{\dfrac{7}{8}}, \qquad \frac{\dfrac{4}{x}}{\dfrac{2}{xy}}, \qquad \frac{\dfrac{1}{2}+\dfrac{3}{4}}{\dfrac{5}{6}-\dfrac{3}{8}}, \qquad \frac{\dfrac{3}{x}+\dfrac{2}{y}}{\dfrac{5}{x}-\dfrac{6}{y^2}}.$$

It is often necessary to *simplify* a complex fraction. Let's examine some techniques for simplifying complex fractions with the four previous examples.

Example 5   Simplify $\dfrac{\dfrac{3}{5}}{\dfrac{7}{8}}$.

Solution   This type of problem creates nothing new since it is merely a divison problem. Thus,

$$\frac{\dfrac{3}{5}}{\dfrac{7}{8}} = \frac{3}{5} \div \frac{7}{8} = \frac{3}{5} \cdot \frac{8}{7} = \frac{24}{35}.$$

Example 6   Simplify $\dfrac{\dfrac{4}{x}}{\dfrac{2}{xy}}$.

Solution $\dfrac{\dfrac{4}{x}}{\dfrac{2}{xy}} = \dfrac{4}{x} \div \dfrac{2}{xy} = \dfrac{\overset{2}{\cancel{4}}}{\cancel{x}} \cdot \dfrac{\cancel{x}y}{\cancel{2}} = 2y.$

Example 7   Simplify $\dfrac{\dfrac{1}{2}+\dfrac{3}{4}}{\dfrac{5}{6}-\dfrac{3}{8}}$.

Let's look at two possible "attacks" for such a problem.

Solution A

$$\frac{\dfrac{1}{2}+\dfrac{3}{4}}{\dfrac{5}{6}-\dfrac{3}{8}}=\frac{\dfrac{2}{4}+\dfrac{3}{4}}{\dfrac{20}{24}-\dfrac{9}{24}}=\frac{\dfrac{5}{4}}{\dfrac{11}{24}}=\frac{5}{4}\cdot\frac{\overset{6}{\cancel{24}}}{11}=\frac{30}{11}.$$

Solution B   The LCD of all four denominators (2, 4, 6, and 8) is 24. Multiply the entire complex fraction by a form of 1, namely, $\dfrac{24}{24}$.

$$\frac{\dfrac{1}{2}+\dfrac{3}{4}}{\dfrac{5}{6}-\dfrac{3}{8}}=\left(\frac{24}{24}\right)\left(\frac{\dfrac{1}{2}+\dfrac{3}{4}}{\dfrac{5}{6}-\dfrac{3}{8}}\right)$$

$$=\frac{24\left(\dfrac{1}{2}+\dfrac{3}{4}\right)}{24\left(\dfrac{5}{6}-\dfrac{3}{8}\right)}$$

$$=\frac{24\left(\dfrac{1}{2}\right)+24\left(\dfrac{3}{4}\right)}{24\left(\dfrac{5}{6}\right)-24\left(\dfrac{3}{8}\right)}$$

$$=\frac{12+18}{20-9}$$

$$=\frac{30}{11}.$$

Example 8   Simplify $\dfrac{\dfrac{3}{x}+\dfrac{2}{y}}{\dfrac{5}{x}-\dfrac{6}{y^2}}$.

Solution A

$$\frac{\dfrac{3}{x}+\dfrac{2}{y}}{\dfrac{5}{x}-\dfrac{6}{y^2}}=\frac{\left(\dfrac{3}{x}\right)\left(\dfrac{y}{y}\right)+\left(\dfrac{2}{y}\right)\left(\dfrac{x}{x}\right)}{\left(\dfrac{5}{x}\right)\left(\dfrac{y^2}{y^2}\right)-\left(\dfrac{6}{y^2}\right)\left(\dfrac{x}{x}\right)}$$

$$=\frac{\dfrac{3y}{xy}+\dfrac{2x}{xy}}{\dfrac{5y^2}{xy^2}-\dfrac{6x}{xy^2}}$$

$$= \frac{\dfrac{3y + 2x}{xy}}{\dfrac{5y^2 - 6x}{xy^2}}$$

$$= \frac{3y + 2x}{xy} \div \frac{5y^2 - 6x}{xy^2}$$

$$= \frac{3y + 2x}{\cancel{xy}} \cdot \frac{\overset{y}{\cancel{xy^2}}}{5y^2 - 6x}$$

$$= \frac{y(3y + 2x)}{5y^2 - 6x}.$$

Solution B   The LCD of all four denominators (x, y, x, and y^2) is xy^2. Multiply the entire complex fraction by a form of 1, namely, $\dfrac{xy^2}{xy^2}$.

$$\frac{\dfrac{3}{x} + \dfrac{2}{y}}{\dfrac{5}{x} - \dfrac{6}{y^2}} = \left(\frac{xy^2}{xy^2}\right) \frac{\dfrac{3}{x} + \dfrac{2}{y}}{\dfrac{5}{x} - \dfrac{6}{y^2}}$$

$$= \frac{xy^2\left(\dfrac{3}{x} + \dfrac{2}{y}\right)}{xy^2\left(\dfrac{5}{x} - \dfrac{6}{y^2}\right)}$$

$$= \frac{xy^2\left(\dfrac{3}{x}\right) + xy^2\left(\dfrac{2}{y}\right)}{xy^2\left(\dfrac{5}{x}\right) - xy^2\left(\dfrac{6}{y^2}\right)}$$

$$= \frac{3y^2 + 2xy}{5y^2 - 6x} \text{ or } \frac{y(3y + 2x)}{5y^2 - 6x}. \qquad \bullet$$

Certainly either approach (Solution A or Solution B) will work with problems such as Examples 7 and 8. Examine Solution B for each of the examples rather carefully. This approach works very effectively with algebraic complex fractions for which the LCD of all the denominators is easy to find.

Let's conclude this section with two more examples involving algebraic complex fractions.

Example 9   Simplify $\dfrac{\dfrac{1}{a} + \dfrac{1}{a - 1}}{\dfrac{1}{a - 1} - \dfrac{1}{a}}$.

Solution   Multiply the entire complex fraction by a form of 1, namely, $\dfrac{a(a-1)}{a(a-1)}$.

$$\frac{\dfrac{1}{a}+\dfrac{1}{a-1}}{\dfrac{1}{a-1}-\dfrac{1}{a}}=\left(\frac{a(a-1)}{a(a-1)}\right)\left(\frac{\dfrac{1}{a}+\dfrac{1}{a-1}}{\dfrac{1}{a-1}-\dfrac{1}{a}}\right)$$

$$=\frac{a(a-1)\left(\dfrac{1}{a}+\dfrac{1}{a-1}\right)}{a(a-1)\left(\dfrac{1}{a-1}-\dfrac{1}{a}\right)}$$

$$=\frac{a(a-1)\left(\dfrac{1}{a}\right)+a(a-1)\left(\dfrac{1}{a-1}\right)}{a(a-1)\left(\dfrac{1}{a-1}\right)-a(a-1)\left(\dfrac{1}{a}\right)}$$

$$=\frac{a-1+a}{a-(a-1)}$$

$$=\frac{2a-1}{a-a+1}$$

$$=\frac{2a-1}{1}\quad\text{or}\quad 2a-1.$$

Example 10   Simplify $1-\dfrac{n}{1-\dfrac{1}{n}}$.

Solution   First simplify the complex fraction $\dfrac{n}{1-\dfrac{1}{n}}$ by multiplying by $\dfrac{n}{n}$.

$$\left(\frac{n}{1-\dfrac{1}{n}}\right)\left(\frac{n}{n}\right)=\frac{n^2}{n-1}.$$

Now we can perform the subtraction.

$$1-\frac{n^2}{n-1}=\left(\frac{n-1}{n-1}\right)\left(\frac{1}{1}\right)-\frac{n^2}{n-1}$$

$$=\frac{n-1}{n-1}-\frac{n^2}{n-1}$$

$$=\frac{n-1-n^2}{n-1}\quad\text{or}\quad\frac{-n^2+n-1}{n-1}.$$

Problem Set 4.4

Perform the indicated operations and express answers in simplest form.

1. $\dfrac{3x}{x^2 - 6x} + \dfrac{2}{x}$

2. $\dfrac{4x}{x^2 + 7x} + \dfrac{3}{x}$

3. $\dfrac{6}{x^2 + 8x} - \dfrac{3}{x}$

4. $\dfrac{-20}{x^2 - 5x} - \dfrac{4}{x}$

5. $\dfrac{x}{x^2 - 9} + \dfrac{4}{x - 3}$

6. $\dfrac{3x}{x^2 - 25} + \dfrac{2}{x + 5}$

7. $\dfrac{4a - 4}{a^2 - 4} - \dfrac{3}{a + 2}$

8. $\dfrac{6a + 4}{a^2 - 1} - \dfrac{5}{a - 1}$

9. $\dfrac{n}{n^2 - 16} - \dfrac{2}{3n + 12}$

10. $\dfrac{5n}{n^2 - 25} - \dfrac{3}{4n + 20}$

11. $\dfrac{3}{x + 1} + \dfrac{x + 5}{x^2 - 1} - \dfrac{3}{x - 1}$

12. $\dfrac{5}{x} - \dfrac{5x - 30}{x^2 + 6x} + \dfrac{x}{x + 6}$

13. $\dfrac{5}{x^2 + 10x + 21} + \dfrac{4}{x^2 + 12x + 27}$

14. $\dfrac{3}{2x^2 + 9x + 4} + \dfrac{7}{2x^2 + 3x + 1}$

15. $\dfrac{2}{a^2 - 3a - 10} - \dfrac{9}{a^2 + 3a - 40}$

16. $\dfrac{8}{a^2 - 3a - 18} - \dfrac{10}{a^2 - 7a - 30}$

17. $\dfrac{2a}{6a^2 + 11a - 10} + \dfrac{a}{2a^2 - 3a - 20}$

18. $\dfrac{3a}{20a^2 - 11a - 3} + \dfrac{1}{12a^2 + 7a - 12}$

19. $\dfrac{5}{x^2 - 1} - \dfrac{2}{x^2 + 6x - 16}$

20. $\dfrac{4}{x^2 + 2} - \dfrac{7}{x^2 + x - 12}$

21. $\dfrac{1}{y^2 + 2y - 35} - \dfrac{3}{y + 7} + \dfrac{2}{y - 5}$

22. $\dfrac{5}{y - 9} + \dfrac{4}{y^2 - 3y - 54} - \dfrac{1}{y + 6}$

23. $x - \dfrac{x^2}{x - 1} + \dfrac{1}{x^2 - 1}$

24. $x - \dfrac{x^2}{x + 7} - \dfrac{x}{x^2 - 49}$

25. $\dfrac{x + 1}{x + 2} + \dfrac{x - 1}{x + 4} + \dfrac{3x - 1}{x^2 + 6x + 8}$

26. $\dfrac{2x - 1}{x + 5} + \dfrac{3x + 2}{x + 2} + \dfrac{x - 1}{(x + 2)(x + 5)}$

27. $\dfrac{n - 1}{n + 4} + \dfrac{n}{n + 6} + \dfrac{2n + 18}{n^2 + 10n + 24}$

28. $\dfrac{n}{n - 6} + \dfrac{n + 3}{n + 8} + \dfrac{12n + 26}{n^2 + 2n - 48}$

29. $\dfrac{t - 3}{2t + 1} + \dfrac{2t^2 + 19t - 46}{2t^2 - 9t - 5} - \dfrac{t + 4}{t - 5}$

30. $\dfrac{t + 3}{3t - 1} + \dfrac{8t^2 + 8t + 2}{3t^2 - 7t + 2} - \dfrac{2t + 3}{t - 2}$

31. $\dfrac{32x + 9}{12x^2 + x - 6} - \dfrac{3}{4x + 3} - \dfrac{x + 5}{3x - 2}$

32. $\dfrac{15x^2 - 10}{5x^2 - 7x + 2} - \dfrac{3x + 4}{x - 1} - \dfrac{2}{5x - 2}$

33. $\dfrac{2n^2}{n^4 - 16} - \dfrac{n}{n^2 - 4} + \dfrac{1}{n + 2}$

34. $\dfrac{n}{n^2 + 1} + \dfrac{n^2 + 3n}{n^4 - 1} - \dfrac{1}{n - 1}$

35. $\dfrac{2x + 5}{x^2 + 3x - 18} - \dfrac{3x - 1}{x^2 + 4x - 12} + \dfrac{5}{x - 2}$

36. $\dfrac{4x - 3}{2x^2 + x - 1} - \dfrac{2x + 7}{3x^2 + x - 2} - \dfrac{3}{3x - 2}$

Simplify each of the following complex fractions.

37. $\dfrac{\dfrac{3}{4} - \dfrac{1}{2}}{\dfrac{3}{8} + \dfrac{1}{4}}$

38. $\dfrac{\dfrac{5}{6} + \dfrac{1}{4}}{\dfrac{7}{8} - \dfrac{1}{12}}$

39. $\dfrac{\dfrac{7}{9} + \dfrac{11}{36}}{\dfrac{1}{6} - \dfrac{5}{12}}$

40. $\dfrac{\dfrac{5}{28} - \dfrac{13}{14}}{\dfrac{3}{8} + \dfrac{1}{2}}$

41. $\dfrac{\dfrac{3}{4x}}{\dfrac{5}{2xy}}$

42. $\dfrac{\dfrac{8}{7x^2y}}{\dfrac{16}{14xy^2}}$

43. $\dfrac{\dfrac{2}{x} + \dfrac{7}{y}}{\dfrac{3}{x} - \dfrac{10}{y}}$

44. $\dfrac{\dfrac{5}{x^2} - \dfrac{3}{x}}{\dfrac{1}{y} + \dfrac{2}{y^2}}$

45. $\dfrac{\dfrac{4}{a^2} - \dfrac{5}{b}}{\dfrac{9}{a} + \dfrac{8}{b^2}}$

46. $\dfrac{\dfrac{3}{a} - \dfrac{7}{b^2}}{\dfrac{5}{ab} - \dfrac{2}{b}}$

47. $\dfrac{\dfrac{1}{x} + 3}{\dfrac{2}{y} + 4}$

48. $\dfrac{1 + \dfrac{1}{x}}{1 - \dfrac{1}{x}}$

49. $\dfrac{2 + \dfrac{3}{n+2}}{4 - \dfrac{1}{n+2}}$

50. $\dfrac{3 - \dfrac{2}{n-4}}{5 + \dfrac{4}{n-4}}$

51. $\dfrac{1 - \dfrac{1}{n+1}}{1 + \dfrac{1}{n-1}}$

52. $\dfrac{\dfrac{1}{n-1} + 1}{\dfrac{1}{n+1} - 1}$

53. $\dfrac{\dfrac{2}{x-y} + \dfrac{3}{x+y}}{\dfrac{5}{x+y} - \dfrac{1}{x^2-y^2}}$

54. $\dfrac{\dfrac{2}{x-3} - \dfrac{3}{x+3}}{\dfrac{5}{x^2-9} - \dfrac{2}{x-3}}$

55. $\dfrac{\dfrac{-2}{x} - \dfrac{4}{x+2}}{\dfrac{3}{x^2+2x} + \dfrac{3}{x}}$

56. $\dfrac{\dfrac{-1}{y-2} + \dfrac{5}{x}}{\dfrac{3}{x} - \dfrac{4}{xy-2x}}$

57. $1 + \dfrac{x}{1 + \dfrac{1}{x}}$

58. $2 - \dfrac{x}{3 - \dfrac{2}{x}}$

59. $\dfrac{\dfrac{a}{1} + 1}{\dfrac{1}{a} + 4}$

60. $\dfrac{\dfrac{3a}{2} - 1}{2 - \dfrac{1}{a}}$

4.5 Dividing Polynomials

In Chapter 3 we saw how the property "$\dfrac{b^n}{b^m} = b^{n-m}$" along with our knowledge of dividing integers was used to divide monomials. For example,

$$\frac{12x^3}{3x} = 4x^2; \qquad \frac{-36x^4y^5}{4xy^2} = -9x^3y^3.$$

In Section 4.3, we used $\dfrac{a}{b} + \dfrac{c}{b} = \dfrac{a+c}{b}$ and $\dfrac{a}{b} - \dfrac{c}{b} = \dfrac{a-c}{b}$ as the basis for adding and subtracting rational expressions. These same equalities, viewed as $\dfrac{a+b}{c} = \dfrac{a}{c} + \dfrac{b}{c}$

and $\dfrac{a-c}{b} = \dfrac{a}{b} - \dfrac{c}{b}$, along with our knowledge of dividing monomials, provide the basis for dividing polynomials by monomials. Consider the following examples.

$$\frac{18x^3 + 24x^2}{6x} = \frac{18x^3}{6x} + \frac{24x^2}{6x} = 3x^2 + 4x;$$

$$\frac{35x^2y^3 - 55x^3y^4}{5xy^2} = \frac{35x^2y^3}{5xy^2} - \frac{55x^3y^4}{5xy^2} = 7xy - 11x^2y^2.$$

To divide a polynomial by a monomial we divide each term of the polynomial by the monomial. As with many skills, once you feel comfortable with the process you may then want to perform some of the steps mentally. Your work could take on the following format:

$$\frac{40x^4y^5 + 72x^5y^7}{8x^2y} = 5x^2y^4 + 9x^3y^6; \qquad \frac{36a^3b^4 - 45a^4b^6}{-9a^2b^3} = -4ab + 5a^2b^3.$$

In Section 4.1 we saw that a fraction like $\dfrac{3x^2 + 11x - 4}{x + 4}$ can be simplified as follows:

$$\frac{3x^2 + 11x - 4}{x + 4} = \frac{(3x - 1)(\cancel{x + 4})}{\cancel{x + 4}} = 3x - 1.$$

The same result can be obtained by using a dividing process similar to long division in arithmetic. The process can be described as follows:

Step 1 Use the conventional long division format and arrange both the dividend and the divisor in descending powers of the variable.

$$x + 4 \overline{)\,3x^2 + 11x - 4}$$

Step 2 Find the first term of the quotient by dividing the first term of the dividend by the first term of the divisor.

$$\begin{array}{r} 3x \\ x + 4 \overline{)\,3x^2 + 11x - 4} \end{array}$$

Step 3 Multiply the entire divisor by the term of the quotient found in Step 2 and place this product so as to be subtracted from the dividend.

$$\begin{array}{r} 3x \\ x + 4 \overline{)\,3x^2 + 11x - 4} \\ 3x^2 + 12x \end{array}$$

Step 4 Subtract.

$$\begin{array}{r} 3x \\ x + 4 \overline{)\,3x^2 + 11x - 4} \\ 3x^2 + 12x \\ \hline -x - 4 \end{array}$$

Remember to add the opposite! ————→

$(3x^2 + 11x - 4) - (3x^2 + 12x) = -x - 4$ ————→

Repeat the process beginning with Step 2; use the polynomial that resulted from the subtraction in Step 4 as a new dividend.

$$\begin{array}{r} 3x - 1 \\ x + 4 \overline{)\,3x^2 + 11x - 4} \\ 3x^2 + 12x \\ \hline -x - 4 \\ -x - 4 \\ \hline \end{array}$$

In the next example let's *think* in terms of the step-by-step procedure described above but arrange our work in a more compact form.

Example 1   Divide $5x^2 + 6x - 8$ by $x + 2$.

Solution

$$
\begin{array}{r}
5x - 4 \\
x + 2 \overline{) 5x^2 + 6x - 8} \\
5x^2 + 10x \\
\hline
-4x - 8 \\
-4x - 8
\end{array}
$$

Think Steps

1. $\dfrac{5x^2}{x} = 5x$.

2. $5x(x + 2) = 5x^2 + 10x$.

3. $(5x^2 + 6x - 8) - (5x^2 + 10x) = -4x - 8$.

4. $\dfrac{-4x}{x} = -4$.

5. $-4(x + 2) = -4x - 8$.   ●

Recall that to check a division problem we can "multiply the divisor times the quotient and add the remainder." This can be expressed as

$$\text{Dividend} = (\text{Divisor})(\text{Quotient}) + \text{Remainder}.$$

Sometimes the remainder is expressed as a fractional part of the divisor. The relationship then becomes

$$\frac{\text{Dividend}}{\text{Divisor}} = \text{Quotient} + \frac{\text{Remainder}}{\text{Divisor}}.$$

Example 2   Divide $2x^2 - 3x + 1$ by $x - 5$.

Solution

$$
\begin{array}{r}
2x + 7 \\
x - 5 \overline{) 2x^2 - 3x + 1} \\
2x^2 - 10x \\
\hline
7x + 1 \\
7x - 35 \\
\hline
36
\end{array}
$$
⟵———————— Remainder

Thus, $\dfrac{2x^2 - 3x + 1}{x - 5} = 2x + 7 + \dfrac{36}{x - 5} \qquad (x \neq 5)$.

Check $(x - 5)(2x + 7) + 36 \overset{?}{=} 2x^2 - 3x + 1$

$2x^2 - 3x - 35 + 36 \overset{?}{=} 2x^2 - 3x + 1$

$2x^2 - 3x + 1 = 2x^2 - 3x + 1.$ ●

Each of the next two examples illustrates another point regarding the division process. Study them carefully and then you should be ready to work the exercises in the next problem set.

Example 3   Divide $t^3 - 8$ by $t - 2$.

Solution

$$
\begin{array}{r}
t^2 + 2t\ \ + 4 \\
t - 2\,\overline{)\,t^3 + 0t^2 + 0t - 8} \\
\underline{t^3 - 2t^2} \\
2t^2 + 0t - 8 \\
\underline{2t^2 - 4t} \\
4t - 8 \\
\underline{4t - 8}
\end{array}
$$

Notice the inserting of a "t squared" and a "t term" with zero coefficients.

(Check this result!) ●

Example 4   Divide $y^3 + 3y^2 - 2y - 1$ by $y^2 + 2y$.

Solution

$$
\begin{array}{r}
y + 1 \\
y^2 + 2y\,\overline{)\,y^3 + 3y^2 - 2y - 1} \\
\underline{y^3 + 2y^2} \\
y^2 - 2y - 1 \\
\underline{y^2 + 2y} \\
-4y - 1
\end{array}
$$

A remainder of $-4y - 1$

(The division process is completed when the degree of the remainder is less than the degree of the divisor.)

Thus, $\dfrac{y^3 + 3y^2 - 2y - 1}{y^2 + 2y} = y + 1 + \dfrac{-4y - 1}{y^2 + 2y}.$ ●

Remark   If the divisor is of the form $x - k$, where the coefficient of the x term is one, then the format of the division process described in this section can be simplified by a procedure called **synthetic division**. This procedure is outlined in Section 9.1.

Problem Set 4.5

Perform the following divisions of polynomials by monomials.

1. $\dfrac{8x^3 + 12x^2}{4x}$

2. $\dfrac{15x^4 - 25x^3}{5x^2}$

3. $\dfrac{-24x^7 + 36x^5}{4x^3}$

4. $\dfrac{-48x^8 - 72x^6}{-8x^4}$

5. $\dfrac{14a^3 - 12a^2 - 8a}{2a}$

6. $\dfrac{30a^5 - 24a^3 + 54a^2}{-6a}$

7. $\dfrac{15x^5 - 17x^3 - 19x}{-x}$

8. $\dfrac{18x^3y^2 + 27x^2y^3}{3xy}$

9. $\dfrac{25x^4y^5 - 45x^3y^7}{-5x^2y^3}$

10. $\dfrac{-20a^3b^2 - 44a^4b^5}{-4a^2b}$

11. $\dfrac{9a^2b^3 - 7a^3b^2 - 5ab^2}{-ab}$

12. $\dfrac{21x^5y^6 + 28x^4y^3 - 35x^5y^4}{7x^2y^3}$

Perform the following divisions.

13. $\dfrac{x^2 - 3x - 54}{x + 6}$

14. $\dfrac{x^2 - 19x + 84}{x - 7}$

15. $\dfrac{2x^2 + 17x + 30}{x + 6}$

16. $\dfrac{3x^2 + 4x - 7}{x - 1}$

17. $(12x^2 + 7x - 10) \div (3x - 2)$

18. $(20x^2 - 39x + 18) \div (5x - 6)$

19. $(y^3 - y^2 - 13y - 3) \div (y + 3)$

20. $(y^3 - 7y^2 + 16y - 12) \div (y - 2)$

21. $\dfrac{3t^3 + 7t^2 - 10t - 4}{3t + 1}$

22. $\dfrac{4t^3 - 17t^2 + 7t + 10}{4t - 5}$

23. $\dfrac{x^2 + 3x - 38}{x + 8}$

24. $\dfrac{x^2 - 2x - 20}{x - 6}$

25. $(3x^2 - 2x + 5) \div (x - 1)$

26. $(4x^2 + 3x - 2) \div (x + 3)$

27. $(6x^2 + 19x + 11) \div (3x + 2)$

28. $(20x^2 + 3x - 1) \div (5x + 2)$

29. $(t^3 + 8) \div (t + 2)$

30. $(x^3 - 27) \div (x - 3)$

31. $\dfrac{x^3 - 8}{x - 4}$

32. $\dfrac{x^3 + 64}{x + 1}$

33. $\dfrac{3x^3 + 2x^2 - 5x - 1}{x^2 + 2x}$

34. $\dfrac{4x^3 - 5x^2 + 2x - 6}{x^2 - 3x}$

35. $\dfrac{5y^3 - 6y^2 - 7y - 2}{y^2 - y}$

36. $\dfrac{8y^3 - y^2 - y + 5}{y^2 + y}$

37. $\dfrac{4a^3 - 2a^2 + 7a - 1}{a^2 - 2a + 3}$

38. $\dfrac{5a^3 + 7a^2 - 2a - 9}{a^2 + 3a - 4}$

39. $\dfrac{3x^2 - 2xy - 8y^2}{x - 2y}$

40. $\dfrac{4a^2 - 8ab + 4b^2}{a - b}$

4.6 Fractional Equations

The fractional equations used in this text are of two basic types. One type has only constants as denominators and the other type contains variables in the denominators.

In Chapter 2 we considered fractional equations involving only constants in the denominators. Let's briefly review our approach to solving such equations as we will be using that same basic technique to solve any type of fractional equation.

Example 1   Solve $\dfrac{x - 2}{3} + \dfrac{x + 1}{4} = \dfrac{1}{6}$.

Solution

$$\frac{x-2}{3} + \frac{x+1}{4} = \frac{1}{6}$$

$$12\left(\frac{x-2}{3} + \frac{x+1}{4}\right) = 12\left(\frac{1}{6}\right)$$ Multiply both sides by 12, which is the LCD of all of the denominators.

$$4(x-2) + 3(x+1) = 2$$

$$4x - 8 + 3x + 3 = 2$$

$$7x - 5 = 2$$

$$7x = 7$$

$$x = 1.$$

The solution set is $\{1\}$. (Check it!) ●

If an equation contains a variable (or variables) in one or more denominators, then we proceed in essentially the same way as in Example 1 above, *except we must avoid any value of the variable that makes a denominator zero.* Consider the following examples.

Example 2   Solve $\dfrac{5}{n} + \dfrac{1}{2} = \dfrac{9}{n}$.

Solution

First, we need to realize that *n cannot equal zero.* (Let's indicate this restriction so that it is not forgotten!) Then we can proceed as follows.

$$\frac{5}{n} + \frac{1}{2} = \frac{9}{n} \qquad (n \neq 0)$$

$$2n\left(\frac{5}{n} + \frac{1}{2}\right) = 2n\left(\frac{9}{n}\right)$$ Multiply both sides by the LCD, which is $2n$.

$$10 + n = 18$$

$$n = 8.$$

The solution set is $\{8\}$. (Check it!) ●

Example 3   Solve $\dfrac{35 - x}{x} = 7 + \dfrac{3}{x}$.

Solution

$$\frac{35 - x}{x} = 7 + \frac{3}{x} \qquad (x \neq 0)$$

$$x\left(\frac{35 - x}{x}\right) = x\left(7 + \frac{3}{x}\right)$$ Multiply both sides by x.

$$35 - x = 7x + 3$$

$$32 = 8x$$

$$4 = x.$$

The solution set is $\{4\}$. ●

Example 4   Solve $\dfrac{3}{a-2} = \dfrac{4}{a+1}$.

Solution

$$\dfrac{3}{a-2} = \dfrac{4}{a+1} \qquad (a \neq 2 \quad \text{and} \quad a \neq -1)$$

$$(a-2)(a+1)\left(\dfrac{3}{a-2}\right) = (a-2)(a+1)\left(\dfrac{4}{a+1}\right) \qquad \begin{array}{l}\text{Multiply both sides by} \\ (a-2)(a+1).\end{array}$$

$$3(a+1) = 4(a-2)$$
$$3a + 3 = 4a - 8$$
$$11 = a.$$

The solution set is $\{11\}$.   ●

Keep in mind that listing the restrictions at the beginning of a problem does not replace *checking* the potential solutions. In Example 4 above, 11 needs to be checked in the original equation.

Example 5   Solve $\dfrac{a}{a-2} + \dfrac{2}{3} = \dfrac{2}{a-2}$.

Solution

$$\dfrac{a}{a-2} + \dfrac{2}{3} = \dfrac{2}{a-2} \qquad (a \neq 2)$$

$$3(a-2)\left(\dfrac{a}{a-2} + \dfrac{2}{3}\right) = 3(a-2)\left(\dfrac{2}{a-2}\right) \qquad \text{Multiply both sides by } 3(a-2).$$

$$3a + 2(a-2) = 6$$
$$3a + 2a - 4 = 6$$
$$5a = 10$$
$$a = 2.$$

Because our initial restriction was $a \neq 2$, we conclude that this equation *has no solution.* Thus, the solution set is $\emptyset$.   ●

Example 5 illustrates the importance of recognizing the restrictions that must be made to exclude division by zero.

Ratio and Proportion

A **ratio** is the comparison of two numbers by division. The fractional form is frequently used to express ratios. For example, the ratio of a to b can be written as $\dfrac{a}{b}$. A statement of equality between two ratios is called a **proportion**. Thus, if $\dfrac{a}{b}$ and $\dfrac{c}{d}$ are two equal ratios, the proportion $\dfrac{a}{b} = \dfrac{c}{d}$ ($b \neq 0$ and $d \neq 0$) can be formed.

An important property of proportions can be deduced as follows:

$$\frac{a}{b} = \frac{c}{d} \quad (b \neq 0 \quad \text{and} \quad d \neq 0)$$

$$bd\left(\frac{a}{b}\right) = bd\left(\frac{c}{d}\right) \qquad \text{Multiply both sides by } bd.$$

$$ad = bc.$$

This is sometimes referred to as the

Cross-Multiplication Property of Proportions

If $\dfrac{a}{b} = \dfrac{c}{d}$, $b \neq 0$ and $d \neq 0$ then $ad = bc$.

Some fractional equations can be treated as proportions and solved by using the cross-multiplication idea, as the next examples illustrate.

Example 6   Solve $\dfrac{5}{x+6} = \dfrac{7}{x-5}$.

Solution
$$\frac{5}{x+6} = \frac{7}{x-5} \qquad (x \neq -6 \quad \text{and} \quad x \neq 5)$$

$$5(x-5) = 7(x+6) \qquad \text{Apply the cross-multiplication property.}$$

$$5x - 25 = 7x + 42$$

$$-67 = 2x$$

$$-\frac{67}{2} = x.$$

The solution set is $\left\{-\dfrac{67}{2}\right\}$.   ●

Example 7   Solve $\dfrac{x}{7} = \dfrac{4}{x+3}$.

Solution
$$\frac{x}{7} = \frac{4}{x+3} \qquad (x \neq -3)$$

$$x(x+3) = 7(4) \qquad \text{Cross-multiplication property.}$$

$$x^2 + 3x = 28$$

$$x^2 + 3x - 28 = 0$$

$$(x+7)(x-4) = 0$$

$$x + 7 = 0 \quad \text{or} \quad x - 4 = 0$$

$$x = -7 \quad \text{or} \quad x = 4.$$

The solution set is $\{-7, 4\}$. (Check these solutions in the original equation.)   ●

Problem Solving

Being able to solve fractional equations broadens our base for working word problems. We are now ready to tackle some word problems that translate into fractional equations.

Problem 1　The sum of a number and its reciprocal is $\dfrac{10}{3}$. Find the number.

Solution　Let n represent the number. Then $\dfrac{1}{n}$ represents its reciprocal.

$$n + \frac{1}{n} = \frac{10}{3}$$

$$3n\left(n + \frac{1}{n}\right) = 3n\left(\frac{10}{3}\right)$$

$$3n^2 + 3 = 10n$$

$$3n^2 - 10n + 3 = 0$$

$$(3n - 1)(n - 3) = 0$$

$$3n - 1 = 0 \quad \text{or} \quad n - 3 = 0$$

$$3n = 1 \quad \text{or} \quad n = 3$$

$$n = \frac{1}{3} \quad \text{or} \quad n = 3.$$

If the number is $\dfrac{1}{3}$, then its reciprocal is $\dfrac{1}{\frac{1}{3}} = 3$. If the number is 3, then its reciprocal is $\dfrac{1}{3}$.

Now let's consider a problem for which we can use the relationship

$$\frac{\text{Dividend}}{\text{Divisor}} = \text{Quotient} + \frac{\text{Remainder}}{\text{Divisor}}$$

as a guideline.

Problem 2　The sum of two numbers is 52. If the larger is divided by the smaller, the quotient is 9 and the remainder is 2. Find the numbers.

Solution　Let n represent the smaller number. Then $52 - n$ represents the larger number. Using the relationship discussed above as a guideline we can proceed as follows.

$$\frac{\text{Dividend}}{\text{Divisor}} = \text{Quotient} + \frac{\text{Remainder}}{\text{Divisor}}$$

$$\frac{52 - n}{n} = 9 + \frac{2}{n}$$

$$n\left(\frac{52 - n}{n}\right) = n\left(9 + \frac{2}{n}\right)$$

$$52 - n = 9n + 2$$

$$50 = 10n$$

$$5 = n.$$

If $n = 5$, then $52 - n$ equals 47. The numbers are 5 and 47.   ●

Some problems can be very conveniently set up and solved using the concepts of ratio and proportion. Let's conclude this section with two such examples.

Problem 3   On a certain map, $1\frac{1}{2}$ inches represents 25 miles. If two cities are $5\frac{1}{4}$ inches apart on the map, find the number of miles between the cities.

Solution   Let m represent the number of miles between the two cities. The following proportion can be set up and solved:

$$\frac{1\frac{1}{2}}{25} = \frac{5\frac{1}{4}}{m}$$

$$\frac{\frac{3}{2}}{25} = \frac{\frac{21}{4}}{m}$$

$$\frac{3}{2}m = 25\left(\frac{21}{4}\right) \qquad \text{Cross-multiplication property}$$

$$\frac{2}{3}\left(\frac{3}{2}m\right) = \frac{\cancel{2}}{\cancel{3}}(25)\left(\frac{\cancel{21}^{7}}{\cancel{4}_{2}}\right) \qquad \text{Multiply both sides by } \frac{2}{3}.$$

$$m = \frac{175}{2} = 87\frac{1}{2}.$$

The distance between the two cities is $87\frac{1}{2}$ miles.   ●

Problem 4   A sum of $750 is to be divided between two people in the ratio of 2 to 3. How much does each person receive?

Solution   Let d represent the amount of money to be received by one person. Then $750 - d$ represents the amount for the other person.

$$\frac{d}{750 - d} = \frac{2}{3}$$
$$3d = 2(750 - d)$$
$$3d = 1500 - 2d$$
$$5d = 1500$$
$$d = 300.$$

If $d = 300$, then $750 - d$ equals 450. Therefore, one person receives $300 and the other person receives $450.   ●

Problem Set 4.6

Solve each of the following equations.

1. $\dfrac{x-1}{4} + \dfrac{x+3}{8} = \dfrac{1}{2}$

2. $\dfrac{x+5}{6} + \dfrac{2x+1}{5} = \dfrac{7}{15}$

3. $\dfrac{2n+1}{7} - \dfrac{3n+2}{4} = \dfrac{5}{2}$

4. $\dfrac{5n-1}{4} - \dfrac{2n-3}{10} = \dfrac{3}{5}$

5. $\dfrac{5}{n} + \dfrac{1}{3} = \dfrac{8}{n}$

6. $\dfrac{7}{n} + \dfrac{2}{5} = \dfrac{11}{n}$

7. $\dfrac{5}{3n} - \dfrac{1}{9} = \dfrac{1}{n}$

8. $\dfrac{9}{n} - \dfrac{1}{4} = \dfrac{7}{n}$

9. $\dfrac{4}{5x} + \dfrac{3}{4} = \dfrac{3}{4x}$

10. $\dfrac{6}{7x} - \dfrac{1}{6} = \dfrac{5}{6x}$

11. $\dfrac{38-n}{n} = 5 + \dfrac{2}{n}$

12. $\dfrac{65-n}{n} = 4 + \dfrac{5}{n}$

13. $\dfrac{x}{46-x} = 5 + \dfrac{4}{46-x}$

14. $\dfrac{x}{79-x} = 5 + \dfrac{1}{79-x}$

15. $n + \dfrac{1}{n} = \dfrac{5}{2}$

16. $n + \dfrac{1}{n} = \dfrac{37}{6}$

17. $n + \dfrac{1}{n} = \dfrac{13}{6}$

18. $n + \dfrac{1}{n} = \dfrac{25}{12}$

19. $\dfrac{3}{2x-1} = \dfrac{5}{3x+2}$

20. $\dfrac{-2}{5x-3} = \dfrac{4}{4x-1}$

21. $\dfrac{x}{x+1} + 3 = \dfrac{4}{x+1}$

22. $\dfrac{x}{x-6} - 3 = \dfrac{6}{x-6}$

23. $\dfrac{a}{a-3} - \dfrac{3}{2} = \dfrac{3}{a-3}$

24. $\dfrac{a}{a+5} - 2 = \dfrac{3a}{a+5}$

25. $\dfrac{x}{x-2} + 1 = \dfrac{8}{x-1}$

26. $\dfrac{x}{x+1} - 2 = \dfrac{3}{x-3}$

27. $-1 + \dfrac{2x}{x+3} = \dfrac{-4}{x+4}$

28. $2 - \dfrac{3x}{x-4} = \dfrac{14}{x+7}$

29. $\dfrac{n}{n+1} + \dfrac{1}{2} = \dfrac{-2}{n+2}$

30. $\dfrac{3n}{n-1} - \dfrac{1}{3} = \dfrac{-40}{3(n-6)}$

31. $\dfrac{s}{2s-1} - 3 = \dfrac{-32}{3(s+5)}$

32. $\dfrac{3s}{s+2} + 1 = \dfrac{35}{2(3s+1)}$

33. $\dfrac{x}{x-4} - \dfrac{2}{x+3} = \dfrac{20}{(x-4)(x+3)}$

34. $\dfrac{2x}{x-2} + \dfrac{15}{(x-2)(x-5)} = \dfrac{3}{x-5}$

Use the cross-multiplication property of proportions to help solve each of the following equations.

35. $\dfrac{3}{x-1} = \dfrac{4}{x+2}$

36. $\dfrac{5}{x+6} = \dfrac{6}{x-3}$

37. $\dfrac{7}{x+4} = \dfrac{3}{x-8}$

38. $\dfrac{-2}{x-5} = \dfrac{1}{x+9}$

39. $\dfrac{5}{2a-1} = \dfrac{-6}{3a+2}$

40. $\dfrac{-3}{4x+5} = \dfrac{2}{5x-7}$

41. $\dfrac{n}{5} = \dfrac{10}{n-5}$

42. $\dfrac{n+6}{27} = \dfrac{1}{n}$

43. $\dfrac{x}{-4} = \dfrac{3}{12x-25}$

44. $\dfrac{3x-7}{10} = \dfrac{2}{x}$

Set up an algebraic equation and solve each of the following problems.

45. The sum of a number and its reciprocal is $\dfrac{29}{10}$. Find the number.

46. The sum of a number and its reciprocal is $\dfrac{53}{14}$. Find the number.

47. The denominator of a rational number is 9 less than three times the numerator. The number in its simplest form is $\dfrac{3}{8}$. Find the rational number.

48. What number must be added to the numerator and denominator of $\dfrac{1}{3}$ to produce a rational number equivalent to $\dfrac{4}{5}$?

49. The sum of two numbers is 69. If the larger is divided by the smaller, the quotient is 10 and the remainder is 3. Find the numbers.

50. One number is 65 larger than another number. If the larger number is divided by the smaller, the quotient is 6 and the remainder is 5. Find the numbers.

Use the concepts of ratio and proportion to help solve the following problems.

51. A blueprint has a scale of 1 inch represents 5 feet. Find the dimensions of a rectangular room that measures $3\frac{1}{2}$ inches by $5\frac{3}{4}$ inches on the blueprint.

52. A sum of $1750 is to be divided between two people in the ratio of 3 to 4. How much does each person receive?

53. The ratio of male students to female students at a certain university is 5 to 7. If there is a total of 16,200 students, find the number of male students and the number of female students.

54. If a home valued at $50,000 is assessed $900 in real estate taxes, at the same rate how much are the taxes on a home valued at $60,000?

55. An inheritance of $300,000 is to be divided between a son and the local heart fund in the ratio of 3 to 1. How much money will the son receive?

56. A 20-foot board is to be cut into two pieces whose lengths are in the ratio of 7 to 3. Find the lengths of the two pieces.

57. The perimeter of a rectangle is 114 centimeters. If the ratio of its width to its length is 7 to 12, find the dimensions of the rectangle.

58. In a certain precinct, 1150 people voted in the last election. If the ratio of female voters to male voters was 3 to 2, how many females and how many males voted?

59. The total value of a house and a lot is $68,000. If the ratio of the value of the house to the value of the lot is 7 to 1, find the value of the house.

60. Together, Laura and Tammy sold $120.75 worth of candy for the school fair. If the ratio of Tammy's sales to Laura's sales was 4 to 3, how much did each sell?

61. The ratio of the complement of an angle to its supplement is 1 to 4. Find the measure of the angle.

62. One angle of a triangle has a measure of 60° and the measures of the other two angles are in a ratio of 2 to 3. Find the measures of the other two angles.

4.7 More Fractional Equations and Applications

Let's begin this section by considering a few more fractional equations. We will continue to solve them using the same basic technique as in the previous section. That is, we will multiply both sides of the equation by the least common denominator of all of the denominators in the equation, keeping in mind the necessary restrictions to avoid division by zero. Some of the denominators in these problems will require factoring before a least common denominator can be determined.

Example 1   Solve $\dfrac{x}{2x - 8} + \dfrac{16}{x^2 - 16} = \dfrac{1}{2}$.

Solution

$$\frac{x}{2x - 8} + \frac{16}{x^2 - 16} = \frac{1}{2}$$

$$\frac{x}{2(x - 4)} + \frac{16}{(x + 4)(x - 4)} = \frac{1}{2}, \quad (x \neq 4 \quad \text{and} \quad x \neq -4)$$

$$2(x - 4)(x + 4)\left(\frac{x}{2(x - 4)} + \frac{16}{(x + 4)(x - 4)}\right) = 2(x + 4)(x - 4)\left(\frac{1}{2}\right)$$

$$x(x + 4) + 2(16) = (x + 4)(x - 4)$$

$$x^2 + 4x + 32 = x^2 - 16$$

$$4x = -48$$

$$x = -12.$$

The solution set is $\{-12\}$. (Perhaps you should check it!)   ●

In Example 1, notice that the restrictions were not indicated until the denominators were expressed in factored form. It is usually easier to determine the necessary restrictions at this step.

Example 2   Solve $\dfrac{3}{n - 5} - \dfrac{2}{2n + 1} = \dfrac{n + 3}{2n^2 - 9n - 5}$.

Solution

$$\frac{3}{n - 5} - \frac{2}{2n + 1} = \frac{n + 3}{2n^2 - 9n - 5}$$

$$\frac{3}{n - 5} - \frac{2}{2n + 1} = \frac{n + 3}{(2n + 1)(n - 5)} \quad \left(n \neq -\frac{1}{2} \quad \text{and} \quad n \neq 5\right)$$

$$(2n + 1)(n - 5)\left(\frac{3}{n - 5} - \frac{2}{2n + 1}\right) = (2n + 1)(n - 5)\left(\frac{n + 3}{(2n + 1)(n - 5)}\right)$$

$$3(2n + 1) - 2(n - 5) = n + 3$$

$$6n + 3 - 2n + 10 = n + 3$$

$$4n + 13 = n + 3$$

$$3n = -10$$

$$n = -\frac{10}{3}.$$

The solution set is $\left\{-\dfrac{10}{3}\right\}$.   ●

Example 3   Solve $2 + \dfrac{4}{x - 2} = \dfrac{8}{x^2 - 2x}$.

Solution

$$2 + \frac{4}{x - 2} = \frac{8}{x^2 - 2x}$$

$$2 + \frac{4}{x - 2} = \frac{8}{x(x - 2)} \qquad (x \neq 0 \quad \text{and} \quad x \neq 2)$$

$$x(x - 2)\left(2 + \frac{4}{x - 2}\right) = x(x - 2)\left(\frac{8}{x(x - 2)}\right)$$

$$2x(x - 2) + 4x = 8$$

$$2x^2 - 4x + 4x = 8$$

$$2x^2 = 8$$

$$x^2 = 4$$

$$x^2 - 4 = 0$$

$$(x + 2)(x - 2) = 0$$

$$x + 2 = 0 \quad \text{or} \quad x - 2 = 0$$

$$x = -2 \quad \text{or} \quad x = 2.$$

Since our initial restriction indicated that $x \neq 2$, then the *only solution* is -2. Thus, the solution set is $\{-2\}$.   ●

In Section 2.4 we discussed using the properties of equality to change the form of various formulas. For example, we considered the simple interest formula "$A = P + Prt$" and changed its form by solving for P as follows.

$$A = P + Prt$$

$$A = P(1 + rt)$$

$$\frac{A}{1 + rt} = P \qquad \text{Multiply both sides by } \frac{1}{1 + rt}.$$

If the formula is in the form of a fractional equation, then the techniques of these last two sections are applicable. Consider the following example.

Example 4   If the original cost of some business property is C dollars and it is depreciated linearly over N years, its value, V, at the end of T years is given by

$$V = C\left(1 - \frac{T}{N}\right).$$

Solve this formula for N in terms of V, C, and T.

Solution

$$V = C\left(1 - \frac{T}{N}\right)$$

$$V = C - \frac{CT}{N}$$

$$N(V) = N\left(C - \frac{CT}{N}\right) \qquad \text{Multiply both sides by } N.$$

$$NV = NC - CT$$
$$NV - NC = -CT$$
$$N(V - C) = -CT$$
$$N = \frac{-CT}{V - C}$$
$$= -\frac{CT}{V - C}.$$

Problem Solving

In Section 2.4 we solved some uniform motion problems. The formula "$d = rt$" was used in the analysis of the problems and we used guidelines involving distance relationships. Now let's consider some uniform motion problems for which guidelines involving either "times" or "rates" are appropriate. These problems will generate fractional equations to solve.

Problem 1   An airplane travels 2050 miles in the same time that a car travels 260 miles. If the rate of the plane is 358 miles per hour greater than the rate of the car, find the rate of each.

Solution   Let r represent the rate of the car. Then $r + 358$ represents the rate of the plane. The fact that the times are equal can be used as a guideline.

$$\underset{\downarrow}{\text{time of plane}} \quad \underset{\downarrow}{\text{equals}} \quad \underset{\downarrow}{\text{time of car}}$$

$$\frac{\text{distance of plane}}{\text{rate of plane}} \quad = \quad \frac{\text{distance of car}}{\text{rate of car}}$$

$$\frac{2050}{r + 358} = \frac{260}{r}$$
$$2050r = 260(r + 358)$$
$$2050r = 260r + 93{,}080$$
$$1790r = 93{,}080$$
$$r = 52.$$

If $r = 52$, then $r + 358$ equals 410. Thus, the rate of the car is 52 m.p.h. and the rate of the plane is 410 m.p.h.

Problem 2   It takes a freight train 2 hours longer to travel 300 miles than it takes an express train to travel 280 miles. The rate of the express is 20 miles per hour greater than the rate of the freight. Find the times and rates of both trains.

Solution   Let t represent the time of the express. Then $t + 2$ represents the time of the freight.

Let's record the information of this problem in a table as follows:

	distance	time	$r = \frac{d}{t}$
express	280	t	$\dfrac{280}{t}$
freight	300	$t + 2$	$\dfrac{300}{t + 2}$

The fact that "the rate of the express is 20 miles per hour greater than the rate of the freight" can be used as a guideline.

rate of express   equals   rate of freight plus 20

$$\dfrac{280}{t} \qquad = \qquad \dfrac{300}{t + 2} + 20$$

$$t(t + 2)\left(\frac{280}{t}\right) = t(t + 2)\left(\frac{300}{t + 2} + 20\right)$$

$$280(t + 2) = 300t + 20t(t + 2)$$

$$280t + 560 = 300t + 20t^2 + 40t$$

$$280t + 560 = 340t + 20t^2$$

$$0 = 20t^2 + 60t - 560$$

$$0 = t^2 + 3t - 28$$

$$0 = (t + 7)(t - 4)$$

$$t + 7 = 0 \quad \text{or} \quad t - 4 = 0$$

$$t = -7 \quad \text{or} \quad t = 4.$$

The negative solution must be discarded, so the time of the express (t) is 4 hours and the time of the freight $(t + 2)$ is 6 hours. The rate of the express $\left(\dfrac{280}{t}\right)$ is $\dfrac{280}{4} = 70$ miles per hour and the rate of the freight $\left(\dfrac{300}{t + 2}\right)$ is $\dfrac{300}{6} = 50$ miles per hour.   ●

Uniform motion problems are a special case of a larger type of problems referred to as "rate-time" problems. For example, if a certain machine can produce 150 items in 10 minutes, then we say that the machine is producing at a rate of $\dfrac{150}{10} = 15$ items per minute. Likewise, if a person can do a certain job in 3 hours, then, assuming a constant rate of work we say that the person is working at a rate of $\dfrac{1}{3}$ of the job per hour. In general, if Q is the quantity of something done in t units of time, then the rate, r, is given by $r = \dfrac{Q}{t}$. The rate is stated in terms of *so much quantity per unit of time*. (In uniform motion problems the "quantity" is distance.) Let's consider some examples of "rate-time" problems.

Problem 3   If Jim can mow a lawn in 50 minutes and his son, Todd, can mow the same lawn in 40 minutes, how long will it take them to mow the lawn if they work together?

Solution   Jim's rate is $\dfrac{1}{50}$ of the lawn per minute and Todd's rate is $\dfrac{1}{40}$ of the lawn per minute.

If we let m represent the number of minutes that they work together, then $\dfrac{1}{m}$ represents their rate when working together. Therefore, since the sum of the individual rates must equal the rate when working together, we can set up and solve the following equation.

$$\frac{1}{50} + \frac{1}{40} = \frac{1}{m}$$

$$200m\left(\frac{1}{50} + \frac{1}{40}\right) = 200m\left(\frac{1}{m}\right)$$

$$4m + 5m = 200$$

$$9m = 200$$

$$m = \frac{200}{9} = 22\frac{2}{9}.$$

It should take them $22\dfrac{2}{9}$ minutes.   ●

Problem 4   Working together, Linda and Kathy can type a term paper in $3\dfrac{3}{5}$ hours. Linda can type the paper by herself in 6 hours. How long would it take Kathy to type the paper by herself?

Solution   Their rate working together is $\dfrac{1}{3\frac{3}{5}} = \dfrac{1}{\frac{18}{5}} = \dfrac{5}{18}$ of the job per hour and Linda's rate

is $\dfrac{1}{6}$ of the job per hour. If we let h represent the number of hours that it would take

Kathy by herself, then her rate is $\dfrac{1}{h}$ of the job per hour. Thus, we have

Linda's rate   Kathy's rate   combined rate
$$\underset{\downarrow}{\frac{1}{6}} \quad + \quad \underset{\downarrow}{\frac{1}{h}} \quad = \quad \underset{\downarrow}{\frac{5}{18}}.$$

Solving this equation yields

$$18h\left(\frac{1}{6} + \frac{1}{h}\right) = 18h\left(\frac{5}{18}\right)$$
$$3h + 18 = 5h$$
$$18 = 2h$$
$$9 = h.$$

It would take Kathy 9 hours to type the paper by herself.

One final example of this section illustrates another approach that some people find meaningful for rate-time problems. This approach has you think in terms of fractional parts of the job. For example, if a person can do a certain job in 5 hours, then at the end of 2 hours he or she has done $\frac{2}{5}$ of the job. (Again a constant rate of work is assumed.) At the end of 4 hours, he or she has finished $\frac{4}{5}$ of the job and, in general, at the end of h hours, he or she has done $\frac{h}{5}$ of the job. Let's see how this works in a problem.

Problem 5   It takes Pat 12 hours to complete a task. After he had been working for 3 hours, he was joined by his brother, Mike, and together they finished the task in 5 hours. How long would it take Mike to do the job by himself?

Solution   Let h represent the number of hours that it would take Mike by himself. Since Pat has been working for 3 hours, he has done $\frac{3}{12} = \frac{1}{4}$ of the job before Mike joins him. Thus, there is $\frac{3}{4}$ of the original job to be done while working together. We can set up and solve the following equation.

fractional part of the remaining $\frac{3}{4}$ of the job that Pat does		fractional part of the remaining $\frac{3}{4}$ of the job that Mike does		
↓		↓		
$\dfrac{5}{12}$	$+$	$\dfrac{5}{h}$	$=$	$\dfrac{3}{4}$

$$12h\left(\frac{5}{12} + \frac{5}{h}\right) = 12h\left(\frac{3}{4}\right)$$
$$5h + 60 = 9h$$
$$60 = 4h$$
$$15 = h.$$

It would take Mike 15 hours to do the entire job by himself.

Problem Set 4.7

Solve each of the following equations.

1. $\dfrac{x}{3x-6}+\dfrac{4}{x^2-4}=\dfrac{1}{3}$

2. $\dfrac{x}{4x-4}+\dfrac{5}{x^2-1}=\dfrac{1}{4}$

3. $\dfrac{3x}{5x+5}-\dfrac{2}{x^2-1}=\dfrac{3}{5}$

4. $\dfrac{5x}{2x+6}-\dfrac{4}{x^2-9}=\dfrac{5}{2}$

5. $\dfrac{2}{n+3}+\dfrac{3}{n-4}=\dfrac{2n-1}{n^2-n-12}$

6. $\dfrac{3}{n-5}+\dfrac{4}{n+7}=\dfrac{2n+11}{n^2+2n-35}$

7. $2+\dfrac{4}{t-1}=\dfrac{4}{t^2-t}$

8. $3+\dfrac{6}{t-3}=\dfrac{6}{t^2-3t}$

9. $\dfrac{5y-4}{6y^2+y-12}-\dfrac{2}{2y+3}=\dfrac{5}{3y-4}$

10. $\dfrac{7y+2}{12y^2+11y-15}-\dfrac{1}{3y+5}=\dfrac{2}{4y-3}$

11. $\dfrac{-2}{3x+2}+\dfrac{x-1}{9x^2-4}=\dfrac{3}{12x-8}$

12. $\dfrac{-1}{2x-5}+\dfrac{2x-4}{4x^2-25}=\dfrac{5}{6x+15}$

13. $3+\dfrac{9}{n-3}=\dfrac{27}{n^2-3n}$

14. $1+\dfrac{1}{n-1}=\dfrac{1}{n^2-n}$

15. $\dfrac{a}{a+2}+\dfrac{3}{a+4}=\dfrac{14}{a^2+6a+8}$

16. $\dfrac{a}{a-5}+\dfrac{2}{a-6}=\dfrac{2}{a^2-11a+30}$

17. $\dfrac{x}{x-4}-\dfrac{2}{x+8}=\dfrac{63}{x^2+4x-32}$

18. $\dfrac{2x}{x+3}-\dfrac{3}{x-6}=\dfrac{29}{x^2-3x-18}$

19. $\dfrac{1}{3x+4}+\dfrac{6}{6x^2+5x-4}=\dfrac{x}{2x-1}$

20. $\dfrac{2}{2x-3}-\dfrac{2}{10x^2-13x-3}=\dfrac{x}{5x+1}$

21. $\dfrac{n}{n+3}+\dfrac{1}{n-4}=\dfrac{11-n}{n^2-n-12}$

22. $\dfrac{2}{n-2}-\dfrac{n}{n+5}=\dfrac{10n+15}{n^2+3n-10}$

23. $\dfrac{2}{n^2+4n}+\dfrac{3}{n^2-3n-28}=\dfrac{5}{n^2-6n-7}$

24. $\dfrac{1}{2x^2-x-1}+\dfrac{3}{2x^2+x}=\dfrac{2}{x^2-1}$

25. $\dfrac{x+1}{2x^2+7x-4}-\dfrac{x}{2x^2-7x+3}=\dfrac{1}{x^2+x-12}$

26. $\dfrac{2n}{6n^2+7n-3}-\dfrac{n-3}{3n^2+11n-4}=\dfrac{5}{2n^2+11n+12}$

27. $\dfrac{2t}{2t^2+9t+10}+\dfrac{1-3t}{3t^2+4t-4}=\dfrac{4}{6t^2+11t-10}$

28. $\dfrac{4t}{4t^2-t-3}+\dfrac{2-3t}{3t^2-t-2}=\dfrac{1}{12t^2+17t+6}$

29. $\dfrac{x}{2x^2+5x}-\dfrac{x}{2x^2+7x+5}=\dfrac{2}{x^2+x}$

30. $\dfrac{x+1}{x^3-9x}-\dfrac{1}{2x^2+x-21}=\dfrac{1}{2x^2+13x+21}$

For Problems 31–44, solve for the indicated variable.

31. $V = C\left(1 - \dfrac{T}{N}\right)$ for T

32. $I = \dfrac{100M}{C}$ for M

33. $\dfrac{1}{R} = \dfrac{1}{S} + \dfrac{1}{T}$ for R

34. $\dfrac{R}{S} = \dfrac{T}{S + T}$ for R

35. $y = \dfrac{3}{4}x - \dfrac{2}{3}$ for x

36. $y = \dfrac{5}{6}x + \dfrac{2}{9}$ for x

37. $\dfrac{7}{y - 3} = \dfrac{3}{x + 1}$ for y

38. $\dfrac{-2}{x - 4} = \dfrac{5}{y - 1}$ for y

39. $\dfrac{y + 5}{x - 2} = \dfrac{3}{7}$ for y

40. $\dfrac{y - 1}{x + 6} = \dfrac{-2}{3}$ for y

41. $\dfrac{y - b}{x} = m$ for y

42. $\dfrac{x}{a} + \dfrac{y}{b} = 1$ for y

43. $y = -\dfrac{a}{b}x + \dfrac{c}{d}$ for x

44. $\dfrac{y - 1}{x - 3} = \dfrac{b - 1}{a - 3}$ for y

Set up an equation and solve each of the following problems.

45. Wendy rides her bicycle 30 miles in the same time that it takes Kim to ride her bicycle 20 miles. If Wendy rides 5 miles per hour faster than Kim, find the rate of each.

46. Kent drives his Mazda 270 miles in the same time that Dave drives his Datsun 250 miles. If Kent averages 4 miles per hour faster than Dave, find their rates.

47. To travel 60 miles, it takes Sue, riding a moped, 2 hours less time than it takes Ann, riding a bicycle, to travel 50 miles. Sue travels 10 miles per hour faster than Ann. Find the times and rates of both girls.

48. Plane A can travel 1400 miles in one hour less time than it takes plane B to travel 2000 miles. The rate of plane B is 50 miles per hour faster than the rate of plane A. Find the times and rates of both planes.

49. Nick jogs for 10 miles and then walks another 10 miles. He jogs $2\dfrac{1}{2}$ miles per hour faster than he walks and the entire distance of 20 miles takes 6 hours. Find the rate that he walks and the rate that he jogs.

50. Debbie rode her bicycle out into the country for a distance of 24 miles. On the way back, she took a much shorter route of 12 miles and made the return trip in one-half hour less time. If her rate out into the country was 4 miles per hour faster than her rate on the return trip, find both rates.

51. Barry can do a certain job in 3 hours, while it takes Roy 5 hours to do the same job. How long would it take them to do the job working together?

52. An inlet pipe can fill a tank in 10 minutes. A drain can empty the tank in 12 minutes. If the tank is empty and both the pipe and drain are open, how long will it take before the tank overflows?

53. If two inlet pipes are both open, they can fill a pool in 1 hour and 12 minutes. One of the pipes can fill the pool by itself in 2 hours. How long would it take the other pipe to fill the pool by itself?

54. It takes Amy twice as long to deliver papers as it does Nancy. How long would it take each if they can deliver the papers together in 40 minutes?

55. Walt can mow a lawn in 1 hour, while his son, Mike, can mow the same lawn in 50 minutes. One day Mike started mowing the lawn by himself and worked for 30 minutes. Then Walt joined him and they finished the lawn. How long did it take them to finish mowing the lawn after Walt started to help?

56. Connie can type 600 words in 5 minutes less than it takes Katie to type 600 words. If Connie types at a rate of 20 words per minute faster than Katie types, find the typing rate of each woman.

57. Al bought some golf balls for $20. The next day they were on sale for $0.50 per ball less and he bought $22.50 worth of balls. If he purchased 5 more balls on the second day than he did on the first day, how many and at what price per ball did he buy each day?

58. Dan agreed to mow a vacant lot for $12. It took him an hour longer than what he had anticipated, so he earned $1 per hour less than he originally calculated. How long had he anticipated that it would take him to mow the lot?

Chapter 4 Summary

(4.1) Any number that can be written in the form $\dfrac{a}{b}$, where a and b are integers and $b \neq 0$, is called a **rational number**.

A **rational expression** is defined as the indicated quotient of two polynomials.

The following properties pertain to rational numbers and rational expressions:

1. $\dfrac{-a}{b} = \dfrac{a}{-b} = -\dfrac{a}{b}$

2. $\dfrac{-a}{-b} = \dfrac{a}{b}$

3. $\dfrac{a \cdot k}{b \cdot k} = \dfrac{a}{b}$ (Fundamental principle of fractions)

(4.2) Multiplication and division of rational expressions are based upon the following:

1. $\dfrac{a}{b} \cdot \dfrac{c}{d} = \dfrac{ac}{bd}$ (Multiplication)

2. $\dfrac{a}{b} \div \dfrac{c}{d} = \dfrac{a}{b} \cdot \dfrac{d}{c} = \dfrac{ad}{bc}$ (Division)

(4.3) Addition and subtraction of rational expressions are based upon the following:

1. $\dfrac{a}{b} + \dfrac{c}{b} = \dfrac{a + c}{b}$ (Addition)

2. $\dfrac{a}{b} - \dfrac{c}{b} = \dfrac{a - c}{b}$ (Subtraction)

(4.4)   The following basic procedure is used to add or subtract rational expressions:

      1. Find the LCD of all denominators.

      2. Change each fraction to an equivalent fraction having the LCD as its denominator.

      3. Add or subtract numerators and place this result over the LCD.

      4. Look for possibilities to simplify the resulting fraction.

Fractional forms that contain rational numbers or rational expressions in the numerators and/or denominators are called **complex fractions**. The fundamental principle of fractions serves as a basis for simplifying complex fractions.

(4.5)   To divide a polynomial by a monomial we divide each term of the polynomial by the monomial.

The procedure for dividing a polynomial by a polynomial, other than a monomial, resembles the long division process in arithmetic. (See examples in Section 4.5.)

(4.6)   To **solve a fractional equation**, it is often easiest to begin by multiplying both sides of the equation by the LCD of all the denominators in the equation. If an equation contains a variable in one or more denominators, then we must be careful to avoid any value of the variable that makes the denominator zero.

A **ratio** is the comparison of two numbers by division. A statement of equality between two ratios is a **proportion**.

Some fractional equations can be treated as proportions and solved by applying the following property, often called the *cross-multiplication* property.

$$\text{If } \frac{a}{b} = \frac{c}{d}, \quad \text{then } ad = bc.$$

(4.7)   The techniques used to solve fractional equations can also be used to change the form of formulas that contain rational expressions and to use them in solving problems.

Chapter 4 Review Problem Set

For Problems 1–6, simplify each of the rational expressions.

1. $\dfrac{26x^2y^3}{39x^4y^2}$

2. $\dfrac{a^2 - 9}{a^2 + 3a}$

3. $\dfrac{n^2 - 3n - 10}{n^2 + n - 2}$

4. $\dfrac{x^4 - 1}{x^3 - x}$

5. $\dfrac{8x^3 - 2x^2 - 3x}{12x^2 - 9x}$

6. $\dfrac{x^4 - 7x^2 - 30}{2x^4 + 7x^2 + 3}$

For Problems 7–10, simplify each complex fraction.

7. $\dfrac{\dfrac{5}{8} - \dfrac{1}{2}}{\dfrac{1}{6} + \dfrac{3}{4}}$

8. $\dfrac{\dfrac{3}{2x} + \dfrac{5}{3y}}{\dfrac{4}{x} - \dfrac{3}{4y}}$

9. $\dfrac{\dfrac{3}{x-2} - \dfrac{4}{x^2-4}}{\dfrac{2}{x+2} + \dfrac{1}{x-2}}$

10. $1 - \dfrac{1}{2 - \dfrac{1}{x}}$

For Problems 11–22, perform the indicated operations and express answers in simplest form.

11. $\dfrac{6xy^2}{7y^3} \div \dfrac{15x^2 y}{5x^2}$

12. $\dfrac{9ab}{3a+6} \cdot \dfrac{a^2 - 4a - 12}{a^2 - 6a}$

13. $\dfrac{n^2 + 10n + 25}{n^2 - n} \cdot \dfrac{5n^3 - 3n^2}{5n^2 + 22n - 15}$

14. $\dfrac{x^2 - 2xy - 3y^2}{x^2 + 9y^2} \div \dfrac{2x^2 + xy - y^2}{2x^2 - xy}$

15. $\dfrac{2x+1}{5} + \dfrac{3x-2}{4}$

16. $\dfrac{3}{2n} + \dfrac{5}{3n} - \dfrac{1}{9}$

17. $\dfrac{3x}{x+7} - \dfrac{2}{x}$

18. $\dfrac{10}{x^2 - 5x} + \dfrac{2}{x}$

19. $\dfrac{3}{n^2 - 5n - 36} + \dfrac{2}{n^2 + 3n - 4}$

20. $\dfrac{3}{2y+3} + \dfrac{5y-2}{2y^2 - 9y - 18} - \dfrac{1}{y-6}$

21. $(18x^2 + 9x - 2) \div (3x + 2)$

22. $(3x^3 + 5x^2 - 6x - 2) \div (x + 4)$

For Problems 23–32, solve each equation.

23. $\dfrac{4x+5}{3} + \dfrac{2x-1}{5} = 2$

24. $\dfrac{3}{4x} + \dfrac{4}{5} = \dfrac{9}{10x}$

25. $\dfrac{a}{a-2} - \dfrac{3}{2} = \dfrac{2}{a-2}$

26. $\dfrac{4}{5y-3} = \dfrac{2}{3y+7}$

27. $n + \dfrac{1}{n} = \dfrac{53}{14}$

28. $\dfrac{1}{2x-7} + \dfrac{x-5}{4x^2 - 49} = \dfrac{4}{6x - 21}$

29. $\dfrac{x}{2x+1} - 1 = \dfrac{-4}{7(x-2)}$

30. $\dfrac{2x}{-5} = \dfrac{3}{4x - 13}$

31. $\dfrac{2n}{2n^2 + 11n - 21} - \dfrac{n}{n^2 + 5n - 14} = \dfrac{3}{n^2 + 5n - 14}$

32. $\dfrac{2}{t^2 - t - 6} + \dfrac{t+1}{t^2 + t - 12} = \dfrac{t}{t^2 + 6t + 8}$

33. Solve $\dfrac{y-6}{x+1} = \dfrac{3}{4}$ for y.

34. Solve $\dfrac{x}{a} - \dfrac{y}{b} = 1$ for y.

For Problems 35–40, set up an equation and solve the problem.

35. A sum of $1400 is to be divided between two people in the ratio of $\dfrac{3}{5}$. How much does each person receive?

36. Working together, Dan and Don can mow a lawn in 12 minutes. Don can mow the lawn by himself in 10 minutes less time than it takes Dan by himself. How long does it take each of them to mow the lawn alone?

37. Car A can travel 250 miles in 3 hours less time than it takes car B to travel 440 miles. The rate of car B is 5 miles per hour faster than car A. Find the rates of both cars.

38. Mark can overhaul an engine in 20 hours and Phil can do the same job by himself in 30 hours. If they both work together for a time and then Mark finishes the job by himself in 5 hours, how long did they work together?

39. Kelly contracted to paint a house for $640. It took him 20 hours longer than he had anticipated, so he earned $1.60 per hour less than he had calculated. How long had he anticipated that it would take him to paint the house?

40. Kent rode his bicycle 66 miles in $4\dfrac{1}{2}$ hours. For the first 40 miles he averaged a certain rate and then for the last 26 miles he reduced his rate by 3 miles per hour. Find his rate for the last 26 miles.

Cumulative Review Problem Set for Chapters 1, 2, 3, and 4

For Problems 1–6, evaluate each algebraic expression for the given values of the variables.

1. $x^2 - 2xy - y^2$ for $x = -2$ and $y = -4$

2. $5(2x - 3) - 3(4x - 5)$ for $x = 5$

3. $\dfrac{4a^2b^3}{12a^3b}$ for $a = 5$ and $b = -8$

4. $\dfrac{\dfrac{1}{x} + \dfrac{1}{y}}{\dfrac{1}{x} - \dfrac{1}{y}}$ for $x = 4$ and $y = 7$

5. $\dfrac{3}{n} + \dfrac{5}{2n} - \dfrac{4}{3n}$ for $n = 25$

6. $\dfrac{4}{x - 1} - \dfrac{2}{x + 2}$ for $x = \dfrac{1}{2}$

For Problems 7–16, perform the indicated operations and express answers in simplified form.

7. $4(3x - 2) - 2(4x - 1) - (2x + 5)$

8. $(3a^2b)(-2ab)(4ab^3)$

9. $(x + 3)(2x^2 - x + 4)$

10. $\dfrac{6xy^2}{14y} \cdot \dfrac{7x^2y}{8x}$

11. $\dfrac{a^2 + 6a - 40}{a^2 - 4a} \div \dfrac{2a^2 + 19a - 10}{a^3 + a^2}$

12. $\dfrac{3x + 4}{6} - \dfrac{5x - 1}{9}$

13. $\dfrac{4}{x^2 + 3x} + \dfrac{5}{x}$

14. $(7x - 3)(4x + 5)$

15. $\dfrac{3n^2 + n}{n^2 + 10n + 16} \cdot \dfrac{2n^2 - 8}{3n^3 - 5n^2 - 2n}$

16. $\dfrac{3}{5x^2 + 3x - 2} - \dfrac{2}{5x^2 - 22x + 8}$

For Problems 17–28, solve each of the equations.

17. $3(x - 2) - 2(3x + 5) = 4(x - 1)$

18. $0.06n + 0.08(n + 50) = 25$

19. $6x^2 - 24 = 0$

20. $a^2 + 14a + 49 = 0$

21. $3n^2 + 14n - 24 = 0$

22. $\dfrac{2}{5x - 2} = \dfrac{4}{6x + 1}$

23. $\dfrac{5}{6x} - \dfrac{2}{3} = \dfrac{7}{10x}$

24. $\dfrac{3}{y + 4} + \dfrac{2y - 1}{y^2 - 16} = \dfrac{-2}{y - 4}$

25. $6x^4 - 23x^2 - 4 = 0$

26. $3n^3 + 3n = 0$

27. $n^2 - 13n - 114 = 0$

28. $n(n + 5) = 36$

For Problems 29–34, solve each of the inequalities.

29. $6 - 2x \geq 10$

30. $4(2x - 1) < 3(x + 5)$

31. $\dfrac{n + 1}{4} + \dfrac{n - 2}{12} > \dfrac{1}{6}$

32. $|2x - 1| < 5$

33. $|3x + 2| > 11$

34. $\dfrac{1}{2}(3x - 1) - \dfrac{2}{3}(x + 4) \leq \dfrac{3}{4}(x - 1)$

5 Exponents and Radicals

It is not uncommon in mathematics to find two separately developed concepts that are closely related to each other. This chapter will first develop the concepts of exponent and root individually and then show how they merge to become even more functional as a unified idea.

5.1 Using Integers as Exponents

Thus far in this text only positive integers have been used as exponents. In Chapter 1 the expression b^n, where b is any real number and n is a positive integer, was defined by

$$b^n = b \cdot b \cdot b \cdots b \qquad (n \text{ factors of } b).$$

Then in Chapter 3 some of the parts of the following property served as a basis for manipulation with polynomials.

Property 5.1	If m and n are positive integers and a and b are real numbers (except $b \neq 0$ whenever it appears in a denominator), then

1. $b^n \cdot b^m = b^{n+m}$

2. $(b^n)^m = b^{mn}$

3. $(ab)^n = a^n b^n$

4. $\left(\dfrac{a}{b}\right)^n = \dfrac{a^n}{b^n}$

5. $\dfrac{b^n}{b^m} = b^{n-m}$ when $n > m$

   $\dfrac{b^n}{b^m} = 1$ when $n = m$

   $\dfrac{b^n}{b^m} = \dfrac{1}{b^{m-n}}$ when $n < m$

We are now ready to extend the concept of an exponent to include the use of zero and the negative integers as exponents.

First, let's consider the use of zero as an exponent. We want to use zero in such a way that the previously listed properties continue to hold. If "$b^n \cdot b^m = b^{n+m}$" is to hold, then $x^4 \cdot x^0 = x^{4+0} = x^4$. In other words, x^0 *acts like* 1 because $x^4 \cdot x^0 = x^4$. This line of reasoning suggests the following definition.

Definition 5.1	If b is a nonzero real number, then $$b^0 = 1.$$

According to Definition 5.1 the following statements are all true.

$$5^0 = 1; \qquad\qquad (-413)^0 = 1;$$
$$\left(\frac{3}{11}\right)^0 = 1; \qquad\qquad n^0 = 1 \quad (n \neq 0);$$
$$(x^3 y^4)^0 = 1 \quad (x \neq 0, y \neq 0).$$

A similar line of reasoning can be used to motivate a definition for the use of negative integers as exponents. Consider the example $x^4 \cdot x^{-4}$. If "$b^n \cdot b^m = b^{n+m}$" is to hold, then $x^4 \cdot x^{-4} = x^{4+(-4)} = x^0 = 1$. Thus, x^{-4} must be the reciprocal of x^4, since their product is 1. That is to say,

$$x^{-4} = \frac{1}{x^4}.$$

This suggests the following general definition.

Definition 5.2	If n is a positive integer and b is a nonzero real number, then $$b^{-n} = \frac{1}{b^n}.$$

According to Definition 5.2 the following statements are true.

$$x^{-5} = \frac{1}{x^5}; \qquad\qquad 2^{-4} = \frac{1}{2^4} = \frac{1}{16};$$

$$10^{-2} = \frac{1}{10^2} = \frac{1}{100} \text{ or } 0.01; \qquad \frac{2}{x^{-3}} = \frac{2}{\dfrac{1}{x^3}} = (2)\left(\frac{x^3}{1}\right) = 2x^3;$$

$$\left(\frac{3}{4}\right)^{-2} = \frac{1}{\left(\dfrac{3}{4}\right)^2} = \frac{1}{\dfrac{9}{16}} = \frac{16}{9}.$$

It can be verified (although it is beyond the scope of this text) that all of the parts of Property 5.1 hold for *all integers*. In fact, the three separate statements for part (5) can be replaced with the following equality.

$$\frac{b^n}{b^m} = b^{n-m} \quad \text{for all integers } n \text{ and } m.$$

Let's restate Property 5.1 as it holds for all integers and include, in parentheses, a "name tag" for each part for easy reference.

Property 5.2

If m and n are integers and a and b are real numbers, except $b \neq 0$ whenever it appears in a denominator, then

1. $b^n \cdot b^m = b^{n+m}$ (Product of two powers)
2. $(b^n)^m = b^{mn}$ (Power of a power)
3. $(ab)^n = a^n b^n$ (Power of a product)
4. $\left(\dfrac{a}{b}\right)^n = \dfrac{a^n}{b^n}$ (Power of a quotient)
5. $\dfrac{b^n}{b^m} = b^{n-m}$ (Quotient of two powers)

Having the use of all integers as exponents allows us to work with a large variety of numerical and algebraic expressions. Let's consider some examples illustrating the use of the various parts of Property 5.2.

Example 1

Simplify each of the following numerical expressions.

(a) $10^{-3} \cdot 10^2$ **(b)** $(2^{-3})^{-2}$ **(c)** $(2^{-1} \cdot 3^2)^{-1}$

(d) $\left(\dfrac{2^{-3}}{3^{-2}}\right)^{-1}$ **(e)** $\dfrac{10^{-2}}{10^{-4}}$

Solution

(a) $10^{-3} \cdot 10^2 = 10^{-3+2}$ Product of two powers.

$$= 10^{-1}$$

$$= \frac{1}{10^1} = \frac{1}{10}.$$

(b) $(2^{-3})^{-2} = 2^{(-2)(-3)}$ Power of a power.

$$= 2^6 = 64.$$

(c) $(2^{-1} \cdot 3^2)^{-1} = (2^{-1})^{-1}(3^2)^{-1}$ Power of a product.

$$= 2^1 \cdot 3^{-2}$$

$$= \frac{2^1}{3^2} = \frac{2}{9}.$$

(d) $\left(\dfrac{2^{-3}}{3^{-2}}\right)^{-1} = \dfrac{(2^{-3})^{-1}}{(3^{-2})^{-1}}$ Power of a quotient.

$\qquad\qquad = \dfrac{2^3}{3^2} = \dfrac{8}{9}.$

(e) $\dfrac{10^{-2}}{10^{-4}} = 10^{-2-(-4)}$ Quotient of two powers.

$\qquad\quad = 10^2 = 100.$ ●

Example 2   Simplify each of the following; express final results without using zero or negative integers as exponents.

$\qquad$ **(a)** $x^2 \cdot x^{-5}$ $\qquad\qquad$ **(b)** $(x^{-2})^4$ $\qquad\qquad$ **(c)** $(x^2 y^{-3})^{-4}$

$\qquad$ **(d)** $\left(\dfrac{a^3}{b^{-5}}\right)^{-2}$ $\qquad\quad$ **(e)** $\dfrac{x^{-4}}{x^{-2}}$

Solution   **(a)** $x^2 \cdot x^{-5} = x^{2+(-5)}$ Product of two powers.

$\qquad\qquad\quad = x^{-3}$

$\qquad\qquad\quad = \dfrac{1}{x^3}.$

$\qquad$ **(b)** $(x^{-2})^4 = x^{4(-2)}$ Power of a power.

$\qquad\qquad\quad = x^{-8}$

$\qquad\qquad\quad = \dfrac{1}{x^8}.$

$\qquad$ **(c)** $(x^2 y^{-3})^{-4} = (x^2)^{-4}(y^{-3})^{-4}$ Power of a product.

$\qquad\qquad\qquad\quad = x^{-4(2)} y^{-4(-3)}$

$\qquad\qquad\qquad\quad = x^{-8} y^{12}$

$\qquad\qquad\qquad\quad = \dfrac{y^{12}}{x^8}.$

$\qquad$ **(d)** $\left(\dfrac{a^3}{b^{-5}}\right)^{-2} = \dfrac{(a^3)^{-2}}{(b^{-5})^{-2}}$ Power of a quotient.

$\qquad\qquad\qquad = \dfrac{a^{-6}}{b^{10}}$

$\qquad\qquad\qquad = \dfrac{1}{a^6 b^{10}}.$

$\qquad$ **(e)** $\dfrac{x^{-4}}{x^{-2}} = x^{-4-(-2)}$ Quotient of two powers.

$\qquad\qquad\quad = x^{-2}$

$\qquad\qquad\quad = \dfrac{1}{x^2}.$

Example 3   Find the indicated products and quotients; express results using positive integral exponents only.

$$\textbf{(a)} \ (3x^2y^{-4})(4x^{-3}y) \qquad \textbf{(b)} \ \frac{12a^3b^2}{-3a^{-1}b^5} \qquad \textbf{(c)} \ \left(\frac{15x^{-1}y^2}{5xy^{-4}}\right)^{-1}$$

Solution

(a) $(3x^2y^{-4})(4x^{-3}y) = 12x^{2+(-3)}y^{-4+1}$

$$= 12x^{-1}y^{-3}$$

$$= \frac{12}{xy^3}.$$

(b) $\dfrac{12a^3b^2}{-3a^{-1}b^5} = -4a^{3-(-1)}b^{2-5}$

$$= -4a^4b^{-3}$$

$$= -\frac{4a^4}{b^3}.$$

(c) $\left(\dfrac{15x^{-1}y^2}{5xy^{-4}}\right)^{-1} = (3x^{-1-1}y^{2-(-4)})^{-1}$ Notice that we are first simplifying inside the parentheses.

$$= (3x^{-2}y^6)^{-1}$$

$$= 3^{-1}x^2y^{-6}$$

$$= \frac{x^2}{3y^6}.$$

$\bullet$

The final examples of this section illustrate the simplifying of numerical and algebraic expressions involving sums and differences. In such cases, Definition 5.2 is used to change from negative to positive exponents so that we can proceed in the usual way.

Example 4   Simplify $2^{-3} + 3^{-1}$.

Solution $2^{-3} + 3^{-1} = \dfrac{1}{2^3} + \dfrac{1}{3^1}$

$$= \frac{1}{8} + \frac{1}{3}$$

$$= \frac{3}{24} + \frac{8}{24}$$

$$= \frac{11}{24}.$$

$\bullet$

Example 5   Simplify $(4^{-1} - 3^{-2})^{-1}$.

Solution $(4^{-1} - 3^{-2})^{-1} = \left(\dfrac{1}{4^1} - \dfrac{1}{3^2}\right)^{-1}$

$$= \left(\dfrac{1}{4} - \dfrac{1}{9}\right)^{-1}$$

$$= \left(\dfrac{9}{36} - \dfrac{4}{36}\right)^{-1}$$

$$= \left(\dfrac{5}{36}\right)^{-1}$$

$$= \dfrac{1}{\left(\dfrac{5}{36}\right)^{1}}$$

$$= \dfrac{1}{\dfrac{5}{36}} = \dfrac{36}{5}.$$

Example 6   Express $a^{-1} + b^{-2}$ as a single fraction involving positive exponents only.

Solution $a^{-1} + b^{-2} = \dfrac{1}{a^1} + \dfrac{1}{b^2}$

$$= \left(\dfrac{1}{a}\right)\left(\dfrac{b^2}{b^2}\right) + \left(\dfrac{1}{b^2}\right)\left(\dfrac{a}{a}\right) \qquad \text{Use } ab^2 \text{ as the LCD.}$$

$$= \dfrac{b^2}{ab^2} + \dfrac{a}{ab^2}$$

$$= \dfrac{b^2 + a}{ab^2}.$$

Problem Set 5.1

Simplify each of the following numerical expressions.

1. 2^{-3} **2.** 3^{-2} **3.** -10^{-3} **4.** 10^{-4}

5. $\dfrac{1}{3^{-3}}$ **6.** $\dfrac{1}{2^{-5}}$ **7.** $\left(\dfrac{1}{2}\right)^{-2}$ **8.** $-\left(\dfrac{1}{3}\right)^{-2}$

9. $\left(-\dfrac{2}{3}\right)^{-3}$ **10.** $\left(\dfrac{5}{6}\right)^{-2}$ **11.** $\left(-\dfrac{1}{5}\right)^{0}$ **12.** $\dfrac{1}{\left(\dfrac{3}{5}\right)^{-2}}$

13. $\dfrac{1}{\left(\dfrac{4}{5}\right)^{-2}}$ **14.** $\left(\dfrac{4}{5}\right)^{0}$ **15.** $2^{5} \cdot 2^{-3}$ **16.** $3^{-2} \cdot 3^{5}$

17. $10^{-6} \cdot 10^{4}$ **18.** $10^{6} \cdot 10^{-9}$ **19.** $10^{-2} \cdot 10^{-3}$ **20.** $10^{-1} \cdot 10^{-5}$

21. $(3^{-2})^{-2}$ **22.** $((-2)^{-1})^{-3}$ **23.** $(4^{2})^{-1}$ **24.** $(3^{-1})^{3}$

25. $(3^{-1} \cdot 2^{2})^{-1}$ **26.** $(2^{3} \cdot 3^{-2})^{-2}$ **27.** $(4^{2} \cdot 5^{-1})^{2}$ **28.** $(2^{-2} \cdot 4^{-1})^{3}$

29. $\left(\dfrac{2^{-2}}{5^{-1}}\right)^{-2}$ **30.** $\left(\dfrac{3^{-1}}{2^{-3}}\right)^{-2}$ **31.** $\left(\dfrac{3^{-2}}{8^{-1}}\right)^{2}$ **32.** $\left(\dfrac{4^{2}}{5^{-1}}\right)^{-1}$

33. $\dfrac{2^{3}}{2^{-3}}$ **34.** $\dfrac{2^{-3}}{2^{3}}$ **35.** $\dfrac{10^{-1}}{10^{4}}$ **36.** $\dfrac{10^{-3}}{10^{-7}}$

37. $3^{-2} + 2^{-3}$ **38.** $2^{-3} + 5^{-1}$ **39.** $\left(\dfrac{2}{3}\right)^{-1} - \left(\dfrac{3}{4}\right)^{-1}$ **40.** $3^{-2} - 2^{3}$

41. $(2^{-4} + 3^{-1})^{-1}$ **42.** $(3^{-2} - 5^{-1})^{-1}$

Simplify each of the following; express final results without using zero or negative integers as exponents.

43. $x^{3} \cdot x^{-7}$ **44.** $x^{-2} \cdot x^{-3}$ **45.** $a^{2} \cdot a^{-3} \cdot a^{-1}$

46. $b^{-3} \cdot b^{5} \cdot b^{-4}$ **47.** $(a^{-3})^{2}$ **48.** $(b^{5})^{-2}$

49. $(x^{3}y^{-4})^{-1}$ **50.** $(x^{4}y^{-2})^{-2}$ **51.** $(ab^{2}c^{-1})^{-3}$

52. $(a^{2}b^{-1}c^{-2})^{-4}$ **53.** $(2x^{2}y^{-1})^{-2}$ **54.** $(3x^{4}y^{-2})^{-1}$

55. $\left(\dfrac{x^{-2}}{y^{-3}}\right)^{-2}$ **56.** $\left(\dfrac{y^{4}}{x^{-1}}\right)^{-3}$ **57.** $\left(\dfrac{2a^{-1}}{3b^{-2}}\right)^{-2}$

58. $\left(\dfrac{3x^{2}y}{4a^{-1}b^{-3}}\right)^{-1}$ **59.** $\dfrac{x^{-5}}{x^{-2}}$ **60.** $\dfrac{a^{-3}}{a^{5}}$

61. $\dfrac{a^{2}b^{-3}}{a^{-1}b^{-2}}$ **62.** $\dfrac{x^{-1}y^{-2}}{x^{3}y^{-1}}$

Find the indicated products and quotients; express results using positive integral exponents only.

63. $(2x^{-1}y^{2})(3x^{-2}y^{-3})$ **64.** $(4x^{-2}y^{3})(-5x^{3}y^{-4})$ **65.** $(-6a^{5}y^{-4})(-a^{-7}y)$

66. $(-8a^{-4}b^{-5})(-6a^{-1}b^{8})$ **67.** $\dfrac{24x^{-1}y^{-2}}{6x^{-4}y^{3}}$ **68.** $\dfrac{56xy^{-3}}{8x^{2}y^{2}}$

69. $\dfrac{-35a^{3}b^{-2}}{7a^{5}b^{-1}}$ **70.** $\dfrac{27a^{-4}b^{-5}}{-3a^{-2}b^{-4}}$ **71.** $\left(\dfrac{14x^{-2}y^{-4}}{7x^{-3}y^{-6}}\right)^{-2}$

72. $\left(\dfrac{24x^{5}y^{-3}}{-8x^{6}y^{-1}}\right)^{-3}$

Express each of the following as a single fraction involving positive exponents only.

73. $x^{-1} + x^{-2}$ **74.** $x^{-2} + x^{-4}$ **75.** $x^{-2} - y^{-1}$

76. $2x^{-1} - 3y^{-3}$ **77.** $3a^{-2} + 2b^{-3}$ **78.** $a^{-2} + a^{-1}b^{-2}$

79. $x^{-1}y - xy^{-1}$ **80.** $x^{2}y^{-1} - x^{-3}y^{2}$

5.2 Roots and Radicals

To **square a number** means to raise it to the second power, that is, to use the number as a factor twice.

$$4^2 = 4 \cdot 4 = 16; \quad \text{(read "four squared equals sixteen")}$$
$$10^2 = 10 \cdot 10 = 100;$$
$$\left(\frac{1}{2}\right)^2 = \frac{1}{2} \cdot \frac{1}{2} = \frac{1}{4};$$
$$(-3)^2 = (-3)(-3) = 9.$$

A **square root of a number** is one of its two equal factors. Thus, 4 is a square root of 16 because $4 \cdot 4 = 16$. Likewise, -4 is also a square root of 16 because $(-4)(-4) = 16$.

In general, a is a square root of b if $a^2 = b$. The following generalizations are a direct consequence of the previous statement.

1. Every positive real number has two square roots; one is positive and the other is negative. They are opposites of each other.

2. Negative real numbers have no real number square roots because any nonzero real number is positive when squared.

3. The square root of 0 is 0.

The symbol $\sqrt{}$, called a **radical sign**, is used to designate the nonnegative square root. The number under the radical sign is called the **radicand**. The entire expression, such as $\sqrt{16}$, is called a **radical**.

$$\sqrt{16} = 4; \qquad \sqrt{16} \text{ indicates the } \textit{nonnegative} \text{ or } \textbf{principal} \text{ square root of 16.}$$
$$-\sqrt{16} = -4; \qquad -\sqrt{16} \text{ indicates the negative square root of 16.}$$
$$\sqrt{0} = 0; \qquad \text{Zero has only one square root. Technically, we could write}$$
$$-\sqrt{0} = -0 = 0.$$
$$\sqrt{-4} \text{ is not a real number;}$$
$$-\sqrt{-4} \text{ is not a real number.}$$

In general, the following definition is useful.

Definition 5.3	If $a \geq 0$ and $b \geq 0$, then $\sqrt{b} = a$ if and only if $a^2 = b$; a is called the *principal square root of b*.

To **cube a number** means to raise it to the third power, that is, to use the number as a factor three times.

$$2^3 = 2 \cdot 2 \cdot 2 = 8; \quad \text{(read "two cubed equals eight")}$$
$$4^3 = 4 \cdot 4 \cdot 4 = 64;$$
$$\left(\frac{2}{3}\right)^3 = \frac{2}{3} \cdot \frac{2}{3} \cdot \frac{2}{3} = \frac{8}{27};$$
$$(-2)^3 = (-2)(-2)(-2) = -8.$$

A **cube root of a number** is one of its three equal factors. Thus, 2 is a cube root of 8 because $2 \cdot 2 \cdot 2 = 8$. (In fact, 2 is the only real number that is a cube root of 8.) Furthermore, -2 is a cube root of -8 because $(-2)(-2)(-2) = -8$. (In fact, -2 is the only real number that is a cube root of -8.)

In general, a is a cube root of b if $a^3 = b$. The following generalizations are a direct consequence of the previous statement.

1. Every positive real number has one positive real number cube root.

2. Every negative real number has one negative real number cube root.

3. The cube root of 0 is 0.

Remark   Technically, every nonzero real number has three cube roots, but only one of them is a real number. The other two roots are classified as complex numbers. We are restricting our work at this time to the set of real numbers.

The symbol $\sqrt[3]{\ }$ is used to designate the cube root of a number. Thus, we can write

$$\sqrt[3]{8} = 2; \qquad \sqrt[3]{\frac{1}{27}} = \frac{1}{3};$$

$$\sqrt[3]{-8} = -2; \qquad \sqrt[3]{-\frac{1}{27}} = -\frac{1}{3}.$$

In general, the following definition is useful.

Definition 5.4 $\sqrt[3]{b} = a$ if and only if $a^3 = b$.

In Definition 5.4, if $b \geq 0$ then $a \geq 0$ whereas if $b < 0$ then $a < 0$. The number a is called the *principal cube root of b* or simply **the cube root of b**.

The concept of root can be extended to fourth roots, fifth roots, sixth roots, and, in general, nth roots. The following generalizations can be made.

If n is an even positive integer, then the following statements are true.

1. Every positive real number has exactly two real nth roots—one positive and one negative. For example, the real fourth roots of 16 are 2 and -2.

2. Negative real numbers do not have real nth roots. For example, there are no real fourth roots of -16.

If n is an odd positive integer greater than one, then the following statements are true.

1. Every real number has exactly one real nth root.

2. The real nth root of a positive number is positive. For example, the fifth root of 32 is 2.

3. The real nth root of a negative number is negative. For example, the fifth root of -32 is -2.

In general, the following definition is useful.

Definition 5.5	$\sqrt[n]{b} = a$ if and only if $a^n = b$.

In Definition 5.5, if n is an even positive integer, then a and b are both nonnegative. If n is an odd positive integer greater than one, then a and b are both nonnegative or both negative. The symbol $\sqrt[n]{\ }$ designates the **principal nth root**. Consider the following examples.

$$\sqrt[4]{81} = 3 \quad \text{because } 3^4 = 81;$$
$$\sqrt[5]{32} = 2 \quad \text{because } 2^5 = 32;$$
$$\sqrt[5]{-32} = -2 \quad \text{because } (-2)^5 = -32.$$

To complete our terminology, the n in the radical $\sqrt[n]{b}$ is called the *index* of the radical. If $n = 2$, we commonly write $\sqrt{b}$ instead of $\sqrt[2]{b}$. In the future as we use symbols such as $\sqrt[n]{b}$, $\sqrt[m]{y}$, and $\sqrt[j]{x}$, we will assume the previous agreements relative to the existence of real roots, without listing the various restrictions, unless a special restriction is needed.

The following property is a direct consequence of Definition 5.5.

Property 5.3	**1.** $(\sqrt[n]{b})^n = b$	n is any positive integer greater than one.
	2. $\sqrt[n]{b^n} = b$	n is any positive integer greater than one if $b \geq 0$; n is an odd positive integer greater than one if $b < 0$.

The following examples illustrate the use of Property 5.3.

$$\sqrt{16^2} = (\sqrt{16})^2 = 4^2 = 16;$$
$$\sqrt[3]{64^3} = (\sqrt[3]{64})^3 = 4^3 = 64;$$
$$\sqrt[3]{(-8)^3} = (\sqrt[3]{-8})^3 = (-2)^3 = -8;$$

but $\sqrt{(-16)^2} \neq (\sqrt{-16})^2$ because $\sqrt{-16}$ is not a real number.

Let's use some examples to lead into the next very useful property of radicals.

$$\sqrt{4 \cdot 9} = \sqrt{36} = 6 \quad \text{and} \quad \sqrt{4} \cdot \sqrt{9} = 2 \cdot 3 = 6;$$
$$\sqrt{16 \cdot 25} = \sqrt{400} = 20 \quad \text{and} \quad \sqrt{16} \cdot \sqrt{25} = 4 \cdot 5 = 20;$$
$$\sqrt[3]{8 \cdot 27} = \sqrt[3]{216} = 6 \quad \text{and} \quad \sqrt[3]{8} \cdot \sqrt[3]{27} = 2 \cdot 3 = 6;$$
$$\sqrt[3]{(-8)(27)} = \sqrt[3]{-216} = -6 \quad \text{and} \quad \sqrt[3]{-8} \cdot \sqrt[3]{27} = (-2)(3) = -6.$$

In general, the following property can be stated.

Property 5.4	$\sqrt[n]{bc} = \sqrt[n]{b}\sqrt[n]{c}$ ($\sqrt[n]{b}$ and $\sqrt[n]{c}$ are real numbers)

Property 5.4 states that *the nth root of a product is equal to the product of the nth roots.*

Simplest Radical Form

The definition of *n*th root, along with Property 5.4, provides the basis for changing radicals to simplest radical form. The concept of **simplest radical form** takes on additional meaning as we encounter more complicated expressions, but for now it simply means that the radicand is not to contain any perfect powers of the index. Let's consider some examples to illustrate this idea.

Example 1   Express each of the following in simplest radical form.

   (a) $\sqrt{8}$ **(b)** $\sqrt{45}$ **(c)** $\sqrt[3]{24}$ **(d)** $\sqrt[3]{54}$

Solution   **(a)** $\sqrt{8} = \sqrt{4 \cdot 2} = \sqrt{4}\sqrt{2} = 2\sqrt{2}.$
                                   ↑
                               4 is a
                               perfect
                               square.

(b) $\sqrt{45} = \sqrt{9 \cdot 5} = \sqrt{9}\sqrt{5} = 3\sqrt{5}.$
                          ↑
                      9 is a
                      perfect
                      square.

(c) $\sqrt[3]{24} = \sqrt[3]{8 \cdot 3} = \sqrt[3]{8}\sqrt[3]{3} = 2\sqrt[3]{3}.$
                          ↑
                      8 is a
                      perfect
                      cube.

(d) $\sqrt[3]{54} = \sqrt[3]{27 \cdot 2} = \sqrt[3]{27}\sqrt[3]{2} = 3\sqrt[3]{2}.$
                          ↑
                      27 is a
                      perfect
                      cube.

The first step in each example is to express the radicand of the given radical as the product of two factors, one of which must be a perfect *n*th power other than 1. Also, observe the radicands of the final radicals. In each case, the radicand *cannot* be expressed as the product of two factors, one of which must be a perfect nth power other than 1. We say that the final radicals $2\sqrt{2}$, $3\sqrt{5}$, $2\sqrt[3]{3}$, and $3\sqrt[3]{2}$ are in **simplest radical form**.

You may vary the steps somewhat in changing to simplest radical form, but the final result should be the same. Consider some different approaches to change $\sqrt{72}$ to simplest form.

$$\sqrt{72} = \sqrt{9}\sqrt{8} = 3\sqrt{8} = 3\sqrt{4}\sqrt{2} = 3 \cdot 2\sqrt{2} = 6\sqrt{2};$$

or $\qquad \sqrt{72} = \sqrt{4}\sqrt{18} = 2\sqrt{18} = 2\sqrt{9}\sqrt{2} = 2 \cdot 3\sqrt{2} = 6\sqrt{2};$

or $\qquad \sqrt{72} = \sqrt{36}\sqrt{2} = 6\sqrt{2}.$

Another variation of the technique for changing radicals to simplest form is to prime factor the radicand and then to look for perfect nth powers in exponential form. The following example illustrates the use of this technique.

Example 2

Express each of the following in simplest radical form.

(a) $\sqrt{50}$ $\qquad\qquad$ **(b)** $3\sqrt{80}$ $\qquad\qquad$ **(c)** $\sqrt[3]{108}$

Solution

(a) $\sqrt{50} = \sqrt{2 \cdot 5 \cdot 5} = \sqrt{5^2}\sqrt{2} = 5\sqrt{2}.$

(b) $3\sqrt{80} = 3\sqrt{2 \cdot 2 \cdot 2 \cdot 2 \cdot 5} = 3\sqrt{2^4}\sqrt{5} = 3 \cdot 2^2\sqrt{5} = 12\sqrt{5}.$

(c) $\sqrt[3]{108} = \sqrt[3]{2 \cdot 2 \cdot 3 \cdot 3 \cdot 3} = \sqrt[3]{3^3}\sqrt[3]{4} = 3\sqrt[3]{4}.$ ●

Another property of nth roots is illustrated by the following examples.

$$\sqrt{\frac{36}{9}} = \sqrt{4} = 2 \quad \text{and} \quad \frac{\sqrt{36}}{\sqrt{9}} = \frac{6}{3} = 2;$$

$$\sqrt[3]{\frac{64}{8}} = \sqrt[3]{8} = 2 \quad \text{and} \quad \frac{\sqrt[3]{64}}{\sqrt[3]{8}} = \frac{4}{2} = 2;$$

$$\sqrt[3]{\frac{-8}{64}} = \sqrt[3]{-\frac{1}{8}} = -\frac{1}{2} \quad \text{and} \quad \frac{\sqrt[3]{-8}}{\sqrt[3]{64}} = \frac{-2}{4} = -\frac{1}{2}.$$

In general, the following property can be stated.

Property 5.5

$$\sqrt[n]{\frac{b}{c}} = \frac{\sqrt[n]{b}}{\sqrt[n]{c}} \qquad \left(\begin{array}{l} \sqrt[n]{b} \text{ and } \sqrt[n]{c} \text{ are real numbers} \\ \text{and } c \neq 0. \end{array} \right)$$

Property 5.5 states that *the nth root of a quotient is equal to the quotient of the nth roots.*

To evaluate radicals such as $\sqrt{\dfrac{4}{25}}$ and $\sqrt[3]{\dfrac{27}{8}}$, for which the numerator and denominator of the fractional radicand are perfect nth powers, you may use Property 5.5 or merely rely on the definition of nth root.

$$\sqrt{\frac{4}{25}} = \frac{\sqrt{4}}{\sqrt{25}} = \frac{2}{5} \quad \text{or} \quad \sqrt{\frac{4}{25}} = \frac{2}{5} \quad \text{because} \quad \frac{2}{5} \cdot \frac{2}{5} = \frac{4}{25};$$

$$\begin{array}{cc} \uparrow & \uparrow \\ \text{Property 5.5} & \text{Definition of } n\text{th root} \\ \downarrow & \downarrow \end{array}$$

$$\sqrt[3]{\frac{27}{8}} = \frac{\sqrt[3]{27}}{\sqrt[3]{8}} = \frac{3}{2} \quad \text{or} \quad \sqrt[3]{\frac{27}{8}} = \frac{3}{2} \quad \text{because} \quad \frac{3}{2} \cdot \frac{3}{2} \cdot \frac{3}{2} = \frac{27}{8}.$$

Radicals such as $\sqrt{\dfrac{28}{9}}$ and $\sqrt[3]{\dfrac{24}{27}}$ in which only the denominators of the radicand are perfect nth powers can be simplified as follows.

$$\sqrt{\frac{28}{9}} = \frac{\sqrt{28}}{\sqrt{9}} = \frac{\sqrt{28}}{3} = \frac{\sqrt{4}\sqrt{7}}{3} = \frac{2\sqrt{7}}{3};$$

$$\sqrt[3]{\frac{24}{27}} = \frac{\sqrt[3]{24}}{\sqrt[3]{27}} = \frac{\sqrt[3]{24}}{3} = \frac{\sqrt[3]{8}\sqrt[3]{3}}{3} = \frac{2\sqrt[3]{3}}{3}.$$

Before considering more examples, let's summarize some ideas pertaining to the simplifying of radicals. A radical is said to be in **simplest radical form** if the following conditions are satisfied.

1. No fraction appears with a radical sign. $\left(\sqrt{\dfrac{3}{4}} \text{ violates this condition.} \right)$

2. No radical appears in the denominator. $\left(\dfrac{\sqrt{2}}{\sqrt{3}} \text{ violates this condition.} \right)$

3. No radicand when expressed in prime factored form contains a factor raised to a power equal to or greater than the index. ($\sqrt{2^3 \cdot 5}$ violates this condition.)

Now let's consider an example in which neither the numerator nor the denominator of the radicand is a perfect nth power.

Example 3   Simplify $\sqrt{\dfrac{2}{3}}$.

Solution $\sqrt{\dfrac{2}{3}} = \dfrac{\sqrt{2}}{\sqrt{3}} = \dfrac{\sqrt{2}}{\sqrt{3}} \cdot \dfrac{\sqrt{3}}{\sqrt{3}} = \dfrac{\sqrt{6}}{3}.$

$$\begin{array}{c} \uparrow \\ \text{form of 1} \end{array}$$

The process used to simplify the radical in Example 3 is referred to as **rationalizing the denominator**. Notice that the denominator becomes a rational number. The process of rationalizing the denominator can often be accomplished in more than one way, as illustrated by the next example.

Example 4   Simplify $\dfrac{\sqrt{5}}{\sqrt{8}}$.

Solution A $\dfrac{\sqrt{5}}{\sqrt{8}} = \dfrac{\sqrt{5}}{\sqrt{8}} \cdot \dfrac{\sqrt{8}}{\sqrt{8}} = \dfrac{\sqrt{40}}{8} = \dfrac{\sqrt{4}\sqrt{10}}{8} = \dfrac{2\sqrt{10}}{8} = \dfrac{\sqrt{10}}{4}.$

Solution B $\dfrac{\sqrt{5}}{\sqrt{8}} = \dfrac{\sqrt{5}}{\sqrt{8}} \cdot \dfrac{\sqrt{2}}{\sqrt{2}} = \dfrac{\sqrt{10}}{\sqrt{16}} = \dfrac{\sqrt{10}}{4}.$

Solution C $\dfrac{\sqrt{5}}{\sqrt{8}} = \dfrac{\sqrt{5}}{\sqrt{4}\sqrt{2}} = \dfrac{\sqrt{5}}{2\sqrt{2}} = \dfrac{\sqrt{5}}{2\sqrt{2}} \cdot \dfrac{\sqrt{2}}{\sqrt{2}} = \dfrac{\sqrt{10}}{4}.$

The three approaches to Example 4 again illustrate the need to think first and then push the pencil. You may find one approach easier than another.

To conclude this section, study the following examples and check the final radicals according to the three conditions previously listed for *simplest radical form*.

Example 5   Simplify each of the following.

(a) $\dfrac{3\sqrt{2}}{5\sqrt{3}}$ (b) $\dfrac{3\sqrt{7}}{2\sqrt{18}}$ (c) $\sqrt[3]{\dfrac{5}{9}}$ (d) $\dfrac{\sqrt[3]{5}}{\sqrt[3]{16}}$

Solution

(a) $\dfrac{3\sqrt{2}}{5\sqrt{3}} = \dfrac{3\sqrt{2}}{5\sqrt{3}} \cdot \dfrac{\sqrt{3}}{\sqrt{3}} = \dfrac{3\sqrt{6}}{5\sqrt{9}} = \dfrac{3\sqrt{6}}{15} = \dfrac{\sqrt{6}}{5}.$

$\uparrow$
form of 1

(b) $\dfrac{3\sqrt{7}}{2\sqrt{18}} = \dfrac{3\sqrt{7}}{2\sqrt{18}} \cdot \dfrac{\sqrt{2}}{\sqrt{2}} = \dfrac{3\sqrt{14}}{2\sqrt{36}} = \dfrac{3\sqrt{14}}{12} = \dfrac{\sqrt{14}}{4}.$

$\uparrow$
form of 1

(c) $\sqrt[3]{\dfrac{5}{9}} = \dfrac{\sqrt[3]{5}}{\sqrt[3]{9}} = \dfrac{\sqrt[3]{5}}{\sqrt[3]{9}} \cdot \dfrac{\sqrt[3]{3}}{\sqrt[3]{3}} = \dfrac{\sqrt[3]{15}}{\sqrt[3]{27}} = \dfrac{\sqrt[3]{15}}{3}.$

$\uparrow$
form of 1

(d) $\dfrac{\sqrt[3]{5}}{\sqrt[3]{16}} = \dfrac{\sqrt[3]{5}}{\sqrt[3]{16}} \cdot \dfrac{\sqrt[3]{4}}{\sqrt[3]{4}} = \dfrac{\sqrt[3]{20}}{\sqrt[3]{64}} = \dfrac{\sqrt[3]{20}}{4}.$

$\uparrow$
form of 1

Problem Set 5.2

Evaluate each of the following. For example, $\sqrt{\dfrac{1}{16}} = \dfrac{1}{4}$.

1. $\sqrt{81}$ **2.** $\sqrt{100}$ **3.** $-\sqrt{49}$ **4.** $-\sqrt{64}$

5. $\sqrt[3]{64}$ **6.** $\sqrt[3]{125}$ **7.** $\sqrt[3]{-27}$ **8.** $\sqrt[3]{-64}$

9. $\sqrt[4]{16}$ **10.** $\sqrt[4]{81}$ **11.** $\sqrt{\dfrac{4}{25}}$ **12.** $\sqrt{\dfrac{36}{49}}$

13. $\sqrt{\dfrac{64}{16}}$ **14.** $\sqrt{\dfrac{256}{64}}$ **15.** $\sqrt{\dfrac{9}{81}}$ **16.** $\sqrt{\dfrac{16}{64}}$

17. $\sqrt[3]{\dfrac{8}{27}}$ **18.** $\sqrt[3]{-\dfrac{27}{8}}$ **19.** $\sqrt[3]{-\dfrac{1}{64}}$ **20.** $\sqrt[3]{\dfrac{64}{27}}$

Change the following radicals to simplest radical form.

21. $\sqrt{12}$ **22.** $\sqrt{18}$ **23.** $\sqrt{24}$ **24.** $\sqrt{32}$

25. $\sqrt{52}$ **26.** $\sqrt{54}$ **27.** $\sqrt{112}$ **28.** $\sqrt{125}$

29. $5\sqrt{8}$ **30.** $6\sqrt{28}$ **31.** $-3\sqrt{44}$ **32.** $-5\sqrt{68}$

33. $\dfrac{1}{2}\sqrt{8}$ **34.** $\dfrac{3}{4}\sqrt{20}$ **35.** $\dfrac{5}{6}\sqrt{48}$ **36.** $\dfrac{3}{8}\sqrt{72}$

37. $\sqrt{\dfrac{17}{9}}$ **38.** $\sqrt{\dfrac{21}{16}}$ **39.** $\sqrt{\dfrac{3}{16}}$ **40.** $\sqrt{\dfrac{12}{25}}$

41. $\sqrt{\dfrac{24}{49}}$ **42.** $\sqrt{\dfrac{75}{81}}$ **43.** $\sqrt{\dfrac{3}{5}}$ **44.** $\sqrt{\dfrac{5}{6}}$

45. $\sqrt{\dfrac{3}{2}}$ **46.** $\sqrt{\dfrac{7}{8}}$ **47.** $\dfrac{\sqrt{7}}{\sqrt{12}}$ **48.** $\dfrac{\sqrt{9}}{\sqrt{8}}$

49. $\dfrac{\sqrt{27}}{\sqrt{18}}$ **50.** $\dfrac{\sqrt{3}}{\sqrt{6}}$ **51.** $\dfrac{\sqrt{42}}{\sqrt{6}}$ **52.** $\dfrac{\sqrt{35}}{\sqrt{7}}$

53. $\dfrac{3\sqrt{2}}{\sqrt{7}}$ **54.** $\dfrac{4\sqrt{3}}{\sqrt{5}}$ **55.** $\dfrac{3\sqrt{12}}{\sqrt{5}}$ **56.** $\dfrac{4\sqrt{27}}{\sqrt{2}}$

57. $\dfrac{3\sqrt{5}}{4\sqrt{3}}$ **58.** $\dfrac{6\sqrt{3}}{7\sqrt{6}}$ **59.** $\dfrac{-6\sqrt{18}}{8\sqrt{50}}$ **60.** $\dfrac{6\sqrt{45}}{-10\sqrt{2}}$

61. $\sqrt[3]{32}$ **62.** $\sqrt[3]{40}$ **63.** $\sqrt[3]{135}$ **64.** $\sqrt[3]{128}$

65. $\dfrac{5}{\sqrt[3]{3}}$ **66.** $\dfrac{4}{\sqrt[3]{9}}$ **67.** $\dfrac{\sqrt[3]{27}}{\sqrt[3]{4}}$ **68.** $\dfrac{\sqrt[3]{8}}{\sqrt[3]{16}}$

69. $\dfrac{\sqrt[3]{5}}{\sqrt[3]{2}}$ **70.** $\dfrac{\sqrt[3]{7}}{\sqrt[3]{3}}$

71. Note the key $\boxed{\sqrt{}}$ on your calculator. It calculates the square root of the displayed number. Use your calculator to find an approximation, to the nearest thousandth, for each of the following.

(a) $\sqrt{2}$ (b) $\sqrt{75}$ (c) $\sqrt{156}$

(d) $\sqrt{691}$ (e) $\sqrt{3249}$ (f) $\sqrt{45123}$

(g) $\sqrt{0.14}$ (h) $\sqrt{0.023}$ (i) $\sqrt{0.8649}$

72. Your calculator may have a key labeled $\boxed{\sqrt[x]{y}}$. It calculates cube roots, fourth roots, fifth roots, and so on. Be sure you can use the $\boxed{\sqrt[x]{y}}$ key on your calculator by evaluating each of the following. (If your calculator does not have a $\boxed{\sqrt[x]{y}}$ key, wait until we study Section 5.6!)

(a) $\sqrt[3]{729}$ (b) $\sqrt[3]{2744}$ (c) $\sqrt[4]{4096}$

(d) $\sqrt[4]{234256}$ (e) $\sqrt[5]{7776}$ (f) $\sqrt[5]{371293}$

73. See how your calculator reacts to each of the following.

(a) $\sqrt{-4}$ (b) $\sqrt[3]{-8}$ (c) $\sqrt[4]{-16}$

5.3 Combining Radicals and Simplifying Radicals Containing Variables

Recall our use of the distributive property as the basis for combining similar terms. For example,

$$3x + 2x = (3 + 2)x = 5x;$$
$$8y - 5y = (8 - 5)y = 3y;$$
$$\frac{2}{3}a^2 + \frac{3}{4}a^2 = \left(\frac{2}{3} + \frac{3}{4}\right)a^2 = \left(\frac{8}{12} + \frac{9}{12}\right)a^2 = \frac{17}{12}a^2.$$

In a like manner, expressions containing radicals can often be simplified by using the distributive property as follows.

$$3\sqrt{2} + 5\sqrt{2} = (3 + 5)\sqrt{2} = 8\sqrt{2};$$
$$7\sqrt[3]{5} - 3\sqrt[3]{5} = (7 - 3)\sqrt[3]{5} = 4\sqrt[3]{5};$$
$$4\sqrt{7} + 5\sqrt{7} + 6\sqrt{11} - 2\sqrt{11} = (4 + 5)\sqrt{7} + (6 - 2)\sqrt{11} = 9\sqrt{7} + 4\sqrt{11}.$$

Notice that *to add or subtract radicals they must have the same index and the same radicand.* Thus, an expression such as $5\sqrt{2} + 7\sqrt{11}$ cannot be simplified.

Simplifying by combining radicals sometimes requires that you first express the given radicals in simplest form and then apply the distributive property. The following examples illustrate this idea.

Example 1   Simplify $3\sqrt{8} + 2\sqrt{18} - 4\sqrt{2}$.

Solution
$$3\sqrt{8} + 2\sqrt{18} - 4\sqrt{2} = 3\sqrt{4}\sqrt{2} + 2\sqrt{9}\sqrt{2} - 4\sqrt{2}$$
$$= 6\sqrt{2} + 6\sqrt{2} - 4\sqrt{2}$$
$$= (6 + 6 - 4)\sqrt{2} = 8\sqrt{2}.$$

Example 2   Simplify $\dfrac{1}{4}\sqrt{45} + \dfrac{1}{3}\sqrt{20}$.

Solution
$$\frac{1}{4}\sqrt{45} + \frac{1}{3}\sqrt{20} = \frac{1}{4}\sqrt{9}\sqrt{5} + \frac{1}{3}\sqrt{4}\sqrt{5}$$
$$= \frac{1}{4}\cdot 3 \cdot \sqrt{5} + \frac{1}{3}\cdot 2 \cdot \sqrt{5}$$
$$= \frac{3}{4}\sqrt{5} + \frac{2}{3}\sqrt{5}$$
$$= \left(\frac{3}{4} + \frac{2}{3}\right)\sqrt{5}$$
$$= \left(\frac{9}{12} + \frac{8}{12}\right)\sqrt{5}$$
$$= \frac{17}{12}\sqrt{5}.$$

Example 3   Simplify $5\sqrt[3]{2} - 2\sqrt[3]{16} - 6\sqrt[3]{54}$.

Solution
$$5\sqrt[3]{2} - 2\sqrt[3]{16} - 6\sqrt[3]{54} = 5\sqrt[3]{2} - 2\sqrt[3]{8}\sqrt[3]{2} - 6\sqrt[3]{27}\sqrt[3]{2}$$
$$= 5\sqrt[3]{2} - 2\cdot 2 \cdot \sqrt[3]{2} - 6\cdot 3 \cdot \sqrt[3]{2}$$
$$= 5\sqrt[3]{2} - 4\sqrt[3]{2} - 18\sqrt[3]{2}$$
$$= (5 - 4 - 18)\sqrt[3]{2}$$
$$= -17\sqrt[3]{2}.$$

Radicals Containing Variables

Before discussing the process of simplifying *radicals containing variables*, there is one technicality that should be called to your attention. Let's look at some examples to illustrate the point.

Consider the radical $\sqrt{x^2}$ as follows:

$$\text{Let } x = 3; \quad \text{then } \sqrt{x^2} = \sqrt{3^2} = \sqrt{9} = 3.$$
$$\text{Let } x = -3; \quad \text{then } \sqrt{x^2} = \sqrt{(-3)^2} = \sqrt{9} = 3.$$

Thus, if $x \geq 0$, then $\sqrt{x^2} = x$, *but if* $x < 0$, *then* $\sqrt{x^2} = -x$. Using the concept of absolute value we can state that *for all real numbers,* $\sqrt{x^2} = |x|$.

Now consider the radical $\sqrt{x^3}$. Since x^3 is negative when x is negative, we need to restrict x to the nonnegative reals when working with $\sqrt{x^3}$. Thus, we can write if $x \geq 0$, then $\sqrt{x^3} = \sqrt{x^2}\sqrt{x} = x\sqrt{x}$, and no absolute value sign is needed.

Finally, let's consider the radical $\sqrt[3]{x^3}$.

$$\text{Let } x = 2; \quad \text{then } \sqrt[3]{x^3} = \sqrt[3]{2^3} = \sqrt[3]{8} = 2.$$

$$\text{Let } x = -2; \quad \text{then } \sqrt[3]{x^3} = \sqrt[3]{(-2)^3} = \sqrt[3]{-8} = -2.$$

Thus, it is correct to write: $\sqrt[3]{x^3} = x$ for all real numbers, and again no absolute value sign is needed.

The previous discussion indicates that technically every radical expression involving variables in the radicand needs to be analyzed individually as to the necessary restrictions imposed on the variables. However, to avoid considering such restrictions on a problem-to-problem basis we shall merely *assume that all variables represent positive real numbers.*

Let's consider the process of simplifying radicals that contain variables in the radicand. Study the following examples and notice the same basic approach as used in Section 5.2.

Example 4

Simplify each of the following.

(a) $\sqrt{8x^3}$ **(b)** $\sqrt{45x^3y^7}$ **(c)** $\sqrt{180a^4b^3}$ **(d)** $\sqrt[3]{40x^4y^8}$

Solution

(a) $\sqrt{8x^3} = \sqrt{4x^2}\sqrt{2x} = 2x\sqrt{2x}.$

$\nwarrow$ $4x^2$ is a perfect square.

(b) $\sqrt{45x^3y^7} = \sqrt{9x^2y^6}\sqrt{5xy} = 3xy^3\sqrt{5xy}.$

$\nwarrow$ $9x^2y^6$ is a perfect square.

(c) If the numerical coefficient of the radicand is quite large, you may want to look at it in prime factored form.

$$\sqrt{180a^4b^3} = \sqrt{2 \cdot 2 \cdot 3 \cdot 3 \cdot 5 \cdot a^4 \cdot b^3}$$
$$= \sqrt{36 \cdot 5 \cdot a^4 \cdot b^3}$$
$$= \sqrt{36a^4b^2}\sqrt{5b}$$
$$= 6a^2b\sqrt{5b}.$$

(d) $\sqrt[3]{40x^4y^8} = \sqrt[3]{8x^3y^6}\sqrt[3]{5xy^2} = 2xy^2\sqrt[3]{5xy^2}.$

$\nwarrow$ $8x^3y^6$ is a perfect cube.

Before considering more examples, let's restate, so as to include radicands containing variables, the conditions necessary for a radical to be in *simplest radical form*.

1. A radicand contains no polynomial factor raised to a power equal to or greater than the index of the radical. ($\sqrt{x^3}$ violates this condition.)

2. No fraction appears within a radical sign. $\left(\sqrt{\dfrac{2x}{3y}} \text{ violates this condition.} \right)$

3. No radical appears in the denominator. $\left(\dfrac{3}{\sqrt[3]{4x}} \text{ violates this condition.} \right)$

Example 5   Express each of the following in simplest radical form.

(a) $\sqrt{\dfrac{2x}{3y}}$ (b) $\dfrac{\sqrt{5}}{\sqrt{12a^3}}$ (c) $\dfrac{\sqrt{8x^2}}{\sqrt{27y^5}}$

(d) $\dfrac{3}{\sqrt[3]{4x}}$ (e) $\dfrac{\sqrt[3]{16x^2}}{\sqrt[3]{9y^5}}$

Solution   (a) $\sqrt{\dfrac{2x}{3y}} = \dfrac{\sqrt{2x}}{\sqrt{3y}} = \dfrac{\sqrt{2x}}{\sqrt{3y}} \cdot \underset{\uparrow}{\dfrac{\sqrt{3y}}{\sqrt{3y}}} = \dfrac{\sqrt{6xy}}{3y}.$

$\text{form of } 1$

(b) $\dfrac{\sqrt{5}}{\sqrt{12a^3}} = \dfrac{\sqrt{5}}{\sqrt{12a^3}} \cdot \underset{\uparrow}{\dfrac{\sqrt{3a}}{\sqrt{3a}}} = \dfrac{\sqrt{15a}}{\sqrt{36a^4}} = \dfrac{\sqrt{15a}}{6a^2}.$

$\text{form of } 1$

(c) $\dfrac{\sqrt{8x^2}}{\sqrt{27y^5}} = \dfrac{\sqrt{4x^2}\sqrt{2}}{\sqrt{9y^4}\sqrt{3y}} = \dfrac{2x\sqrt{2}}{3y^2\sqrt{3y}} = \dfrac{2x\sqrt{2}}{3y^2\sqrt{3y}} \cdot \dfrac{\sqrt{3y}}{\sqrt{3y}}$

$= \dfrac{2x\sqrt{6y}}{(3y^2)(3y)} = \dfrac{2x\sqrt{6y}}{9y^3}.$

(d) $\dfrac{3}{\sqrt[3]{4x}} = \dfrac{3}{\sqrt[3]{4x}} \cdot \dfrac{\sqrt[3]{2x^2}}{\sqrt[3]{2x^2}} = \dfrac{3\sqrt[3]{2x^2}}{\sqrt[3]{8x^3}} = \dfrac{3\sqrt[3]{2x^2}}{2x}.$

(e) $\dfrac{\sqrt[3]{16x^2}}{\sqrt[3]{9y^5}} = \dfrac{\sqrt[3]{16x^2}}{\sqrt[3]{9y^5}} \cdot \dfrac{\sqrt[3]{3y}}{\sqrt[3]{3y}} = \dfrac{\sqrt[3]{48x^2y}}{\sqrt[3]{27y^6}} = \dfrac{\sqrt[3]{8}\sqrt[3]{6x^2y}}{3y^2} = \dfrac{2\sqrt[3]{6x^2y}}{3y^2}.$

●

Notice that in **(c)** we did some simplifying first before rationalizing the denominator, whereas in **(b)** we proceeded immediately to rationalize the denominator. This is an individual choice and you should probably do it both ways a few times to help determine your preference.

Problem Set 5.3

Use the distributive property to help simplify each of the following. For example,

$$3\sqrt{8} + 5\sqrt{2} - \sqrt{32} = 3\sqrt{4}\sqrt{2} + 5\sqrt{2} - \sqrt{16}\sqrt{2}$$
$$= 6\sqrt{2} + 5\sqrt{2} - 4\sqrt{2}$$
$$= (6 + 5 - 4)\sqrt{2}$$
$$= 7\sqrt{2}.$$

All variables represent positive real numbers.

1. $5\sqrt{12} + 2\sqrt{3}$

2. $3\sqrt{18} - 5\sqrt{2}$

3. $9\sqrt{8} - 3\sqrt{18}$

4. $5\sqrt{12} + 2\sqrt{48}$

5. $3\sqrt{20} - 6\sqrt{45}$

6. $4\sqrt{50} - 9\sqrt{32}$

7. $3\sqrt{12} + 2\sqrt{3} + \sqrt{48}$

8. $2\sqrt{20} - 4\sqrt{5} - 3\sqrt{45}$

9. $2\sqrt{28} - 3\sqrt{63} + 8\sqrt{7}$

10. $-5\sqrt{24} + 2\sqrt{54} + 4\sqrt{6}$

11. $\dfrac{1}{5}\sqrt{2} + \dfrac{1}{3}\sqrt{8}$

12. $\dfrac{5}{6}\sqrt{7} - \dfrac{1}{2}\sqrt{28}$

13. $\dfrac{5}{6}\sqrt{48} - \dfrac{3}{4}\sqrt{12}$

14. $\dfrac{2}{5}\sqrt{40} + \dfrac{1}{6}\sqrt{90}$

15. $\dfrac{\sqrt{20}}{3} + \dfrac{\sqrt{45}}{4} - \dfrac{\sqrt{80}}{2}$

16. $\dfrac{2\sqrt{8}}{3} - \dfrac{3\sqrt{18}}{5} - \dfrac{\sqrt{50}}{2}$

17. $4\sqrt[3]{2} + 2\sqrt[3]{16} - \sqrt[3]{54}$

18. $4\sqrt[3]{3} + \sqrt[3]{24} - 5\sqrt[3]{81}$

19. $-2\sqrt[3]{24} + 5\sqrt[3]{3} - 6\sqrt[3]{81}$

20. $-3\sqrt[3]{16} - 5\sqrt[3]{54} + 2\sqrt[3]{2}$

21. $\sqrt{18x} + \sqrt{8x} - \sqrt{50x}$

22. $2\sqrt{4x} - 3\sqrt{9x} + 5\sqrt{16x}$

23. $\sqrt{27n} - 3\sqrt{12n} - 2\sqrt{3n}$

24. $3\sqrt{8n} + 4\sqrt{18n} - \sqrt{72n}$

25. $-\sqrt{2x^3} + \sqrt{8x^3} - 2\sqrt{32x^3}$

26. $\sqrt{4ab} - 2\sqrt{16ab} - 7\sqrt{25ab}$

Express each of the following in simplest radical form. All variables represent positive real numbers.

27. $\sqrt{8x}$

28. $\sqrt{18y}$

29. $\sqrt{12y^2}$

30. $\sqrt{27x^3}$

31. $\sqrt{20x^2y}$

32. $\sqrt{45xy^2}$

33. $\sqrt{49x^3y^5}$

34. $\sqrt{64x^4y^7}$

35. $\sqrt{50a^5b^6}$

36. $\sqrt{72ab^5}$

37. $\sqrt{75x^6y^8}$

38. $\sqrt{48x^4y^{10}}$

39. $3\sqrt{32a^3}$

40. $5\sqrt{50a^7}$

41. $\dfrac{3}{7}\sqrt{45xy^6}$

42. $\dfrac{4}{9}\sqrt{54x^4y^3}$

43. $\sqrt{\dfrac{3x}{2y}}$

44. $\sqrt{\dfrac{2x}{5y}}$

45. $\sqrt{\dfrac{7}{8x^2}}$

46. $\sqrt{\dfrac{5}{12x^4}}$

47. $\dfrac{3}{\sqrt{12x}}$

48. $\dfrac{5}{\sqrt{18y}}$

49. $\dfrac{\sqrt{5y}}{\sqrt{18x^3}}$

50. $\dfrac{\sqrt{7x}}{\sqrt{8y^5}}$

51. $\dfrac{\sqrt{2x^3}}{\sqrt{9y}}$ **52.** $\dfrac{\sqrt{18y^3}}{\sqrt{16x}}$ **53.** $\dfrac{\sqrt{12a^2b}}{\sqrt{5a^3b^3}}$ **54.** $\dfrac{\sqrt{24a^2b^3}}{\sqrt{7ab^6}}$

55. $\sqrt[3]{16x^2}$ **56.** $\sqrt[3]{24y}$ **57.** $\sqrt[3]{54x^3}$ **58.** $\sqrt[3]{16x^4}$

59. $\sqrt[3]{81x^5y^6}$ **60.** $\sqrt[3]{56x^6y^8}$ **61.** $\sqrt[3]{\dfrac{5}{2x}}$ **62.** $\sqrt[3]{\dfrac{7}{9x^2}}$

63. $\dfrac{\sqrt[3]{2y}}{\sqrt[3]{3x}}$ **64.** $\dfrac{\sqrt[3]{3y}}{\sqrt[3]{16x^4}}$ **65.** $\dfrac{5}{\sqrt[3]{9xy^2}}$ **66.** $\dfrac{\sqrt[3]{12xy}}{\sqrt[3]{3x^2y^5}}$

67. $\sqrt{4x+4y}$ (Hint: $\sqrt{4x+4y}=\sqrt{4(x+y)}$)

68. $\sqrt{8x+12y}$ **69.** $\sqrt{27x+18y}$ **70.** $\sqrt{16x+48y}$

Miscellaneous Problems

71. (a) Use 1.414 as a rational approximation for $\sqrt{2}$ and find the value of $\dfrac{1}{\sqrt{2}}$ to the nearest one-thousandth.

(b) Change $\dfrac{1}{\sqrt{2}}$ to simplest radical form.

(c) Use 1.414 as a rational approximation for $\sqrt{2}$ and find the value of $\dfrac{\sqrt{2}}{2}$ to the nearest one-thousandth.

(d) Use 1.414 as a rational approximation for $\sqrt{2}$ and find the value of the following to the nearest one-thousandth.

(1) $\dfrac{5}{\sqrt{2}}$ **(2)** $\dfrac{3}{4\sqrt{2}}$ **(3)** $\sqrt{8}+\sqrt{18}$

(4) $3\sqrt{32}-4\sqrt{50}$ **(5)** $\dfrac{3}{\sqrt{8}}$ **(6)** $\dfrac{-5}{\sqrt{18}}$

5.4 Products and Quotients Involving Radicals

As we have seen, Property 5.4 ($\sqrt[n]{bc}=\sqrt[n]{b}\sqrt[n]{c}$) is used to express one radical as the product of two radicals and also to express the product of two radicals as one radical. In fact, we have used the property for both purposes within the framework of simplifying radicals. For example,

$$\frac{\sqrt{3}}{\sqrt{32}}=\frac{\sqrt{3}}{\sqrt{16}\sqrt{2}}=\frac{\sqrt{3}}{4\sqrt{2}}=\frac{\sqrt{3}}{4\sqrt{2}}\cdot\frac{\sqrt{2}}{\sqrt{2}}=\frac{\sqrt{6}}{8}.$$

$\uparrow$ $\sqrt[n]{bc}=\sqrt[n]{b}\sqrt[n]{c}$ $\qquad$ $\uparrow$ $\sqrt[n]{b}\sqrt[n]{c}=\sqrt[n]{bc}$

The following examples illustrate the use of Property 5.4 to multiply radicals and express the product in simplest form.

Example 1   Multiply and simplify where possible.

(a) $(2\sqrt{3})(3\sqrt{5})$ (b) $(3\sqrt{8})(5\sqrt{2})$

(c) $(7\sqrt{6})(3\sqrt{8})$ (d) $(2\sqrt[3]{6})(5\sqrt[3]{4})$

Solution   (a) $(2\sqrt{3})(3\sqrt{5}) = 2 \cdot 3 \cdot \sqrt{3} \cdot \sqrt{5} = 6\sqrt{15}.$

(b) $(3\sqrt{8})(5\sqrt{2}) = 3 \cdot 5 \cdot \sqrt{8} \cdot \sqrt{2} = 15\sqrt{16} = 15 \cdot 4 = 60.$

(c) $(7\sqrt{6})(3\sqrt{8}) = 7 \cdot 3 \cdot \sqrt{6} \cdot \sqrt{8} = 21\sqrt{48} = 21\sqrt{16}\sqrt{3} = 21 \cdot 4 \cdot \sqrt{3}$
$$= 84\sqrt{3}.$$

(d) $(2\sqrt[3]{6})(5\sqrt[3]{4}) = 2 \cdot 5 \cdot \sqrt[3]{6} \cdot \sqrt[3]{4} = 10\sqrt[3]{24}$
$$= 10\sqrt[3]{8}\sqrt[3]{3}$$
$$= 10 \cdot 2 \cdot \sqrt[3]{3}$$
$$= 20\sqrt[3]{3}. \qquad \bullet$$

Recall the use of the distributive property when finding the product of a monomial and a polynomial. For example, $3x^2(2x + 7) = 3x^2(2x) + 3x^2(7) = 6x^3 + 21x^2$. In a similar manner, the distributive property, along with Property 5.4, provides the basis for finding certain special products involving radicals. The following examples illustrate this idea.

Example 2   Multiply and simplify where possible.

(a) $\sqrt{3}(\sqrt{6} + \sqrt{12})$ (b) $2\sqrt{2}(4\sqrt{3} - 5\sqrt{6})$

(c) $\sqrt{6x}(\sqrt{8x} + \sqrt{12xy})$ (d) $\sqrt[3]{2}(5\sqrt[3]{4} - 3\sqrt[3]{16})$

Solution   (a) $\sqrt{3}(\sqrt{6} + \sqrt{12}) = \sqrt{3}\sqrt{6} + \sqrt{3}\sqrt{12}$
$$= \sqrt{18} + \sqrt{36}$$
$$= \sqrt{9}\sqrt{2} + 6$$
$$= 3\sqrt{2} + 6.$$

(b) $2\sqrt{2}(4\sqrt{3} - 5\sqrt{6}) = (2\sqrt{2})(4\sqrt{3}) - (2\sqrt{2})(5\sqrt{6})$
$$= 8\sqrt{6} - 10\sqrt{12}$$
$$= 8\sqrt{6} - 10\sqrt{4}\sqrt{3}$$
$$= 8\sqrt{6} - 20\sqrt{3}.$$

(c) $\sqrt{6x}(\sqrt{8x} + \sqrt{12xy}) = (\sqrt{6x})(\sqrt{8x}) + (\sqrt{6x})(\sqrt{12xy})$

$$= \sqrt{48x^2} + \sqrt{72x^2y}$$
$$= \sqrt{16x^2}\sqrt{3} + \sqrt{36x^2}\sqrt{2y}$$
$$= 4x\sqrt{3} + 6x\sqrt{2y}.$$

(d) $\sqrt[3]{2}(5\sqrt[3]{4} - 3\sqrt[3]{16}) = (\sqrt[3]{2})(5\sqrt[3]{4}) - (\sqrt[3]{2})(3\sqrt[3]{16})$

$$= 5\sqrt[3]{8} - 3\sqrt[3]{32}$$
$$= 5 \cdot 2 - 3\sqrt[3]{8}\sqrt[3]{4}$$
$$= 10 - 6\sqrt[3]{4}.$$
●

The distributive property also plays a central role in finding the product of two binomials. For example, $(x + 2)(x + 3) = x(x + 3) + 2(x + 3) = x^2 + 3x + 2x + 6 = x^2 + 5x + 6$. Finding the product of two binomial expressions involving radicals can be handled in a similar fashion, as the next examples illustrate.

Example 3　　Find the following products and simplify.

(a) $(\sqrt{3} + \sqrt{5})(\sqrt{2} + \sqrt{6})$　　　　**(b)** $(2\sqrt{2} - \sqrt{7})(3\sqrt{2} + 5\sqrt{7})$

(c) $(\sqrt{8} + \sqrt{6})(\sqrt{8} - \sqrt{6})$　　　　**(d)** $(\sqrt{x} + \sqrt{y})(\sqrt{x} - \sqrt{y})$

Solution　　**(a)** $(\sqrt{3} + \sqrt{5})(\sqrt{2} + \sqrt{6}) = \sqrt{3}(\sqrt{2} + \sqrt{6}) + \sqrt{5}(\sqrt{2} + \sqrt{6})$

$$= \sqrt{3}\sqrt{2} + \sqrt{3}\sqrt{6} + \sqrt{5}\sqrt{2} + \sqrt{5}\sqrt{6}$$
$$= \sqrt{6} + \sqrt{18} + \sqrt{10} + \sqrt{30}$$
$$= \sqrt{6} + 3\sqrt{2} + \sqrt{10} + \sqrt{30}.$$

(b) $(2\sqrt{2} - \sqrt{7})(3\sqrt{2} + 5\sqrt{7}) = 2\sqrt{2}(3\sqrt{2} + 5\sqrt{7}) - \sqrt{7}(3\sqrt{2} + 5\sqrt{7})$

$$= (2\sqrt{2})(3\sqrt{2}) + (2\sqrt{2})(5\sqrt{7})$$
$$\quad -(\sqrt{7})(3\sqrt{2}) - (\sqrt{7})(5\sqrt{7})$$
$$= 12 + 10\sqrt{14} - 3\sqrt{14} - 35$$
$$= -23 + 7\sqrt{14}.$$

(c) $(\sqrt{8} + \sqrt{6})(\sqrt{8} - \sqrt{6}) = \sqrt{8}(\sqrt{8} - \sqrt{6}) + \sqrt{6}(\sqrt{8} - \sqrt{6})$

$$= \sqrt{8}\sqrt{8} - \sqrt{8}\sqrt{6} + \sqrt{6}\sqrt{8} - \sqrt{6}\sqrt{6}$$
$$= 8 - \sqrt{48} + \sqrt{48} - 6$$
$$= 2.$$

(d) $(\sqrt{x} + \sqrt{y})(\sqrt{x} - \sqrt{y}) = \sqrt{x}(\sqrt{x} - \sqrt{y}) + \sqrt{y}(\sqrt{x} - \sqrt{y})$

$$= \sqrt{x}\sqrt{x} - \sqrt{x}\sqrt{y} + \sqrt{y}\sqrt{x} - \sqrt{y}\sqrt{y}$$
$$= x - \sqrt{xy} + \sqrt{xy} - y$$
$$= x - y.$$
●

Notice parts **(c)** and **(d)** of Example 3. They fit the special product pattern $(a + b)(a - b) = a^2 - b^2$. Furthermore, in each case the final product is in rational form. This suggests a way of rationalizing the denominator of an expression containing a binomial denominator with radicals. Consider the following example.

Example 4

Simplify $\dfrac{4}{\sqrt{5} + \sqrt{2}}$ by rationalizing the denominator.

Solution

$$\frac{4}{\sqrt{5} + \sqrt{2}} = \frac{4}{\sqrt{5} + \sqrt{2}} \cdot \left(\frac{\sqrt{5} - \sqrt{2}}{\sqrt{5} - \sqrt{2}}\right) \quad \text{a form of 1}$$

$$= \frac{4(\sqrt{5} - \sqrt{2})}{(\sqrt{5} + \sqrt{2})(\sqrt{5} - \sqrt{2})}$$

$$= \frac{4(\sqrt{5} - \sqrt{2})}{5 - 2}$$

$$= \frac{4(\sqrt{5} - \sqrt{2})}{3} \quad \text{or} \quad \frac{4\sqrt{5} - 4\sqrt{2}}{3}.$$

either answer
is acceptable

The next examples further illustrate the process of rationalizing and simplifying expressions containing binomial denominators.

Example 5

For each of the following, rationalize the denominator and simplify.

(a) $\dfrac{\sqrt{3}}{\sqrt{6} - 9}$ **(b)** $\dfrac{7}{3\sqrt{5} + 2\sqrt{3}}$ **(c)** $\dfrac{\sqrt{x} + 2}{\sqrt{x} - 3}$ **(d)** $\dfrac{2\sqrt{x} - 3\sqrt{y}}{\sqrt{x} + \sqrt{y}}$

Solution

(a) $\dfrac{\sqrt{3}}{\sqrt{6} - 9} = \dfrac{\sqrt{3}}{\sqrt{6} - 9} \cdot \dfrac{\sqrt{6} + 9}{\sqrt{6} + 9}$

$$= \frac{\sqrt{3}(\sqrt{6} + 9)}{(\sqrt{6} - 9)(\sqrt{6} + 9)}$$

$$= \frac{\sqrt{18} + 9\sqrt{3}}{6 - 81}$$

$$= \frac{3\sqrt{2} + 9\sqrt{3}}{-75}$$

$$= \frac{3(\sqrt{2} + 3\sqrt{3})}{(-3)(25)}$$

$$= -\frac{\sqrt{2} + 3\sqrt{3}}{25} \quad \text{or} \quad \frac{-\sqrt{2} - 3\sqrt{3}}{25}.$$

(b) $\dfrac{7}{3\sqrt{5}+2\sqrt{3}} = \dfrac{7}{3\sqrt{5}+2\sqrt{3}} \cdot \dfrac{3\sqrt{5}-2\sqrt{3}}{3\sqrt{5}-2\sqrt{3}}$

$ = \dfrac{7(3\sqrt{5}-2\sqrt{3})}{(3\sqrt{5}+2\sqrt{3})(3\sqrt{5}-2\sqrt{3})}$

$ = \dfrac{7(3\sqrt{5}-2\sqrt{3})}{45-12}$

$ = \dfrac{7(3\sqrt{5}-2\sqrt{3})}{33} \quad \text{or} \quad \dfrac{21\sqrt{5}-14\sqrt{3}}{33}.$

(c) $\dfrac{\sqrt{x}+2}{\sqrt{x}-3} = \dfrac{\sqrt{x}+2}{\sqrt{x}-3} \cdot \dfrac{\sqrt{x}+3}{\sqrt{x}+3}$

$ = \dfrac{(\sqrt{x}+2)(\sqrt{x}+3)}{(\sqrt{x}-3)(\sqrt{x}+3)}$

$ = \dfrac{x+3\sqrt{x}+2\sqrt{x}+6}{x-9}$

$ = \dfrac{x+5\sqrt{x}+6}{x-9}.$

(d) $\dfrac{2\sqrt{x}-3\sqrt{y}}{\sqrt{x}+\sqrt{y}} = \dfrac{2\sqrt{x}-3\sqrt{y}}{\sqrt{x}+\sqrt{y}} \cdot \dfrac{\sqrt{x}-\sqrt{y}}{\sqrt{x}-\sqrt{y}}$

$ = \dfrac{(2\sqrt{x}-3\sqrt{y})(\sqrt{x}-\sqrt{y})}{(\sqrt{x}+\sqrt{y})(\sqrt{x}-\sqrt{y})}$

$ = \dfrac{2x-2\sqrt{xy}-3\sqrt{xy}+3y}{x-y}$

$ = \dfrac{2x-5\sqrt{xy}+3y}{x-y}.$

Problem Set 5.4

Multiply and simplify where possible.

1. $\sqrt{6}\sqrt{8}$ 2. $\sqrt{12}\sqrt{5}$ 3. $(3\sqrt{2})(2\sqrt{5})$

4. $(5\sqrt{3})(3\sqrt{7})$ 5. $(4\sqrt{7})(5\sqrt{7})$ 6. $(3\sqrt{10})(2\sqrt{10})$

7. $(4\sqrt{3})(6\sqrt{8})$ 8. $(5\sqrt{8})(3\sqrt{7})$ 9. $(3\sqrt[3]{4})(5\sqrt[3]{2})$

10. $(2\sqrt[3]{9})(4\sqrt[3]{6})$

Find the following products by applying the distributive property. Express answers in simplest radical form. All variables represent nonnegative real numbers.

11. $\sqrt{3}(\sqrt{5}+\sqrt{7})$ 12. $\sqrt{5}(\sqrt{6}-\sqrt{2})$

13. $3\sqrt{5}(2\sqrt{7} - 3\sqrt{11})$

14. $2\sqrt{3}(5\sqrt{2} + 4\sqrt{10})$

15. $5\sqrt{2}(2\sqrt{12} + 3\sqrt{6})$

16. $3\sqrt{6}(2\sqrt{8} - 3\sqrt{12})$

17. $2\sqrt[3]{2}(3\sqrt[3]{6} - 4\sqrt[3]{5})$

18. $2\sqrt[3]{3}(5\sqrt[3]{4} + \sqrt[3]{6})$

19. $3\sqrt[3]{3}(4\sqrt[3]{9} + 5\sqrt[3]{7})$

20. $4\sqrt[3]{2}(3\sqrt[3]{4} - 2\sqrt[3]{16})$

21. $\sqrt{2x}(2\sqrt{y} - 3\sqrt{5})$

22. $2\sqrt{x}(3\sqrt{2} + 5\sqrt{y})$

23. $3\sqrt{x}(5\sqrt{xy} + 4\sqrt{x})$

24. $\sqrt{xy}(2\sqrt{xy} - 3\sqrt{x})$

25. $\sqrt{3x}(\sqrt{6xy} - \sqrt{8y})$

26. $\sqrt{6y}(\sqrt{8x} + \sqrt{10y^2})$

27. $(\sqrt{3} + 2)(\sqrt{3} + 5)$

28. $(\sqrt{2} - 3)(\sqrt{2} + 4)$

29. $(\sqrt{2} + \sqrt{3})(\sqrt{5} - \sqrt{7})$

30. $(3\sqrt{5} - 2\sqrt{3})(2\sqrt{7} + \sqrt{2})$

31. $(4\sqrt{2} + \sqrt{3})(3\sqrt{2} + 2\sqrt{3})$

32. $(2\sqrt{6} + 3\sqrt{5})(3\sqrt{6} + 4\sqrt{5})$

33. $(2\sqrt{6} - 3\sqrt{8})(4\sqrt{3} - 2\sqrt{5})$

34. $(6\sqrt{2} - 2\sqrt{6})(5\sqrt{10} - \sqrt{7})$

35. $(\sqrt{7} + 3)(\sqrt{7} - 3)$

36. $(\sqrt{10} + 4)(\sqrt{10} - 4)$

37. $(6 + 2\sqrt{5})(6 - 2\sqrt{5})$

38. $(7 - 3\sqrt{2})(7 + 3\sqrt{2})$

39. $(\sqrt{3} + \sqrt{7})(\sqrt{3} - \sqrt{7})$

40. $(\sqrt{11} + \sqrt{6})(\sqrt{11} - \sqrt{6})$

41. $(3\sqrt{2} + 2\sqrt{3})(3\sqrt{2} - 2\sqrt{3})$

42. $(5\sqrt{5} + 2\sqrt{3})(5\sqrt{5} - 2\sqrt{3})$

43. $(\sqrt{a} + \sqrt{b})(\sqrt{a} - \sqrt{b})$

44. $(\sqrt{2x} + \sqrt{3y})(\sqrt{2x} - \sqrt{3y})$

45. $(2\sqrt{x} + 3\sqrt{y})(2\sqrt{x} - 3\sqrt{y})$

46. $(3\sqrt{x} + 5\sqrt{y})(3\sqrt{x} - 5\sqrt{y})$

For each of the following, rationalize the denominator and simplify. All variables represent positive real numbers.

47. $\dfrac{3}{\sqrt{5} + 2}$

48. $\dfrac{7}{\sqrt{10} - 3}$

49. $\dfrac{5}{\sqrt{6} - 1}$

50. $\dfrac{3}{\sqrt{7} + 2}$

51. $\dfrac{2}{\sqrt{5} + \sqrt{3}}$

52. $\dfrac{4}{\sqrt{7} - \sqrt{3}}$

53. $\dfrac{\sqrt{6}}{\sqrt{7} - \sqrt{3}}$

54. $\dfrac{\sqrt{8}}{\sqrt{2} + \sqrt{11}}$

55. $\dfrac{\sqrt{5}}{2\sqrt{3} - 1}$

56. $\dfrac{\sqrt{7}}{3\sqrt{2} + 4}$

57. $\dfrac{3}{4\sqrt{3} - 2\sqrt{5}}$

58. $\dfrac{5}{5\sqrt{2} - 3\sqrt{5}}$

59. $\dfrac{\sqrt{2}}{2\sqrt{5} + 3\sqrt{7}}$

60. $\dfrac{\sqrt{6}}{5\sqrt{3} + 4\sqrt{5}}$

61. $\dfrac{3}{\sqrt{x} - 2}$

62. $\dfrac{4}{\sqrt{x} + 5}$

63. $\dfrac{\sqrt{x}}{\sqrt{x} - 1}$

64. $\dfrac{\sqrt{x}}{\sqrt{x} + 2}$

65. $\dfrac{\sqrt{x} + 3}{\sqrt{x} + 2}$

66. $\dfrac{\sqrt{x} - 4}{\sqrt{x} - 1}$

67. $\dfrac{\sqrt{x}}{\sqrt{x} + \sqrt{y}}$

68. $\dfrac{2\sqrt{x}}{\sqrt{x} - \sqrt{y}}$

69. $\dfrac{2\sqrt{x} + \sqrt{y}}{3\sqrt{x} - 2\sqrt{y}}$

70. $\dfrac{3\sqrt{x} - 2\sqrt{y}}{2\sqrt{x} + 5\sqrt{y}}$

5.5 Equations Involving Radicals

Equations containing radicals with variables in a radicand are often referred to as **radical equations**. In this section we shall discuss techniques for solving such equations containing one or more radicals. To solve radical equations we need the following property of equality.

Property 5.6	Let a and b be real numbers and n a positive integer. $$\text{If } a = b, \quad \text{then } a^n = b^n.$$

Property 5.6 states that we can *raise both sides of an equation to a positive integral power. However,* raising both sides of an equation to a positive integral power sometimes produces results that do not satisfy the original equation. Let's consider two examples to illustrate this point.

Example 1   Solve $\sqrt{2x - 5} = 7$.

Solution

$$\sqrt{2x - 5} = 7$$
$$(\sqrt{2x - 5})^2 = 7^2 \qquad \text{Square both sides.}$$
$$2x - 5 = 49$$
$$2x = 54$$
$$x = 27.$$

Check $\sqrt{2x - 5} = 7$
$$\sqrt{2(27) - 5} \overset{?}{=} 7$$
$$\sqrt{49} \overset{?}{=} 7$$
$$7 = 7.$$

The solution set for $\sqrt{2x - 5} = 7$ is $\{27\}$.

Example 2   Solve $\sqrt{3a + 4} = -4$.

Solution

$$\sqrt{3a + 4} = -4$$
$$(\sqrt{3a + 4})^2 = (-4)^2 \qquad \text{Square both sides.}$$
$$3a + 4 = 16$$
$$3a = 12$$
$$a = 4.$$

Check $\sqrt{3a + 4} = -4$

$$\sqrt{3(4) + 4} \overset{?}{=} -4$$

$$\sqrt{16} \overset{?}{=} -4$$

$$4 \neq -4.$$

Since 4 does not check, the original equation *has no real number solution*. Thus, the solution set is ∅.   ●

In general, raising both sides of an equation to a positive integral power produces an equation that has all of the solutions of the original equation, but it may also have some extra solutions that will not satisfy the original equation. Such extra solutions are called **extraneous solutions**. Therefore, when using Property 5.6 you *must* check each potential solution in the original equation.

Let's consider some examples to illustrate different situations that arise when solving radical equations.

Example 3   Solve $\sqrt{2t - 4} = t - 2$.

Solution

$$\sqrt{2t - 4} = t - 2$$

$$(\sqrt{2t - 4})^2 = (t - 2)^2 \qquad \text{Square both sides.}$$

$$2t - 4 = t^2 - 4t + 4$$

$$0 = t^2 - 6t + 8$$

$$0 = (t - 2)(t - 4) \qquad \text{Factor the right side.}$$

$t - 2 = 0$ or $t - 4 = 0$ \qquad Apply: if $ab = 0$, then $a = 0$ or $b = 0$.

  $t = 2$ or $t = 4$.

Check $\sqrt{2t - 4} = t - 2$ \qquad\qquad $\sqrt{2t - 4} = t - 2$

$\sqrt{2(2) - 4} \overset{?}{=} 2 - 2$ or $\sqrt{2(4) - 4} \overset{?}{=} 4 - 2$

$\qquad \sqrt{0} \overset{?}{=} 0$ \qquad\qquad\qquad $\sqrt{4} \overset{?}{=} 2$

$\qquad\quad 0 = 0$ \qquad\qquad\qquad\qquad $2 = 2.$

The solution set is $\{2, 4\}$.   ●

Example 4   Solve $\sqrt{y} + 6 = y$.

Solution

$$\sqrt{y} + 6 = y$$

$$\sqrt{y} = y - 6$$

$$(\sqrt{y})^2 = (y - 6)^2 \qquad \text{Square both sides.}$$

$$y = y^2 - 12y + 36$$

$$0 = y^2 - 13y + 36$$

$$0 = (y - 4)(y - 9) \qquad \text{Factor the right side.}$$

$y - 4 = 0$ or $y - 9 = 0$ \qquad Apply: if $ab = 0$, then $a = 0$ or $b = 0$.

  $y = 4$ or $y = 9$.

Check $\quad\sqrt{y+6}=y\qquad\qquad\sqrt{y+6}=y$

$\qquad\qquad\sqrt{4+6}\overset{?}{=}4\qquad$ or $\qquad\sqrt{9+6}\overset{?}{=}9$

$\qquad\qquad\quad 2+6\overset{?}{=}4\qquad\qquad\qquad 3+6\overset{?}{=}9$

$\qquad\qquad\qquad\quad 8\neq 4.\qquad\qquad\qquad\qquad 9=9.$

The only solution is 9; the solution set is $\{9\}$.

In Example 4, note that we changed the form of the original equation $\sqrt{y+6}=y$ to $\sqrt{y}=y-6$ before squaring both sides. Squaring both sides of $\sqrt{y+6}=y$ produces $y+12\sqrt{y}+36=y^2$, a more complex equation that still contains a radical. So, again it pays to think ahead a few steps before carrying out the details. Finally, let's consider an example involving cube root.

Example 5   Solve $\sqrt[3]{n^2-1}=2$.

Solution

$$\sqrt[3]{n^2-1}=2$$
$$(\sqrt[3]{n^2-1})^3=2^3\qquad\text{Cube both sides.}$$
$$n^2-1=8$$
$$n^2-9=0$$
$$(n+3)(n-3)=0$$
$$n+3=0\quad\text{ or }\quad n-3=0$$
$$n=-3\quad\text{ or }\qquad n=3.$$

Check $\qquad\sqrt[3]{n^2-1}=2\qquad\qquad\sqrt[3]{n^2-1}=2$

$\qquad\qquad\sqrt[3]{(-3)^2-1}\overset{?}{=}2\quad$ or $\quad\sqrt[3]{3^2-1}\overset{?}{=}2$

$\qquad\qquad\qquad\sqrt[3]{8}\overset{?}{=}2\qquad\qquad\qquad\sqrt[3]{8}\overset{?}{=}2$

$\qquad\qquad\qquad\quad 2=2.\qquad\qquad\qquad\quad 2=2.$

The solution set is $\{-3, 3\}$.

Problem Set 5.5

Solve each of the following equations. Don't forget to check each of your potential solutions.

1. $\sqrt{2x}=4$

2. $\sqrt{3x}=6$

3. $\sqrt{3x}=-3$

4. $\sqrt{2x}=-8$

5. $2\sqrt{n}=5$

6. $3\sqrt{n}=4$

7. $\sqrt{2y-1}=3$

8. $\sqrt{3y-2}=4$

9. $\sqrt{4x-3}=-1$

10. $\sqrt{5x+2}=-3$

11. $\sqrt{5x-3}-6=0$

12. $\sqrt{4x-5}-1=0$

13. $-\sqrt{x-4}=5$

14. $-\sqrt{x+6}=2$

15. $\sqrt{5x-2}=\sqrt{2x+7}$

16. $\sqrt{3x-1}=\sqrt{4x-5}$

17. $\sqrt{6x-5}=\sqrt{3x+5}$

18. $-\sqrt{2x-1}=-\sqrt{x+2}$

19. $3\sqrt{t+1}=4$

20. $2\sqrt{t-5}=1$

21. $\sqrt{x^2+3}=2$

22. $\sqrt{x^2+7}-4=0$

23. $\sqrt{x^2-2x+1}=5$

24. $\sqrt{x^2+11x+19}=1$

25. $\sqrt{n^2-2n-4}=n$

26. $\sqrt{x^2-x+1}=x+1$

27. $\sqrt{x^2+2x+1}=x+3$

28. $\sqrt{x^2+3x+7}=x+2$

29. $\sqrt{2x-1}=x-2$

30. $\sqrt{-4x+17}=x-3$

31. $\sqrt{n+2}=n+2$

32. $\sqrt{n+4}=n+4$

33. $2\sqrt{y}=y-3$

34. $\sqrt{3y}=y-6$

35. $\sqrt{-x-6}=x$

36. $4\sqrt{x+5}=x$

37. $\sqrt[3]{2x+1}=3$

38. $\sqrt[3]{2x-5}=1$

39. $\sqrt[3]{3x-1}=-4$

40. $\sqrt[3]{2x+3}=-3$

41. $\sqrt[3]{n^2+1}=1$

42. $\sqrt[3]{n^2-1}=-1$

43. $\sqrt[3]{3x-1}=\sqrt[3]{2-5x}$

44. $\sqrt[3]{2x+5}=\sqrt[3]{4-x}$

45. $\sqrt{x+4}=\sqrt{x-1}+1$ (Hint: Apply Property 5.6 twice.)

46. $\sqrt{x+19}-\sqrt{x+28}=-1$

47. $\sqrt{2x-1}-\sqrt{x+3}=1$

48. $\sqrt{3x+1}+\sqrt{2x+4}=3$

49. $\sqrt{n-3}+\sqrt{n+5}=2\sqrt{n}$

50. $\sqrt{n-4}+\sqrt{n+4}=2\sqrt{n-1}$

5.6 The Merging of Exponents and Roots

Recall that the basic properties of positive integral exponents provided the basis for a definition for the use of negative integers as exponents. In this section, the properties of integral exponents are used to form definitions for the use of rational numbers as exponents. These definitions will tie together the concepts of *exponent* and *root*.

    Let's consider the following comparisons.

From our study of radicals we know that

If $(b^n)^m=b^{mn}$ is to hold when n equals a rational number of the form $\dfrac{1}{p}$, where p is a positive integer greater than one, then

$(\sqrt{5})^2=5;$

$\left(5^{\frac{1}{2}}\right)^2=5^{2\left(\frac{1}{2}\right)}=5^1=5;$

$(\sqrt[3]{8})^3=8;$

$\left(8^{\frac{1}{3}}\right)^3=8^{3\left(\frac{1}{3}\right)}=8^1=8;$

$(\sqrt[4]{21})^4=21.$

$\left(21^{\frac{1}{4}}\right)^4=21^{4\left(\frac{1}{4}\right)}=21^1=21.$

It would seem reasonable to make the following definition.

Definition 5.6	If b is a real number, n is a positive integer greater than one, and $\sqrt[n]{b}$ exists, then $$b^{\frac{1}{n}} = \sqrt[n]{b}.$$

Definition 5.6 states that $b^{\frac{1}{n}}$ *means the nth root of b.* We shall assume that b and n are chosen so that $\sqrt[n]{b}$ exists. For example, $(-25)^{\frac{1}{2}}$ is not meaningful at this time because $\sqrt{-25}$ is not a real number.

Consider the following examples that illustrate the use of Definition 5.6.

$$25^{\frac{1}{2}} = \sqrt{25} = 5; \qquad\qquad 16^{\frac{1}{4}} = \sqrt[4]{16} = 2;$$

$$8^{\frac{1}{3}} = \sqrt[3]{8} = 2; \qquad\qquad \left(\frac{36}{49}\right)^{\frac{1}{2}} = \sqrt{\frac{36}{49}} = \frac{6}{7};$$

$$(-27)^{\frac{1}{3}} = \sqrt[3]{-27} = -3.$$

The following definition provides the basis for the use of *all* rational numbers as exponents.

Definition 5.7	If $\dfrac{m}{n}$ is a rational number, where n is a positive integer greater than one, and b is a real number such that $\sqrt[n]{b}$ exists, then $$b^{\frac{m}{n}} = \sqrt[n]{b^m} = (\sqrt[n]{b})^m.$$

In Definition 5.7, notice that the denominator of the exponent is the index of the radical and the numerator of the exponent is either the exponent of the radicand or the exponent of the root.

Whether we use the form $\sqrt[n]{b^m}$ or $(\sqrt[n]{b})^m$ for computational purposes depends somewhat on the magnitude of the problem. Let's use both forms on two problems to illustrate this point.

$$8^{\frac{2}{3}} = \sqrt[3]{8^2} \qquad\qquad 8^{\frac{2}{3}} = (\sqrt[3]{8})^2$$
$$\phantom{8^{\frac{2}{3}}} = \sqrt[3]{64} \quad \text{or} \quad \phantom{8^{\frac{2}{3}}} = 2^2$$
$$\phantom{8^{\frac{2}{3}}} = 4. \qquad\qquad\qquad = 4.$$

$$27^{\frac{2}{3}} = \sqrt[3]{27^2} \qquad\qquad 27^{\frac{2}{3}} = (\sqrt[3]{27})^2$$
$$\phantom{27^{\frac{2}{3}}} = \sqrt[3]{729} \quad \text{or} \quad \phantom{27^{\frac{2}{3}}} = 3^2$$
$$\phantom{27^{\frac{2}{3}}} = 9. \qquad\qquad\qquad = 9.$$

To compute $8^{\frac{2}{3}}$, either form seems to work about as well as the other one. However, to compute $27^{\frac{2}{3}}$, it should be obvious that $(\sqrt[3]{27})^2$ is much easier to handle than $\sqrt[3]{27^2}$.

Example 1   Simplify each of the following numerical expressions.

(a) $25^{\frac{3}{2}}$ (b) $16^{\frac{3}{4}}$ (c) $(32)^{-\frac{2}{5}}$ (d) $(-64)^{\frac{2}{3}}$ (e) $-8^{\frac{1}{3}}$

Solution   (a) $25^{\frac{3}{2}} = (\sqrt{25})^3 = 5^3 = 125.$

(b) $16^{\frac{3}{4}} = (\sqrt[4]{16})^3 = 2^3 = 8.$

(c) $(32)^{-\frac{2}{5}} = \dfrac{1}{(32)^{\frac{2}{5}}} = \dfrac{1}{(\sqrt[5]{32})^2} = \dfrac{1}{2^2} = \dfrac{1}{4}.$

(d) $(-64)^{\frac{2}{3}} = (\sqrt[3]{-64})^2 = (-4)^2 = 16.$

(e) $-8^{\frac{1}{3}} = -\sqrt[3]{8} = -2.$ ●

It can be shown that the basic laws of exponents stated in Property 5.2 are true for all rational exponents. Therefore, without restating Property 5.2 we shall henceforth use it for rational as well as integral exponents.

Some problems can be handled better in exponential form and others in radical form. Thus, we must be able to switch forms with a certain amount of ease. Let's consider some examples illustrating the switch from one form to the other.

Example 2   Write each of the following expressions in radical form.

(a) $x^{\frac{3}{4}}$ (b) $3y^{\frac{2}{5}}$

(c) $x^{\frac{1}{4}}y^{\frac{3}{4}}$ (d) $(x + y)^{\frac{2}{3}}$

Solution   (a) $x^{\frac{3}{4}} = \sqrt[4]{x^3}.$ (b) $3y^{\frac{2}{5}} = 3\sqrt[5]{y^2}.$

(c) $x^{\frac{1}{4}}y^{\frac{3}{4}} = (xy^3)^{\frac{1}{4}} = \sqrt[4]{xy^3}.$ (d) $(x + y)^{\frac{2}{3}} = \sqrt[3]{(x + y)^2}.$ ●

Example 3   Write each of the following using positive rational exponents.

(a) $\sqrt{xy}$ (b) $\sqrt[4]{a^3b}$ (c) $4\sqrt[3]{x^2}$ (d) $\sqrt[5]{(x + y)^4}$

Solution   (a) $\sqrt{xy} = (xy)^{\frac{1}{2}} = x^{\frac{1}{2}}y^{\frac{1}{2}}.$ (b) $\sqrt[4]{a^3b} = (a^3b)^{\frac{1}{4}} = a^{\frac{3}{4}}b^{\frac{1}{4}}.$

(c) $4\sqrt[3]{x^2} = 4x^{\frac{2}{3}}.$ (d) $\sqrt[5]{(x + y)^4} = (x + y)^{\frac{4}{5}}.$ ●

The basic properties of exponents provide the basis for simplifying algebraic expressions containing rational exponents, as these next examples illustrate.

Example 4   Simplify each of the following. Express final results using positive exponents only.

(a) $\left(3x^{\frac{1}{2}}\right)\left(4x^{\frac{2}{3}}\right)$ (b) $\left(5a^{\frac{1}{3}}b^{\frac{1}{2}}\right)^2$ (c) $\dfrac{12y^{\frac{1}{3}}}{6y^{\frac{1}{2}}}$ (d) $\left(\dfrac{3x^{\frac{2}{5}}}{2y^{\frac{2}{3}}}\right)^4$

Solution

(a) $\left(3x^{\frac{1}{2}}\right)\left(4x^{\frac{2}{3}}\right) = 3 \cdot 4 \cdot x^{\frac{1}{2}} \cdot x^{\frac{2}{3}}$

$= 12x^{\frac{1}{2}+\frac{2}{3}}$ $(b^n \cdot b^m = b^{n+m})$

$= 12x^{\frac{3}{6}+\frac{4}{6}}$

$= 12x^{\frac{7}{6}}.$

(b) $\left(5a^{\frac{1}{3}}b^{\frac{1}{2}}\right)^2 = 5^2 \cdot \left(a^{\frac{1}{3}}\right)^2 \cdot \left(b^{\frac{1}{2}}\right)^2$ $((ab)^n = a^n b^n)$

$= 25a^{\frac{2}{3}}b.$ $((b^n)^m = b^{mn})$

(c) $\dfrac{12y^{\frac{1}{3}}}{6y^{\frac{1}{2}}} = 2y^{\frac{1}{3}-\frac{1}{2}}$ $\left(\dfrac{b^n}{b^m} = b^{n-m}\right)$

$= 2y^{\frac{2}{6}-\frac{3}{6}}$

$= 2y^{-\frac{1}{6}}$

$= \dfrac{2}{y^{\frac{1}{6}}}.$

(d) $\left(\dfrac{3x^{\frac{2}{5}}}{2y^{\frac{2}{3}}}\right)^4 = \dfrac{\left(3x^{\frac{2}{5}}\right)^4}{\left(2y^{\frac{2}{3}}\right)^4}$ $\left(\left(\dfrac{a}{b}\right)^n = \dfrac{a^n}{b^n}\right)$

$= \dfrac{3^4 \cdot \left(x^{\frac{2}{5}}\right)^4}{2^4 \cdot \left(y^{\frac{2}{3}}\right)^4}$ $((ab)^n = a^n b^n)$

$= \dfrac{81x^{\frac{8}{5}}}{16y^{\frac{8}{3}}}.$ $((b^n)^m = b^{mn})$ ●

The link between exponents and roots also provides a basis for multiplying and dividing some radicals even if they have a different index. The general procedure is one of changing from radical form to exponential form, applying the properties of exponents, and then changing back to radical form. Let's consider three examples to illustrate this process.

Example 5   Perform the indicated operations and express the answer in simplest radical form.

(a) $\sqrt{2}\sqrt[3]{2}$ (b) $\dfrac{\sqrt{5}}{\sqrt[3]{5}}$ (c) $\dfrac{\sqrt{4}}{\sqrt[3]{2}}$

Solution

(a) $\sqrt{2}\,\sqrt[3]{2} = 2^{\frac{1}{2}} \cdot 2^{\frac{1}{3}}$

$\qquad = 2^{\frac{1}{2}+\frac{1}{3}}$

$\qquad = 2^{\frac{3}{6}+\frac{2}{6}}$

$\qquad = 2^{\frac{5}{6}}$

$\qquad = \sqrt[6]{2^5} = \sqrt[6]{32}.$

(b) $\dfrac{\sqrt{5}}{\sqrt[3]{5}} = \dfrac{5^{\frac{1}{2}}}{5^{\frac{1}{3}}}$

$\qquad = 5^{\frac{1}{2}-\frac{1}{3}}$

$\qquad = 5^{\frac{3}{6}-\frac{2}{6}}$

$\qquad = 5^{\frac{1}{6}} = \sqrt[6]{5}.$

(c) $\dfrac{\sqrt{4}}{\sqrt[3]{2}} = \dfrac{4^{\frac{1}{2}}}{2^{\frac{1}{3}}}$

$\qquad = \dfrac{(2^2)^{\frac{1}{2}}}{2^{\frac{1}{3}}}$

$\qquad = \dfrac{2^1}{2^{\frac{1}{3}}}$

$\qquad = 2^{1-\frac{1}{3}}$

$\qquad = 2^{\frac{2}{3}} = \sqrt[3]{2^2} = \sqrt[3]{4}.$

Problem Set 5.6

Simplify each of the following numerical expressions.

1. $64^{\frac{1}{2}}$

2. $100^{\frac{1}{2}}$

3. $64^{\frac{1}{3}}$

4. $32^{\frac{1}{5}}$

5. $(-8)^{\frac{1}{3}}$

6. $\left(-\dfrac{8}{27}\right)^{\frac{1}{3}}$

7. $-16^{\frac{1}{2}}$

8. $-27^{\frac{1}{3}}$

9. $49^{-\frac{1}{2}}$

10. $\left(\dfrac{1}{4}\right)^{-\frac{1}{2}}$

11. $\left(\dfrac{1}{8}\right)^{-\frac{1}{3}}$

12. $\left(-\dfrac{27}{8}\right)^{-\frac{1}{3}}$

13. $4^{\frac{5}{2}}$

14. $27^{\frac{2}{3}}$

15. $64^{\frac{2}{3}}$

16. $16^{\frac{3}{2}}$

17. $(-8)^{\frac{4}{3}}$

18. $(-1)^{\frac{5}{3}}$

19. $-27^{\frac{2}{3}}$

20. $-4^{\frac{3}{2}}$

21. $\left(\dfrac{8}{27}\right)^{\frac{4}{3}}$

22. $\left(\dfrac{1}{32}\right)^{\frac{3}{5}}$

23. $\left(\dfrac{1}{27}\right)^{-\frac{2}{3}}$

24. $\left(-\dfrac{1}{8}\right)^{-\frac{2}{3}}$

25. $64^{-\frac{5}{6}}$

26. $32^{-\frac{3}{5}}$

27. $-16^{\frac{5}{2}}$

28. $-16^{\frac{5}{4}}$

29. $81^{\frac{3}{4}}$

30. $125^{\frac{4}{3}}$

Write each of the following in radical form; for example, $3x^{\frac{2}{3}} = 3\sqrt[3]{x^2}$.

31. $x^{\frac{2}{3}}$

32. $x^{\frac{3}{5}}$

33. $2x^{\frac{1}{2}}$

34. $3x^{\frac{3}{4}}$

35. $(3y)^{\frac{1}{2}}$ **36.** $(2xy)^{\frac{1}{3}}$ **37.** $(x - 2y)^{\frac{1}{3}}$ **38.** $(2x - y)^{\frac{1}{2}}$

39. $(2a + b)^{\frac{3}{5}}$ **40.** $(3a + 2b)^{\frac{4}{7}}$ **41.** $x^{\frac{1}{3}}y^{\frac{2}{3}}$ **42.** $x^{\frac{2}{5}}y^{\frac{3}{5}}$

43. $-3x^{\frac{1}{4}}y^{\frac{3}{4}}$ **44.** $-2x^{\frac{2}{3}}y^{\frac{1}{3}}$

Write each of the following using positive rational exponents; for example, $\sqrt{ab} = (ab)^{\frac{1}{2}} = a^{\frac{1}{2}}b^{\frac{1}{2}}$.

45. $\sqrt{2x}$ **46.** $\sqrt{3xy}$ **47.** $2\sqrt{x}$ **48.** $3\sqrt{ab}$

49. $\sqrt[3]{x^2 y}$ **50.** $\sqrt[3]{xy^2}$ **51.** $\sqrt[4]{ab^3}$ **52.** $\sqrt[5]{a^2 b^3}$

53. $\sqrt[3]{(2x - y)^2}$ **54.** $\sqrt[5]{(3x + 2y)^3}$ **55.** $2x\sqrt{y}$ **56.** $3y\sqrt[5]{x^2}$

57. $-\sqrt[3]{a + b}$ **58.** $-\sqrt[4]{(a - b)^3}$

Simplify each of the following. Express final results using positive exponents only. For example,

$$\left(2x^{\frac{1}{2}}\right)\left(3x^{\frac{1}{3}}\right) = 6x^{\frac{5}{6}}.$$

59. $\left(3x^{\frac{1}{4}}\right)\left(5x^{\frac{1}{3}}\right)$ **60.** $\left(2x^{\frac{2}{5}}\right)\left(6x^{\frac{1}{4}}\right)$ **61.** $\left(y^{\frac{3}{4}}\right)\left(y^{-\frac{1}{2}}\right)$ **62.** $\left(y^{\frac{2}{3}}\right)\left(y^{-\frac{1}{4}}\right)$

63. $\left(2x^{\frac{1}{3}}\right)\left(x^{-\frac{1}{2}}\right)$ **64.** $\left(x^{\frac{2}{5}}\right)\left(4x^{-\frac{1}{2}}\right)$ **65.** $\left(4x^{\frac{1}{4}}y^{\frac{1}{2}}\right)^3$ **66.** $\left(5x^{\frac{1}{2}}y\right)^2$

67. $(9x^2 y^4)^{\frac{1}{2}}$ **68.** $(8x^6 y^3)^{\frac{1}{3}}$ **69.** $\dfrac{18x^{\frac{1}{2}}}{9x^{\frac{1}{3}}}$ **70.** $\dfrac{24x^{\frac{3}{5}}}{6x^{\frac{1}{3}}}$

71. $\dfrac{56a^{\frac{1}{6}}}{8a^{\frac{1}{4}}}$ **72.** $\dfrac{48b^{\frac{1}{3}}}{12b^{\frac{3}{4}}}$ **73.** $\left(\dfrac{2x^{\frac{1}{3}}}{3y^{\frac{1}{4}}}\right)^4$ **74.** $\left(\dfrac{6x^{\frac{2}{5}}}{7y^{\frac{2}{3}}}\right)^2$

75. $\left(\dfrac{x^2}{y^3}\right)^{-\frac{1}{2}}$ **76.** $\left(\dfrac{a^3}{b^{-2}}\right)^{-\frac{1}{3}}$

Perform the indicated operations and express answers in simplest radical form. (See Example 5.)

77. $\sqrt{2}\sqrt[4]{2}$ **78.** $\sqrt[3]{3}\sqrt{3}$ **79.** $\sqrt[3]{5}\sqrt{5}$

80. $\sqrt[4]{6}\sqrt{6}$ **81.** $\dfrac{\sqrt{2}}{\sqrt[3]{2}}$ **82.** $\dfrac{\sqrt[3]{3}}{\sqrt[4]{3}}$

83. $\dfrac{\sqrt{9}}{\sqrt[3]{3}}$ **84.** $\dfrac{\sqrt[3]{8}}{\sqrt[4]{4}}$ **85.** $\dfrac{\sqrt[3]{16}}{\sqrt[6]{4}}$

86. If your calculator has $\boxed{y^x}$ and $\boxed{1/x}$ keys, then they can be used to evaluate cube roots, fourth roots, fifth roots, and so on. For example, on some calculators $\sqrt[3]{4913}$ can be evaluated as follows:

$$4913 \ \boxed{y^x} \ 3 \ \boxed{1/x} = 17.$$

Use your calculator to evaluate each of the following.

(a) $\sqrt[3]{1728}$ **(b)** $\sqrt[3]{5832}$ **(c)** $\sqrt[4]{2401}$

(d) $\sqrt[4]{65536}$ **(e)** $\sqrt[5]{161051}$ **(f)** $\sqrt[5]{6436343}$

87. Definition 5.7 stated that

$$b^{\frac{m}{n}} = \sqrt[n]{b^m} = (\sqrt[n]{b})^m.$$

Use your calculator to verify each of the following.

(a) $\sqrt[3]{27^2} = (\sqrt[3]{27})^2$ (b) $\sqrt[3]{8^5} = (\sqrt[3]{8})^5$ (c) $\sqrt[4]{16^3} = (\sqrt[4]{16})^3$

(d) $\sqrt[3]{16^2} = (\sqrt[3]{16})^2$ (e) $\sqrt[5]{9^4} = (\sqrt[5]{9})^4$ (f) $\sqrt[3]{12^4} = (\sqrt[3]{12})^4$

88. Use your calculator to evaluate each of the following.

(a) $16^{\frac{5}{2}}$ (b) $25^{\frac{7}{2}}$ (c) $16^{\frac{9}{4}}$

(d) $27^{\frac{5}{3}}$ (e) $343^{\frac{2}{3}}$ (f) $512^{\frac{4}{3}}$

89. Use your calculator to estimate each of the following to the nearest thousandth.

(a) $7^{\frac{4}{3}}$ (b) $10^{\frac{4}{5}}$ (c) $12^{\frac{2}{5}}$

(d) $19^{\frac{2}{5}}$ (e) $7^{\frac{3}{4}}$ (f) $10^{\frac{5}{4}}$

90. (a) Since $\dfrac{4}{5} = 0.8$, we can evaluate $10^{\frac{4}{5}}$ by evaluating $10^{0.8}$. This can be done using a shorter sequence of "calculator steps." Evaluate parts **(b)**, **(c)**, **(d)**, **(e)**, and **(f)** of Problem 89 taking advantage of decimal exponents.

(b) What problem is created when we try to evaluate $7^{\frac{4}{3}}$ by changing the exponent to decimal form?

5.7 Scientific Notation (Optional)

Many applications of mathematics involve the use of very large and very small numbers. For example,

1. A light year—the distance that a ray of light travels in one year—is approximately 5,900,000,000,000 miles;

2. The national debt is approximately 950,000,000,000 dollars;

3. In the metric system, a millimicron equals 0.000000001 of a meter;

4. The weight of an oxygen molecule is approximately 0.00000000000000000000000053 of a gram.

Working with numbers of this type in standard form is quite cumbersome. It is much more convenient to represent very small and very large numbers in **scientific notation**, sometimes called **scientific form**. A number is expressed in scientific notation when it is written as a product of a number between 1 and 10 and an integral power of 10. Symbolically, a number written in scientific notation has the form

$$(N)(10)^k$$

where N is a number between 1 and 10, written in decimal form, and k is an integer. Consider the following examples which show a comparison between ordinary notation and scientific notation.

Ordinary Notation	Scientific Notation
2.14	$(2.14)(10)^0$
31.78	$(3.178)(10)^1$
412.9	$(4.129)(10)^2$
8,000,000	$(8)(10)^6$
0.14	$(1.4)(10)^{-1}$
0.0379	$(3.79)(10)^{-2}$
0.00000049	$(4.9)(10)^{-7}$

To switch from ordinary notation to scientific notation, you can use the following procedure:

Write the given number as the product of a number between 1 and 10 and a power of 10. The exponent of 10 is determined by counting the number of places that the decimal point was moved in going from the original number to the number between 1 and 10. This exponent is a) negative if the original number is less than 1, b) positive if the original number is greater than 10, and c) 0 if the original number itself is between 1 and 10.

Thus, we can write

$$0.00467 = (4.67)(10)^{-3};$$
$$87,000 = (8.7)(10)^4;$$
$$3.1416 = (3.1416)(10)^0.$$

To switch from scientific notation to ordinary notation you can use the following procedure:

Move the decimal point the number of places indicated by the exponent of 10. The decimal point is moved to the right if the exponent is positive and to the left if it is negative.

Thus, we can write

$$(4.78)(10)^4 = 47,800;$$
$$(8.4)(10)^{-3} = 0.0084.$$

Scientific notation can frequently be used to simplify numerical calculations. We merely change the numbers to scientific notation and use the appropriate properties of exponents. Consider the following examples.

Example 1

Perform the indicated operations.

(a) $(0.00024)(20,000)$

(b) $\dfrac{7,800,000}{0.0039}$

(c) $\dfrac{(0.00069)(0.0034)}{(0.0000017)(0.023)}$

(d) $\sqrt{0.000004}$

Solution

(a) $(0.00024)(20,000) = (2.4)(10)^{-4}(2)(10)^4$

$= (2.4)(2)(10)^{-4}(10)^4$

$= (4.8)(10)^0$

$= (4.8)(1)$

$= 4.8.$

(b) $\dfrac{7,800,000}{0.0039} = \dfrac{(7.8)(10)^6}{(3.9)(10)^{-3}}$

$= (2)(10)^9$

$= 2,000,000,000.$

(c) $\dfrac{(0.00069)(0.0034)}{(0.0000017)(0.023)} = \dfrac{(6.9)(10)^{-4}(3.4)(10)^{-3}}{(1.7)(10)^{-6}(2.3)(10)^{-2}}$

$= \dfrac{\overset{3}{\cancel{(6.9)}}\overset{2}{\cancel{(3.4)}}(10)^{-7}}{\cancel{(1.7)}\cancel{(2.3)}(10)^{-8}}$

$= (6)(10)^1$

$= 60.$

(d) $\sqrt{0.000004} = \sqrt{(4)(10)^{-6}}$

$= ((4)(10)^{-6})^{\frac{1}{2}}$

$= (4^{\frac{1}{2}})((10)^{-6})^{\frac{1}{2}}$

$= (2)(10)^{-3}$

$= 0.002.$ ●

Many calculators are equipped to display numbers in scientific notation. The display panel shows the number between 1 and 10 and the appropriate exponent of 10. For example, evaluating $(3,800,000)^2$ yields

$$\boxed{1.444 \qquad 13}\,.$$

Thus, $(3,800,000)^2 = (1.444)(10)^{13} = 14,440,000,000,000.$
Similarly, the answer for $(0.000168)^2$ is displayed as

$$\boxed{2.8224 \qquad -08}\,.$$

Thus, $(0.000168)^2 = (2.8224)(10)^{-8} = 0.000000028224.$
Calculators vary as to the number of digits displayed in the number between 1 and 10 when using scientific notation. For example, we used two different calculators to

estimate $(6729)^6$ and obtained the following results:

| 9.2833 | 22 | ;

| 9.283316768 | 22 | .

Obviously, you need to know the capabilities of your calculator when working with problems in scientific notation.

Many calculators also allow the entry of a number in scientific notation. Such calculators are equipped with an "enter the exponent" key (often labeled as | EE | or | E EX |). Thus, a number such as $(3.14)(10)^8$ might be entered as follows:

Enter	Press	Display
3.14	EE	3.14 00
8		3.14 08

Furthermore, it may be that your calculator will perform the switch from ordinary notation to scientific notation with a routine such as the following.

Enter	Press	Display
4721	EE	4721 00
	=	4.721 03

Be sure that you know the routine that will accomplish this switch on your calculator.

It should be evident from this brief discussion that even when using a calculator, you need to have a thorough understanding of scientific notation.

Problem Set 5.7

Write each of the following in scientific notation. For example, $27,800 = (2.78)(10)^4$.

1. 89	**2.** 117	**3.** 4290
4. 812,000	**5.** 6,120,000	**6.** 72,400,000
7. 40,000,000	**8.** 500,000,000	**9.** 376.4
10. 9126.21	**11.** 0.347	**12.** 0.2165
13. 0.0214	**14.** 0.0037	**15.** 0.00005
16. 0.00000082	**17.** 0.00000000194	**18.** 0.00000000003

Write each of the following in ordinary notation. For example, $(3.18)(10)^2 = 318$.

19. $(2.3)(10)^1$	**20.** $(1.62)(10)^2$	**21.** $(4.19)(10)^3$
22. $(7.631)(10)^4$	**23.** $(5)(10)^8$	**24.** $(7)(10)^9$

25. $(3.14)(10)^{10}$ **26.** $(2.04)(10)^{12}$ **27.** $(4.3)(10)^{-1}$

28. $(5.2)(10)^{-2}$ **29.** $(9.14)(10)^{-4}$ **30.** $(8.76)(10)^{-5}$

31. $(5.123)(10)^{-8}$ **32.** $(6)(10)^{-9}$

Use scientific notation and the properties of exponents to help perform the following operations.

33. $(0.0037)(0.00002)$ **34.** $(0.00003)(0.00025)$ **35.** $(0.00007)(11,000)$

36. $(0.000004)(120,000)$ **37.** $\dfrac{360,000,000}{0.0012}$ **38.** $\dfrac{66,000,000,000}{0.022}$

39. $\dfrac{0.000064}{16,000}$ **40.** $\dfrac{0.00072}{0.0000024}$ **41.** $\dfrac{(60,000)(0.006)}{(0.0009)(400)}$

42. $\dfrac{(0.00063)(960,000)}{(3,200)(0.0000021)}$ **43.** $\dfrac{(0.0045)(60000)}{(1800)(0.00015)}$ **44.** $\dfrac{(0.00016)(300)(0.028)}{0.064}$

45. $\sqrt{9,000,000}$ **46.** $\sqrt{0.00000009}$ **47.** $\sqrt[3]{8000}$

48. $\sqrt[3]{0.001}$ **49.** $(90,000)^{\frac{3}{2}}$ **50.** $(8000)^{\frac{2}{3}}$

51. Sometimes it is more convenient to express a number as a product of a power of 10 and a number that is not between 1 and 10. For example, suppose that we want to calculate $\sqrt{640,000}$. We can proceed as follows:

$$\sqrt{640,000} = \sqrt{(64)(10)^4}$$
$$= ((64)(10)^4)^{\frac{1}{2}}$$
$$= (64)^{\frac{1}{2}}(10^4)^{\frac{1}{2}}$$
$$= (8)(10)^2$$
$$= 8(100) = 800.$$

Compute each of the following.

(a) $\sqrt{49,000,000}$ **(b)** $\sqrt{0.0025}$ **(c)** $\sqrt{14400}$

(d) $\sqrt{0.000121}$ **(e)** $\sqrt[3]{27000}$ **(f)** $\sqrt[3]{0.000064}$

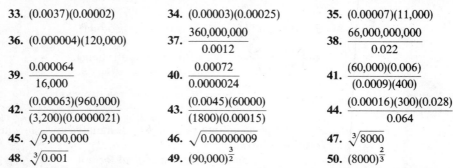

52. Use your calculator to evaluate each of the following. Express final answers in ordinary notation.

(a) $(27,000)^2$ **(b)** $(450,000)^2$ **(c)** $(14,800)^2$

(d) $(1700)^3$ **(e)** $(900)^4$ **(f)** $(60)^5$

(g) $(0.0213)^2$ **(h)** $(0.000213)^2$ **(i)** $(0.000198)^2$

(j) $(0.000009)^3$

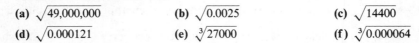

53. Use your calculator to estimate each of the following. Express final answers in scientific notation with the number between 1 and 10 rounded to the nearest thousandth.

(a) $(4576)^4$ **(b)** $(719)^{10}$ **(c)** $(28)^{12}$

(d) $(8619)^6$ **(e)** $(314)^5$ **(f)** $(145,723)^2$

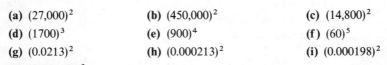

54. Use your calculator to estimate each of the following. Express final answers in ordinary notation rounded to the nearest thousandth.

(a) $(1.09)^5$ **(b)** $(1.08)^{10}$ **(c)** $(1.14)^7$

(d) $(1.12)^{20}$ **(e)** $(0.785)^4$ **(f)** $(0.492)^5$

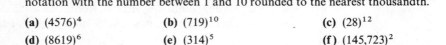

Chapter 5 Summary

(5.1)   The following properties form the basis for manipulating with exponents.

    1. $b^n \cdot b^m = b^{n+m}$ (Product of two powers)

    2. $(b^n)^m = b^{mn}$ (Power of a power)

    3. $(ab)^n = a^n b^n$ (Power of a product)

    4. $\left(\dfrac{a}{b}\right)^n = \dfrac{a^n}{b^n}$ (Power of a quotient)

    5. $\dfrac{b^n}{b^m} = b^{n-m}$ (Quotient of two powers)

(5.2) and (5.3)   The **principal nth root of b** is designated by $\sqrt[n]{b}$, where n is the **index** and b is the **radicand**.

A radical expression is in **simplest radical form** if:

    1. A radicand contains no polynomial factor raised to a power equal to or greater than the index of the radical.

    2. No fraction appears within a radical sign.

    3. No radical appears in the denominator.

The following properties are used to express radicals in simplest radical form:

$$\sqrt[n]{bc} = \sqrt[n]{b}\,\sqrt[n]{c}; \qquad\qquad \sqrt[n]{\frac{b}{c}} = \frac{\sqrt[n]{b}}{\sqrt[n]{c}}.$$

Simplifying by combining radicals sometimes requires that you first express the given radicals in simplest form and then apply the distributive property.

(5.4)   The distributive property and the property $\sqrt[n]{b}\,\sqrt[n]{c} = \sqrt[n]{bc}$ are used to find products of expressions involving radicals.

The special product pattern $(a + b)(a - b) = a^2 - b^2$ suggests a procedure for **rationalizing the denominator** of an expression containing a binomial denominator with radicals.

(5.5)   Equations containing radicals with variables in a radicand are called **radical equations**. The property "if $a = b$, then $a^n = b^n$" forms the basis for solving radical equations. Raising both sides of an equation to a positive integral power may produce **extraneous solutions**, that is, solutions that do not satisfy the original equation. Therefore, you **must check** each potential solution.

(5.6)   If b is a real number, n is a positive integer greater than one, and $\sqrt[n]{b}$ exists, then

$$b^{\frac{1}{n}} = \sqrt[n]{b}.$$

Thus, $b^{\frac{1}{n}}$ means *the nth root of b*.

If $\dfrac{m}{n}$ is a rational number, where n is a positive integer greater than one, and b is a real number such that $\sqrt[n]{b}$ exists, then

$$b^{\frac{m}{n}} = \sqrt[n]{b^m} = (\sqrt[n]{b})^m.$$

Both $\sqrt[n]{b^m}$ and $(\sqrt[n]{b})^m$ can be used for computational purposes.

We need to be able to switch back and forth between **exponential** and **radical form**. The link between exponents and roots provides a basis for multiplying and dividing some radicals even if they have different indices.

Chapter 5 Review Problem Set

Evaluate each of the following numerical expressions.

1. 4^{-3}

2. $\left(\dfrac{2}{3}\right)^{-2}$

3. $(3^2 \cdot 3^{-3})^{-1}$

4. $\sqrt[3]{-8}$

5. $\sqrt[4]{\dfrac{16}{81}}$

6. $4^{\frac{5}{2}}$

7. $(-1)^{-\frac{2}{3}}$

8. $\left(\dfrac{8}{27}\right)^{\frac{2}{3}}$

9. $-16^{\frac{3}{2}}$

10. $\dfrac{2^3}{2^{-2}}$

11. $(4^{-2} \cdot 4^2)^{-1}$

12. $\left(\dfrac{3^{-1}}{3^2}\right)^{-1}$

Express each of the following radicals in simplest radical form.

13. $\sqrt{54}$

14. $\sqrt{48x^3 y}$

15. $\dfrac{4\sqrt{3}}{\sqrt{6}}$

16. $\sqrt{\dfrac{5}{12x^3}}$

17. $\sqrt[3]{56}$

18. $\dfrac{\sqrt[3]{2}}{\sqrt[3]{9}}$

19. $\sqrt[3]{\dfrac{9}{5}}$

20. $\sqrt{\dfrac{3x^3}{7}}$

21. $\sqrt[3]{108x^4 y^8}$

22. $\dfrac{3}{4}\sqrt{150}$

23. $\dfrac{2}{3}\sqrt{45xy^3}$

24. $\dfrac{\sqrt{8x^2}}{\sqrt{2x}}$

Multiply and simplify.

25. $(3\sqrt{8})(4\sqrt{5})$

26. $(5\sqrt[3]{2})(6\sqrt[3]{4})$

27. $3\sqrt{2}(4\sqrt{6} - 2\sqrt{7})$

28. $(\sqrt{x} + 3)(\sqrt{x} - 5)$

29. $(2\sqrt{5} - \sqrt{3})(2\sqrt{5} + \sqrt{3})$

30. $(3\sqrt{2} + \sqrt{6})(5\sqrt{2} - 3\sqrt{6})$

31. $(2\sqrt{a} + \sqrt{b})(3\sqrt{a} - 4\sqrt{b})$

32. $(4\sqrt{8} - \sqrt{2})(\sqrt{8} + 3\sqrt{2})$

Rationalize the denominator and simplify.

33. $\dfrac{4}{\sqrt{7} - 1}$

34. $\dfrac{\sqrt{3}}{\sqrt{8} + \sqrt{5}}$

35. $\dfrac{3}{2\sqrt{3} + 3\sqrt{5}}$

36. $\dfrac{3\sqrt{2}}{2\sqrt{6} - \sqrt{10}}$

Simplify each of the following, expressing the final results using positive exponents.

37. $(x^{-3}y^4)^{-2}$

38. $\left(\dfrac{2a^{-1}}{3b^4}\right)^{-3}$

39. $\left(4x^{\frac{1}{2}}\right)\left(5x^{\frac{1}{5}}\right)$

40. $\dfrac{42a^{\frac{3}{4}}}{6a^{\frac{1}{3}}}$

41. $\left(\dfrac{x^3}{y^4}\right)^{-\frac{1}{3}}$

42. $\left(\dfrac{6x^{-2}}{2x^4}\right)^{-2}$

Use the distributive property to help simplify each of the following.

43. $3\sqrt{45} - 2\sqrt{20} - \sqrt{80}$

44. $4\sqrt[3]{24} + 3\sqrt[3]{3} - 2\sqrt[3]{81}$

45. $3\sqrt{24} - \dfrac{2\sqrt{54}}{5} + \dfrac{\sqrt{96}}{4}$

46. $-2\sqrt{12x} + 3\sqrt{27x} - 5\sqrt{48x}$

Express each of the following as a single fraction involving positive exponents only.

47. $x^{-2} + y^{-1}$

48. $a^{-2} - 2a^{-1}b^{-1}$

Solve each of the following equations.

49. $\sqrt{7x - 3} = 4$

50. $\sqrt{2y + 1} = \sqrt{5y - 11}$

51. $\sqrt{2x} = x - 4$

52. $\sqrt{5n - 1} = -7$

53. $\sqrt[3]{2x - 1} = 3$

54. $\sqrt{t^2 + 9t - 1} = 3$

55. $\sqrt{x^2 + 3x - 6} = x$

56. $\sqrt{x + 1} - \sqrt{2x} = -1$

6 Quadratic Equations and Inequalities

Solving equations is one of the central themes throughout this text. Let's pause for a moment and reflect back upon the different types of equations that we have solved.

Type of equation	Examples
First degree equations in one variable	$3x + 2x = x - 4$; $5(x + 4) = 12$; $\dfrac{x + 2}{3} + \dfrac{x - 1}{4} = 2.$
Second degree equations in one variable *that are factorable*	$x^2 + 5x = 0$; $x^2 + 5x + 6 = 0$; $x^2 - 9 = 0$; $x^2 - 10x + 25 = 0.$
Fractional equations	$\dfrac{2}{x} + \dfrac{3}{x} = 4$; $\dfrac{5}{a - 1} = \dfrac{6}{a - 2}$; $\dfrac{2}{x^2 - 9} + \dfrac{3}{x + 3} = \dfrac{4}{x - 3}.$
Radical equations	$\sqrt{x} = 2$; $\sqrt{3x - 2} = 5$; $\sqrt{5y + 1} = \sqrt{3y + 4}.$

As indicated in the above chart, we have solved second degree equations in one variable, but only those that are factorable. In this chapter we will expand our work to include more general types of second degree equations as well as inequalities in one variable.

6.1 Quadratic Equations

A second degree equation in one variable contains the variable with an exponent of two, but no higher power. Such equations are also called **quadratic equations**. The following are examples of quadratic equations.

$$x^2 = 36; \qquad y^2 + 4y = 0; \qquad x^2 + 5x - 2 = 0;$$
$$3n^2 + 2n - 1 = 0; \qquad 5x^2 + x + 2 = 3x^2 - 2x - 1.$$

A quadratic equation in the variable x can also be defined as any equation that can be written in the form

$$ax^2 + bx + c = 0$$

where a, b, and c are real numbers and $a \neq 0$. The form $ax^2 + bx + c = 0$ is called the **standard form** of a quadratic equation.

In previous chapters you solved quadratic equations (the term *quadratic* was not used at that time) by factoring and applying the property "if $ab = 0$ then $a = 0$ or $b = 0$." Let's review a few examples of that type.

Example 1   Solve $3n^2 + 14n - 5 = 0$.

Solution

$$3n^2 + 14n - 5 = 0$$
$$(3n - 1)(n + 5) = 0 \qquad \text{Factor left side.}$$
$$3n - 1 = 0 \quad \text{or} \quad n + 5 = 0 \qquad \text{Apply: if } ab = 0, \text{ then } a = 0 \text{ or } b = 0.$$
$$3n = 1 \quad \text{or} \quad n = -5$$
$$n = \frac{1}{3} \quad \text{or} \quad n = -5.$$

The solution set is $\left\{ -5, \frac{1}{3} \right\}$.

Example 2   Solve $x^2 + 3kx - 10k^2 = 0$ for x.

Solution

$$x^2 + 3kx - 10k^2 = 0$$
$$(x + 5k)(x - 2k) = 0 \qquad \text{Factor left side.}$$
$$x + 5k = 0 \quad \text{or} \quad x - 2k = 0 \qquad \text{Apply: if } ab = 0, \text{ then } a = 0 \text{ or } b = 0.$$
$$x = -5k \quad \text{or} \quad x = 2k.$$

The solution set is $\{ -5k, 2k \}$.

Example 3   Solve $2\sqrt{x} = x - 8$.

Solution

$$2\sqrt{x} = x - 8$$
$$(2\sqrt{x})^2 = (x - 8)^2 \qquad \text{Square both sides.}$$
$$4x = x^2 - 16x + 64$$
$$0 = x^2 - 20x + 64$$
$$0 = (x - 16)(x - 4) \qquad \text{Factor right side.}$$
$$x - 16 = 0 \quad \text{or} \quad x - 4 = 0 \qquad \text{Apply: if } ab = 0, \text{ then } a = 0 \text{ or } b = 0.$$
$$x = 16 \quad \text{or} \quad x = 4.$$

Check $2\sqrt{x} = x - 8$ $2\sqrt{x} = x - 8$

$\qquad 2\sqrt{16} \overset{?}{=} 16 - 8 \quad$ or $\quad 2\sqrt{4} \overset{?}{=} 4 - 8$

$\qquad\qquad 2(4) \overset{?}{=} 8 \qquad\qquad\qquad 2(2) \overset{?}{=} -4$

$\qquad\qquad\quad 8 = 8 \qquad\qquad\qquad\qquad 4 \neq -4.$

The solution set is $\{16\}$.   ●

Two comments about Example 3 should be made. First, remember that applying the property "if $a = b$, then $a^n = b^n$" might produce extraneous solutions. Therefore, we *must* check all potential solutions. Secondly, the equation $2\sqrt{x} = x - 8$ is said to be of *quadratic form* because it can be written as $2x^{\frac{1}{2}} = \left(x^{\frac{1}{2}}\right)^2 - 8$. More will be said about the phrase *quadratic form* later.

Let's consider quadratic equations of the form $x^2 = a$, where x is the variable and a is any nonnegative real number. We can solve $x^2 = a$ as follows:

$$x^2 = a$$
$$x^2 - a = 0$$
$$x^2 - (\sqrt{a})^2 = 0 \qquad (a = (\sqrt{a})^2)$$
$$(x - \sqrt{a})(x + \sqrt{a}) = 0 \qquad \text{Factor left side.}$$
$$x - \sqrt{a} = 0 \quad \text{or} \quad x + \sqrt{a} = 0 \qquad \text{Apply: if } ab = 0, \text{ then } a = 0 \text{ or } b = 0.$$
$$x = \sqrt{a} \quad \text{or} \quad x = -\sqrt{a}.$$

The solutions are $\sqrt{a}$ and $-\sqrt{a}$.

The previous result can be stated as a general property and used to solve certain types of quadratic equations.

Property 6.1

For any nonnegative real number a,

$$\text{if } x^2 = a, \quad \text{then } x = \sqrt{a} \quad \text{or} \quad x = -\sqrt{a}.$$

(The statement "$x = \sqrt{a}$ or $x = -\sqrt{a}$" can be written as $x = \pm\sqrt{a}$.)

Property 6.1, along with our knowledge of square root, makes it very easy to solve quadratic equations of the form $x^2 = a$.

Example 4   Solve $x^2 = 45$.

Solution

$$x^2 = 45$$
$$x = \pm\sqrt{45}$$
$$x = \pm 3\sqrt{5}. \qquad (\sqrt{45} = \sqrt{9}\sqrt{5} = 3\sqrt{5})$$

The solution set is $\{-3\sqrt{5}, 3\sqrt{5}\}$.   ●

Example 5   Solve $7n^2 = 12$.

Solution $7n^2 = 12$

$$n^2 = \frac{12}{7}$$

$$n = \pm\sqrt{\frac{12}{7}} \qquad \left(\sqrt{\frac{12}{7}} = \frac{\sqrt{12}}{\sqrt{7}} \cdot \frac{\sqrt{7}}{\sqrt{7}} = \frac{\sqrt{84}}{7}\right)$$

$$n = \pm\frac{2\sqrt{21}}{7}. \qquad \left(\quad = \frac{\sqrt{4}\sqrt{21}}{7} = \frac{2\sqrt{21}}{7}\right)$$

The solution set is $\left\{-\dfrac{2\sqrt{21}}{7}, \dfrac{2\sqrt{21}}{7}\right\}$.   ●

The final two examples of this section further illustrate the use of Property 6.1. As shown in Example 7, it may be necessary to do a bit of changing of form before Property 6.1 can be applied.

Example 6   Solve $(3n + 1)^2 = 25$.

Solution $(3n + 1)^2 = 25$

$$(3n + 1) = \pm\sqrt{25}$$

$$3n + 1 = \pm 5$$

$$3n + 1 = 5 \quad \text{or} \quad 3n + 1 = -5$$

$$3n = 4 \quad \text{or} \quad 3n = -6$$

$$n = \frac{4}{3} \quad \text{or} \quad n = -2.$$

The solution set is $\left\{-2, \dfrac{4}{3}\right\}$.   ●

Example 7   Solve $3(2x - 3)^2 + 8 = 44$.

Solution $3(2x - 3)^2 + 8 = 44$

$$3(2x - 3)^2 = 36$$

$$(2x - 3)^2 = 12$$

$$2x - 3 = \pm\sqrt{12}$$

$$2x - 3 = \pm 2\sqrt{3}$$

$$2x - 3 = 2\sqrt{3} \quad \text{or} \quad 2x - 3 = -2\sqrt{3}$$

$$2x = 3 + 2\sqrt{3} \quad \text{or} \quad 2x = 3 - 2\sqrt{3}$$

$$x = \frac{3 + 2\sqrt{3}}{2} \quad \text{or} \quad x = \frac{3 - 2\sqrt{3}}{2}.$$

The solution set is $\left\{\dfrac{3 - 2\sqrt{3}}{2}, \dfrac{3 + 2\sqrt{3}}{2}\right\}$.   ●

It should be noted that quadratic equations of the form $x^2 = a$, where a is a *negative* number, have no real number solutions. For example, $x^2 = -9$ has no real number solution because any real number squared is nonnegative. In a like manner, we can recognize by inspection that an equation such as $(x + 3)^2 = -17$ has no real number solutions.

Back to the Pythagorean Theorem

Our work with radicals, Property 6.1, and the Pythagorean Theorem merge to form a basis for solving a variety of problems pertaining to right triangles.

Problem 1 Find c in the accompany figure.

Solution Applying the Pythagorean Theorem we obtain

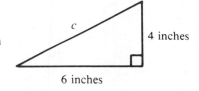

c

4 inches

6 inches

$$c^2 = 6^2 + 4^2$$
$$= 36 + 16$$
$$= 52.$$

Therefore,

$$c = \pm\sqrt{52} = \pm 2\sqrt{13}.$$

Disregarding the negative solution, we find that $c = 2\sqrt{13}$ inches.   ●

There are two special kinds of right triangles that are used extensively in later mathematics courses. An **isosceles right triangle** is a right triangle having both legs of the same length. Let's consider a problem involving an isosceles right triangle.

Problem 2 Find the length of each leg of an isosceles right triangle having a hypotenuse of length 5 meters.

Solution Let's sketch an isosceles right triangle and let x represent the length of each leg. Then we can apply the Pythagorean Theorem as follows.

$$x^2 + x^2 = 5^2$$
$$2x^2 = 25$$
$$x^2 = \frac{25}{2}$$
$$x = \pm\sqrt{\frac{25}{2}} = \pm\frac{5}{\sqrt{2}} = \pm\frac{5\sqrt{2}}{2}.$$

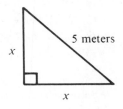

5 meters

x

x

Each leg is $\dfrac{5\sqrt{2}}{2}$ meters long.   ●

The other special kind of right triangle that is used frequently is one that contains acute angles of 30° and 60°. In such a right triangle, often referred to as a **30°–60° right triangle**, *the side opposite the 30° angle is equal in length to one-half of the length of the hypotenuse.* This relationship, along with the Pythagorean Theorem, provides us with another problem-solving technique.

Problem 3   Suppose that the side opposite the 60° angle in a 30°–60° right triangle is 9 feet long. Find the length of the other leg and the length of the hypotenuse.

Solution   Let x represent the length of the side opposite the 30° angle. Then $2x$ represents the length of the hypotenuse, as indicated in the accompanying diagram. Thus, we can apply the Pythagorean Theorem as follows.

$$x^2 + 9^2 = (2x)^2$$
$$x^2 + 81 = 4x^2$$
$$81 = 3x^2$$
$$27 = x^2$$
$$\pm\sqrt{27} = x$$
$$\pm 3\sqrt{3} = x.$$

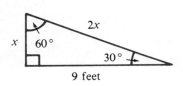

The other leg is $3\sqrt{3}$ feet long and the hypotenuse is $2(3\sqrt{3}) = 6\sqrt{3}$ feet long. ●

Problem Set 6.1

Solve each of the following quadratic equations by factoring and applying the property "if $ab = 0$, then $a = 0$ or $b = 0$." If necessary, return to Chapter 3 and review the factoring techniques that we have studied.

1. $x^2 - 7x = 0$	**2.** $x^2 + 8x = 0$	**3.** $x^2 = -9x$
4. $x^2 = 13x$	**5.** $4y^2 + 12y = 0$	**6.** $5y^2 - 20y = 0$
7. $3n^2 - 7n = 0$	**8.** $7n^2 + 11n = 0$	**9.** $x^2 - x - 20 = 0$
10. $x^2 + 7x - 8 = 0$	**11.** $x^2 - 7x - 18 = 0$	**12.** $x^2 - 14x + 33 = 0$
13. $3x^2 + 19x = -6$	**14.** $5x^2 + 37x = -14$	**15.** $12n^2 + 7n - 10 = 0$
16. $8n^2 - 10n - 63 = 0$	**17.** $13y - 15 = -6y^2$	**18.** $22y - 8 = -21y^2$
19. $9x^2 - 6x + 1 = 0$	**20.** $25x^2 + 30x + 9 = 0$	

Solve each of the following radical equations. Don't forget, you *must* check potential solutions.

21. $3\sqrt{2x} = x + 4$	**22.** $3\sqrt{x} = x + 2$	**23.** $\sqrt{x} = x - 2$
24. $\sqrt{2x} = x - 4$	**25.** $\sqrt{5x + 10} = x$	**26.** $\sqrt{3x + 6} = x$

Solve each of the following equations for x by factoring and applying the property "if $ab = 0$, then $a = 0$ or $b = 0$."

27. $x^2 + 3kx = 0$

28. $x^2 - 5kx = 0$

29. $x^2 = 16kx$

30. $x^2 = 25k^2x$

31. $x^2 - 7kx + 6k^2 = 0$

32. $x^2 + 5kx + 4k^2 = 0$

33. $x^2 - 6kx + 9k^2 = 0$

34. $x^2 + 12kx + 36k^2 = 0$

Use Property 6.1 to help solve each of the following quadratic equations. Express irrational solutions in simplest radical form.

35. $x^2 = 25$

36. $x^2 = 1$

37. $9x^2 = 9$

38. $9x^2 = 36$

39. $x^2 = 5$

40. $x^2 = 15$

41. $n^2 - 20 = 0$

42. $n^2 - 32 = 0$

43. $4t^2 = 48$

44. $2t^2 = 36$

45. $8 - 3x^2 = 0$

46. $0 = 24 - 5x^2$

47. $7y^2 = 40$

48. $9y^2 = 12$

49. $8x^2 = 45$

50. $12x^2 = 75$

51. $0 = -24x^2 + 25$

52. $-27x^2 + 64 = 0$

53. $(x + 1)^2 = 4$

54. $(x - 2)^2 = 1$

55. $(x - 3)^2 = -9$

56. $(x + 5)^2 = 16$

57. $(3x - 2)^2 = 1$

58. $(4x + 3)^2 = 4$

59. $(x + 2)^2 - 7 = 0$

60. $(x - 3)^2 + 5 = 0$

61. $(n + 4)^2 = 8$

62. $(n - 4)^2 = 12$

63. $18 - (2n - 1)^2 = 0$

64. $20 - (3n + 5)^2 = 0$

65. $(4x - 3)^2 = 32$

66. $(5x - 1)^2 = 27$

67. $2(x + 6)^2 - 9 = 63$

68. $3(x + 7)^2 + 4 = 79$

69. $3(4x - 1)^2 + 1 = 16$

70. $2(5x - 2)^2 + 5 = 25$

For Problems 71–76, a and b represent the lengths of the legs of a right triangle, and c represents the length of the hypotenuse. Express answers in simplest radical form.

71. Find c if $a = 2$ centimeters and $b = 4$ centimeters.

72. Find c if $a = 5$ inches and $b = 7$ inches.

73. Find a if $c = 10$ feet and $b = 4$ feet.

74. Find a if $c = 6$ meters and $b = 3$ meters.

75. Find b if $c = 25$ meters and $a = 24$ meters.

76. Find b if $c = 12$ feet and $a = 6$ feet.

For Problems 77–80, use the following *isosceles right triangle*. Express answers in simplest radical form.

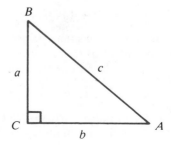

77. If $b = 5$ inches, find c.

78. If $a = 10$ centimeters, find c.

79. If $c = 7$ feet, find a and b.

80. If $c = 12$ meters, find a and b.

For Problems 81–86, use the accompanying figure. Express answers in simplest radical form.

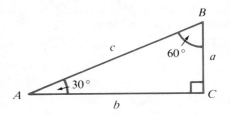

81. If $a = 2$ inches, find b and c.

82. If $a = 5$ inches, find b and c.

83. If $c = 16$ centimeters, find a and b.

84. If $c = 7$ centimeters, find a and b.

85. If $b = 8$ feet, find a and c.

86. If $b = 12$ meters, find a and c.

6.2 Completing the Square

Thus far we have solved quadratic equations by factoring and applying the property "if $ab = 0$, then $a = 0$ or $b = 0$," or by applying the property "if $x^2 = a$, then $x = \pm\sqrt{a}$." In this section we shall examine another method called **completing the square**, which will give us the power to solve *any* quadratic equation.

A factoring technique studied in Chapter 3 relied on recognizing **perfect square trinomials**. In each of the following, the perfect square trinomial on the right side is the result of squaring the binomial on the left side.

$$(x + 4)^2 = x^2 + 8x + 16; \qquad (x - 6)^2 = x^2 - 12x + 36;$$
$$(x + 7)^2 = x^2 + 14x + 49; \qquad (x - 9)^2 = x^2 - 18x + 81.$$

Notice that in each of the square trinomials *the constant term is equal to the square of one-half of the coefficient of the x-term.* This relationship allows us to form a perfect square trinomial by adding a proper constant term. For example, suppose that we want to form a perfect square trinomial from $x^2 + 10x$. Since $\frac{1}{2}(10) = 5$ and $5^2 = 25$, the perfect square trinomial $x^2 + 10x + 25$ can be formed. Let's use the previous ideas to help solve some quadratic equations.

Example 1   Solve $x^2 + 10x - 2 = 0$.

Solution

$$x^2 + 10x - 2 = 0$$
$$x^2 + 10x = 2$$
$$x^2 + 10x + 25 = 2 + 25$$ We add 25 to the left side to form a perfect square trinomial; 25 must also be added to the right side.
$$(x + 5)^2 = 27.$$

(Now we can proceed as in the last section.)

$$x + 5 = \pm\sqrt{27}$$
$$x + 5 = \pm 3\sqrt{3}$$
$$x + 5 = 3\sqrt{3} \quad \text{or} \quad x + 5 = -3\sqrt{3}$$
$$x = -5 + 3\sqrt{3} \quad \text{or} \quad x = -5 - 3\sqrt{3}.$$

The solution set is $\{-5 - 3\sqrt{3}, -5 + 3\sqrt{3}\}$.   ●

Notice from Example 1 that the method of completing the square to solve a quadratic equation is merely what the name implies. A perfect square trinomial is formed, then the equation can be changed to the necessary form for applying the property "if $x^2 = a$, then $x = \pm\sqrt{a}$." Let's consider another example.

Example 2   Solve $x^2 - 3x + 1 = 0$.

Solution

$$x^2 - 3x + 1 = 0$$
$$x^2 - 3x = -1$$
$$x^2 - 3x + \frac{9}{4} = -1 + \frac{9}{4} \qquad \left(\frac{1}{2}(3) = \frac{3}{2} \quad \text{and} \quad \left(\frac{3}{2}\right)^2 = \frac{9}{4}\right)$$
$$\left(x - \frac{3}{2}\right)^2 = \frac{5}{4}$$
$$x - \frac{3}{2} = \pm\sqrt{\frac{5}{4}}$$
$$x - \frac{3}{2} = \pm\frac{\sqrt{5}}{2}$$
$$x - \frac{3}{2} = \frac{\sqrt{5}}{2} \quad \text{or} \quad x - \frac{3}{2} = -\frac{\sqrt{5}}{2}$$
$$x = \frac{3}{2} + \frac{\sqrt{5}}{2} \quad \text{or} \quad x = \frac{3}{2} - \frac{\sqrt{5}}{2}$$
$$x = \frac{3 + \sqrt{5}}{2} \quad \text{or} \quad x = \frac{3 - \sqrt{5}}{2}.$$

The solution set is $\left\{\dfrac{3 - \sqrt{5}}{2}, \dfrac{3 + \sqrt{5}}{2}\right\}$.   ●

In Example 2, notice that since the coefficient of the x-term is odd, we are forced into the realm of fractions. The use of common fractions rather than decimals makes our previous work with radicals applicable.

The relationship for a perfect square trinomial that states that *the constant term is equal to the square of one-half of the coefficient of the x-term* holds only if the coefficient of x^2 is 1. Thus, a slight adjustment needs to be made when solving quadratic equations having a coefficient of x^2 other than 1. The next example shows how to make this adjustment.

Example 3   Solve $2x^2 + 12x - 5 = 0$.

Solution

$$2x^2 + 12x - 5 = 0$$
$$2x^2 + 12x = 5$$
$$x^2 + 6x = \frac{5}{2} \qquad \text{Multiply both sides by } \frac{1}{2}.$$
$$x^2 + 6x + 9 = \frac{5}{2} + 9$$
$$x^2 + 6x + 9 = \frac{23}{2}$$
$$(x + 3)^2 = \frac{23}{2}$$
$$x + 3 = \pm\sqrt{\frac{23}{2}}$$
$$x + 3 = \pm\frac{\sqrt{46}}{2} \qquad \left(\sqrt{\frac{23}{2}} = \frac{\sqrt{23}}{\sqrt{2}} \cdot \frac{\sqrt{2}}{\sqrt{2}} = \frac{\sqrt{46}}{2} \right)$$
$$x + 3 = \frac{\sqrt{46}}{2} \quad \text{or} \quad x + 3 = -\frac{\sqrt{46}}{2}$$
$$x = -3 + \frac{\sqrt{46}}{2} \quad \text{or} \quad x = -3 - \frac{\sqrt{46}}{2}$$
$$x = \frac{-6 + \sqrt{46}}{2} \quad \text{or} \quad x = \frac{-6 - \sqrt{46}}{2}.$$

The solution set is $\left\{ \dfrac{-6 - \sqrt{46}}{2}, \dfrac{-6 + \sqrt{46}}{2} \right\}$.   ●

As mentioned earlier, the method of completing the square can be used to solve *any* quadratic equation. To illustrate, let's use it to solve an equation that could also be solved by factoring.

Example 4   Solve $x^2 - 2x - 8 = 0$ by completing the square.

Solution

$$x^2 - 2x - 8 = 0$$
$$x^2 - 2x = 8$$
$$x^2 - 2x + 1 = 8 + 1$$
$$(x - 1)^2 = 9$$
$$x - 1 = \pm 3$$
$$x - 1 = 3 \quad \text{or} \quad x - 1 = -3$$
$$x = 4 \quad \text{or} \quad x = -2.$$

The solution set is $\{-2, 4\}$.   ●

No claim is made that using the method of completing the square with an equation such as the one in Example 4 is easier than the factoring technique. However, you should recognize that the method of completing the square will work with any quadratic equation.

The next example illustrates that the method of completing the square will also identify quadratic equations that have no real number solutions.

Example 5   Solve $x^2 + 4x + 6 = 0$.

Solution

$$x^2 + 4x + 6 = 0$$
$$x^2 + 4x = -6$$
$$x^2 + 4x + 4 = -6 + 4$$
$$(x + 2)^2 = -2.$$

We can stop here and reason as follows: any value of x will yield a nonnegative value for $(x + 2)^2$; thus, $(x + 2)^2$ cannot equal -2. Therefore, the original equation, $x^2 + 4x + 6 = 0$, *has no real number solutions.*   ●

Problem Set 6.2

Solve each of the following quadratic equations by using (a) the factoring method, and (b) the method of completing the square.

1. $x^2 + 4x - 12 = 0$ **2.** $x^2 + 4x - 32 = 0$ **3.** $x^2 - 10x + 16 = 0$

4. $x^2 - 10x + 24 = 0$ **5.** $x^2 + 3x = 10$ **6.** $x^2 + 3x = 28$

7. $3n^2 - 5n - 2 = 0$ **8.** $2n^2 + 5n + 2 = 0$ **9.** $4n^2 + 14n - 8 = 0$

10. $9n^2 + 21n - 18 = 0$

Use the method of completing the square to solve each of the following quadratic equations.

11. $x^2 + 6x - 1 = 0$ **12.** $x^2 + 8x + 14 = 0$ **13.** $x^2 + 8x - 2 = 0$

14. $x^2 + 2x - 5 = 0$ **15.** $y^2 + 6y - 15 = 0$ **16.** $y^2 + 4y - 8 = 0$

17. $x^2 + 2x + 6 = 0$ **18.** $n^2 + 4n - 24 = 0$ **19.** $n^2 = 17 - 2n$

20. $x^2 + 6x + 10 = 0$ **21.** $x^2 - 3x - 1 = 0$ **22.** $x^2 = 5x + 2$

23. $x^2 + 5x + 1 = 0$ **24.** $x^2 + 7x + 3 = 0$ **25.** $-y^2 - y + 8 = 0$

26. $-y^2 - 3y + 12 = 0$ **27.** $3x^2 + 12x - 2 = 0$ **28.** $3n^2 - 6n + 2 = 0$

29. $2t^2 - 4t + 1 = 0$ **30.** $3x^2 + 5x - 1 = 0$

Solve each of the following quadratic equations using whatever method seems most appropriate.

31. $x^2 = 16x$ **32.** $(x - 3)^2 = 12$ **33.** $x^2 + 5x - 14 = 0$

34. $x^2 + 8x - 48 = 0$ **35.** $x^2 + 6x - 3 = 0$ **36.** $x^2 + 12x - 4 = 0$

37. $6n^2 + n - 2 = 0$ **38.** $6n^2 + 23n + 21 = 0$ **39.** $2n^2 - 2n - 1 = 0$

40. $3n^2 - 6n - 2 = 0$ **41.** $6x^2 - 13x = 28$ **42.** $3x^2 + 29x = -66$

43. $t(t - 26) = -160$ **44.** $n(n + 8) = 240$ **45.** $5(x + 2)^2 + 1 = 16$

46. $2x^2 + 4x = -7$ **47.** $3x^2 + 6x = 1$ **48.** $-2n^2 + 8n = 3$

49. $-12n^2 + 7n - 1 = 0$ **50.** $(x + 2)(x - 7) = 10$

51. Use the method of completing the square to solve $ax^2 + bx + c = 0$ for x, where $a, b,$ and c are real numbers and $a \neq 0$.

Miscellaneous Problems

Solve each of the following for the indicated variable. Assume that all letters represent positive numbers.

52. $A = \pi r^2$ for r **53.** $s = \frac{1}{2}gt^2$ for t

54. $\dfrac{x^2}{a^2} + \dfrac{y^2}{b^2} = 1$ for x **55.** $\dfrac{x^2}{a^2} - \dfrac{y^2}{b^2} = 1$ for y

Solve each of the following equations for x.

56. $x^2 - 5ax + 6a^2 = 0$ **57.** $x^2 + 8ax + 15a^2 = 0$

58. $6x^2 + ax - 2a^2 = 0$ **59.** $10x^2 - 31ax - 14a^2 = 0$

60. $9x^2 - 12bx + 4b^2 = 0$ **61.** $4x^2 + 4bx + b^2 = 0$

6.3 The Quadratic Formula

As we saw in the last section, the method of completing the square can be used to solve *any* quadratic equation. Furthermore, the equation $ax^2 + bx + c = 0$, where $a, b,$ and c are real numbers with $a \neq 0$, can be used to represent *any* quadratic equation. These two ideas merge to produce the **quadratic formula**, a formula that can be used to solve *any* quadratic equation. The merging is accomplished by using the method of completing the square to solve the equation $ax^2 + bx + c = 0$ as follows.

$$ax^2 + bx + c = 0$$

$$ax^2 + bx = -c$$

$$x^2 + \frac{b}{a}x = -\frac{c}{a} \qquad \text{Multiply both sides by } \frac{1}{a}.$$

$$x^2 + \frac{b}{a}x + \frac{b^2}{4a^2} = -\frac{c}{a} + \frac{b^2}{4a^2} \qquad \textit{Complete the square} \text{ by adding } \frac{b^2}{4a^2} \text{ to}$$

$$\left(x + \frac{b}{2a}\right)^2 = \frac{b^2 - 4ac}{4a^2} \qquad \begin{array}{l} \text{both sides.} \\ \text{The right side is combined into a single} \\ \text{fraction.} \end{array}$$

$$x + \frac{b}{2a} = \pm\sqrt{\frac{b^2 - 4ac}{4a^2}}$$

$$x + \frac{b}{2a} = \pm\frac{\sqrt{b^2 - 4ac}}{\sqrt{4a^2}}$$

$$x + \frac{b}{2a} = \pm\frac{\sqrt{b^2 - 4ac}}{2a}$$

$$x + \frac{b}{2a} = \frac{\sqrt{b^2 - 4ac}}{2a} \qquad \text{or} \qquad x + \frac{b}{2a} = -\frac{\sqrt{b^2 - 4ac}}{2a}$$

$$x = -\frac{b}{2a} + \frac{\sqrt{b^2 - 4ac}}{2a} \qquad \text{or} \qquad x = -\frac{b}{2a} - \frac{\sqrt{b^2 - 4ac}}{2a}$$

$$x = \frac{-b + \sqrt{b^2 - 4ac}}{2a} \qquad \text{or} \qquad x = \frac{-b - \sqrt{b^2 - 4ac}}{2a}$$

The quadratic formula is usually stated as follows:

Quadratic Formula

$$x = \frac{-b \pm \sqrt{b^2 - 4ac}}{2a}, \qquad a \neq 0$$

This formula can be used to solve any quadratic equation by expressing the equation in the standard form, $ax^2 + bx + c = 0$, and substituting the values for a, b, and c into the formula. Let's consider some examples.

Example 1   Solve $x^2 + 5x + 2 = 0$.

Solution   The given equation is in standard form, so $a = 1$, $b = 5$, and $c = 2$. Substituting these values into the formula and simplifying, we obtain

$$x = \frac{-b \pm \sqrt{b^2 - 4ac}}{2a}$$

$$x = \frac{-5 \pm \sqrt{5^2 - 4(1)(2)}}{2(1)}$$

$$x = \frac{-5 \pm \sqrt{25 - 8}}{2}$$

$$x = \frac{-5 \pm \sqrt{17}}{2}.$$

The solution set is $\left\{ \dfrac{-5 - \sqrt{17}}{2}, \dfrac{-5 + \sqrt{17}}{2} \right\}$. ●

Example 2

Solve $x^2 - 2x - 4 = 0$.

Solution

We need to think of $x^2 - 2x - 4 = 0$ as $x^2 + (-2x) + (-4) = 0$ to determine the values $a = 1$, $b = -2$, and $c = -4$. Substituting these values into the quadratic formula and simplifying, we obtain

$$x = \frac{-b \pm \sqrt{b^2 - 4ac}}{2a}$$

$$x = \frac{-(-2) \pm \sqrt{(-2)^2 - 4(1)(-4)}}{2(1)}$$

$$x = \frac{2 \pm \sqrt{4 + 16}}{2}$$

$$x = \frac{2 \pm \sqrt{20}}{2}$$

$$x = \frac{2 \pm 2\sqrt{5}}{2} = \frac{\cancel{2}(1 \pm \sqrt{5})}{\cancel{2}}.$$

The solution set is $\{1 - \sqrt{5}, 1 + \sqrt{5}\}$. ●

Example 3

Solve $2x^2 + 4x - 3 = 0$.

Solution

Substituting $a = 2$, $b = 4$, and $c = -3$ into the quadratic formula and simplifying, we obtain

$$x = \frac{-b \pm \sqrt{b^2 - 4ac}}{2a}$$

$$x = \frac{-4 \pm \sqrt{4^2 - 4(2)(-3)}}{2(2)}$$

$$x = \frac{-4 \pm \sqrt{16 + 24}}{4}$$

$$x = \frac{-4 \pm \sqrt{40}}{4}$$

$$x = \frac{-4 \pm 2\sqrt{10}}{4} = \frac{-2 \pm \sqrt{10}}{2}.$$

The solution set is $\left\{ \dfrac{-2 + \sqrt{10}}{2}, \dfrac{-2 - \sqrt{10}}{2} \right\}$.

Example 4   Solve $n(3n - 10) = 25$.

Solution   First, we need to change the equation to the standard form $an^2 + bn + c = 0$.

$$n(3n - 10) = 25$$
$$3n^2 - 10n = 25$$
$$3n^2 - 10n - 25 = 0.$$

Now substituting $a = 3$, $b = -10$, and $c = -25$ into the quadratic formula we obtain

$$n = \frac{-b \pm \sqrt{b^2 - 4ac}}{2a}$$

$$n = \frac{-(-10) \pm \sqrt{(-10)^2 - 4(3)(-25)}}{2(3)}$$

$$n = \frac{10 \pm \sqrt{100 + 300}}{2(3)}$$

$$n = \frac{10 \pm \sqrt{400}}{6}$$

$$n = \frac{10 \pm 20}{6}$$

$$n = \frac{10 + 20}{6} \quad \text{or} \quad n = \frac{10 - 20}{6}$$

$$n = 5 \quad \text{or} \quad n = -\frac{5}{3}.$$

The solution set is $\left\{ -\dfrac{5}{3}, 5 \right\}$.

In Example 4, notice that the variable n is used. The quadratic formula is usually stated in terms of x, but certainly it can be applied to quadratic equations in other variables.

Quadratic equations that have no real number solutions can be easily identified when using the quadratic formula, as the next example illustrates.

Example 5   Solve $x^2 - 3x + 7 = 0$.

Solution   Substituting $a = 1$, $b = -3$, and $c = 7$ into the quadratic formula, we obtain

$$x = \frac{-b \pm \sqrt{b^2 - 4ac}}{2a}$$

$$x = \frac{-(-3) \pm \sqrt{(-3)^2 - 4(1)(7)}}{2(1)}$$

$$x = \frac{3 \pm \sqrt{9 - 28}}{2}$$

$$x = \frac{3 \pm \sqrt{-19}}{2}.$$

Since $\sqrt{-19}$ is not a real number, we can conclude that the original equation *has no real number solutions*; thus, the solution set is $\varnothing$.   ●

The quadratic formula makes it easy to determine the nature of the roots of a quadratic equation without completely solving the equation. The number

$$b^2 - 4ac,$$

which appears under the radical sign in the quadratic formula, is called the *discriminant* of the quadratic equation. It is the indicator as to the kind of roots that the equation possesses. For example, suppose that we start to solve the equation $x^2 - 4x - 7 = 0$ as follows:

$$x = \frac{-b \pm \sqrt{b^2 - 4ac}}{2a}$$

$$x = \frac{-(-4) \pm \sqrt{(-4)^2 - 4(1)(-7)}}{2(1)}$$

$$x = \frac{4 \pm \sqrt{16 + 28}}{2}$$

$$x = \frac{4 \pm \sqrt{44}}{2}.$$

Now we should be able to look ahead and realize that we will obtain two real solutions for the equation. In other words, the discriminant, 44, is the indicator as to the type of roots that we will obtain.

The following general statements can be made relative to the roots of a quadratic equation of the form $ax^2 + bx + c = 0$.

1. If $b^2 - 4ac < 0$, then the equation has no real solutions.

2. If $b^2 - 4ac = 0$, then the equation has one real solution.

3. If $b^2 - 4ac > 0$, then the equation has two real solutions.

The following examples illustrate each of these situations. (You may want to solve the equations completely to verify the conclusions.)

Equation	Discriminant	Nature of roots
$x^2 - 3x + 7 = 0$	$b^2 - 4ac = (-3)^2 - 4(1)(7)$ $= 9 - 28$ $= -19.$	no real solutions
$9x^2 - 12x + 4 = 0$	$b^2 - 4ac = (-12)^2 - 4(9)(4)$ $= 144 - 144$ $= 0.$	one real solution
$2x^2 + 5x - 3 = 0$	$b^2 - 4ac = (5)^2 - 4(2)(-3)$ $= 25 + 24$ $= 49.$	two real solutions

There is another very useful relationship involving the roots of a quadratic equation and the numbers a, b, and c of the general form $ax^2 + bx + c = 0$. Suppose that we let x_1 and x_2 be the two roots generated by the quadratic formula. Thus, we have

$$x_1 = \frac{-b + \sqrt{b^2 - 4ac}}{2a} \quad \text{and} \quad x_2 = \frac{-b - \sqrt{b^2 - 4ac}}{2a}.$$

Now let's consider the sum and product of the two roots.

Sum $x_1 + x_2 = \dfrac{-b + \sqrt{b^2 - 4ac}}{2a} + \dfrac{-b - \sqrt{b^2 - 4ac}}{2a} = \dfrac{-2b}{2a} = -\dfrac{b}{a}.$

Product $(x_1)(x_2) = \left(\dfrac{-b + \sqrt{b^2 - 4ac}}{2a}\right)\left(\dfrac{-b - \sqrt{b^2 - 4ac}}{2a}\right) = \dfrac{b^2 - (b^2 - 4ac)}{4a^2}$

$$= \frac{b^2 - b^2 + 4ac}{4a^2}$$

$$= \frac{4ac}{4a^2} = \frac{c}{a}.$$

These relationships provide another way of checking potential solutions when solving quadratic equations. For example, back in Example 3 we solved the equation $2x^2 + 4x - 3 = 0$ and obtained solutions of $\dfrac{-2 + \sqrt{10}}{2}$ and $\dfrac{-2 - \sqrt{10}}{2}$. Let's check these potential solutions by using the sum and product relationships.

Check for Example 3

Sum of roots $\left(\dfrac{-2+\sqrt{10}}{2}\right) + \left(\dfrac{-2-\sqrt{10}}{2}\right) = -\dfrac{4}{2} = -2$ and $-\dfrac{b}{a} = -2.$

Product of roots $\left(\dfrac{-2+\sqrt{10}}{2}\right)\left(\dfrac{-2-\sqrt{10}}{2}\right) = -\dfrac{6}{4} = -\dfrac{3}{2}$ and $\dfrac{c}{a} = -\dfrac{3}{2}.$

Notice that for Example 3 it was much easier to check by using the sum and product relationships than it would have been by substituting back into the original equation. Therefore, from now on, let's keep in mind that we do have an alternate way of checking potential solutions for a quadratic equation.

Remark A point of clarification should be made at this time. It was stated earlier that if $b^2 - 4ac < 0$, then the quadratic equation has *no real solutions*. Furthermore, it was stated that if $b^2 - 4ac = 0$, then the quadratic equation has *one real solution*. A discussion followed of the *sum and product* relationships of the *two roots* of a quadratic equation. Later, you will find that every quadratic equation does have *two* roots and they can be described as follows:

1. If $b^2 - 4ac < 0$, then the equation has *two complex number solutions*.
2. If $b^2 - 4ac = 0$, then the equation has *two equal real number solutions*. (This is sometimes referred to as *one real solution with a multiplicity of two*.)
3. If $b^2 - 4ac > 0$, then the equation has *two unequal real number solutions*.

Problem Set 6.3

1. Find the discriminant of each of the following quadratic equations and determine whether the equation has (1) no real solutions, (2) one real solution, or (3) two real solutions.

   (a) $x^2 + 4x - 12 = 0$ (b) $x^2 - 3x - 10 = 0$

   (c) $x^2 - 6x + 11 = 0$ (d) $4x^2 + 4x + 1 = 0$

   (e) $3x^2 + 4x - 2 = 0$ (f) $6x^2 + 5x + 3 = 0$

   (g) $15x^2 + 7x - 2 = 0$ (h) $4x^2 + x - 6 = 0$

2. Solve each of the equations in Problem 1.

3. Without solving, find the sum and product of the roots of each of the following quadratic equations.

   (a) $x^2 + 9x + 8 = 0$ (b) $x^2 - 13x + 30 = 0$

   (c) $2x^2 - 9x - 18 = 0$ (d) $3x^2 + 7x - 20 = 0$

   (e) $x^2 - 2x - 2 = 0$ (f) $x^2 - 4x - 1 = 0$

   (g) $3x^2 - 4x - 1 = 0$ (h) $2x^2 - 6x - 1 = 0$

4. Solve each of the equations in Problem 3, and check your solutions by using your answers to Problem 3.

Use the quadratic formula to solve each of the following quadratic equations. Check your solutions by using the *sum and product relationships*.

5. $x^2 + 3x - 2 = 0$

6. $x^2 + 5x - 3 = 0$

7. $x^2 + 4x - 1 = 0$

8. $x^2 + 2x - 1 = 0$

9. $a^2 - 6a = 2$

10. $a^2 - 8a = 4$

11. $x^2 + 4x + 9 = 0$

12. $x^2 - 18x + 80 = 0$

13. $x^2 + 19x + 70 = 0$

14. $n^2 - 5n + 8 = 0$

15. $-y^2 + 7y = 4$

16. $-y^2 = -9y + 5$

17. $2x^2 + 5x - 2 = 0$

18. $2x^2 + x - 4 = 0$

19. $3x^2 + 5x + 1 = 0$

20. $4x^2 = -7x - 2$

21. $2a^2 + 1 = 6a$

22. $3a^2 - 9a + 4 = 0$

23. $4n^2 - 3n + 5 = 0$

24. $5n^2 - 3n - 4 = 0$

25. $2x^2 - 17x + 30 = 0$

26. $3x^2 + 19x + 20 = 0$

27. $-5y^2 + 4y + 2 = 0$

28. $-2n^2 + 6n - 9 = 0$

29. $6x^2 = x + 2$

30. $12x^2 - 5x = 3$

31. $3x^2 + 14x = 0$

32. $4x^2 - 11x = 0$

33. $0 = 3 - 4x^2$

34. $3x^2 = 5$

35. $4t^2 + 12t - 6 = 0$

36. $25t^2 - 30t + 9 = 0$

37. $n^2 - 4n - 192 = 0$

38. $n^2 + 32n + 252 = 0$

39. $6x^2 + 11x - 255 = 0$

40. $12x^2 - 73x + 110 = 0$

41. $9x^2 - 12x + 2 = 0$

42. $6x^2 - 12x - 8 = 0$

43. $-2x^2 + 4x + 1 = 0$

44. $-6x^2 + 2x + 1 = 0$

45. The solution set for $x^2 - 4x - 37 = 0$ is $\{2 - \sqrt{41}, 2 + \sqrt{41}\}$. Using a calculator, we found a rational approximation, to the nearest thousandth, for each of these solutions.

$$2 - \sqrt{41} = -4.403; \qquad 2 + \sqrt{41} = 8.403.$$

Thus, the solution set is $\{-4.403, 8.403\}$, with answers rounded to the nearest thousandth.

Solve each of the following equations expressing solutions to the nearest thousandth.

(a) $x^2 - 6x - 10 = 0$

(b) $x^2 - 16x - 24 = 0$

(c) $x^2 + 6x - 44 = 0$

(d) $x^2 + 10x - 46 = 0$

(e) $3x^2 - 7x - 2 = 0$

(f) $2x^2 - 3x - 10 = 0$

(g) $4x^2 - 6x + 1 = 0$

(h) $5x^2 - 9x + 1 = 0$

(i) $2x^2 - 11x - 5 = 0$

(j) $3x^2 - 12x - 10 = 0$

6.4 More Quadratic Equations and Applications

Which method should be used to solve a particular quadratic equation? There is no definite answer to that question; it depends upon the *type* of equation and your personal preference. In the following examples reasons will be stated for choosing a specific technique. However, keep in mind that usually this is a decision *you* must make as the need arises. So become familiar with the strengths and weaknesses of each method.

Example 1   Solve $2x^2 - 3x - 1 = 0$.

Solution   Because of the leading coefficient of 2 and the constant term of -1, there are very few factoring possibilities to consider. Therefore, with such problems, first try the factoring approach. Unfortunately, this particular polynomial is not factorable using integers. Thus, let's use the quadratic formula to solve the equation.

$$x = \frac{-b \pm \sqrt{b^2 - 4ac}}{2a}$$

$$x = \frac{-(-3) \pm \sqrt{(-3)^2 - 4(2)(-1)}}{2(2)}$$

$$x = \frac{3 \pm \sqrt{9 + 8}}{4}$$

$$x = \frac{3 \pm \sqrt{17}}{4}.$$

Check   We can use the *sum and product of roots* relationships for checking purposes.

Sum of roots $\dfrac{3 + \sqrt{17}}{4} + \dfrac{3 - \sqrt{17}}{4} = \dfrac{6}{4} = \dfrac{3}{2}$ and $-\dfrac{b}{a} = -\dfrac{-3}{2} = \dfrac{3}{2}.$

Product of roots

$$\left(\frac{3 + \sqrt{17}}{4}\right)\left(\frac{3 - \sqrt{17}}{4}\right) = \frac{9 - 17}{16} = -\frac{8}{16} = -\frac{1}{2} \quad \text{and} \quad \frac{c}{a} = \frac{-1}{2} = -\frac{1}{2}.$$

The solution set is $\left\{\dfrac{3 - \sqrt{17}}{4}, \dfrac{3 + \sqrt{17}}{4}\right\}.$ ●

Example 2   Solve $\dfrac{3}{n} + \dfrac{10}{n + 6} = 1$.

Solution $\dfrac{3}{n} + \dfrac{10}{n + 6} = 1, \quad n \neq 0 \quad \text{and} \quad n \neq -6.$

$$n(n + 6)\left(\frac{3}{n} + \frac{10}{n + 6}\right) = 1(n)(n + 6) \qquad \text{Multiply both sides by } n(n + 6), \text{ which is the LCD.}$$

$$3(n + 6) + 10n = n(n + 6)$$
$$3n + 18 + 10n = n^2 + 6n$$
$$13n + 18 = n^2 + 6n$$
$$0 = n^2 - 7n - 18.$$

This is an easy one to consider the possibilities for factoring, and it does factor as follows.

$$0 = (n - 9)(n + 2)$$
$$n - 9 = 0 \quad \text{or} \quad n + 2 = 0$$
$$n = 9 \quad \text{or} \qquad n = -2.$$

Check   Substituting 9 and -2 back into the original equation, we obtain

$$\frac{3}{n} + \frac{10}{n + 6} = 1 \qquad\qquad \frac{3}{n} + \frac{10}{n + 6} = 1$$

$$\frac{3}{9} + \frac{10}{9 + 6} \overset{?}{=} 1 \qquad\qquad \frac{3}{-2} + \frac{10}{-2 + 6} \overset{?}{=} 1$$

$$\frac{1}{3} + \frac{10}{15} \overset{?}{=} 1 \quad \text{or} \qquad -\frac{3}{2} + \frac{10}{4} \overset{?}{=} 1$$

$$\frac{1}{3} + \frac{2}{3} \overset{?}{=} 1 \qquad\qquad -\frac{3}{2} + \frac{5}{2} \overset{?}{=} 1$$

$$1 = 1. \qquad\qquad\qquad \frac{2}{2} = 1.$$

The solution set is $\{-2, 9\}$.

Two comments should be made about Example 2. First, notice the indication of the initial restrictions $n \neq 0$ and $n \neq -6$. Remember the need for doing this when solving fractional equations. Secondly, the *sum and product of roots* relationships were not used for checking purposes in this problem. Those relationships would only check the validity of our work from the step $0 = n^2 - 7n - 18$ to the finish. In other words, an error made in changing the original equation to quadratic form would not be detected by checking the sum and product of potential roots. Thus, with such a problem the only *absolute check* is to substitute the potential solutions back into the *original equation*.

Example 3   Solve $x^2 + 22x + 112 = 0$.

Solution   The size of the constant term makes the factoring approach a little cumbersome for this problem. Furthermore, since the leading coefficient is 1 and the coefficient of the x-term is even, the method of completing the square will work rather effectively as follows.

$$x^2 + 22x + 112 = 0$$
$$x^2 + 22x = -112$$
$$x^2 + 22x + 121 = -112 + 121$$
$$(x + 11)^2 = 9$$
$$x + 11 = \pm\sqrt{9}$$
$$x + 11 = \pm 3$$
$$x + 11 = 3 \quad \text{or} \quad x + 11 = -3$$
$$x = -8 \quad \text{or} \quad x = -14.$$

Check

Sum of roots $-8 + (-14) = -22$ and $-\dfrac{b}{a} = -22.$

Product of roots $(-8)(-14) = 112$ and $\dfrac{c}{a} = 112.$

The solution set is $\{-14, -8\}$.   ●

An equation such as $x^4 - 20x^2 + 96 = 0$ is not a quadratic equation, but it can be solved using the techniques that we use on quadratic equations. That is to say, we can factor the polynomial and apply the property "if $ab = 0$, then $a = 0$ or $b = 0$" as follows.

$$x^4 - 20x^2 + 96 = 0$$
$$(x^2 - 12)(x^2 - 8) = 0$$
$$x^2 - 12 = 0 \quad \text{or} \quad x^2 - 8 = 0$$
$$x^2 = 12 \quad \text{or} \quad x^2 = 8$$
$$x = \pm\sqrt{12} = \pm 2\sqrt{3} \quad \text{or} \quad x = \pm\sqrt{8} = \pm 2\sqrt{2}.$$

The solution set is $\{-2\sqrt{3}, -2\sqrt{2}, 2\sqrt{2}, 2\sqrt{3}\}$.

Remark If we substitute y for x^2 and y^2 for x^4, the equation $x^4 - 20x^2 + 96 = 0$ becomes the quadratic equation $y^2 - 20y + 96 = 0$. Thus, we say that $x^4 - 20x^2 + 96 = 0$ is of *quadratic form*. We could solve the quadratic equation $y^2 - 20y + 96 = 0$ and then use the equation $y = x^2$ to determine the solutions for x.

Applications

Before concluding this section with some word problems that can be solved using quadratic equations, let's restate the suggestions for solving word problems made in an earlier chapter.

Suggestions for Solving Word Problems:

1. Read the problem carefully, making certain that you understand the meanings of all the words. Be especially alert for any technical terms used in the statement of the problem.

2. Read the problem a second time (perhaps even a third time) to get an overview of the situation being described and to determine the known facts, as well as what is to be found.

3. Sketch any figure, diagram, or chart that might be helpful in analyzing the problem.

4. Choose a meaningful variable to represent an unknown quantity in the problem (perhaps *l*, if the length of a rectangle is an unknown quantity) and represent any other unknowns in terms of that variable.

5. Look for a *guideline* that can be used to set up an equation. A guideline might be a formula such as "$A = lw$" or a relationship such as "the fractional part of a job done by Bill plus the fractional part of the job done by Mary equals the total job."

6. Form an equation containing the variable that translates the conditions of the guideline from English to algebra.

7. Solve the equation and use the solutions to determine all facts requested in the problem.

8. Check all answers back into the *original statement of the problem*.

Keeping these suggestions in mind, let's now consider some word problems.

Problem 1   A page for a magazine contains 70 square inches of type. The height of a page is twice the width. If the margin around the type is to be 2 inches uniformly, what are the dimensions of a page?

Solution   Let x represent the width of a page. Then $2x$ represents the height of a page. Now let's draw and label a model of a page.

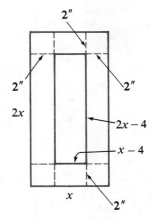

width of   height of   area of
typed   typed   typed
material   material   material
$$\searrow \qquad \searrow \qquad \searrow$$
$$(x - 4)(2x - 4) = 70$$
$$2x^2 - 12x + 16 = 70$$
$$2x^2 - 12x - 54 = 0$$
$$x^2 - 6x - 27 = 0$$
$$(x - 9)(x + 3) = 0$$
$$x - 9 = 0 \quad \text{or} \quad x + 3 = 0$$
$$x = 9 \quad \text{or} \quad x = -3.$$

The negative solution has to be disregarded; thus, the page is 9 inches wide and its height is $2(9) = 18$ inches.   ●

Let's consider some applications of the business world that can be analyzed using our knowledge of quadratic equations. For example, if P dollars is invested at r rate of interest compounded annually for t years, then the amount of money, A, accumulated at the end of t years is given by the formula

$$A = P(1 + r)^t.$$

This compound interest formula serves as a guideline for the next problem.

Problem 2

Suppose that $100 is invested at a certain rate of interest compounded annually for 2 years. If the accumulated value at the end of 2 years is $121, find the rate of interest.

Solution

Let r represent the rate of interest. Substituting the known values into the compound interest formula yields

$$A = P(1 + r)^t$$
$$121 = 100(1 + r)^2.$$

Solving this equation, we obtain

$$\frac{121}{100} = (1 + r)^2$$

$$\pm\sqrt{\frac{121}{100}} = (1 + r)$$

$$\pm\frac{11}{10} = 1 + r$$

$$1 + r = \frac{11}{10} \quad \text{or} \quad 1 + r = -\frac{11}{10}$$

$$r = -1 + \frac{11}{10} \quad \text{or} \quad r = -1 - \frac{11}{10}$$

$$r = \frac{1}{10} \quad \text{or} \quad r = -\frac{21}{10}.$$

The negative solution must be disregarded, so $r = \dfrac{1}{10}$ is the only solution. Changing $\dfrac{1}{10}$ to a percent, the rate of interest is 10%. ●

Problem 3

A businesswoman bought a parcel of land on speculation, for $120,000. She subdivided the land into lots and when she had sold all but 18 lots at a profit of $6000 per lot, the entire cost of the land had been regained. How many lots were sold and at what price per lot?

Solution

Let x represent the number of lots sold. Then $x + 18$ represents the total number of lots. Therefore, $\dfrac{120,000}{x}$ represents the selling price per lot and $\dfrac{120,000}{x + 18}$ represents the cost per lot. The following equation represents the situation.

selling price equals cost per lot plus $6000
   per lot
      ↓   ↓   ↓

$$\frac{120,000}{x} = \frac{120,000}{x + 18} + 6000.$$

Solving this equation, we obtain

$$x(x + 18)\left(\frac{120,000}{x}\right) = \left(\frac{120,000}{x + 18} + 6000\right)(x)(x + 18)$$
$$120,000(x + 18) = 120,000x + 6000x(x + 18)$$
$$120,000x + 2,160,000 = 120,000x + 6000x^2 + 108,000x$$
$$0 = 6000x^2 + 108,000x - 2,160,000$$
$$0 = x^2 + 18x - 360.$$

The method of completing the square works very well with this equation.

$$x^2 + 18x = 360$$
$$x^2 + 18x + 81 = 441$$
$$(x + 9)^2 = 441$$
$$x + 9 = \pm\sqrt{441}$$
$$x + 9 = \pm 21$$
$$x + 9 = 21 \quad \text{or} \quad x + 9 = -21$$
$$x = 12 \quad \text{or} \quad x = -30.$$

The negative solution is discarded; thus, 12 lots were sold at $\dfrac{120,000}{x} = \dfrac{120,000}{12}$ = \$10,000 per lot.   ●

Problem 4   Barry bought a number of shares of stock for \$600. A week later the value of the stock increased \$3 per share and he sold all but 10 shares and regained his original investment of \$600. How many shares did he sell and at what price per share?

Solution   Let s represent the number of shares Barry sold. Then $s + 10$ represents the number of shares purchased. Therefore $\dfrac{600}{s}$ represents the selling price per share and

$\dfrac{600}{s + 10}$ represents the cost per share.

selling price cost per
  per share   share
     ↓   ↓

$$\frac{600}{s} = \frac{600}{s + 10} + 3.$$

Solving this equation yields

$$s(s + 10)\left(\frac{600}{s}\right) = \left(\frac{600}{s + 10} + 3\right)(s)(s + 10)$$

$$600(s + 10) = 600s + 3s(s + 10)$$

$$600s + 6000 = 600s + 3s^2 + 30s$$

$$0 = 3s^2 + 30s - 6000$$

$$0 = s^2 + 10s - 2000.$$

Using the quadratic formula, we obtain

$$s = \frac{-10 \pm \sqrt{10^2 - 4(1)(-2000)}}{2(1)}$$

$$s = \frac{-10 \pm \sqrt{100 + 8000}}{2}$$

$$s = \frac{-10 \pm \sqrt{8100}}{2}$$

$$s = \frac{-10 \pm 90}{2}$$

$$s = \frac{-10 + 90}{2} \quad \text{or} \quad s = \frac{-10 - 90}{2}$$

$$s = 40 \quad\quad \text{or} \quad\quad s = -50.$$

The negative solution is discarded and we know that 40 shares were sold at $\dfrac{600}{s} = \dfrac{600}{40}$ = \$15 per share. ●

   This next problem set contains a large variety of word problems. Not only are there some business applications similar to those discussed in this section, but there are also more problems of the types discussed back in Chapters 3 and 4. Try to give them your best shot without referring back to examples in earlier chapters.

Problem Set 6.4

Solve each of the following quadratic equations using the method that seems most appropriate to you. Check your solutions by using the *sum and product of roots* relationships.

1. $x^2 - 3x - 7 = 0$

2. $x^2 - 4x - 6 = 0$

3. $2x^2 + x - 28 = 0$

4. $n^2 + 24n + 135 = 0$

5. $n^2 + 22n + 105 = 0$

6. $3x^2 + 23x - 36 = 0$

7. $-3y^2 + 2y + 4 = 0$

8. $-2x^2 + 5x + 2 = 0$

9. $12x^2 + 23x - 9 = 0$

10. $2x^2 - 4x + 7 = 0$

11. $5x^2 - 89x + 70 = 0$

12. $-3x^2 + 54x - 45 = 0$

13. $3x^2 - 2x + 8 = 0$

14. $20y^2 + 17y - 10 = 0$

15. $5t^2 + 5t - 1 = 0$

16. $4t^2 + 4t - 1 = 0$

17. $x^2 + 20x = 25$

18. $x^2 - 18x = 9$

Solve each of the following equations.

19. $n + \dfrac{1}{n} = \dfrac{25}{12}$

20. $n + \dfrac{1}{n} = \dfrac{29}{10}$

21. $\dfrac{2}{x} + \dfrac{5}{x + 2} = 1$

22. $\dfrac{3}{x} + \dfrac{7}{x - 1} = 1$

23. $\dfrac{15}{x + 4} + \dfrac{6}{x} = \dfrac{11}{2}$

24. $\dfrac{6}{x} + \dfrac{40}{x + 5} = 7$

25. $\dfrac{240}{x + 3} = \dfrac{240}{x} - 4$

26. $\dfrac{12}{t} + \dfrac{18}{t + 8} = \dfrac{9}{2}$

27. $\dfrac{3}{t + 2} + \dfrac{4}{t - 2} = 2$

28. $\dfrac{5}{n - 3} - \dfrac{3}{n + 3} = 1$

29. $x^4 - 7x^2 + 6 = 0$

30. $x^4 - 25x^2 + 100 = 0$

31. $x^4 - 8x^2 + 12 = 0$

32. $x^4 - 11x^2 + 28 = 0$

33. $6x^4 - 11x^2 + 4 = 0$

34. $12x^4 - 19x^2 + 5 = 0$

35. $4x^4 + 13x^2 - 12 = 0$

36. $3x^4 - 2x^2 - 5 = 0$

Set up an equation and solve each of the following problems.

37. Find two positive numbers having a sum of 23 and a product of 126.

38. Find two consecutive whole numbers such that the sum of their squares is 113.

39. The sum of two numbers is 12 and the sum of their squares is 74. Find the numbers.

40. Two positive integers differ by 6, and their product is 667. Find the numbers.

41. The length of a rectangular floor is 3 meters longer than its width. If a diagonal of the rectangle is 15 meters long, find the length and width of the floor.

42. The sum of the lengths of the two legs of a right triangle is 14 inches. If the length of the hypotenuse is 10 inches, find the length of each leg.

43. A 5-inch-by-7-inch picture is surrounded by a frame of uniform width. The area of the picture and frame together is 80 square inches. Find the width of the frame.

44. A rectangular plot of ground measuring 12 meters by 20 meters is surrounded by a sidewalk of uniform width. The area of the sidewalk is 68 square meters. Find the width of the walk.

45. A rectangular piece of cardboard is 2 units longer than it is wide. From each of its corners a square piece 2 units on a side is cut out. The flaps are then turned up to form an open box having a volume of 70 cubic units. Find the length and width of the original piece of cardboard.

46. The perimeter of a rectangle is 44 inches and its area is 112 square inches. Find the length and width of the rectangle.

47. Find two numbers such that their sum is 6 and their product is 7.

48. Find two numbers such that their sum is 10 and their product is 22.

49. Larry drove 156 miles in one hour more than it took Mike to drive 108 miles. Mike drove at an average rate of 2 miles per hour faster than Larry. How fast did each one travel?

50. Charlotte traveled 250 miles in one hour more time than it took Lorraine to travel 180 miles. Charlotte drove 5 miles per hour faster than Lorraine. How fast did each one travel?

51. On a 135-mile bicycle excursion, Maria averaged 5 miles per hour faster for the first 60 miles than she did for the last 75 miles. The entire trip took 8 hours. Find her rate for the first 60 miles.

52. It takes Terry 2 hours longer to do a certain job than it takes Tom. They worked together for 3 hours; then Tom left and Terry finished the job in 1 hour. How long would it take each of them to do the job alone?

53. Arlene can mow the lawn in 40 minutes less time with the power mower than with the push mower. One day the power mower broke down after she had been mowing for 30 minutes. She finished the lawn with the push mower in 20 minutes. How long does it take Arlene to mow the entire lawn with the power mower?

54. Tony bought a number of shares of stock for $720. A month later the value of the stock increased by $8 per share and he sold all but 20 shares and regained his original investment plus a profit of $80. How many shares did he sell and at what price per share?

55. A group of students agreed to each chip in the same amount to pay for a party that would cost $100. Then they found 5 more students interested in the party and in sharing the expenses. This decreased the amount each had to pay by $1. How many students were involved in the party and how much did they each have to pay?

56. A retailer bought a number of special mugs for $48. Two of the mugs were broken in the store, but by selling each of the other mugs $3 above the original cost per mug she made a total profit of $22. How many mugs did she buy and at what price per mug did she sell them?

57. The formula $D = \dfrac{n(n-3)}{2}$ yields the number of diagonals, D, in a polygon of n sides. Find the number of sides of a polygon having 54 diagonals.

58. The formula $S = \dfrac{n(n+1)}{2}$ yields the sum, S, of the first n natural numbers $1, 2, 3, 4, \ldots$. How many consecutive natural numbers starting with 1 will give a sum of 1275?

59. At a point 16 yards from the base of a tower, the distance to the top of the tower is 4 yards more than the height of the tower. Find the height of the tower.

60. Suppose that $500 is invested at a certain rate of interest compounded annually for 2 years. If the accumulated value at the end of 2 years is $594.05, find the rate of interest.

61. Suppose that $10,000 is invested at a certain rate of interest compounded annually for 2 years. If the accumulated value at the end of 2 years is $12,544, find the rate of interest.

Miscellaneous Problems

Solve each of the following equations.

62. $x - 9\sqrt{x} + 18 = 0$ (Hint: Let $y = \sqrt{x}$.)

63. $x - 4\sqrt{x} + 3 = 0$ **64.** $x + \sqrt{x} - 2 = 0$

65. $x^{\frac{2}{3}} + x^{\frac{1}{3}} - 6 = 0$ (Hint: Let $y = x^{\frac{1}{3}}$.)

66. $6x^{\frac{2}{3}} - 5x^{\frac{1}{3}} - 6 = 0$ **67.** $x^{-2} + 4x^{-1} - 12 = 0$

68. $12x^{-2} - 17x^{-1} - 5 = 0$

6.5 Quadratic Inequalities

The equation $ax^2 + bx + c = 0$ has been referred to as the standard form of a quadratic equation in one variable. Similarly, the following forms are used to express **quadratic inequalities** in one variable.

$$ax^2 + bx + c > 0;$$
$$ax^2 + bx + c \geq 0;$$
$$ax^2 + bx + c < 0;$$
$$ax^2 + bx + c \leq 0.$$

The number line can be used very effectively to solve quadratic inequalities for which the quadratic polynomial is factorable. Let's consider some examples illustrating use of the number line.

Example 1 Solve and graph the solutions for $x^2 + 2x - 8 > 0$.

Solution First, let's factor the polynomial.

$$x^2 + 2x - 8 > 0$$
$$(x + 4)(x - 2) > 0$$

On a number line, we now indicate that at $x = -4$ and $x = 2$ the product $(x + 4)(x - 2)$ equals zero.

The numbers -4 and 2 divide the number line into three intervals: the numbers less than -4, the numbers between -4 and 2, and the numbers greater than 2. We can choose a **test number** from each of these intervals and see how it affects the factors $(x + 4)$ and $(x - 2)$ and, consequently, the product of these factors. For example, if $x < -4$ (try $x = -5$), then $(x + 4)$ is negative and $(x - 2)$ is negative; so their product is positive. If $-4 < x < 2$ (try $x = 0$), then $(x + 4)$ is positive and $(x - 2)$ is negative; so their product is negative. If $x > 2$ (try $x = 3$), then $(x + 4)$ is positive and $(x - 2)$ is positive; so their product is positive. This information can be conveniently arranged using the number line as follows.

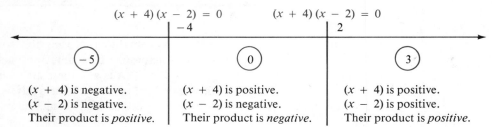

Thus, the given inequality, $x^2 + 2x - 8 > 0$, is satisfied for values of x such that $x < -4$ or $x > 2$. These solutions can be indicated on a number line as follows.

Numbers such as -4 and 2 in the preceding example, for which the given polynomial or algebraic expression equals zero or is undefined, are referred to as **critical numbers**. Let's consider some additional examples making use of critical numbers and test numbers.

Example 2

Solve and graph the solutions for $x^2 + 2x - 3 \leq 0$.

Solution

First, we factor the polynomial.

$$x^2 + 2x - 3 \leq 0$$
$$(x + 3)(x - 1) \leq 0$$

Second, we locate the values for which $(x + 3)(x - 1)$ equals zero.

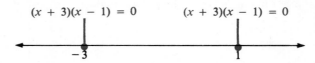

We put dots at -3 and 1 as a reminder that these two numbers are to be included in the solution set, since the given statement includes equality. Now let's choose a test number from each of the three intervals and observe the sign behavior of the factors.

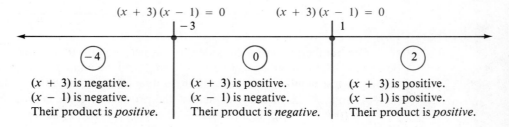

Therefore, we see that $x^2 + 2x - 3 \leq 0$ if x is between -3 and 1, inclusive. These solutions can be shown on a number line as follows.

Examples 1 and 2 illustrate a systematic approach for solving quadratic inequalities for which the polynomial is factorable. This same approach will work with inequalities such as $(x - 1)(x + 2)(x - 4) > 0$. (A few of these are included in the next set of exercises.) Furthermore, this same type of **number-line analysis** can be used for inequalities involving indicated quotients. Let's consider some examples.

Example 3   Solve and graph the solutions for $\dfrac{x+1}{x-5} > 0$.

Solution   First, we indicate that at $x = -1$ the quotient $\dfrac{x+1}{x-5}$ equals zero and that at $x = 5$

the quotient $\dfrac{x+1}{x-5}$ is undefined. Then we can choose test numbers in each of the

intervals and observe the sign behavior of $(x + 1)$ and of $(x - 5)$.

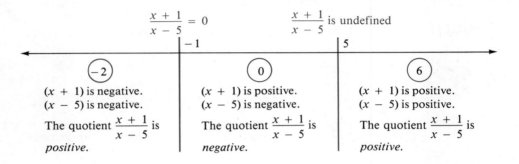

Therefore, $\dfrac{x+1}{x-5} > 0$ if $x < -1$ or $x > 5$. These solutions can be shown on a

number line as follows.

Example 4   Solve and graph the solutions for $\dfrac{x+2}{x+4} \le 0$.

Solution   First, we observe that the indicated quotient $\dfrac{x+2}{x+4}$ equals zero when $x = -2$ and

is undefined when $x = -4$. Then using test numbers -5, -3, and 0, we can study the
sign behavior of $(x + 2)$ and $(x + 4)$.

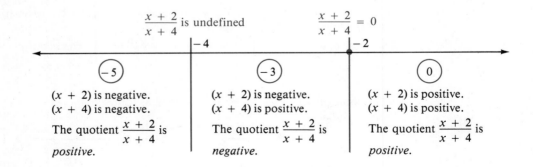

Thus, the solution set for $\dfrac{x + 2}{x + 4} \leq 0$ is graphed as follows.

The final example illustrates that sometimes we need to change form before using the number-line approach.

Example 5

Solve and graph the solutions for $\dfrac{x}{x + 2} \leq 3$.

Solution

First, let's change the form of the given inequality.

$$\frac{x}{x + 2} \leq 3$$

$$\frac{x}{x + 2} - 3 \leq 0 \qquad \text{Add } -3 \text{ to both sides.}$$

$$\frac{x - 3(x + 2)}{x + 2} \leq 0 \qquad \text{Express the left side over a common denominator.}$$

$$\frac{x - 3x - 6}{x + 2} \leq 0$$

$$\frac{-2x - 6}{x + 2} \leq 0$$

Now we can proceed as in the previous examples. If $x = -3$, then $\dfrac{-2x - 6}{x + 2}$ equals zero; and if $x = -2$, then $\dfrac{-2x - 6}{x + 2}$ is undefined. Then choosing test numbers such as -4, $-2\dfrac{1}{2}$, and 0, we can study the sign behavior of $(-2x - 6)$ and $(x + 2)$.

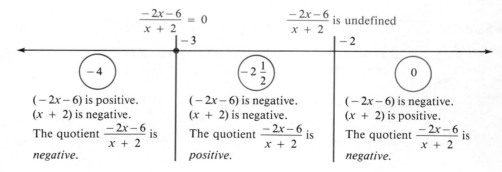

The solution set for $\dfrac{x}{x + 2} \leq 3$ is graphed as follows.

Problem Set 6.5

Solve each of the following inequalities, and indicate the solutions on a number line graph.

1. $(x - 1)(x - 4) > 0$

2. $(x + 5)(x + 2) < 0$

3. $(2x - 3)(3x + 4) \leq 0$

4. $(5x + 1)(x - 3) \geq 0$

5. $x^2 - x - 12 < 0$

6. $x^2 - x - 20 > 0$

7. $8 \geq 9x - x^2$

8. $x^2 + 8x + 12 \leq 0$

9. $3x^2 - 7x + 2 > 0$

10. $1 < 5x - 4x^2$

11. $6x^2 - x < 2$

12. $12x^2 + 8x > 15$

13. $x(x - 6) > 16$

14. $x(x + 5) < 24$

15. $x^2 - 4x + 4 \geq 0$

16. $x^2 + 6x + 9 \leq 0$

17. $(x - 1)(x - 3)(x + 2) > 0$

18. $(x + 4)(x + 2)(x - 2) \leq 0$

19. $(x - 4)^2(x - 1) \leq 0$

20. $(x + 1)(x - 3)^2 > 0$

Solve each of the following inequalities and indicate the solutions on a number line graph.

21. $\dfrac{x - 2}{x + 3} > 0$

22. $\dfrac{x + 4}{x - 1} < 0$

23. $\dfrac{x + 1}{x + 6} \leq 0$

24. $\dfrac{x - 3}{x - 7} \geq 0$

25. $\dfrac{-x + 3}{x - 5} \geq 0$

26. $\dfrac{-x - 4}{x + 1} \leq 0$

27. $\dfrac{x}{x - 1} > 2$

28. $\dfrac{2x}{x + 3} > 4$

29. $\dfrac{x + 2}{x + 4} \leq 3$

30. $\dfrac{x - 1}{x - 5} \leq 2$

31. $\dfrac{2x - 1}{x + 2} \geq 0$

32. $\dfrac{3x + 1}{x - 3} \leq 0$

33. $\dfrac{x - 1}{x - 2} < -1$

34. $\dfrac{x + 2}{x - 3} > -2$

35. $\dfrac{2x - 1}{x + 2} \geq -1$

36. $\dfrac{3x + 2}{x + 4} \leq 2$

6.6 Complex Numbers

Each of the following quadratic equations has been used as an example in a previous section of this chapter.

$$x^2 = -9; \qquad (x + 3)^2 = -17; \qquad (x + 2)^2 = -2.$$

For each one the solution set was found to be $\emptyset$. That is, none of these equations yielded any *real number solutions*. In each example, our conclusion of "no real number solution" was based on the fact that squaring any real number produces a nonnegative result.

In this section we will examine a set of numbers that contains some numbers whose squares are negative real numbers. Then in a following section we shall see that this set of numbers, called the **complex numbers**, provides solutions for not only equations such as $x^2 = -9$, but also for any quadratic equation in one variable with real number coefficients.

To provide a solution for the equation $x^2 + 1 = 0$, we use the number i, such that

$$i^2 = -1.$$

The number i is not a real number and is often called the **imaginary unit**, but the number i^2 is the real number -1.

The imaginary unit i is used to define a complex number as follows:

Definition 6.1

A **complex number** is any number that can be expressed in the form

$$a + bi$$

where a and b are real numbers.

The form $a + bi$ is called the **standard form** of a complex number. The real number a is called the **real part** of the complex number and b is called the **imaginary part**. (Note that b is a real number even though it is called the imaginary part.) The following examples illustrate this terminology.

1. The number $7 + 5i$ is a complex number having a real part of 7 and an imaginary part of 5.

2. The number $\frac{2}{3} + i\sqrt{2}$ is a complex number having a real part of $\frac{2}{3}$ and an imaginary part of $\sqrt{2}$. (It is easy to mistake $\sqrt{2i}$ for $\sqrt{2}\,i$. Thus, it is common to write $i\sqrt{2}$ instead of $\sqrt{2}\,i$ to avoid any difficulties with the radical sign.)

3. The number $-4 - 3i$ can be written in the standard form $-4 + (-3i)$ and therefore is a complex number having a real part of -4 and an imaginary part of -3. (The form $-4 - 3i$ is often used knowing that it means $-4 + (-3i)$.)

4. The number $-9i$ can be written as $0 + (-9i)$; thus, it is a complex number having a real part of 0 and an imaginary part of -9. (Complex numbers, such as $-9i$, for which $a = 0$ and $b \neq 0$ are called **pure imaginary numbers**.)

5. The real number 4 can be written as $4 + 0i$ and is thus a complex number having a real part of 4 and an imaginary part of 0.

From Example 5 we see that the set of real numbers is a subset of the set of complex numbers. The following diagram indicates the organizational format of the complex numbers.

Complex numbers ($a + bi$, where a and b are real numbers)

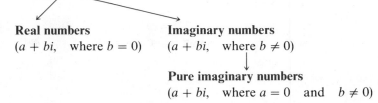

Real numbers
($a + bi$, where $b = 0$)

Imaginary numbers
($a + bi$, where $b \neq 0$)

Pure imaginary numbers
($a + bi$, where $a = 0$ and $b \neq 0$)

Two complex numbers $a + bi$ and $c + di$ are said to be *equal* if and only if $a = c$ and $b = d$.

Adding and Subtracting Complex Numbers

To **add complex numbers**, we simply add their real parts and add their imaginary parts. Thus,

$$(a + bi) + (c + di) = (a + c) + (b + d)i.$$

The following examples illustrate addition of two complex numbers.

1. $(4 + 3i) + (5 + 9i) = (4 + 5) + (3 + 9)i = 9 + 12i.$

2. $(-6 + 4i) + (8 - 7i) = (-6 + 8) + (4 - 7)i$
$$= 2 - 3i.$$

3. $\left(\dfrac{1}{2} + \dfrac{3}{4}i\right) + \left(\dfrac{2}{3} + \dfrac{1}{5}i\right) = \left(\dfrac{1}{2} + \dfrac{2}{3}\right) + \left(\dfrac{3}{4} + \dfrac{1}{5}\right)i$
$$= \left(\dfrac{3}{6} + \dfrac{4}{6}\right) + \left(\dfrac{15}{20} + \dfrac{4}{20}\right)i$$
$$= \dfrac{7}{6} + \dfrac{19}{20}i.$$

The set of complex numbers is closed with respect to addition; that is, the sum of two complex numbers is a complex number. Furthermore, the commutative and associative properties of addition hold for all complex numbers. The addition identity element is $0 + 0i$ (or simply the real number 0). The additive inverse of $a + bi$ is $-a - bi$, because

$$(a + bi) + (-a - bi) = 0.$$

To **subtract complex numbers**, $c + di$ from $a + bi$, add the additive inverse of $c + di$. Thus,

$$(a + bi) - (c + di) = (a + bi) + (-c - di)$$
$$= (a - c) + (b - d)i.$$

In other words, we subtract the real parts and subtract the imaginary parts, as the next examples illustrate.

1. $(9 + 8i) - (5 + 3i) = (9 - 5) + (8 - 3)i$
$$= 4 + 5i.$$

2. $(3 - 2i) - (4 - 10i) = (3 - 4) + (-2 - (-10))i$
$$= -1 + 8i.$$

Products and Quotients of Complex Numbers

Since $i^2 = -1$, i is a square root of negative one; so we let $i = \sqrt{-1}$. It should also be evident that $-i$ is a square root of negative one since

$$(-i)^2 = (-i)(-i) = i^2 = -1.$$

Thus, in the set of complex numbers, -1 has two square roots, i and $-i$. These are symbolically expressed as

$$\sqrt{-1} = i \quad \text{and} \quad -\sqrt{-1} = -i.$$

Let us extend our definition so that in the set of complex numbers every negative real number has two square roots. We simply define $\sqrt{-b}$, where b is a positive real number, to be the number whose square is $-b$. Thus,

$$(\sqrt{-b})^2 = -b, \quad \text{for } b > 0.$$

Furthermore, since $(i\sqrt{b})(i\sqrt{b}) = i^2(b) = -1(b) = -b$ we see that

$$\sqrt{-b} = i\sqrt{b}.$$

In other words, a square root of any negative real number can be represented as the product of a real number and the imaginary unit i. Consider the following examples.

$$\sqrt{-4} = i\sqrt{4} = 2i;$$
$$\sqrt{-17} = i\sqrt{17};$$
$$\sqrt{-24} = i\sqrt{24} = i\sqrt{4}\sqrt{6} = 2i\sqrt{6}.$$

(Notice that we simplified the radical $\sqrt{24}$ to $2\sqrt{6}$.)

We should also observe that $-\sqrt{-b}$, where $b > 0$, is a square root of $-b$ since

$$(-\sqrt{-b})^2 = (-i\sqrt{b})^2 = i^2(b) = -1(b) = -b.$$

Thus, in the set of complex numbers, $-b$ (where $b > 0$) has two square roots, $i\sqrt{b}$ and $-i\sqrt{b}$. These are symbolically expressed as

$$\sqrt{-b} = i\sqrt{b} \quad \text{and} \quad -\sqrt{-b} = -i\sqrt{b}.$$

We must be very careful with the use of the symbol $\sqrt{-b}$, where $b > 0$. Some relationships that are true in the set of real numbers involving the square root symbol do not hold if the square root symbol does not represent a real number. For example, $\sqrt{a}\sqrt{b} = \sqrt{ab}$ *does not hold* if a and b are both negative numbers.

Correct $\sqrt{-4}\sqrt{-9} = (2i)(3i) = 6i^2 = 6(-1) = -6.$

Incorrect $\sqrt{-4}\sqrt{-9} = \sqrt{(-4)(-9)} = \sqrt{36} = 6.$

To avoid difficulty with this idea, you should rewrite all expressions of the form $\sqrt{-b}$, where $b > 0$, in the form $i\sqrt{b}$ before doing *any computations*. The following examples further illustrate this point.

1. $\sqrt{-5}\sqrt{-7} = (i\sqrt{5})(i\sqrt{7}) = i^2\sqrt{35} = (-1)\sqrt{35} = -\sqrt{35}.$
2. $\sqrt{-2}\sqrt{-8} = (i\sqrt{2})(i\sqrt{8}) = i^2\sqrt{16} = (-1)(4) = -4.$
3. $\sqrt{-6}\sqrt{-8} = (i\sqrt{6})(i\sqrt{8}) = i^2\sqrt{48} = (-1)\sqrt{16}\sqrt{3} = -4\sqrt{3}.$
4. $\dfrac{\sqrt{-75}}{\sqrt{-3}} = \dfrac{i\sqrt{75}}{i\sqrt{3}} = \dfrac{\sqrt{75}}{\sqrt{3}} = \sqrt{\dfrac{75}{3}} = \sqrt{25} = 5.$
5. $\dfrac{\sqrt{-48}}{\sqrt{12}} = \dfrac{i\sqrt{48}}{\sqrt{12}} = i\sqrt{\dfrac{48}{12}} = i\sqrt{4} = 2i.$

Since complex numbers have a *binomial form*, we find the *product* of two complex numbers in the same way as we find the product of two binomials. Then by replacing i^2 with -1, we are able to simplify and express the final result in standard form. Consider the following examples.

6. $(2 + 3i)(4 + 5i) = 2(4 + 5i) + 3i(4 + 5i)$
$= 8 + 10i + 12i + 15i^2$
$= 8 + 22i + 15i^2$
$= 8 + 22i + 15(-1) = -7 + 22i.$

7. $(-3 + 6i)(2 - 4i) = -3(2 - 4i) + 6i(2 - 4i)$
$= -6 + 12i + 12i - 24i^2$
$= -6 + 24i - 24(-1)$
$= -6 + 24i + 24 = 18 + 24i.$

8. $(1 - 7i)^2 = (1 - 7i)(1 - 7i)$
$= 1(1 - 7i) - 7i(1 - 7i)$
$= 1 - 7i - 7i + 49i^2$
$= 1 - 14i + 49(-1)$
$= 1 - 14i - 49 = -48 - 14i.$

9. $(2 + 3i)(2 - 3i) = 2(2 - 3i) + 3i(2 - 3i)$
$= 4 - 6i + 6i - 9i^2$
$= 4 - 9(-1)$
$= 4 + 9 = 13.$

Example **9.** illustrates an important situation. The complex numbers $2 + 3i$ and $2 - 3i$ are called conjugates of each other. In general, the two complex numbers $a + bi$ and $a - bi$ are called **conjugates** of each other and *the product of a complex number and its conjugate is always a real number*. This can be shown as follows:

$$
\begin{aligned}
(a + bi)(a - bi) &= a(a - bi) + bi(a - bi) \\
&= a^2 - abi + abi - b^2 i^2 \\
&= a^2 - b^2(-1) \\
&= a^2 + b^2.
\end{aligned}
$$

Conjugates are used to *simplify expressions* such as $\dfrac{3i}{5 + 2i}$ that *indicate the quotient* of two complex numbers. To eliminate i in the denominator and change the indicated quotient to the standard form of a complex number, we can multiply both the numerator and the denominator by the conjugate of the denominator as follows:

$$
\begin{aligned}
\frac{3i}{5 + 2i} &= \frac{3i(5 - 2i)}{(5 + 2i)(5 - 2i)} \\
&= \frac{15i - 6i^2}{25 - 4i^2} \\
&= \frac{15i - 6(-1)}{25 - 4(-1)} \\
&= \frac{15i + 6}{29} \\
&= \frac{6}{29} + \frac{15}{29}i.
\end{aligned}
$$

The following examples further illustrate the process of *dividing* complex numbers.

10. $\dfrac{2 - 3i}{4 - 7i} = \dfrac{(2 - 3i)(4 + 7i)}{(4 - 7i)(4 + 7i)}$ ($4 + 7i$ is the conjugate of $4 - 7i$.)

$$
\begin{aligned}
&= \frac{8 + 14i - 12i - 21i^2}{16 - 49i^2} \\
&= \frac{8 + 2i - 21(-1)}{16 - 49(-1)} \\
&= \frac{8 + 2i + 21}{16 + 49} \\
&= \frac{29 + 2i}{65} \\
&= \frac{29}{65} + \frac{2}{65}i.
\end{aligned}
$$

11. $\dfrac{4 - 5i}{2i} = \dfrac{(4 - 5i)(-2i)}{(2i)(-2i)}$ ($-2i$ is the conjugate of $2i$.)

$= \dfrac{-8i + 10i^2}{-4i^2}$

$= \dfrac{-8i + 10(-1)}{-4(-1)}$

$= \dfrac{-8i - 10}{4}$

$= -\dfrac{5}{2} - 2i.$

For a problem such as Example 11 in which the denominator is a pure imaginary number, we can change to standard form by choosing a multiplier other than the conjugate. Consider the following alternate approach for Example 11.

$\dfrac{4 - 5i}{2i} = \dfrac{(4 - 5i)(i)}{(2i)(i)}$

$= \dfrac{4i - 5i^2}{2i^2}$

$= \dfrac{4i - 5(-1)}{2(-1)}$

$= \dfrac{4i + 5}{-2}$

$= -\dfrac{5}{2} - 2i.$

Problem Set 6.6

Label each of the following statements true or false.

1. Every real number is a complex number.
2. Every complex number is a real number.
3. Every complex number is a pure imaginary number.
4. The real part of the complex number $6i$ is 0.
5. The imaginary part of the complex number 7 is 0.
6. The sum of two complex numbers is always a complex number.
7. The sum of two pure imaginary numbers is always a pure imaginary number.
8. The sum of two complex numbers is sometimes a real number.

Add or subtract as indicated.

9. $(5 + 2i) + (8 + 6i)$

10. $(3 + 6i) + (7 + 12i)$

11. $(-9 + 3i) + (5 + 5i)$

12. $(6 - 7i) + (-8 + 4i)$

13. $(8 + 6i) - (5 + 2i)$

14. $(14 + 7i) - (13 + i)$

15. $(-6 + 4i) - (4 + 6i)$

16. $(9 - 2i) - (10 + 3i)$

17. $(-7 - 3i) + (-4 + 4i)$

18. $(8 - 12i) + (-7 - 10i)$

19. $(6 - 7i) - (7 - 6i)$

20. $(-10 - 2i) - (11 + 3i)$

21. $(-2 - 3i) - (-1 - i)$

22. $(-3 - 6i) - (-4 + 9i)$

23. $\left(\frac{1}{3} + \frac{2}{5}i\right) + \left(\frac{1}{2} + \frac{1}{4}i\right)$

24. $\left(\frac{1}{2} + \frac{2}{3}i\right) + \left(\frac{5}{6} + \frac{3}{4}i\right)$

25. $\left(\frac{5}{8} + \frac{1}{2}i\right) - \left(\frac{7}{8} + \frac{1}{5}i\right)$

26. $\left(\frac{4}{9} - \frac{2}{5}i\right) - \left(\frac{2}{3} - \frac{5}{6}i\right)$

Write each of the following in terms of i and simplify. For example,
$$\sqrt{-20} = i\sqrt{20} = i\sqrt{4}\sqrt{5} = 2i\sqrt{5}.$$

27. $\sqrt{-9}$

28. $\sqrt{-49}$

29. $\sqrt{-19}$

30. $\sqrt{-31}$

31. $\sqrt{-\frac{4}{9}}$

32. $\sqrt{-\frac{25}{36}}$

33. $\sqrt{-8}$

34. $\sqrt{-18}$

35. $\sqrt{-27}$

36. $\sqrt{-32}$

37. $\sqrt{-54}$

38. $\sqrt{-40}$

39. $3\sqrt{-36}$

40. $5\sqrt{-64}$

41. $4\sqrt{-18}$

42. $6\sqrt{-8}$

Write each of the following in terms of i, perform the indicated operations, and simplify. For example,
$$\sqrt{-9}\sqrt{-16} = (i\sqrt{9})(i\sqrt{16}) = (3i)(4i) = 12i^2 = 12(-1) = -12.$$

43. $\sqrt{-4}\sqrt{-16}$

44. $\sqrt{-25}\sqrt{-9}$

45. $\sqrt{-2}\sqrt{-3}$

46. $\sqrt{-3}\sqrt{-7}$

47. $\sqrt{-5}\sqrt{-4}$

48. $\sqrt{-7}\sqrt{-9}$

49. $\sqrt{-6}\sqrt{-10}$

50. $\sqrt{-2}\sqrt{-12}$

51. $\sqrt{-8}\sqrt{-7}$

52. $\sqrt{-12}\sqrt{-5}$

53. $\frac{\sqrt{-36}}{\sqrt{-4}}$

54. $\frac{\sqrt{-64}}{\sqrt{-16}}$

55. $\frac{\sqrt{-54}}{\sqrt{-9}}$

56. $\frac{\sqrt{-18}}{\sqrt{-3}}$

Find each of the following products and express answers in standard form.

57. $(3i)(7i)$

58. $(-5i)(8i)$

59. $(4i)(3 - 2i)$

60. $(5i)(2 + 6i)$

61. $(3 + 2i)(4 + 6i)$

62. $(7 + 3i)(8 + 4i)$

63. $(4 + 5i)(2 - 9i)$

64. $(1 + i)(2 - i)$

65. $(-2 - 3i)(4 + 6i)$

66. $(-3 - 7i)(2 + 10i)$

67. $(6 - 4i)(-1 - 2i)$

68. $(7 - 3i)(-2 - 8i)$

69. $(3 + 4i)^2$

70. $(4 - 2i)^2$

71. $(-1 - 2i)^2$

72. $(-2 + 5i)^2$

73. $(8 - 7i)(8 + 7i)$ **74.** $(5 + 3i)(5 - 3i)$

75. $(-2 + 3i)(-2 - 3i)$ **76.** $(-6 - 7i)(-6 + 7i)$

Find each of the following quotients, expressing answers in standard form.

77. $\dfrac{4i}{3 - 2i}$ **78.** $\dfrac{3i}{6 + 2i}$ **79.** $\dfrac{2 + 3i}{3i}$ **80.** $\dfrac{3 - 5i}{4i}$

81. $\dfrac{3}{2i}$ **82.** $\dfrac{7}{4i}$ **83.** $\dfrac{3 + 2i}{4 + 5i}$ **84.** $\dfrac{2 + 5i}{3 + 7i}$

85. $\dfrac{4 + 7i}{2 - 3i}$ **86.** $\dfrac{3 + 9i}{4 - i}$ **87.** $\dfrac{3 - 7i}{-2 + 4i}$ **88.** $\dfrac{4 - 10i}{-3 + 7i}$

89. $\dfrac{-1 - i}{-2 - 3i}$ **90.** $\dfrac{-4 + 9i}{-3 - 6i}$

6.7 Quadratic Equations Revisited: Complex Solutions

In Section 6.6 we used $i = \sqrt{-1}$ to provide a solution for the equation $x^2 + 1 = 0$. Since $i^2 = -1$, it should be obvious that i is indeed a solution for $x^2 + 1 = 0$. Furthermore, since $(-i)^2 = i^2 = -1$, we see that $-i$ is also a solution. Thus, the solution set for $x^2 + 1 = 0$ is $\{-i, i\}$.

It was further stated in Section 6.6 that this new set of numbers, called the complex numbers, would provide solutions for any quadratic equation in one variable having real number coefficients. To find complex solutions for quadratic equations we simply continue to use the techniques of factoring, completing the square, and the quadratic formula. Furthermore, the restriction that a is a nonnegative real number can be lifted from Property 6.1 which stated "if $x^2 = a$, then $x = \pm\sqrt{a}$." In other words, Property 6.1 holds for all real number values for a. Let's use this property in two examples.

Example 1   Solve $x^2 = -9$.

Solution

$$x^2 = -9$$
$$x = \pm\sqrt{-9}$$
$$x = \pm 3i. \qquad\qquad (\sqrt{-9} = i\sqrt{9} = 3i)$$

Check

$$\begin{array}{ccc}
x^2 = -9 & & x^2 = -9 \\
(3i)^2 \overset{?}{=} -9 & \text{or} & (-3i)^2 \overset{?}{=} -9 \\
9i^2 \overset{?}{=} -9 & & 9i^2 \overset{?}{=} -9 \\
9(-1) \overset{?}{=} -9 & & 9(-1) \overset{?}{=} -9 \\
-9 = -9. & & -9 = -9.
\end{array}$$

Thus, the solution set is $\{-3i, 3i\}$.

Example 2   Solve $(x - 3)^2 = -10$.

Solution
$$(x - 3)^2 = -10$$
$$x - 3 = \pm\sqrt{-10}$$
$$x - 3 = \pm i\sqrt{10}$$
$$x = 3 \pm i\sqrt{10}.$$

Check

$$(x - 3)^2 = -10$$
$$(3 + i\sqrt{10} - 3)^2 \overset{?}{=} -10$$
$$(i\sqrt{10})^2 \overset{?}{=} -10 \quad \text{or}$$
$$10i^2 \overset{?}{=} -10$$
$$10(-1) \overset{?}{=} -10$$
$$-10 = -10.$$

$$(x - 3)^2 = -10$$
$$(3 - i\sqrt{10} - 3)^2 \overset{?}{=} -10$$
$$(-i\sqrt{10})^2 \overset{?}{=} -10$$
$$10i^2 \overset{?}{=} -10$$
$$10(-1) \overset{?}{=} -10$$
$$-10 = -10.$$

Thus, the solution set is $\{3 - i\sqrt{10}, 3 + i\sqrt{10}\}$.   ●

The next example illustrates the use of *completing the square* to find the complex solutions.

Example 3   Solve $x^2 + 4x + 7 = 0$.

Solution
$$x^2 + 4x + 7 = 0$$
$$x^2 + 4x = -7$$
$$x^2 + 4x + 4 = -7 + 4$$
$$(x + 2)^2 = -3$$
$$x + 2 = \pm\sqrt{-3}$$
$$x + 2 = \pm i\sqrt{3}$$
$$x = -2 \pm i\sqrt{3}.$$

Check   Remember that the sum of the roots of the equation $ax^2 + bx + c = 0$ is $-\dfrac{b}{a}$ and the product of the roots is $\dfrac{c}{a}$. These relationships can continue to be used for checking purposes.

Sum of roots $(-2 + i\sqrt{3}) + (-2 - i\sqrt{3}) = -4$ and $-\dfrac{b}{a} = -\dfrac{4}{1} = -4.$

Product of roots $(-2 + i\sqrt{3})(-2 - i\sqrt{3}) = 4 - 3i^2$
$$= 4 - 3(-1) = 7 \quad \text{and} \quad \frac{c}{a} = \frac{7}{1} = 7.$$

The solution set is $\{-2 - i\sqrt{3}, -2 + i\sqrt{3}\}$.   ●

Now let's use the quadratic formula to find the complex solutions of a quadratic equation.

Example 4   Solve $x^2 - 2x + 19 = 0$.

Solution $x^2 - 2x + 19 = 0$

$$x = \frac{-(-2) \pm \sqrt{(-2)^2 - 4(1)(19)}}{2(1)}$$

$$x = \frac{2 \pm \sqrt{4 - 76}}{2}$$

$$x = \frac{2 \pm \sqrt{-72}}{2}$$

$$x = \frac{2 \pm 6i\sqrt{2}}{2} \qquad (\sqrt{-72} = i\sqrt{72} = i\sqrt{36}\sqrt{2} = 6i\sqrt{2})$$

$$x = 1 \pm 3i\sqrt{2}.$$

Check

Sum of roots $(1 + 3i\sqrt{2}) + (1 - 3i\sqrt{2}) = 2$ and $-\dfrac{b}{a} = -\dfrac{-2}{1} = 2.$

Product of roots $(1 + 3i\sqrt{2})(1 - 3i\sqrt{2}) = 1 - (9i^2)(2) = 1 - 18i^2$

$$= 1 - 18(-1) = 1 + 18 = 19 \quad \text{and} \quad \frac{c}{a} = \frac{19}{1} = 19.$$

The solution set is $\{1 - 3i\sqrt{2}, 1 + 3i\sqrt{2}\}$.   ●

Our next and final example reminds us that the set of real numbers is a subset of the set of complex numbers. In other words, the complex solutions of a quadratic equation may turn out to be real numbers.

Example 5   Solve $x^2 + 3x - 10 = 0$.

Solution   We can factor $x^2 + 3x - 10$ and proceed as follows:

$$x^2 + 3x - 10 = 0$$
$$(x + 5)(x - 2) = 0$$
$$x + 5 = 0 \quad \text{or} \quad x - 2 = 0$$
$$x = -5 \quad \text{or} \qquad x = 2.$$

The solution set is $\{-5, 2\}$.   ●

To summarize our work in this chapter, consider the following approach to solving a quadratic equation.

1. Use the property "if $x^2 = a$, then $x = \pm\sqrt{a}$," if the equation is in an appropriate form.

2. If the quadratic expression is obviously factorable using integers, factor it and apply the property "if $ab = 0$, then $a = 0$ or $b = 0$."

3. If neither **1.** nor **2.** applies, use either the quadratic formula or the process of completing the square.

In Section 6.3 we used the *discriminant*, $b^2 - 4ac$, to determine the types of roots of a particular quadratic equation. Now that we have the entire set of complex numbers to be used as possible solutions, the following general statements can be made relative to the roots of a quadratic equation of the form $ax^2 + bx + c = 0$, where a, b, and c are real numbers and $a \neq 0$.

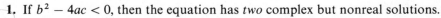

1. If $b^2 - 4ac < 0$, then the equation has *two* complex but nonreal solutions.
2. If $b^2 - 4ac = 0$, then the equation has *two* equal real solutions.
3. If $b^2 - 4ac > 0$, then the equation has *two* unequal real solutions.

The following examples illustrate each of these situations. (You may want to solve the equations completely to verify the conclusions.)

Equation	Discriminant	Nature of roots
$x^2 - 3x + 9 = 0$	$b^2 - 4ac = (-3)^2 - 4(1)(9)$ $= 9 - 36$ $= -27.$	two complex but nonreal solutions
$4x^2 - 20x + 25 = 0$	$b^2 - 4ac = (-20)^2 - 4(4)(25)$ $= 400 - 400$ $= 0.$	two equal real solutions
$3x^2 + 2x - 1 = 0$	$b^2 - 4ac = 2^2 - 4(3)(-1)$ $= 4 + 12$ $= 16.$	two unequal real solutions

Notice that in each of the last two examples the value of the discriminant is a perfect square. Whenever this happens, we know that the roots are rational and the equation can be solved by the factoring method.

Problem Set 6.7

Solve each of the following quadratic equations and check your solutions.

1. $x^2 = -36$
2. $x^2 = -81$
3. $(x + 4)^2 = -9$
4. $(x - 5)^2 = -16$
5. $(x - 2)^2 = -5$
6. $(x + 4)^2 = -7$
7. $(x - 1)^2 + 12 = 0$
8. $(x - 2)^2 + 24$
9. $4 - (3x - 2)^2 = 0$
10. $16 - (4x + 3)^2 = 0$

11. $x^2 + 2x = -5$ **12.** $x^2 - 4x = -20$

13. $y^2 - 6y + 10 = 0$ **14.** $y^2 - 8y + 17 = 0$

15. $a^2 + 3a - 28 = 0$ **16.** $a^2 + 4a - 32 = 0$

17. $n^2 = 2n + 1$ **18.** $n^2 = -4n - 1$

19. $2x^2 - 3x + 4 = 0$ **20.** $3x^2 - 2x + 5 = 0$

21. $3x^2 - 2x + 1 = 0$ **22.** $2x^2 + 5x + 5 = 0$

23. $6x^2 + 13x - 5 = 0$ **24.** $8x^2 + 10x - 3 = 0$

25. $40x^2 - 10x + 10 = 0$ **26.** $25x^2 - 10x + 15 = 0$

27. $x^2 - 6x + 25 = 0$ **28.** $x^2 - 8x + 25 = 0$

29. $4x^2 - 28x + 49 = 0$ **30.** $9x^2 + 12x + 4 = 0$

31. Find the discriminant of each of the following quadratic equations and determine whether the equation has (1) two complex but nonreal solutions, (2) two equal real solutions, or (3) two unequal real solutions.

   (a) $x^2 + 6x - 7 = 0$ **(b)** $x^2 + 4x + 5 = 0$

   (c) $x^2 - 14x + 49 = 0$ **(d)** $x^2 - x - 1 = 0$

   (e) $2x^2 + 5x + 7 = 0$ **(f)** $3x^2 - 2x - 1 = 0$

   (g) $5x^2 + 3x - 9 = 0$ **(h)** $6x^2 - 4x - 7 = 0$

   (i) $16x^2 - 40x + 25 = 0$ **(j)** $x^2 = 7x - 2$

32. Determine k so that the solutions of $x^2 - 2x + k = 0$ are complex but nonreal. (Hint: Use the discriminant.)

33. Determine k so that $4x^2 - kx + 1 = 0$ has two equal real solutions.

34. Determine k so that $3x^2 - kx - 2 = 0$ has real solutions.

Chapter 6 Summary

(6.1) The *standard form* for a **quadratic equation** in one variable is

$$ax^2 + bx + c = 0,$$

where a, b, and c are real numbers and $a \neq 0$.

Some quadratic equations can be solved by *factoring* and applying the property "if $ab = 0$, then $a = 0$ or $b = 0$."

Don't forget that applying the property "if $a = b$, then $a^n = b^n$" might produce extraneous solutions. Therefore, we *must check* all potential solutions.

Some quadratic equations can be solved by applying the property "if $x^2 = a$, then $x = \sqrt{a}$ or $x = -\sqrt{a}$."

If $x^2 = a$ where a is a negative number, the equation has no real number solutions.

(6.2) To solve a quadratic equation of the form $x^2 + bx = k$ by **completing the square**, we (1) add $\left(\dfrac{b}{2}\right)^2$ to both sides, (2) factor the left side, and (3) apply the property "if $x^2 = a$, then $x = \pm\sqrt{a}$."

(6.3) Any quadratic equation of the form $ax^2 + bx + c = 0$ can be solved by the **quadratic formula** which is usually stated as

$$x = \frac{-b \pm \sqrt{b^2 - 4ac}}{2a}.$$

The *discriminant*, $b^2 - 4ac$, can be used to determine the nature of the roots of a quadratic equation as follows:

1. If $b^2 - 4ac < 0$, then the equation has no real solutions.
2. If $b^2 - 4ac = 0$, then the equation has two equal real solutions.
3. If $b^2 - 4ac > 0$, then the equation has two unequal real solutions.

If x_1 and x_2 are roots of a quadratic equation, then the following relationships exist:

$$x_1 + x_2 = -\frac{b}{a} \quad \text{and} \quad (x_1)(x_2) = \frac{c}{a}.$$

These *sum and product relationships* can be used to check potential solutions of quadratic equations.

(6.4) To review the strengths and weaknesses of the three basic methods for solving a quadratic equation (factoring, completing the square, the quadratic formula), go back over the examples in this section.

Keep the following suggestions in mind as you solve word problems:

1. Read the problem carefully.
2. Sketch any figure, diagram, or chart that might help organize and analyze the problem.
3. Choose a meaningful variable.
4. Look for a guideline that can be used to set up an equation.
5. Form an equation that translates the guideline from English to algebra.
6. Solve the equation and use the solutions to determine all facts requested in the problem.
7. Check all answers back into the original statement of the problem.

(6.5) The number line, along with *critical numbers* and *test numbers*, provides a good basis for solving **quadratic inequalities** for which the polynomial is factorable. This same basic approach can be used to solve inequalities, such as $\dfrac{3x + 1}{x - 4} > 0$, that indicate quotients.

(6.6) A number of the form $a + bi$, where a and b are real numbers and i is the imaginary unit defined by $i = \sqrt{-1}$, is a **complex number**.

Two complex numbers $a + bi$ and $c + di$ are said to be *equal* if and only if $a = c$ and $b = d$.

Addition and subtraction of complex numbers are described as follows:

$$(a + bi) + (c + di) = (a + c) + (b + d)i;$$
$$(a + bi) - (c + di) = (a - c) + (b - d)i.$$

A square root of any negative real number can be represented as the product of a real number and the imaginary unit i. That is,

$$\sqrt{-b} = i\sqrt{b}, \quad \text{where } b \text{ is a positive real number.}$$

The product of two complex numbers is defined to conform with the product of two binomials.

The **conjugate** of $a + bi$ is $a - bi$. The product of a complex number and its conjugate is a real number. Therefore, conjugates are used to simplify expressions, such as $\dfrac{4 + 3i}{5 - 2i}$, that indicate the quotient of two complex numbers.

(6.7) To find complex solutions for quadratic equations we continue to use the techniques of factoring, completing the square, the quadratic formula, and applying the property "if $x^2 = a$, then $x = \pm\sqrt{a}$."

If the discriminant, $b^2 - 4ac$, is less than zero, then the equation has two nonreal complex solutions.

Chapter 6 Review Problem Set

For Problems 1–14, solve each of the equations.

1. $x^2 - 17x = 0$

2. $(x - 2)^2 = 36$

3. $x^2 - 4x - 21 = 0$

4. $x^2 + 2x - 9 = 0$

5. $4\sqrt{x} = x - 5$

6. $3n^2 + 10n - 8 = 0$

7. $n^2 - 10n = 200$

8. $3a^2 + a - 5 = 0$

9. $2a^2 + 4a - 5 = 0$

10. $t(t + 5) = 36$

11. $\dfrac{3}{x} + \dfrac{2}{x + 3} = 1$

12. $(x - 4)(x - 2) = 80$

13. $2x^4 - 23x^2 + 56 = 0$

14. $\dfrac{3}{n - 2} = \dfrac{n + 5}{4}$

For Problems 15–18, find the discriminant of each equation and determine whether the equation has (1) no real solutions, (2) two equal real solutions, or (3) two unequal real solutions.

15. $4x^2 - 20x + 25 = 0$ **16.** $2x^2 + 3x + 7 = 0$

17. $3x^2 - x - 5 = 0$ **18.** $5x^2 - 2x = 4$

For Problems 19–22, solve each inequality and indicate the solution set on a number line graph.

19. $x^2 + 3x - 10 > 0$ **20.** $2x^2 + x - 21 \leq 0$

21. $\dfrac{x - 4}{x + 6} \geq 0$ **22.** $\dfrac{2x - 1}{x + 1} > 4$

Set up an equation and solve each of the following problems.

23. Find two numbers such that their sum is 6 and their product is 2.

24. Sherry bought a number of shares of stock for $250. Six months later the value of the stock increased by $5 per share and she sold all but 5 shares and regained her original investment plus a profit of $50. How many shares did she sell and at what price per share?

25. Dave traveled 270 miles in one hour more time than it took Sandy to travel 260 miles. Sandy drove 7 miles per hour faster than Dave. How fast did each one travel?

26. The area of a square is numerically equal to twice its perimeter. Find the length of a side of the square.

27. Find two consecutive even whole numbers such that the sum of their squares is 164.

28. The perimeter of a rectangle is 38 inches and its area is 84 square inches. Find the length and width of the rectangle.

29. It takes Billy 2 hours longer to do a certain job than it takes Janet. They worked together for 2 hours; then Janet left and Billy finished the job in 1 hour. How long would it take each of them to do the job alone?

30. A company has a rectangular parking lot 40 meters wide and 60 meters long. They plan to increase the area of the lot by 1100 square meters by adding a strip of equal width to one side and one end. Find the width of the strip to be added.

Problems 31–41 pertain to Sections 6.6 and 6.7. For Problems 31–36, perform the indicated operations and express the answers in the standard form of a complex number.

31. $(-7 + 3i) + (9 - 5i)$ **32.** $(4 - 10i) - (7 - 9i)$

33. $5i(3 - 6i)$ **34.** $(5 - 7i)(6 + 8i)$

35. $(-2 - 3i)(4 - 8i)$ **36.** $\dfrac{4 + 3i}{6 - 2i}$

For Problems 37–41, solve each of the equations.

37. $(2x - 1)^2 = -64$ **38.** $x^2 - 6x = -34$

39. $x^2 - x + 3 = 0$ **40.** $2x^2 - 5x + 6 = 0$

41. $x^2 + 4x + 9 = 0$

7 Coordinate Geometry and Graphing Techniques

René Descartes, a French mathematician of the seventeenth century, was able to transform geometric problems into an algebraic setting and then use the tools of algebra to solve the problems. This merging of algebraic and geometric ideas is the foundation of a branch of mathematics called **analytic geometry** or more commonly **coordinate geometry**. Basically, there are two kinds of problems in coordinate geometry:

1. Given an algebraic equation, find its geometric graph.
2. Given a set of conditions pertaining to a geometric graph, find its algebraic equation.

We will discuss problems of both types in this chapter.

7.1 The Rectangular Coordinate System

Consider two number lines, one vertical and one horizontal, perpendicular to each other at the point associated with zero on both lines (Figure 7.1). These number lines are referred to individually as the **horizontal** and **vertical axes** or together as the **coordinate axes**. They partition the plane into four regions called **quadrants**. The quadrants are numbered counterclockwise from I through IV as indicated in Figure 7.1. The point of intersection of the two axes is called the **origin**.

It is now possible to set up a one-to-one correspondence between **ordered pairs** of real numbers and the points in a plane. To each ordered pair of real numbers there

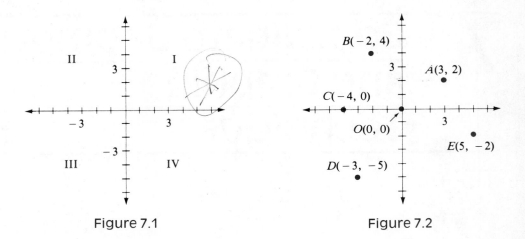

Figure 7.1   Figure 7.2

corresponds a unique point in the plane, and to each point in the plane there corresponds a unique ordered pair of real numbers. We have illustrated a part of this correspondence in Figure 7.2. The ordered pair (3, 2) means that the point A is located 3 units to the right and two units up from the origin. [The ordered pair (0, 0) is associated with the origin 0.] The ordered pair $(-3, -5)$ means that the point D is located three units to the left and five units down from the origin.

In general, the real numbers a and b in the ordered pair (a, b) are associated with a point; they are referred to as the **coordinates of the point**. The first number, a, called the **abscissa**, is the directed distance of the point from the vertical axis measured parallel to the horizontal axis. The second number, b, called the **ordinate**, is the directed distance of the point from the horizontal axis measured parallel to the vertical axis [Figure 7.3(a)]. Thus, in the first quadrant all points have a positive abscissa and a positive ordinate. In the second quadrant all points have a negative abscissa and a positive ordinate. We have indicated the sign situations for all four quadrants in Figure 7.3(b). This system of associating points in a plane with pairs of real numbers is called the **rectangular coordinate system** or the **Cartesian coordinate system**.

For most problems in coordinate geometry it is customary to label the horizontal axis the **x-axis** and the vertical axis the **y-axis**. Then, ordered pairs representing points in the xy-plane are of the form (x, y); that is, x is the first coordinate, and y is the second coordinate.

Figure 7.3

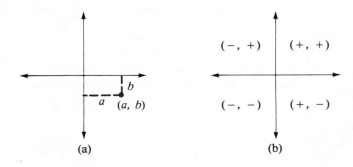

(a)   (b)

Distance

As we work with the rectangular coordinate system, it is sometimes necessary to express the length of certain line segments. In other words, we need to be able to find the *distance between* two points. Let's first consider two specific examples and then develop the general distance formula.

Example 1   Find the distance between the points $A(2, 2)$ and $B(5, 2)$, and between the points $C(-2, 5)$ and $D(-2, -4)$.

Solution   Let's plot the points and draw $\overline{AB}$ and $\overline{CD}$ as in Figure 7.4. Since $\overline{AB}$ is parallel to the x-axis, its length can be expressed as $|5 - 2|$ or $|2 - 5|$. (The absolute value symbol is used to ensure a nonnegative value.) Thus, the length of $\overline{AB}$ is 3 units. Likewise, since $\overline{CD}$ is parallel to the y-axis, the length of $\overline{CD}$ is $|5 - (-4)| = |-4 - 5| = 9$ units.

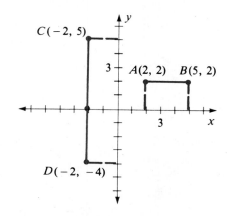

Figure 7.4   ●

Example 2   Find the distance between the points $A(2, 3)$ and $B(5, 7)$.

Solution   Let's plot the points and form a right triangle as indicated in Figure 7.5. Notice that the coordinates of point D are $(5, 3)$. Because $\overline{AD}$ is parallel to the horizontal axis, its length is easily determined to be 3 units. Likewise, $\overline{DB}$ is parallel to the vertical axis and its length is 4 units. Letting d represent the length of $\overline{AB}$ and applying the Pythagorean Theorem, we obtain

$$d^2 = 3^2 + 4^2$$
$$d^2 = 9 + 16$$
$$d^2 = 25$$
$$d = \pm\sqrt{25} = \pm 5.$$

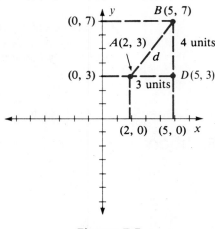

Figure 7.5

Since *distance between* is a nonnegative value, the length of $\overline{AB}$ is 5 units.   ●

The approach in Example 2 can be used to develop a general distance formula for finding the distance between any two points in a coordinate plane. The development proceeds as follows:

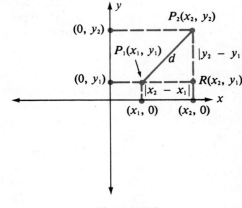

1. Let $P_1(x_1, y_1)$ and $P_2(x_2, y_2)$ represent any two points in a coordinate plane.

2. Form a right triangle as indicated in Figure 7.6. The coordinates of the vertex of the right angle, point R, are (x_2, y_1).

Figure 7.6

The length of $\overline{P_1 R}$ is $|x_2 - x_1|$, and the length of $\overline{RP_2}$ is $|y_2 - y_1|$. (The absolute value symbol is used to ensure a nonnegative value.) Letting d represent the length of $\overline{P_1 P_2}$ and applying the Pythagorean Theorem, yields

$$d^2 = |x_2 - x_1|^2 + |y_2 - y_1|^2.$$

Since $|a|^2 = a^2$, the distance formula can be stated as follows:

Distance Formula

$$d = \sqrt{(x_2 - x_1)^2 + (y_2 - y_1)^2}.$$

It makes no difference which point you call P_1 or P_2 when using the distance formula. Remember, if you forget the formula, don't panic, but merely form a right triangle and apply the Pythagorean Theorem as we did in Example 2.

Let's consider a few examples illustrating the use of the distance formula.

Example 3   Find the distance between $(-1, 4)$ and $(1, 2)$.

Solution   Let $(-1, 4)$ be P_1 and $(1, 2)$ be P_2. Using the distance formula, we obtain

$$d = \sqrt{[1 - (-1)]^2 + (2 - 4)^2}$$
$$= \sqrt{2^2 + (-2)^2}$$
$$= \sqrt{4 + 4}$$
$$= \sqrt{8} = 2\sqrt{2}. \qquad \text{(Express the answer in simplest radical form.)}$$

The distance between the two points is $2\sqrt{2}$ units.

In Example 3, notice the simplicity of the approach when the distance formula is used. No diagram is needed; we merely *plug in* the values and do the computation.

Example 4   Verify that the points $(-3, 6)$, $(3, 4)$, and $(1, -2)$ are vertices of an isosceles triangle. (An isosceles triangle has two sides of the same length.)

Solution   Let's plot the points and draw the triangle (Figure 7.7). Using the distance formula, we can find the lengths d_1, d_2, and d_3 as follows:

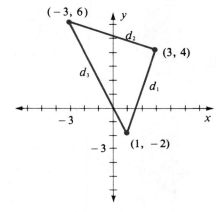

$$d_1 = \sqrt{[3 - 1]^2 + [4 - (-2)]^2}$$
$$= \sqrt{2^2 + 6^2} = \sqrt{40} = 2\sqrt{10};$$
$$d_2 = \sqrt{[-3 - 3]^2 + [6 - 4]^2}$$
$$= \sqrt{(-6)^2 + 2^2} = \sqrt{40} = 2\sqrt{10};$$
$$d_3 = \sqrt{[-3 - 1]^2 + [6 - (-2)]^2}$$
$$= \sqrt{(-4)^2 + 8^2} = \sqrt{80} = 4\sqrt{5}.$$

Since $d_1 = d_2$, we know that the triangle is an isosceles triangle.

Figure 7.7   ●

Slope of a Line

In coordinate geometry, the concept of **slope** is used to discuss the steepness of lines. The slope of a line is the ratio of the vertical change compared to the horizontal change as we move from one point on a line to another point. This is illustrated in Figure 7.8 using points P_1 and P_2. A precise definition for slope can be given by considering the coordinates of the points P_1, P_2, and R as indicated in Figure 7.9. The horizontal

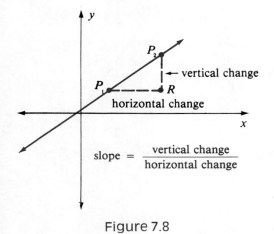

Figure 7.8

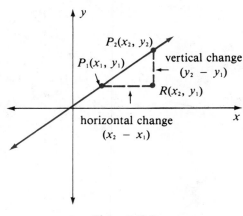

Figure 7.9

change as we move from P_1 to P_2 is $x_2 - x_1$, and the vertical change is $y_2 - y_1$. Thus, the following definition for slope is given.

Definition 7.1	If points P_1 and P_2 with coordinates (x_1, y_1) and (x_2, y_2), respectively, are any two different points on a line, then the **slope** of the line (denoted by m) is $$m = \frac{y_2 - y_1}{x_2 - x_1}, \qquad x_2 \neq x_1.$$

Since $\dfrac{y_2 - y_1}{x_2 - x_1} = \dfrac{y_1 - y_2}{x_1 - x_2}$, how we designate P_1 and P_2 is not important. Let's use Definition 7.1 to find the slopes of some lines.

Example 5   Find the slope of the line determined by each of the following pairs of points, and graph the lines.

    (a) $(-1, 1)$ and $(3, 2)$
    (b) $(4, -2)$ and $(-1, 5)$
    (c) $(2, -3)$ and $(-3, -3)$

Solutions   **(a)** Let $(-1, 1)$ be P_1, and let $(3, 2)$ be P_2 (Figure 7.10).

$$m = \frac{y_2 - y_1}{x_2 - x_1} = \frac{2 - 1}{3 - (-1)} = \frac{1}{4}.$$

    (b) Let $(4, -2)$ be P_1, and let $(-1, 5)$ be P_2 (Figure 7.11).

$$m = \frac{y_2 - y_1}{x_2 - x_1} = \frac{5 - (-2)}{-1 - 4} = \frac{7}{-5} = -\frac{7}{5}.$$

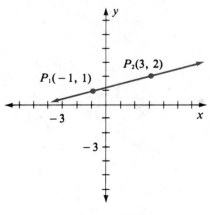

Figure 7.10

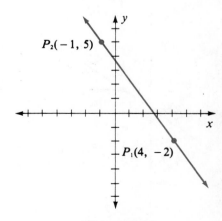

Figure 7.11

(c) Let $(2, -3)$ be P_1, and let $(-3, -3)$ be P_2 (Figure 7.12).

$$m = \frac{y_2 - y_1}{x_2 - x_1} = \frac{-3 - (-3)}{-3 - 2} = \frac{0}{-5} = 0.$$

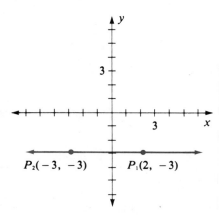

Figure 7.12

The three parts of Example 5 illustrate the three basic possibilities for slope; that is, the slope of a line can be positive, negative, or zero. A line having a positive slope, as in Figure 7.10, rises as we move from left to right. A line having a negative slope, as in Figure 7.11, falls as we move from left to right. A horizontal line, as in Figure 7.12, has a slope of zero. Finally, we need to realize that *the concept of slope is undefined for vertical lines.* This is due to the fact that for any vertical line the horizontal change as we move from one point on the line to another is zero. Thus, the ratio $\dfrac{y_2 - y_1}{x_2 - x_1}$ will have a denominator of zero and will therefore be undefined. So in Definition 7.1 the restriction $x_2 \neq x_1$ is made.

One final idea pertaining to the concept of slope needs to be emphasized. The slope of a line is a *ratio*, the ratio of vertical change compared to horizontal change. A slope of $\dfrac{2}{3}$ means that for every 2 units of vertical change there must be a corresponding 3 units of horizontal change. Thus, starting at some point on a line having a slope of $\dfrac{2}{3}$, we could locate other points on the line as follows:

$$\frac{2}{3} = \frac{4}{6} \qquad \text{By moving 4 units } up \text{ and 6 units to the } right$$

$$\frac{2}{3} = \frac{8}{12} \qquad \text{By moving 8 units } up \text{ and 12 units to the } right$$

$$\frac{2}{3} = \frac{-2}{-3} \qquad \text{By moving 2 units } down \text{ and 3 units to the } left$$

Likewise, if a line has a slope of $-\frac{3}{4}$, then by starting at some point on the line we could locate other points on the line as follows:

$$-\frac{3}{4} = \frac{-3}{4}$$ By moving 3 units *down* and 4 units to the *right*

$$-\frac{3}{4} = \frac{3}{-4}$$ By moving 3 units *up* and 4 units to the *left*

$$-\frac{3}{4} = \frac{-9}{12}$$ By moving 9 units *down* and 12 units to the *right*

$$-\frac{3}{4} = \frac{15}{-20}$$ By moving 15 units *up* and 20 units to the *left*

Problem Set 7.1

For Problems 1–10, find the distance between each of the pairs of points. Express answers in simplest radical form.

1. $(2, 1), (10, 7)$ **2.** $(-2, -1), (7, 11)$ **3.** $(-1, 3), (2, -2)$

4. $(1, -1), (3, -4)$ **5.** $(-5, 2), (-1, 6)$ **6.** $(6, -4), (9, -7)$

7. $(-2, -4), (4, 0)$ **8.** $(-3, 3), (0, -3)$ **9.** $(-2, 3), (-2, -7)$

10. $(1, -6), (-5, -6)$

11. Verify that the points $(0, 3)$, $(2, -3)$, and $(-4, -5)$ are vertices of an isosceles triangle.

12. Verify that the points $(-3, 1)$, $(5, 7)$, and $(8, 3)$ are vertices of a right triangle. (Hint: If $a^2 + b^2 = c^2$, then it is a right triangle with the right angle opposite side c.)

13. Verify that $(3, 1)$ is the midpoint of the line segment joining $(-2, 6)$ and $(8, -4)$.

14. Verify that the points $(7, 12)$ and $(11, 18)$ divide the line segment joining $(3, 6)$ and $(15, 24)$ into three segments of equal length.

For Problems 15–25, find the slope of the line determined by each pair of points.

15. $(3, 1), (7, 4)$ **16.** $(-4, 1), (-1, 3)$ **17.** $(-3, -4), (3, 0)$

18. $(5, -4), (13, 2)$ **19.** $(-2, 3), (-7, 4)$ **20.** $(2, 1), (3, -6)$

21. $(-3, 6), (7, 6)$ **22.** $(-2, -5), (4, -5)$ **23.** $(a, 0), (0, b)$

24. $(a, b), (c, d)$ **25.** $(-2, 3), (-2, -7)$

26. Find x if the line through $(-2, 4)$ and $(x, 6)$ has a slope of $\frac{2}{9}$.

27. Find y if the line through $(1, y)$ and $(4, 2)$ has a slope of $\frac{5}{3}$.

28. Find x if the line through $(x, 4)$ and $(2, -5)$ has a slope of $-\frac{9}{4}$.

29. Find y if the line through $(5, 2)$ and $(-3, y)$ has a slope of $-\frac{7}{8}$.

For Problems 30–37, you are given one point on a line and the slope of the line. Find the coordinates of three other points on the line.

30. $(3, 2)$, $m = \dfrac{1}{3}$ **31.** $(4, 7)$, $m = \dfrac{3}{7}$ **32.** $(-2, 5)$, $m = 4$

33. $(-3, -4)$, $m = 1$ **34.** $(4, -6)$, $m = -\dfrac{1}{4}$ **35.** $(3, -2)$, $m = -\dfrac{5}{6}$

36. $(2, 6)$, $m = -\dfrac{5}{3}$ **37.** $(-4, 1)$, $m = -2$

For Problems 38–45, find the coordinates of two points on the given line and then use those coordinates to find the slope of the line.

38. $3x + 2y = 6$ **39.** $x - 2y = 4$ **40.** $3x - y = 6$

41. $5x - 2y = 10$ **42.** $2x - 3y = 7$ **43.** $3x + 5y = 9$

44. $-4x + 5y = 13$ **45.** $-2x + y = -7$

46. The concept of slope is used for highway construction. The "grade" of a highway expressed as a percent means the number of feet that the highway changes in elevation for each 100 feet of horizontal change.

    (a) A certain highway has a 2% grade. How many feet does it rise in a horizontal distance of 1 mile? (1 mile = 5280 feet)

    (b) The grade of a highway up a hill is 30%. How much change in horizontal distance is there if the vertical height of the hill is 75 feet?

47. Slope is often expressed as the ratio of "rise to run" in construction of steps.

    (a) If the ratio of rise to run is to be $\dfrac{3}{5}$ for some steps and the rise is 19 centimeters, find the measure of the run to the nearest centimeter.

    (b) If the ratio of rise to run is to be $\dfrac{2}{3}$ for some steps and the run is 28 centimeters, find the rise to the nearest centimeter.

48. Suppose that a county ordinance requires a $2\dfrac{1}{4}\%$ "fall" for a sewage pipe from the house to the main pipe at the street. How much vertical drop must there be for a horizontal distance of 45 feet? Express the answer to the nearest tenth of a foot.

7.2 Graphing Techniques—Linear Equations and Inequalities

As you continue to study mathematics, you will find that the ability to "quickly" sketch the graph of an equation is an important skill. Thus, throughout pre-calculus and calculus courses, various curve-sketching techniques are discussed. A good portion of this chapter will be devoted to expanding your repertoire of graphing techniques.

First, let's briefly review some basic ideas by considering the solutions for the equation $y = x + 2$. A **solution** of an equation in two variables is an ordered pair of real numbers that satisfy the equation. When the variables x and y are used, the ordered pairs are of the form (x, y). We see that $(1, 3)$ is a solution for $y = x + 2$, because if x is replaced by 1 and y by 3, the true numerical statement $3 = 1 + 2$ is obtained. Likewise, $(-2, 0)$ is a solution because $0 = -2 + 2$ is a true statement. An infinite number of pairs of real numbers that satisfy $y = x + 2$ can be found by arbitrarily choosing values for x and then determining corresponding values for y. Let's use a table to record some of the solutions for $y = x + 2$.

Choose x	Determine y from $y = x + 2$	Solutions for $y = x + 2$
0	2	$(0, 2)$
1	3	$(1, 3)$
3	5	$(3, 5)$
5	7	$(5, 7)$
-2	0	$(-2, 0)$
-4	-2	$(-4, -2)$
-6	-4	$(-6, -4)$

Plotting the points associated with the ordered pairs from the table above produces Figure 7.13(a). The straight line containing the points [Figure 7.13(b)] is called the **graph of the equation $y = x + 2$**.

Figure 7.13

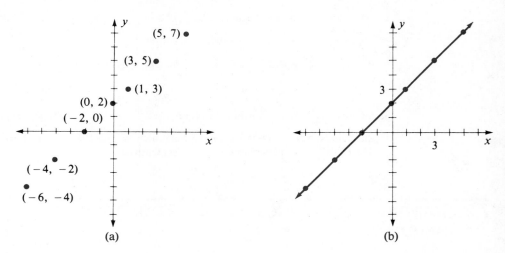

(a)   (b)

Graphing Linear Equations

Probably the most valuable graphing technique is the ability to recognize the kind of graph that a certain type of equation produces. For example, from previous mathematics courses you probably remember that any equation of the form $Ax + By = C$, where A, B, and C are constants (A and B not both zero) and x and y are variables, is a *linear equation* and *its graph is a straight line*. Two comments about this description of a linear equation should be made. First, the choice of x and y as variables is arbitrary; any two letters could be used to represent the variables. For example, an equation such as $3r + 2s = 9$ is also a linear equation in two variables. In order to avoid constantly changing the labeling of the coordinate axes when graphing equations, we will use the same two variables, x and y, in all equations. Second, the statement "any equation of the form $Ax + By = C$" technically means any equation of the form $Ax + By = C$ or *equivalent* to that form. For example, the equation $y = 2x - 1$ is equivalent to $-2x + y = -1$ and therefore is linear and produces a straight line graph.

Before graphing some linear equations, let's define in general the *intercepts* of a graph.

> The x-coordinates of the points that a graph has in common with the x-axis are called the **x-intercepts** of the graph. (To compute the x-intercepts, let $y = 0$ and solve for x.)

> The y-coordinates of the points that a graph has in common with the y-axis are called the **y-intercepts** of the graph. (To compute the y-intercepts, let $x = 0$ and solve for y.)

Knowing that any equation of the form $Ax + By = C$ produces a straight line graph and that two points determine a straight line makes graphing linear equations a simple process. We can find two points and draw the line determined by them. The two points involving the intercepts are usually easy to find, and it's generally a good idea to plot a third point to serve as a check.

Example 1

Graph $3x - 2y = 6$.

Solution

First, let's find the intercepts. If $x = 0$, then

$$3(0) - 2y = 6$$
$$-2y = 6$$
$$y = -3.$$

Therefore, the point $(0, -3)$ is on the line.
If $y = 0$, then

$$3x - 2(0) = 6$$
$$3x = 6$$
$$x = 2.$$

Thus, the point $(2, 0)$ is also on the line.
Now let's find a check point. If $x = -2$, then

$$3(-2) - 2y = 6$$
$$-6 - 2y = 6$$
$$-2y = 12$$
$$y = -6.$$

So the point $(-2, -6)$ is also on the line.

In Figure 7.14, the three points are plotted and the graph of $3x - 2y = 6$ is drawn.

Figure 7.14

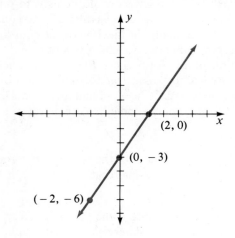

Notice in Example 1 that we did not solve the given equation for y in terms of x or for x in terms of y. Since we know it is a straight line, there is no need for an extensive table of values; thus, there is no need to change the form of the original equation. Furthermore, the point $(-2, -6)$ served as a check point. If it had not been on the line determined by the two intercepts, then we would have known that an error had been made in finding the intercepts.

Example 2   Graph $y = -2x$.

Solution   If $x = 0$, then $y = -2(0) = 0$; so the origin $(0, 0)$ is on the line. Since both intercepts are determined by the point $(0, 0)$, another point is necessary to determine the line. Then a third point should be found as a check point.

The graph of $y = -2x$ is shown in Figure 7.15.

x	y
0	0
1	-2
-1	2

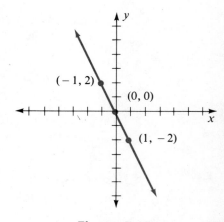

Figure 7.15

Example 2 illustrates the general concept that for the form $Ax + By = C$, if $C = 0$, then the line contains the origin. Stated another way, the graph of any equation of the form $y = kx$, where k is any real number, is a straight line containing the origin.

Example 3

Graph $x = 2$.

Solution

Because we are considering linear equations *in two variables*, the equation $x = 2$ is equivalent to $x + 0(y) = 2$. Any value of y can be used, but the x-value must always be 2. Therefore, some of the solutions are $(2, 0), (2, 1), (2, 2), (2, -1)$, and $(2, -2)$. The graph of $x = 2$ is the vertical line shown in Figure 7.16.

Figure 7.16

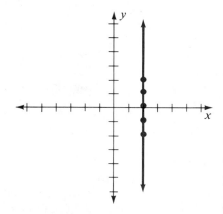

In general, the graph of any equation of the form $Ax + By = C$, where $A = 0$ or $B = 0$ (but not both), is a line parallel to one of the axes. More specifically, any equation of the form $x = a$, where a is any real number, is a line parallel to the y-axis (a vertical line) having an x-intercept of a. Any equation of the form $y = b$, where b is a real number, is a line parallel to the x-axis (a horizontal line) having a y-intercept of b.

Graphing Linear Inequalities

Linear inequalities in two variables are of the form $Ax + By > C$ or $Ax + By < C$, where A, B, and C are real numbers. (Combined linear equality and inequality statements are of the form $Ax + By \geq C$ or $Ax + By \leq C$.) Graphing linear inequalities is almost as easy as graphing linear equations. The following discussion leads into a simple, step-by-step process.

Let's consider the following equation and related inequalities.

$$x + y = 2, \qquad x + y > 2, \qquad x + y < 2.$$

The straight line in Figure 7.17 is the graph of $x + y = 2$. The line divides the plane into two half-planes, one above the line and one below the line. For each point in the half-plane *above* the line, the ordered pair associated with the point satisfies the inequality $x + y > 2$. For example, the ordered pair $(3, 4)$ produces the true statement $3 + 4 > 2$. Likewise, for each point in the half-plane *below* the line, the ordered pair associated

Figure 7.17

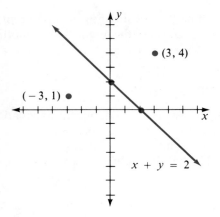

with the point satisfies the inequality $x + y < 2$. For example, $(-3, 1)$ produces the true statement $-3 + 1 < 2$.

Now let's use the ideas from the previous discussion to graph some inequalities.

Example 4   Graph $x - 2y > 4$.

Solution   First, graph $x - 2y = 4$ as a dashed line since equality is not included in the expression $x - 2y > 4$ (Figure 7.18).

Second, since *all* of the points in a specific half-plane satisfy either $x - 2y > 4$ or $x - 2y < 4$, let's try a **test point**. For example, try the origin. Then $x - 2y > 4$ becomes $0 - 2(0) > 4$, which is a false statement. Because the ordered pairs in the half-plane containing the origin do not satisfy $x - 2y > 4$, the ordered pairs in the other half-plane must satisfy it. Therefore, the graph of $x - 2y > 4$ is the half-plane below the line, as indicated by the shaded portion in Figure 7.19.

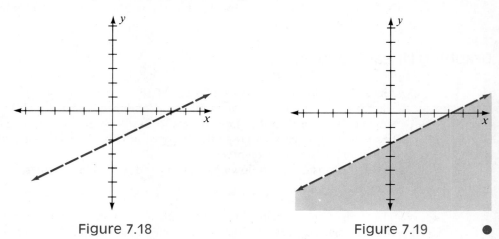

Figure 7.18   Figure 7.19   ●

The following steps are suggested for graphing a linear inequality.

1. Graph the corresponding equality. Use a solid line if equality is included in the original statement and a dashed line if equality is not included.

2. Choose a test point not on the line, and substitute its coordinates into the inequality. (The origin is a convenient point if it is not on the line.)

3. The graph of the original inequality is

   (a) the half-plane containing the test point if the inequality is satisfied by that point; or

   (b) the half-plane not containing the test point if the inequality is not satisfied by the point.

Example 5   Graph $2x + 3y \geq -6$.

Solution

1. Graph $2x + 3y = -6$ as a solid line (Figure 7.20).

2. Choose the origin as a test point. Then $2x + 3y \geq -6$ becomes $2(0) + 3(0) \geq -6$, which is true.

3. Since the test point satisfies the given inequality, all points in the same half-plane as the test point satisfy it. Thus, the graph of $2x + 3y \geq -6$ is the line and the half-plane above the line (Figure 7.20).

Figure 7.20

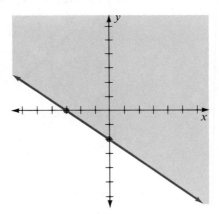

Problem Set 7.2

Graph each of the following linear equations.

1. $x - 2y = 4$	**2.** $2x + y = -4$	**3.** $3x + 2y = 6$
4. $2x - 3y = 6$	**5.** $4x = 5y + 20$	**6.** $4y = 20 - 5x$
7. $x - y = 3$	**8.** $-x + y = 4$	**9.** $y = 3x - 1$
10. $y = -2x + 3$	**11.** $y = -x$	**12.** $y = 4x$
13. $x = 0$	**14.** $y = -1$	**15.** $y = \dfrac{2}{3}x$

16. $y = -\dfrac{1}{2}x$

Graph each of the following linear inequalities.

17. $x + 2y > 4$
18. $2x - y < -4$
19. $3x - 2y < 6$
20. $2x + 3y < 6$
21. $2x + 5y \leq 10$
22. $4x + 5y \leq 20$
23. $y > -x - 1$
24. $y < 3x - 2$
25. $y \leq -x$
26. $y \geq x$
27. $x + 2y < 0$
28. $3x - y > 0$
29. $x > -1$
30. $y < 3$

Miscellaneous Problems

From our work with absolute value we know that $|x + y| = 4$ is equivalent to $x + y = 4$ or $x + y = -4$. Therefore, the graph of $|x + y| = 4$ is the two lines $x + y = 4$ and $x + y = -4$. Graph each of the following.

31. $|x - y| = 2$
32. $|2x + y| = 1$
33. $|x - 2y| \leq 4$
34. $|3x - 2y| \geq 6$
35. $|2x + 3y| > 6$
36. $|5x + 2y| < 10$

By the definition of absolute value, the equation $y = |x| + 2$ becomes $y = x + 2$ for $x \geq 0$ and $y = -x + 2$ for $x < 0$. Therefore, the graph of $y = |x| + 2$ is as follows.

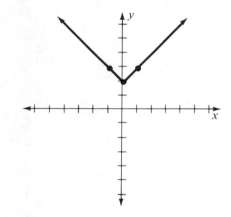

Graph each of the following.

37. $y = |x| - 1$
38. $y = |x - 2|$
39. $|y| = x$
40. $|y| = |x|$
41. $y = 2|x|$
42. $|x| + |y| = 4$

7.3 Determining the Equation of a Line

As we stated earlier, there are basically two types of problems in coordinate geometry:

1. Given an algebraic equation, find its geometric graph.
2. Given a set of conditions pertaining to a geometric figure, find its algebraic equation.

Problems of type **1** were our primary concern in the previous section. At this time we want to introduce some problems of type **2** that specifically deal with straight lines. In other words, given certain facts about a line, we need to be able to determine its algebraic equation. Let's consider some examples.

Example 1

Find the equation of the line having a slope of $\frac{2}{3}$ and containing the point (1, 2).

Solution

First, let's draw the line and record the given information. Then we will choose a point (x, y) that represents any point on the line other than the given point (1, 2). (See Figure 7.21.)

The slope determined by (1, 2) and (x, y) is $\frac{2}{3}$. Thus,

$$\frac{y - 2}{x - 1} = \frac{2}{3}$$
$$2(x - 1) = 3(y - 2)$$
$$2x - 2 = 3y - 6$$
$$2x - 3y = -4.$$

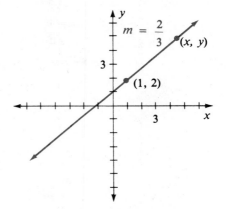

Figure 7.21

Example 2

Find the equation of the line containing (3, 2) and $(-2, 5)$.

Solution

First, let's draw the line determined by the given points (Figure 7.22). Since two points are known, the slope can be found:

$$m = \frac{y_2 - y_1}{x_2 - x_1} = \frac{3}{-5} = -\frac{3}{5}.$$

Now we can use the same approach as in Example 1. We form an equation using a variable point (x, y), one of the two given points, and the slope of $-\frac{3}{5}$.

$$\frac{y - 5}{x + 2} = \frac{3}{-5} \qquad \left(-\frac{3}{5} = \frac{3}{-5}\right)$$
$$3(x + 2) = -5(y - 5)$$
$$3x + 6 = -5y + 25$$
$$3x + 5y = 19.$$

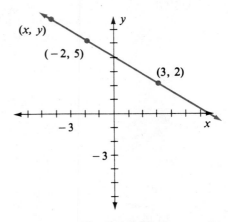

Figure 7.22

Example 3   Find the equation of the line that has a slope of $\frac{1}{4}$ and a y-intercept of 2.

Solution   A y-intercept of 2 means that the point $(0, 2)$ is on the line (Figure 7.23). Choosing a variable point (x, y), we can proceed as in the previous examples.

$$\frac{y - 2}{x - 0} = \frac{1}{4}$$

$$1(x - 0) = 4(y - 2)$$

$$x = 4y - 8$$

$$x - 4y = -8.$$

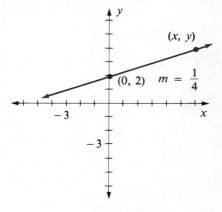

Figure 7.23   ●

Perhaps it would be helpful for you to pause a moment and look back over Examples 1, 2, and 3. Notice that the same basic approach is used in all three situations—that is, choosing a variable point (x, y) and using it to determine the equation that satisfies the conditions given in the problem. We should also recognize that the approach taken in the previous examples can be generalized to produce some special forms of equations of straight lines.

Point-Slope Form

Example 4   Find the equation of the line having a slope of m and containing the point (x_1, y_1).

Solution   If we choose (x, y) to represent any other point on the line (Figure 7.24), then the slope of the line is given by

$$m = \frac{y - y_1}{x - x_1}, \qquad x \neq x_1$$

from which we have

$$y - y_1 = m(x - x_1).$$

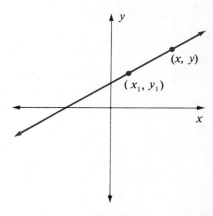

Figure 7.24   ●

The equation

$$y - y_1 = m(x - x_1)$$

is referred to as the **point-slope form** of the equation of a straight line. Instead of using the approach used in Example 1, we can use the point-slope form to write the equation of a line with a given slope and containing a given point. For example, the equation of the line having a slope of $\frac{3}{5}$ and containing the point (2, 4) can be determined as follows:

Substituting (2, 4) for (x_1, y_1) and $\frac{3}{5}$ for m in

$$y - y_1 = m(x - x_1),$$

we have

$$y - 4 = \frac{3}{5}(x - 2)$$

$$5(y - 4) = 3(x - 2)$$
$$5y - 20 = 3x - 6$$
$$-14 = 3x - 5y.$$

Slope-Intercept Form

Now consider the equation of the line having a slope of m and a y-intercept of b (Figure 7.25). A y-intercept of b means that the line contains the point (0, b); therefore, we can use the point-slope form as follows:

$$y - y_1 = m(x - x_1)$$
$$y - b = m(x - 0)$$
$$y - b = mx$$
$$y = mx + b.$$

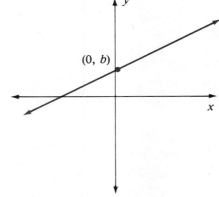

Figure 7.25

The equation

$$y = mx + b$$

is referred to as the **slope-intercept form** of the equation of a straight line. It can be used for two primary purposes, as the next two examples illustrate.

Example 5   Find the equation of the line that has a slope of $\frac{1}{4}$ and a y-intercept of 2.

Solution   This is a restatement of Example 3, but this time we will use the slope-intercept form ($y = mx + b$) of a line to write its equation. Since $m = \frac{1}{4}$ and $b = 2$, these values can be substituted into $y = mx + b$.

$$y = mx + b$$

$$y = \frac{1}{4}x + 2$$

$4y = x + 8$ Multiply both sides by 4.

$x - 4y = -8$ (Same result as in Example 3.)   ●

Example 6   Find the slope of the line when the equation is $3x + 2y = 6$.

Solution   We can solve the equation for y in terms of x and then compare the resulting equation to the slope-intercept form to determine the slope. Thus,

$$3x + 2y = 6$$

$$2y = -3x + 6$$

$$y = -\frac{3}{2}x + 3$$

$$y = -\frac{3}{2}x + 3. \qquad y = mx + b.$$

The slope of the line is $-\frac{3}{2}$. Furthermore, the y-intercept is 3.   ●

In general, if the equation of a nonvertical line is written in slope-intercept form ($y = mx + b$), *the coefficient of x is the slope of the line and the constant term is the y-intercept.* (Remember that the concept of slope is not defined for a vertical line.)

Parallel and Perpendicular Lines

Two important relationships between lines and their slopes can be used to solve certain kinds of problems. It can be shown that nonvertical parallel lines have the same slope, and that two nonvertical lines are perpendicular if the product of their slopes is -1.

(Details for verifying these facts are left to another course.) In other words, if two lines have slopes m_1 and m_2, respectively, then

1. the two lines are parallel if and only if $m_1 = m_2$;
2. the two lines are perpendicular if and only if $(m_1)(m_2) = -1$.

The following examples illustrate the use of these properties.

Example 7   (a) Verify that the graphs of $2x + 3y = 7$ and $4x + 6y = 11$ are parallel lines.
          (b) Verify that the graphs of $8x - 12y = 3$ and $3x + 2y = 2$ are perpendicular lines.

Solutions   (a) Let's change each equation to slope-intercept form.

$$2x + 3y = 7 \longrightarrow 3y = -2x + 7$$
$$y = -\frac{2}{3}x + \frac{7}{3}.$$
$$4x + 6y = 11 \longrightarrow 6y = -4x + 11$$
$$y = -\frac{4}{6}x + \frac{11}{6}$$
$$y = -\frac{2}{3}x + \frac{11}{6}.$$

Both lines have a slope of $-\frac{2}{3}$ and different y-intercepts. Therefore, the two lines are parallel.

(b) Solving each equation for y in terms of x, we obtain

$$8x - 12y = 3 \longrightarrow -12y = -8x + 3$$
$$y = \frac{8}{12}x - \frac{3}{12}$$
$$y = \frac{2}{3}x - \frac{1}{4}.$$
$$3x + 2y = 2 \longrightarrow 2y = -3x + 2$$
$$y = -\frac{3}{2}x + 1.$$

Because $\left(\frac{2}{3}\right)\left(-\frac{3}{2}\right) = -1$ (the product of the two slopes is -1), the lines are perpendicular.   ●

Remark The statement "the product of two slopes is -1" is equivalent to saying that the two slopes are negative reciprocals of each other; that is, $m_1 = -\dfrac{1}{m_2}$.

Example 8   Find the equation of the line that contains the point $(1, 4)$ and is parallel to the line determined by $x + 2y = 5$.

Solution   First, let's draw a figure to help in our analysis of the problem (Figure 7.26). Since the line through $(1, 4)$ is to be parallel to the line determined by $x + 2y = 5$, it must have the same slope. So let's find the slope by changing $x + 2y = 5$ to the slope-intercept form.

$$x + 2y = 5$$
$$2y = -x + 5$$
$$y = -\frac{1}{2}x + \frac{5}{2}.$$

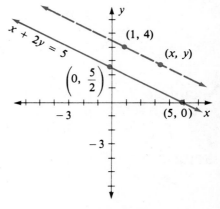

Figure 7.26

The slope of both lines is $-\frac{1}{2}$. Now we can choose a variable point (x, y) on the line through $(1, 4)$ and proceed as we did in earlier examples.

$$\frac{y - 4}{x - 1} = \frac{1}{-2}$$
$$1(x - 1) = -2(y - 4)$$
$$x - 1 = -2y + 8$$
$$x + 2y = 9.$$

Example 9   Find the equation of the line that contains the point $(-1, -2)$ and is perpendicular to the line determined by $2x - y = 6$.

Solution   First, let's draw a figure to help in our analysis of the problem (Figure 7.27). Because the line through $(-1, -2)$ is to be perpendicular to the line determined by $2x - y = 6$, its slope must be the negative reciprocal of the slope of $2x - y = 6$. So let's find the slope of $2x - y = 6$ by changing it to the slope-intercept form.

$$2x - y = 6$$
$$-y = -2x + 6$$
$$y = 2x - 6.$$

(The slope is 2.)

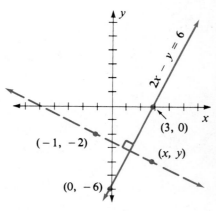

Figure 7.27

The slope of the desired line is $-\dfrac{1}{2}$ (the negative reciprocal of 2), and we can proceed as before using a variable point (x, y).

$$\frac{y + 2}{x + 1} = \frac{1}{-2}$$
$$1(x + 1) = -2(y + 2)$$
$$x + 1 = -2y - 4$$
$$x + 2y = -5.$$

Two forms of equations of straight lines are used extensively. They are referred to as the **standard form** and the **slope-intercept form** and can be described as follows:

Standard Form

$$Ax + By = C,$$

where B and C are integers and A is a nonnegative integer (A and B not both zero).

Slope-Intercept Form

$$y = mx + b,$$

where m is a real number representing the slope and b is a real number representing the y-intercept.

Problem Set 7.3

For Problems 1–8, write the equation of each line having the indicated slope and containing the indicated point. Express final equations in standard form.

1. $m = \dfrac{3}{4}, (2, 5)$ **2.** $m = \dfrac{2}{5}, (3, 7)$ **3.** $m = 2, (-1, 4)$

4. $m = -3, (-2, 4)$ **5.** $m = -\dfrac{3}{5}, (-2, -3)$ **6.** $m = -\dfrac{1}{4}, (4, -1)$

7. $m = 0, (3, -4)$ **8.** $m = \dfrac{5}{2}, (-2, -6)$

For Problems 9–18, write the equation of each line containing the indicated pair of points. Express final equations in standard form.

9. $(1, 2), (4, 4)$ **10.** $(3, 2), (10, 4)$ **11.** $(2, 3), (5, 9)$

12. $(-3, 1), (-1, 7)$ **13.** $(4, 2), (-1, 3)$ **14.** $(-4, 2), (2, -3)$

15. $(-3, 1), (-2, -1)$ **16.** $(-6, 4), (-4, -2)$ **17.** $(2, 7), (2, 5)$

18. $(3, -1), (4, -1)$

For Problems 19–26, write the equation of each line having the indicated slope (m) and y-intercept (b). Express final equations in standard form.

19. $m = \dfrac{1}{5}, b = 2$ 20. $m = \dfrac{4}{9}, b = 4$ 21. $m = 4, b = -1$

22. $m = 1, b = -3$ 23. $m = -\dfrac{2}{7}, b = 4$ 24. $m = -\dfrac{5}{3}, b = 2$

25. $m = -4, b = -2$ 26. $m = -3, b = -1$

For Problems 27–32, write the equation of each line satisfying the given conditions. Express final equations in standard form.

27. x-intercept of 3 and y-intercept of -2

28. y-intercept of 0 and slope of $-\dfrac{4}{5}$

29. contains the point $(-2, 4)$ and is parallel to the x-axis

30. contains the point $(3, -4)$ and is parallel to the y-axis

31. contains the point $(2, 7)$ and is perpendicular to the x-axis

32. contains the points $(-1, 4)$ and $(-1, -2)$

For Problems 33–42, determine whether each pair of lines is (a) parallel, (b) perpendicular, or (c) a pair of intersecting lines that are not perpendicular.

33. $y = \dfrac{3}{4}x + 7$ 34. $y = 3x - 2$

 $y = \dfrac{3}{4}x - 1$ $y = -\dfrac{1}{3}x + 5$

35. $3x + 2y = 5$ 36. $5x + 3y = 9$
 $2x - 3y = 11$ $10x + 6y = 1$

37. $4x + y = 9$ 38. $2x - 9y = 11$
 $3x - y = 5$ $2x + 9y = 5$

39. $5x - 3y = 10$ 40. $4x + 5y = 6$
 $15x - 9y = 4$ $5x - 4y = 3$

41. $x = 2$ 42. $y = 3x$
 $y = 4$ $y = -2x$

For Problems 43–50, write the equation of each line satisfying the given conditions. Express final equations in standard form.

43. containing $(2, 7)$ and parallel to $x + 3y = 6$

44. containing $(-1, 3)$ and parallel to $x - 2y = 4$

45. containing $(-2, -4)$ and parallel to $3x - 4y = 7$

46. containing the origin and parallel to $5x - 2y = 10$

47. containing $(1, 4)$ and perpendicular to $2x + 3y = -6$

48. containing $(2, -3)$ and perpendicular to $3x - 4y = -8$

49. containing the origin and perpendicular to $5x + 7y = 10$

50. containing $(3, 7)$ and perpendicular to $y = -3x$

For Problems 51–54, use the slope-intercept equation, $y = mx + b$, to help write the equation of each line having the indicated slope and containing the given point. Express final equations in slope-intercept form.

51. $m = \dfrac{3}{4}, (4, 1)$

52. $m = \dfrac{2}{3}, (6, -1)$

53. $m = \dfrac{2}{5}, (3, -4)$

54. $m = -\dfrac{3}{7}, (-5, -1)$

55. The slope-intercept form of a line can also be used for graphing purposes. Suppose that we want to graph the equation $y = \dfrac{1}{4}x + 2$. Since the y-intercept is 2, the point $(0, 2)$ is on the line. Furthermore, since the slope is $\dfrac{1}{4}$, another point can be located by moving 1 unit *up* and 4 units to the *right*. Thus, the point $(4, 3)$ is also on the line. The two points $(0, 2)$ and $(4, 3)$ determine the line.

Use the slope-intercept form to graph the following lines.

(a) $y = \dfrac{3}{4}x + 1$

(b) $y = \dfrac{2}{3}x - 2$

(c) $y = -\dfrac{2}{5}x - 1$

(d) $y = -\dfrac{1}{2}x + 3$

(e) $x + 2y = 5$

(f) $2x - y = 7$

(g) $-y = -4x + 7$

(h) $3x = 2y$

(i) $7y = -2x$

(j) $y = -3$

(k) $x = 2$

56. Some real-world situations can be described by the use of linear equations in two variables. If two pairs of values are known, then the equation can be determined by using the approach used in Example 2 of this section. For each of the following cases, assume that the relationship can be expressed as a linear equation in two variables, and use the given information to determine the equation. Express the equation in standard form.

(a) A company produces 10 fiberglass shower stalls for $2015 and 15 stalls for $3015. Let y be the cost and x the number of stalls.

(b) A company can produce 6 boxes of candy for $8 and 10 boxes of candy for $13. Let y represent the cost and x the number of boxes of candy.

(c) Two banks on opposite corners of a town square have signs displaying the current temperature. One bank displays the temperature in Celsius degrees and the other in degrees Fahrenheit. One day a temperature of $10°C$ was displayed at the same time as a temperature of $50°F$. On another day, a temperature of $-5°C$ was displayed at the same time as a temperature of $23°F$. Let y represent the temperature in Fahrenheit and x the temperature in Celsius.

Miscellaneous Problems

57. The equation of a line containing the two points (x_1, y_1) and (x_2, y_2) is $\dfrac{y - y_1}{x - x_1} = \dfrac{y_2 - y_1}{x_2 - x_1}$.

This form is often referred to as the *two-point form* of the equation of a straight line. Using the two-point form, write the equation of each of the lines containing the indicated pair of points. Express final equations in standard form.

(a) $(1, 1)$ and $(5, 2)$

(b) $(2, 4)$ and $(-2, -1)$

(c) $(-3, 5)$ and $(3, 1)$

(d) $(-5, 1)$ and $(2, -7)$

58. Let $Ax + By = C$ and $A'x + B'y = C'$ represent two lines. Change each of these equations to slope-intercept form, and then verify each of the following properties.

(a) If $\dfrac{A}{A'} = \dfrac{B}{B'} \neq \dfrac{C}{C'}$, then the lines are parallel.

(b) If $AA' = -BB'$, then the lines are perpendicular.

59. The properties in Problem 58 provide us with another way to write the equation of a line parallel or perpendicular to a given line and containing a given point not on the line. For example, suppose that we want the equation of the line perpendicular to $3x + 4y = 6$ and containing the point (1, 2). The form $4x - 3y = k$, where k is a constant, represents a family of lines perpendicular to $3x + 4y = 6$ because we have satisfied the condition $AA' = -BB'$. Therefore, to find the specific line of the family containing (1, 2), we substitute 1 for x and 2 for y to determine k.

$$4x - 3y = k$$
$$4(1) - 3(2) = k$$
$$-2 = k.$$

Thus, the equation of the desired line is $4x - 3y = -2$.

Use the properties from Problem 58 to write the equation of each of the following lines.

(a) contains (1, 8) and is parallel to $2x + 3y = 6$

(b) contains $(-1, 4)$ and is parallel to $x - 2y = 4$

(c) contains $(2, -7)$ and is perpendicular to $3x - 5y = 10$

(d) contains $(-1, -4)$ and is perpendicular to $2x + 5y = 12$

7.4 Graphing Parabolas

Let's return to the problem of sketching a graph of an algebraic equation in two variables. Suppose that we want the graph of the equation $y = x^2$. First we set up a table indicating some of the solutions. Then we can plot the points associated with these solutions, as indicated in Figure 7.28(a). Finally, by connecting these points with a

x	$y = x^2$	Solutions (x, y)
0	0	(0, 0)
1	1	(1, 1)
2	4	(2, 4)
3	9	(3, 9)
-1	1	$(-1, 1)$
-2	4	$(-2, 4)$
-3	9	$(-3, 9)$

Figure 7.28

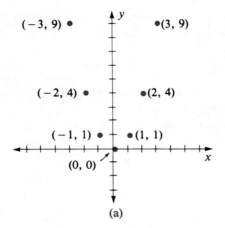

(a)

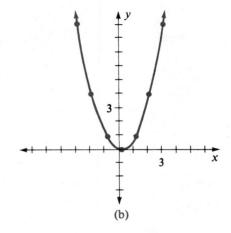

(b)

smooth curve, we produce Figure 7.28(b). The curve shown in Figure 7.28(b) is called a **parabola**. The graph of any equation of the form $y = ax^2 + bx + c$, where a, b, and c are real numbers and $a \neq 0$, is a parabola. As we work with parabolas, the vocabulary introduced in Figure 7.29 will be used.

Figure 7.29

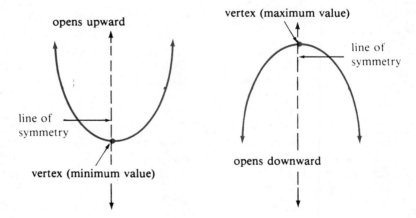

The parabola in Figure 7.28(b) is said to be **symmetric with respect to the y-axis**. That is to say, the y-axis is a line of symmetry. Each half of the curve is a mirror image of the other half through the y-axis. Notice that in the table of values for the equation $y = x^2$, for each ordered pair (x, y) the ordered pair $(-x, y)$ is also a solution. A general test for y-axis symmetry can be stated as follows.

---

y-axis symmetry

The graph of an equation is symmetric with respect to the y-axis if replacing x with $-x$ results in an equivalent equation.

The equation $y = x^2$ exhibits y-axis symmetry because replacing x with $-x$ produces $y = (-x)^2 = x^2$. Likewise, the equations $y = x^2 + 6$, $y = x^4$, and $y = x^4 + 2x^2$ exhibit y-axis symmetry.

Graphing parabolas requires being able to find the vertex, determine which way the parabola opens, and locate two points on opposite sides of the line of symmetry. Some of this information can be found by comparing the parabolas produced by various types of equations such as $y = x^2 + k$, $y = ax^2$, $y = (x - h)^2$, and $y = a(x - h)^2 + k$ to the **basic parabola** produced by the equation $y = x^2$. Let's first consider some equations of the form $y = x^2 + k$, where k is a constant.

Example 1   Graph $y = x^2 + 1$.

Solution   Let's set up a table of values to compare y-values for $y = x^2 + 1$ to corresponding y-values for $y = x^2$.

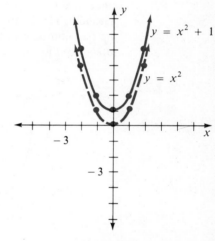

Figure 7.30

X	Y = X²	Y = X² + 1
0	0	1
1	1	2
2	4	5
−1	1	2
−2	4	5

It should be evident that y-values for $y = x^2 + 1$ are one *larger* than corresponding y-values for $y = x^2$. For example, if $x = 2$ then $y = 4$ for the equation $y = x^2$, but if $x = 2$ then $y = 5$ for the equation $y = x^2 + 1$. Thus, the graph of $y = x^2 + 1$ is the same as the graph of $y = x^2$ but *moved up 1 unit* (Figure 7.30).

Example 2   Graph $y = x^2 - 2$.

Solution   The y-values for $y = x^2 - 2$ are 2 *less than* the corresponding y-values for $y = x^2$, as indicated in the following table.

X	Y = X²	Y = X² - 2
0	0	−2
1	1	−1
2	4	2
−1	1	−1
−2	4	2

Thus, the graph of $y = x^2 - 2$ is the same as the graph of $y = x^2$, but *moved down 2 units* (Figure 7.31).

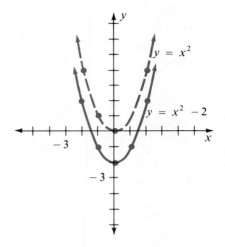

Figure 7.31   ●

In general, the graph of a quadratic equation of the form $y = x^2 + k$ is the same as the graph of $y = x^2$ but moved up or down k units depending on whether k is positive or negative.

Now, let's consider some quadratic equations of the form $y = ax^2$, where a is a nonzero constant.

Example 3   Graph $y = 2x^2$.

Solution   Let's again use a table to make some comparisons of y-values.

x	$y = x^2$	$y = 2x^2$
0	0	0
1	1	2
2	4	8
−1	1	2
−2	4	8

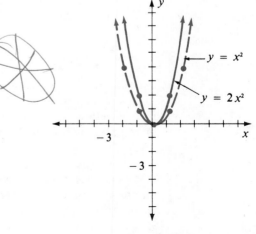

Obviously, the y-values for $y = 2x^2$ are *twice* the corresponding y-values for $y = x^2$. Thus, the parabola associated with $y = 2x^2$ has the same vertex (the origin) as the graph of $y = x^2$, but it is narrower (Figure 7.32).

Figure 7.32   ●

Example 4

Graph $y = \frac{1}{2}x^2$.

Solution

The following table indicates some comparisons of y-values.

x	$y = x^2$	$y = \frac{1}{2}x^2$
0	0	0
1	1	$\frac{1}{2}$
2	4	2
-1	1	$\frac{1}{2}$
-2	4	2

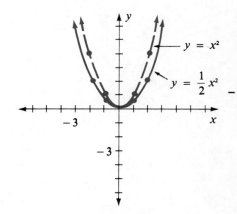

The y-values for $y = \frac{1}{2}x^2$ are *one-half* of the corresponding y-values for $y = x^2$. Therefore, the graph of $y = \frac{1}{2}x^2$ is wider than the basic parabola (Figure 7.33).

Figure 7.33

Example 5

Graph $y = -x^2$.

Solution

x	$y = x^2$	$y = -x^2$
0	0	0
1	1	-1
2	4	-4
-1	1	-1
-2	4	-4

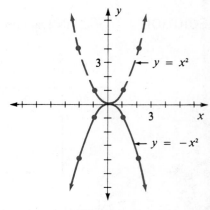

The y-values for $y = -x^2$ are the *opposites* of the corresponding y-values for $y = x^2$. Thus, the graph of $y = -x^2$ is a reflection across the x-axis of the basic parabola (Figure 7.34).

Figure 7.34

In general, the graph of a quadratic equation of the form $y = ax^2$ has its vertex at the origin and opens upward if a is positive and downward if a is negative. The parabola is narrower than the basic parabola if $|a| > 1$ and wider if $|a| < 1$.

Let's continue our investigation of quadratic equations by considering those of the form $y = (x - h)^2$, where h is a nonzero constant.

Example 6   Graph $y = (x - 2)^2$.

Solution   A fairly extensive table of values illustrates a pattern.

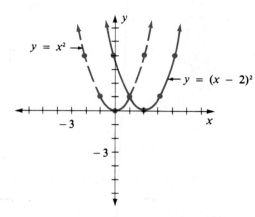

x	$y = x^2$	$y = (x - 2)^2$
-2	4	16
-1	1	9
0	0	4
1	1	1
2	4	0
3	9	1
4	16	4
5	25	9

Figure 7.35

Notice that $y = (x - 2)^2$ and $y = x^2$ take on the same y-values, *but* for different values of x. More specifically, if $y = x^2$ achieves a certain y-value at x = a constant, then $y = (x - 2)^2$ achieves that same y-value at x = the *constant plus two*. In other words, the graph of $y = (x - 2)^2$ is the same as the graph of $y = x^2$ but *moved 2 units to the right* (Figure 7.35). ●

Example 7   Graph $y = (x + 3)^2$.

Solution

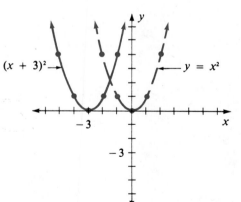

x	$y = x^2$	$y = (x + 3)^2$
-3	9	0
-2	4	1
-1	1	4
0	0	9
1	1	16
2	4	25
3	9	36

Figure 7.36

If $y = x^2$ achieves a certain y-value at x = a constant, then $y = (x + 3)^2$ achieves that same y-value at x = that *constant minus* 3. Therefore, the graph of $y = (x + 3)^2$ is the same as the graph of $y = x^2$ but *moved 3 units to the left* (Figure 7.36). ●

In general, the graph of a quadratic equation of the form $y = (x - h)^2$ is the same as the graph of $y = x^2$ but moved to the right h units if h is positive or moved to the left h units if h is negative. For example,

$$y = (x - 4)^2 \qquad \longleftarrow \text{ Moved to the } right \text{ 4 units}$$
$$y = (x + 2)^2 = [x - (-2)]^2. \qquad \longleftarrow \text{ Moved to the } left \text{ 2 units}$$

The following diagram summarizes our work thus far in graphing quadratic equations.

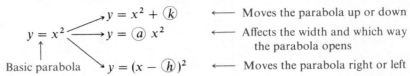

$y = x^2 + \boxed{k}$ $\longleftarrow$ Moves the parabola up or down

$y = x^2 \longrightarrow y = \boxed{a}\, x^2$ $\longleftarrow$ Affects the width and which way the parabola opens

Basic parabola $\longrightarrow y = (x - \boxed{h})^2$ $\longleftarrow$ Moves the parabola right or left

Equations of the form $y = x^2 + k$ and $y = ax^2$ are symmetrical about the y-axis. The final two examples of this section illustrate putting these ideas together to graph a quadratic equation of the form $y = a(x - h)^2 + k$.

Example 8   Graph $y = 2(x - 3)^2 + 1$.

Solution $y = 2(x - 3)^2 + 1$

Narrows the parabola and opens it *upward* / Moves the parabola 3 units to the *right* / Moves the parabola 1 unit *up*

The parabola is drawn in Figure 7.37. Two points in addition to the vertex are located to determine the parabola.

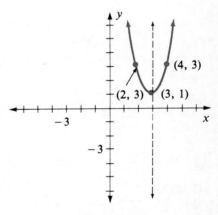

Figure 7.37

Example 9   Graph $y = -\dfrac{1}{2}(x + 1)^2 - 2$.

Solution $y = -\dfrac{1}{2}(x + 1)^2 - 2$

Widens the parabola and opens it *downward* / Moves the parabola 1 unit to the *left* / Moves the parabola 2 units *down*

The parabola is drawn in Figure 7.38.

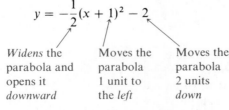

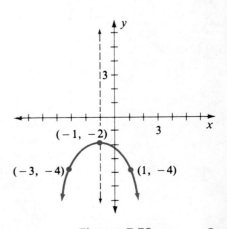

Figure 7.38

Problem Set 7.4

Graph each of the following parabolas.

1. $y = x^2 + 3$ **2.** $y = x^2 - 4$ **3.** $y = x^2 - 1$ **4.** $y = x^2 + 5$

5. $y = 3x^2$ **6.** $y = 5x^2$ **7.** $y = -2x^2$ **8.** $y = -3x^2$

9. $y = \dfrac{1}{4}x^2$ **10.** $y = \dfrac{2}{3}x^2$ **11.** $y = -\dfrac{1}{3}x^2$ **12.** $y = -\dfrac{1}{2}x^2$

13. $y = (x - 1)^2$ **14.** $y = (x + 2)^2$ **15.** $y = (x + 4)^2$

16. $y = (x - 5)^2$ **17.** $y = 2x^2 + 3$ **18.** $y = 3x^2 - 4$

19. $y = -2x^2 - 1$ **20.** $y = -\dfrac{1}{2}x^2 + 4$ **21.** $y = (x - 2)^2 + 4$

22. $y = (x + 3)^2 - 2$ **23.** $y = 3(x - 1)^2 - 2$ **24.** $y = 2(x + 4)^2 + 5$

25. $y = \dfrac{1}{2}(x - 3)^2 + 6$ **26.** $y = \dfrac{1}{3}(x - 2)^2 - 4$ **27.** $y = -2(x + 4)^2 - 1$

28. $y = -3(x - 4)^2 + 2$ **29.** $y = -(x + 1)^2 + 1$ **30.** $y = -\dfrac{1}{2}(x + 3)^2 - 5$

7.5 More Parabolas and Some Circles

We are now ready to graph quadratic equations of the form $y = ax^2 + bx + c$, where a, b, and c are real numbers and $a \neq 0$. The general approach is one of changing equations of the form $y = ax^2 + bx + c$ to the form $y = a(x - h)^2 + k$. Then we can proceed to graph them as we did in the previous section. The process of *completing the square* serves as the basis for making the change in the form of the equations. Let's consider some examples to illustrate the details.

Example 1   Graph $y = x^2 + 6x + 8$.

Solution

$y = x^2 + 6x + 8$

$y = (x^2 + 6x \qquad) + 8$ ⟵ Add 9, which is the square of one-half of the coefficient of x.

$y = (x^2 + 6x + 9) + 8 - 9$ ⟵ Subtract 9 to compensate for the 9 which was added.

$y = (x + 3)^2 - 1.$ $x^2 + 6x + 9 = (x + 3)^2$

The graph of $y = (x + 3)^2 - 1$ is the basic parabola moved three units to the left and one unit down (Figure 7.39).

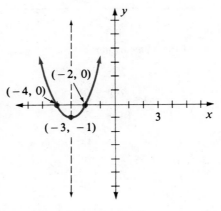

Figure 7.39

Example 2   Graph $y = x^2 - 3x - 1$.

Solution

$y = x^2 - 3x - 1$

$y = (x^2 - 3x \quad) - 1$

$y = \left(x^2 - 3x + \dfrac{9}{4}\right) - 1 - \dfrac{9}{4}$ Add and subtract $\dfrac{9}{4}$, which is the square of one-half of the coefficient of x.

$y = \left(x - \dfrac{3}{2}\right)^2 - \dfrac{13}{4}$.

The graph of $y = \left(x - \dfrac{3}{2}\right)^2 - \dfrac{13}{4}$ is the basic parabola moved $1\dfrac{1}{2}$ units to the right and $3\dfrac{1}{4}$ units down (Figure 7.40).

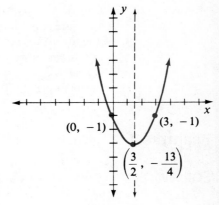

Figure 7.40

If the coefficient of x^2 is not 1, then a slight adjustment has to be made before the process of completing the square is applied. The next two examples illustrate this situation.

Example 3   Graph $y = 2x^2 + 8x + 9$.

Solution
$$y = 2x^2 + 8x + 9$$
$$y = 2(x^2 + 4x \qquad) + 9 \qquad \text{Factor a 2 from the first two terms on the right side.}$$
$$y = 2(x^2 + 4x + 4) + 9 - 8 \qquad \text{Add 4 inside the parentheses, which is the square of}$$
one-half of the coefficient of x.

Subtract 8 to compensate for the 4 added inside the parentheses times the factor of 2.

$$y = 2(x + 2)^2 + 1.$$

See Figure 7.41 for the graph of $y = 2(x + 2)^2 + 1$.   ●

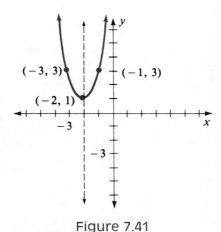

$(-3, 3)$ $(-1, 3)$

$(-2, 1)$

-3

-3

Figure 7.41   ●

Example 4   Graph $y = -3x^2 + 6x - 5$.

Solution
$$y = -3x^2 + 6x - 5.$$
$$y = -3(x^2 - 2x \qquad) - 5 \qquad \text{Factor } -3 \text{ from the first two terms on the}$$
right side.

$$y = -3(x^2 - 2x + 1) - 5 + 3 \qquad \text{Add 1 inside the parentheses to complete}$$
the square.

Add 3 to compensate for the 1 added inside the parentheses times the factor of -3.

$$y = -3(x - 1)^2 - 2.$$

The graph of $y = -3(x - 1)^2 - 2$ is drawn in Figure 7.42.

Figure 7.42

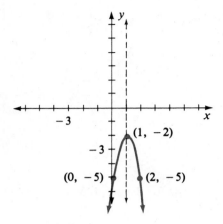

Circles

Applying the distance formula (developed in Section 7.1), $d = \sqrt{(x_2 - x_1)^2 + (y_2 - y_1)^2}$, to the definition of a circle produces what is known as the **standard equation of a circle**. We start with a precise definition of a circle.

Definition 7.2	A **circle** is the set of all points in a plane equidistant from a given fixed point called the **center**. A line segment determined by the center and any point on the circle is called a **radius**.

Let's consider a circle having a radius of length r and a center at (h, k) on a coordinate system (Figure 7.43). For any point P on the circle with coordinates (x, y), the length of a radius (denoted by r) can be expressed as

$$r = \sqrt{(x - h)^2 + (y - k)^2}.$$

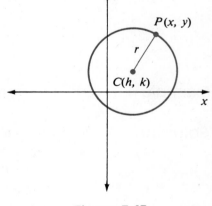

Figure 7.43

Thus, squaring both sides of the equation, we obtain the **standard form of the equation of a circle**,

$$(x - h)^2 + (y - k)^2 = r^2.$$

The standard form of the equation of a circle can be used to solve the two basic kinds of problems:

1. Given the coordinates of the center and the length of a radius of a circle, find its equation.

2. Given the equation of a circle, find its center and the length of a radius.

Let's look at some examples of such problems.

Example 5   Write the equation of a circle having its center at $(3, -5)$ and a radius of length 6 units.

Solution   Let's substitute 3 for h, -5 for k, and 6 for r into the standard form.

$(x - h)^2 + (y - k)^2 = r^2$ becomes $(x - 3)^2 + (y + 5)^2 = 6^2$,

which can be simplified as follows:

$$(x - 3)^2 + (y + 5)^2 = 6^2$$
$$x^2 - 6x + 9 + y^2 + 10y + 25 = 36$$
$$x^2 + y^2 - 6x + 10y - 2 = 0.$$ ●

Notice that in Example 5 we simplified the equation to the form $x^2 + y^2 + Dx + Ey + F = 0$, where D, E, and F are integers. This is another form commonly used in work with circles.

Example 6   Graph $x^2 + y^2 + 4x - 6y + 9 = 0$.

Solution   This equation is of the form $x^2 + y^2 + Dx + Ey + F = 0$, so its graph is a circle. We can change the given equation into the form $(x - h)^2 + (y - k)^2 = r^2$ by completing the square on x and on y as follows:

$$x^2 + y^2 + 4x - 6y + 9 = 0$$
$$(x^2 + 4x \quad) + (y^2 - 6y \quad) = -9$$
$$(x^2 + 4x + 4) + (y^2 - 6y + 9) = -9 + 4 + 9.$$

Add 4 to complete the square on x Add 9 to complete the square on y Add 4 and 9 to compensate for the 4 and 9 added on the left side

Then

$$(x + 2)^2 + (y - 3)^2 = 4$$
$$(x - (-2))^2 + (y - 3)^2 = 2^2.$$
$$\uparrow \qquad\quad \uparrow \quad\ \uparrow$$
$$h \qquad\qquad k \quad\ r$$

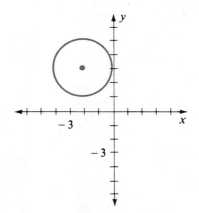

Figure 7.44

The center of the circle is at $(-2, 3)$, and the length of a radius is 2 (Figure 7.44).   ●

As illustrated by Examples 5 and 6, both forms, $(x - h)^2 + (y - k)^2 = r^2$ and $x^2 + y^2 + Dx + Ey + F = 0$, play an important role in solving problems dealing with circles.

Finally, we need to recognize that the standard form of a circle having its center at the origin is $x^2 + y^2 = r^2$. This is the result of letting $h = 0$ and $k = 0$ in the general standard form.

$$(x - h)^2 + (y - k)^2 = r^2$$
$$(x - 0)^2 + (y - 0)^2 = r^2$$
$$x^2 + y^2 = r^2.$$

Thus, by inspection we can recognize that $x^2 + y^2 = 9$ is a circle with its center at the origin; the length of the radius is 3 units. Furthermore, the equation of a circle having its center at the origin and a radius of length 6 units is $x^2 + y^2 = 36$.

Problem Set 7.5

Graph each of the following parabolas.

1. $y = (x - 4)^2 + 5$
2. $y = (x + 3)^2 - 2$
3. $y = 2(x - 1)^2 + 6$
4. $y = -2(x + 4)^2 - 5$
5. $y = x^2 - 4x + 7$
6. $y = x^2 - 6x + 13$
7. $y = x^2 + 8x + 14$
8. $y = x^2 + 2x + 6$
9. $y = x^2 + 3x + 1$
10. $y = x^2 - 5x + 3$
11. $y = x^2 - x - 1$
12. $y = x^2 + 7x + 14$
13. $y = 2x^2 + 4x + 7$
14. $y = 3x^2 - 6x + 5$
15. $y = 3x^2 + 24x + 49$
16. $y = 4x^2 - 24x + 32$
17. $y = -2x^2 + 8x - 5$
18. $y = -2x^2 - 4x - 5$
19. $y = -x^2 - 6x - 7$
20. $y = -x^2 + 8x - 21$
21. $y = 2x^2 + 3x + 1$
22. $y = 2x^2 - x + 2$
23. $y = 3x^2 - x - 1$
24. $y = 3x^2 + 2x + 1$
25. $y = -2x^2 + x - 2$
26. $y = -3x^2 - 7x - 2$

Find the center and the length of a radius, and then graph each of the following circles.

27. $x^2 + y^2 = 1$
28. $x^2 + y^2 = 16$
29. $x^2 + y^2 = 10$
30. $x^2 + y^2 = 12$

Write the equation of each of the following circles. Express the final equations in the form $x^2 + y^2 + Dx + Ey + F = 0$.

31. center at $(2, 4)$ and $r = 3$
32. center at $(5, 1)$ and $r = 1$
33. center at $(-3, 4)$ and $r = 7$
34. center at $(-2, -6)$ and $r = 4$
35. center at $(0, 2)$ and $r = 2$
36. center at $(4, 0)$ and $r = 5$
37. center at $(0, 0)$ and $r = 8$
38. center at $(0, 0)$ and $r = \sqrt{5}$

Find the center and the length of a radius of each of the following circles.

39. $x^2 + y^2 - 2x + 4y - 4 = 0$ **40.** $x^2 + y^2 + 2x + 6y + 9 = 0$

41. $x^2 + y^2 + 6x - 8y = 0$ **42.** $x^2 - 8x + y^2 - 14y + 61 = 0$

43. $x^2 - 6x + y^2 - 10y + 24 = 0$ **44.** $x^2 + 4x + y^2 - 12y + 35 = 0$

45. Find the equation of the circle that passes through the origin and has its center at $(-3, 0)$.

46. Find the equation of the circle that passes through the origin and has its center at $(3, 4)$.

47. If an object is thrown upward with an initial velocity of 32 feet per second, then its height (h) at a time of t seconds is given by the following equation:

$$h = 32t - 16t^2.$$

Label the horizontal axis t and the vertical axis h, and graph the equation $h = 32t - 16t^2$.

48. The area of a rectangle having a perimeter of 12 meters is given by the equation

$$A = l(6 - l)$$

where l represents the length in meters and A the area in square meters. Label the horizontal axis l and the vertical axis A, and graph the equation $A = l(6 - l)$.

Miscellaneous Problems

49. The points (x, y) and (y, x) are mirror images of each other across the line $y = x$. Therefore, by interchanging x and y in the equation $y = ax^2 + bx + c$, we obtain the equation of its mirror image across the line $y = x$—namely, $x = ay^2 + by + c$. Thus to graph $x = y^2 + 2$, we can first graph $y = x^2 + 2$ and then reflect it across the line $y = x$ as indicated in the following figure.

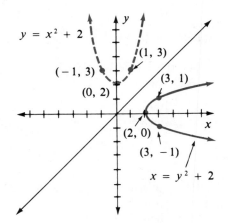

Graph each of the following parabolas.

(a) $x = y^2$ **(b)** $x = -y^2$ **(c)** $x = y^2 - 1$

(d) $x = -y^2 + 3$ **(e)** $x = -2y^2$ **(f)** $x = 3y^2$

(g) $x = y^2 + 4y + 7$ **(h)** $x = y^2 - 2y - 3$

50. Expanding $(x - h)^2 + (y - k)^2 = r^2$, we obtain $x^2 - 2hx + h^2 + y^2 - 2ky + k^2 - r^2 = 0$. Comparing this result to the form $x^2 + y^2 + Dx + Ey + F = 0$, we see that $D = -2h$, $E = -2k$, and $F = h^2 + k^2 - r^2$. Therefore, the center and length of a radius of a circle can be found by using $h = \dfrac{D}{-2}$, $k = \dfrac{E}{-2}$, and $r = \sqrt{h^2 + k^2 - F}$. Use these relationships to find the center and length of a radius of each of the following circles.

(a) $x^2 + y^2 - 2x - 8y + 8 = 0$ **(b)** $x^2 + y^2 + 4x - 14y + 49 = 0$

(c) $x^2 + y^2 + 12x + 8y - 12 = 0$ **(d)** $x^2 + y^2 - 16x + 20y + 115 = 0$

(e) $x^2 + y^2 - 12y - 45 = 0$ **(f)** $x^2 + y^2 + 14x = 0$

7.6 Ellipses and Hyperbolas

In the previous section we found that the graph of the equation $x^2 + y^2 = 36$ is a circle of radius 6 units with its center at the origin. More generally, it is true that any equation of the form $Ax^2 + By^2 = C$, where $A = B$ and A, B, and C are nonzero constants having the same sign, is a circle with the center at the origin. For example, $3x^2 + 3y^2 = 12$ is equivalent to $x^2 + y^2 = 4$ (divide both sides of the given equation by 3) and thus it is a circle of radius 2 units with its center at the origin.

The general equation $Ax^2 + By^2 = C$ can be used to describe other geometric figures if the restrictions on A and B are changed. For example, if A, B, and C are of the same sign but $A \neq B$, then the graph of the equation $Ax^2 + By^2 = C$ is an ellipse. Let's consider two examples.

Example 1   Graph $4x^2 + 25y^2 = 100$.

Solution   Let's find the x- and y-intercepts.
Let $x = 0$. Then

$$4(0)^2 + 25y^2 = 100$$
$$25y^2 = 100$$
$$y^2 = 4$$
$$y = \pm 2.$$

Thus, the points $(0, 2)$ and $(0, -2)$ are on the graph.
Let $y = 0$. Then

$$4x^2 + 25(0)^2 = 100$$
$$4x^2 = 100$$
$$x^2 = 25$$
$$x = \pm 5.$$

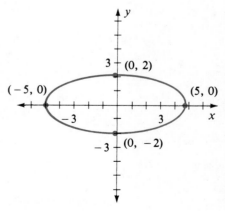

Figure 7.45

Thus, the points (5, 0) and (−5, 0) are also on the graph. Plotting the four points we have and knowing that it is an ellipse, we can make a pretty good sketch of the figure (Figure 7.45). ●

In Figure 7.45 the line segment with endpoints at (−5, 0) and (5, 0) is called the **major axis** of the ellipse. The shorter line segment with endpoints at (0, −2) and (0, 2) is called the **minor axis**. Establishing the endpoints of the major and minor axes provides a basis for sketching an ellipse.

Example 2   Graph $9x^2 + 4y^2 = 36$.

Solution   Again let's find the x- and y-intercepts.
Let $x = 0$. Then

$$9(0)^2 + 4y^2 = 36$$
$$4y^2 = 36$$
$$y^2 = 9$$
$$y = \pm 3.$$

Thus, the points (0, 3) and (0, −3) are on the graph.
Let $y = 0$. Then

$$9x^2 + 4(0)^2 = 36$$
$$9x^2 = 36$$
$$x^2 = 4$$
$$x = \pm 2.$$

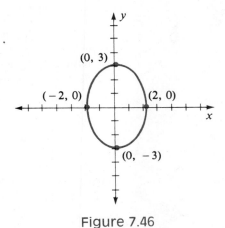

Figure 7.46

Thus, the points (2, 0) and (−2, 0) are also on the graph. The ellipse is sketched in Figure 7.46. ●

In Figure 7.46, the major axis has endpoints at (0, −3) and (0, 3), and the minor axis has endpoints at (−2, 0) and (2, 0). The ellipses in Figures 7.45 and 7.46 are symmetrical about the x-axis and about the y-axis. In other words, both the x-axis and the y-axis serve as lines of symmetry.

Hyperbolas

The graph of an equation of the form $Ax^2 + By^2 = C$, where A, B, and C are nonzero constants and A and B are of unlike signs, is a **hyperbola**. The next two examples illustrate the graphing of hyperbolas.

Example 3   Graph $x^2 - y^2 = 9$.

Solution   If we let $y = 0$, we obtain

$$x^2 - 0^2 = 9$$
$$x^2 = 9$$
$$x = \pm 3.$$

Thus, the points $(3, 0)$ and $(-3, 0)$ are on the graph.
If we let $x = 0$, we obtain

$$0^2 - y^2 = 9$$
$$-y^2 = 9$$
$$y^2 = -9.$$

Since $y^2 = -9$ has no real number solutions, there are no points of the y-axis on this graph. That is to say, the graph does not intersect the y-axis.

Now let's solve the given equation for y so as to have a more convenient form for finding other solutions.

$$x^2 - y^2 = 9$$
$$-y^2 = 9 - x^2$$
$$y^2 = x^2 - 9$$
$$y = \pm \sqrt{x^2 - 9}.$$

Since the radicand, $x^2 - 9$, must be nonnegative, the values chosen for x must be greater than or equal to 3, or less than or equal to -3. With this in mind, we can form the following table of values.

x	y	
3	0	intercepts
-3	0	
4	$\pm\sqrt{7}$	
-4	$\pm\sqrt{7}$	other points
5	± 4	
-5	± 4	

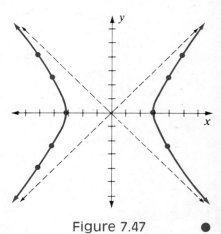

These points are plotted and the hyperbola is drawn in Figure 7.47. (This graph is also symmetrical about both axes.)

Figure 7.47   ●

Notice the dashed lines in Figure 7.47; they are called **asymptotes**. Each branch of the hyperbola approaches one of these lines, but does not intersect it. Therefore, being able to sketch the asymptotes of a hyperbola is very helpful in graphing the hyperbola. Fortunately, the equations of the asymptotes are easy to determine. They can be found by replacing the constant term in the given equation of the hyperbola with 0 and solving for y. (The reason this works will become evident in a later course.) Thus, for the hyperbola in Example 3 we obtain

$$x^2 - y^2 = 0$$
$$y^2 = x^2$$
$$y = \pm x.$$

So, the two lines $y = x$ and $y = -x$ are the asymptotes indicated by the dashed lines in Figure 7.47.

Example 4   Graph $y^2 - 5x^2 = 4$.

Solution   If we let $x = 0$, we obtain

$$y^2 - 5(0)^2 = 4$$
$$y^2 = 4$$
$$y = \pm 2.$$

The points $(0, 2)$ and $(0, -2)$ are on the graph.
  If we let $y = 0$, we obtain

$$0^2 - 5x^2 = 4$$
$$-5x^2 = 4$$
$$x^2 = -\frac{4}{5}.$$

Since $x^2 = -\dfrac{4}{5}$ has no real number solutions, we know that this hyperbola does not intersect the x-axis. Solving the given equation for y yields

$$y^2 - 5x^2 = 4$$
$$y^2 = 5x^2 + 4$$
$$y = \pm\sqrt{5x^2 + 4}.$$

The following table shows some additional solutions for the equation.

x	y	
0	2	
0	−2	intercepts
1	±3	other points
−1	±3	
2	$\pm\sqrt{24}$	
−2	$\pm\sqrt{24}$	

The equations of the asymptotes are determined as follows.

$$y^2 - 5x^2 = 0$$
$$y^2 = 5x^2$$
$$y = \pm\sqrt{5}x.$$

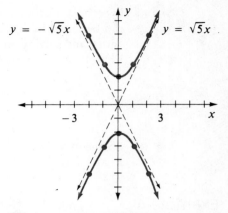

Figure 7.48

Sketching the asymptotes and plotting the points determined by the table of values helps to determine the hyperbola in Figure 7.48. (Notice that this hyperbola is also symmetrical about the x-axis and the y-axis.)   ●

The graphing of ellipses and hyperbolas whose centers are not at the origin can be done in much the same way as we handled circles in Section 7.5. The final two examples of this section illustrate these ideas.

Example 5   Graph $4x^2 + 24x + 9y^2 - 36y + 36 = 0$.

Solution   Let's complete the square on x and y as follows:

$$4x^2 + 24x + 9y^2 - 36y + 36 = 0$$
$$4(x^2 + 6x + \underline{\hspace{2em}}) + 9(y^2 - 4y + \underline{\hspace{2em}}) = -36$$
$$4(x^2 + 6x + 9) + 9(y^2 - 4y + 4) = -36 + 36 + 36$$
$$4(x + 3)^2 + 9(y - 2)^2 = 36$$
$$4(x - (-3))^2 + 9(y - 2)^2 = 36.$$

Since 4 and 9 are of the same sign, but not equal, the graph is an ellipse. The center of the ellipse is at $(-3, 2)$.

The equation $4(x + 3)^2 + 9(y - 2)^2 = 36$ can be used to find the endpoints of the major and minor axes as follows:

Let $y = 2$. Then

$$4(x + 3)^2 + 9(2 - 2)^2 = 36$$
$$4(x + 3)^2 = 36$$
$$(x + 3)^2 = 9$$
$$x + 3 = \pm 3$$
$$x + 3 = 3 \quad \text{or} \quad x + 3 = -3$$
$$x = 0 \quad \text{or} \quad x = -6.$$

The endpoints of the major axis are at (0, 2) and (−6, 2).

Let $x = -3$. Then

$$4(-3 + 3)^2 + 9(y - 2)^2 = 36$$
$$9(y - 2)^2 = 36$$
$$(y - 2)^2 = 4$$
$$y - 2 = \pm 2$$
$$y - 2 = 2 \quad \text{or} \quad y - 2 = -2$$
$$y = 4 \quad \text{or} \quad y = 0.$$

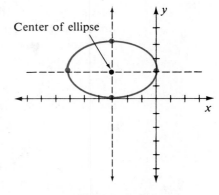

Center of ellipse

Figure 7.49

The endpoints of the minor axis are at $(-3, 4)$ and $(-3, 0)$. The ellipse is sketched in Figure 7.49. ●

Example 6

Graph $4x^2 - 8x - y^2 - 4y - 16 = 0$.

Solution

Completing the square on x and on y, we obtain

$$4x^2 - 8x - y^2 - 4y - 16 = 0$$
$$4(x^2 - 2x + \underline{\quad}) - (y^2 + 4y + \underline{\quad}) = 16$$
$$4(x^2 - 2x + 1) - (y^2 + 4y + 4) = 16 + 4 - 4$$
$$4(x - 1)^2 - (y + 2)^2 = 16$$
$$4(x - 1)^2 - 1(y - (-2))^2 = 16.$$

Since 4 and −1 are of opposite signs, the graph is a hyperbola. The center of the hyperbola is at $(1, -2)$.

Now using the equation $4(x - 1)^2 - (y + 2)^2 = 16$, we can proceed as follows:

Let $y = -2$. Then

$$4(x - 1)^2 - (-2 + 2)^2 = 16$$
$$4(x - 1)^2 = 16$$
$$(x - 1)^2 = 4$$
$$x - 1 = \pm 2$$
$$x - 1 = 2 \quad \text{or} \quad x - 1 = -2$$
$$x = 3 \quad \text{or} \quad x = -1.$$

Thus, the hyperbola intersects the horizontal line $y = -2$ at $(3, -2)$ and at $(-1, -2)$.

Let $x = 1$. Then

$$4(1 - 1)^2 - (y + 2)^2 = 16$$
$$-(y + 2)^2 = 16$$
$$(y + 2)^2 = -16.$$

Since $(y + 2)^2 = -16$ has no real number solutions, we know that the hyperbola does not intersect the vertical line $x = 1$.

Replacing the constant term of $4(x - 1)^2 - (y + 2)^2 = 16$ with 0 and solving for y produces the equations of the asymptotes as follows:

$$4(x - 1)^2 - (y + 2)^2 = 0.$$

The left side can be factored using the pattern of the difference of squares.

$$(2(x - 1) + (y + 2))(2(x - 1) - (y + 2)) = 0$$
$$(2x - 2 + y + 2)(2x - 2 - y - 2) = 0$$
$$(2x + y)(2x - y - 4) = 0$$
$$2x + y = 0 \quad \text{or} \quad 2x - y - 4 = 0$$
$$y = -2x \quad \text{or} \quad 2x - 4 = y.$$

So, the equations of the asymptotes are $y = -2x$ and $y = 2x - 4$. Sketching the asymptotes and plotting the two points $(3, -2)$ and $(-1, -2)$, we can draw the hyperbola as in Figure 7.50.

Figure 7.50

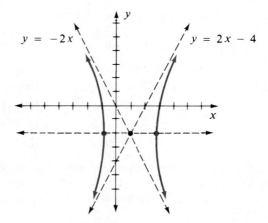

As a way of summarizing, let's focus our attention on the continuity pattern used in these last two sections. In Section 7.5 we used the definition of a circle to generate a standard form for the equation of a circle. Then, in Section 7.6 ellipses and hyperbolas were discussed, not from a definition viewpoint, but by considering variations of the general equation of a circle with its center at the origin ($Ax^2 + By^2 = C$, where A, B, and C are of the same sign and $A = B$). In a subsequent mathematics course, parabolas, ellipses, and hyperbolas will also be developed from a definition viewpoint. That is to say, first each concept will be defined and then the definition will be used to generate a standard form of its equation.

One final comment should be made relative to the material in these last two sections. Parabolas, circles, ellipses, and hyperbolas can be formed by intersecting a plane with a conical surface as shown in Figure 7.51; these curves are often called **conic sections**.

Figure 7.51

circle

ellipse

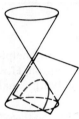

parabola

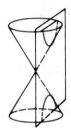

hyperbola

Problem Set 7.6

Graph each of the following.

1. $25x^2 + 4y^2 = 100$ **2.** $4x^2 + 9y^2 = 36$ **3.** $x^2 + 4y^2 = 16$

4. $9x^2 + y^2 = 36$ **5.** $3x^2 + 4y^2 = 12$ **6.** $2x^2 + 5y^2 = 50$

7. $x^2 - y^2 = 4$ **8.** $x^2 - y^2 = 1$ **9.** $y^2 - 3x^2 = 9$

10. $y^2 - 4x^2 = 25$ **11.** $3x^2 - 2y^2 = 12$ **12.** $4y^2 - 5x^2 = 16$

13. $2y^2 - x^2 = 10$ **14.** $3x^2 - 4y^2 = 15$

15. The graphs of equations of the form $xy = k$, where k is a nonzero constant, are also hyperbolas, sometimes referred to as rectangular hyperbolas. Graph each of the following.

    (a) $xy = 4$ **(b)** $xy = 2$

    (c) $xy = -2$ **(d)** $xy = -5$

16. What is the graph of $xy = 0$? Defend your answer.

17. We have graphed various equations of the form $Ax^2 + By^2 = C$, where C is a nonzero constant. Now graph each of the following.

    (a) $x^2 + y^2 = 0$ **(b)** $2x^2 + 3y^2 = 0$

    (c) $x^2 - y^2 = 0$ **(d)** $4x^2 - 9y^2 = 0$

18. A flashlight produces a *cone of light* that can be cut by the plane of a wall to illustrate the conic sections. Try shining a flashlight against a wall at different angles to produce a circle, an ellipse, a parabola, and one branch of a hyperbola. You may find it difficult to distinguish between a parabola and a branch of a hyperbola.

For Problems 19–26, graph each of the ellipses and hyperbolas.

19. $9x^2 - 36x + 4y^2 - 24y + 36 = 0$ **20.** $4x^2 + 8x + 16y^2 - 64y + 4 = 0$

21. $x^2 - 4x - y^2 + 6y - 14 = 0$ **22.** $-4x^2 + 32x + 9y^2 - 18y - 91 = 0$

23. $4x^2 - 24x + y^2 + 4y + 24 = 0$ **24.** $x^2 + 8x + 9y^2 + 36y + 16 = 0$

25. $x^2 + 4x - 9y^2 + 54y - 113 = 0$ **26.** $-4x^2 + 24x + 16y^2 + 64y - 36 = 0$

7.7 More on Graphing

As you saw in the previous sections, it is very helpful to recognize that a certain type of equation produces a particular kind of graph. However, we also need to develop some general graphing techniques to be used with equations for which we do not recognize the graph.

In Section 7.4 the following test for y-axis symmetry was stated.

---

y-axis symmetry

The graph of an equation is symmetric with respect to the y-axis if replacing x with $-x$ results in an equivalent equation.

Figure 7.52 illustrates the reasoning that justifies this test. The points $A(x, y)$ and $B(-x, y)$ are y-axis reflections of each other. Thus, if an equation is satisfied by both (x, y) and $(-x, y)$, it exhibits y-axis symmetry. Likewise, as indicated in Figure 7.52, $A(x, y)$ and $C(x, -y)$ are x-axis reflections of each other. Therefore, the following test for x-axis symmetry can be stated.

Figure 7.52

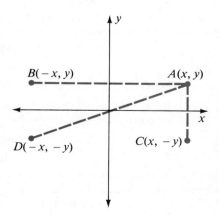

x-axis symmetry

The graph of an equation is symmetric with respect to the x-axis if replacing y with $-y$ results in an equivalent equation.

Again in Figure 7.52, we have illustrated that $A(x, y)$ and $D(-x, -y)$ are origin reflections of each other. That is, the origin is the midpoint of the line segment $\overline{DA}$. Thus, the following test for origin symmetry can be stated.

Origin symmetry

The graph of an equation is symmetric with respect to the origin if replacing x with $-x$ and y with $-y$ results in an equivalent equation.

Following is a list of graphing suggestions. The order of the suggestions indicates the order in which we usually attack a "new" graphing problem—that is, one that we do not recognize from its equation.

1. Determine the type of symmetry that the equation exhibits.
2. Find the intercepts.
3. Solve the equation for y in terms of x or for x in terms of y if it is not already in such a form.
4. Determine the necessary restrictions so as to ensure real number solutions. (This will be explained in a moment.)
5. Set up a table of ordered pairs that satisfy the equation. The type of symmetry and the restrictions will affect your choice of values in the table.
6. Plot the points associated with the ordered pairs, and connect them with a smooth curve. Then, if appropriate, reflect this curve according to the symmetry shown by the graph.

The following examples illustrate the use of these suggestions.

Example 1   Graph $y = x^4$.

Solution   Replacing x with $-x$ produces $y = (-x)^4 = x^4$. Therefore, this graph has *y-axis symmetry*.

If $x = 0$, then $y = 0$; so the graph contains the origin.

Because of the y-axis symmetry, we can limit the table of values to positive values of x.

x	y
$\dfrac{1}{2}$	$\dfrac{1}{16}$
1	1
$\dfrac{3}{2}$	$\dfrac{81}{16}$
2	16

Plotting these points and connecting them with a smooth curve, we obtain the portion of the curve in the first quadrant, as shown in Figure 7.53(a). Then reflecting this portion of the curve across the y-axis produces the complete graph, as shown in Figure 7.53(b).

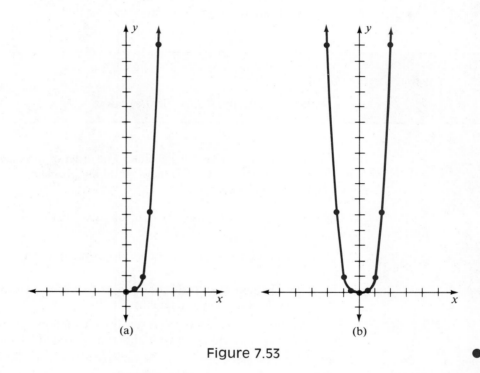

(a)   (b)

Figure 7.53   ●

Remark   The curve in Figure 7.53(b) is not a parabola even though its general shape resembles one. This curve is much flatter than a parabola at the bottom, and it rises much more rapidly.

Example 2   Graph $y = x^3$.

Solution   Replacing x with $-x$ and y with $-y$ produces $-y = (-x)^3 = -x^3$. This equation is equivalent to $y = x^3$, since multiplying both sides of $-y = -x^3$ by -1 produces $y = x^3$. Therefore, this graph has *origin symmetry*.

If $x = 0$, then $y = 0$; so the graph contains the origin.

Because of origin symmetry, we can limit the table of values to positive values of x.

x	y
$\frac{1}{2}$	$\frac{1}{8}$
1	1
$\frac{3}{2}$	$\frac{27}{8}$
2	8

Plotting these points and connecting them with a smooth curve, we obtain Figure 7.54(a). Then reflecting this portion of the curve through the origin produces the complete graph, shown in Figure 7.54(b).

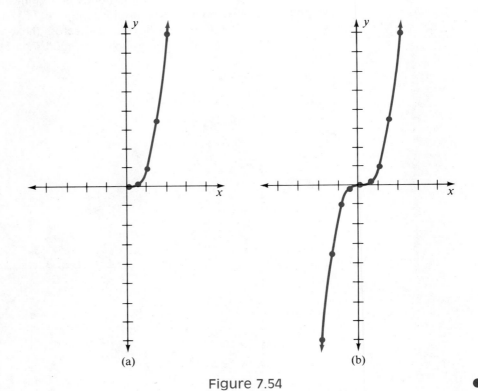

(a)

(b)

Figure 7.54

Remark From the symmetry tests we can see that if a curve has both x-axis and y-axis symmetry, it must have origin symmetry. However, it is possible for a curve to have origin symmetry and not be symmetrical to either axis. Figure 7.54(b) is an example of such a curve.

Example 3   Graph $y^2 = x^3 + 1$.

Solution   Replacing y with $-y$ produces $(-y)^2 = x^3 + 1$, which is equivalent to $y^2 = x^3 + 1$. Therefore, this graph is *symmetric with respect to the x-axis*.

If $x = 0$, then $y^2 = 1$; so $y = \pm 1$ and the y-intercepts are 1 and -1. Thus, the points $(0, 1)$ and $(0, -1)$ are on the graph.

If $y = 0$, then $x^3 + 1 = 0$; so $x^3 = -1$ and $x = -1$. (The other two solutions for $x^3 + 1 = 0$ are complex numbers.) Thus, the point $(-1, 0)$ is on the graph.

Solving $y^2 = x^3 + 1$ for y produces $y = \pm\sqrt{x^3 + 1}$. Using $y = \sqrt{x^3 + 1}$, we can set up the following table of values.

x	y
$-\dfrac{1}{2}$	$\dfrac{\sqrt{14}}{4}$
$\dfrac{1}{2}$	$\dfrac{3\sqrt{2}}{4}$
1	$\sqrt{2}$
2	3
3	$2\sqrt{7}$

Plotting these points and connecting them with a smooth curve produces Figure 7.55(a). Then reflecting this portion of the curve across the x-axis produces the complete graph shown in Figure 7.55(b).

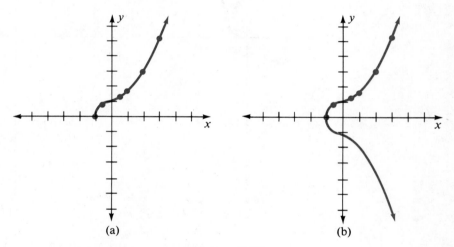

(a)   (b)

Figure 7.55

Example 4   Graph $x^2y^2 = 9$.

Solution   All three tests for symmetry are satisfied; therefore, this graph is *symmetric with respect to both axes and to the origin.*

It should be evident from the equation that neither x nor y can equal zero. Therefore, no points of either axis are on the graph.

Solving $x^2y^2 = 9$ for y produces $y = \pm\sqrt{\dfrac{9}{x^2}} = \pm\dfrac{3}{x}$. Using $y = \dfrac{3}{x}$, we can set up the following table of values.

x	y
$\dfrac{1}{2}$	6
1	3
2	$\dfrac{3}{2}$
3	1
4	$\dfrac{3}{4}$

Plotting these points and connecting them with a smooth curve produces Figure 7.56(a). Then reflecting this portion of the curve through the x-axis, y-axis, and the origin produces the complete graph shown in Figure 7.56(b).

Figure 7.56

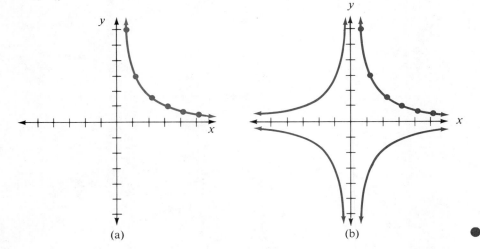

(a) (b)

Another graphing consideration is that of *restricting a variable* to ensure real number solutions. The following example illustrates this point.

Example 5 Graph $y = \sqrt{x - 1}$.

Solution The radicand, $x - 1$, must be nonnegative. Therefore,

$$x - 1 \geq 0$$
$$x \geq 1.$$

The restriction, $x \geq 1$, indicates that we need not look for a y-intercept. The x-intercept can be found as follows.

If $y = 0$, then

$$0 = \sqrt{x - 1}$$
$$0 = x - 1$$
$$1 = x.$$

The point $(1, 0)$ is on the graph.

Now, keeping the restriction in mind, we can determine the following table of values.

x	y
2	1
3	$\sqrt{2}$
4	$\sqrt{3}$
5	2
10	3

Plotting these points and connecting them with a smooth curve produces the curve in Figure 7.57.

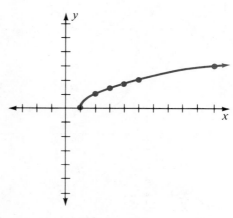

Figure 7.57

Problem Set 7.7 has a section of graphing problems that includes a mixture of "old" and "new" graphs. That is, some of the graphs are straight lines, circles, parabolas, ellipses, and hyperbolas, but others are curves that you probably will not recognize from their equations. Each time you face a "new" curve, remember the suggestions offered in this section. You will be exposed to more graphing ideas in Chapters 8 and 9.

Problem Set 7.7

For each of the following points, determine the points that are symmetric with respect to the (a) x-axis, (b) y-axis, and (c) origin.

1. $(4, 3)$ 2. $(-2, 5)$ 3. $(-6, -1)$

4. $(3, -7)$ 5. $(0, 4)$ 6. $(-5, 0)$

Determine the type of symmetry (x-axis, y-axis, origin) possessed by each of the following graphs. Do *not* sketch the graph.

7. $y = x^2 - 6$ 8. $x = y^2 + 1$

9. $x^3 = y^2$ 10. $x^2 y^2 = 4$

11. $x^2 + 2y^2 = 6$

12. $3x^2 - y^2 + 4x = 6$

13. $x^2 - 2x + y^2 - 3y - 4 = 0$

14. $xy = 4$

15. $y = x$

16. $2x - 3y = 15$

17. $y = x^3 + 2$

18. $y = x^4 + x^2$

19. $5x^2 - y^2 + 2y - 1 = 0$

20. $x^2 + y^2 - 2y - 4 = 0$

Graph each of the following equations. Remember that some of them are straight lines, circles, parabolas, ellipses, and hyperbolas, but some of them are "new" curves.

21. $y = x^2$

22. $y = -x^2$

23. $y = x^2 + 2$

24. $y = -x^2 - 1$

25. $xy = 4$

26. $xy = -2$

27. $y = -x^3$

28. $y = x^3 + 2$

29. $y^2 = x^3$

30. $y^3 = x^2$

31. $y^2 - x^2 = 4$

32. $x^2 - 2y^2 = 8$

33. $y = -\sqrt{x}$

34. $y = \sqrt{x + 1}$

35. $x^2 y = 4$

36. $xy^2 = 4$

37. $x^2 + 2y^2 = 8$

38. $2x^2 + y^2 = 4$

39. $y = \dfrac{4}{x^2 + 1}$

40. $y = \dfrac{-2}{x^2 + 1}$

41. $y = -2x - 2$

42. $-2x + y = -4$

43. $y = x^3 - x$

44. $y^2 = x^3 + 8$

45. $x^2 - 6x + y^2 + 6y + 9 = 0$

46. $x^2 + 6x + y^2 - 8y = 0$

47. $y = x^4 - x^2$

48. $y = -x^4$

49. $x = y^2 - 1$

50. $x = y^2 + 2$

Chapter 7 Summary

(7.1) The **Cartesian (or rectangular) coordinate system** is used to graph ordered pairs of real numbers. The first number, a, of the ordered pair (a, b) is called the **abscissa**, and the second number, b, is called the **ordinate**; together they are referred to as the **coordinates** of a point.

Two basic kinds of problems exist in coordinate geometry:

1. Given an algebraic equation, find its geometric graph.

2. Given a set of conditions pertaining to a geometric figure, find its algebraic equation.

The distance between any two points (x_1, y_1) and (x_2, y_2) is given by the **distance**

formula,

$$d = \sqrt{(x_2 - x_1)^2 + (y_2 - y_1)^2}.$$

The **slope** (denoted by m) of a line determined by the points (x_1, y_1) and (x_2, y_2) is given by the slope formula,

$$m = \frac{y_2 - y_1}{x_2 - x_1}, \qquad x_2 \neq x_1.$$

(7.2)　A **solution** of an equation in two variables is an ordered pair of real numbers that satisfies the equation.

Any equation of the form $Ax + By = C$, where A, B, and C are constants (A and B not both zero) and x and y are variables, is a **linear equation**, and its graph is a **straight line**.

Any equation of the form $Ax + By = C$ for which $C = 0$ is a straight line containing the **origin**.

Any equation of the form $x = a$, where a is a constant, is a line parallel to the y-axis having an x-intercept of a.

Any equation of the form $y = b$, where b is a constant, is a line parallel to the x-axis having a y-intercept of b.

Linear inequalities in two variables are of the form $Ax + By > C$ or $Ax + By < C$. The following steps are suggested for graphing a linear inequality.

1. First, graph the corresponding equality. Use a solid line if equality is included in the original statement and a dashed line if equality is not included.

2. Choose a test point not on the line, and substitute its coordinates into the inequality.

3. The graph of the original inequality is
   (a) the half-plane containing the test point if the inequality is satisfied by that point, or
   (b) the half-plane not containing the test point if the inequality is not satisfied by that point.

(7.3)　Look back at Examples 1, 2, and 3 of this section to review the *general approach* to writing the equation of a line given certain facts about the line.

The equation

$$y - y_1 = m(x - x_1)$$

is called the **point-slope form** of the equation of a line.

The equation

$$y = mx + b$$

is called the **slope-intercept form** of the equation of a line. If the equation of a nonvertical line is written in this form, the coefficient of x is the slope of the line and the constant term is the y-intercept.

If two lines have slopes m_1 and m_2, respectively, then

   1. the two lines are parallel if and only if $m_1 = m_2$;

   2. the two lines are perpendicular if and only if $(m_1)(m_2) = -1$.

(7.4) and (7.5)   The graph of any quadratic equation of the form $y = ax^2 + bx + c$, where a, b, and c are real numbers and $a \neq 0$, is a **parabola**.

The following diagram summarizes the graphing of parabolas:

$$y = x^2 + \textcircled{k} \quad \longleftarrow \text{ Moves the parabola up or down}$$

$$y = x^2$$

$$y = \textcircled{a}x^2 \quad \longleftarrow \text{ Affects the width and which way the parabola opens}$$

$$\uparrow$$

Basic parabola

$$y = (x - \textcircled{h})^2 \quad \longleftarrow \text{ Moves the parabola right or left}$$

The **standard form of the equation of a circle** with its center at (h, k) and a radius of length r is

$$(x - h)^2 + (y - k)^2 = r^2.$$

The standard form of the equation of a circle with its center at the origin and a radius of length r is

$$x^2 + y^2 = r^2.$$

(7.6)   The graph of an equation of the form

$$Ax^2 + By^2 = C,$$

where A, B, and C are nonzero constants of the same sign and $A \neq B$, is an **ellipse**.

The graph of an equation of the form

$$Ax^2 + By^2 = C,$$

where A, B, and C are nonzero constants and A and B are of unlike signs, is a **hyperbola**.

Circles, ellipses, parabolas, and hyperbolas are often referred to as **conic sections**.

(7.7)   Probably the most effective graphing technique is to be able to recognize the kind of graph that is produced by a certain type of equation. However, for determining "new" curves, we offer the following suggestions.

   1. Determine the type of symmetry that the equation exhibits.

   2. Find the x- and y-intercepts.

   3. Solve the equation for y in terms of x or for x in terms of y.

   4. Determine the necessary restrictions to ensure real number solutions.

5. Set up a table of ordered pairs that satisfy the equation.

6. Plot the points associated with the ordered pairs.

7. Connect the points with a smooth curve. Then, if appropriate, reflect this portion of the curve according to the symmetry possessed by the graph.

Chapter 7 Review Problem Set

1. Find the slope of the line determined by each pair of points.

   (a) $(3, 4), (-2, -2)$ (b) $(-2, 3), (4, -1)$

2. Find the slope of each of the following lines.

   (a) $4x + y = 7$ (b) $2x - 7y = 3$

3. Find the lengths of the sides of a triangle whose vertices are at $(2, 3)$, $(5, -1)$, and $(-4, -5)$.

For Problems 4–8, write the equation of each of the lines satisfying the stated conditions. Express final equations in standard form.

4. containing the points $(-1, 2)$ and $(3, -5)$

5. having a slope of $-\dfrac{3}{7}$ and a y-intercept of 4

6. containing the point $(-1, -6)$ and having a slope of $\dfrac{2}{3}$

7. containing the point $(2, 5)$ and parallel to the line $x - 2y = 4$

8. containing the point $(-2, -6)$ and perpendicular to the line $3x + 2y = 12$

For Problems 9–20, graph each of these equations.

9. $2x - y = 6$ 10. $y = -2x^2 - 1$

11. $x^2 + y^2 = 1$ 12. $4x^2 + y^2 = 16$

13. $y = -4x$ 14. $2x^2 - y^2 = 8$

15. $y = x^2 + 4x - 1$ 16. $y = 4x^2 - 8x + 2$

17. $xy^2 = -1$ 18. $x^2 - 4x + 2y^2 - 4y - 2 = 0$

19. $y^2 + 4y - 4x^2 - 24x - 36 = 0$ 20. $y = -2x^2 + 12x - 16$

For Problems 21 and 22, graph each of the inequalities.

21. $x + 2y \geq 4$ 22. $2x - 3y < 6$

23. Find the center and length of a radius of the circle $x^2 + y^2 + 6x - 8y + 16 = 0$.

24. Find the coordinates of the vertex of the parabola $y = x^2 + 8x + 10$.

25. Find the equations of the asymptotes of the hyperbola $9x^2 - 4y^2 = 72$.

26. Find the length of the major axis of the ellipse $4x^2 + y^2 = 9$.

8 Functions

One of the fundamental concepts of mathematics is that of a function. Functions are used to unify mathematics and also to serve as a meaningful way of applying mathematics to many real-world problems. Functions provide a means of studying quantities that vary with one another, that is, when a change in one quantity causes a corresponding change in another.

This chapter will (1) introduce the basic ideas pertaining to the function concept, (2) review some concepts from Chapter 7 regarding functions, and (3) discuss some applications using functions.

8.1 Relations and Functions

Mathematically, a function is a special kind of **relation**, so we will begin our discussion with a simple definition of a relation.

Definition 8.1	A *relation* is a set of ordered pairs.

Thus, a set of ordered pairs such as $\{(1, 2), (3, 7), (8, 14)\}$ is a relation. The set of all first components of the ordered pairs is the **domain** of the relation and the set of all second components is the **range** of the relation. The relation $\{(1, 2), (3, 7), (8, 14)\}$ has a domain of $\{1, 3, 8\}$ and a range of $\{2, 7, 14\}$.

The ordered pairs referred to in Definition 8.1 may be generated by various means, such as a graph or a chart. However, one of the most common ways of generating ordered pairs is by the use of equations. Since the solution set of an equation in two variables is a set of ordered pairs, such an equation describes a relation. Each of the following equations describes a relation between the variables x and y. We have listed *some* of the infinitely many ordered pairs (x, y) of each relation.

1. $x^2 + y^2 = 4$: $(1, \sqrt{3}), (1, -\sqrt{3}), (0, 2), (0, -2)$
2. $y^2 = x^3$: $(0, 0), (1, 1)(1, -1), (4, 8), (4, -8)$
3. $y = x + 2$: $(0, 2), (1, 3), (2, 4), (-1, 1), (5, 7)$
4. $y = \dfrac{1}{x - 1}$: $(0, -1), (2, 1), \left(3, \dfrac{1}{2}\right), \left(-1, -\dfrac{1}{2}\right), \left(-2, -\dfrac{1}{3}\right)$
5. $y = x^2$: $(0, 0), (1, 1), (2, 4), (-1, 1), (-2, 4)$

Now we direct your attention to the ordered pairs associated with equations **3.**, **4.**, and **5.**. Note that in each case no two ordered pairs have the same first component. Such a set of ordered pairs is called a **function**.

Definition 8.2	A *function* is a relation in which no two ordered pairs have the same first component.

Stated another way, Definition 8.2 means that a function is a relation for which each member of the domain is assigned *one and only one* member of the range. Thus, it is easy to determine that each of the following sets of ordered pairs is a function.

$$f = \{(x, y) | y = x + 2\};$$
$$g = \left\{(x, y) \Big| y = \frac{1}{x - 1}\right\};$$
$$h = \{(x, y) | y = x^2\}.$$

In each case there is one and only one value of y (an element of the range) associated with each value of x (an element of the domain).

Notice that we named the previous functions f, g, and h. It is common to name functions by means of a single letter and the letters f, g, and h are often used. We would suggest more meaningful choices when functions are used to portray real-world situations. For example, if a problem involves a profit function, then naming the function p or even P would seem natural.

The symbol for a function can be used along with a variable that represents an element in the domain to represent the associated element in the range. For example, suppose that we have a function f specified in terms of the variable x. The symbol, $f(x)$, (read "f of x" or "the value of f at x") represents the element in the range associated with the element x from the domain. The function $f = \{(x, y) | y = x + 2\}$ can be written as $f = \{(x, f(x)) | f(x) = x + 2\}$ and this is usually shortened to read "f is the function determined by the equation $f(x) = x + 2$."

Remark Be careful with the symbolism $f(x)$. As we stated above, it means the value of the function f at x. It does not mean f times x.

This *function notation* is very convenient when computing and expressing various values of the function. For example, the value of the function $f(x) = 3x - 5$ at $x = 1$ is

$$f(1) = 3(1) - 5 = -2.$$

Likewise, the functional values for $x = 2$, $x = -1$, and $x = 5$ are

$$f(2) = 3(2) - 5 = 1;$$
$$f(-1) = 3(-1) - 5 = -8; \quad \text{and}$$
$$f(5) = 3(5) - 5 = 10.$$

Thus, this function f contains the ordered pairs $(1, -2), (2, 1), (-1, -8), (5, 10)$, and in general all ordered pairs of the form $(x, f(x))$, where $f(x) = 3x - 5$ and x is any real number.

Figure 8.1

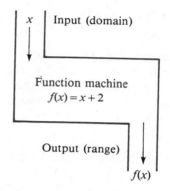

It may be helpful for you to mentally picture the concept of a function in terms of a "function machine" as illustrated in Figure 8.1. Each time that a value of x is put into the machine, the equation $f(x) = x + 2$ is used to generate one and only one value for $f(x)$ to be ejected from the machine. For example, if 3 is put into this machine, then $f(3) = 3 + 2 = 5$, and 5 is ejected. Thus, the ordered pair $(3, 5)$ is one element of the function. Now let's look at some examples to help pull together some of the ideas about functions.

Example 1

Determine whether the relation $\{(x, y)|y^2 = x\}$ is a function and specify its domain and range.

Solution

Because $y^2 = x$ is equivalent to $y = \pm\sqrt{x}$, to each value of x there are assigned *two* values for y. Therefore, this relation is not a function.

The expression $\sqrt{x}$ requires that x be nonnegative; therefore, the domain (D) is

$$D = \{x|x \geq 0\}.$$

To each nonnegative real number, the relation assigns two real numbers, $\sqrt{x}$ and $-\sqrt{x}$. Thus, the range (R) is

$$R = \{y|y \text{ is a real number}\}.$$

Example 2   For the function $f(x) = x^2$,

(a) specify its domain,   (b) determine its range, and
(c) evaluate $f(-2)$, $f(0)$, and $f(4)$.

Solution   (a) Any real number can be squared; therefore, the domain (D) is

$$D = \{x|x \text{ is a real number}\}.$$

(b) Squaring a real number always produces a nonnegative result. Thus, the range (R) is

$$R = \{f(x)|f(x) \geq 0\}.$$

(c) $f(-2) = (-2)^2 = 4;$
$$f(0) = (0)^2 = 0;$$
$$f(4) = (4)^2 = 16.$$ ●

For our purposes in this text, if the domain of a function is not specifically indicated or determined by a real-world application, then we assume the domain to be all **real number** replacements for the variable, which represents an element in the domain, that will produce **real number** functional values. Consider the following examples.

Example 3   Specify the domain for each of the following.

(a) $f(x) = \dfrac{1}{x - 1}$ (b) $f(t) = \dfrac{1}{t^2 - 4}$ (c) $f(s) = \sqrt{s - 3}$

Solution   (a) We can replace x with any real number except 1, because 1 makes the denominator zero. Thus, the domain is given by

$$D = \{x|x \neq 1\}.$$

(b) We need to eliminate any value of t that will make the denominator zero. Thus, let's solve the equation $t^2 - 4 = 0$.

$$t^2 - 4 = 0$$
$$t^2 = 4$$
$$t = \pm 2.$$

The domain is the set

$$D = \{t|t \neq -2 \text{ and } t \neq 2\}.$$

(c) The radicand, $s - 3$, must be nonnegative.

$$s - 3 \geq 0$$
$$s \geq 3.$$

The domain is the set

$$D = \{s|s \geq 3\}.$$ ●

Example 4   If $f(x) = -2x + 7$ and $g(x) = x^2 - 5x + 6$, find $f(3)$, $f(-4)$, $g(2)$, and $g(-1)$.

Solution
$$f(x) = -2x + 7$$
$$f(3) = -2(3) + 7 = 1;$$
$$f(-4) = -2(-4) + 7 = 15.$$

$$g(x) = x^2 - 5x + 6$$
$$g(2) = 2^2 - 5(2) + 6 = 0;$$
$$g(-1) = (-1)^2 - 5(-1) + 6 = 12.$$ ●

In Example 4, notice that we are working with two different functions in the same problem. Thus, different names, f and g, are used.

Example 5   If $f(x) = x^2 + 2x - 3$, find $\dfrac{f(a + h) - f(a)}{h}$.

Solution
$$f(a + h) = (a + h)^2 + 2(a + h) - 3$$
$$= a^2 + 2ah + h^2 + 2a + 2h - 3;$$

and

$$f(a) = a^2 + 2a - 3.$$

Therefore,

$$f(a + h) - f(a) = (a^2 + 2ah + h^2 + 2a + 2h - 3) - (a^2 + 2a - 3)$$
$$= a^2 + 2ah + h^2 + 2a + 2h - 3 - a^2 - 2a + 3$$
$$= 2ah + h^2 + 2h;$$

and

$$\frac{f(a + h) - f(a)}{h} = \frac{2ah + h^2 + 2h}{h}$$
$$= \frac{\not{h}(2a + h + 2)}{\not{h}}$$
$$= 2a + h + 2.$$ ●

Functions and functional notation provide the basis for describing many real-world relationships. The next example illustrates this point.

Example 6   Suppose a factory determines that the overhead for producing a quantity of a certain item is $500 and the cost for each item is $25. Express the total expenses as a function of the number of items produced and compute the expenses for producing 12, 25, 50, 75, and 100 items.

Solution   Let n represent the number of items produced. Then $25n + 500$ represents the total expenses. Let's use E to represent the *expense function*, so that we have

$$E(n) = 25n + 500, \text{ where } n \text{ is a whole number,}$$

from which we obtain

$$E(12) = 25(12) + 500 = 800;$$
$$E(25) = 25(25) + 500 = 1125;$$
$$E(50) = 25(50) + 500 = 1750;$$
$$E(75) = 25(75) + 500 = 2375;$$
$$E(100) = 25(100) + 500 = 3000.$$

So the total expenses for producing 12, 25, 50, 75, and 100 items are $800, $1125, $1750, $2375, and $3000, respectively. ●

Problem Set 8.1

1. **Specify the domain and the range for each of the following relations. Also state whether or not each relation is a function.**

   (a) $\{(1, 4), (2, 6), (3, 12), (4, 17)\}$

   (b) $\{(0, 0), (1, 1), (2, 8), (3, 27)\}$

   (c) $\{(0, 0), (1, 1), (1, -1), (-1, 1)\}$

   (d) $\{(0, 2), (1, 2), (2, 2), (3, 2), (4, 2)\}$

   (e) $\{(1, 3), (1, 4), (1, -1), (1, -2)\}$

   (f) $\{(4, 2), (4, -2), (9, 3), (9, -3), (16, 4), (16, -4)\}$

   (g) $\{(x, y) | x + 2y = 4\}$

   (h) $\{(x, y) | y = x^3\}$

   (i) $\{(x, y) | x^2 + y^2 = 1\}$

   (j) $\{(x, y) | x^2 - y^2 = 1\}$

Specify the domain for each of the following functions (Problems 2–23).

2. $f(x) = x^2 + 2x - 1$

3. $f(x) = 3x - 7$

4. $f(x) = \dfrac{1}{x}$

5. $f(x) = \dfrac{3}{x + 6}$

6. $g(x) = \dfrac{4x}{3x + 2}$

7. $g(x) = \dfrac{7x}{5x - 2}$

8. $h(x) = \dfrac{3}{(x + 2)(x - 3)}$

9. $h(x) = \dfrac{-4}{(x - 1)(x + 5)}$

10. $f(x) = \dfrac{1}{x^2 + 5x + 6}$

11. $f(x) = \dfrac{2}{x^2 - 2x - 15}$

12. $f(x) = \dfrac{-3}{x^2 + 2x}$

13. $f(x) = \dfrac{-5}{x^2 - 6x}$

14. $f(t) = \dfrac{2}{t^2 + 4}$

15. $f(t) = \dfrac{5}{t^2 + 9}$

16. $g(x) = \dfrac{-3}{x^2 - 9}$ **17.** $g(x) = \dfrac{-4}{x^2 - 25}$

18. $f(x) = \sqrt{x + 1}$ **19.** $f(x) = \sqrt{2x - 3}$

20. $f(s) = \sqrt{s - 2} + 6$ **21.** $f(s) = \sqrt{s + 4} - 3$

22. $f(x) = \sqrt{x^2 - 1}$ **23.** $f(x) = \sqrt{x^2 - 9}$

24. If $f(x) = 3x + 6$, find $f(0)$, $f(2)$, $f(-3)$, and $f(a)$.

25. If $f(x) = -5x + 8$, find $f(-1)$, $f(-2)$, $f(3)$, and $f(5)$.

26. If $f(x) = \dfrac{2}{3}x + \dfrac{3}{4}$, find $f(2)$, $f\left(\dfrac{1}{2}\right)$, $f\left(-\dfrac{1}{3}\right)$, and $f(k)$.

27. If $g(x) = x^2 + 3x - 1$, find $g(1)$, $g(-1)$, $g(3)$, and $g(-4)$.

28. If $g(x) = 2x^2 - 5x - 7$, find $g(-1)$, $g(2)$, $g(-3)$, and $g(4)$.

29. If $h(x) = -x^2 - 3$, find $h(1)$, $h(-1)$, $h(-3)$, and $h(5)$.

30. If $h(x) = -2x^2 - x + 4$, find $h(-2)$, $h(-3)$, $h(4)$, and $h(5)$.

31. If $f(x) = \sqrt{x - 1}$, find $f(1)$, $f(5)$, $f(13)$, and $f(26)$.

32. If $f(x) = \sqrt{2x + 1}$, find $f(3)$, $f(4)$, $f(10)$, and $f(12)$.

33. If $f(x) = \dfrac{3}{x - 2}$, find $f(3)$, $f(0)$, $f(-1)$, and $f(-5)$.

34. If $f(x) = \dfrac{-4}{x + 3}$, find $f(1)$, $f(-1)$, $f(3)$, and $f(-6)$.

35. If $f(x) = 2x^2 - 7$ and $g(x) = x^2 + x - 1$, find $f(-2)$, $f(3)$, $g(-4)$, and $g(5)$.

36. If $f(x) = 5x^2 - 2x + 3$ and $g(x) = -x^2 + 4x - 5$, find $f(-2)$, $f(3)$, $g(-4)$, and $g(6)$.

37. If $f(x) = |3x - 2|$ and $g(x) = |x| + 2$, find $f(1)$, $f(-1)$, $g(2)$, and $g(-3)$.

38. If $f(x) = 3|x| - 1$ and $g(x) = -|x| + 1$, find $f(-2)$, $f(3)$, $g(-4)$, and $g(5)$.

39. If $f(x) = 3x + 2$, find $\dfrac{f(a + h) - f(a)}{h}$.

40. If $f(x) = -4x + 3$, find $\dfrac{f(a + h) - f(a)}{h}$.

41. If $f(x) = x^2 + 3$, find $\dfrac{f(a + h) - f(a)}{h}$.

42. If $f(x) = x^2 - 6x + 4$, find $\dfrac{f(a + h) - f(a)}{h}$.

43. If $f(x) = 2x^2 + x - 1$, find $\dfrac{f(a + h) - f(a)}{h}$.

44. Suppose that the cost function for producing a certain item is given by $C(n) = 3n + 5$, where n represents the number of items produced. Compute $C(150)$, $C(500)$, $C(750)$, and $C(1500)$.

45. The height of a projectile fired vertically into the air (neglecting air resistance) at an initial velocity of 64 feet per second is a function of the time (t) and is given by the equation

$$h(t) = 64t - 16t^2.$$

Compute $h(1)$, $h(2)$, $h(3)$, and $h(4)$.

46. The profit function for selling n items is given by $P(n) = -n^2 + 500n - 61,500$. Compute $P(200)$, $P(230)$, $P(250)$, and $P(260)$.

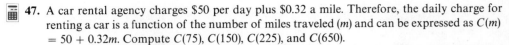 **47.** A car rental agency charges \$50 per day plus \$0.32 a mile. Therefore, the daily charge for renting a car is a function of the number of miles traveled (m) and can be expressed as $C(m) = 50 + 0.32m$. Compute $C(75)$, $C(150)$, $C(225)$, and $C(650)$.

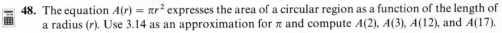 **48.** The equation $A(r) = \pi r^2$ expresses the area of a circular region as a function of the length of a radius (r). Use 3.14 as an approximation for π and compute $A(2)$, $A(3)$, $A(12)$, and $A(17)$.

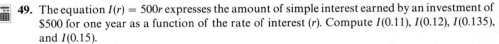 **49.** The equation $I(r) = 500r$ expresses the amount of simple interest earned by an investment of \$500 for one year as a function of the rate of interest (r). Compute $I(0.11)$, $I(0.12)$, $I(0.135)$, and $I(0.15)$.

8.2 Special Functions and Their Graphs

In Section 7.1 we used phrases such as "the graph of the solution set of the equation $y = x - 1$," or simply "the graph of the equation $y = x - 1$" is a line containing the points $(0, -1)$ and $(1, 0)$. Because the equation $y = x - 1$ (which can be written as $f(x) = x - 1$) can be used to specify a function, the line previously referred to is also called the **graph of the function specified by the equation** or simply the **graph of the function**. Generally speaking, the graph of any equation that determines a function is also called the graph of the function. Thus, the graphing techniques discussed in Chapter 7 will continue to play an important role as we graph functions.

As we use the function concept in our study of mathematics, it is helpful to classify certain types of functions and become familiar with their equations, characteristics, and graphs. This section will discuss two special types of functions—**linear** and **quadratic functions**. These functions are merely an outgrowth of our earlier study of linear and quadratic equations.

Linear Functions

Any function that can be written in the form

$$f(x) = ax + b,$$

where a and b are real numbers, is called a **linear function**. The following are examples of linear functions.

$$f(x) = -3x + 6; \qquad f(x) = 2x + 4; \qquad f(x) = -\frac{1}{2}x - \frac{3}{4}.$$

Graphing linear functions is quite easy because the graph of every linear function is a straight line. Therefore, all we need to do is determine two points of the graph and draw the line determined by those two points. You may want to continue using a third point as a check point.

Example 1   Graph the function $f(x) = -3x + 6$.

Solution   Because $f(0) = 6$, the point $(0, 6)$ is on the graph. Likewise, because $f(1) = 3$, the point $(1, 3)$ is on the graph. Plotting these two points and drawing the line determined by the two points produces Figure 8.2.

Figure 8.2

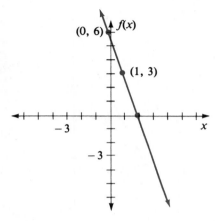

Note that in Figure 8.2 we labeled the vertical axis $f(x)$. It could also be labeled y, since $f(x) = -3x + 6$ and $y = -3x + 6$ mean the same thing. We will continue to use $f(x)$ in this chapter to help you adjust to the function notation.

Example 2   Graph the function $f(x) = x$.

Solution   The equation $f(x) = x$ can be written as $f(x) = 1x + 0$; thus, it is a linear function. Since $f(0) = 0$ and $f(2) = 2$, the points $(0, 0)$ and $(2, 2)$ determine the line in Figure 8.3. The function $f(x) = x$ is often called the **identity function**.

Figure 8.3

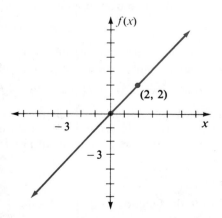

As you graph functions by using function notation, it is often helpful to think of the ordinate of every point on the graph as the value of the function at a specific value of x. Geometrically, this functional value is the directed distance of the point from the x-axis as illustrated in Figure 8.4, with the function $f(x) = 2x - 4$. For example, consider the graph of the function $f(x) = 2$. The function $f(x) = 2$ means that every functional value is 2, or geometrically, that every point of the graph is 2 units above the x-axis. Thus, the graph is the horizontal line indicated in Figure 8.5.

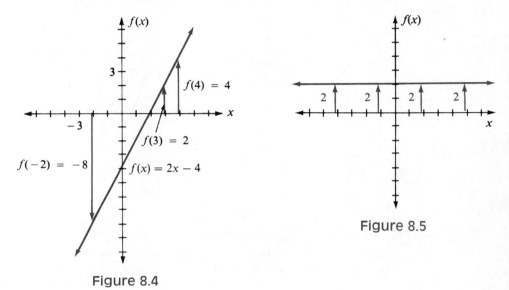

Figure 8.4

Figure 8.5

Any linear function of the form $f(x) = ax + b$, where $a = 0$, is called a **constant function** and its graph is a *horizontal line*.

Quadratic Functions

Any function that can be written in the form

$$f(x) = ax^2 + bx + c,$$

where a, b, and c are real numbers with $a \neq 0$, is called a **quadratic function**. The following are examples of quadratic functions.

$$f(x) = 3x^2; \qquad f(x) = -2x^2 + 5x; \qquad f(x) = 4x^2 - 7x + 1.$$

The techniques discussed in Chapter 7 relative to graphing quadratic equations of the form $y = ax^2 + bx + c$ provide the basis for graphing quadratic functions. Let's review some work from Chapter 7 with an example.

Example 3   Graph the function $f(x) = 2x^2 - 4x + 5$.

Solution

$$f(x) = 2x^2 - 4x + 5$$
$$= 2(x^2 - 2x + \quad) + 5 \qquad \text{(Recall the process of completing the square!)}$$
$$= 2(x^2 - 2x + 1) + 5 - 2$$
$$= 2(x - 1)^2 + 3.$$

From this form we can obtain the following information about the parabola.

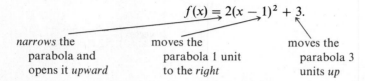

$$f(x) = 2(x - 1)^2 + 3.$$

narrows the
parabola and
opens it *upward*

moves the
parabola 1 unit
to the *right*

moves the
parabola 3
units *up*

Thus, the parabola can be drawn as in Figure 8.6.

Figure 8.6

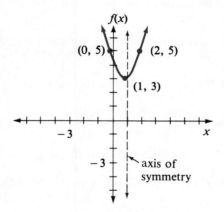

In general, if we complete the square on

$$f(x) = ax^2 + bx + c,$$

we obtain

$$f(x) = a\left(x^2 + \frac{b}{a}x + \underline{\quad}\right) + c$$

$$= a\left(x^2 + \frac{b}{a}x + \frac{b^2}{4a^2}\right) + c - \frac{b^2}{4a}$$

$$= a\left(x + \frac{b}{2a}\right)^2 + \frac{4ac - b^2}{4a}.$$

Therefore, the parabola associated with $f(x) = ax^2 + bx + c$ has its vertex at $\left(-\dfrac{b}{2a}, \dfrac{4ac - b^2}{4a}\right)$ and the equation of its axis of symmetry is $x = -\dfrac{b}{2a}$. These facts are illustrated in Figure 8.7.

Figure 8.7

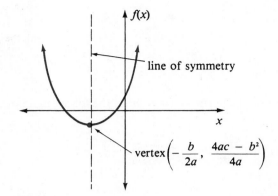

By using the information from Figure 8.7 we now have another way of graphing quadratic functions of the form $f(x) = ax^2 + bx + c$, as indicated by the following steps.

1. Determine whether the parabola opens upward (if $a > 0$) or downward (if $a < 0$).

2. Find $-\dfrac{b}{2a}$, which is the x-coordinate of the vertex.

3. Find $f\left(-\dfrac{b}{2a}\right)$, which is the y-coordinate of the vertex. $\left(\text{You could also find the } y\text{-coordinate by evaluating } \dfrac{4ac - b^2}{4a}.\right)$

4. Locate another point on the parabola and also locate its image across the line of symmetry, $x = -\dfrac{b}{2a}$.

The three points found in Steps **2.**, **3.**, and **4.** should determine the general shape of the parabola. Let's try two examples and use these steps.

Example 4   Graph $f(x) = 3x^2 - 6x + 5$.

Solution   **Step 1**   Because $a = 3$, the parabola opens upward.

Step 2 $-\dfrac{b}{2a} = -\dfrac{-6}{6} = 1.$

Step 3 $f\left(-\dfrac{b}{2a}\right) = f(1) = 3 - 6 + 5 = 2.$
Thus, the vertex is at $(1, 2)$.

Step 4   Letting $x = 2$, we obtain $f(2) = 12 - 12 + 5 = 5$. Thus $(2, 5)$ is on the graph and so is its reflection $(0, 5)$ across the line of symmetry $x = 1$.

The three points $(1, 2)$, $(2, 5)$, and $(0, 5)$ are used to graph the parabola in Figure 8.8.

Figure 8.8

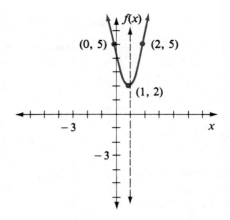

Example 5   Graph $f(x) = -x^2 - 4x - 7$.

Solution   **Step 1**   Since $a = -1$, the parabola opens downward.

Step 2 $-\dfrac{b}{2a} = -\dfrac{-4}{-2} = -2$.

Step 3 $f\left(-\dfrac{b}{2a}\right) = f(-2) = -(-2)^2 - 4(-2) - 7 = -3$.
Thus, the vertex is at $(-2, -3)$.

Step 4   Letting $x = 0$, we obtain $f(0) = -7$. Thus $(0, -7)$ is on the graph and so is its reflection $(-4, -7)$ across the line of symmetry $x = -2$.

The three points $(-2, -3)$, $(0, -7)$, and $(-4, -7)$ are used to draw the parabola in Figure 8.9.

Figure 8.9

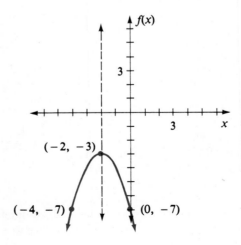

In summary, to graph a quadratic function we basically have two methods:

1. we can express the function in the form $f(x) = a(x - h)^2 + k$ and use the values of a, h, and k to determine the parabola; *or*

2. we can express the function in the form $f(x) = ax^2 + bx + c$ and use the approach demonstrated in Examples 4 and 5.

Other Functions

To graph a new function, that is, one we are unfamiliar with, we can use some of the graphing suggestions offered in Chapter 7. Let's restate those suggestions in terms of functional vocabulary and notation.

1. Determine the domain of the function.

2. Find the *y*-intercept (we are labeling the *y*-axis with $f(x)$) by evaluating $f(0)$. Find the *x*-intercept by finding the value of x such that $f(x) = 0$.

3. Set up a table of ordered pairs that satisfy the equation specifying the function.

4. Plot the points associated with the ordered pairs.

5. Connect the points with a smooth curve.

Let's consider some examples in terms of these suggestions.

Example 6   Graph $f(x) = \sqrt{x}$.

Solution   First, since the radicand must be nonnegative, the domain is

$$D = \{x \mid x \geq 0\}.$$

Therefore, we will use only nonnegative values for x. Second, we see that $f(0) = 0$; so both intercepts are 0. That is, the origin $(0, 0)$ is a point of the graph. Third, let's set up a table of values.

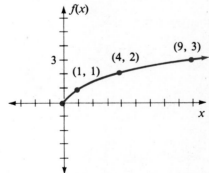

x	$f(x)$
0	0
1	1
4	2
9	3

Finally, plotting these points and connecting them with a smooth curve, we obtain Figure 8.10.

Figure 8.10   ●

Sometimes a new function is defined in terms of old functions. In such cases, the definition plays an important role in the study of the new function. Consider the following example.

Example 7   Graph the function $f(x) = |x|$.

Solution   The concept of absolute value is defined for all real numbers as

$|x| = x$ if $x \geq 0$,

$|x| = -x$ if $x < 0$.

Therefore, the absolute value function can be expressed as

$$f(x) = |x| = \begin{cases} x & \text{if } x \geq 0 \\ -x & \text{if } x < 0 \end{cases}.$$

The graph of $f(x) = x$ for $x \geq 0$ is the ray in the first quadrant and the graph of $f(x) = -x$ for $x < 0$ is the half-line in the second quadrant as indicated in Figure 8.11.

Figure 8.11

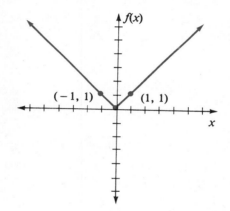

Problem Set 8.2

Graph each of the following linear functions.

1. $f(x) = 2x + 4$ 2. $f(x) = 3x - 1$ 3. $f(x) = -3x + 3$

4. $f(x) = -2x + 4$ 5. $f(x) = 3x$ 6. $f(x) = -2x$

7. $f(x) = 2$ 8. $f(x) = -1$

Graph each of the following quadratic functions.

9. $f(x) = 2x^2$ 10. $f(x) = -4x^2$

11. $f(x) = -2(x + 3)^2 - 1$ 12. $f(x) = 3(x + 2)^2 - 1$

13. $f(x) = x^2 + 8x + 17$ 14. $f(x) = x^2 + 2x - 4$

15. $f(x) = -x^2 + 2x - 6$ 16. $f(x) = -x^2 - 2x - 2$

17. $f(x) = 2x^2 + 12x + 19$ 18. $f(x) = 3x^2 - 12x + 11$

19. $f(x) = -3x^2 + 24x - 46$ 20. $f(x) = -2x^2 - 4x + 2$

21. $f(x) = x^2 + 3x - 1$ 22. $f(x) = x^2 + 5x + 2$

23. $f(x) = -2x^2 + 5x + 1$ 24. $f(x) = -3x^2 + 2x - 1$

Graph each of the following as variations of the basic square root curve, $f(x) = \sqrt{x}$. (See Figure 8.10 on Page 354.)

25. $f(x) = \sqrt{x} + 1$

27. $f(x) = \sqrt{x - 1}$

29. $f(x) = -\sqrt{x}$

31. $f(x) = \sqrt{x + 2} - 1$

33. $f(x) = 2\sqrt{x}$

26. $f(x) = \sqrt{x} - 2$

28. $f(x) = \sqrt{x + 3}$

30. $f(x) = -\sqrt{x - 2}$

32. $f(x) = -\sqrt{x + 1} - 2$

34. $f(x) = -3\sqrt{x}$

Graph each of the following as variations of the basic absolute value curve, $f(x) = |x|$. (See Figure 8.11 on Page 355.)

35. $f(x) = |x| + 3$

37. $f(x) = |x - 1|$

39. $f(x) = -|x + 1|$

41. $f(x) = -|x| + 1$

43. $f(x) = -|x - 3| + 2$

45. $f(x) = 3|x|$

47. $f(x) = 2|x - 1| + 1$

36. $f(x) = |x| - 2$

38. $f(x) = |x + 3|$

40. $f(x) = -|x - 2|$

42. $f(x) = -|x| - 2$

44. $f(x) = -|x + 2| + 4$

46. $f(x) = -2|x|$

48. $f(x) = -3|x + 2| - 1$

49. (a) Graph $f(x) = x^3$.

(b) Graph each of the following as variations of the basic cubic curve from Part **(a)**.

(i) $f(x) = x^3 + 1$

(iii) $f(x) = (x - 2)^3$

(v) $f(x) = -x^3$

(vii) $f(x) = 2x^3$

(ii) $f(x) = x^3 - 2$

(iv) $f(x) = (x + 1)^3$

(vi) $f(x) = -(x - 2)^3$

(viii) $f(x) = (x - 1)^3 + 2$

50. (a) Graph $f(x) = \dfrac{1}{x}$.

(b) Graph each of the following as variations of the basic curve from Part **(a)**.

(i) $f(x) = \dfrac{1}{x} + 2$

(iv) $f(x) = \dfrac{1}{x + 3}$

(ii) $f(x) = \dfrac{1}{x} - 1$

(v) $f(x) = -\dfrac{1}{x}$

(iii) $f(x) = \dfrac{1}{x - 2}$

(vi) $f(x) = \dfrac{2}{x}$

8.3 Problem Solving and the Composition of Functions

As we have seen, the vertex of the graph of a quadratic function is either the lowest or the highest point on the graph. Thus, the vocabulary *minimum value* or *maximum value* of a function is often used in applications of the parabola. The x-value of the vertex indicates where the minimum or maximum occurs and $f(x)$ yields the minimum or maximum value of the function. Let's consider some examples that use these ideas.

Example 1   A farmer has 120 rods of fencing and wants to enclose a rectangular plot of land that requires fencing on only three sides, since it is bounded by a river on one side. Find the length and width of the plot that will maximize the area.

Solution   Let x represent the width; then $120 - 2x$ represents the length, as indicated in Figure 8.12.

Figure 8.12

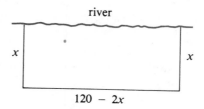

river

x x

$120 - 2x$

The function $A(x) = x(120 - 2x)$ represents the area of the plot in terms of the width x. Since

$$A(x) = x(120 - 2x)$$
$$= 120x - 2x^2$$
$$= -2x^2 + 120x,$$

we have a quadratic function with $a = -2$, $b = 120$, and $c = 0$. Therefore, the x-value where the maximum value of the function is obtained is

$$-\frac{b}{2a} = -\frac{120}{2(-2)} = 30.$$

If $x = 30$, then $120 - 2x = 120 - 2(30) = 60$.

Thus, the farmer should make the plot 30 rods wide and 60 rods long to maximize the area at $(30)(60) = 1800$ square rods.   ●

Example 2   Find two numbers whose sum is 30, such that the sum of their squares is a minimum.

Solution   Let x represent one of the numbers; then $30 - x$ represents the other number. By expressing the sum of the squares as a function of x we obtain

$$f(x) = x^2 + (30 - x)^2,$$

which can be simplified to

$$f(x) = x^2 + 900 - 60x + x^2$$
$$= 2x^2 - 60x + 900.$$

This is a quadratic function with $a = 2$, $b = -60$, and $c = 900$. Therefore, the x-value where the minimum occurs is

$$-\frac{b}{2a} = -\frac{-60}{4} = 15.$$

If $x = 15$, then $30 - x = 30 - (15) = 15$. Thus, the two numbers should both be 15. ●

Example 3   A golf pro-shop operator finds that she can sell 30 sets of golf clubs at $500 per set in a year. Furthermore, she predicts that for each $25 decrease in price, three extra sets of golf clubs could be sold. At what price should she sell the clubs to maximize gross income?

Solution   Sometimes in analyzing such a problem it helps to start setting up a table as follows:

	Number of sets	Price per set	Income
3 additional sets	30	$500	$15,000
can be sold for →	33	$475	$15,675
a $25 decrease	36	$450	$16,200
in price			

Let x represent the number of $25 decreases in price. Then the income can be expressed as a function of x as follows:

$$f(x) = (30 + 3x)(500 - 25x).$$

$$\underset{\substack{\text{number} \\ \text{of sets}}}{\uparrow} \qquad \underset{\text{price per set}}{\uparrow}$$

Simplifying this, we obtain

$$f(x) = 15{,}000 - 750x + 1500x - 75x^2$$
$$= -75x^2 + 750x + 15{,}000.$$

Completing the square we obtain

$$f(x) = -75x^2 + 750x + 15{,}000$$
$$= -75(x^2 - 10x + \underline{\quad}) + 15{,}000$$
$$= -75(x^2 - 10x + 25) + 15{,}000 + 1875$$
$$= -75(x - 5)^2 + 16{,}875.$$

From this form we know that the vertex of the parabola is at $(5, 16875)$. So 5 decreases of $25, that is, a $125 reduction in price, will give a maximum income of $16,875. The golf clubs should be sold at $375 per set. ●

The Composition of Functions

The basic operations of addition, subtraction, multiplication, and division can be performed on functions. However, for our purposes in this text, there is an additional operation, called **composition**, that we will use in the next section. Here is the definition and an illustration of this operation.

Definition 8.3	The *composition* of functions f and g is defined by $$(f \circ g)(x) = f(g(x)),$$ for all x in the domain of g such that $g(x)$ is in the domain of f.

The left side, $(f \circ g)(x)$, of the equation in Definition 8.3 can be read as "the composition of f and g" and the right side, $f(g(x))$, can be read as "f of g of x." It may also be helpful for you to mentally picture Definition 8.3 as two function machines *hooked together* to produce another function (often called a **composite function**) as illustrated in Figure 8.13. Notice that what comes out of the function g is substituted into the function f. Thus, composition is sometimes called the substitution of functions.

Figure 8.13

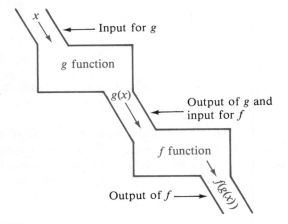

Figure 8.13 also vividly illustrates the fact that $f \circ g$ is defined *for all x in the domain of g such that $g(x)$ is in the domain of f.* In other words, what comes out of g must be capable of being fed into f. Let's consider some examples.

Example 4

If $f(x) = x^2$ and $g(x) = x - 3$, find $(f \circ g)(x)$ and determine its domain.

Solution

Applying Definition 8.3 we obtain

$$(f \circ g)(x) = f(g(x))$$
$$= f(x - 3)$$
$$= (x - 3)^2.$$

Because g and f are both defined for all real numbers, so is $f \circ g$. ●

Example 5   If $f(x) = \sqrt{x}$ and $g(x) = x - 4$, find $(f \circ g)(x)$ and determine its domain.

Solution   Applying Definition 8.3 we obtain

$$(f \circ g)(x) = f(g(x))$$
$$= f(x - 4)$$
$$= \sqrt{x - 4}.$$

The domain of g is all real numbers but the domain of f is only the nonnegative real numbers. Thus $g(x)$, which is $x - 4$, has to be nonnegative. So

$$x - 4 \geq 0$$
$$x \geq 4,$$

and the domain of $f \circ g$ is $D = \{x | x \geq 4\}$.   ●

Definition 8.3, with f and g interchanged, defines the composition of g and f as $(g \circ f)(x) = g(f(x))$.

Example 6   If $f(x) = x^2$ and $g(x) = x - 3$, find $(g \circ f)(x)$ and determine its domain.

Solution
$$(g \circ f)(x) = g(f(x))$$
$$= g(x^2)$$
$$= x^2 - 3.$$

Since f and g are both defined for all real numbers, the domain of $g \circ f$ is the set of all real numbers.   ●

The results of Examples 4 and 6 demonstrate an important idea, namely, that the composition of functions is *not a commutative operation*. In other words, it is not true that $f \circ g = g \circ f$ for all functions f and g. However, as we will see in the next section, there is a special class of functions where $f \circ g = g \circ f$.

Example 7   If $f(x) = 2x + 3$ and $g(x) = \sqrt{x - 1}$, determine each of the following:

(a) $(f \circ g)(x)$ **(b)** $(g \circ f)(x)$ **(c)** $(f \circ g)(5)$ **(d)** $(g \circ f)(7)$

Solution
(a) $(f \circ g)(x) = f(g(x))$
$$= f(\sqrt{x - 1})$$
$$= 2\sqrt{x - 1} + 3, \qquad D = \{x | x \geq 1\}$$
(b) $(g \circ f)(x) = g(f(x))$
$$= g(2x + 3)$$
$$= \sqrt{2x + 3 - 1}$$
$$= \sqrt{2x + 2}, \qquad D = \{x | x \geq -1\}$$

(c) By using the composite function formed in part **(a)**, we obtain

$$(f \circ g)(5) = 2\sqrt{5 - 1} + 3$$
$$= 2\sqrt{4} + 3$$
$$= 2(2) + 3 = 7.$$

(d) By using the composite function formed in part **(b)** we obtain

$$(g \circ f)(7) = \sqrt{2(7) + 2} = \sqrt{16} = 4.$$ ●

Problem Set 8.3

1. Suppose that the equation $p(x) = -2x^2 + 280x - 1000$, where x represents the number of items sold, describes the profit function for a certain business. How many items should be sold to maximize the profit?

2. Suppose that the cost function for a particular item is given by the equation $C(x) = 2x^2 - 320x + 12{,}920$, where x represents the number of items. How many items should be produced to minimize the cost?

3. The height of a projectile fired vertically into the air (neglecting air resistance) at an initial velocity of 96 feet per second is a function of the time and is given by the equation $f(x) = 96x - 16x^2$, where x represents the time. Find the highest point reached by the projectile.

4. Find two numbers whose sum is 30, such that the sum of the square of one number plus ten times the other number is a minimum.

5. Find two numbers whose sum is 50 and whose product is a maximum.

6. Two hundred and forty meters of fencing is available to enclose a rectangular playground. What should be the dimensions of the playground to maximize the area?

7. A motel advertises that they will provide dinner, a dance, and drinks for $50 per couple for a New Year's Eve party. They must have a guarantee of 30 couples. Furthermore, they will agree that for each couple in excess of 30, they will reduce the price per couple for all attending by $0.50. How many couples will it take to maximize the motel's revenue?

8. A Cable TV company has 1000 subscribers who each pay $15 per month. Based on a survey, they feel that for each decrease of $0.25 on the monthly rate, they could obtain 20 additional subscribers. At what rate will maximum revenue be obtained and how many subscribers will it take at that rate?

Determine $(f \circ g)(x)$ and $(g \circ f)(x)$ for each of the following. Also specify the domain for each.

9. $f(x) = 2x, g(x) = 3x - 1$

10. $f(x) = 4x + 1, g(x) = 3x$

11. $f(x) = 5x - 3, g(x) = 2x + 1$

12. $f(x) = 3 - 2x, g(x) = -4x$

13. $f(x) = 3x + 4, g(x) = x^2 + 1$

14. $f(x) = 3, g(x) = -3x^2 - 1$

15. $f(x) = 3x - 4, g(x) = x^2 + 3x - 4$

16. $f(x) = 2x^2 - x - 1, g(x) = x + 4$

17. $f(x) = \dfrac{1}{x}, g(x) = 2x + 7$

18. $f(x) = \dfrac{1}{x^2}, g(x) = x$

19. $f(x) = \sqrt{x - 2}, g(x) = 3x - 1$

20. $f(x) = \dfrac{1}{x}, g(x) = \dfrac{1}{x^2}$

21. $f(x) = \dfrac{1}{x-1}, \; g(x) = \dfrac{2}{x}$ **22.** $f(x) = \dfrac{4}{x+2}, \; g(x) = \dfrac{3}{2x}$

23. If $f(x) = 3x - 2$ and $g(x) = x^2 + 1$, find $(f \circ g)(-1)$ and $(g \circ f)(3)$.

24. If $f(x) = x^2 - 2$ and $g(x) = x + 4$, find $(f \circ g)(2)$ and $(g \circ f)(-4)$.

25. If $f(x) = 2x - 3$ and $g(x) = x^2 - 3x - 4$, find $(f \circ g)(-2)$ and $(g \circ f)(1)$.

26. If $f(x) = \dfrac{1}{x}$ and $g(x) = 2x + 1$, find $(f \circ g)(1)$ and $(g \circ f)(2)$.

27. If $f(x) = \sqrt{x}$ and $g(x) = 3x - 1$, find $(f \circ g)(4)$ and $(g \circ f)(4)$.

28. If $f(x) = x + 5$ and $g(x) = |x|$, find $(f \circ g)(-4)$ and $(g \circ f)(-4)$.

For each of the following, show that $(f \circ g)(x) = x$ and $(g \circ f)(x) = x$.

29. $f(x) = 2x, \; g(x) = \dfrac{1}{2}x$ **30.** $f(x) = \dfrac{3}{4}x, \; g(x) = \dfrac{4}{3}x$

31. $f(x) = x - 2, \; g(x) = x + 2$ **32.** $f(x) = 2x + 1, \; g(x) = \dfrac{x-1}{2}$

33. $f(x) = 3x + 4, \; g(x) = \dfrac{x-4}{3}$ **34.** $f(x) = 4x - 3, \; g(x) = \dfrac{x+3}{4}$

8.4 Inverse Functions

Graphically, the distinction between a relation and a function can be easily recognized. In Figure 8.14 we have sketched four graphs. Which of these are graphs of functions and which are graphs of relations that are not functions? Think in terms of "to each member of the domain there is assigned one and only one member of the range;" this is the basis for what is known as the **vertical line test for functions**. Because each value of x produces only one value of $f(x)$, any vertical line drawn through a **graph of a function must not intersect the graph in more than one point**. Therefore, parts (a) and (c) of Figure 8.14 are graphs of functions and parts (b) and (d) are graphs of relations that are not functions.

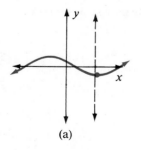

(a)

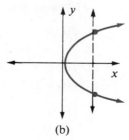

(b)

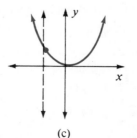

(c)

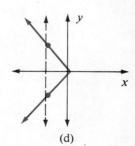

(d)

Figure 8.14

There is also a useful distinction made between two basic types of functions. Consider the graphs of the two functions $f(x) = 2x - 1$ and $f(x) = x^2$ in Figure 8.15. In part (a) any *horizontal line* will intersect the graph in no more than one point. Therefore, every value of $f(x)$ has only one value of x associated with it. Any function that has the additional property of having only one value of x associated with each value of $f(x)$ is called a **one-to-one function**. The function $f(x) = x^2$ is not a one-to-one function because the horizontal line in part (b) of Figure 8.15 intersects the parabola in two points.

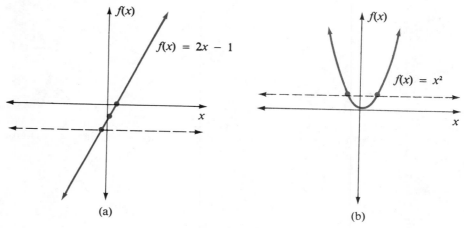

Figure 8.15

In terms of ordered pairs, a one-to-one function does not contain any ordered pairs having the same second component. For example,

$$f = \{(1, 3), (2, 6), (4, 12)\}$$

is a one-to-one function, but

$$g = \{(1, 2), (2, 5), (-2, 5)\}$$

is not a one-to-one function.

If the components of each ordered pair of a given one-to-one function are interchanged, the resulting function and the given function are called **inverses** of each other. Thus,

$$\{(1, 3), (2, 6), (4, 12)\} \quad \text{and} \quad \{(3, 1), (6, 2), (12, 4)\}$$

are **inverse functions**. The inverse of a function f is denoted by f^{-1} (read "f inverse" or "the inverse of f"). If (a, b) is an ordered pair of f, then (b, a) is an ordered pair of f^{-1}. The domain and range of f^{-1} are the range and domain, respectively, of f.

Remark Do not confuse the -1 in f^{-1} with a negative exponent. The symbol f^{-1} does not mean $\dfrac{1}{f^1}$, but refers to the inverse function of function f.

Graphically, two functions that are inverses of each other are mirror images with reference to the line $y = x$. This is due to the fact that ordered pairs (a, b) and (b, a) are mirror images with respect to the line $y = x$ as illustrated in Figure 8.16.

Figure 8.16

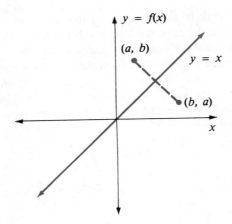

Therefore, if the graph of a function f is known, as in Figure 8.17(a), then the graph of f^{-1} can be determined by reflecting f across the line $y = x$ (Figure 8.17(b)).

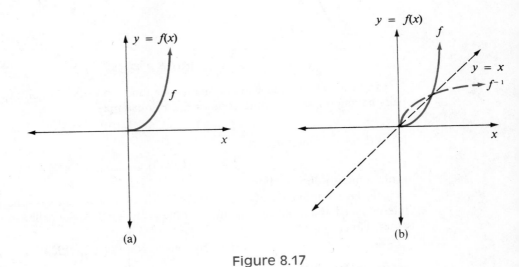

(a)   (b)

Figure 8.17

Another useful way of viewing inverse functions is in terms of composition. Basically, inverse functions *undo* each other and this can be more formally stated as follows. If f and g are inverses of each other, then

1. $(f \circ g)(x) = f(g(x)) = x$ for all x in domain of g; and
2. $(g \circ f)(x) = g(f(x)) = x$ for all x in domain of f.

As we will see in a moment, this relationship of inverse functions can be used to verify whether two functions are indeed inverses of each other.

Finding Inverse Functions

The idea of inverse functions *undoing each other* provides the basis for a rather informal approach to finding the inverse of a function. Consider the function

$$f(x) = 2x + 1.$$

To each x this function assigns *twice x plus* 1. To *undo* this function, we could *subtract* 1 *and divide by* 2. So, the inverse should be

$$f^{-1}(x) = \frac{x - 1}{2}.$$

Now let's verify that f and f^{-1} are inverses of each other.

$$(f \circ f^{-1})(x) = f(f^{-1}(x)) = f\left(\frac{x - 1}{2}\right) = 2\left(\frac{x - 1}{2}\right) + 1 = x$$

and

$$(f^{-1} \circ f)(x) = f^{-1}(f(x)) = f^{-1}(2x + 1) = \frac{2x + 1 - 1}{2} = x.$$

Thus, the inverse of f is given by

$$f^{-1}(x) = \frac{x - 1}{2}.$$

Let's consider another example of finding an inverse function by the *undoing* process.

Example 1   Find the inverse of $f(x) = 3x - 5$.

Solution   To each x, the function f assigns *three times x minus* 5. To *undo* this we can *add* 5 and then *divide by* 3. So, the inverse should be

$$f^{-1}(x) = \frac{x + 5}{3}.$$

To verify that f and f^{-1} are inverses we can show that

$$(f \circ f^{-1})(x) = f(f^{-1}(x)) = f\left(\frac{x + 5}{3}\right) = 3\left(\frac{x + 5}{3}\right) - 5 = x$$

and

$$(f^{-1} \circ f)(x) = f^{-1}(f(x)) = f^{-1}(3x - 5) = \frac{3x - 5 + 5}{3} = x.$$

Thus, f and f^{-1} are inverses and we can write

$$f^{-1}(x) = \frac{x + 5}{3}.$$

The following is a more formal technique for finding the inverse of a function.

1. Let $y = f(x)$.
2. Solve for x in terms of y.
3. Interchange x and y.
4. $f^{-1}(x)$ is determined by the final equation.

The following examples illustrate this procedure.

Example 2   Find the inverse of $f(x) = 3x - 5$.

Solution   First, let $y = f(x)$, so that the equation becomes

$$y = 3x - 5.$$

Second, solving this equation for x produces

$$y = 3x - 5$$
$$y + 5 = 3x$$
$$\frac{y + 5}{3} = x.$$

Third, interchanging x and y produces

$$\frac{x + 5}{3} = y.$$

Thus, the inverse of f is

$$f^{-1}(x) = \frac{x + 5}{3}.$$

Example 3   Find the inverse of $f(x) = \frac{2}{3}x + \frac{3}{5}$.

Solution   Let $y = f(x)$, so the equation becomes

$$y = \frac{2}{3}x + \frac{3}{5}.$$

Solving for x produces

$$15y = \left(\frac{2}{3}x + \frac{3}{5}\right)15$$
$$15y = 10x + 9$$

$$15y - 9 = 10x$$

$$\frac{15y - 9}{10} = x.$$

Interchanging variables we obtain

$$\frac{15x - 9}{10} = y$$

where the inverse of f

$$f^{-1}(x) = \frac{15x - 9}{10}$$

is obtained.

To verify that f and f^{-1} are inverses, we can show that

$$(f \circ f^{-1})(x) = x \quad \text{and} \quad (f^{-1} \circ f)(x) = x. \text{ (You may complete this.)}$$

●

Problem Set 8.4

Identify each of the following as (a) a function or (b) a relation that is not a function. Use the vertical line test.

1.

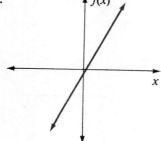

2.

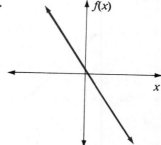

3.

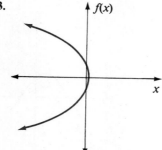

4.

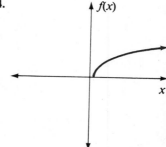

5.

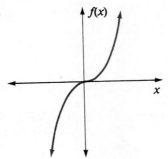

6.

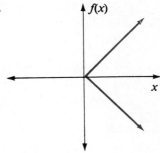

Identify each of the following functions as (*a*) a one-to-one function or (*b*) a function that is not one-to-one. Use the horizontal line test.

7.

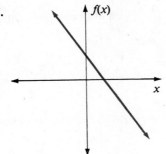

8.

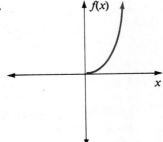

9.

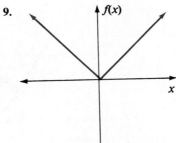

10.

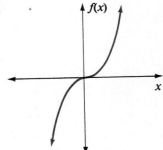

11.

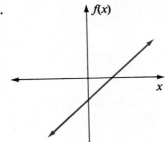

12.

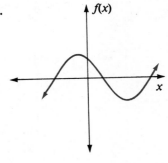

For each of the following functions, (a) list the domain and range, (b) form the inverse function, and (c) list the domain and range of the inverse.

13. $f = \{(1, 5), (2, 9), (5, 21)\}$

14. $f = \{(1, 1), (4, 2), (9, 3), (16, 4)\}$

15. $f = \{(0, 0), (2, 8), (-1, -1), (-2, -8)\}$

16. $f = \{(-1, 1), (-2, 4), (-3, 9), (-4, 16)\}$

For each of the following functions, (a) find the inverse by using the *undoing* process and (b) verify that the two functions are inverses by showing that $(f \circ f^{-1})(x) = x$ and $(f^{-1} \circ f)(x) = x$.

17. $f(x) = 2x + 3$

18. $f(x) = 4x - 5$

19. $f(x) = \dfrac{1}{2}x$

20. $f(x) = -\dfrac{1}{3}x$

21. $f(x) = 5x + 9$

22. $f(x) = 3x + 7$

23. $f(x) = \dfrac{1}{3}x - 4$

24. $f(x) = \dfrac{1}{2}x + 5$

25. $f(x) = \dfrac{2}{3}x + 5$

26. $f(x) = \dfrac{3}{4}x - 1$

For each of the following functions, (a) find the inverse by using the process illustrated in Examples 2 and 3, and (b) verify that the two functions are inverses by showing that $(f \circ f^{-1})(x) = x$ and $(f^{-1} \circ f)(x) = x$.

27. $f(x) = 4x$

28. $f(x) = \dfrac{1}{2}x$

29. $f(x) = 2x + 9$

30. $f(x) = 3x - 11$

31. $f(x) = -\dfrac{2}{3}x$

32. $f(x) = -\dfrac{4}{3}x$

33. $f(x) = -3x - 4$

34. $f(x) = -5x + 6$

35. $f(x) = \dfrac{3}{4}x - \dfrac{5}{6}$

36. $f(x) = \dfrac{2}{3}x - \dfrac{1}{4}$

For each of the following functions (a) find the inverse, and (b) graph the given function and its inverse on the same set of axes.

37. $f(x) = 6x$

38. $f(x) = \dfrac{3}{5}x$

39. $f(x) = -\dfrac{1}{4}x$

40. $f(x) = -2x$

41. $f(x) = 2x - 1$

42. $f(x) = 3x + 4$

43. $f(x) = -4x + 3$

44. $f(x) = -5x - 1$

45. Find the inverse of $f(x) = x^2$, where $x \geq 0$, and graph both f and f^{-1} on the same set of axes.

46. Find the inverse of $f(x) = x^2 + 1$, where $x \geq 0$, and graph both f and f^{-1} on the same set of axes.

47. Explain why every nonconstant linear function has an inverse.

Miscellaneous Problems

48. The composition idea can also be used to find the inverse of a function. For example, to find the inverse of $f(x) = 5x + 3$, we could proceed as follows:

$$f(f^{-1}(x)) = 5(f^{-1}(x)) + 3 \quad \text{and} \quad f(f^{-1}(x)) = x.$$

Therefore, equating the two expressions for $f(f^{-1}(x))$ we obtain

$$5(f^{-1}(x)) + 3 = x$$
$$5(f^{-1}(x)) = x - 3$$
$$f^{-1}(x) = \frac{x-3}{5}.$$

Use this approach to find the inverse of each of the following functions.

(a) $f(x) = 2x + 1$ **(b)** $f(x) = 3x - 2$ **(c)** $f(x) = -4x + 5$

(d) $f(x) = -x + 1$ **(e)** $f(x) = 2x$ **(f)** $f(x) = -5x$

8.5 Direct and Inverse Variations

"The distance a car travels at a fixed rate *varies directly* as the time." "At a constant temperature, the volume of an enclosed gas *varies inversely* as the pressure." Such statements illustrate two basic types of functional relationships, called **direct** and **inverse variation**, which are widely used, especially in the physical sciences. These relationships can be expressed by equations that specify functions. The purpose of this section is to investigate these special functions.

The statement *y varies directly as x* means

$$y = kx$$

where k is a nonzero constant, called the **constant of variation**. The phrase "y is directly proportional to x," is also used to indicate direct variation; k is then referred to as the **constant of proportionality**.

Remark Notice that the equation $y = kx$ defines a function and could be written as $f(x) = kx$ by using function notation. However, in this section it is more convenient to avoid the function notation and use variables that are meaningful in terms of the physical entities involved in the problem.

Statements indicating direct variation may also involve powers of x. For example, "y varies directly as the square of x" can be written as

$$y = kx^2.$$

In general, "y varies directly as the nth power of $x(n > 0)$" means

$$y = kx^n.$$

There are basically three types of problems in dealing with direct variation, namely, (1) translating an English statement into an equation expressing the direct variation, (2) finding the constant of variation from given values of the variables, and (3) finding additional values of the variables once the constant of variation has been determined. Let's consider an example of each of these types of problems.

Example 1   Translate the statement "the tension on a spring varies directly as the distance it is stretched" into an equation using k as the constant of variation.

Solution   Letting t represent the tension and d the distance, the equation becomes

$$t = kd.$$

●

Example 2   If A varies directly as the square of s, and $A = 28$ when $s = 2$, find the constant of variation.

Solution   Since A varies directly as the square of s, we have

$$A = ks^2.$$

Substituting $A = 28$ and $s = 2$, we obtain

$$28 = k(2)^2.$$

Solving this equation for k yields

$$28 = 4k$$
$$7 = k.$$

The constant of variation is 7.

●

Example 3   If y is directly proportional to x and if $y = 6$ when $x = 9$, find the value of y when $x = 24$.

Solution   The statement "y is directly proportional to x" translates into

$$y = kx.$$

Letting $y = 6$ and $x = 9$, the constant of variation becomes

$$6 = k(9)$$
$$6 = 9k$$
$$\frac{6}{9} = k$$
$$\frac{2}{3} = k.$$

So, the specific equation is $y = \frac{2}{3}x$.

Now, letting $x = 24$, we obtain

$$y = \frac{2}{3}(24) = 16.$$

The required value of y is 16.

Inverse Variation

The second basic type of variation, called **inverse variation**, is defined as follows. The statement y *varies inversely as* x means

$$y = \frac{k}{x}$$

where k is a nonzero constant and is again referred to as the constant of variation. The phrase "y is inversely proportional to x" is also used to express inverse variation. As with direct variation, statements indicating inverse variation may involve powers of x. For example, "y varies inversely as the square of x" can be written as

$$y = \frac{k}{x^2}.$$

In general, "y varies inversely as the nth power of $x(n > 0)$" means

$$y = \frac{k}{x^n}.$$

The following examples illustrate the three basic kinds of problems that we run across involving inverse variation.

Example 4   Translate the statement "the length of a rectangle of fixed area varies inversely as the width" into an equation using k as the constant of variation.

Solution   Letting l represent the length and w the width, the equation is

$$l = \frac{k}{w}.$$

Example 5   If y is inversely proportional to x and $y = 4$ when $x = 12$, find the constant of variation.

Solution   Since y is inversely proportional to x, we have

$$y = \frac{k}{x}.$$

Substituting $y = 4$ and $x = 12$, we obtain

$$4 = \frac{k}{12}.$$

Solving this equation for k yields

$$k = 48.$$

The constant of variation is 48.   ●

Example 6   Suppose the number of days it takes to complete a construction job varies inversely as the number of people assigned to the job. If it takes 7 people 8 days to do the job, how long would it take 14 people to complete the job?

Solution   Let d represent the number of days and p the number of people. The phrase "number of days … varies inversely as the number of people" translates into

$$d = \frac{k}{p}.$$

Letting $d = 8$ when $p = 7$, the constant of variation becomes

$$8 = \frac{k}{7}$$

$$k = 56.$$

So, the specific equation is

$$d = \frac{56}{p}.$$

Now, letting $p = 14$, we obtain

$$d = \frac{56}{14}$$

$$d = 4.$$

It should take 14 people 4 days to complete the job.   ●

The terms *direct* and *inverse*, as applied to variation, refer to the relative behavior of the variables involved in the equation. That is to say, in direct variation ($y = kx$) an assignment of *increasing absolute values* for x produces *increasing absolute values* for y. Whereas, in inverse variation $\left(y = \dfrac{k}{x} \right)$ an assignment of *increasing absolute values* for x produces *decreasing absolute values* for y.

Joint Variation

Variation may involve more than two variables. The following table illustrates some variation statements and their equivalent algebraic equations using k as the constant of variation.

Variation statement	Algebraic equation
1. y *varies jointly* as x and z	$y = kxz$
2. y *varies jointly* as x, z, and w	$y = kxzw$
3. V *varies jointly* as h and the square of r	$V = khr^2$
4. h varies directly as V and inversely as w	$h = \dfrac{kV}{w}$
5. y is directly proportional to x and inversely proportional to the square of z	$y = \dfrac{kx}{z^2}$
6. y varies jointly as w and z, and inversely as x	$y = \dfrac{kwz}{x}$

Statements **1.**, **2.**, and **3.** illustrate the concept of **joint variation**. Statements **4.** and **5.** show that both direct and inverse variation may occur in the same problem. Statement **6.** combines joint variation with inverse variation.

The two final examples that follow illustrate problems involving some of these possible variation situations.

Example 7   The length of a rectangular box with a fixed height varies directly as the volume and inversely as the width. If the length is 12 centimeters when the volume is 960 cubic centimeters and the width is 8 centimeters, find the length when the volume is 700 centimeters and the width is 5 centimeters.

Solution   Using l for length, V for volume, and w for width, the phrase "length varies directly as the volume and inversely as the width" translates into

$$l = \frac{kV}{w}.$$

Substituting $l = 12$, $V = 960$, and $w = 8$ the constant of variation becomes

$$12 = \frac{k(960)}{8}$$

$$12 = 120k$$

$$\frac{1}{10} = k.$$

So the specific equation is

$$l = \frac{\frac{1}{10}V}{w} = \frac{V}{10w}.$$

Now, letting $V = 700$ and $w = 5$, we obtain

$$l = \frac{700}{10(5)} = \frac{700}{50} = 14.$$

The length is 14 centimeters.

Example 8

Suppose that y varies jointly as x and z, and inversely as w. If $y = 154$ when $x = 6$, $z = 11$, and $w = 3$, find the constant of variation.

Solution

The statement "y varies jointly as x and z, and inversely as w" translates into

$$y = \frac{kxz}{w}.$$

Substituting $y = 154$, $x = 6$, $z = 11$, and $w = 3$, we obtain

$$154 = \frac{k(6)(11)}{3}$$

$$154 = 22k$$

$$7 = k.$$

The constant of variation is 7.

Problem Set 8.5

Translate each of the following statements of variation into an equation using k as the constant of variation.

1. y varies directly as the square of x.

2. y varies inversely as the cube of x.

3. V varies jointly as l and w.

4. c varies directly as g and inversely as the cube of t.

5. At a constant temperature, the volume (V) of a gas varies inversely as the pressure (P).

6. The volume (V) of a sphere is directly proportional to the cube of its radius (r).

7. The intensity of illumination (I) received from a source of light is inversely proportional to the square of the distance (d) from the source.

8. The surface area (S) of a cube varies directly as the square of the length of an edge (e).

9. The volume (V) of a gas varies directly as the absolute temperature (T) and inversely as the pressure (P).

10. The volume (V) of a cone varies jointly as its height and the square of its radius.

Find the constant of variation for each of the following stated conditions.

11. y varies directly as x, and $y = 24$ when $x = 3$.

12. y varies directly as x, and $y = 9$ when $x = 12$.

13. A varies directly as the square of r, and $A = 154$ when $r = 7$.

14. V varies directly as the cube of e, and $V = 27$ when $e = 3$.

15. A varies jointly as b and h, and $A = 52$ when $b = 8$ and $h = 13$.

16. V varies jointly as B and h, and $V = 88$ when $B = 24$ and $h = 11$.

17. y varies inversely as x, and $y = 6$ when $x = 3$.

18. r varies inversely as t, and $r = 24$ when $t = \dfrac{5}{2}$.

19. y varies directly as x and inversely as z, and $y = 8$ when $x = 12$ and $z = 6$.

20. y is directly proportional to the square of x and inversely proportional to the cube of z, and $y = 18$ when $x = 9$ and $z = 3$.

Solve each of the following problems.

21. If y is directly proportional to x, and $y = 10$ when $x = 15$, find the value of y when $x = 24$.

22. If y is inversely proportional to the square of x, and $y = 4$ when $x = 3$, find y when $x = 6$.

23. The time required for a car to travel a certain distance varies inversely as the rate at which it travels. If it takes 4 hours at 50 miles per hour to travel the distance, how long will it take at 40 miles per hour?

24. The volume of a gas at a constant temperature varies inversely as the pressure. What is the volume of a gas under a pressure of 25 pounds if the gas occupies 15 cubic centimeters under a pressure of 20 pounds?

25. The distance that a freely falling body falls varies directly as the square of the time it falls. If a body falls 144 feet in 3 seconds, how far will it fall in 5 seconds?

26. The volume (V) of a gas varies directly as the temperature (T) and inversely as the pressure (P). If $V = 48$ when $T = 320$ and $P = 20$, find V when $T = 280$ and $P = 30$.

27. The simple interest earned by a certain amount of money varies jointly as the rate of interest and the time (in years) that the money is invested. If $120 is earned for the money invested at 12% for 2 years, how much is earned if the money is invested at 14% for 3 years?

28. The period (the time required for one complete oscillation) of a simple pendulum varies directly as the square root of its length. If a pendulum 12 feet long has a period of 4 seconds, find the period of a pendulum of length 3 feet.

29. The volume of a cylinder varies jointly as its altitude and the square of the radius of its base. If the volume of a cylinder is 1386 cubic centimeters when the radius of the base is 7 centimeters and its altitude is 9 centimeters, find the volume of a cylinder that has a base of radius 14 centimeters; the altitude of the cylinder is 5 centimeters.

30. The electrical resistance of a wire varies directly as its length and inversely as the square of its diameter. If the resistance of 200 meters of wire having a diameter of $\dfrac{1}{2}$ centimeter is 1.5 ohms, find the resistance of 400 meters of wire with a diameter of $\dfrac{1}{4}$ centimeter.

In the previous problems numbers were chosen to make computations reasonable without the use of a calculator. However, many times variation-type problems involve messy computations and the calculator becomes a very useful tool. Use your calculator to help solve the following problems.

31. The simple interest earned by a certain amount of money varies jointly as the rate of interest and the time (in years) that the money is invested.

(a) If some money invested at 11% for 2 years earns $385, how much would the same amount earn at 12% for 1 year?

(b) If some money invested at 12% for 3 years earns $819, how much would the same amount earn at 14% for 2 years?

(c) If some money invested at 14% for 4 years earns $1960, how much would the same amount earn at 15% for 2 years?

32. The period (the time required for one complete oscillation) of a simple pendulum varies directly as the square root of its length. If a pendulum 9 inches long has a period of 2.4 seconds, find the period of a pendulum of length 12 inches. Express answer to the nearest one-tenth of a second.

33. The volume of a cylinder varies jointly as its altitude and the square of the radius of its base. If the volume of a cylinder is 549.5 cubic meters when the radius of the base is 5 meters and its altitude is 7 meters, find the volume of a cylinder that has a base of radius 9 meters and an altitude of 14 meters.

34. If y is directly proportional to x and inversely proportional to the square of z, and if $y = 0.336$ when $x = 6$ and $z = 5$, find the constant of variation.

35. If y is inversely proportional to the square root of x, and $y = 0.08$ when $x = 225$, find y when $x = 625$.

Chapter 8 Summary

(8.1)   A **relation** is a set of ordered pairs; a **function** is a relation in which no two ordered pairs have the same first component. The **domain** of a relation (or function) is the set of all first components and the **range** is the set of all second components.

Single symbols such as f, g, and h are commonly used to name functions. The symbol $f(x)$ represents the element in the range associated with x from the domain. Thus, if $f(x) = 3x + 7$, then $f(1) = 3(1) + 7 = 10$.

(8.2)   Three special types of functions are summarized as follows:

Equation	Type of function	Graph
1. $f(x) = b$, where b is a real number	constant	horizontal line
2. $f(x) = ax + b$, where a and b are real numbers and $a \neq 0$	linear	straight line
3. $f(x) = ax^2 + bx + c$, where a, b, and c are real numbers and $a \neq 0$	quadratic	parabola

The following suggestions are given for graphing functions you are unfamiliar with.

1. Determine the domain of the function.
2. Find the intercepts.
3. Set up a table of values.
4. Plot the points associated with the ordered pairs.
5. Connect the points with a smooth curve.

A useful graphing tool is the ability to recognize variations of certain basic curves.

(8.3) Some applications involving maximum or minimum values can be solved by using our knowledge of parabolas that are generated by quadratic functions.

The **composition** of two functions f and g is defined by

$$(f \circ g)(x) = f(g(x))$$

for all x in the domain of g such that $g(x)$ is in the domain of f.

(8.4) **A one-to-one function** is a function such that no two ordered pairs have the same second component.

If the components of each ordered pair of a given one-to-one function are interchanged, the resulting function and the given function are **inverses** of each other. The inverse of a function f is denoted by f^{-1}.

Graphically, two functions that are inverses of each other are mirror images with reference to the line $y = x$.

Two functions f and f^{-1} can be shown to be inverses of each other by verifying that

1. $(f^{-1} \circ f)(x) = x$ for all x in the domain of f.
2. $(f \circ f^{-1})(x) = x$ for all x in the domain of f^{-1}.

A technique for finding the inverse of a function can be described as follows.

1. Let $y = f(x)$.
2. Solve the equation for x in terms of y.
3. Interchange x and y.
4. $f^{-1}(x)$ is determined by the final equation.

(8.5) The equation $y = kx$ (k is a nonzero constant) defines a function called a **direct variation**. The equation $y = \dfrac{k}{x}$ defines a function called **inverse variation**. In both cases, k is called the **constant of variation**.

Chapter 8 Review Problem Set

For Problems 1–4, specify the domain of each function.

1. $f = \{(1, 3), (2, 5), (4, 9)\}$

2. $f(x) = \dfrac{4}{x - 5}$

3. $f(x) = \dfrac{3}{x^2 + 4x}$

4. $f(x) = \sqrt{x^2 - 25}$

5. If $f(x) = x^2 - 2x - 1$, find $f(2)$, $f(-3)$, and $f(a)$.

6. If $f(x) = 2x^2 + x - 7$, find $\dfrac{f(a + h) - f(a)}{h}$.

For Problems 7–16, graph each of the functions.

7. $f(x) = 4$

8. $f(x) = -3x + 2$

9. $f(x) = x^2 + 2x + 2$

10. $f(x) = |x| + 4$

11. $f(x) = -|x - 2|$

12. $f(x) = \sqrt{x - 2} - 3$

13. $f(x) = \dfrac{1}{x^2}$

14. $f(x) = -\dfrac{1}{2}x^2$

15. $f(x) = -3x^2 + 6x - 2$

16. $f(x) = -\sqrt{x + 1} - 2$

17. Find the coordinates of the vertex and the equation of the line of symmetry for each of the following parabolas.

(a) $f(x) = x^2 + 10x - 3$

(b) $f(x) = -2x^2 - 14x + 9$

For Problems 18–20, determine $(f \circ g)(x)$ and $(g \circ f)(x)$ for each pair of functions.

18. $f(x) = 2x - 3$ and $g(x) = 3x - 4$

19. $f(x) = x - 4$ and $g(x) = x^2 - 2x + 3$

20. $f(x) = x^2 - 5$ and $g(x) = -2x + 5$

For Problems 21–23, find the inverse (f^{-1}) of the given function.

21. $f(x) = 6x - 1$

22. $f(x) = \dfrac{2}{3}x + 7$

23. $f(x) = -\dfrac{3}{5}x - \dfrac{2}{7}$

24. If y varies directly as x and inversely as z, and if $y = 21$ when $x = 14$ and $z = 6$, find the constant of variation.

25. If y varies jointly as x and the square root of z, and if $y = 60$ when $x = 2$ and $z = 9$, find y when $x = 3$ and $z = 16$.

26. The weight of a body above the surface of the earth varies inversely as the square of its distance from the center of the earth. Assuming the radius of the earth to be 4000 miles, how much would a man weigh 1000 miles above the earth's surface if he weighs 200 pounds on the surface?

27. Find two numbers whose sum is 40 and whose product is a maximum.

28. Find two numbers whose sum is 50 such that the square of one number plus six times the other number is a minimum.

29. Suppose that 50 students are able to raise $250 for a party by each one contributing $5. Furthermore, they figure that for each additional student they can find to contribute, the cost per student will decrease by a nickel. How many additional students do they need to find so as to maximize the amount they will have for a party?

30. The surface area of a cube varies directly as the square of the length of an edge. If the surface area of a cube having edges 8 inches long is 384 square inches, find the surface area of a cube having edges 10 inches long.

9 Polynomial and Rational Functions

Earlier in this text we solved linear and quadratic equations as well as graphed linear and quadratic functions. In this chapter we will expand our equation-solving processes and graphing techniques to include more general polynomial equations and functions. Then our knowledge of polynomial functions will allow us to work with rational functions. The function concept will again serve as a unifying thread throughout the chapter. To facilitate our study in this chapter, we will first review the concept of dividing polynomials and will introduce a special division technique called synthetic division.

9.1 Synthetic Division

In Section 4.5 we discussed the process of dividing polynomials by using the following format.

$$
\require{enclose}
\begin{array}{r}
x^2 - 2x + 4 \\
3x + 1 \enclose{longdiv}{3x^3 - 5x^2 + 10x + 1} \\
\underline{3x^3 + x^2 } \\
-6x^2 + 10x + 1 \\
\underline{-6x^2 - 2x } \\
12x + 1 \\
\underline{12x + 4} \\
-3
\end{array}
$$

We also suggested writing the final result as

$$\frac{3x^3 - 5x^2 + 10x + 1}{3x + 1} = x^2 - 2x + 4 + \frac{-3}{3x + 1}.$$

Multiplying both sides of this equation by $3x + 1$ produces

$$3x^3 - 5x^2 + 10x + 1 = (3x + 1)(x^2 - 2x + 4) + (-3),$$

which is of the familiar form,

$$\text{Dividend} = (\text{Divisor})(\text{Quotient}) + \text{Remainder}.$$

This result is commonly called the **Division Algorithm for Polynomials** and can be stated in general terms as follows.

Division Algorithm for Polynomials

If $f(x)$ and $g(x)$ are polynomials and $g(x) \neq 0$, then there exist unique polynomials $q(x)$ and $r(x)$ such that

$$f(x) = g(x)q(x) + r(x)$$

$$\underset{\text{Dividend}}{} \quad \underset{\text{Divisor}}{} \quad \underset{\text{Quotient}}{} \quad \underset{\text{Remainder}}{}$$

where $r(x) = 0$ or the degree of $r(x)$ is less than the degree of $g(x)$.

If the divisor is of the form $x - c$, where c is a constant, then the typical long division algorithm can be conveniently simplified into a process called **synthetic division**. First, let's consider an example using the usual algorithm. Then, in a step-by-step fashion, we will list some short-cuts to use that will lead us into the synthetic division procedure. Consider the division problem $(2x^4 + x^3 - 17x^2 + 13x + 2) \div (x - 2)$.

$$
\begin{array}{r}
2x^3 + 5x^2 - 7x - 1 \\
x - 2\overline{)\,2x^4 + x^3 - 17x^2 + 13x + 2} \\
\underline{2x^4 - 4x^3} \\
5x^3 - 17x^2 \\
\underline{5x^3 - 10x^2} \\
-7x^2 + 13x \\
\underline{-7x^2 + 14x} \\
-x + 2 \\
\underline{-x + 2}
\end{array}
$$

Notice that because the dividend $(2x^4 + x^3 - 17x^2 + 13x + 2)$ is written in descending powers of x, the quotient $(2x^3 + 5x^2 - 7x - 1)$ is also in descending powers of x. In other words, the numerical coefficients are the key. So let's rewrite the above problem in terms of its coefficients.

$$
\begin{array}{r}
2 \quad\; 5 \;-\; 7 \;-1 \\
1-2)\overline{2 \quad\; 1 \;-17 \quad 13 \quad 2} \\
\textcircled{2} \;-4 \\
\overline{5 \;\;\textcircled{-17}} \\
\textcircled{5} \;-10 \\
\overline{-7 \;\;\textcircled{13}} \\
\overline{\textcircled{-7} \;\; 14} \\
-1 \;\textcircled{2} \\
\overline{\textcircled{-1} \;\; 2}
\end{array}
$$

Now observe that the numbers circled are simply repetitions of the numbers directly above them in the format. Thus, the circled numbers could be omitted, and the format would be as follows. (Disregard the arrows for the moment.)

$$
\begin{array}{r}
2 \quad\; 5 \;-\; 7 \;-1 \\
1-2)\overline{2 \quad\; 1 \;-17 \quad 13 \quad 2} \\
-4 \\
\overline{5} \\
\overline{-10} \\
\overline{-7} \\
14 \\
\overline{-1} \\
2
\end{array}
$$

Next, moving some numbers up as indicated by the arrows and omitting writing 1 as the coefficient of x in the divisor yields the following more compact form.

$$
\begin{array}{r}
2 \quad\; 5 \;-\; 7 \;-1 \hspace{3em} (1)\\
-2)\overline{2 \quad\; 1 \;-17 \quad 13 \quad 2} \hspace{2em} (2)\\
-4 \;-10 \quad 14 \quad 2 \hspace{2em} (3)\\
\overline{5 \;-\; 7 \;-1} \hspace{3em} (4)
\end{array}
$$

Notice that line (4) reveals all of the coefficients of the quotient [line (1)] except for the first coefficient, 2. Thus, we can omit line (1), begin line (4) with the first coefficient, and then use the following form.

$$
\begin{array}{r}
-2)\overline{2 \quad\; 1 \;-17 \quad 13 \quad 2} \hspace{2em} (5)\\
-4 \;-10 \quad 14 \quad 2 \hspace{2em} (6)\\
\overline{2 \quad\; 5 \;-\; 7 \;-1 \quad 0} \hspace{2em} (7)
\end{array}
$$

Line (7) contains the coefficients of the quotient; the 0 indicates the remainder.

Finally, changing the constant in the divisor to 2 (instead of -2), which will change the signs of the numbers in line (6), allows us to add the corresponding entries in lines (5) and (6) rather than subtract them. Thus, the final synthetic division form for this problem is as follows.

$$
\begin{array}{r}
2)\overline{2 \quad 1 \;-17 \quad 13 \quad 2} \\
4 \quad\; 10 \;-14 \;-2 \\
\overline{2 \quad 5 \;-\; 7 \;-\; 1 \quad 0}
\end{array}
$$

Now we will consider another problem, following a step-by-step procedure for setting up and carrying out the synthetic division process. Suppose that we want to do the following division problem.

$$x + 4 \overline{) 2x^3 + 5x^2 - 13x - 2}$$

1. Write the coefficients of the dividend as follows.

$$\overline{)2 \quad 5 \quad -13 \quad -2}$$

2. In the divisor, use -4 instead of 4 so that later we can add rather than subtract.

$$-4 \overline{)2 \quad 5 \quad -13 \quad -2}$$

3. Bring down the first coefficient of the dividend.

$$-4 \overline{)2 \quad 5 \quad -13 \quad -2}$$
$$\overline{}$$
$$2$$

4. Multiply that first coefficient times the divisor, which yields $2(-4) = -8$. This result is added to the second coefficient of the dividend.

$$-4 \overline{)2 \quad\quad 5 \quad -13 \quad -2}$$
$$\quad\quad\quad -8$$
$$\overline{}$$
$$2 \quad -3$$

5. Multiply $(-3)(-4)$, which yields 12; this result is added to the third coefficient of the dividend.

$$-4 \overline{)2 \quad\quad 5 \quad -13 \quad -2}$$
$$\quad\quad\quad -8 \quad\quad 12$$
$$\overline{}$$
$$2 \quad -3 \quad -1$$

6. Multiply $(-1)(-4)$, which yields 4; this result is added to the last term of the dividend.

$$-4 \overline{)2 \quad\quad 5 \quad -13 \quad -2}$$
$$\quad\quad\quad -8 \quad\quad 12 \quad\quad 4$$
$$\overline{}$$
$$2 \quad -3 \quad -1 \quad 2$$

The last row indicates a quotient of $2x^2 - 3x - 1$ and a remainder of 2.

Let's consider three more examples showing only the final compact form for synthetic division.

Example 1   Find the quotient and remainder for $(2x^3 - 5x^2 + 6x + 4) \div (x - 2)$.

Solution
$$2 \overline{)2 \quad -5 \quad\quad 6 \quad\quad 4}$$
$$\quad\quad\quad 4 \quad -2 \quad\quad 8$$
$$\overline{}$$
$$2 \quad -1 \quad\quad 4 \quad 12$$

Therefore, the quotient is $2x^2 - x + 4$, and the remainder is 12.   ●

Example 2   Find the quotient and remainder for $(4x^4 - 2x^3 + 6x - 1) \div (x - 1)$.

Solution

$$1\overline{)4 \quad -2 \quad 0 \quad 6 \quad -1}$$
$$ \quad 4 \quad 2 \quad 2 \quad 8$$
$$4 \quad 2 \quad 2 \quad 8 \quad 7$$

Notice that a zero has been inserted as the coefficient of the missing x^2 term.

Thus, the quotient is $4x^3 + 2x^2 + 2x + 8$, and the remainder is 7.

Example 3   Find the quotient and remainder for $(x^3 + 8x^2 + 13x - 6) \div (x + 3)$.

Solution

$$-3\overline{)1 \quad 8 \quad 13 \quad -6}$$
$$ \quad -3 \quad -15 \quad 6$$
$$1 \quad 5 \quad -2 \quad 0$$

Thus, the quotient is $x^2 + 5x - 2$, and the remainder is 0.

In Example 3, because the remainder is 0, we can say that $x + 3$ is a factor of $x^3 + 8x^2 + 13x - 6$. We will use this idea a bit later when solving polynomial equations.

Problem Set 9.1

Use synthetic division to determine the quotient and the remainder for each of the following.

1. $(3x^2 + x - 4) \div (x - 1)$

2. $(2x^2 - 5x - 3) \div (x - 3)$

3. $(x^2 + 2x - 10) \div (x - 4)$

4. $(x^2 - 10x + 15) \div (x - 8)$

5. $(4x^2 + 5x - 4) \div (x + 2)$

6. $(5x^2 + 18x - 8) \div (x + 4)$

7. $(x^3 - 2x^2 - x + 2) \div (x - 2)$

8. $(x^3 - 5x^2 + 2x + 8) \div (x + 1)$

9. $(3x^4 - x^3 + 2x^2 - 7x - 1) \div (x + 1)$

10. $(2x^3 - 5x^2 - 4x + 6) \div (x - 2)$

11. $(x^3 - 7x - 6) \div (x + 2)$

12. $(x^3 + 6x^2 - 5x - 1) \div (x - 1)$

13. $(x^4 + 4x^3 - 7x - 1) \div (x - 3)$

14. $(2x^4 + 3x^2 + 3) \div (x + 2)$

15. $(x^3 + 6x^2 + 11x + 6) \div (x + 3)$

16. $(x^3 - 4x^2 - 11x + 30) \div (x - 5)$

17. $(x^5 - 1) \div (x - 1)$

18. $(x^5 - 1) \div (x + 1)$

19. $(x^5 + 1) \div (x - 1)$

20. $(x^5 + 1) \div (x + 1)$

21. $(2x^5 + 3x^4 - 4x^3 - x^2 + 5x - 2) \div (x + 2)$

22. $(x^5 + 3x^4 - 5x^3 - 3x^2 + 3x - 4) \div (x + 4)$

23. $(3x^5 - 8x^4 + 5x^3 + 2x^2 - 9x + 4) \div (x - 2)$

24. $(4x^5 - 6x^4 + 2x^3 + 2x^2 - 5x + 2) \div (x - 1)$

25. $(2x^3 + 3x^2 - 2x + 3) \div \left(x + \dfrac{1}{2}\right)$

26. $(9x^3 - 6x^2 + 3x - 4) \div \left(x - \dfrac{1}{3}\right)$

27. $(4x^4 - 5x^2 + 1) \div \left(x - \dfrac{1}{2}\right)$

28. $(3x^4 - 2x^3 + 5x^2 - x - 1) \div \left(x + \dfrac{1}{3}\right)$

9.2 The Remainder and Factor Theorems

Let's consider the division algorithm (stated in the previous section) when the dividend, $f(x)$, is divided by a linear polynomial of the form $x - c$. Then the division algorithm

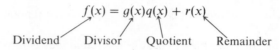

$$f(x) = g(x)q(x) + r(x)$$

Dividend   Divisor   Quotient   Remainder

becomes

$$f(x) = (x - c)q(x) + r(x).$$

Because the degree of the remainder, $r(x)$, must be less than the degree of the divisor, $x - c$, the remainder is a constant. Therefore, letting R represent the remainder, we have

$$f(x) = (x - c)q(x) + R.$$

If the functional value at c is found, we obtain

$$\begin{aligned} f(c) &= (c - c)q(c) + R \\ &= 0 \cdot q(c) + R \\ &= R. \end{aligned}$$

In other words, if a polynomial is divided by a linear polynomial of the form $x - c$, then the remainder is given by the value of the polynomial at c. Let's state this result more formally as the Remainder Theorem.

Remainder Theorem
Property 9.1 If the polynomial $f(x)$ is divided by $x - c$, then the remainder is equal to $f(c)$.

Example 1   If $f(x) = x^3 + 2x^2 - 5x - 1$, find $f(2)$ by (a) using synthetic division and the Remainder Theorem and (b) evaluating $f(2)$ directly.

Solution

(a)

$$2 \overline{)\begin{array}{rrr} 1 & 2 & -5 & -1 \\ & 2 & 8 & 6 \\ \hline 1 & 4 & 3 & \boxed{5} \end{array}} \quad \longleftarrow R = f(2)$$

(b) $f(2) = 2^3 + 2(2)^2 - 5(2) - 1 = 8 + 8 - 10 - 1 = 5$ ●

Example 2   If $f(x) = x^4 + 7x^3 + 8x^2 + 11x + 5$, find $f(-6)$ by (a) using synthetic division and the Remainder Theorem and (b) evaluating $f(-6)$ directly.

Solution

(a)

$$-6 \overline{)\begin{array}{rrrrr} 1 & 7 & 8 & 11 & 5 \\ & -6 & -6 & -12 & 6 \\ \hline 1 & 1 & 2 & -1 & \boxed{11} \end{array}} \quad \longleftarrow R = f(-6)$$

(b) $f(-6) = (-6)^4 + 7(-6)^3 + 8(-6)^2 + 11(-6) + 5$

$\qquad = 1296 - 1512 + 288 - 66 + 5$

$\qquad = 11$ ●

In Example 2, notice that the computations involved in finding $f(-6)$ using synthetic division and the Remainder Theorem are much easier than those required to evaluate $f(-6)$ directly. This is not always the case, but often using synthetic division is easier than evaluating $f(c)$ directly.

Example 3   Find the remainder when $x^3 + 3x^2 - 13x - 15$ is divided by $x + 1$.

Solution   Letting $f(x) = x^3 + 3x^2 - 13x - 15$ and writing $x + 1$ as $x - (-1)$, we can apply the Remainder Theorem.

$$f(-1) = (-1)^3 + 3(-1)^2 - 13(-1) - 15 = 0$$

Thus, the remainder is 0. ●

Example 3 illustrates an important special case of the Remainder Theorem—the situation in which the remainder is *zero*. Thus, we can say that $x + 1$ is a factor of $x^3 + 3x^2 - 13x - 15$.

Factor Theorem

A general factor theorem can be formulated by considering the equation

$$f(x) = (x - c)q(x) + R.$$

If $x - c$ is a factor of $f(x)$, then the remainder R, which is also $f(c)$, must be zero.

Conversely, if $R = f(c) = 0$, then $f(x) = (x - c)q(x)$; in other words, $x - c$ is a factor of $f(x)$. The Factor Theorem can be stated as follows.

Factor Theorem

Property 9.2   A polynomial $f(x)$ has a factor $x - c$ if and only if $f(c) = 0$.

Example 4   Is $x - 1$ a factor of $x^3 + 5x^2 + 2x - 8$?

Solution   Letting $f(x) = x^3 + 5x^2 + 2x - 8$ and computing $f(1)$, we obtain

$$f(1) = 1^3 + 5(1)^2 + 2(1) - 8 = 0.$$

By the Factor Theorem, therefore, $x - 1$ is a factor of $f(x)$.   ●

Example 5   Is $x + 3$ a factor of $2x^3 + 5x^2 - 6x - 7$?

Solution   Using synthetic division, we obtain the following.

$$
\begin{array}{r|rrrr}
-3 & 2 & 5 & -6 & -7 \\
   &   & -6 & 3 & 9 \\
\hline
   & 2 & -1 & -3 & ② \\
\end{array}
\qquad \longleftarrow R = f(-3)
$$

Since $R \neq 0$, we know that $x + 3$ is not a factor of the given polynomial.   ●

In Examples 4 and 5 we were only concerned with determining whether a linear polynomial of the form $x - c$ was a factor of another polynomial. For such problems, it is reasonable to compute $f(c)$ either directly or by synthetic division, whichever way seems easier for a particular problem. However, if more information is required, such as the complete factorization of the given polynomial, then the use of synthetic division becomes appropriate, as the next two examples illustrate.

Example 6   Show that $x - 1$ is a factor of $x^3 - 2x^2 - 11x + 12$, and find the other linear factors of the polynomial.

Solution   Let's use synthetic division to divide $x^3 - 2x^2 - 11x + 12$ by $x - 1$.

$$
\begin{array}{r|rrrr}
1 & 1 & -2 & -11 & 12 \\
  &   & 1 & -1 & -12 \\
\hline
  & 1 & -1 & -12 & 0 \\
\end{array}
$$

The last line indicates a quotient of $x^2 - x - 12$ and a remainder of 0. The remainder of 0 means that $x - 1$ is a factor. Furthermore, we can write

$$x^3 - 2x^2 - 11x + 12 = (x - 1)(x^2 - x - 12).$$

The quadratic polynomial $x^2 - x - 12$ can be factored as $(x - 4)(x + 3)$ using our conventional factoring techniques. Thus, we obtain

$$x^3 - 2x^2 - 11x + 12 = (x - 1)(x - 4)(x + 3).$$ ●

Example 7   Show that $x + 4$ is a factor of $f(x) = x^3 - 5x^2 - 22x + 56$, and complete the factorization of $f(x)$.

Solution   Using synthetic division to divide $x^3 - 5x^2 - 22x + 56$ by $x + 4$, we obtain

$$
\begin{array}{r|rrrr}
-4 & 1 & -5 & -22 & 56 \\
   &   & -4 & 36 & -56 \\
\hline
   & 1 & -9 & 14 & 0 \\
\end{array}
$$

The last line indicates a quotient of $x^2 - 9x + 14$ and a remainder of 0. The remainder of 0 means that $x + 4$ is a factor. Furthermore, we can write

$$x^3 - 5x^2 - 22x + 56 = (x + 4)(x^2 - 9x + 14)$$

and then complete the factoring to obtain

$$f(x) = x^3 - 5x^2 - 22x + 56 = (x + 4)(x - 7)(x - 2).$$ ●

The Factor Theorem also plays a significant role in determining some general factorization ideas, as the last example of this section illustrates.

Example 8   Verify that $x + 1$ is a factor of $x^n + 1$ for all odd positive integral values of n.

Solution   Letting $f(x) = x^n + 1$ and computing $f(-1)$, we obtain

$$f(-1) = (-1)^n + 1$$
$$= -1 + 1 \quad \text{(Any odd power of } -1 \text{ is } -1.)$$
$$= 0.$$

Since $f(-1) = 0$, we know that $x + 1$ is a factor of $f(x)$. ●

Problem Set 9.2

For Problems 1–10, find $f(c)$ by (a) using synthetic division and the Remainder Theorem and (b) evaluating $f(c)$ directly.

1. $f(x) = x^2 + x - 8$ and $c = 2$
2. $f(x) = x^3 + x^2 - 2x - 4$ and $c = -1$
3. $f(x) = 3x^3 + 4x^2 - 5x + 3$ and $c = -4$
4. $f(x) = 2x^4 + x^2 + 6$ and $c = 1$
5. $f(x) = x^4 - 2x^3 - 3x^2 + 5x - 1$ and $c = -2$
6. $f(x) = 2x^4 + x^3 - 4x^2 - x + 1$ and $c = 2$

7. $f(t) = 6t^3 - 35t^2 + 8t - 10$ and $c = 6$

8. $f(t) = 2t^5 - 1$ and $c = -2$

9. $f(n) = 3n^4 - 2n^3 + 4n - 1$ and $c = 3$

10. $f(n) = -2n^4 + 4n - 5$ and $c = -3$

For Problems 11–18, find $f(c)$ *either* by using synthetic division and the Remainder Theorem *or* by evaluating $f(c)$ directly.

11. $f(x) = 5x^6 - x^3 - 1$ and $c = -1$

12. $f(x) = 2x^3 - 3x^2 - 5x + 4$ and $c = 4$

13. $f(x) = x^4 - 8x^3 + 9x^2 - 15x + 2$ and $c = 7$

14. $f(t) = 5t^3 - 8t^2 + 9t - 4$ and $c = -5$

15. $f(n) = -2n^4 + 2n^2 - n - 5$ and $c = -2$

16. $f(x) = 4x^7 + 3$ and $c = 3$

17. $f(x) = 2x^3 - 5x^2 + 4x - 3$ and $c = \dfrac{1}{2}$

18. $f(x) = 3x^3 + 4x^2 - 5x - 7$ and $c = -\dfrac{1}{3}$

For Problems 19–28, use the Factor Theorem to help answer some questions about factors.

19. Is $x - 2$ a factor of $3x^2 - 4x - 4$?

20. Is $x + 3$ a factor of $6x^2 + 13x - 15$?

21. Is $x + 2$ a factor of $x^3 + x^2 - 7x - 10$?

22. Is $x - 3$ a factor of $2x^3 - 3x^2 - 10x + 3$?

23. Is $x - 1$ a factor of $3x^3 + 5x^2 - x - 2$?

24. Is $x + 4$ a factor of $x^3 - 4x^2 + 2x - 8$?

25. Is $x - 2$ a factor of $x^3 - 8$?

26. Is $x + 2$ a factor of $x^3 + 8$?

27. Is $x - 3$ a factor of $x^4 - 81$?

28. Is $x + 3$ a factor of $x^4 - 81$?

For Problems 29–34, use synthetic division to show that $g(x)$ is a factor of $f(x)$, and complete the factorization of $f(x)$.

29. $g(x) = x + 2$, $f(x) = x^3 + 7x^2 + 4x - 12$

30. $g(x) = x - 1$, $f(x) = 3x^3 + 19x^2 - 38x + 16$

31. $g(x) = x - 3$, $f(x) = 6x^3 - 17x^2 - 5x + 6$

32. $g(x) = x + 2$, $f(x) = 12x^3 + 29x^2 + 8x - 4$

33. $g(x) = x + 1$, $f(x) = x^3 - 2x^2 - 7x - 4$

34. $g(x) = x - 5$, $f(x) = 2x^3 + x^2 - 61x + 30$

For Problems 35–38, find the value(s) of k that makes the second polynomial a factor of the first.

35. $x^3 - kx^2 + 5x + k$; $x - 2$

36. $k^2x^4 + 3kx^2 - 4$; $x - 1$

37. $x^3 + 4x^2 - 11x + k$; $x + 2$

38. $kx^3 + 19x^2 + x - 6$; $x + 3$

39. Show that $x + 2$ is a factor of $x^{12} - 4096$.

40. Argue that $f(x) = 2x^4 + x^2 + 3$ has no factor of the form $x - c$, where c is a real number.

41. Verify that $x - 1$ is a factor of $x^n - 1$ for all positive integral values of n.

42. Verify that $x + 1$ is a factor of $x^n - 1$ for all even positive integral values of n.

43. (a) Verify that $x - y$ is a factor of $x^n - y^n$ for all positive integral values of n.

   (b) Verify that $x + y$ is a factor of $x^n - y^n$ for all even positive integral values of n.

   (c) Verify that $x + y$ is a factor of $x^n + y^n$ for all odd positive integral values of n.

Miscellaneous Problems

The Remainder and Factor Theorems are true for any complex value of c. Therefore, for Problems 44–46, find $f(c)$ by (a) using synthetic division and the Remainder Theorem and (b) evaluating $f(c)$ directly.

44. $f(x) = x^3 - 5x^2 + 2x + 1$ and $c = i$

45. $f(x) = x^2 + 4x - 2$ and $c = 1 + i$

46. $f(x) = x^3 + 2x^2 + x - 2$ and $c = 2 - 3i$

47. Show that $x - 2i$ is a factor of $f(x) = x^4 + 6x^2 + 8$.

48. Show that $x + 3i$ is a factor of $f(x) = x^4 + 14x^2 + 45$.

49. Consider changing the form of the polynomial $f(x) = x^3 + 4x^2 - 3x + 2$ as follows.

$$f(x) = x^3 + 4x^2 - 3x + 2$$
$$= x(x^2 + 4x - 3) + 2$$
$$= x[x(x + 4) - 3] + 2$$

The final form $f(x) = x[x(x + 4) - 3] + 2$ is called the **nested form** of the polynomial. It is particularly well suited for evaluating functional values of f either by hand or with a calculator. For each of the following, find the indicated functional values using the nested form of the given polynomial.

   (a) $f(4)$, $f(-5)$, and $f(7)$ for $f(x) = x^3 + 5x^2 - 2x + 1$

   (b) $f(3)$, $f(6)$, and $f(-7)$ for $f(x) = 2x^3 - 4x^2 - 3x + 2$

   (c) $f(4)$, $f(5)$, and $f(-3)$ for $f(x) = -2x^3 + 5x^2 - 6x - 7$

   (d) $f(5)$, $f(6)$, and $f(-3)$ for $f(x) = x^4 + 3x^3 - 2x^2 + 5x - 1$

9.3 Polynomial Equations

We have solved a large variety of linear equations of the form $ax + b = 0$ and quadratic equations of the form $ax^2 + bx + c = 0$. Linear and quadratic equations are special cases of a general class of equations referred to as **polynomial equations**. The equation

$$a_n x^n + a_{n-1} x^{n-1} + \cdots + a_1 x + a_0 = 0,$$

where the coefficients $a_0, a_1, \ldots, a_n$ are real numbers and n is a positive integer, is called a **polynomial equation of degree n**. The following are examples of polynomial equations.

$$\sqrt{2}x - 6 = 0 \qquad \text{Degree 1}$$

$$\frac{3}{4}x^2 - \frac{2}{3}x + 5 = 0 \qquad \text{Degree 2}$$

$$4x^3 - 3x^2 - 7x - 9 = 0 \qquad \text{Degree 3}$$

$$5x^4 - x + 6 = 0 \qquad \text{Degree 4}$$

Remark   The most general polynomial equation would allow complex numbers as coefficients. However, for our purposes in this text we will restrict the coefficients to real numbers. Such equations are often referred to as **polynomial equations over the reals**.

In general, solving polynomial equations of degree greater than 2 can be very difficult and often requires mathematics beyond the scope of this text. However, there are some general properties pertaining to the solving of polynomial equations with which you should be familiar, and furthermore, there are certain types of polynomial equations that we can solve using the techniques available to us at this time.

Let's begin by listing some polynomial equations and corresponding solution sets that we have already encountered in this text.

Equation	Solution Set
$3x + 4 = 7$	$\{1\}$
$x^2 + x - 6 = 0$	$\{-3, 2\}$
$2x^3 - 3x^2 - 2x + 3 = 0$	$\left\{-1, 1, \frac{3}{2}\right\}$
$x^4 - 16 = 0$	$\{-2, 2, -2i, 2i\}$

Notice that in each of the above examples the number of solutions corresponds to the degree of the equation. The first-degree equation has one solution, the second-degree equation has two solutions, the third-degree equation has three solutions, and the fourth-degree equation has four solutions.

Now consider the equation

$$(x - 4)^2(x + 5)^3 = 0.$$

It can be written as

$$(x - 4)(x - 4)(x + 5)(x + 5)(x + 5) = 0,$$

which implies that

$$x - 4 = 0 \quad \text{or} \quad x - 4 = 0 \quad \text{or} \quad x + 5 = 0 \quad \text{or} \quad x + 5 = 0 \quad \text{or} \quad x + 5 = 0.$$

Therefore,

$$x = 4 \quad \text{or} \quad x = 4 \quad \text{or} \quad x = -5 \quad \text{or} \quad x = -5 \quad \text{or} \quad x = -5.$$

We state that the solution set of the original equation is $\{-5, 4\}$, but we also say that the equation has a solution of 4 with a **multiplicity of two**, and a solution of -5 with a **multiplicity of three**. Furthermore, notice that the sum of the multiplicities is 5, which agrees with the degree of the equation.

The following general property can be stated.

Property 9.3	A polynomial equation of degree n has n solutions, where any solution of multiplicity p is counted p times.

Finding Rational Solutions

Although solving polynomial equations of a degree greater than 2 can, in general, be very difficult, **rational solutions of polynomial equations with integral coefficients** can be found using techniques presented in this chapter. The following property restricts the potential rational solutions of such equations.

Property 9.4	**Rational Root Theorem** Consider the polynomial equation $$a_n x^n + a_{n-1} x^{n-1} + \cdots + a_1 x + a_0 = 0,$$ where the coefficients $a_0, a_1, \ldots, a_n$ are *integers*. If the rational number $\dfrac{c}{d}$, reduced to lowest terms, is a solution of the equation, then c is a factor of the constant term a_0 and d is a factor of the leading coefficient a_n.

The "why" behind the Rational Root Theorem is based on some simple factoring ideas, as indicated by the following outline of a proof for the theorem.

Outline of Proof If $\dfrac{c}{d}$ is to be a solution, then

$$a_n \left(\frac{c}{d}\right)^n + a_{n-1} \left(\frac{c}{d}\right)^{n-1} + \cdots + a_1 \left(\frac{c}{d}\right) + a_0 = 0.$$

Multiplying both sides of this equation by d^n and adding $-a_0 d^n$ to both sides yields

$$a_n c^n + a_{n-1} c^{n-1} d + \cdots + a_1 c d^{n-1} = -a_0 d^n.$$

Because c is a factor of the left side of this equation, c must also be a factor of $-a_0 d^n$. Furthermore, because $\dfrac{c}{d}$ is in reduced form, c and d have no common factors other than

-1 or 1. Thus, c is a factor of a_0. In the same way, from the equation

$$a_{n-1}c^{n-1}d + \cdots + a_1 cd^{n-1} + a_0 d^n = -a_n c^n$$

we can conclude that d is a factor of the left side and therefore d is also a factor of a_n.

The Rational Root Theorem, synthetic division, the Factor Theorem, and some previous knowledge pertaining to solving linear and quadratic equations merge to form a basis for finding rational solutions. Let's consider some examples.

Example 1   Find all rational solutions of $3x^3 + 8x^2 - 15x + 4 = 0$.

Solution   If $\dfrac{c}{d}$ is a rational solution, then c must be a factor of 4 and d must be a factor of 3. Therefore, the possible values for c and d are as follows:

for c: $\pm 1, \pm 2, \pm 4,$
for d: $+1, \pm 3.$

Thus, the possible values for $\dfrac{c}{d}$ are

$$\pm 1, \pm \frac{1}{3}, \pm 2, \pm \frac{2}{3}, \pm 4, \pm \frac{4}{3}.$$

Using synthetic division, we obtain

$$
\begin{array}{r|rrrr}
1 & 3 & 8 & -15 & 4 \\
  &   & 3 & 11 & -4 \\
\hline
  & 3 & 11 & -4 & 0
\end{array}
$$

which shows that $x - 1$ is a factor of the given polynomial; therefore, 1 is a rational solution of the equation. Furthermore, the synthetic division result also indicates that we can factor the given polynomial as follows.

$$3x^3 + 8x^2 - 15x + 4 = 0$$
$$(x - 1)(3x^2 + 11x - 4) = 0$$

The quadratic factor can be factored further using our previous techniques; we can proceed as follows.

$$(x - 1)(3x^2 + 11x - 4) = 0$$
$$(x - 1)(3x - 1)(x + 4) = 0$$
$$x - 1 = 0 \quad \text{or} \quad 3x - 1 = 0 \quad \text{or} \quad x + 4 = 0$$
$$x = 1 \quad \text{or} \quad x = \frac{1}{3} \quad \text{or} \quad x = -4$$

Thus, the entire solution set consists of rational numbers and can be listed as $\left\{ -4, \dfrac{1}{3}, 1 \right\}$.

In Example 1, we were fortunate that a rational solution was the result of our first use of synthetic division. This does not happen frequently; we often need to conduct an organized search, as the next example illustrates.

Example 2   Find all rational solutions of $3x^3 + 7x^2 - 22x - 8 = 0$.

Solution   If $\dfrac{c}{d}$ is a rational solution, then c must be a factor of -8 and d must be a factor of 3. Therefore, the possible values for c and d are as follows:

for c: $\pm1, \pm2, \pm4, \pm8$,

for d: $\pm1, \pm3$.

Thus, the possible values for $\dfrac{c}{d}$ are

$$\pm1, \ \pm\frac{1}{3}, \ \pm2, \ \pm\frac{2}{3}, \ \pm4, \ \pm\frac{4}{3}, \ \pm8, \ \pm\frac{8}{3}.$$

Let's begin our search for rational solutions, trying the integers first.

$$
\begin{array}{r|rrrr}
1 & 3 & 7 & -22 & -8 \\
  &   & 3 & 10 & -12 \\
\hline
  & 3 & 10 & -12 & \boxed{-20}
\end{array}
$$
⟵ This remainder indicates that $x - 1$ is not a factor and thus 1 is not a solution.

$$
\begin{array}{r|rrrr}
-1 & 3 & 7 & -22 & -8 \\
   &   & -3 & -4 & 26 \\
\hline
   & 3 & 4 & -26 & \boxed{18}
\end{array}
$$
⟵ This remainder indicates that -1 is not a solution.

$$
\begin{array}{r|rrrr}
2 & 3 & 7 & -22 & -8 \\
  &   & 6 & 26 & 8 \\
\hline
  & 3 & 13 & 4 & 0
\end{array}
$$

Now we know that $x - 2$ is a factor; we can proceed as follows.

$$3x^3 + 7x^2 - 22x - 8 = 0$$
$$(x - 2)(3x^2 + 13x + 4) = 0$$
$$(x - 2)(3x + 1)(x + 4) = 0$$

$x - 2 = 0$ or $3x + 1 = 0$ or $x + 4 = 0$

$x = 2$ or $3x = -1$ or $x = -4$

$x = 2$ or $x = -\dfrac{1}{3}$ or $x = -4$

The solution set is $\left\{-4, -\dfrac{1}{3}, 2\right\}$.

In Examples 1 and 2 we were solving third-degree equations. Therefore, after finding one linear factor by synthetic division we were able to factor the remaining quadratic factor in the usual way. However, if the given equation is of degree 4 or more, we may need to find more than one linear factor by synthetic division, as the next example illustrates.

Example 3   Solve $x^4 - 6x^3 + 22x^2 - 30x + 13 = 0$.

Solution   The possible values for $\dfrac{c}{d}$ are as follows:

for $\dfrac{c}{d}$: $\pm 1, \pm 13$.

By synthetic division we find that

$$\begin{array}{r|rrrrr} 1 & 1 & -6 & 22 & -30 & 13 \\ & & 1 & -5 & 17 & -13 \\ \hline & 1 & -5 & 17 & -13 & 0 \end{array}$$

which indicates that $x - 1$ is a factor of the given polynomial. The bottom line of the synthetic division indicates that the given polynomial can be factored as follows.

$$x^4 - 6x^3 + 22x^2 - 30x + 13 = 0$$
$$(x - 1)(x^3 - 5x^2 + 17x - 13) = 0$$

Therefore,

$$x - 1 = 0 \quad \text{or} \quad x^3 - 5x^2 + 17x - 13 = 0.$$

Now we can use the same approach to look for rational solutions of the expression $x^3 - 5x^2 + 17x - 13 = 0$. The possible values for $\dfrac{c}{d}$ are as follows:

for $\dfrac{c}{d}$: $\pm 1, \pm 13$.

By synthetic division we find that

$$\begin{array}{r|rrrr} 1 & 1 & -5 & 17 & -13 \\ & & 1 & -4 & 13 \\ \hline & 1 & -4 & 13 & 0 \end{array}$$

which indicates that $x - 1$ is a factor of $x^3 - 5x^2 + 17x - 13$ and that the other factor is $x^2 - 4x + 13$.

Now we can solve the original equation as follows.

$$x^4 - 6x^3 + 22x^2 - 30x + 13 = 0$$
$$(x - 1)(x^3 - 5x^2 + 17x - 13) = 0$$
$$(x - 1)(x - 1)(x^2 - 4x + 13) = 0$$
$$x - 1 = 0 \quad \text{or} \quad x - 1 = 0 \quad \text{or} \quad x^2 - 4x + 13 = 0$$
$$x = 1 \qquad \text{or} \qquad x = 1 \quad \text{or} \quad x^2 - 4x + 13 = 0$$

Using the quadratic formula on $x^2 - 4x + 13 = 0$ produces

$$x = \frac{4 \pm \sqrt{16 - 52}}{2} = \frac{4 \pm \sqrt{-36}}{2}$$

$$= \frac{4 + 6i}{2} = 2 \pm 3i$$

Thus, the original equation has a rational solution of 1 with a multiplicity of 2 and has two complex solutions, $2 + 3i$ and $2 - 3i$. The solution set would be listed as $\{1, 2 \pm 3i\}$.

●

Example 3 illustrates two general properties. First, notice that the coefficient of x^4 is 1, and thus the possible rational solutions must be integers. *In general, the possible rational solutions of $x^n + a_{n-1}, x^{n-1} + \cdots + a_1x + a_0 = 0$ are the integral factors of a_0.* Second, notice that the complex solutions of Example 3 are conjugates of each other. The following general property can be stated.

Property 9.5	Nonreal complex solutions of polynomial equations with real coefficients, if they exist, must occur in conjugate pairs.

Remark The justification for Property 9.5 is based on some properties of conjugate complex numbers. We will not show the details of such a proof in this text.

Each of Properties 9.3, 9.4, and 9.5 yields some information about the solutions of a polynomial equation. Before stating the final property of this section, which will give us some additional information, we need to consider two ideas.

First, in a polynomial that is arranged in descending powers of x, if two successive terms differ in sign, then there is said to be a **variation in sign**. (Terms with zero coefficients are disregarded when sign variations are counted.) For example, the polynomial

$$3x^3 - 2x^2 + 4x + 7$$

has *two* sign variations, whereas the polynomial

$$x^5 - 4x^3 + x - 5$$

has *three* variations.

Second, the solutions of

$$a_n(-x)^n + a_{n-1}(-x)^{n-1} + \cdots + a_1(-x) + a_0 = 0$$

are the opposites of the solutions of

$$a_nx^n + a_{n-1}x^{n-1} + \cdots + a_1x + a_0 = 0.$$

In other words, if a new equation is formed by replacing x with $-x$ in a given equation, then the solutions of the newly formed equation are the opposites of the solutions of the given equation. For example, the solution set of $x^2 + 7x + 12 = 0$ is

$\{-4, -3\}$, and the solution set of $\{-x)^2 + 7(-x) + 12 = 0$, which simplifies to $x^2 - 7x + 12 = 0$, is $\{3, 4\}$.

Now we can state a property that often helps us to determine the nature of the solutions of a polynomial equation without actually solving the equation.

Descartes' Rule of Signs

Property 9.6

Let $a_n x^n + a_{n-1} x^{n-1} + \cdots + a_1 x + a_0 = 0$ be a polynomial equation with real coefficients.

1. The number of *positive real solutions* of the given equation is either equal to the number of variations in sign of the polynomial or else less than the number of variations by a positive even integer.

2. The number of *negative real solutions* of the given equation is either equal to the number of variations in sign of the polynomial $a_n(-x)^n + a_{n-1}(-x)^{n-1} + \cdots + a_1(-x) + a_0$ or else less than the number of variations by a positive even integer.

Along with Properties 9.3 and 9.5, Property 9.6 allows us to acquire some information about the solutions of a polynomial equation without actually solving the equation. Let's consider some equations and see how much we know about their solutions, without solving them.

1. $x^3 + 3x^2 + 5x + 4 = 0$

   (a) No variations of sign in $x^3 + 3x^2 + 5x + 4$ means that there are *no positive solutions*.

   (b) Replacing x with $-x$ in the given polynomial produces $(-x)^3 + 3(-x)^2 + 5(-x) + 4$, which simplifies to $-x^3 + 3x^2 - 5x + 4$ and contains three variations of sign; thus, there are *three or one negative solutions*.

   Conclusion: The given equation has three negative real solutions or else one negative real solution and two nonreal complex solutions.

2. $2x^4 + 3x^2 - x - 1 = 0$

   (a) There is one variation of the sign in the given polynomial; thus, the equation has *one positive solution*.

   (b) Replacing x with $-x$ produces $2(-x)^4 + 3(-x)^2 - (-x) - 1$, which simplifies to $2x^4 + 3x^2 + x - 1$ and contains one variation of sign. Thus, the given equation has *one negative solution*.

   Conclusion: The given equation has one positive, one negative, and two nonreal complex solutions.

3. $3x^4 + 2x^2 + 5 = 0$

   (a) No variations of sign in the given polynomial means that there are *no positive solutions*.

   (b) Replacing x with $-x$ produces $3(-x)^4 + 2(-x)^2 + 5$, which simplifies to $3x^4 + 2x^2 + 5$ and contains no variations of sign. Thus, there are *no negative solutions*.

Conclusion: The given equation contains four nonreal complex solutions. These solutions will appear in conjugate pairs.

4. $2x^5 - 4x^3 + 2x - 5 = 0$

(a) The fact that there are three variations of sign in the given polynomial implies that the *number of positive solutions is three or one.*

(b) Replacing x with $-x$ produces $2(-x)^5 - 4(-x)^3 + 2(-x) - 5$, which simplifies to $-2x^5 + 4x^3 - 2x - 5$ and contains two variations of sign. Thus, the *number of negative solutions is two or zero.*

Conclusion: The given equation has either three positive and two negative solutions; three positive and two nonreal complex solutions; one positive, two negative, and two nonreal complex solutions; or one positive and four nonreal complex solutions.

It should be evident from the previous discussions that sometimes we can truly pinpoint the nature of the solutions of a polynomial equation. However, for some equations (such as the last example) the best we can do with the properties discussed in this section is to restrict the possibilities for the nature of the solutions.

It might be helpful for you to again look over Examples 1, 2, and 3 of this section and show that the solution sets do satisfy Properties 9.3, 9.5, and 9.6.

Problem Set 9.3

Use the Rational Root Theorem and the Factor Theorem to help solve each of the following equations. Verify that the number of solutions for each equation agrees with Property 9.3, taking into account multiplicity of solutions.

1. $x^3 + x^2 - 4x - 4 = 0$

2. $x^3 - 2x^2 - 11x + 12 = 0$

3. $6x^3 + x^2 - 10x + 3 = 0$

4. $8x^3 - 2x^2 - 41x - 10 = 0$

5. $3x^3 + 13x^2 - 52x + 28 = 0$

6. $15x^3 + 14x^2 - 3x - 2 = 0$

7. $x^3 - 2x^2 - 7x - 4 = 0$

8. $x^3 - x^2 - 8x + 12 = 0$

9. $x^4 - 4x^3 - 7x^2 + 34x - 24 = 0$

10. $x^4 + 4x^3 - x^2 - 16x - 12 = 0$

11. $x^3 - 10x - 12 = 0$

12. $x^3 - 4x^2 + 8 = 0$

13. $3x^4 - x^3 - 8x^2 + 2x + 4 = 0$

14. $2x^4 + 3x^3 - 11x^2 - 9x + 15 = 0$

15. $6x^4 - 13x^3 - 19x^2 + 12x = 0$

16. $x^3 - x^2 + x - 1 = 0$

17. $x^4 - 3x^3 + 2x^2 + 2x - 4 = 0$

18. $x^4 + x^3 - 3x^2 - 17x - 30 = 0$

19. $2x^5 - 5x^4 + x^3 + x^2 - x + 6 = 0$

20. $4x^4 + 12x^3 + x^2 - 12x + 4 = 0$

Verify that the following equations have no rational solutions.

21. $x^4 - x^3 - 8x^2 - 3x + 1 = 0$

22. $x^4 + 3x - 2 = 0$

23. $2x^4 - 3x^3 + 6x^2 - 24x + 5 = 0$

24. $3x^4 - 4x^3 - 10x^2 + 3x - 4 = 0$

25. $x^5 - 2x^4 + 3x^3 + 4x^2 + 7x - 1 = 0$

26. $x^5 + 2x^4 - 2x^3 + 5x^2 - 2x - 3 = 0$

27. The Rational Root Theorem pertains to polynomial equations with integral coefficients. However, if the coefficients are nonintegral rational numbers, we can first apply the multiplication property of equality to produce an equivalent equation with integral coefficients. Solve each of the following equations.

(a) $\frac{1}{10}x^3 + \frac{1}{2}x^2 + \frac{1}{5}x - \frac{4}{5} = 0$

(b) $\frac{1}{10}x^3 + \frac{1}{5}x^2 - \frac{1}{2}x - \frac{3}{5} = 0$

(c) $x^3 + \frac{9}{2}x^2 - x - 12 = 0$

(d) $x^3 - \frac{5}{6}x^2 - \frac{22}{3}x + \frac{5}{2} = 0$

For Problems 28–37, use Descartes' Rule of Signs (Property 9.6) to help you list the possibilities for the nature of the solutions for each of the equations. *Do not solve the equations.*

28. $6x^2 + 7x - 20 = 0$

29. $8x^2 - 14x + 3 = 0$

30. $2x^3 + x - 3 = 0$

31. $4x^3 + 3x + 7 = 0$

32. $3x^3 - 2x^2 + 6x + 5 = 0$

33. $4x^3 + 5x^2 - 6x - 2 = 0$

34. $x^5 - 3x^4 + 5x^3 - x^2 + 2x - 1 = 0$

35. $2x^5 + 3x^3 - x + 1 = 0$

36. $x^5 + 32 = 0$

37. $2x^6 + 3x^4 - 2x^2 - 1 = 0$

Miscellaneous Problems

38. Use the Rational Root Theorem to argue that $\sqrt{2}$ is not a rational number. (Hint: The solutions of $x^2 - 2 = 0$ are $\pm\sqrt{2}$.)

39. Use the Rational Root Theorem to argue that $\sqrt{12}$ is not a rational number.

40. Defend this statement: "Every polynomial equation of odd degree with real coefficients has at least one real number solution."

41. The following synthetic division shows that 2 is a solution of $x^4 + x^3 + x^2 - 9x - 10 = 0$.

$$
\begin{array}{r|rrrrr}
2 & 1 & 1 & 1 & -9 & -10 \\
  &   & 2 & 6 & 14 & 10 \\
\hline
  & 1 & 3 & 7 & 5 & 0 \longleftarrow
\end{array}
$$

Notice that the new quotient row (indicated by the arrow) consists entirely of nonnegative numbers. This indicates that searching for solutions greater than 2 would be a waste of time, since larger divisors would continue to increase each of the numbers (except the one on the far left) in the new quotient row. (Try 3 as a divisor!) Thus, we say that 2 is an **upper bound** for the real number solutions of the given equation.

   Now consider the following synthetic division, which shows that -1 is also a solution of $x^4 + x^3 + x^2 - 9x - 10 = 0$.

$$
\begin{array}{r|rrrrr}
-1 & 1 & 1 & 1 & -9 & -10 \\
   &   & -1 & 0 & -1 & 10 \\
\hline
   & 1 & 0 & 1 & -10 & 0 \longleftarrow
\end{array}
$$

The new quotient row (indicated by the arrow) shows that there is no need to look for solutions less than -1, because any divisor less than -1 would increase the absolute value of each number (except the one on the far left) in the new quotient row. (Try -2 as a divisor!) Thus, we say that -1 is a **lower bound** for the real number solutions of the given equation.

The following general property can be stated: If $a_n x^n + a_{n-1} x^{n-1} + \cdots + a_1 x + a_0 = 0$ is a polynomial equation with real coefficients, where $a_n > 0$, and if the polynomial is divided synthetically by $x - c$, then

1. if $c > 0$ and all numbers in the new quotient row of the synthetic division are nonnegative, c is an upper bound of the solutions of the given equation;

2. if $c < 0$ and the numbers in the new quotient row alternate in sign (with 0 considered either positive or negative, as needed), c is a lower bound of the solutions of the given equation.

Find the smallest positive integer and the largest negative integer that are upper and lower bounds, respectively, for the real number solutions of each of the following equations. Keep in mind that the integers that serve as bounds do not necessarily have to be solutions of the equation.

(a) $x^3 - 3x^2 + 25x - 75 = 0$

(b) $x^3 + x^2 - 4x - 4 = 0$

(c) $x^4 + 4x^3 - 7x^2 - 22x + 24 = 0$

(d) $3x^3 + 7x^2 - 22x - 8 = 0$

(e) $x^4 - 2x^3 - 9x^2 + 2x + 8 = 0$

9.4 Graphing Polynomial Functions

The terms that classify functions are analogous to the linear equation–quadratic equation–polynomial equation vocabulary. In Chapter 8 we defined a linear function in terms of the equation

$$f(x) = ax + b$$

and a quadratic function in terms of the equation

$$f(x) = ax^2 + bx + c.$$

Both of these are special cases of a general class of functions called polynomial functions. Any function of the form

$$f(x) = a_n x^n + a_{n-1} x^{n-1} + \cdots + a_1 x + a_0,$$

is called a **polynomial function of degree** n, where a_n is a nonzero real number, $a_{n-1}, \ldots,$ a_1, a_0 are real numbers, and n is a nonnegative integer. The following are examples of polynomial functions.

$$f(x) = 5x^3 - 2x^2 + x - 4 \qquad \text{Degree 3}$$
$$f(x) = -2x^4 - 5x^3 + 3x^2 + 4x - 1 \qquad \text{Degree 4}$$
$$f(x) = 3x^5 + 2x^2 - 3 \qquad \text{Degree 5}$$

Remark Our previous work with polynomial equations is sometimes presented as "finding zeros of polynomial functions." The **solutions**, or **roots**, of a polynomial equation are also called the **zeros** of the polynomial function. For example, -2 and 2 are solutions of $x^2 - 4 = 0$, and they are zeros of $f(x) = x^2 - 4$. That is, $f(-2) = 0$ and $f(2) = 0$.

For a complete discussion of graphing polynomial functions, we would need some tools from calculus. However, the graphing techniques that we have discussed in this text will allow us to graph certain kinds of polynomial functions. For example, polynomial functions of the form

$$f(x) = ax^n$$

are quite easy to graph. We know from our previous work that if $n = 1$, then functions such as $f(x) = 2x$, $f(x) = -3x$, and $f(x) = \frac{1}{2}x$ are lines through the origin having slopes of 2, -3, and $\frac{1}{2}$, respectively.

Furthermore, if $n = 2$, we know that the graphs of functions of the form $f(x) = ax^2$ are parabolas symmetrical with respect to the y-axis and having their vertices at the origin.

We have also previously graphed the special case of $f(x) = ax^n$, where $a = 1$ and $n = 3$—namely, the function $f(x) = x^3$. This graph is shown in Figure 9.1.

Figure 9.1

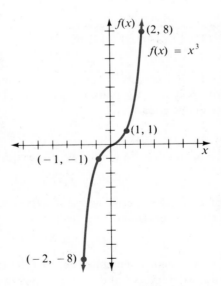

The graphs of functions of the form $f(x) = ax^3$ for which $a \neq 1$ are slight variations of $f(x) = x^3$ and can be easily determined by plotting a few points. The graphs of $f(x) = \frac{1}{2}x^3$ and $f(x) = -x^3$ appear in Figure 9.2.

Figure 9.2

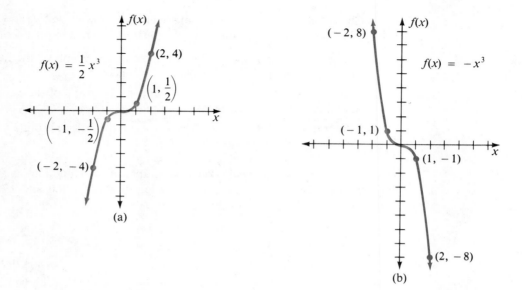

(a)

(b)

There are two general patterns that emerge from studying functions of the form $f(x) = x^n$. If n is odd and greater than 3, the graphs closely resemble Figure 9.1. The graph of $f(x) = x^5$ is shown in Figure 9.3. Notice that the curve "flattens out" a little more around the origin than it does in the graph of $f(x) = x^3$; it increases and decreases more rapidly because of the larger exponent. If n is even and greater than 2, the graphs of $f(x) = x^n$ are not parabolas. They resemble the basic parabola, but they are flatter at the bottom, and steeper on the sides. Figure 9.4 shows the graph of $f(x) = x^4$.

Graphs of functions of the form $f(x) = ax^n$, where n is an integer greater than 2 and $a \neq 1$, are variations of those shown in Figures 9.1 and 9.4. If n is odd, the curve is symmetrical about the origin. If n is even, the graph is symmetrical about the y-axis.

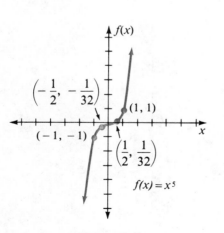

Figure 9.3

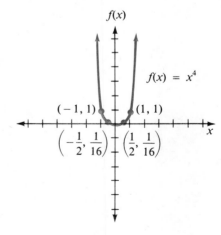

Figure 9.4

In Chapter 8 we found that functions such as $f(x) = x^2 + 2$, $f(x) = \sqrt{x - 1}$, and $f(x) = -|x|$ could easily be graphed by comparing them to a basic function as follows.

1. The graph of $f(x) = x^2 + 2$ is the same as the graph of the basic parabola $f(x) = x^2$ except *moved up 2 units.*

2. The graph of $f(x) = \sqrt{x - 1}$ is the same as the graph of the basic square root function $f(x) = \sqrt{x}$ except *moved to the right 1 unit.*

3. The graph of $f(x) = -|x|$ is the graph of the basic absolute value function $f(x) = |x|$ *reflected across the x-axis.*

In a similar fashion, we can easily sketch the graphs of such functions as $f(x) = x^3 + 2$, $f(x) = (x - 1)^5$, and $f(x) = -x^4$ as indicated in Figures 9.5, 9.6, and 9.7, respectively.

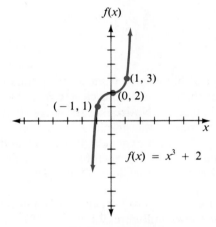

Figure 9.5

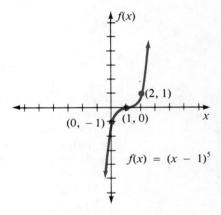

Figure 9.6

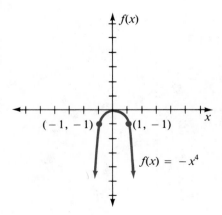

Figure 9.7

Graphing Polynomial Functions in Factored Form

As the degree of the polynomial increases, the graphs often become more complicated. We do know, however, that polynomial functions produce smooth continuous curves with a number of turning points, as illustrated in Figures 9.8 and 9.9. Some typical graphs of polynomial functions of odd degree are shown in Figure 9.8. As the graphs suggest, every polynomial function of odd degree has at least one *real zero*—that is, at least one real number c such that $f(c) = 0$. Geometrically, the zeros of the function are the x-intercepts of the graph. Figure 9.9 illustrates some possible graphs of polynomial functions of even degree.

The **turning points** are the places where the function either changes from increasing to decreasing or from decreasing to increasing. Using calculus we are able to verify that

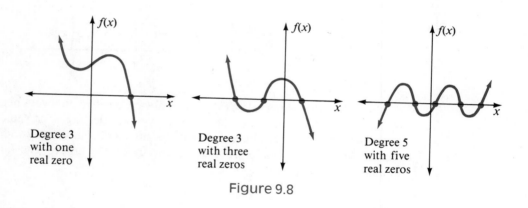

Degree 3 with one real zero

Degree 3 with three real zeros

Degree 5 with five real zeros

Figure 9.8

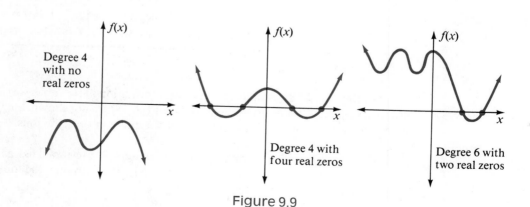

Degree 4 with no real zeros

Degree 4 with four real zeros

Degree 6 with two real zeros

Figure 9.9

a polynomial function of degree n *has at most* n−1 *turning points.* Now let's illustrate how this information along with some other techniques can be used to graph polynomial functions that are expressed in factored form.

Example 1   Graph $f(x) = (x + 2)(x - 1)(x - 3)$.

Solution   First, let's find the x-intercepts (zeros of the function) by setting each factor equal to zero and solving for x.

$$x + 2 = 0 \quad \text{or} \quad x - 1 = 0 \quad \text{or} \quad x - 3 = 0$$
$$x = -2 \quad \text{or} \quad x = 1 \quad \text{or} \quad x = 3$$

Thus, the points $(-2, 0), (1, 0)$, and $(3, 0)$ are on the graph. Second, the points associated with the x-intercepts divide the x-axis into four intervals as follows.

In each of these intervals, $f(x)$ is either always positive or always negative. That is to say, the graph is either above or below the x-axis. Selecting a test value for x in each of the intervals will determine whether x is positive or negative. Any additional points that are easily obtained improve the accuracy of the graph. The following table summarizes these results.

Interval	Test value	Sign of $f(x)$	Location of graph
$x < -2$	$f(-3) = -24$	negative	below x-axis
$-2 < x < 1$	$f(0) = 6$	positive	above x-axis
$1 < x < 3$	$f(2) = -4$	negative	below x-axis
$x > 3$	$f(4) = 18$	positive	above x-axis

Additional values: $f(-1) = 8$

Making use of the x-intercepts and the information in the table, we can sketch the graph in Figure 9.10. The points $(-3, -24)$ and $(4, 18)$ are not shown, but they are used to indicate a rapid decrease and increase of the curve in those regions.

Remark   In Figure 9.10 turning points of the graph are indicated at $(2, -4)$ and $(-1, 8)$. Keep in mind that these are only approximations. Again, the tools of calculus are needed to find the exact turning points.

Figure 9.10

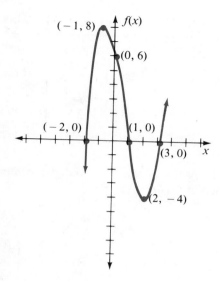

Example 2   Graph $f(x) = -x^4 + 3x^3 - 2x^2$.

Solution   The polynomial can be factored as follows.

$$f(x) = -x^4 + 3x^3 - 2x^2$$
$$= -x^2(x^2 - 3x + 2)$$
$$= -x^2(x - 1)(x - 2)$$

Now we can find the x-intercepts.

$$-x^2 = 0 \quad \text{or} \quad x - 1 = 0 \quad \text{or} \quad x - 2 = 0$$
$$x = 0 \quad \text{or} \quad x = 1 \quad \text{or} \quad x = 2$$

The points $(0, 0)$, $(1, 0)$, and $(2, 0)$ are on the graph and divide the x-axis into four intervals as follows.

$$
\begin{array}{c|c|c|c}
x < 0 & 0 < x < 1 & 1 < x < 2 & x > 2 \\
\end{array}
$$

$$0 \qquad 1 \qquad 2$$

In the following table some points are determined and the sign behavior of $f(x)$ is summarized.

Interval	Test value	Sign of $f(x)$	Location of graph
$x < 0$	$f(-1) = -6$	negative	below x-axis
$0 < x < 1$	$f\left(\dfrac{1}{2}\right) = -\dfrac{3}{16}$	negative	below x-axis
$1 < x < 2$	$f\left(\dfrac{3}{2}\right) = \dfrac{9}{16}$	positive	above x-axis
$x > 2$	$f(3) = -18$	negative	below x-axis

Making use of the table and the x-intercepts, we can determine the graph, as illustrated in Figure 9.11.

Figure 9.11

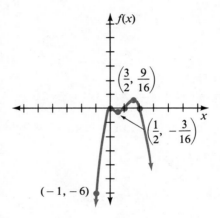

Example 3   Graph $f(x) = x^3 + 3x^2 - 4$.

Solution   Using the Rational Root Theorem, synthetic division, and the Factor Theorem, we can factor the given polynomial as follows.

$$f(x) = x^3 + 3x^2 - 4$$
$$= (x - 1)(x^2 + 4x + 4)$$
$$= (x - 1)(x + 2)^2$$

Now we can find the x-intercepts.

$$x - 1 = 0 \quad \text{or} \quad (x + 2)^2 = 0$$
$$x = 1 \quad \text{or} \quad x = -2$$

The points $(-2, 0)$ and $(1, 0)$ are on the graph and divide the x-axis into three intervals as follows.

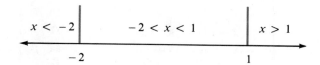

In the following table some points are determined and the sign behavior of $f(x)$ is summarized.

Interval	Test value	Sign of f(x)	Location of graph
$x < -2$	$f(-3) = -4$	negative	below x-axis
$-2 < x < 1$	$f(0) = -4$	negative	below x-axis
$x > 1$	$f(2) = 16$	positive	above x-axis

Additional values: $f(-1) = -2$
$f(-4) = -20$

As a result of the table and the x-intercepts, we can sketch the graph, as shown in Figure 9.12.

Figure 9.12

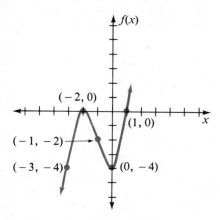

Problem Set 9.4

Graph each of the following polynomial functions.

1. $f(x) = x^3 - 3$
2. $f(x) = (x + 1)^3$
3. $f(x) = (x - 2)^3 + 1$
4. $f(x) = -(x - 3)^3$
5. $f(x) = x^4 - 2$
6. $f(x) = (x + 3)^4$
7. $f(x) = (x + 1)^4 + 3$
8. $f(x) = -x^5$

9. $f(x) = (x - 1)^5 + 2$

10. $f(x) = -(x - 2)^4$

11. $f(x) = (x - 1)(x + 1)(x - 3)$

12. $f(x) = (x - 2)(x + 1)(x + 3)$

13. $f(x) = (x + 4)(x + 1)(1 - x)$

14. $f(x) = x(x + 2)(2 - x)$

15. $f(x) = -x(x + 3)(x - 2)$

16. $f(x) = -x^2(x - 1)(x + 1)$

17. $f(x) = (x + 3)(x + 1)(x - 1)(x - 2)$

18. $f(x) = (2x - 1)(x - 2)(x - 3)$

19. $f(x) = (x - 1)^2(x + 2)$

20. $f(x) = (x + 2)^3(x - 4)$

21. $f(x) = (x + 1)^2(x - 1)^2$

22. $f(x) = x(x - 2)^2(x + 1)$

Graph each of the following polynomial functions by first factoring the given polynomial. You may need to use some factoring techniques from Chapter 3 as well as the Rational Root Theorem and the Factor Theorem.

23. $f(x) = x^3 + x^2 - 2x$

24. $f(x) = -x^3 - x^2 + 6x$

25. $f(x) = -x^4 - 3x^3 - 2x^2$

26. $f(x) = x^4 - 6x^3 + 8x^2$

27. $f(x) = x^3 - x^2 - 4x + 4$

28. $f(x) = x^3 + 2x^2 - x - 2$

29. $f(x) = x^3 - 13x + 12$

30. $f(x) = x^3 - x^2 - 9x + 9$

31. $f(x) = x^3 - 2x^2 - 11x + 12$

32. $f(x) = 2x^3 - 3x^2 - 3x + 2$

33. $f(x) = -x^3 + 6x^2 - 11x + 6$

34. $f(x) = x^4 - 5x^2 + 4$

For each of the following, (a) find the y-intercepts, (b) find the x-intercepts, and (c) find the intervals of x for which $f(x) > 0$ and those for which $f(x) < 0$. *Do not sketch the graph.*

35. $f(x) = (x - 5)(x + 4)(x - 3)$

36. $f(x) = (x + 3)(x - 6)(8 - x)$

37. $f(x) = (x - 4)^2(x + 3)^3$

38. $f(x) = (x + 3)^4(x - 1)^3$

39. $f(x) = (x + 2)^2(x - 1)^3(x - 2)$

40. $f(x) = x(x - 6)^2(x + 4)$

41. $f(x) = (x + 2)^5(x - 4)^2$

Miscellaneous Problems

42. A polynomial function with real coefficients is continuous everywhere; that is, its graph has no holes or breaks. This is the basis for the following property: *If $f(x)$ is a polynomial with real coefficients, and if $f(a)$ and $f(b)$ are of opposite sign, then there is at least one real zero between a and b.* This property, along with our previous knowledge of polynomial functions, provides the basis for locating and approximating irrational solutions of a polynomial equation.

Consider the equation $x^3 + 2x - 4 = 0$. Applying Decartes' Rule of Signs, we can determine that this equation has one positive real solution and two nonreal complex solutions. (You may want to confirm this!) The Rational Root Theorem indicates that the only possible positive rational solutions are 1, 2, and 4. Using a little more compact format for synthetic division, we obtain the following results when testing for 1 and 2 as possible solutions.

	1	0	2	−4
1	1	1	3	−1
2	1	2	6	8

Since $f(1) = -1$ and $f(2) = 8$, there must be an irrational solution between 1 and 2. Furthermore, since -1 is closer to 0 than is 8, our guess is that the solution is closer to 1 than to 2. Let's start looking at 1.0, 1.1, 1.2, etc., until we can place the solution between two numbers.

	1	0	2	-4
1.0	1	1	3	-1
1.1	1	1.1	3.21	-0.469
1.2	1	1.2	3.44	0.128

A calculator is very helpful at this time.

Since $f(1.1) = -0.469$ and $f(1.2) = 0.128$, the irrational solution must be between 1.1 and 1.2. Furthermore, since 0.128 is closer to 0 than is -0.469, our guess is that the solution is closer to 1.2 than to 1.1. Let's start looking at 1.15, 1.16, etc.

	1	0	2	-4
1.15	1	1.15	3.3225	-0.179
1.16	1	1.16	3.3456	-0.119
1.17	1	1.17	3.3689	-0.058
1.18	1	1.18	3.3924	0.003

Since $f(1.17) = -0.058$ and $f(1.18) = 0.003$, the irrational solution must be between 1.17 and 1.18. Therefore, we can use 1.2 as a rational approximation to the nearest tenth.

For each of the following equations, (a) verify that the equation has exactly one irrational solution and (b) find an approximation, to the nearest tenth, of that solution.

(a) $x^3 + x - 6 = 0$

(d) $x^3 - x^2 - x - 1 = 0$

(b) $x^3 - 6x - 4 = 0$

(e) $x^3 - 24x - 32 = 0$

(c) $x^3 - 27x + 18 = 0$

(f) $x^3 - 5x^2 + 3 = 0$

9.5 Graphing Rational Functions

A function of the form

$$f(x) = \frac{p(x)}{q(x)}, \quad q(x) \neq 0,$$

where $p(x)$ and $q(x)$ are polynomial functions, is called a **rational function**. The following are examples of rational functions:

$$f(x) = \frac{2}{x - 1}, \qquad f(x) = \frac{x}{x - 2},$$

$$f(x) = \frac{x^2}{x^2 - x - 6}, \qquad f(x) = \frac{x^3 - 8}{x + 4}$$

In each of these examples the domain of the rational function is the set of all real numbers except for those that make the denominator zero. For example, the domain of $f(x) = \dfrac{2}{x - 1}$ is the set of all real numbers except 1. As we will soon see, these exclusions from the domain are important numbers from a graphing standpoint; they represent breaks in an otherwise continuous curve.

Let's set the stage for graphing rational functions by considering in detail the function $f(x) = \dfrac{1}{x}$. First, note that at $x = 0$ the function is undefined. Second, let's consider a rather extensive table of values to find some number trends and to build a basis for defining the concept of an asymptote.

x	$f(x) = \dfrac{1}{x}$	
1	1	
2	0.5	These values indicate that the value of $f(x)$ is positive and approaches zero from above as x gets larger and larger.
10	0.1	
100	0.01	
1000	0.001	
0.5	2	
0.1	10	These values indicate that $f(x)$ is positive and is getting larger and larger as x approaches zero from the right.
0.01	100	
0.001	1000	
0.0001	10000	
-0.5	-2	
-0.1	-10	These values indicate that $f(x)$ is negative and is getting smaller and smaller as x approaches zero from the left.
-0.01	-100	
-0.001	-1000	
-0.0001	-10000	
-1	-1	
-2	-0.5	These values indicate that $f(x)$ is negative and approaches zero from below as x gets smaller and smaller.
-10	-0.1	
-100	-0.01	
-1000	-0.001	

A sketch of $f(x) = \dfrac{1}{x}$, drawn using a few points from this table and the patterns discussed, is shown in Figure 9.13. Notice that the graph approaches but does not touch either axis. We say that the y-axis [or $f(x)$-axis] is a **vertical asymptote** and the x-axis is a **horizontal asymptote**. In general, the following definitions can be given.

Figure 9.13

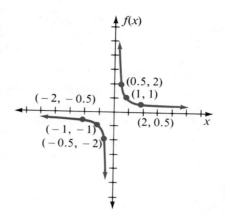

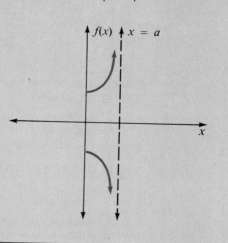

Vertical Asymptote

A line $x = a$ is a vertical asymptote for the graph of a function f if

1. $f(x)$ either increases or decreases without bound as x approaches a from the right

or

2. $f(x)$ either increases or decreases without bound as x approaches a from the left.

Horizontal Asymptote

A line $y = b$ [or $f(x) = b$] is a horizontal asymptote for the graph of a function f if

1. $f(x)$ approaches b from above or below as x gets infinitely small

   or

2. $f(x)$ approaches b from above or below as x gets infinitely large.

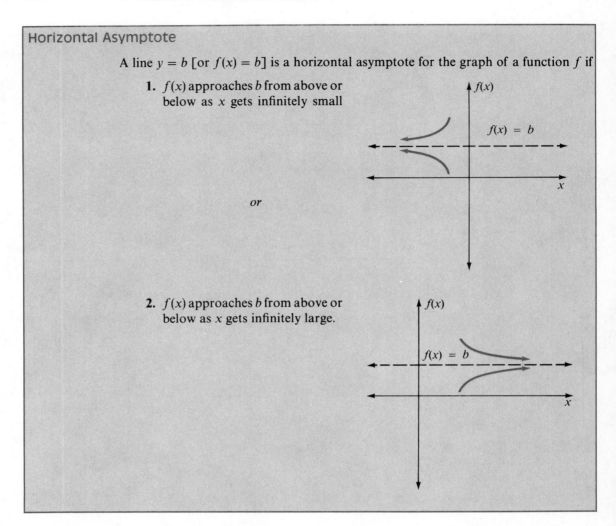

Remark   We know that the equation $f(x) = \dfrac{1}{x}$ exhibits origin symmetry because $f(-x) = -f(x)$. Thus, the graph in Figure 9.13 could have been drawn by first determining the part of the curve in the first quadrant and then reflecting that curve through the origin.

Following are some suggestions for graphing rational functions of the type being considered in this section.

1. Check for y-axis and origin symmetry.
2. Find any vertical asymptote by setting the denominator equal to zero and solving for x.
3. Find any horizontal asymptote by studying the behavior of $f(x)$ as x gets infinitely large or as x gets infinitely small.
4. Study the behavior of the graph when it is close to the asymptotes.

5. Plot as many points as necessary to determine the shape of the graph. The number needed may be affected by whether or not the graph has any kind of symmetry.

Keep these suggestions in mind as you study the following examples.

Example 1   Graph $f(x) = \dfrac{-2}{x-1}$.

Solution   Since $x = 1$ makes the denominator zero, the line $x = 1$ is a vertical asymptote, indicated in Figure 9.14 with a dashed line.

Figure 9.14

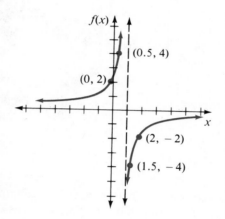

Now let's look for a horizontal asymptote by checking some large and some small values of x.

x	$f(x)$
10	$-\dfrac{2}{9}$
100	$-\dfrac{2}{99}$
1000	$-\dfrac{2}{999}$
-10	$\dfrac{2}{11}$
-100	$\dfrac{2}{101}$
-1000	$\dfrac{2}{1001}$

This portion of the table shows that as x gets very large, the value of $f(x)$ approaches zero from below.

This portion shows that as x gets very small, the value of $f(x)$ approaches zero from above.

Therefore, the x-axis is a horizontal asymptote.

Finally, let's check the behavior of the graph near the vertical asymptote.

x	$f(x)$
2	-2
1.5	-4
1.1	-20
1.01	-200
1.001	-2000
0	2
0.5	4
0.9	20
0.99	200
0.999	2000

As x approaches 1 from the right side, the value of $f(x)$ gets smaller and smaller.

As x approaches 1 from the left side, the value of $f(x)$ gets larger and larger.

The graph of $f(x) = \dfrac{-2}{x-1}$ is shown in Figure 9.14.

Example 2   Graph $f(x) = \dfrac{x}{x+2}$.

Solution   Since $x = -2$ makes the denominator zero, the line $x = -2$ is a vertical asymptote. To study the behavior of $f(x)$ as x gets very large or very small, let's change the form of the rational expression by dividing numerator and denominator by x.

$$f(x) = \frac{x}{x+2} = \frac{\dfrac{x}{x}}{\dfrac{x+2}{x}} = \frac{1}{\dfrac{x}{x} + \dfrac{2}{x}} = \frac{1}{1 + \dfrac{2}{x}}$$

Figure 9.15

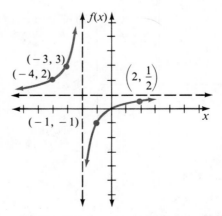

Now we can see that as x gets larger and larger, the value of $f(x)$ approaches 1 from below; and as x gets smaller and smaller, the value of $f(x)$ approaches 1 from above. (Perhaps you should check these claims by plugging in some values for x.) Thus, the line $f(x) = 1$ is a horizontal asymptote. Drawing the asymptotes (dashed lines) and plotting a few points allows us to complete the graph in Figure 9.15. ●

In the next two examples, pay special attention to the role of symmetry. It will allow us to direct our efforts toward quadrants I and IV and then reflect those portions of the curve across the vertical axis to complete the graph.

Example 3 Graph $f(x) = \dfrac{2x^2}{x^2 + 4}$.

Solution First, notice that $f(-x) = f(x)$; therefore, this graph is symmetrical with respect to the vertical axis. Second, the denominator $x^2 + 4$ cannot equal zero for any real number value of x. Thus, there is no vertical asymptote. Third, dividing both numerator and denominator of the rational expression by x^2 produces

$$f(x) = \frac{2x^2}{x^2 + 4} = \frac{\dfrac{2x^2}{x^2}}{\dfrac{x^2 + 4}{x^2}} = \frac{2}{\dfrac{x^2}{x^2} + \dfrac{4}{x^2}} = \frac{2}{1 + \dfrac{4}{x^2}}.$$

Now we can see that as x gets larger and larger, the value of $f(x)$ approaches 2 from below. Therefore, the line $f(x) = 2$ is a horizontal asymptote. So we can plot a few points using positive values for x, sketch this part of the curve, and then reflect across the $f(x)$-axis to obtain the complete graph, as shown in Figure 9.16.

Figure 9.16

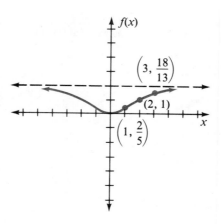

Example 4   Graph $f(x) = \dfrac{3}{x^2 - 4}$.

Solution   First, notice that $f(-x) = f(x)$; therefore, this graph is symmetrical about the y-axis. So by setting the denominator equal to zero and solving for x, we obtain

$$x^2 - 4 = 0$$
$$x^2 = 4$$
$$x = \pm 2.$$

The lines $x = 2$ and $x = -2$ are vertical asymptotes. Next, we can see that $\dfrac{3}{x^2 - 4}$ approaches zero from above as x gets larger and larger. Finally, we can plot a few points using positive values for x (other than 2), sketch this part of the curve, and then reflect it across the $f(x)$-axis to obtain the complete graph, shown in Figure 9.17.

Figure 9.17

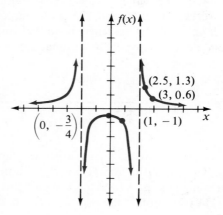

Problem Set 9.5

Graph each of the following rational functions.

1. $f(x) = \dfrac{-1}{x}$

2. $f(x) = \dfrac{1}{x^2}$

3. $f(x) = \dfrac{3}{x + 1}$

4. $f(x) = \dfrac{-1}{x - 3}$

5. $f(x) = \dfrac{2}{(x - 1)^2}$

6. $f(x) = \dfrac{-3}{(x + 2)^2}$

7. $f(x) = \dfrac{x}{x - 3}$

8. $f(x) = \dfrac{2x}{x - 1}$

9. $f(x) = \dfrac{-3x}{x + 2}$

10. $f(x) = \dfrac{-x}{x + 1}$

11. $f(x) = \dfrac{1}{x^2 - 1}$

12. $f(x) = \dfrac{-2}{x^2 - 4}$

13. $f(x) = \dfrac{-2}{(x + 1)(x - 2)}$

14. $f(x) = \dfrac{3}{(x + 2)(x - 4)}$

15. $f(x) = \dfrac{2}{x^2 + x - 2}$

16. $f(x) = \dfrac{-1}{x^2 + x - 6}$

17. $f(x) = \dfrac{x + 2}{x}$

18. $f(x) = \dfrac{2x - 1}{x}$

19. $f(x) = \dfrac{4}{x^2 + 2}$

20. $f(x) = \dfrac{4x^2}{x^2 + 1}$

21. $f(x) = \dfrac{2x^4}{x^4 + 1}$

22. $f(x) = \dfrac{x^2 - 4}{x^2}$

Miscellaneous Problems

23. The rational function $f(x) = \dfrac{(x - 2)(x + 3)}{x - 2}$ has a domain of all real numbers except 2 and can be simplified to $f(x) = x + 3$. Thus, its graph is a straight line with a hole at (2, 5). Graph each of the following functions.

(a) $f(x) = \dfrac{(x + 4)(x - 1)}{x + 4}$

(b) $f(x) = \dfrac{x^2 - 5x + 6}{x - 2}$

(c) $f(x) = \dfrac{x - 1}{x^2 - 1}$

(d) $f(x) = \dfrac{x + 2}{x^2 + 6x + 8}$

Chapter 9 Summary

(9.1) If the divisor is of the form $x - c$, where c is a constant, then the typical long-division format for dividing polynomials can be simplified to a process called **synthetic division**. This process can be reviewed by studying the examples of this section.

The Division Algorithm for Polynomials states that if $f(x)$ and $g(x)$ are polynomials and $g(x) \neq 0$, then there exist unique polynomials $q(x)$ and $r(x)$ such that

$$f(x) = g(x)q(x) + r(x),$$

where $r(x) = 0$ or the degree of $r(x)$ is less than the degree of $g(x)$.

(9.2) The Remainder Theorem states that if a polynomial $f(x)$ is divided by $x - c$, then the remainder is equal to $f(c)$. Thus, a polynomial can be evaluated for a given number either by direct substitution or by using synthetic division.

The Factor Theorem states that a polynomial $f(x)$ has a factor $x - c$ if and only if $f(c) = 0$.

(9.3)   The following concepts and properties provide a basis for solving polynomial equations:

1. Synthetic division.

2. The Factor Theorem.

3. Property 9.3: A polynomial equation of degree n has n solutions, where any solution of multiplicity p is counted p times.

4. The Rational Root Theorem: Consider the polynomial equation

$$a_n x^n + a_{n-1} x^{n-1} + \cdots + a_1 x + a_0 = 0,$$

where the *coefficients are integers*. If the rational number $\dfrac{c}{d}$, reduced to lowest terms, is a solution of the equation, then c is a factor of the constant term a_0 and d is a factor of the leading coefficient a_n.

5. Property 9.5: Nonreal complex solutions of polynomial equations with real coefficients, if they exist, must occur in conjugate pairs.

6. Descartes' Rule of Signs: Let $a_n x^n + a_{n-1} x^{n-1} + \cdots + a_1 x + a_0 = 0$ be a polynomial equation with real coefficients.

    (a) The number of *positive real solutions* is either equal to the number of sign variations or else less than the number of sign variations by a positive even integer.

    (b) The number of *negative real solutions* is either equal to the number of sign variations in

    $$a_n(-x)^n + a_{n-1}(-x)^{n-1} + \cdots + a_1(-x) + a_0,$$

    or else less than the number of sign variations by a positive even integer.

(9.4)   The following steps may be used to graph a polynomial function that is expressed in factored form,

1. Find the x-intercepts, which are also called the **zeros of the polynomial**.

2. Use a test value in each of the intervals determined by the x-intercepts to find out whether the function is positive or negative over that interval.

3. Plot any additional points that are needed to determine the graph.

(9.5)   The following steps may be used to graph a rational function.

1. Check for vertical axis and origin symmetry.

2. Find any vertical asymptotes by setting the denominator equal to zero and solving for x.

3. Find any horizontal asymptotes by studying the behavior of $f(x)$ as x gets very large or very small. This may require changing the form of the original rational expression.

4. Study the behavior of the graph when it is close to the asymptotic lines.

5. Plot as many points as necessary to determine the graph. The number needed may be affected by whether or not the graph has any kind of symmetry.

Chapter 9 Review Problem Set

For Problems 1–4, use synthetic division to determine the quotient and the remainder.

1. $(3x^3 - 4x^2 + 6x - 2) \div (x - 1)$

2. $(5x^3 + 7x^2 - 9x + 10) \div (x + 2)$

3. $(-2x^4 + x^3 - 2x^2 - x - 1) \div (x + 4)$

4. $(-3x^4 - 5x^2 + 9) \div (x - 3)$

For Problems 5–8, find $f(c)$ either by using synthetic division and the Remainder Theorem or by evaluating $f(c)$ directly.

5. $f(x) = 4x^5 - 3x^3 + x^2 - 1$ and $c = 1$

6. $f(x) = 4x^3 - 7x^2 + 6x - 8$ and $c = -3$

7. $f(x) = -x^4 + 9x^2 - x - 2$ and $c = -2$

8. $f(x) = x^4 - 9x^3 + 9x^2 - 10x + 16$ and $c = 8$

For Problems 9–12, use the Factor Theorem to help answer some questions about factors.

9. Is $x + 2$ a factor of $2x^3 + x^2 - 7x - 2$?

10. Is $x - 3$ a factor of $x^4 + 5x^3 - 7x^2 - x + 3$?

11. Is $x - 4$ a factor of $x^5 - 1024$?

12. Is $x + 1$ a factor of $x^5 + 1$?

For Problems 13–16, use the Rational Root Theorem and the Factor Theorem to help solve each of the equations.

13. $x^3 - 3x^2 - 13x + 15 = 0$

14. $8x^3 + 26x^2 - 17x - 35 = 0$

15. $x^4 - 5x^3 + 34x^2 - 82x + 52 = 0$

16. $x^3 - 4x^2 - 10x + 4 = 0$

For Problems 17 and 18, use Descartes' Rule of Signs (Property 9.6) to help list the possibilities for the nature of the solutions. *Do not solve the equations.*

17. $4x^4 - 3x^3 + 2x^2 + x + 4 = 0$

18. $x^5 + 3x^3 + x + 7 = 0$

For Problems 19–22, graph each of the polynomial functions.

19. $f(x) = -(x - 2)^3 + 3$

20. $f(x) = (x + 3)(x - 1)(3 - x)$

21. $f(x) = x^4 - 4x^2$

22. $f(x) = x^3 - 4x^2 + x + 6$

For Problems 23 and 24, graph each of the rational functions. Be sure to identify the asymptotes.

23. $f(x) = \dfrac{2x}{x - 3}$

24. $f(x) = \dfrac{-3}{x^2 + 1}$

10 Exponential and Logarithmic Functions

In this chapter we will (1) extend the meaning of an exponent, (2) work with some exponential functions, (3) consider the concept of a logarithm, (4) work with some logarithmic functions, and (5) use the concepts of exponent and logarithm to expand problem-solving skills. Your calculator will be a valuable tool throughout this chapter.

10.1 Exponents and Exponential Functions

In Chapter 1, the expression b^n was defined to mean n factors of b, where n is any positive integer and b is any real number. For example,

$$2^3 = 2 \cdot 2 \cdot 2 = 8, \qquad \left(\frac{1}{3}\right)^4 = \left(\frac{1}{3}\right)\left(\frac{1}{3}\right)\left(\frac{1}{3}\right)\left(\frac{1}{3}\right) = \frac{1}{81},$$

$$(-4)^2 = (-4)(-4) = 16, \qquad -(0.5)^3 = -[(0.5)(0.5)(0.5)] = -0.125.$$

Then in Chapter 5, by defining $b^0 = 1$ and $b^{-n} = \dfrac{1}{b^n}$, where n is any positive integer and b is any nonzero real number, we extended the concept of an exponent to include all integers. Examples include

$$(0.76)^0 = 1, \qquad 2^{-3} = \frac{1}{2^3} = \frac{1}{8},$$

$$\left(\frac{2}{3}\right)^{-2} = \frac{1}{\left(\frac{2}{3}\right)^2} = \frac{1}{\frac{4}{9}} = \frac{9}{4}, \qquad (0.4)^{-1} = \frac{1}{(0.4)^1} = \frac{1}{0.4} = 2.5.$$

In Chapter 5, we also provided for the use of all rational numbers as exponents by defining

$$b^{m/n} = \sqrt[n]{b^m} = (\sqrt[n]{b})^m,$$

where n is a positive integer greater than 1 and b is a real number such that $\sqrt[n]{b}$ exists. Some examples are

$$27^{2/3} = (\sqrt[3]{27})^2 = 9, \qquad 16^{1/4} = \sqrt[4]{16^1} = 2,$$

$$\left(\frac{1}{9}\right)^{1/2} = \sqrt{\frac{1}{9}} = \frac{1}{3}, \qquad 32^{-1/5} = \frac{1}{32^{1/5}} = \frac{1}{\sqrt[5]{32}} = \frac{1}{2}.$$

Formally extending the concept of an exponent to include the use of irrational numbers requires some ideas from calculus and is therefore beyond the scope of this text. However, we can take a brief glimpse at the general idea involved. Consider the number $2^{\sqrt{3}}$. By using the nonterminating and nonrepeating decimal representation $1.73205\ldots$ for $\sqrt{3}$, we can form the sequence of numbers $2^1, 2^{1.7}, 2^{1.73}, 2^{1.732}, 2^{1.7320}, 2^{1.73205}, \ldots$. It should seem reasonable that each successive power gets closer to $2^{\sqrt{3}}$. This is precisely what happens if b^n, where n is irrational, is properly defined using the concept of a limit. Furthermore, this will ensure that an expression such as 2^x will yield exactly one value for each value of x.

So from now on we can use any real number as an exponent, and the basic properties stated in Chapter 5 can be extended to include all real numbers as exponents. Let's restate those properties with the restriction that the bases a and b must be positive numbers, to avoid expressions such as $(-4)^{1/2}$, which do not represent real numbers.

Property 10.1	If a and b are positive real numbers and m and n are any real numbers, then
	1. $b^n \cdot b^m = b^{n+m}$ Product of two powers
	2. $(b^n)^m = b^{mn}$ Power of a power
	3. $(ab)^n = a^n b^n$ Power of a product
	4. $\left(\dfrac{a}{b}\right)^n = \dfrac{a^n}{b^n}$ Power of a quotient
	5. $\dfrac{b^n}{b^m} = b^{n-m}$ Quotient of two powers

Another property that can be used to solve certain types of equations involving exponents can be stated as follows.

Property 10.2	If $b > 0$, $b \neq 1$, and m and n are real numbers, then $b^n = b^m$ if and only if $n = m$.

The following three examples illustrate the use of Property 10.2.

Example 1   Solve $2^x = 32$.

Solution $2^x = 32$

$2^x = 2^5$ $(32 = 2^5)$

$x = 5$ Apply Property 10.2.

The solution set is $\{5\}$.   ●

Example 2 $2^{3x} = \dfrac{1}{64}$.

Solution $2^{3x} = \dfrac{1}{64}$

$2^{3x} = \dfrac{1}{2^6}$

$2^{3x} = 2^{-6}$

$3x = -6$ Apply Property 10.2.

$x = -2$

The solution set is $\{-2\}$.   ●

Example 3   Solve $\left(\dfrac{1}{5}\right)^{x-4} = \dfrac{1}{125}$.

Solution $\left(\dfrac{1}{5}\right)^{x-4} = \dfrac{1}{125}$

$\left(\dfrac{1}{5}\right)^{x-4} = \left(\dfrac{1}{5}\right)^3$

$x - 4 = 3$ Apply Property 10.2.

$x = 7$

The solution set is $\{7\}$.   ●

If b is any positive number, then the expression b^x designates exactly one real number for every real value of x. Therefore, the equation $f(x) = b^x$ defines a function whose domain is the set of real numbers. Furthermore, if we include the additional restriction $b \neq 1$, then any equation of the form $f(x) = b^x$ describes a one-to-one function and is called an **exponential function**. This leads to the following definition.

> **Definition 10.1**
>
> If $b > 0$ and $b \neq 1$, then the function f defined by
> $$f(x) = b^x,$$
> where x is any real number, is called the *exponential function with base b*.

Remark   The function $f(x) = 1^x$ is a constant function and therefore is not a one-to-one function. Remember from Chapter 8 that one-to-one functions have inverses; this becomes a key issue in a later section.

Now let's consider graphing some exponential functions.

Example 4   Graph the function $f(x) = 2^x$.

Solution   Let's set up a table of values, keeping in mind that the domain is the set of real numbers and that the equation $f(x) = 2^x$ exhibits no symmetry. Plotting these points and connecting them with a smooth curve produces Figure 10.1.

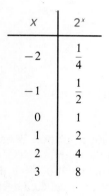

x	2^x
-2	$\dfrac{1}{4}$
-1	$\dfrac{1}{2}$
0	1
1	2
2	4
3	8

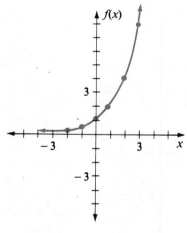

Figure 10.1

In the table for Example 4 we chose integral values for x to keep the computation simple. However, with the use of a calculator we could easily acquire functional values by using nonintegral exponents. Consider the following additional values for $f(x) = 2^x$.

$$f(0.5) \approx 1.41, \qquad f(1.7) \approx 3.25,$$
$$f(-0.5) \approx 0.71, \qquad f(-2.6) \approx 0.16$$

Use your calculator to check these results. Also notice that the points generated by these values do fit the graph in Figure 10.1.

Example 5   Graph $f(x) = \left(\dfrac{1}{2}\right)^x$.

Solution   Again, let's set up a table of values, plot the points, and connect them with a smooth curve. The graph is shown in Figure 10.2.

x	$\left(\frac{1}{2}\right)^x$
-2	4
-1	2
0	1
1	$\frac{1}{2}$
2	$\frac{1}{4}$
3	$\frac{1}{8}$

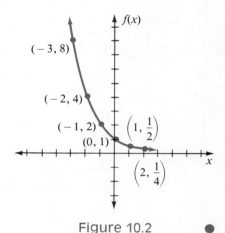

Figure 10.2

Remark Since $\left(\dfrac{1}{2}\right)^x = \dfrac{1}{2^x} = 2^{-x}$, the graphs of $f(x) = 2^x$ and $f(x) = \left(\dfrac{1}{2}\right)^x$ are reflections of each other across the y-axis. Therefore, Figure 10.2 could have been drawn by reflecting Figure 10.1 across the y-axis.

The graphs in Figures 10.1 and 10.2 illustrate a general behavior pattern of exponential functions. That is to say, if $b > 1$, then the graph of $f(x) = b^x$ "goes up to the right" and the function is called an **increasing function**. If $0 < b < 1$, then the graph of $f(x) = b^x$ "goes down to the right" and the function is called a **decreasing function**. These facts are illustrated in Figure 10.3. Notice that $b^0 = 1$ for any $b > 0$; thus, all graphs of $f(x) = b^x$ contain the point $(0, 1)$.

As you graph exponential functions, don't forget to use previous graphing experiences. For example,

1. The graph of $f(x) = 2^x + 3$ is the graph of $f(x) = 2^x$ *moved up 3 units*.
2. The graph of $f(x) = 2^{x-4}$ is the graph of $f(x) = 2^x$ *moved to the right 4 units*.
3. The graph of $f(x) = -2^x$ is the graph of $f(x) = 2^x$ *reflected across the x-axis*.
4. The graph of $f(x) = 2^x + 2^{-x}$ is symmetrical with respect to the y-axis because $f(-x) = 2^{-x} + 2^x = f(x)$.

Furthermore, if you are faced with an exponential function that is not of the basic form $f(x) = b^x$ or a variation thereof, don't forget the graphing suggestions offered in Chapter 7. Let's consider one such example.

Figure 10.3

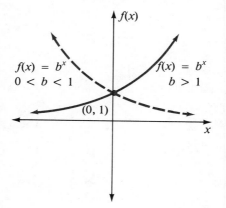

Example 6   Graph $f(x) = 2^{-x^2}$.

Solution   Since $f(-x) = 2^{-(-x)^2} = 2^{-x^2} = f(x)$, we know that this curve is symmetrical with respect to the y-axis. Therefore, let's set up a table of values using nonnegative values for x. Plotting these points, connecting them with a smooth curve, and reflecting this portion of the curve across the y-axis produces the graph in Figure 10.4.

x	2^{-x^2}
0	1
$\frac{1}{2}$	0.84
1	0.5
$\frac{3}{2}$	0.21
2	0.06

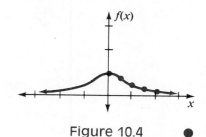

Figure 10.4

Problem Set 10.1

Solve each of the following equations.

1. $3^x = 27$

2. $2^x = 64$

3. $\left(\frac{1}{2}\right)^x = \frac{1}{8}$

4. $\left(\frac{1}{2}\right)^n = 4$

5. $3^{-x} = \frac{1}{81}$

6. $3^{x+1} = 9$

7. $5^{2n-1} = 125$

8. $2^{3-n} = 8$

9. $\left(\frac{2}{3}\right)^t = \frac{9}{4}$

10. $\left(\frac{3}{4}\right)^n = \frac{64}{27}$

11. $4^{3x-1} = 256$

12. $16^x = 64$

13. $4^n = 8$

14. $27^{4x} = 9^{x+1}$

15. $32^x = 16^{1-x}$

16. $\left(\dfrac{1}{8}\right)^{-2t} = 2^{t+3}$

Graph each of the following exponential functions.

17. $f(x) = 3^x$

18. $f(x) = \left(\dfrac{1}{3}\right)^x$

19. $f(x) = 4^x$

20. $f(x) = \left(\dfrac{1}{4}\right)^x$

21. $f(x) = \left(\dfrac{2}{3}\right)^x$

22. $f(x) = \left(\dfrac{3}{2}\right)^x$

23. $f(x) = 2^x + 1$

24. $f(x) = 2^x - 3$

25. $f(x) = 2^{x-1}$

26. $f(x) = 2^{x+2}$

27. $f(x) = -3^x$

28. $f(x) = -2^x$

29. $f(x) = 2^{-x+1}$

30. $f(x) = 2^{-x-2}$

31. $f(x) = 2^x + 2^{-x}$

32. $f(x) = 2^{x^2}$

33. $f(x) = 3^{1-x^2}$

34. $f(x) = 2^{|x|}$

35. $f(x) = 2^{-|x|}$

36. $f(x) = 2^x - 2^{-x}$

10.2 Applications of Exponential Functions

Many real-world situations exhibiting growth or decay can be represented by equations that describe exponential functions. For example, suppose an economist predicts an annual inflation rate of 5% per year for the next 10 years. This means that an item that presently costs \$8 will cost $8(105\%) = 8(1.05) = \$8.40$ a year from now. The same item will cost $[8(105\%)](105\%) = 8(1.05)^2 = \$8.82$ in 2 years. In general, the equation

$$P = P_0(1.05)^t$$

yields the predicted price P of an item in t years if the present cost is P_0 and the annual inflation rate is 5%. Using this equation, we can look at some future prices based on the prediction of a 5% inflation rate.

1. A \$3.27 container of cocoa mix will cost $\$3.27(1.05)^3 = \$3.79$ in 3 years.
2. A \$4.07 jar of coffee will cost $\$4.07(1.05)^5 = \$5.19$ in 5 years.
3. A \$9500 car will cost $\$9500(1.05)^7 = \$13,367$ (to the nearest dollar) in 7 years.

Suppose it is estimated that the value of a car depreciates 15% per year for the first five years. Therefore, a car costing \$9500 will be worth $\$9500(100\% - 15\%) = \$9500(85\%) = \$9500(0.85) = \$8075$ in 1 year. In 2 years the value of the car will have depreciated to $9500(0.85)^2 = \$6864$ (to the nearest dollar). The equation

$$V = V_0(0.85)^t$$

yields the value V of a car in t years if the initial cost is V_0 and the value depreciates 15% per year. Therefore, we can estimate some car values to the nearest dollar as follows.

1. A $6900 car will be worth $6900(0.85)^3 = \$4237$ in 3 years.
2. A $10,900 car will be worth $10,900(0.85)^4 = \$5690$ in 4 years.
3. A $13,000 car will be worth $13,000(0.85)^5 = \$5768$ in 5 years.

Compound Interest

Compound interest provides another illustration of exponential growth. Suppose that $500, called the **principal**, is invested at an interest rate of 8% *compounded annually*. The interest earned the first year is $500(0.08) = \$40$, and this amount is added to the original $500 to form a new principal of $540 for the second year. The interest earned during the second year is $540(0.08) = \$43.20$, and this amount is added to $540 to form a new principal of $583.20 for the third year. Each year a new principal is formed by reinvesting the interest earned during that year.

In general, suppose that a sum of money P (the principal) is invested at an interest rate of r percent compounded annually. The interest earned the first year is Pr, and the new principal for the second year is $P + Pr$, or $P(1 + r)$. Note that the new principal for the second year can be found by multiplying the original principal P by $(1 + r)$. In like fashion, the new principal for the third year can be found by multiplying the previous principal $P(1 + r)$ by $1 + r$, thus obtaining $P(1 + r)^2$. If this process is continued, after t years the total amount of money accumulated, A, is given by

$$A = P(1 + r)^t.$$

Consider the following examples of investments made at a certain rate of interest compounded annually.

1. $750 invested for 5 years at 9% compounded annually produces

$$A = \$750(1.09)^5 = \$1153.97.$$

2. $1000 invested for 10 years at 11% compounded annually produces

$$A = \$1000(1.11)^{10} = \$2839.42.$$

3. $5000 invested for 20 years at 12% compounded annually produces

$$A = \$5000(1.12)^{20} = \$48,231.47.$$

If money invested at a certain rate of interest is compounded more than once a year, then the basic formula $A = P(1 + r)^t$ can be adjusted according to the number of compounding periods in a year. For example, for **semiannual compounding**, the formula becomes $A = P\left(1 + \dfrac{r}{2}\right)^{2t}$; for **quarterly compounding**, the formula becomes $A = P\left(1 + \dfrac{r}{4}\right)^{4t}$. In general, if n represents the number of compounding periods in a year, the formula becomes

$$A = P\left(1 + \frac{r}{n}\right)^{nt}.$$

The following examples illustrate the use of the formula.

1. $750 invested for 5 years at 9% compounded semiannually produces

$$A = \$750\left(1 + \frac{0.09}{2}\right)^{2(5)} = \$750(1.045)^{10} = \$1164.73.$$

2. $1000 invested for 10 years at 11% compounded quarterly produces

$$A = \$1000\left(1 + \frac{0.11}{4}\right)^{4(10)} = \$1000(1.0275)^{40} = \$2959.87.$$

3. $5000 invested for 20 years at 12% compounded monthly produces

$$A = \$5000\left(1 + \frac{0.12}{12}\right)^{12(20)} = \$5000(1.01)^{240} = \$54,462.77.$$

You may find it interesting to compare these results with those obtained earlier for annual compounding.

The Number e

We find an interesting situation when we consider the compound interest formula for $P = \$1, r = 100\%$, and $t = 1$ year. The formula becomes $A = 1\left(1 + \frac{1}{n}\right)^n$. The following table shows some values, rounded to seven decimals, for different values of n.

n	$\left(1 + \frac{1}{n}\right)^n$
5	2.4883200
10	2.5937425
100	2.7048138
1000	2.7169236
10,000	2.7181459
100,000	2.7182818

The table suggests that as n increases, the value of $\left(1 + \frac{1}{n}\right)^n$ gets closer to some fixed number. This does happen, and the fixed number is called e. To five decimal places,

$$e = 2.71828.$$

Exponential expressions using e as a base are found in many real-world applications. Before considering some of those applications, let's take a look at the graph of the basic exponential function using e as a base—namely, the function $f(x) = e^x$.

Example 1   Graph $f(x) = e^x$.

Solution   For graphing purposes, let's use 2.72 as an approximation for e and express the functional values to the nearest tenth. The table below can easily be obtained by using a calculator. The graph of $f(x) = e^x$ is shown in Figure 10.5.

x	$(2.72)^x$
0	1
1	2.7
2	7.4
-1	0.4
-2	0.1

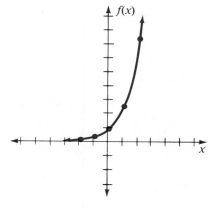

Figure 10.5   ●

Let's return to the concept of compound interest. If the number of compounding periods in a year is increased indefinitely, we arrive at the concept of **continuous compounding**. Mathematically, this can be accomplished by applying the limit concept to the expression $P\left(1 + \dfrac{r}{n}\right)^{nt}$. We will not show the details here, but the following result is obtained. The formula

$$A = Pe^{rt}$$

yields the accumulated value (A) of a sum of money (P) that has been invested for t years at a rate of r percent compounded continuously. The following examples illustrate the use of this formula. (We are using 2.718 as an approximation for e in these calculations.)

1. $750 invested for 5 years at 9% compounded continuously produces

$$A = \$750(2.718)^{(0.09)(5)} = \$750(2.718)^{0.45} = \$1176.18.$$

2. \$1000 invested for 10 years at 11% compounded continuously produces
$$A = \$1000(2.718)^{(0.11)(10)} = \$1000(2.718)^{1.1} = \$3003.82.$$

3. \$5000 invested for 20 years at 12% compounded continuously produces
$$A = \$5000(2.718)^{(0.12)(20)} = \$5000(2.718)^{2.4} = \$55,102.17.$$

Again, you may find it interesting to compare these results to those obtained earlier using a different number of compounding periods.

Law of Exponential Growth

The ideas behind compounding continuously carry over to other growth situations. The law of exponential growth

$$Q(t) = Q_0 e^{kt}$$

is used as a mathematical model for numerous growth-and-decay applications. In this equation, $Q(t)$ represents the quantity of a given substance at any time t, Q_0 is the initial amount of the substance (when $t = 0$), and k is a constant that depends on the particular application. If $k < 0$, then $Q(t)$ decreases as t increases, and the model is referred to as the **law of decay**.

Let's consider some growth-and-decay applications.

Example 2   Suppose that in a certain culture the equation $Q(t) = 15,000e^{0.3t}$ expresses the number of bacteria present as a function of the time t, where t is expressed in hours. How many bacteria are present at the end of 3 hours?

Solution   Using $Q(t) = 15,000e^{0.3t}$, we obtain

$$Q(3) = 15,000(2.718)^{0.3(3)}$$
$$= 15,000(2.718)^{0.9}$$
$$= 36,891 \quad \text{(to the nearest whole number)}$$

Example 3   Suppose the number of bacteria present in a certain culture after t minutes is given by the equation $Q(t) = Q_0 e^{0.05t}$, where Q_0 represents the initial number of bacteria. If 5000 bacteria are present after 20 minutes, how many bacteria were present initially?

Solution   Since $Q(20) = 5000$, we obtain

$$5000 = Q_0(2.718)^{(0.05)(20)}$$
$$5000 = Q_0(2.718)^1$$
$$\frac{5000}{2.718} = Q_0$$
$$1840 = Q_0 \quad \text{(to the nearest whole number)}$$

Therefore, there were approximately 1840 bacteria present initially.

Example 4   The number of grams of a certain radioactive substance present after t seconds is given by the equation $Q(t) = 200e^{-0.3t}$. How many grams remain after 7 seconds?

Solution   Using $Q(t) = 200e^{-0.3t}$, we obtain

$$Q(7) = 200e^{(-0.3)(7)}$$
$$= 200e^{-2.1}$$
$$= 200(2.718)^{-2.1}$$
$$= 24.5 \quad \text{(to the nearest tenth)}$$

Thus, approximately 24.5 grams remain after 7 seconds.   ●

The concept of half-life plays an important role in many decay type problems. The **half-life** of a substance is the time required for one-half of the substance to disintegrate or disappear. The half-life of radium is approximately 1600 years. Thus, if 2000 grams of radium are present now, in 1600 years 1000 grams will be present.

Example 5   Suppose that a certain substance has a half-life of 5 years. If there are presently 100 grams of the substance, then the equation $Q(t) = 100(2)^{-t/5}$ yields the amount remaining after t years. How much remains after (a) 10 years and (b) 18 years?

Solution   (a) $Q(10) = 100(2)^{-10/5}$

$$= 100(2)^{-2}$$
$$= 100\left(\frac{1}{4}\right)$$
$$= 25$$

Therefore, 25 grams remain after 10 years.

(b) $Q(18) = 100(2)^{-18/5}$

$$= 100(2)^{-3.6}$$
$$= 8 \quad \text{(to the nearest whole number)}$$

Therefore, approximately 8 grams remain after 18 years.   ●

Problem Set 10.2

1. Assuming that the rate of inflation is 7% per year, the equation $P = P_0(1.07)^t$ yields the predicted price P in t years of an item that presently costs P_0. Find the predicted price of each of the following items for the indicated number of years from now.

(a) a $0.55 can of soup in 3 years

(b) a $3.43 container of cocoa mix in 5 years

(c) a $1.76 jar of coffee creamer in 4 years

(d) a $0.44 can of beans in 10 years

(e) a $9000 car in 5 years (to the nearest dollar)

(f) a $50,000 house in 8 years (to the nearest dollar)

(g) a $500 TV set in 7 years (to the nearest dollar)

2. Suppose it is estimated that the value of a car depreciates 20% per year for the first 5 years. The equation $A = P_0(0.8)^t$ yields the value (A) of a car after t years if the original price is P_0. Find the value (to the nearest dollar) of each of the following cars after the indicated time.

(a) a $9000 car after 4 years

(b) a $5295 car after 2 years

(c) a $6395 car after 5 years

(d) a $15,595 car after 3 years

For Problems 3–12, use the formula $A = P\left(1 + \dfrac{r}{n}\right)^{nt}$ to find the total amount of money accumulated at the end of the indicated time period for each of the following investments.

3. $250 for 5 years at 9% compounded annually

4. $350 for 7 years at 11% compounded annually

5. $300 for 6 years at 8% compounded semiannually

6. $450 for 10 years at 10% compounded semiannually

7. $600 for 12 years at 12% compounded quarterly

8. $750 for 15 years at 9% compounded quarterly

9. $1000 for 5 years at 12% compounded monthly

10. $1250 for 8 years at 9% compounded monthly

11. $600 for 10 years at $8\frac{1}{2}$% compounded annually

12. $1500 for 15 years at $9\frac{1}{4}$% compounded semiannually

For Problems 13–18, use the formula $A = Pe^{rt}$ to find the total amount of money accumulated at the end of the indicated time period by compounding continuously. Use 2.718 as an approximation for e.

13. $500 for 5 years at 8%

14. $750 for 7 years at 9%

15. $800 for 10 years at 10%

16. $1000 for 10 years at $9\frac{1}{2}$%

17. $1500 for 12 years at 9%

18. $2000 for 20 years at 11%

19. Complete the following chart, which will illustrate what happens to $1000 in 10 years based on different rates of interest and different numbers of compounding periods. Round your answers to the nearest dollar.

$1000 for 10 years

	8%	10%	12%	14%
compounded annually				
compounded semiannually				
compounded quarterly				
compounded monthly				
compounded continuously				

20. Complete the following chart, which will illustrate what happens to $1000 invested at 12% for different lengths of time and different numbers of compounding periods. Round all of your answers to the nearest dollar.

$1000 at 12%

	1 year	5 years	10 years	20 years
compounded annually				
compounded semiannually				
compounded quarterly				
compounded monthly				
compounded continuously				

21. Complete the following chart, which will illustrate what happens to $1000 invested at various rates of interest for different lengths of time, always compounded continuously.

$1000 Compounded Continuously

	8%	10%	12%	14%
5 years				
10 years				
15 years				
20 years				
25 years				

For Problems 22–27, graph each of the exponential functions.

22. $f(x) = e^x + 1$

23. $f(x) = e^x - 2$

24. $f(x) = 2e^x$

25. $f(x) = -e^x$

26. $f(x) = e^{2x}$

27. $f(x) = e^{-x}$

28. Suppose that for a certain culture, the equation $Q(t) = 1000e^{0.4t}$ expresses the number of bacteria present as a function of the time t, where t is expressed in hours. How many bacteria are present at the end of 2 hours? 3 hours? 5 hours?

29. The number of bacteria present at a given time under certain conditions is given by the equation $Q(t) = 5000e^{0.05t}$, where t is expressed in minutes. How many bacteria are present at the end of 10 minutes? 30 minutes? 1 hour?

30. The number of bacteria present in a certain culture after t hours is given by the equation $Q(t) = Q_0 e^{0.3t}$, where Q_0 represents the initial number of bacteria. If 6640 bacteria are present after 4 hours, how many bacteria were present initially?

31. The number of grams of a certain radioactive substance present after t seconds is given by the equation $Q(t) = 1500e^{-0.4t}$. How many grams remain after 5 seconds? 10 seconds? 20 seconds?

32. The atmospheric pressure, measured in pounds per square inch, is a function of the altitude above sea level. The equation $P(a) = 14.7e^{-0.21a}$, where a is the altitude measured in miles, can be used to approximate atmospheric pressure. Find the atmospheric pressure at each of the following locations.

(a) Mount McKinley in Alaska—altitude of 3.85 miles

(b) Denver—the "mile-high" city

(c) Asheville, North Carolina—altitude of 1985 feet

(d) Phoenix, Arizona—altitude of 1090 feet

33. Suppose that the present population of a city is 75,000. Using the equation $P(t) = 75000e^{0.01t}$ to estimate future growth, estimate the population (a) 10 years from now, (b) 15 years from now, and (c) 25 years from now.

34. Suppose that a certain substance has a half-life of 20 years. If there are presently 2500 milligrams of the substance, then the equation $Q(t) = 2500(2)^{-t/20}$ yields the amount remaining after t years. How much remains after 40 years? 50 years?

35. Assume that the half-life of a certain radioactive substance is 1200 years and that the amount of the substance remaining after t years is given by the equation $Q(t) = 500(2)^{-t/1200}$. If the substance is measured in grams, how many grams are present now? How many grams will remain after 1500 years?

36. The half-life of radium is approximately 1600 years. If the initial amount is Q_0 milligrams, then the quantity $Q(t)$ remaining after t years is given by $Q(t) = Q_0(2)^{kt}$. Find the value of k.

37. The half-life of a certain substance is 25 years. If the initial amount is Q_0 grams, then the quantity $Q(t)$ remaining after t years is given by $Q(t) = Q_0(2)^{kt}$. Find the value of k.

Miscellaneous Problems

Graph each of the following exponential functions.

38. $f(x) = x(2^x)$

39. $f(x) = \dfrac{e^x + e^{-x}}{2}$

40. $f(x) = \dfrac{2}{e^x + e^{-x}}$

41. $f(x) = \dfrac{e^x - e^{-x}}{2}$

42. $f(x) = \dfrac{2}{e^x - e^{-x}}$

10.3 Logarithms

In Sections 10.1 and 10.2, we gave meaning to exponential expressions of the form b^n, where b is any positive real number and n is any real number; used exponential expressions of the form b^n to define exponential functions; and used exponential functions to help solve problems. In the next three sections we will follow the same basic pattern with respect to a new concept—that of a logarithm. Let's begin with the following definition.

| Definition 10.2 | If r is any positive real number, then the unique exponent t such that $b^t = r$ is called the *logarithm of r with base b* and is denoted by $\log_b r$. |

According to Definition 10.2, the logarithm of 16 base 2 is the exponent t such that $2^t = 16$; thus, we can write $\log_2 16 = 4$. Likewise, we can write $\log_{10} 1000 = 3$ because $10^3 = 1000$. In general, Definition 10.2 can be remembered in terms of the statement

$$\log_b r = t \text{ is equivalent to } b^t = r.$$

Therefore, we can easily switch back and forth between exponential and logarithmic forms of equations, as the next examples illustrate.

$$\log_2 8 = 3 \text{ is equivalent to } 2^3 = 8.$$
$$\log_{10} 100 = 2 \text{ is equivalent to } 10^2 = 100.$$
$$\log_3 81 = 4 \text{ is equivalent to } 3^4 = 81.$$
$$\log_{10} 0.001 = -3 \text{ is equivalent to } 10^{-3} = 0.001.$$
$$2^7 = 128 \text{ is equivalent to } \log_2 128 = 7.$$
$$5^3 = 125 \text{ is equivalent to } \log_5 125 = 3.$$
$$\left(\frac{1}{2}\right)^4 = \frac{1}{16} \text{ is equivalent to } \log_{1/2}\left(\frac{1}{16}\right) = 4.$$
$$10^{-2} = 0.01 \text{ is equivalent to } \log_{10} 0.01 = -2.$$

Some logarithms can be determined by changing to exponential form and using the properties of exponents, as the next two examples illustrate.

Example 1   Evaluate $\log_{10} 0.0001$.

Solution   Let $\log_{10} 0.0001 = x$. Then by changing to exponential form, we have $10^x = 0.0001$, which can be solved as follows.

$$10^x = 0.0001$$

$$10^x = 10^{-4} \qquad \left(0.0001 = \frac{1}{10000} = \frac{1}{10^4} = 10^{-4}\right)$$

$$x = -4$$

Thus, we have $\log_{10} 0.0001 = -4$.   ●

Example 2   Evaluate $\log_9 \left(\dfrac{\sqrt[5]{27}}{3}\right)$.

Solution   Let $\log_9 \left(\dfrac{\sqrt[5]{27}}{3}\right) = x$. Then by changing to exponential form, we have $9^x = \dfrac{\sqrt[5]{27}}{3}$, which can be solved as follows.

$$9^x = \frac{(27)^{1/5}}{3}$$

$$(3^2)^x = \frac{(3^3)^{1/5}}{3}$$

$$3^{2x} = \frac{3^{3/5}}{3}$$

$$3^{2x} = 3^{-2/5}$$

$$2x = -\frac{2}{5}$$

$$x = -\frac{1}{5}$$

Therefore, we have $\log_9 \left(\dfrac{\sqrt[5]{27}}{3} \right) = -\dfrac{1}{5}$.

Some equations involving logarithms can also be solved by changing to exponential form and using our knowledge of exponents.

Example 3   Solve $\log_8 x = \dfrac{2}{3}$.

Solution   Changing $\log_8 x = \dfrac{2}{3}$ to exponential form, we obtain

$$8^{2/3} = x.$$

Therefore,

$$x = (\sqrt[3]{8})^2$$
$$= 2^2$$
$$= 4.$$

The solution set is $\{4\}$.

Example 4   Solve $\log_b \left(\dfrac{27}{64} \right) = 3$.

Solution   Changing $\log_b \left(\dfrac{27}{64} \right) = 3$ to exponential form, we obtain

$$b^3 = \frac{27}{64}.$$

Therefore,

$$b = \sqrt[3]{\frac{27}{64}}$$
$$= \frac{3}{4}.$$

The solution set is $\left\{ \dfrac{3}{4} \right\}$.

Properties of Logarithms

There are some properties of logarithms that are a direct consequence of Definition 10.2 and the properties of exponents. For example, the following property is obtained by writing the exponential equations $b^1 = b$ and $b^0 = 1$ in logarithmic form.

Property 10.3	For $b > 0$ and $b \neq 1$, $$\log_b b = 1 \quad \text{and} \quad \log_b 1 = 0.$$

Therefore, according to Property 10.3 we can write these equations.

$$\log_{10} 10 = 1, \quad \log_4 4 = 1,$$
$$\log_{10} 1 = 0, \quad \log_5 1 = 0$$

Also from Definition 10.2 we know that $\log_b r$ is the exponent t such that $b^t = r$. Therefore, raising b to the $\log_b r$ power must produce r. This fact is stated in Property 10.4.

Property 10.4	For $b > 0$, $b \neq 1$, and $r > 0$, $$b^{\log_b r} = r.$$

Therefore, according to Property 10.4 we can write the following.

$$10^{\log_{10} 72} = 72 \qquad 3^{\log_3 85} = 85 \qquad e^{\log_e 7} = 7$$

Because a logarithm is by definition an exponent, it would seem reasonable to predict that some properties of logarithms correspond to the basic exponential properties. This is an accurate prediction; these properties provide a basis for computational work with logarithms. Let's state the first of these properties and show how it can be verified using our knowledge of exponents.

Property 10.5	For positive numbers b, r, and s, where $b \neq 1$, $$\log_b rs = \log_b r + \log_b s.$$

To verify Property 10.5, we can proceed as follows. Let $m = \log_b r$ and $n = \log_b s$. Change each of these equations to exponential form.

$$m = \log_b r \text{ becomes } r = b^m.$$
$$n = \log_b s \text{ becomes } s = b^n.$$

Thus, the product rs becomes

$$rs = b^m \cdot b^n = b^{m+n}.$$

Now, by changing $rs = b^{m+n}$ back to logarithmic form, we obtain

$$\log_b rs = m + n.$$

Replacing m with $\log_b r$ and replacing n with $\log_b s$ yields

$$\log_b rs = \log_b r + \log_b s.$$

The following two examples illustrate the use of Property 10.5.

Example 5   If $\log_2 5 = 2.3222$ and $\log_2 3 = 1.5850$, evaluate $\log_2 15$.

Solution   Because $15 = 5 \cdot 3$, we can apply Property 10.5 as follows:

$$\log_2 15 = \log_2 (5 \cdot 3)$$
$$= \log_2 5 + \log_2 3$$
$$= 2.3222 + 1.5850 = 3.9072.$$

Example 6   Given that $\log_{10} 178 = 2.2504$ and $\log_{10} 89 = 1.9494$, evaluate $\log_{10} (178 \cdot 89)$.

Solution
$$\log_{10} (178 \cdot 89) = \log_{10} 178 + \log_{10} 89$$
$$= 2.2504 + 1.9494 = 4.1998.$$

Since $\dfrac{b^m}{b^n} = b^{m-n}$, we would expect a corresponding property pertaining to logarithms. Property 10.6 is that property. It can be verified by using an approach similar to the one used to verify Property 10.5. This verification is left for you to do as an exercise in the next problem set.

Property 10.6   For positive numbers b, r, and s, where $b \neq 1$,

$$\log_b \left(\frac{r}{s}\right) = \log_b r - \log_b s.$$

Property 10.6 can be used to change a division problem into an equivalent subtraction problem, as the next two examples illustrate.

Example 7   If $\log_5 36 = 2.2265$ and $\log_5 4 = 0.8614$, evaluate $\log_5 9$.

Solution   Since $9 = \dfrac{36}{4}$, we can use Property 10.6 as follows:

$$\log_5 9 = \log_5 \left(\frac{36}{4}\right)$$
$$= \log_5 36 - \log_5 4$$
$$= 2.2265 - 0.8614 = 1.3651.$$

Example 8   Evaluate $\log_{10}\left(\dfrac{379}{86}\right)$ given that $\log_{10} 379 = 2.5786$ and $\log_{10} 86 = 1.9345$.

Solution

$$\log_{10}\left(\frac{379}{86}\right) = \log_{10} 379 - \log_{10} 86$$

$$= 2.5786 - 1.9345$$

$$= 0.6441$$

●

Another property of exponents states that $(b^n)^m = b^{mn}$. The corresponding property of logarithms is stated in Property 10.7. Again, we will leave the verification of this property as an exercise for you to do in the next set of problems.

Property 10.7   If r is a positive real number, b is a positive real number other than 1, and p is any real number, then

$$\log_b r^p = p(\log_b r).$$

The next two examples illustrate the use of Property 10.7.

Example 9   Evaluate $\log_2 22^{1/3}$ given that $\log_2 22 = 4.4598$.

Solution

$$\log_2 22^{1/3} = \frac{1}{3}\log_2 22 \qquad \text{Property 10.7}$$

$$= \frac{1}{3}(4.4598)$$

$$= 1.4866$$

●

Example 10   Evaluate $\log_{10} (8540)^{3/5}$ given that $\log_{10} 8540 = 3.9315$.

Solution

$$\log_{10} (8540)^{3/5} = \frac{3}{5}\log_{10} 8540$$

$$= \frac{3}{5}(3.9315)$$

$$= 2.3589$$

●

Used together, the properties of logarithms allow us to change the forms of various logarithmic expressions. For example, an expression such as $\log_b \sqrt{\dfrac{xy}{z}}$ can be rewritten in terms of sums and differences of simpler logarithmic quantities as follows.

$$\log_b \sqrt{\frac{xy}{z}} = \log_b \left(\frac{xy}{z}\right)^{1/2}$$

$$= \frac{1}{2} \log_b \left(\frac{xy}{z}\right) \qquad \text{Property 10.7}$$

$$= \frac{1}{2}(\log_b xy - \log_b z) \qquad \text{Property 10.6}$$

$$= \frac{1}{2}(\log_b x + \log_b y - \log_b z) \qquad \text{Property 10.5}$$

The properties of logarithms along with the link between logarithmic form and exponential form provide the basis for solving certain types of equations involving logarithms. Our final example of this section illustrates this idea.

Example 11   Solve $\log_{10} x + \log_{10} (x + 9) = 1$.

Solution
$$\log_{10} x + \log_{10} (x + 9) = 1$$
$$\log_{10} [x(x + 9)] = 1 \qquad \text{Property 10.5}$$
$$10^1 = x(x + 9) \qquad \text{Change to exponential form.}$$
$$10 = x^2 + 9x$$
$$0 = x^2 + 9x - 10$$
$$0 = (x + 10)(x - 1)$$
$$x + 10 = 0 \qquad \text{or} \qquad x - 1 = 0$$
$$x = -10 \quad \text{or} \qquad x = 1$$

Since the left-hand number of the original equation is meaningful only if $x > 0$ and $x + 9 > 0$, the solution -10 must be discarded. Thus, the solution set is $\{1\}$.   ●

Problem Set 10.3

Write each of the following in logarithmic form. For example, $2^4 = 16$ becomes $\log_2 16 = 4$.

1. $3^2 = 9$ 2. $2^5 = 32$ 3. $5^3 = 125$

4. $10^1 = 10$ 5. $2^{-4} = \dfrac{1}{16}$ 6. $\left(\dfrac{2}{3}\right)^{-3} = \dfrac{27}{8}$

7. $10^{-2} = 0.01$ 8. $10^5 = 100{,}000$

Write each of the following in exponential form. For example, $\log_2 8 = 3$ becomes $2^3 = 8$.

9. $\log_2 64 = 6$ 10. $\log_3 27 = 3$ 11. $\log_{10} 0.1 = -1$

12. $\log_5 \left(\dfrac{1}{25}\right) = -2$ 13. $\log_2 \left(\dfrac{1}{16}\right) = -4$ 14. $\log_{10} 0.00001 = -5$

Evaluate each of the following.

15. $\log_6 36$

16. $\log_3 243$

17. $\log_5 \left(\dfrac{1}{5} \right)$

18. $\log_4 \left(\dfrac{1}{64} \right)$

19. $\log_{10} 10$

20. $\log_{10} 1$

21. $\log_3 \sqrt{3}$

22. $\log_5 \sqrt[3]{25}$

23. $\log_3 \left(\dfrac{\sqrt{27}}{3} \right)$

24. $\log_{1/2} \left(\dfrac{\sqrt[4]{8}}{2} \right)$

25. $\log_{1/4} \left(\dfrac{\sqrt[4]{32}}{2} \right)$

26. $\log_2 \left(\dfrac{\sqrt[3]{16}}{4} \right)$

27. $10^{\log_{10} 7}$

28. $5^{\log_5 13}$

29. $\log_2 (\log_5 5)$

30. $\log_6 (\log_2 64)$

Solve each of the following equations.

31. $\log_5 x = 2$

32. $\log_{10} x = 3$

33. $\log_8 t = \dfrac{5}{3}$

34. $\log_4 m = \dfrac{3}{2}$

35. $\log_b 3 = \dfrac{1}{2}$

36. $\log_b 2 = \dfrac{1}{2}$

37. $\log_{10} x = 0$

38. $\log_{10} x = 1$

Given that $\log_{10} 2 = 0.3010$ and $\log_{10} 7 = 0.8451$, evaluate each of the following by using Properties 10.5–10.7.

39. $\log_{10} 14$

40. $\log_{10} \left(\dfrac{7}{2} \right)$

41. $\log_{10} 4$

42. $\log_{10} 49$

43. $\log_{10} 343$

44. $\log_{10} 32$

45. $\log_{10} \sqrt{2}$

46. $\log_{10} \sqrt[3]{7}$

47. $\log_{10} (7)^{4/3}$

48. $\log_{10} (2)^{3/5}$

49. $\log_{10} 28$

50. $\log_{10} 56$

51. $\log_{10} 98$

52. $\log_{10} 20$

53. $\log_{10} 200$

54. $\log_{10} 70$

55. $\log_{10} 1400$

56. $\log_{10} 4900$

Express each of the following as the sum or difference of simpler logarithmic quantities. (Assume that all variables represent positive real numbers.) For example,

$$\log_b \left(\frac{x^3}{y^2} \right) = \log_b x^3 - \log_b y^2$$
$$= 3 \log_b x - 2 \log_b y.$$

57. $\log_b xyz$

58. $\log_b \left(\dfrac{x^2}{y} \right)$

59. $\log_b x^2 y^3$

60. $\log_b x^{2/3} y^{3/4}$

61. $\log_b \sqrt{xy}$

62. $\log_b \sqrt[3]{x^2 z}$

63. $\log_b \sqrt{\dfrac{x}{y}}$

64. $\log_b \left(x \sqrt{\dfrac{x}{y}} \right)$

Express each of the following as a single logarithm. (Assume that all variables represent positive real numbers.) For example,

$$3 \log_b x + 5 \log_b y = \log_b x^3 y^5.$$

65. $\log_b x + \log_b y - \log_b z$

66. $2 \log_b x - 4 \log_b y$

67. $(\log_b x - \log_b y) - \log_b z$

68. $\log_b x - (\log_b y - \log_b z)$

69. $\log_b x + \dfrac{1}{2} \log_b y$

70. $2 \log_b x + 4 \log_b y - 3 \log_b z$

71. $2 \log_b x + \dfrac{1}{2} \log_b (x - 1) - 4 \log_b (2x + 5)$

72. $\dfrac{1}{2} \log_b x - 3 \log_b x + 4 \log_b y$

Solve each of the following equations.

73. $\log_{10} 5 + \log_{10} x = 1$

74. $\log_{10} x + \log_{10} 25 = 2$

75. $\log_{10} 20 - \log_{10} x = 1$

76. $\log_{10} x + \log_{10} (x - 3) = 1$

77. $\log_{10} (x + 2) - \log_{10} x = 1$

78. $\log_{10} x + \log_{10} (x - 21) = 2$

79. $\log_{10} (x - 4) + \log_{10} (x - 1) = 1$

80. $\log_{10} (x + 2) + \log_{10} (x - 1) = 1$

81. Verify Property 10.6.

82. Verify Property 10.7.

10.4 Logarithmic Functions

The concept of a logarithm can now be used to define a logarithmic function as follows.

Definition 10.3	If $b > 0$ and $b \neq 1$, then the function defined by $$f(x) = \log_b x,$$ where x is any positive real number, is called the *logarithmic function with base b*.

We can obtain the graph of a specific logarithmic function in various ways. For example, the equation $y = \log_2 x$ can be changed to the exponential equation $2^y = x$, from which a table of values can be determined. The next set of exercises asks you to graph some logarithmic functions using this approach.

The graph of a logarithmic function can also be obtained by setting up a table of values directly from the logarithmic equation. Example 1 illustrates this approach.

Example 1   Graph $f(x) = \log_2 x$.

Solution   Let's choose some values for x for which the corresponding values for $\log_2 x$ are easily determined. (Remember that logarithms are only defined for the positive real numbers.)

x	f(x)
$\frac{1}{8}$	-3
$\frac{1}{4}$	-2
$\frac{1}{2}$	-1
1	0
2	1
4	2
8	3

$$\log_2 \frac{1}{8} = -3 \text{ because } 2^{-3} = \frac{1}{2^3} = \frac{1}{8}$$

$$\log_2 1 = 0 \text{ because } 2^0 = 1$$

Plotting these points and connecting them with a smooth curve produces Figure 10.6.

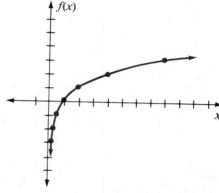

Figure 10.6

Now suppose that we consider two functions f and g as follows:

$f(x) = b^x$ Domain: all real numbers
                         Range: positive real numbers

$g(x) = \log_b x$ Domain: positive real numbers
                         Range: all real numbers

Furthermore, suppose that we consider the composition of f and g and the composition of g and f.

$$(f \circ g)(x) = f(g(x)) = f(\log_b x) = b^{\log_b x} = x$$
$$(g \circ f)(x) = g(f(x)) = g(b^x) = \log_b b^x = x \log_b b = x(1) = x$$

Because the domain of f is the range of g, the range of f is the domain of g, $f(g(x)) = x$, and $g(f(x)) = x$, the two functions f and g *are inverses of each other*.

Remember from Chapter 8 that the graph of a function and the graph of its inverse are reflections of each other through the line $y = x$. Thus, the graph of a logarithmic function can also be determined by reflecting the graph of its inverse exponential function through the line $y = x$. This idea is illustrated in Figure 10.7, where the graph of $y = 2^x$ has been reflected across the line $y = x$ to produce the graph of $y = \log_2 x$.

The general behavior patterns of exponential functions were illustrated back in Figure 10.3. We can now reflect each of these graphs through the line $y = x$ and observe the general behavior patterns of logarithmic functions, shown in Figure 10.8.

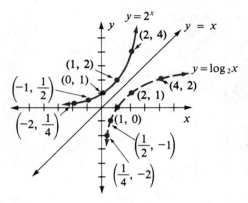

Figure 10.7

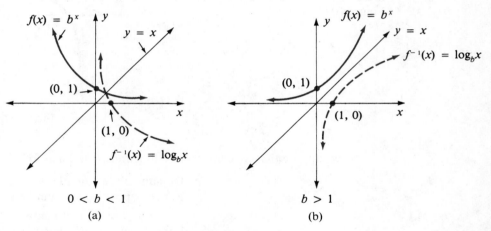

$0 < b < 1$

(a)

$b > 1$

(b)

Figure 10.8

Common Logarithms — Base 10

The properties of logarithms discussed in Section 10.3 are true for any valid base. However, since the Hindu-Arabic numeration system that we use is a base 10 system, logarithms to base 10 have historically been used for computational purposes. Base 10 logarithms are called **common logarithms**.

Originally, common logarithms were developed as an aid to numerical calculations. Today they are seldom used for that purpose because the calculator and computer can more effectively handle messy computational problems. Thus, in this section we will restrict our discussion to evaluating common logarithms, and then in an optional section at the end of this chapter we will illustrate a few of their computational characteristics.

As we know from earlier work, the definition of a logarithm provides the basis for evaluating $\log_{10} x$ for values of x that are integral powers of 10. Consider the following examples.

$$\log_{10} 1000 = 3 \text{ because } 10^3 = 1000.$$
$$\log_{10} 100 = 2 \text{ because } 10^2 = 100.$$
$$\log_{10} 10 = 1 \text{ because } 10^1 = 10.$$
$$\log_{10} 1 = 0 \text{ because } 10^0 = 1.$$

$$\log_{10} 0.1 = -1 \text{ because } 10^{-1} = \frac{1}{10} = 0.1.$$

$$\log_{10} 0.01 = -2 \text{ because } 10^{-2} = \frac{1}{10^2} = 0.01.$$

$$\log_{10} 0.001 = -3 \text{ because } 10^{-3} = \frac{1}{10^3} = 0.001.$$

When one is working exclusively with base 10 logarithms, it is customary to omit writing the numeral 10 to designate the base. Thus, the expression $\log_{10} x$ is written as $\log x$ and a statement such as $\log_{10} 1000 = 3$ becomes $\log 1000 = 3$. We will follow this practice from now on in this chapter, but don't forget that the base is understood to be 10.

To find the common logarithm of a positive number that is not an integral power of 10, we can use an appropriately equipped calculator or a table such as the one that appears inside the back cover of this text. Using a calculator equipped with a common logarithm function (ordinarily this key is labeled $\boxed{\log}$), we obtain the following results rounded to four decimal places.

$$\log 1.75 = 0.2430$$
$$\log 23.8 = 1.3766$$
$$\log 134 = 2.1271$$
$$\log 0.192 = -0.7167$$
$$\log 0.0246 = -1.6091$$

Be sure that you know how to use a calculator to obtain these results.

Using a table to find a common logarithm is relatively easy, but it does require a little more effort than pushing a button as you do with a calculator. Let's consider a small part of the table that appears in the back of the book.

Each number in the column headed n represents the first two significant digits of a number between 1 and 10; each of the column headings 0 through 9 represents the third significant digit. To find the logarithm of a number such as 1.75, we look at the intersection of the row containing 1.7 and the column headed 5. Thus, we obtain $\log 1.75 = 0.2430$. Similarly, we can find that $\log 2.09 = 0.3201$ and $\log 2.40 = 0.3802$. Keep in mind that these values are rounded to four decimal places. Also, don't lose sight of the meaning of such logarithmic statements. That is to say, $\log 1.75 = 0.2430$ means that 10 raised to the 0.2430 power is approximately 1.75. (Try it on your calculator!)

Table of Common Logarithms

n	0	1	2	3	4	5	6	7	8	9
1.0	0.0000	0.0043	0.0086	0.0128	0.0170	0.0212	0.0253	0.0294	0.0334	0.0374
1.1	0.0414	0.0453	0.0492	0.0531	0.0569	0.0607	0.0645	0.0682	0.0719	0.0755
1.2	0.0792	0.0828	0.0864	0.0899	0.0934	0.0969	0.1004	0.1038	0.1072	0.1106
1.3	0.1139	0.1173	0.1206	0.1239	0.1271	0.1303	0.1335	0.1367	0.1399	0.1430
1.4	0.1461	0.1492	0.1523	0.1553	0.1584	0.1614	0.1644	0.1673	0.1703	0.1732
1.5	0.1761	0.1790	0.1818	0.1847	0.1875	0.1903	0.1931	0.1959	0.1987	0.2014
1.6	0.2041	0.2068	0.2095	0.2122	0.2148	0.2175	0.2201	0.2227	0.2253	0.2279
1.7	0.2304	0.2330	0.2355	0.2380	0.2405	0.2430	0.2455	0.2480	0.2504	0.2529
1.8	0.2553	0.2577	0.2601	0.2625	0.2648	0.2672	0.2695	0.2718	0.2742	0.2765
1.9	0.2788	0.2810	0.2833	0.2856	0.2878	0.2900	0.2923	0.2945	0.2967	0.2989
2.0	0.3010	0.3032	0.3054	0.3075	0.3096	0.3118	0.3139	0.3160	0.3181	0.3201
2.1	0.3222	0.3243	0.3263	0.3284	0.3304	0.3324	0.3345	0.3365	0.3385	0.3404
2.2	0.3424	0.3444	0.3464	0.3483	0.3502	0.3522	0.3541	0.3560	0.3579	0.3598
2.3	0.3617	0.3636	0.3655	0.3674	0.3692	0.3711	0.3729	0.3747	0.3766	0.3784
2.4	0.3802	0.3820	0.3838	0.3856	0.3874	0.3892	0.3909	0.3927	0.3945	0.3962

(*complete table in back of text*)

Now suppose that we want to use the table to find the logarithm of a positive number greater than 10 or less than 1. This can be accomplished by representing the number in scientific notation and applying the property $\log rs = \log r + \log s$. For example, to find $\log 134$, we can proceed as follows.

$$\log 134 = \log (1.34 \cdot 10^2)$$
$$= \log 1.34 + \log 10^2$$
$$= \log 1.34 + 2 \log 10$$
$$= .1271 + 2 = 2.1271$$

From the table — By inspection, since $\log 10 = 1$

The decimal part (.1271) of the logarithm 2.1271 is called the **mantissa**, and the integral part (2) is called the **characteristic**. We can find the characteristic of a common logarithm by inspection (since it is the exponent of 10 when the number is written in scientific notation), and the mantissa we can get from a table. Let's consider two more examples.

$$\log 23.8 = \log (2.38 \cdot 10^1)$$
$$= \log 2.38 + \log 10^1$$
$$= .3766 + 1$$

From the table — Exponent of 10

$$= 1.3766$$

$$\log 0.192 = \log (1.92 \cdot 10^{-1})$$
$$= \log 1.92 + \log 10^{-1}$$
$$= .2833 + (-1)$$

From the   Exponent of 10
table

$$= .2833 + (-1)$$

Notice that in the last example we expressed the logarithm of 0.192 as .2833 + (−1); we did not add .2833 and −1. This is normal procedure when a table of common logarithms is used, because the mantissas given in the table are positive numbers. However, you should recognize that adding .2833 and −1 produces −0.7167, which agrees with our earlier result obtained with a calculator.

You should realize that the common logarithm table in this text restricts us to reading directly from the table an approximation for the common logarithm of any number between 1.00 and 9.99 with *three* significant digits. A reasonable approximation can also be obtained from the table for the common logarithms of numbers with *four* significant digits by using a process called *linear interpolation*. This process is discussed in a later section of this chapter.

Antilogarithms

It is often necessary to be able to find a number when given the common logarithm of the number; that is, given log x we need to be able to determine x. In this situation, x is referred to as the **antilogarithm** (abbreviated **antilog**) of log x. Many calculators are equipped to find antilogarithms two ways. Let's consider some examples.

Example 2   Determine antilog 0.2430.

Solution A   The phrase "determine the antilog of 0.2430" means to find a value for x such that log x = 0.2430. Changing log x = 0.2430 to exponential form, we obtain $10^{0.2430} = x$. Now on our calculator we can enter 10, press $\boxed{y^x}$, enter .2430, press $\boxed{=}$, and obtain 1.7498467, which equals 1.75 when rounded to three significant digits.

Solution B   Considered in terms of functions, the antilog function is the inverse of the log function. Therefore, on some calculators we can enter .2430, press $\boxed{\text{INV}}$, press $\boxed{\text{log}}$, and obtain 1.7498467, which rounds to 1.75.   ●

Example 3   Determine antilog −1.6091.

Solution   Using the inverse routine, we can enter −1.6091, press $\boxed{\text{INV}}$, press $\boxed{\text{log}}$, and obtain .02459801, which equals 0.0246 when rounded to three significant digits.   ●

The common logarithm table at the back of the book can also be used to determine antilogarithms. The procedure is illustrated in the next three examples.

Example 4

Determine antilog 1.3365.

Solution

Finding an antilogarithm simply involves reversing the process used before for finding a logarithm. Thus, antilog 1.3365 means that 1 is the characteristic and .3365 is the mantissa. We look for .3365 in the body of the common logarithm table, and we find that it is located at the intersection of the 2.1 row and the 7 column. Therefore, the antilogarithm is

$$2.17 \cdot 10^1 = 21.7.$$

Example 5

Determine antilog $[0.1523 + (-2)]$.

Solution

The mantissa, .1523, is located at the intersection of the 1.4 row and 2 column. The characteristic is -2, and therefore the antilogarithm is

$$1.42 \cdot 10^{-2} = 0.0142.$$

Example 6

Determine antilog -2.6038.

Solution

The mantissas given in a logarithm table are *positive* numbers. Thus, we need to express -2.6038 in terms of a positive mantissa, which can be done by adding and subtracting 3 as follows:

$$(-2.6038 + 3) - 3 = 0.3962 + (-3).$$

Now we can look for .3962, which we find at the intersection of the 2.4 row and the 9 column. Therefore, the antilogarithm is

$$2.49 \cdot 10^{-3} = 0.00249.$$

Natural Logarithms

In many practical applications of logarithms, the number e (remember $e \approx 2.71828$) is used as a base. Logarithms with a base of e are called **natural logarithms**, and the symbol **ln x** is commonly used instead of $\log_e x$.

Natural logarithms can be found with an appropriately equipped calculator or with a table of natural logarithms. Using a calculator with a natural logarithm function (ordinarily a key labeled $\boxed{\ln x}$), we can obtain the following results rounded to four decimal places.

$$\ln 3.21 = 1.1663$$
$$\ln 47.28 = 3.8561$$

$$\ln 842 = 6.7358$$
$$\ln 0.21 = -1.5606$$
$$\ln 0.0046 = -5.3817$$
$$\ln 10 = 2.3026$$

Be sure that you can use a calculator to obtain these results. Keep in mind the significance of a statement such as ln 3.21 = 1.1663. We are claiming that e raised to the 1.1663 power is approximately 3.21. Using 2.71828 as an approximation for e, we obtain

$$(2.71828)^{1.1663} = 3.2100908.$$

Furthermore, since $f(x) = \ln x$ and $g(x) = e^x$ are inverses of each other, we can enter 1.1663, press $\boxed{\text{INV}}$, press $\boxed{\ln x}$, and obtain 3.2100933. (The last three digits of this result differ from our previous result of 3.2100908 because a better approximation for e is used in this sequence of operations.)

Table 10.1 contains the natural logarithms for numbers between 0.1 and 10, inclusive, at intervals of 0.1. Reading directly from the table, we obtain ln 1.6 = 0.4700, ln 4.8 = 1.5686, and ln 9.2 = 2.2192.

Table 10.1   Natural Logarithms

n	$\ln n$	n	$\ln n$	n	$\ln n$	n	$\ln n$
0.1	−2.3026	2.6	0.9555	5.1	1.6292	7.6	2.0281
0.2	−1.6094	2.7	0.9933	5.2	1.6487	7.7	2.0412
0.3	−1.2040	2.8	1.0296	5.3	1.6677	7.8	2.0541
0.4	−0.9163	2.9	1.0647	5.4	1.6864	7.9	2.0669
0.5	−0.6931	3.0	1.0986	5.5	1.7047	8.0	2.0794
0.6	−0.5108	3.1	1.1314	5.6	1.7228	8.1	2.0919
0.7	−0.3567	3.2	1.1632	5.7	1.7405	8.2	2.1041
0.8	−0.2231	3.3	1.1939	5.8	1.7579	8.3	2.1163
0.9	−0.1054	3.4	1.2238	5.9	1.7750	8.4	2.1282
1.0	0.0000	3.5	1.2528	6.0	1.7918	8.5	2.1401
1.1	0.0953	3.6	1.2809	6.1	1.8083	8.6	2.1518
1.2	0.1823	3.7	1.3083	6.2	1.8245	8.7	2.1633
1.3	0.2624	3.8	1.3350	6.3	1.8405	8.8	2.1748
1.4	0.3365	3.9	1.3610	6.4	1.8563	8.9	2.1861
1.5	0.4055	4.0	1.3863	6.5	1.8718	9.0	2.1972
1.6	0.4700	4.1	1.4110	6.6	1.8871	9.1	2.2083
1.7	0.5306	4.2	1.4351	6.7	1.9021	9.2	2.2192
1.8	0.5878	4.3	1.4586	6.8	1.9169	9.3	2.2300
1.9	0.6419	4.4	1.4816	6.9	1.9315	9.4	2.2407
2.0	0.6931	4.5	1.5041	7.0	1.9459	9.5	2.2513
2.1	0.7419	4.6	1.5261	7.1	1.9601	9.6	2.2618
2.2	0.7885	4.7	1.5476	7.2	1.9741	9.7	2.2721
2.3	0.8329	4.8	1.5686	7.3	1.9879	9.8	2.2824
2.4	0.8755	4.9	1.5892	7.4	2.0015	9.9	2.2925
2.5	0.9163	5.0	1.6094	7.5	2.0149	10	2.3026

When using Table 10.1 to find the natural logarithm of a positive number less than 0.1 or greater than 10, we can use the property $\ln rs = \ln r + \ln s$ as follows.

$$\ln 0.0084 = \ln (8.4 \cdot 10^{-3})$$
$$= \ln 8.4 + \ln 10^{-3}$$
$$= \ln 8.4 + (-3)(\ln 10)$$
$$= 2.1282 - 3(2.3026)$$

From the   From the
table   table

$$= 2.1282 - 6.9078 = -4.7796.$$

$$\ln 190 = \ln (1.9 \cdot 10^2)$$
$$= \ln 1.9 + \ln 10^2$$
$$= \ln 1.9 + 2 \ln 10$$
$$= 0.6419 + 2(2.3026)$$

From the   From the
table   table

$$= 5.2471.$$

Problem Set 10.4

For Problems 1–6, graph each of the logarithmic functions.

1. $f(x) = \log_3 x$ **2.** $f(x) = \log_4 x$ **3.** $f(x) = \log_{10} x$

4. $f(x) = \log_5 x$ **5.** $f(x) = \log_{1/2} x$ **6.** $f(x) = \log_{1/3} x$

7. Graph $y = \log_3 x$ by changing the equation to exponential form.

8. Graph $y = \log_4 x$ by changing the equation to exponential form.

9. Graph $f(x) = \log_{1/2} x$ by reflecting the graph of $g(x) = \left(\dfrac{1}{2}\right)^x$ across the line $y = x$.

10. Graph $f(x) = \log_{1/3} x$ by reflecting the graph of $g(x) = \left(\dfrac{1}{3}\right)^x$ across the line $y = x$.

For Problems 11–14, graph each of the logarithmic functions. Don't forget that the graph of $f(x) = g(x) + 1$ is the graph of $g(x)$ moved up 1 unit.

11. $f(x) = 1 + \log_{10} x$ **12.** $f(x) = -2 + \log_{10} x$

13. $f(x) = \log_{10} (x - 1)$ **14.** $f(x) = \log_{10} (x + 2)$

For Problems 15–24, use a calculator to find each of the *common logarithms*. Express answers to four decimal places.

15. log 9.45   **16.** log 1.07   **17.** log 34.62

18. log 578.1   **19.** log 4721.4   **20.** log 52,698

21. log 0.612   **22.** log 0.08134   **23.** log 0.0047

24. log 0.000076

For Problems 25–36, find each *common logarithm* by using the table at the back of the book.

25. log 8.72   **26.** log 6.04   **27.** log 56.9

28. log 48   **29.** log 708   **30.** log 14,100

31. log 0.492   **32.** log 0.023   **33.** log 0.00528

34. log 0.000415   **35.** log 763,000   **36.** log 9,180,000

For Problems 37–46, use a calculator to find each antilogarithm. Express answers to five significant digits of accuracy. For example,

$$\text{antilog } 3.2147 = 1639.4569$$
$$= 1639.5 \quad \text{(to five significant digits)}$$

37. antilog 1.5263

38. antilog 2.7185

39. antilog 3.9335

40. antilog 4.9547

41. antilog 0.5517

42. antilog 1.9006

43. antilog −0.1452

44. antilog −1.3148

45. antilog −2.6542

46. antilog −2.1928

For Problems 47–60, find each antilogarithm by using the table at the back of the book.

47. antilog 0.5502

48. antilog 0.9624

49. antilog 1.4829

50. antilog 1.9170

51. antilog 2.9926

52. antilog 3.4533

53. antilog 5.8062

54. antilog 4.6812

55. antilog (−1 + .7340)

56. antilog (−2 + .7774)

57. antilog (−3 + .8639)

58. antilog (−4 + .0969)

59. antilog −0.7471

60. antilog −1.1232

For Problems 61–72, use a calculator to find each *natural logarithm*. Express answers to four decimal places.

61. ln 2

62. ln 9

63. ln 21.4

64. ln 87.6

65. ln 412

66. ln 384.2

67. ln 0.32

68. ln 0.417

69. ln 0.0715

70. ln 0.006285

71. ln 0.0008

72. ln 52,173

For Problems 73–82, find each *natural logarithm* by using the table on page 451.

73. ln 3.7

74. ln 2.8

75. ln 78

76. ln 140

77. ln 620

78. ln 9800

79. ln 0.42

80. ln 0.051

81. ln 0.0085

82. ln 0.00056

10.5 Exponential Equations, Logarithmic Equations, and Problem Solving

In Section 5.1 we solved exponential equations such as $3^x = 81$ by expressing both sides of the equation as a power of 3 and then applying the property "if $b^n = b^m$, then $n = m$." However, if we try this same approach with an equation such as $3^x = 5$, we face the difficulty of expressing 5 as a power of 3. We can solve this type of problem by using the properties of logarithms and the following property of equality.

Property 10.8	If $x > 0$, $y > 0$, $b > 0$, and $b \neq 1$, then $x = y$ if and only if $\log_b x = \log_b y$.

Property 10.8 is stated in terms of any valid base b; however, for most applications either common logarithms or natural logarithms are used. Let's consider some examples.

Example 1   Solve $3^x = 5$ to the nearest hundredth.

Solution   Using common logarithms, we can proceed as follows.

$$3^x = 5$$

$\log 3^x = \log 5$ Property 10.8

$x \log 3 = \log 5$ $\log r^p = p \log r$

$$x = \frac{\log 5}{\log 3}$$

$$x = \frac{0.6990}{0.4771} = 1.47$$ (to the nearest hundredth)

Check   Using a calculator to raise 3 to the 1.47 power, we obtain $3^{1.47} = 5.0276871$. Thus, we say that to the nearest hundredth the solution set for $3^x = 5$ is $\{1.47\}$.   ●

A Word of Caution The expression $\dfrac{\log 5}{\log 3}$ means that we must *divide*—not subtract—the logarithms. That is, $\dfrac{\log 5}{\log 3}$ *does not* mean $\log\left(\dfrac{5}{3}\right)$.

Example 2   Solve $e^{x+1} = 19$ to the nearest hundredth.

Solution   Since base e is used in the exponential expression, let's use natural logarithms to help solve this equation.

$$e^{x+1} = 5$$

$\ln e^{x+1} = \ln 5$ Property 10.8

$(x + 1) \ln e = \ln 5$ $\ln r^p = p \ln r$

$(x + 1)(1) = \ln 5$ $\ln e = 1$

$x = \ln 5 - 1$

$x = 1.6094 - 1$

$x = 0.6094 = 0.61$ (to the nearest hundredth)

The solution set is $\{0.61\}$. Check it!   ●

Example 3   Solve $2^{3x-2} = 3^{2x+1}$ to the nearest hundredth.

Solution
$$2^{3x-2} = 3^{2x+1}$$
$$\log 2^{3x-2} = \log 3^{2x+1}$$
$$(3x - 2)\log 2 = (2x + 1)\log 3$$
$$3x \log 2 - 2 \log 2 = 2x \log 3 + \log 3$$
$$3x \log 2 - 2x \log 3 = \log 3 + 2 \log 2$$
$$x(3 \log 2 - 2 \log 3) = \log 3 + 2 \log 2$$
$$x = \frac{\log 3 + 2 \log 2}{3 \log 2 - 2 \log 3}$$

At this point we could evaluate x, but instead let's use the properties of logarithms to simplify the expression for x.

$$x = \frac{\log 3 + \log 2^2}{\log 2^3 - \log 3^2}$$
$$= \frac{\log 3 + \log 4}{\log 8 - \log 9}$$
$$= \frac{\log 12}{\log \left(\dfrac{8}{9}\right)}$$

Now, evaluating x, we obtain

$$x \approx \frac{1.0792}{-0.0512} = -21.08 \qquad \text{(to the nearest hundredth)}.$$

The solution set is $\{-21.08\}$. Check it!   ●

Logarithmic Equations

In Example 11 of Section 10.3 we solved the logarithmic equation

$$\log_{10} x + \log_{10}(x + 9) = 1$$

by simplifying the left side of the equation to $\log_{10}[x(x + 9)]$ and then changing the equation to exponential form to complete the solution. Now, using Property 10.8, we can solve such a logarithmic equation another way and also expand our equation-solving capabilities. Let's consider some examples.

Example 4   Solve $\log x + \log(x - 15) = 2$.

Solution   Since $\log 100 = 2$, the given equation becomes

$$\log x + \log(x - 15) = \log 100.$$

Now, simplifying the left side and applying Property 10.8, we can proceed as follows.

$$\log (x)(x - 15) = \log 100$$
$$x(x - 15) = 100$$
$$x^2 - 15x - 100 = 0$$
$$(x - 20)(x + 5) = 0$$
$$x - 20 = 0 \quad \text{or} \quad x + 5 = \quad 0$$
$$x = 20 \quad \text{or} \quad x = -5$$

The domain of a logarithmic function must contain only positive numbers, so x and $x - 15$ must be positive in this problem. Therefore, the solution of -5 is discarded, and the solution set is $\{20\}$.   ●

Example 5   Solve $\ln (x + 2) = \ln (x - 4) + \ln 3$.

Solution $$\ln (x + 2) = \ln (x - 4) + \ln 3$$
$$\ln (x + 2) = \ln [3(x - 4)]$$
$$x + 2 = 3(x - 4)$$
$$x + 2 = 3x - 12$$
$$14 = 2x$$
$$7 = x$$

The solution set is $\{7\}$.   ●

Example 6   Solve $\log_b (x + 2) + \log_b (2x - 1) = \log_b x$.

Solution $$\log_b (x + 2) + \log_b (2x - 1) = \log_b x$$
$$\log_b [(x + 2)(2x - 1)] = \log_b x$$
$$(x + 2)(2x - 1) = x$$
$$2x^2 + 3x - 2 = x$$
$$2x^2 + 2x - 2 = 0$$
$$x^2 + x - 1 = 0$$

Using the quadratic formula we obtain

$$x = \frac{-1 \pm \sqrt{1 + 4}}{2}$$
$$= \frac{-1 \pm \sqrt{5}}{2}.$$

Since $x + 2$, $2x - 1$, and x have to be positive, the solution of $\dfrac{-1 - \sqrt{5}}{2}$ has to be discarded; the solution set is $\left\{\dfrac{-1 + \sqrt{5}}{2}\right\}$.   ●

Problem Solving

In Section 10.2 we used the compound interest formula

$$A = P\left(1 + \frac{r}{n}\right)^{nt}$$

to determine the amount of money (A) accumulated at the end of t years if P dollars is invested at rate of interest r compounded n times per year. Now let's use this formula to solve other types of problems that deal with compound interest.

Example 7   How long will it take for $500 to double itself if it is invested at 12% compounded quarterly?

Solution   "To double itself" means that the $500 must grow into $1000. Thus,

$$1000 = 500\left(1 + \frac{0.12}{4}\right)^{4t}$$
$$= 500(1 + 0.03)^{4t}$$
$$= 500(1.03)^{4t}.$$

Multiplying both sides of $1000 = 500(1.03)^{4t}$ by $\frac{1}{500}$ yields

$$2 = (1.03)^{4t}.$$

Therefore,

$$\log 2 = \log (1.03)^{4t} \qquad \text{Property 10.8}$$
$$= 4t \log 1.03. \qquad \log r^p = p \log r$$

Solving for t, we obtain

$$\log 2 = 4t \log 1.03$$
$$\frac{\log 2}{\log 1.03} = 4t$$
$$\frac{\log 2}{4 \log 1.03} = t \qquad \text{Multiply both sides by } \frac{1}{4}.$$
$$\frac{0.3010}{4(0.0128)} = t$$
$$\frac{0.3010}{0.0512} = t$$
$$5.9 = t \qquad \text{(to the nearest tenth)}.$$

Therefore, we are claiming that $500 invested at 12% interest compounded quarterly will double itself in approximately 5.9 years.

Check $500 invested at 12% compounded quarterly for 5.9 years will produce

$$A = \$500\left(1 + \frac{0.12}{4}\right)^{4(5.9)}$$
$$= \$500(1.03)^{23.6}$$
$$= \$1004.45.$$

●

Example 8   What rate of interest (to the nearest tenth of a percent) is needed for an investment of $1000 to yield $4000 in 10 years if the interest is compounded annually?

Solution   Substituting the known facts into the compound interest formula, we obtain

$$4000 = 1000(1 + r)^{10}.$$

Multiplying both sides of this equation by $\frac{1}{1000}$ yields

$$4 = (1 + r)^{10}.$$

Therefore,

$$\log 4 = \log (1 + r)^{10} \qquad \text{Property 10.8}$$
$$= 10 \log (1 + r). \qquad \log r^p = p \log r$$

Multiplying both sides by $\frac{1}{10}$ produces

$$\frac{\log 4}{10} = \log (1 + r).$$

Using log 4 = 0.6021, we obtain

$$\frac{0.6021}{10} = \log (1 + r)$$
$$0.0602 = \log (1 + r).$$

Finding the antilog of 0.0602 and solving for r, we obtain

$$1.149 = 1 + r$$
$$0.149 = r$$
$$14.9\% = r.$$

Therefore, a rate of interest of approximately 14.9% is needed.

Check $1000 invested at 14.9% compounded annually for 10 years will produce

$$A = \$1000(1.149)^{10} = \$4010.52.$$

●

Example 9   How long will it take $100 to triple itself if it is invested at 8% interest compounded continuously?

Solution   "To triple itself" means that the $100 must grow into $300. Using the formula for interest that is compounded continuously, we can proceed as follows.

$$A = Pe^{rt}$$
$$\$300 = \$100e^{(0.08)t}$$
$$3 = e^{0.08t}$$

$\ln 3 = \ln e^{0.08t}$	Property 10.8
$\ln 3 = 0.08t \ln e$	$\ln r^p = p \ln r$
$\ln 3 = 0.08t$	$\ln e = 1$

$$\frac{\ln 3}{0.08} = t$$

$$\frac{1.0986}{0.08} = t$$

$$13.7 = t \quad \text{(to the nearest tenth)}$$

Therefore, in approximately 13.7 years, $100 will triple itself at 8% interest compounded continuously.

Check $100 invested at 8% compounded continuously for 13.7 years produces

$$A = Pe^{rt}$$
$$= \$100e^{0.08(13.7)}$$
$$= \$100(2.718)^{1.096}$$
$$= \$299.18.$$

Example 10   Suppose that the number of bacteria present in a certain culture after t minutes is given by the equation $Q(t) = Q_0 e^{0.04t}$, where Q_0 represents the initial number of bacteria. How long would it take for the bacteria count to grow from 500 to 2000?

Solution   Substituting into $Q(t) = Q_0 e^{0.04t}$ and solving for t, we obtain the following.

$$2000 = 500e^{0.04t}$$
$$4 = e^{0.04t}$$
$$\ln 4 = \ln e^{0.04t}$$
$$\ln 4 = 0.04t \ln e$$
$$\ln 4 = 0.04t$$
$$\frac{\ln 4}{0.04} = t$$
$$34.7 = t \quad \text{(to the nearest tenth)}$$

It should take approximately 34.7 minutes.

The basic approach of applying Property 10.8 and using either common or natural logarithms can also be used to evaluate a logarithm to some base other than 10 or e. The next example illustrates this process.

Example 11   Evaluate $\log_3 41$.

Solution   Let $x = \log_3 41$. Changing to exponential form, we obtain

$$3^x = 41.$$

Now we can apply Property 10.8 and proceed as follows.

$$\log 3^x = \log 41$$
$$x \log 3 = \log 41$$
$$x = \frac{\log 41}{\log 3}$$
$$x = \frac{1.6128}{0.4771}$$
$$x = 3.3804 \qquad \text{(to four decimal places)}$$

Therefore, we are claiming that 3 raised to the 3.3804 power will produce approximately 41. Check it!   ●

Using the method of Example 11 to evaluate $\log_a r$ produces the following formula, often referred to as the **change of base formula for logarithms**.

Property 10.9	If a, b, and r are positive numbers, with $a \neq 1$ and $b \neq 1$, then $$\log_a r = \frac{\log_b r}{\log_b a}.$$

By using Property 10.9 we can easily determine a relationship between logarithms of different bases. For example, suppose that in Property 10.9 we let $a = 10$ and $b = e$.

$$\log_a r = \frac{\log_b r}{\log_b a}$$

becomes

$$\log_{10} r = \frac{\log_e r}{\log_e 10}$$
$$\log_e r = (\log_e 10)(\log_{10} r)$$
$$\log_e r = (2.3026)(\log_{10} r).$$

Thus, the natural logarithm of any positive number is approximately equal to the common logarithm of the number times 2.3026.

Problem Set 10.5

Solve each of the following exponential equations, expressing approximate solutions to the nearest hundredth.

1. $2^x = 9$

2. $3^x = 20$

3. $5^t = 23$

4. $4^t = 12$

5. $2^{x+1} = 7$

6. $3^{x-2} = 11$

7. $7^{2t-1} = 35$

8. $5^{3t+1} = 9$

9. $e^x = 4.1$

10. $e^x = 30$

11. $e^{x-1} = 8.2$

12. $e^{x-2} = 13.1$

13. $2e^x = 12.4$

14. $3e^x - 1 = 17$

15. $3^{x-1} = 2^{x+3}$

16. $5^{2x+1} = 7^{x+3}$

17. $5^{x-1} = 2^{2x+1}$

18. $3^{2x+1} = 2^{3x+2}$

Solve each of the following logarithmic equations, expressing irrational solutions in lowest radical form.

19. $\log x + \log (x + 3) = 1$

20. $\log x + \log (x + 21) = 2$

21. $\log (2x - 1) - \log (x - 3) = 1$

22. $\log (3x - 1) = 1 + \log (5x - 2)$

23. $\log (x - 2) = 1 - \log (x + 3)$

24. $\log (x + 1) = \log 3 - \log (2x - 1)$

25. $\log (x + 1) - \log (x + 2) = \log \dfrac{1}{x}$

26. $\log (x + 2) - \log (2x + 1) = \log x$

27. $\ln (3t - 4) - \ln (t + 1) = \ln 2$

28. $\ln (2t + 5) = \ln 3 + \ln (t - 1)$

29. $\log (x^2) = (\log x)^2$

30. $\log \sqrt{x} = \sqrt{\log x}$

Approximate each of the following logarithms to three decimal places. (Example 11 and/or Property 10.9 should be of some help.)

31. $\log_3 14$

32. $\log_4 94$

33. $\log_5 2.1$

34. $\log_6 0.345$

35. $\log_7 176$

36. $\log_8 296$

37. $\log_9 14.32$

38. $\log_7 0.024$

Solve each of the following problems.

39. How long will it take $1000 to double itself if it is invested at 9% interest compounded semiannually?

40. How long will it take $750 to be worth $1000 if it is invested at 12% interest compounded quarterly?

41. How long will it take $500 to triple itself if it is invested at 9% interest compounded continuously?

42. How long will it take $2000 to double itself if it is invested at 13% interest compounded continuously?

43. What rate of interest (to the nearest tenth of a percent) compounded annually is needed for an investment of $200 to grow to $350 in 5 years?

44. What rate of interest (to the nearest tenth of a percent) compounded continuously is needed for an investment of $500 to grow to $900 in 10 years?

45. A piece of machinery valued at $30,000 depreciates at a rate of 10% yearly. How long will it take for it to reach a value of $15,000?

46. For a certain strain of bacteria, the number of bacteria present after t hours is given by the equation $Q = Q_0 e^{0.34t}$, where Q_0 represents the initial number of bacteria. How long will it take 400 bacteria to increase to 4000 bacteria?

47. The number of grams of a certain radioactive substance present after t hours is given by the equation $Q = Q_0 e^{-0.45t}$, where Q_0 represents the initial number of grams. How long would it take 2500 grams to be reduced to 1250 grams?

48. The equation $P(a) = 14.7e^{-0.21a}$, where a is the altitude above sea level measured in miles, yields the atmospheric pressure in pounds per square inch. If the atmospheric pressure at Cheyenne, Wyoming is approximately 11.53 pounds per square inch, find that city's altitude above sea level. Express your answer to the nearest hundred feet.

49. Suppose that the equation $P(t) = P_0 e^{0.02t}$, where P_0 represents an initial population and t is the time in years, is used to predict population growth. How long would it take a city of 50,000 to double its population?

50. For a certain culture the equation $Q(t) = Q_0 e^{0.4t}$, where Q_0 is an initial number of bacteria and t is time measured in hours, yields the number of bacteria as a function of time. How long will it take 500 bacteria to increase to 2000?

51. Radon is formed by the radioactive decay of radium. It has a half-life of approximately 4 days, and the equation $Q(t) = Q_0 e^{-4k}$ yields the amount of radon that remains after t days if the initial amount is Q_0. Solve the equation $\frac{1}{2}Q_0 = Q_0 e^{-4k}$ for k. Express the answer to the nearest hundredth.

52. Polonium is also formed from the radioactive decay of radium and has a half-life of approximately 140 days. The equation $Q(t) = Q_0 e^{-140k}$ yields the amount of polonium that remains after t days if the initial amount is Q_0. Solve the equation $\frac{1}{2}Q_0 = Q_0 e^{-140k}$ for k, and express the answer to the nearest thousandth.

Miscellaneous Problems

53. Use the approach of Example 11 to develop Property 10.9.

54. Let $r = b$ in Property 10.9, and verify that $\log_a b = \dfrac{1}{\log_b a}$.

55. Solve the equation $\dfrac{5^x - 5^{-x}}{2} = 3$. Express your answer to the nearest hundredth.

56. Solve the equation $y = \dfrac{10^x + 10^{-x}}{2}$ for x in terms of y.

57. Solve the equation $y = \dfrac{e^x - e^{-x}}{2}$ for x in terms of y.

10.6 Computation with Common Logarithms (Optional)

As we mentioned earlier, the calculator has replaced common logarithms for most computational purposes. Nevertheless, a brief look at how common logarithms were used will give you a better insight into the meaning of logarithms and their properties.

Although the computations in this section should be done by using logarithms and the table of common logarithms at the back of the book, it is recommended that you check each problem by using a calculator. This will provide you with some additional practice with your calculator and will establish the validity of our work with logarithms.

Linear Interpolation

Let's begin by expanding our use of the common logarithm table at the back of the book. Suppose that we try to determine log 2.744 from the table. Because the table contains only logarithms of numbers with at most three significant digits, we have a problem. However, by a process called **linear interpolation** we can extend the capabilities of the table to include numbers of four significant digits.

First, let's consider a geometric basis of linear interpolation, and then we will suggest a systematic procedure for carrying out the necessary calculations. A portion of the graph of $y = \log_{10} x$, with the curvature exaggerated to help illustrate the principle involved, is shown in Figure 10.9. The line segment joining points P and Q is used to approximate the curve from P to Q. The actual value of log 2.744 is the ordinate of point C—that is, the length of $\overline{AC}$. This cannot be determined from the table. Instead we will use the ordinate of point B (the length of $\overline{AB}$) as an approximation for log 2.744.

Now consider Figure 10.10, where line segments $\overline{DB}$ and $\overline{EQ}$ are drawn perpendicular to $\overline{PE}$. The right triangles formed, $\triangle PDB$ and $\triangle PEQ$, are similar, and therefore the lengths of their corresponding sides are proportional. Thus, we can write

$$\frac{PD}{PE} = \frac{DB}{EQ}.$$

Also from Figure 10.10 we see that

$$PD = 2.744 - 2.74 = 0.004,$$
$$PE = 2.75 - 2.74 = 0.01,$$
$$EQ = 0.4393 - 0.4378 = 0.0015.$$

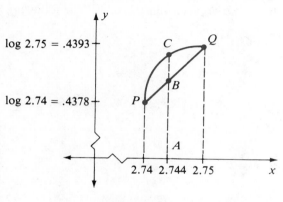

Figure 10.9

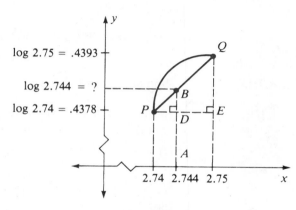

Figure 10.10

Therefore, the above proportion becomes

$$\frac{0.004}{0.01} = \frac{DB}{0.0015}.$$

Solving this proportion for DB yields

$$(0.01)(DB) = (0.004)(0.0015)$$
$$DB = 0.0006.$$

Since $AB = AD + DB$, we have

$$AB = 0.4378 + 0.0006$$
$$= 0.4384.$$

Thus, we obtain log 2.744 = 0.4384.

   Now let's suggest an abbreviated format for carrying out the calculations necessary to find log 2.744.

$$
\begin{array}{cc}
x & \log x \\
4\left\{\begin{matrix} 2.740 \\ 2.744 \end{matrix}\right\}10 & k\left\{\begin{matrix} 0.4378 \\ ? \end{matrix}\right\}0.0015 \\
2.750 & 0.4393
\end{array}
$$

Notice that we have used 4 and 10 for the differences between values of x instead of 0.004 and 0.01, because the ratio $\dfrac{0.004}{0.01}$ equals $\dfrac{4}{10}$. Setting up a proportion and solving for k yields

$$\frac{4}{10} = \frac{k}{0.0015}$$
$$10k = 4(0.0015) = 0.0060$$
$$k = 0.0006.$$

Thus, log 2.744 = 0.4378 + 0.0006 = 0.4384.

   The process of linear interpolation can also be used to approximate an antilogarithm when the mantissa is between two values in the table. The following example illustrates this procedure.

Example 1   Find antilog 1.6157.

Solution   From the table we see that the mantissa, .6157, is between .6149 and .6160. We can carry out the interpolation as follows.

$$
\begin{array}{cc}
h\left\{\begin{matrix} 4.120 \\ ? \end{matrix}\right\}0.010 & 8\left\{\begin{matrix} .6149 \\ .6157 \end{matrix}\right\}11 \\
4.130 & .6160
\end{array}
$$

$$\frac{h}{0.010} = \frac{8}{11}$$

$$11h = 8(0.010) = 0.080$$

$$h = \frac{1}{11}(0.080) = 0.007 \qquad \text{(to the nearest thousandth)}$$

Thus, antilog .6157 = 4.120 + 0.007 = 4.127. Therefore,

$$\text{antilog } 1.6157 = \text{antilog } (.6157 + 1)$$
$$= 4.127 \cdot 10^1$$
$$= 41.27.$$

Computation with Common Logarithms

Let's first restate the basic properties of logarithms in terms of common logarithms. Remember that we are writing $\log x$ instead of $\log_{10} x$.

If x and y are positive real numbers, then

1. $\log xy = \log x + \log y$,

2. $\log \dfrac{x}{y} = \log x - \log y$, and

3. $\log x^p = p \log x \qquad$ (where p is any real number).

The following two properties of equality pertaining to logarithms will also be used.

4. If $x = y$ (where x and y are positive), then $\log x = \log y$.

5. If $\log x = \log y$, then $x = y$.

Now let's illustrate how common logarithms can be used for computational purposes.

Example 2   Evaluate $\dfrac{(571.4)(8.236)}{71.68}$.

Solution   Let $N = \dfrac{(571.4)(8.236)}{71.68}$.

$$\log N = \log \frac{(571.4)(8.236)}{71.68}$$
$$= \log 571.4 + \log 8.236 - \log 71.68$$
$$= 2.7569 + 0.9157 - 1.8554$$
$$= 1.8172$$

Therefore,

$N = $ antilog 1.8172

$\quad = $ antilog $(0.8172 + 1)$

$\quad = 6.564 \cdot 10^1$

$\quad = 65.64.$

Check   Using a calculator, we obtain

$$N = \frac{(571.4)(8.236)}{71.68} = 65.653605. \qquad \bullet$$

When a table of logarithms is used, it is sometimes necessary to change the form of writing a logarithm so that the mantissa is positive. The next example illustrates this idea.

Example 3   Find the quotient $\dfrac{1.73}{5.08}$.

Solution   Let $N = \dfrac{1.73}{5.08}$. Therefore,

$\log N = \log \dfrac{1.73}{5.08}$

$\qquad = \log 1.73 - \log 5.08$

$\qquad = 0.2380 - 0.7059 = -0.4679.$

Now, by adding 1 and subtracting 1, which changes the form but not the value, we obtain

$\log N = -0.4679 + 1 - 1$

$\qquad = 0.5321 - 1$

$\qquad = 0.5321 + (-1).$

Therefore,

$N = $ antilog $[0.5321 + (-1)]$

$\quad = 3.405 \cdot 10^{-1} = 0.3405.$

Check   Using a calculator, we obtain

$$N = \frac{1.73}{5.08} = 0.34055118. \qquad \bullet$$

Sometimes it is necessary to change the form of a logarithm so that a subsequent calculation will produce an integer for the characteristic part of the logarithm. Example 4 illustrates this process.

Example 4   Evaluate $\sqrt[4]{0.0767}$.

Solution   Let $N = \sqrt[4]{0.0767} = (0.0767)^{1/4}$. Therefore,

$$\log N = \log (0.0767)^{1/4}$$

$$= \frac{1}{4} \log 0.0767$$

$$= \frac{1}{4} [0.8848 + (-2)] = \frac{1}{4}(-2 + 0.8848).$$

At this stage we recognize that applying the distributive property will produce a nonintegral characteristic—namely, $-\frac{1}{2}$. Therefore, let's add 4 and subtract 4 inside the parentheses, which will change the form as follows.

$$\log N = \frac{1}{4}(-2 + 0.8848 + 4 - 4)$$

$$= \frac{1}{4}(4 - 2 + 0.8848 - 4)$$

$$= \frac{1}{4}(2.8848 - 4)$$

Now, applying the distributive property, we obtain

$$\log N = \frac{1}{4}(2.8848) - \frac{1}{4}(4)$$

$$= 0.7212 - 1$$

$$= 0.7212 + (-1).$$

Therefore,

$$N = \text{antilog} [0.7212 + (-1)]$$

$$= 5.262 \cdot 10^{-1} = 0.5262.$$

Check   Using a calculator, we obtain

$$N = \sqrt[4]{0.0767} = 0.52625816.$$

Problem Set 10.6

Use the table at the back of the book and linear interpolation to find each of the following common logarithms.

1. log 4.327	**2.** log 27.43	**3.** log 128.9
4. log 3526	**5.** log 0.8761	**6.** log 0.07692
7. log 0.005186	**8.** log 0.0002558	

Use the table at the back of the book and linear interpolation to find each of the following antilogarithms to four significant digits.

9. antilog 0.4690

10. antilog 1.7971

11. antilog 2.1925

12. antilog 3.7225

13. antilog $[0.5026 + (-1)]$

14. antilog $[0.9397 + (-2)]$

Use common logarithms and linear interpolation to help evaluate each of the following. Express your answers with four significant digits. Check your answers using a calculator.

15. $(294)(71.2)$

16. $(192.6)(4.017)$

17. $\dfrac{23.4}{4.07}$

18. $\dfrac{718.5}{8.248}$

19. $(17.3)^5$

20. $(48.02)^3$

21. $\dfrac{(108)(76.2)}{13.4}$

22. $\dfrac{(126.3)(24.32)}{8.019}$

23. $\sqrt[5]{0.821}$

24. $\sqrt[4]{645.3}$

25. $(79.3)^{3/5}$

26. $(176.8)^{3/4}$

27. $\sqrt{\dfrac{(7.05)(18.7)}{0.521}}$

28. $\sqrt[3]{\dfrac{(41.3)(0.271)}{8.05}}$

Chapter 10 Summary

(10.1) If a and b are positive real numbers and m and n are any real numbers, then

1. $b^n \cdot b^m = b^{n+m}$ Product of two powers

2. $(b^n)^m = b^{mn}$ Power of a power

3. $(ab)^n = a^n b^n$ Power of a product

4. $\left(\dfrac{a}{b}\right)^n = \dfrac{a^n}{b^n}$ Power of a quotient

5. $\dfrac{b^n}{b^m} = b^{n-m}$ Quotient of two powers

If $b > 0$, $b \neq 1$, and m and n are real numbers, then $b^n = b^m$ if and only if $n = m$.

A function defined by an equation of the form

$$f(x) = b^x \qquad (b > 0 \text{ and } b \neq 1)$$

is called an **exponential function**.

(10.2)   A general formula for any principal, P, compounded n times per year for any number t, at a rate r is

$$A = P\left(1 + \frac{r}{n}\right)^{nt}$$

where A represents the total amount of money accumulated at the end of the t years.

The value of $\left(1 + \dfrac{1}{n}\right)^n$, as n gets infinitely large, approaches the number e, where e equals 2.71828 to five decimal places.

The formula

$$A = Pe^{rt}$$

yields the accumulated value, A, of a sum of money, P, that has been invested for t years at a rate of r percent **compounded continuously**.

The equation

$$Q(t) = Q_0 e^{kt}$$

is used as a mathematical model for many growth-and-decay applications.

(10.3)   If r is any positive real number, then the unique exponent t such that $b^t = r$ is called the **logarithm of r with base b** and is denoted by $\log_b r$. For $b \geq 0$, $b \neq 1$, and $r > 0$,

   1. $\log_b b = 1$
   2. $\log_b 1 = 0$
   3. $r = b^{\log_b r}$.

The following properties of logarithms are derived from the definition of a logarithm and the properties of exponents. For positive real numbers b, r, and s, where $b \neq 1$,

   1. $\log_b rs = \log_b r + \log_b s,$
   2. $\log_b \left(\dfrac{r}{s}\right) = \log_b r - \log_b s,$ and
   3. $\log_b r^p = p \log_b r$ (where p is any real number).

(10.4)   A function defined by an equation of the form

$$f(x) = \log_b x \qquad (b > 0 \text{ and } b \neq 1)$$

is called a **logarithmic function**. The equation $y = \log_b x$ is equivalent to $x = b^y$.

The two functions $f(x) = b^x$ and $g(x) = \log_b x$ are inverses of each other.

Logarithms with a base of 10 are called **common logarithms**. The expression $\log_{10} x$ is commonly written as $\log x$.

Many calculators are equipped with a common logarithm function. Often a key labeled $\boxed{\log}$ is used to find common logarithms.

Tables of common logarithms, such as the one found in the back of this book, can be used to find logarithms as follows:

$$\log 87.6 = \log (8.76 \cdot 10^1)$$
$$= \log 8.76 + \log 10^1$$
$$= .9425 + 1$$

From the   Exponent
table   of 10

$$= 1.9425$$

The decimal part (.9425) of the logarithm 1.9425 is called the **mantissa**, and the integral part (1) is called the **characteristic**.

Calculators and tables can also be used to find x given log x. The number x is referred to as an **antilogarithm**.

Natural logarithms are logarithms having a base of e, where e is an irrational number whose decimal approximation to eight digits is 2.7182818. Natural logarithms are denoted by $\log_e x$ or ln x.

Many calculators are also equipped with a natural logarithm function. Often a key labeled $\boxed{\ln x}$ is used for this purpose.

(10.5)   Equations that contain a variable in an exponent are called **exponential equations**. The solution sets for some exponential equations can be found by using common and natural logarithms.

The formula

$$\log_a r = \frac{\log_b r}{\log_b a}$$

is often called the **change of base formula**.

Chapter 10 Review Problem Set

Evaluate each of the following.

1. $8^{5/3}$

2. $-25^{3/2}$

3. $(-27)^{4/3}$

4. $\log_6 216$

5. $\log_7 \left(\dfrac{1}{49}\right)$

6. $\log_2 \sqrt[3]{2}$

7. $\log_2 \left(\dfrac{\sqrt[4]{32}}{2}\right)$

8. $\log_{10} 0.00001$

9. ln e

10. $7^{\log_7 12}$

Solve each of the following equations. Express approximate solutions to the nearest hundredth.

11. $\log_{10} 2 + \log_{10} x = 1$

12. $\log_3 x = -2$

13. $4^x = 128$

14. $3^t = 42$

15. $\log_2 x = 3$

16. $\left(\dfrac{1}{27}\right)^{3x} = 3^{2x-1}$

17. $2e^x = 14$

18. $2^{2x+1} = 3^{x+1}$

19. $\ln(x + 4) - \ln(x + 2) = \ln x$

20. $\log x + \log(x - 15) = 2$

21. $\log(\log x) = 2$

22. $\log(7x - 4) - \log(x - 1) = 1$

23. $\ln(2t - 1) = \ln 4 + \ln(t - 3)$

24. $64^{2t+1} = 8^{-t+2}$

For Problems 25–28, if $\log 3 = 0.4771$ and $\log 7 = 0.8451$, evaluate each of the following.

25. $\log\left(\dfrac{7}{3}\right)$

26. $\log 21$

27. $\log 27$

28. $\log 7^{2/3}$

29. Express each of the following as the sum or difference of simpler logarithmic quantities. Assume that all variables represent positive real numbers.

(a) $\log_b\left(\dfrac{x}{y^2}\right)$

(b) $\log_b \sqrt[4]{xy^2}$

(c) $\log_b\left(\dfrac{\sqrt{x}}{y^3}\right)$

30. Express each of the following as a single logarithm. Assume that all variables represent positive real numbers.

(a) $3\log_b x + 2\log_b y$

(b) $\dfrac{1}{2}\log_b y - 4\log_b x$

(c) $\dfrac{1}{2}(\log_b x + \log_b y) - 2\log_b z$

For Problems 31–34, approximate each of the logarithms to three decimal places.

31. $\log_2 3$

32. $\log_3 2$

33. $\log_4 191$

34. $\log_2 0.23$

For Problems 35–42, graph each of the functions.

35. $f(x) = \left(\dfrac{3}{4}\right)^x$

36. $f(x) = 2^{x+2}$

37. $f(x) = e^{x-1}$

38. $f(x) = -1 + \log x$

39. $f(x) = 3^x - 3^{-x}$

40. $f(x) = e^{-x^2/2}$

41. $f(x) = \log_2(x - 3)$

42. $f(x) = 3\log_3 x$

For Problems 43–45, use the compound interest formula $A = P\left(1 + \dfrac{r}{n}\right)^{nt}$ to find the total amount of money accumulated at the end of the indicated time period for each of the investments.

43. $750 for 10 years at 11% compounded quarterly

44. $1250 for 15 years at 9% compounded monthly

45. $2500 for 20 years at 9.5% compounded semiannually

46. How long will it take $100 to double itself if it is invested at 14% interest compounded annually?

47. How long will it take $1000 to be worth $3500 if it is invested at 10.5% interest compounded quarterly?

48. What rate of interest (to the nearest tenth of a percent) compounded continuously is needed for an investment of $500 to grow to $1000 in 8 years?

49. Suppose that the present population of a city is 50,000. Using the equation $P(t) = P_0 e^{0.02t}$, where P_0 represents an initial population, to estimate future populations, estimate the population of that city in 10 years, 15 years, and 20 years.

50. The number of bacteria present in a certain culture after t hours is given by the equation $Q = Q_0 e^{0.29t}$, where Q_0 represents the initial number of bacteria. How long will it take 500 bacteria to increase to 2000 bacteria?

11 Systems of Equations and Inequalities

"Find two numbers whose sum is 125 and whose difference is 21." Such a problem translates naturally into the two equations $x + y = 125$ and $x - y = 21$, where x represents the larger number and y the smaller number. The two equations considered together form a system of linear equations, and the original problem can be solved by solving the system of equations. Throughout most of this chapter we will be considering systems of linear equations and their applications.

11.1 Systems of Two Linear Equations in Two Variables

In Chapter 7 we stated that any equation of the form $Ax + By = C$, where A, B, and C are real numbers (A and B not both zero) is a **linear equation** in the two variables x and y, and its graph is a straight line. Two linear equations in two variables considered together form a **system of two linear equations in two variables**, as illustrated by the following:

$$\begin{pmatrix} x + y = 6 \\ x - y = 2 \end{pmatrix}, \quad \begin{pmatrix} 3x + 2y = 1 \\ 5x - 2y = 23 \end{pmatrix}, \quad \begin{pmatrix} 4x - 5y = 21 \\ 3x + 7y = -38 \end{pmatrix}.$$

To solve a system, such as the three above, means to find all of the ordered pairs that satisfy both equations in the system. For example, if we graph the two equations $x + y = 6$ and $x - y = 2$ on the same set of axes, as in Figure 11.1, then the ordered pair associated with the point of intersection of the two lines is the **solution of the system**. Thus, we say that $\{(4, 2)\}$ is the solution set of the system

$$\begin{pmatrix} x + y = 6 \\ x - y = 2 \end{pmatrix}.$$

Figure 11.1

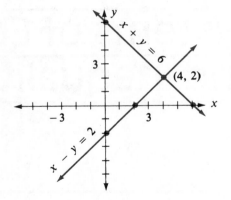

To check, we substitute 4 for x and 2 for y in the two equations.

$$x + y \text{ becomes } 4 + 2 = 6. \qquad \text{(a true statement)}$$
$$x - y \text{ becomes } 4 - 2 = 2. \qquad \text{(a true statement)}$$

Because the graph of a linear equation in two variables is a straight line, there are only three possible situations that can arise when a system of two linear equations in two variables is solved. These situations are illustrated in Figure 11.2.

Case 1 The graphs of the two equations are two lines intersecting in *one* point. There is *one solution*, and the system is called a **consistent system**.

Case 2 The graphs of the two equations are parallel lines. There is *no solution*, and the system is called an **inconsistent system**.

Case 3 The graphs of the two equations are the same line, and there are *infinitely many solutions* to the system. Any pair of real numbers that satisfies one of the equations will also satisfy the other equation, and we say that the equations are **dependent**.

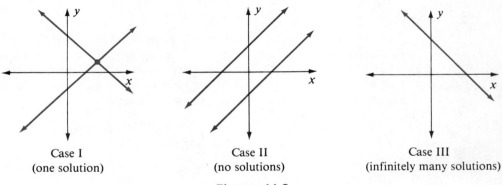

Case I
(one solution)

Case II
(no solutions)

Case III
(infinitely many solutions)

Figure 11.2

Thus, when we solve a system of two linear equations in two variables, we know what to expect. The system will have *no* solutions, *one* ordered pair as a solution, or *infinitely many* ordered pairs as solutions.

It should be evident that solving specific systems of equations by graphing requires accurate graphs. In fact, unless the solutions are integers, it is quite difficult to obtain exact solutions from a graph. Therefore, we will consider some other methods for solving systems of equations in this chapter. One such method, called the **elimination-by-addition method**, depends on the following property.

Property 11.1	If an ordered pair (p, q) satisfies the two equations $A_1 x + B_1 y = C_1$ and $A_2 x + B_2 y = C_2$, then it will also satisfy the equation $$k(A_1 x + B_1 y) + l(A_2 x + B_2 y) = kC_1 + lC_2,$$ where k and l are real numbers.

Property 11.1 states that "you can multiply each equation in a system by a constant and then add the resulting equations." The objective is to choose multipliers so that adding the resulting equations will eliminate one of the variables. Let's consider some examples.

Example 1

Solve the system $\begin{pmatrix} 3x + 2y = 1 \\ 5x - 2y = 23 \end{pmatrix}$.

Solution

Notice that the numerical coefficients of the y-terms in the two equations are additive inverses of each other. Thus, we can immediately add the two equations and *eliminate the ys.* (Technically, we are letting $k = 1$ and $l = 1$ in Property 11.1.)

$$3x + 2y = 1$$
$$\underline{5x - 2y = 23}$$
$$8x = 24$$
$$x = 3$$

Now we can substitute 3 for x in one of the original equations; let's choose $3x + 2y = 1$.

$$3(3) + 2y = 1$$
$$2y = -8$$
$$y = -4$$

To check, substitute 3 for x and -4 for y in both of the original equations.

$$3(3) + 2(-4) = 1 \qquad \text{(a true statement)}$$
$$5(3) - 2(-4) = 23 \qquad \text{(a true statement)}$$

Thus, the solution set is $\{(3, -4)\}$.

Example 2   Solve the system $\begin{pmatrix} 4x - 5y = 21 \\ 3x + 7y = -38 \end{pmatrix}$.

Solution A

$4x - 5y =\ \ \ 21$	Multiply both sides by 3.	$12x - 15y =\ \ 63$
$3x + 7y = -38$	Multiply both sides by -4.	$-12x - 28y = 152$

$$-43y = 215$$
$$y = -5$$

Now we can substitute -5 for y in one of the *original equations*; let's use $4x - 5y = 21$.

$$4x - 5(-5) = 21$$
$$4x + 25 = 21$$
$$4x = -4$$
$$x = -1$$

The solution set is $\{(-1, -5)\}$.

Solution B

$4x - 5y =\ \ \ 21$	Multiply both sides by 7.	$28x - 35y =\ \ \ \ 147$
$3x + 7y = -38$	Multiply both sides by 5.	$15x + 35y = -190$

$$43x\ \ \ \ \ \ \ \ = -43$$
$$x =\ \ \ -1$$

Substituting -1 for x in $4x - 5y = 21$, we obtain

$$4(-1) - 5y = 21$$
$$-4 - 5y = 21$$
$$-5y = 25$$
$$y = -5.$$

The solution set is $\{(-1, -5)\}$.   ●

Look back over Solutions A and B of Example 2. Notice especially the first steps, in which the original equations are multiplied by a constant. In Solution A the multipliers, 3 and -4, were chosen so that adding the resulting equations would eliminate the x-variable. In Solution B we chose multipliers 7 and 5 so that the y-variable was eliminated when the resulting equations were added. Either approach works; choose the one that involves the easiest computation.

Example 3   Solve the system $\begin{pmatrix} 3x - 2y = 5 \\ 2x + 7y = 9 \end{pmatrix}$.

Solution

$3x - 2y = 5$	Multiply both sides by -2.	$-6x +\ \ 4y = -10$
$2x + 7y = 9$	Multiply both sides by 3.	$6x + 21y =\ \ \ 27$

$$25y =\ \ \ 17$$
$$y =\ \ \ \frac{17}{25}$$

Since substituting $\dfrac{17}{25}$ for y in one of the original equations will produce some messy calculations, let's solve for x by eliminating the ys.

$$3x - 2y = 5 \qquad \underrightarrow{\text{Multiply both sides by 7.}} \qquad 21x - 14y = 35$$

$$2x + 7y = 9 \qquad \underrightarrow{\text{Multiply both sides by 2.}} \qquad \dfrac{4x + 14y = 18}{25x \qquad\;\; = 53}$$

$$x = \dfrac{53}{25}$$

The solution set is $\left\{\left(\dfrac{53}{25}, \dfrac{17}{25}\right)\right\}$. (Perhaps you should check these results!)   ●

Example 4

Solve the system $\left(\begin{array}{l} \dfrac{1}{2}x + \dfrac{2}{3}y = -4 \\[2mm] \dfrac{1}{4}x - \dfrac{3}{2}y = 20 \end{array}\right)$.

Solution

First, let's multiply each equation by an appropriate constant (the least common denominator of the fractions involved in the equation) to obtain integral coefficients.

$$\dfrac{1}{2}x + \dfrac{2}{3}y = -4 \qquad \underrightarrow{\text{Multiply both sides by 6.}} \qquad 3x + 4y = -24$$

$$\dfrac{1}{4}x - \dfrac{3}{2}y = 20 \qquad \underrightarrow{\text{Multiply both sides by 4.}} \qquad x - 6y = \;\;\; 80$$

Now we can proceed as in previous examples.

$$3x + 4y = -24 \qquad \underrightarrow{\text{Do not change.}} \qquad 3x + \;\;4y = \;\;-24$$

$$x - 6y = \;\;\; 80 \qquad \underrightarrow{\text{Multiply both sides by} -3.} \qquad \dfrac{-3x + 18y = -240}{22y = -264}$$

$$y = \;\; -12$$

Substituting -12 for y in $\dfrac{1}{2}x + \dfrac{2}{3}y = -4$, we obtain

$$\dfrac{1}{2}x + \dfrac{2}{3}(-12) = -4$$

$$\dfrac{1}{2}x - 8 = -4$$

$$\dfrac{1}{2}x = 4$$

$$x = 8.$$

Check
$$\frac{1}{2}(8) + \frac{2}{3}(-12) \overset{?}{=} -4$$
$$4 + (-8) \overset{?}{=} -4$$
$$-4 = -4 \quad \text{(a true statement)}$$
$$\frac{1}{4}(8) - \frac{3}{2}(-12) \overset{?}{=} 20$$
$$2 - (-18) \overset{?}{=} 20$$
$$20 = 20 \quad \text{(a true statement)}$$

The solution set is $\{(8, -12)\}$.   ●

Problem Solving

Many word problems that were solved earlier in this text by using one variable and one equation can also be solved by using a system of two linear equations in two variables. In fact, in many of these problems you may find it more natural to use two variables. Let's consider some examples.

Problem 1   The sum of two numbers is 42, and their difference is 14. Find the numbers.

Solution   Let l represent the larger number, and let s represent the smaller number. The first sentence of Problem 1 translates into the following system.

$$\begin{pmatrix} l + s = 42 \\ l - s = 14 \end{pmatrix} \qquad \begin{array}{l} \text{sum of 42} \\ \text{difference of 14} \end{array}$$

Solving this system, we obtain

$$\begin{array}{r} l + s = 42 \\ \underline{l - s = 14} \\ 2l = 56 \\ l = 28. \end{array}$$

Substituting 28 for l in $l + s = 42$ yields

$$28 + s = 42$$
$$s = 14.$$

The numbers are 14 and 28.   ●

Problem 2   Jackie invested $1000, part at 12% and the remainder at 14%. Find the amount invested at each rate if the yearly income from the two investments is $136.

Solution   Let x represent the amount invested at 12%. Let y represent the amount invested at 14%. The problem translates into the following system.

$$\begin{pmatrix} x + \quad y = 1000 \\ 0.12x + 0.14y = 136 \end{pmatrix}$$ The two investments total $1000.
The yearly interest from the two investments was $136.

The system can be solved as follows.

$x + \quad y = 1000$ <u>Multiply both sides by -12.</u> $-12x - 12y = -12{,}000$

$0.12x + 0.14y = \ 136$ <u>Multiply both sides by 100.</u> $\underline{12x + 14y = \quad 13{,}600}$

$$2y = \quad 1600$$
$$y = \quad 800$$

Substituting 800 for y in $x + y = 1000$ yields

$x + 800 = 1000$
$x = 200.$

Thus, we are claiming that $200 was invested at 12% and $800 at 14%. Do these figures check according to the original statement of the problem? The problem states that "Jackie invested $1000"; since $200 + $800 = $1000, that part checks. Furthermore, 12% of $200 is $24, 14% of $800 is $112, and $24 + $112 = $136, so that also checks. Therefore, we know that our answers are correct.   ●

Problem Set 11.1

Use the graphing approach to determine whether each of the following systems is consistent, inconsistent, or dependent. If the system is consistent, find the solution set from the graph and check it.

1. $\begin{pmatrix} x + y = 11 \\ x - y = 5 \end{pmatrix}$

2. $\begin{pmatrix} 2x - \ y = -4 \\ 3x + 2y = 1 \end{pmatrix}$

3. $\begin{pmatrix} 3x + 2y = 6 \\ 6x + 4y = 7 \end{pmatrix}$

4. $\begin{pmatrix} 3x + 4y = 1 \\ 5x - 2y = 19 \end{pmatrix}$

5. $\begin{pmatrix} 2x - \ y = 3 \\ 3x - 4y = 17 \end{pmatrix}$

6. $\begin{pmatrix} 4x - 2y = 3 \\ 2x - \ y = 10 \end{pmatrix}$

7. $\begin{pmatrix} \dfrac{1}{2}x - \dfrac{2}{3}y = 1 \\ 3x - 4y = 6 \end{pmatrix}$

8. $\begin{pmatrix} 6x + 5y = 30 \\ \dfrac{2}{5}x + \dfrac{1}{3}y = 2 \end{pmatrix}$

9. $\begin{pmatrix} x + 5y = 9 \\ 3x - \ y = 11 \end{pmatrix}$

10. $\begin{pmatrix} 5x - \ y = -26 \\ x + 4y = -1 \end{pmatrix}$

Use the elimination-by-addition method (Property 11.1) to solve the following systems.

11. $\begin{pmatrix} x + y = 9 \\ 2x - y = 3 \end{pmatrix}$

12. $\begin{pmatrix} -x + 2y = 16 \\ x - 3y = -23 \end{pmatrix}$

13. $\begin{pmatrix} 3x + 2y = -18 \\ 5x - y = -17 \end{pmatrix}$

14. $\begin{pmatrix} x - 5y = 46 \\ 3x + 4y = -33 \end{pmatrix}$

15. $\begin{pmatrix} 4x + 3y = -7 \\ 3x - 2y = 16 \end{pmatrix}$

16. $\begin{pmatrix} 5x - 3y = -34 \\ 2x + 7y = -30 \end{pmatrix}$

17. $\begin{pmatrix} 2x + 3y = 3 \\ 4x - 9y = -4 \end{pmatrix}$

18. $\begin{pmatrix} 4x - 5y = 3 \\ 8x + 15y = -24 \end{pmatrix}$

19. $\begin{pmatrix} 7a - 2b = 1 \\ 4a + 5b = 2 \end{pmatrix}$

20. $\begin{pmatrix} 6a - 3b = 4 \\ 5a + 2b = -1 \end{pmatrix}$

21. $\begin{pmatrix} 5a - 4b = -8 \\ 6a - 5b = -12 \end{pmatrix}$

22. $\begin{pmatrix} 4a - 3b = 18 \\ 3a - 2b = 15 \end{pmatrix}$

23. $\begin{pmatrix} 2x - y = 9 \\ 7x + 4y = 1 \end{pmatrix}$

24. $\begin{pmatrix} 3x + y = 2 \\ 11x - 3y = 5 \end{pmatrix}$

25. $\begin{pmatrix} \dfrac{1}{4}x - \dfrac{2}{3}y = -3 \\ \dfrac{1}{3}x + \dfrac{1}{3}y = 7 \end{pmatrix}$

26. $\begin{pmatrix} \dfrac{2}{3}x + \dfrac{1}{4}y = -1 \\ \dfrac{1}{2}x - \dfrac{1}{3}y = -7 \end{pmatrix}$

27. $\begin{pmatrix} \dfrac{x}{2} - \dfrac{y}{3} = \dfrac{1}{4} \\ \dfrac{x}{4} + \dfrac{y}{5} = \dfrac{1}{2} \end{pmatrix}$

28. $\begin{pmatrix} \dfrac{2x}{3} + \dfrac{3y}{4} = \dfrac{1}{6} \\ \dfrac{3x}{5} - \dfrac{2y}{3} = -\dfrac{1}{2} \end{pmatrix}$

29. $\begin{pmatrix} \dfrac{3x}{2} - \dfrac{2y}{7} = -1 \\ 4x + y = 2 \end{pmatrix}$

30. $\begin{pmatrix} \dfrac{4x}{5} - \dfrac{3y}{2} = \dfrac{1}{5} \\ -2x + y = -1 \end{pmatrix}$

Solve each of the following problems by using a system of equations.

31. The sum of two numbers is -2, and their difference is 10. Find the numbers.

32. The sum of two numbers is 25, and their difference is 9. Find the numbers.

33. Joan invested part of $500 at 12% and the remainder at 13%. Her yearly income from the two investments is $62. How much did she invest at each rate?

34. Leon invested $950, part at 12% and the remainder at 14%. His yearly income from the two investments is $128. How much did he invest at each rate?

35. Al bought 55 stamps for $10.25. Some of them were 17-cent stamps, and the rest were 20-cent stamps. How many of each kind did he buy?

36. The income from a student production was $10,000. The price of a student ticket was $3, and nonstudent tickets were sold at $5 each. Altogether, 3000 tickets were sold. How many tickets of each kind were sold?

37. Cindy has $5.10 in dimes and quarters. If she has 30 coins, how many coins of each kind does she have?

38. A deposit slip lists $700 in cash to be deposited. There are 100 bills, some of them five-dollar bills and the remainder ten-dollar bills. How many bills of each denomination are to be deposited?

39. For moving purposes, the Hendersons bought 25 cardboard boxes for $97.50. There were two kinds of boxes; the large ones cost $7.50 per box, and the small ones were $3 per box. How many boxes of each kind did they buy?

40. The cost of 3 tennis balls and 2 golf balls is $7. The cost of 6 tennis balls and 3 golf balls is $12. Find the cost of 1 tennis ball and the cost of 1 golf ball.

41. A library is buying a total of 35 books for $462. Some of the books cost $12 each, and the remainder cost $14 per book. How many books of each price are they buying?

42. A 10% salt solution is to be mixed with a 20% salt solution to produce 20 gallons of a 17.5% salt solution. How many gallons of the 10% solution and how many gallons of the 20% solution will be needed?

43. Sue bought 3 packages of cookies and 2 bags of potato chips for $3.65. Later she bought 2 more packages of cookies and 5 additional bags of potato chips for $4.23. Find the price of a package of cookies.

Miscellaneous Problems

44. There is another way of telling whether a system of two linear equations in two unknowns is consistent, inconsistent, or dependent without taking the time to graph each equation. It can be shown that any system of the form

$$\begin{pmatrix} a_1x + b_1y = c_1 \\ a_2x + b_2y = c_2 \end{pmatrix}$$

has one and only one solution if

$$\frac{a_1}{a_2} \neq \frac{b_1}{b_2} \qquad \text{(consistent)};$$

has no solution if

$$\frac{a_1}{a_2} = \frac{b_1}{b_2} \neq \frac{c_1}{c_2} \qquad \text{(inconsistent)};$$

and has infinitely many solutions if

$$\frac{a_1}{a_2} = \frac{b_1}{b_2} = \frac{c_1}{c_2} \qquad \text{(dependent)}.$$

Determine whether each of the following systems is consistent, inconsistent, or dependent.

(a) $\begin{pmatrix} 4x - 3y = 7 \\ 9x + 2y = 5 \end{pmatrix}$
  $\qquad$ **(b)** $\begin{pmatrix} 5x - y = 6 \\ 10x - 2y = 19 \end{pmatrix}$

(c) $\begin{pmatrix} 5x - 4y = 11 \\ 4x + 5y = 12 \end{pmatrix}$
  $\qquad$ **(d)** $\begin{pmatrix} x + 2y = 5 \\ x - 2y = 9 \end{pmatrix}$

(e) $\begin{pmatrix} x - 3y = 5 \\ 3x - 9y = 15 \end{pmatrix}$
  $\qquad$ **(f)** $\begin{pmatrix} 4x + 3y = 7 \\ 2x - y = 10 \end{pmatrix}$

(g) $\begin{pmatrix} 3x + 2y = 4 \\ y = -\frac{3}{2}x - 1 \end{pmatrix}$
  $\qquad$ **(h)** $\begin{pmatrix} y = \frac{4}{3}x - 2 \\ 4x - 3y = 6 \end{pmatrix}$

11.2 The Substitution Method

There is another method of solving systems of equations called the **substitution method**. Like the elimination method, it produces exact solutions and can be used on any system of two linear equations involving two variables. However, we will find that some systems lend themselves more to the substitution method than do others. Let's consider a few examples to illustrate the use of the substitution method.

Example 1 Solve the system $\left(\begin{array}{c} x + y = 16 \\ y = x + 2 \end{array}\right)$.

Solution Since the second equation states that y equals $x + 2$, we can substitute $x + 2$ for y in the first equation.

$$x + y = 16 \quad \underrightarrow{\text{Substitute } x + 2 \text{ for } y.} \quad x + (x + 2) = 16$$

Now we have an equation with one variable that can be solved in the usual way.

$$\begin{aligned} x + (x + 2) &= 16 \\ 2x + 2 &= 16 \\ 2x &= 14 \\ x &= 7 \end{aligned}$$

Substituting 7 for x in one of the two original equations (we chose the second one) yields

$$y = 7 + 2 = 9.$$

To check, we can substitute 7 for x and 9 for y in both of the original equations.

$$\begin{aligned} 7 + 9 &= 16 \qquad \text{(a true statement)} \\ 9 &= 7 + 2 \qquad \text{(a true statement)} \end{aligned}$$

The solution set is $\{(7, 9)\}$. ●

Example 2 Solve the system $\left(\begin{array}{c} x = 3y - 25 \\ 4x + 5y = 19 \end{array}\right)$.

Solution In this case the first equation states that x equals $3y - 25$. Therefore, we can substitute $3y - 25$ for x in the second equation.

$$4x + 5y = 19 \quad \underrightarrow{\text{Substitute } 3y - 25 \text{ for } x.} \quad 4(3y - 25) + 5y = 19$$

Solving this equation yields

$$\begin{aligned} 4(3y - 25) + 5y &= 19 \\ 12y - 100 + 5y &= 19 \\ 17y &= 119 \\ y &= 7. \end{aligned}$$

Substituting 7 for y in the first equation produces

$$x = 3(7) - 25$$
$$= 21 - 25 = -4.$$

The solution set is $\{(-4, 7)\}$. Check it.   ●

Note that the key idea behind the substitution method is the elimination of a variable, but the elimination is accomplished by a substitution rather than by adding the equations. The substitution method is especially convenient if at least one of the equations is of the form *y equals* or *x equals*. In Example 1, the second equation is of the form *y equals*; in Example 2, the first equation is of the form *x equals*.

Which Method to Use

Both the elimination and the substitution method can be used to obtain exact solutions for any system of two linear equations in two unknowns. As the examples thus far in this section and those of the previous section have indicated, deciding which method to use on a particular system often involves determining whether the system lends itself to one or the other of the methods as a result of the original format of the equations. Let's reemphasize that point with two more examples.

Example 3   Solve the system $\begin{pmatrix} 2x - 5y = 7 \\ 3x + 4y = 9 \end{pmatrix}$.

Solution   Since neither equation is of the form *y equals* or *x equals* and, furthermore, changing the form of either equation in preparation for the substitution method would produce a fractional form, we are probably better off using the elimination method.

$2x - 5y = 7$ $\xrightarrow{\text{Multiply both sides by } -3.}$ $-6x + 15y = -21$

$3x + 4y = 9$ $\xrightarrow{\text{Multiply both sides by 2.}}$ $\underline{6x + 8y = 18}$

$$23y = -3$$
$$y = -\frac{3}{23}$$

Since substituting $-\dfrac{3}{23}$ for y in either of the original equations will produce some messy calculations, let's solve for x by eliminating the y's.

$2x - 5y = 7$ $\xrightarrow{\text{Multiply both sides by 4.}}$ $8x - 20y = 28$

$3x + 4y = 9$ $\xrightarrow{\text{Multiply both sides by 5.}}$ $\underline{15x + 20y = 45}$

$$23x = 73$$
$$x = \frac{73}{23}$$

Check
$$2x - 5y = 7$$

$$2\left(\frac{73}{23}\right) - 5\left(-\frac{3}{23}\right) \overset{?}{=} 7$$

$$\frac{146}{23} + \frac{15}{23} \overset{?}{=} 7$$

$$\frac{161}{23} \overset{?}{=} 7$$

$$7 = 7$$

$$3x + 4y = 9$$

$$3\left(\frac{73}{23}\right) + 4\left(-\frac{3}{23}\right) \overset{?}{=} 9$$

$$\frac{219}{23} - \frac{12}{23} \overset{?}{=} 9$$

$$\frac{207}{23} \overset{?}{=} 9$$

$$9 = 9$$

The solution set is $\left\{\left(\frac{73}{23}, -\frac{3}{23}\right)\right\}$.

Example 4   Solve the system $\begin{pmatrix} 6x + 5y = -3 \\ y = -2x - 7 \end{pmatrix}$.

Solution   Since the second equation is of the form *y equals*, let's use the substitution method. From the second equation, we can substitute $-2x - 7$ for y in the first equation.

$$6x + 5y = -3 \quad \xrightarrow{\text{Substitute } -2x - 7 \text{ for } y.} \quad 6x + 5(-2x - 7) = -3$$

Solving this equation yields

$$6x + 5(-2x - 7) = -3$$
$$6x - 10x - 35 = -3$$
$$-4x - 35 = -3$$
$$-4x = 32$$
$$x = -8.$$

Substituting -8 for x in the second equation yields

$$y = -2(-8) - 7$$
$$= 16 - 7 = 9.$$

The solution set is $\{(-8, 9)\}$.

In Section 11.1 we discussed the fact that whether a system of two linear equations in two unknowns has no solution, one solution, or infinitely many solutions can be determined by graphing the equations of the system. That is, the two lines may be parallel (no solution), they may intersect in one point (one solution), or they may coincide (infinitely many solutions).

From a practical viewpoint, the systems having one solution deserve most of our attention. However, we do need to be able to deal with the other situations because they do occur occasionally. Let's use two examples to illustrate what happens when we hit a *no solution* or an *infinitely many solutions* situation when using either the elimination or the substitution method.

Example 5   Solve the system $\left(\begin{array}{c} y = 3x - 1 \\ -9x + 3y = 4 \end{array}\right)$.

Solution   Using the substitution method, we can proceed as follows.

$$-9x + 3y = 4 \quad \underrightarrow{\text{Substitute } 3x - 1 \text{ for } y.} \quad -9x + 3(3x - 1) = 4$$

Solving this equation yields

$$-9x + 3(3x - 1) = 4$$
$$-9x + 9x - 3 = 4$$
$$-3 = 4.$$

The *false numerical statement*, $-3 = 4$, implies that the system has *no solution*. (You may want to graph the two lines to verify our conclusion!)   ●

Example 6   Solve the system $\left(\begin{array}{c} 5x + y = 2 \\ 10x + 2y = 4 \end{array}\right)$.

Solution   Using the elimination method, we can proceed as follows.

$$
\begin{array}{lll}
5x + y = 2 & \underrightarrow{\text{Multiply both sides by } -2.} & -10x - 2y = -4 \\
10x + 2y = 4 & \underrightarrow{\text{Leave alone.}} & \underline{10x + 2y = 4} \\
& & 0 + 0 = 0
\end{array}
$$

The *true numerical statement*, $0 + 0 = 0$, implies that the system has *infinitely many solutions*. Any ordered pair that satisfies one of the equations will also satisfy the other equation. The solution set can be expressed as $\{(x, y) \mid 5x + y = 2\}$.   ●

The original forms of the equations of a system may be the result of a direct translation of a word problem. That is to say, some word problems translate rather naturally into systems that lend themselves to being solved by the elimination method, whereas other problems translate into systems that can be more effectively solved by the substitution method. Let's consider two examples.

Problem 1   One solution contains 30% alcohol and a second solution contains 70% alcohol. How many liters of each solution should be mixed to make 10 liters containing 40% alcohol?

Solution   Let x represent the number of liters of the 30% solution to be used. Let y represent the number of liters of the 70% solution to be used. The problem translates into the following system.

$\left(\begin{array}{c} x + y = 10 \\ 0.3x + 0.7y = 4 \end{array}\right)$ The total amount of solution is to be 10 liters.

The amount of pure alcohol in the 30% solution plus the amount of pure alcohol in the 70% solution equals the total amount of pure alcohol, which is 40% of 10, or 4 liters.

Then solve this system by *elimination*.

$x + y = 10$	Multiply both sides by -3. $\longrightarrow$	$-3x - 3y = -30$
$0.3x + 0.7y = 4$	Multiply both sides by 10. $\longrightarrow$	$3x + 7y = 40$

$$4y = 10$$
$$y = 2.5$$

Substituting 2.5 for y in $x + y = 10$ yields

$$x + y = 10$$
$$x + 2.5 = 10$$
$$x = 7.5.$$

We need to mix 7.5 liters of the 30% solution and 2.5 liters of the 70% solution. (You should check these results to make sure that they satisfy the conditions stated in the problem.) ●

The two-variable expression $10t + u$ can be used to represent any two-digit number. The t represents the tens digit, and the u represents the units digit. For example, if $t = 5$ and $u = 2$, then $10t + u$ becomes $10(5) + 2 = 52$. We use this general representation for a two-digit number in the next problem.

Problem 2   The tens digit of a two-digit number is 2 more than twice the units digit. The number with the digits reversed is 45 less than the original number. Find the original number.

Solution   Let u represent the units digit of the original number. Let t represent the tens digit of the original number. Then $10t + u$ represents the original number, and $10u + t$ represents the number with the digits reversed. The problem translates into the following system.

$$\left(\begin{array}{l} t = 2u + 2 \\ 10u + t = 10t + u - 45 \end{array} \right)$$ The tens digit is 2 more than twice the units digit.
The number with the digits reversed is 45 less than the original number.

Simplifying the second equation, we find that the system becomes

$$\left(\begin{array}{l} t = 2u + 2 \\ -9t + 9u = -45 \end{array} \right).$$

From the first equation we can *substitute* $2u + 2$ for t in the second equation and solve.

$$-9t + 9u = -45$$
$$-9(2u + 2) + 9u = -45$$
$$-18u - 18 + 9u = -45$$
$$-9u = -27$$
$$u = 3$$

Substituting 3 for u in $t = 2u + 2$, we obtain

$$t = 2u + 2$$
$$= 2(3) + 2 = 8.$$

The tens digit is 8 and the units digit is 3, so the number is 83. (You should check to see if 83 satisfies the original conditions stated in the problem.) ●

Problem Set 11.2

Solve each of the following systems by the substitution method.

1. $\begin{pmatrix} x + y = 10 \\ x = y + 8 \end{pmatrix}$

2. $\begin{pmatrix} y = x + 5 \\ x + y = -11 \end{pmatrix}$

3. $\begin{pmatrix} y = 3x - 13 \\ 3x + 4y = -22 \end{pmatrix}$

4. $\begin{pmatrix} 4x + 5y = 22 \\ x = 2y + 1 \end{pmatrix}$

5. $\begin{pmatrix} 7x + 3y = 16 \\ y = -5x \end{pmatrix}$

6. $\begin{pmatrix} y = -4x \\ 9x + 2y = 3 \end{pmatrix}$

7. $\begin{pmatrix} y = 3x - 5 \\ 2x + 3y = 6 \end{pmatrix}$

8. $\begin{pmatrix} 3x - 4y = 9 \\ x = 4y - 1 \end{pmatrix}$

9. $\begin{pmatrix} t + u = 11 \\ t = u + 7 \end{pmatrix}$

10. $\begin{pmatrix} u = t - 2 \\ t + u = 12 \end{pmatrix}$

11. $\begin{pmatrix} y = \dfrac{3}{4}x - 5 \\ 5x - 4y = 9 \end{pmatrix}$

12. $\begin{pmatrix} y = \dfrac{2}{5}x - 1 \\ 3x + 5y = 4 \end{pmatrix}$

13. $\begin{pmatrix} 5x - y = 9 \\ x = \dfrac{1}{2}y - 3 \end{pmatrix}$

14. $\begin{pmatrix} 7x - 3y = -2 \\ x = \dfrac{3}{4}y + 1 \end{pmatrix}$

Solve each of the following systems by either the substitution or the elimination method— whichever seems more appropriate.

15. $\begin{pmatrix} 3x - 2y = -5 \\ 5x + 2y = 13 \end{pmatrix}$

16. $\begin{pmatrix} -2x + 4y = 0 \\ 2x - 3y = -1 \end{pmatrix}$

17. $\begin{pmatrix} 9x - 2y = 28 \\ y = -3x + 1 \end{pmatrix}$

18. $\begin{pmatrix} x = 4y + 13 \\ 3x + 6y = -33 \end{pmatrix}$

19. $\begin{pmatrix} y = 2x - 23 \\ y = -5x + 33 \end{pmatrix}$

20. $\begin{pmatrix} x = -4y + 30 \\ x = 3y - 33 \end{pmatrix}$

21. $\begin{pmatrix} 3a - b = 9 \\ 5a + 7b = 1 \end{pmatrix}$

22. $\begin{pmatrix} 4a - 3b = 2 \\ 5a - b = 3 \end{pmatrix}$

23. $\begin{pmatrix} 2x - 3y = 4 \\ y = \dfrac{2}{3}x - \dfrac{4}{3} \end{pmatrix}$

24. $\begin{pmatrix} 3x - 2y = 7 \\ 5x + 7y = 1 \end{pmatrix}$

25. $\begin{pmatrix} 4x + 7y = 2 \\ 9x - 2y = 1 \end{pmatrix}$

26. $\begin{pmatrix} 5x - 2y = 1 \\ 10x - 4y = 7 \end{pmatrix}$

27. $\begin{pmatrix} y = \dfrac{2x}{3} - \dfrac{3}{4} \\ 2x + 3y = 11 \end{pmatrix}$

28. $\begin{pmatrix} \dfrac{x}{2} - \dfrac{y}{3} = 1 \\ \dfrac{2x}{5} + \dfrac{y}{2} = 2 \end{pmatrix}$

29. $\begin{pmatrix} \dfrac{x}{6} + \dfrac{y}{4} = 1 \\ \dfrac{x}{2} - \dfrac{2y}{3} = 1 \end{pmatrix}$

30. $\begin{pmatrix} 5x - 3y = 7 \\ x = \dfrac{3y}{4} - \dfrac{1}{3} \end{pmatrix}$

31. $\begin{pmatrix} y = \dfrac{2}{3}x - 4 \\ 5x - 3y = 9 \end{pmatrix}$

32. $\begin{pmatrix} -2x + 5y = -16 \\ x = \dfrac{3}{4}y + 1 \end{pmatrix}$

33. $\begin{pmatrix} 2(x - 1) - 3(y + 2) = 30 \\ 3(x + 2) + 2(y - 1) = -4 \end{pmatrix}$

34. $\begin{pmatrix} 5(x + 1) - (y + 3) = -6 \\ 2(x - 2) + 3(y - 1) = 0 \end{pmatrix}$

35. $\begin{pmatrix} -2(x + 2) + 4(y - 3) = -34 \\ 3(x + 4) - 5(y + 2) = 23 \end{pmatrix}$

36. $\begin{pmatrix} -(x - 6) + 6(y + 1) = 58 \\ 3(x + 1) - 4(y - 2) = -15 \end{pmatrix}$

37. $\begin{pmatrix} \dfrac{x - 2}{4} + \dfrac{y + 1}{3} = 2 \\ \dfrac{x + 1}{7} + \dfrac{y - 3}{2} = \dfrac{1}{2} \end{pmatrix}$

38. $\begin{pmatrix} \dfrac{x + 1}{5} - \dfrac{y + 2}{2} = 3 \\ \dfrac{x - 3}{2} + \dfrac{y - 1}{7} = -\dfrac{1}{2} \end{pmatrix}$

39. $\begin{pmatrix} \dfrac{x - y}{4} - \dfrac{2x - y}{3} = -\dfrac{1}{4} \\ \dfrac{2x + y}{3} + \dfrac{x + y}{2} = \dfrac{17}{6} \end{pmatrix}$

40. $\begin{pmatrix} \dfrac{3x + y}{2} + \dfrac{x - 2y}{5} = 8 \\ \dfrac{x - y}{3} - \dfrac{x + y}{6} = \dfrac{10}{3} \end{pmatrix}$

Solve each of the following problems by setting up and solving a system of equations.

41. Laurie invested some money at 12% and some money at 14%. She invested $500 more at 14% than she did at 12%. Her total yearly interest from the two investments is $135. How much did Laurie invest at each rate?

42. Eric invested a total of $2000, part at 10% and the remainder at 11%. His yearly income from the two investments is $212. How much did Eric invest at each rate?

43. The units digit of a two-digit number is 1 more than twice the tens digit. The sum of the digits is 10. Find the number.

44. The sum of the digits of a two-digit number is 8. The tens digit is three times the units digit. Find the number.

45. A motel rents double rooms at $28 per day and single rooms at $19 per day. If a total of 50 rooms were rented one day for $1265, how many of each kind were rented?

46. Jo invest $600, part at 13% and the balance at 15%. The income on the 15% investment was $8 more than twice the income on the 13% investment. How much did Jo invest at each rate?

47. The sum of the digits of a two-digit number is 13. If the digits are reversed, the new number is 45 less than the original number. Find the original number.

48. The units digit of a two-digit number is 3 larger than the tens digit. If the digits are reversed, the new number is 20 less than twice the original number. Find the original number.

49. One solution contains 50% alcohol, and another solution contains 80% alcohol. How many liters of each solution should be mixed to make 10.5 liters of a 70% solution?

50. A man bought 2 pounds of coffee and 1 pound of butter for a total of $7.75. A month later the prices had not changed (this makes the problem fictitious), and he bought 3 pounds of coffee and 2 pounds of butter for $12.50. Find the price per pound of both the coffee and the butter.

51. If the numerator of a certain fraction is increased by 5 and the denominator is decreased by 1, the resulting fraction is $\dfrac{8}{3}$. However, if the numerator of the original fraction is doubled and the denominator is increased by 7, the resulting fraction is $\dfrac{6}{11}$. Find the original fraction.

52. If a certain two-digit number is divided by the sum of its digits, the quotient is 2. If the digits are reversed, the new number is 9 less than five times the original number. Find the original number.

53. Suppose that a fulcrum is placed so that weights of 40 pounds and 80 pounds are in balance. When 20 pounds is added to the 40-pound weight, the 80-pound weight must be moved $1\dfrac{1}{2}$ feet further from the fulcrum to preserve the balance. Find the original distance between the two weights.

54. A fulcrum is placed so that weights of 60 pounds and 100 pounds are in balance. If 20 pounds are subtracted from the 100-pound weight, then the 60-pound weight must be moved 1 foot closer to the fulcrum to preserve the balance. Find the original distance between the 60-pound and 100-pound weights.

55. Suppose that we have a rectangular-shaped book cover. If the width is increased by 2 centimeters and the length is decreased by 1 centimeter, the area is increased by 28 square centimeters. However, if the width is decreased by 1 centimeter and the length is increased by 2 centimeters, the area is increased by 10 square centimeters. Find the dimensions of the book cover.

56. A blueprint indicates a master bedroom in the shape of a rectangle. If the width is increased by 2 feet and the length remains the same, the area is increased by 36 square feet. However, if the width is increased by 1 foot and the length is increased by 2 feet, the area is increased by 48 square feet. Find the dimensions of the room as indicated on the blueprint.

Miscellaneous Problems

57. A system such as

$$\left(\begin{aligned} \frac{3}{x} + \frac{2}{y} &= 2 \\ \frac{2}{x} - \frac{3}{y} &= \frac{1}{4} \end{aligned} \right)$$

is not a system of linear equations but can be transformed into a linear system by changing variables. For example, if we substitute u for $\dfrac{1}{x}$ and v for $\dfrac{1}{y}$, the above system becomes

$$\begin{pmatrix} 3u + 2v = 2 \\ 2u - 3v = \dfrac{1}{4} \end{pmatrix}.$$

This "new" system can be solved by either elimination or substitution (we will leave the details for you), producing $u = \dfrac{1}{2}$ and $v = \dfrac{1}{4}$. Since $u = \dfrac{1}{x}$ and $v = \dfrac{1}{y}$, we have

$$\frac{1}{x} = \frac{1}{2} \quad \text{and} \quad \frac{1}{y} = \frac{1}{4}.$$

Solving these equations yields

$$x = 2 \quad \text{and} \quad y = 4.$$

The solution set of the original system is $\{(2, 4)\}$. Solve each of the following systems.

(a) $\begin{pmatrix} \dfrac{1}{x} + \dfrac{2}{y} = \dfrac{7}{12} \\ \dfrac{3}{x} - \dfrac{2}{y} = \dfrac{5}{12} \end{pmatrix}$

(b) $\begin{pmatrix} \dfrac{2}{x} + \dfrac{3}{y} = \dfrac{19}{15} \\ -\dfrac{2}{x} + \dfrac{1}{y} = -\dfrac{7}{15} \end{pmatrix}$

(c) $\begin{pmatrix} \dfrac{3}{x} - \dfrac{2}{y} = \dfrac{13}{6} \\ \dfrac{2}{x} + \dfrac{3}{y} = 0 \end{pmatrix}$

(d) $\begin{pmatrix} \dfrac{4}{x} + \dfrac{1}{y} = 11 \\ \dfrac{3}{x} - \dfrac{5}{y} = -9 \end{pmatrix}$

(e) $\begin{pmatrix} \dfrac{5}{x} - \dfrac{2}{y} = 23 \\ \dfrac{4}{x} + \dfrac{3}{y} = \dfrac{32}{2} \end{pmatrix}$

(f) $\begin{pmatrix} \dfrac{2}{x} - \dfrac{7}{y} = \dfrac{9}{10} \\ \dfrac{5}{x} + \dfrac{4}{y} = \dfrac{41}{20} \end{pmatrix}$

58. Solve the following system for x and y.

$$\begin{pmatrix} a_1 x + b_1 y = c_1 \\ a_2 x + b_2 y = c_2 \end{pmatrix}$$

11.3 Systems of Three Linear Equations in Three Variables

Consider a linear equation in three variables x, y, and z, such as $3x - 2y + z = 7$. Any **ordered triple** (x, y, z) that makes the equation a true numerical statement is said to be a solution of the equation. For example, the ordered triple $(2, 1, 3)$ is a solution because $3(2) - 2(1) + 3 = 7$. However, the ordered triple $(5, 2, 4)$ is not a solution because $3(5) - 2(2) + 4 \neq 7$. There are infinitely many solutions in the solution set.

Remark The idea of a *linear* equation is generalized to include equations of more than two variables. Thus, an equation such as $5x - 2y + 9z = 8$ is called a *linear equation in three variables*, the equation $5x - 7y + 2z - 11w = 1$ is called a *linear equation in four variables*, and so on.

To *solve* a system of three linear equations in three variables, such as

$$\begin{pmatrix} 3x - y + 2z = 13 \\ 4x + 2y + 5z = 30 \\ 5x - 3y - z = 3 \end{pmatrix},$$

means to find all of the ordered triples that satisfy all three equations. In other words, the solution set of the system is the intersection of the solution sets of all three equations in the system.

The graph of a linear equation in three variables is a plane, not a line. In fact, graphing equations in three variables requires the use of a three-dimensional coordinate system. Thus, using a graphing approach to solving systems of three linear equations in three variables is not at all practical. However, a simple graphic analysis does provide us with some idea as to what we can expect as we begin solving such systems.

In general, because each linear equation in three variables produces a plane, a system of three such equations produces three planes. There are various ways in which three planes can be related. For example, they may be mutually parallel, or two of the planes may be parallel and the third intersects each of the two. (You may want to analyze all of the other possibilities for the three planes!) However, for our purposes at this time we need to realize that from a solution-set viewpoint, a system of three linear equations in three variables produces one of the following possibilities.

1. There is *one ordered triple* that satisfies all three equations. The three planes have a common point of intersection, as indicated in Figure 11.3.

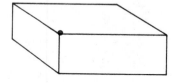

Figure 11.3

2. There are *infinitely many ordered triples* in the solution set, all of which are coordinates of points on a line common to the planes. This can happen if the

Figure 11.4

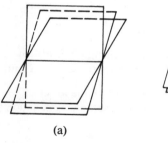

(a)

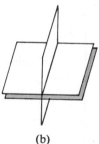

(b)

three planes have a common line of intersection, as in Figure 11.4(a), or if two of the planes coincide and the third plane intersects them, as in Figure 11.4(b).

3. There are *infinitely many ordered triples* in the solution set, all of which are coordinates of points on a plane. This can happen if the three planes coincide, as illustrated in Figure 11.5.

Figure 11.5

4. The *solution set is empty*; thus we write Ø. This can happen in various ways, as illustrated in Figure 11.6. Notice that in each situation there are no points common to all three planes.

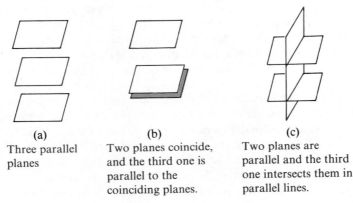

(a)	(b)	(c)	(d)
Three parallel planes	Two planes coincide, and the third one is parallel to the coinciding planes.	Two planes are parallel and the third one intersects them in parallel lines.	No two planes are parallel, but two of them intersect in a line which is parallel to the third plane.

Figure 11.6

Now that we know what possibilities exist, let's consider finding the solution sets for some systems. One approach is basically the *elimination method* used previously with systems of two equations and two unknowns. We will comment on the details of this approach as we solve some systems.

Example 1   Solve the system

$$\begin{pmatrix} 3x - y + 2z = 13 \\ 4x + 2y + 5z = 30 \\ 5x - 3y - z = 3 \end{pmatrix}$$

(1)
(2)
(3)

Solution   We can begin by eliminating a variable from equations (1) and (2). We can easily eliminate y by multiplying equation (1) by 2 and adding the result to equation (2).

$$\begin{array}{r} 6x - 2y + 4z = 26 \quad \longleftarrow \text{ This is the result of} \\ \underline{4x + 2y + 5z = 30} \quad\quad \text{multiplying equation (1) by 2.} \\ 10x \quad\quad + 9z = 56 \end{array}$$

We now have one equation with two variables. To get another equation with the same two variables, we need to eliminate y from a different pair of the original equations. This can be done by multiplying equation (1) by -3 and adding the result to equation (3).

$$\begin{array}{rl} -9x + 3y - 6z = -39 & \longleftarrow \text{This is the result of} \\ \underline{5x - 3y - z = 3} & \text{multiplying equation (1) by } -3. \\ -4x - 7z = -36 & \end{array}$$

Now we can solve the system

$$\begin{pmatrix} 10x + 9z = 56 \\ -4x - 7z = -36 \end{pmatrix} \qquad\qquad \begin{array}{l} (4) \\ (5) \end{array}$$

by using the elimination method.

$$\begin{array}{lll} 10x + 9z = 56 & \xrightarrow{\text{Multiply both sides by 2.}} & 20x + 18z = 112 \\ -4x - 7z = -36 & \xrightarrow{\text{Multiply both sides by 5.}} & \underline{-20x - 35z = -180} \\ & & -17z = -68 \\ & & z = 4 \end{array}$$

Substitute 4 for z in either equation (4) or (5). If we choose equation (4), we obtain

$$\begin{aligned} 10x + 9z &= 56 \\ 10x + 9(4) &= 56 \\ 10x &= 20 \\ x &= 2. \end{aligned}$$

Substitute 4 for z and 2 for x in either equation (1), (2), or (3). Choosing equation (1), we obtain

$$\begin{aligned} 3x - y + 2z &= 13 \\ 3(2) - y + 2(4) &= 13 \\ 6 - y + 8 &= 13 \\ -y &= -1 \\ y &= 1. \end{aligned}$$

Check   We must substitute the ordered triple (2, 1, 4) into all three of the original equations.

$$\begin{array}{ll} 3x - y + 2z = 13 & 3(2) - 1 + 2(4) \overset{?}{=} 13 \\ & 6 - 1 + 8 \overset{?}{=} 13 \\ & 13 = 13 \\[2mm] 4x + 2y + 5z = 30 & 4(2) + 2(1) + 5(4) \overset{?}{=} 30 \\ & 8 + 2 + 20 \overset{?}{=} 30 \\ & 30 = 30 \end{array}$$

$$5x - 3y - z = 3 \longrightarrow 5(2) - 3(1) - 4 \overset{?}{=} 3$$
$$10 - 3 - 4 \overset{?}{=} 3$$
$$3 = 3$$

The solution set is $\{(2, 1, 4)\}$.

Example 2   Solve the system

$$\begin{pmatrix} 4x + 3y - 2z = -11 \\ 3x - 7y + 3z = 10 \\ 9x - 8y + 5z = 9 \end{pmatrix}.$$ (1)
(2)
(3)

Solution   A glance at the coefficients in the system indicates that eliminating either the x's or the z's would be the easiest. Let's eliminate the x's.

$$-12x - 9y + 6z = 33 \qquad \text{Multiply equation (1) by } -3.$$
$$\underline{12x - 28y + 12z = 40} \qquad \text{Multiply equation (2) by 4.}$$
$$-37y + 18z = 73$$

$$-9x + 21y - 9z = -30 \qquad \text{Multiply equation (2) by } -3.$$
$$\underline{9x - 8y + 5z = 9} \qquad \text{Leave equation (3) alone.}$$
$$13y - 4z = -21$$

Now we need to solve the system

$$\begin{pmatrix} -37y + 18z = 73 \\ 13y - 4z = -21 \end{pmatrix}.$$ (4)
(5)

$$-74y + 36z = 146 \qquad \text{Multiply equation (4) by 2.}$$
$$\underline{117y - 36z = -189} \qquad \text{Multiply equation (5) by 9.}$$
$$43y = -43$$
$$y = -1$$

Substituting -1 for y in equation (5) produces

$$13y - 4z = -21$$
$$13(-1) - 4z = -21$$
$$-4z = -8$$
$$z = 2.$$

Substituting -1 for y and 2 for z in equation (1) produces

$$4x + 3y - 2z = -11$$
$$4x + 3(-1) - 2(2) = -11$$
$$4x - 3 - 4 = -11$$
$$4x = -4$$
$$x = -1.$$

The solution set is $\{(-1, -1, 2)\}$. Check it!

Example 3

Solve the system

$$\begin{pmatrix} x - 2y + 3z = 1 \\ 3x - 5y - 2z = 4 \\ 2x - 4y + 6z = 7 \end{pmatrix}.$$

(1)
(2)
(3)

Solution

A glance at the coefficients indicates that the x's would be easy to eliminate.

$$\begin{array}{ll} -3x + 6y - 9z = -3 & \text{Multiply equation (1) by } -3. \\ \underline{3x - 5y - 2z = 4} & \text{Leave equation (2) alone.} \\ y - 11z = 1 & \end{array}$$

$$\begin{array}{ll} -2x + 4y - 6z = -2 & \text{Multiply equation (1) by } -2. \\ \underline{2x - 4y + 6z = 7} & \text{Leave equation (3) alone.} \\ 0 = 5 & \end{array}$$

The false statement, $0 = 5$, indicates that the system is inconsistent, and therefore the solution set is empty. If you were to graph this system, equations (1) and (3) would produce parallel planes, the situation depicted in Figure 11.6(c). ●

Example 4

Solve the system

$$\begin{pmatrix} 2x - y + 4z = 1 \\ 3x + 2y - z = 5 \\ 5x - 6y + 17z = -1 \end{pmatrix}.$$

(1)
(2)
(3)

Solution

Let's eliminate the y's.

$$\begin{array}{ll} 4x - 2y + 8z = 2 & \text{Multiply equation (1) by 2.} \\ \underline{3x + 2y - z = 5} & \text{Leave equation (2) alone.} \\ 7x + 7z = 7 & \end{array}$$

$$\begin{array}{ll} -12x + 6y - 24z = -6 & \text{Multiply equation (1) by } -6. \\ \underline{5x - 6y + 17z = -1} & \text{Leave equation (3) alone.} \\ -7x - 7z = -7 & \end{array}$$

Now we need to solve the system

$$\begin{pmatrix} 7x + 7z = 7 \\ -7x - 7z = -7 \end{pmatrix}.$$

Adding these two equations produces

$$\begin{array}{l} 7x + 7z = 7 \\ \underline{-7x - 7z = -7} \\ 0 + 0 = 0. \end{array}$$

The true numerical statement, $0 + 0 = 0$, indicates that the system has *infinitely many solutions*, the situation illustrated in Figure 11.4(a). ●

Remark   It can be shown that the solutions for the system in Example 4 are of the form $(t, 3 - 2t, 1 - t)$, where t is any real number. For example, if we let $t = 2$, then we get the ordered triple $(2, -1, -1)$, and this triple will satisfy all three of the original equations. For our purposes at this time we will simply indicate that such a system has *infinitely many solutions*.

Problem Set 11.3

Solve each of the following systems. If the solution set is $\emptyset$ or if it contains infinitely many solutions, so indicate.

1. $\begin{pmatrix} x - 2y + 3z = 7 \\ 2x + y + 5z = 17 \\ 3x - 4y - 2z = 1 \end{pmatrix}$

2. $\begin{pmatrix} 2x - y + z = 0 \\ 3x - 2y + 4z = 4 \\ 5x + y - 6z = -32 \end{pmatrix}$

3. $\begin{pmatrix} 3x + 2y - z = -11 \\ 2x - 3y + 4z = 11 \\ 5x + y - 2z = -17 \end{pmatrix}$

4. $\begin{pmatrix} 3x + 2y - 2z = 14 \\ 2x - 5y + 3z = 7 \\ 4x - 3y + 7z = 5 \end{pmatrix}$

5. $\begin{pmatrix} x + y - z = 2 \\ 3x - 4y + 2z = 5 \\ 2x + 2y - 2z = 7 \end{pmatrix}$

6. $\begin{pmatrix} x + 2y - 3z = 2 \\ 3x - z = -8 \\ 2x - 3y + 5z = -9 \end{pmatrix}$

7. $\begin{pmatrix} 2x + 3y - 4z = -10 \\ 4x - 5y + 3z = 2 \\ 2y + z = 8 \end{pmatrix}$

8. $\begin{pmatrix} x - y + 2z = 4 \\ 2x - 2y + 4z = 7 \\ 3x - 3y + 6z = 1 \end{pmatrix}$

9. $\begin{pmatrix} 2x - y + 3z = -14 \\ 4x + 2y - z = 12 \\ 6x - 3y + 4z = -22 \end{pmatrix}$

10. $\begin{pmatrix} x - 2y + z = -4 \\ 2x + 4y - 3z = -1 \\ -3x - 6y + 7z = 4 \end{pmatrix}$

11. $\begin{pmatrix} 2x - y + 3z = 1 \\ 4x + 7y - z = 7 \\ x + 4y - 2z = 3 \end{pmatrix}$

12. $\begin{pmatrix} 3x - 2y + 4z = 6 \\ 9x + 4y - z = 0 \\ 6x - 8y - 3z = 3 \end{pmatrix}$

13. $\begin{pmatrix} 5x - 3y + z = 1 \\ 2x - 5y = -2 \\ 3x - 2y - 4z = -27 \end{pmatrix}$

14. $\begin{pmatrix} 4x - y + z = 5 \\ 3x + y + 2z = 4 \\ x - 2y - z = 1 \end{pmatrix}$

15. $\begin{pmatrix} x + 3y - 2z = 19 \\ 3x - y - z = 7 \\ -2x + 5y + z = 2 \end{pmatrix}$

16. $\begin{pmatrix} 4x - y + 3z = -12 \\ 2x + 3y - z = 8 \\ 6x + y + 2z = -8 \end{pmatrix}$

17. $\begin{pmatrix} x + y + z = 1 \\ 2x - 3y + 6z = 1 \\ -x + y + z = 0 \end{pmatrix}$

18. $\begin{pmatrix} 3x + 2y - 2z = -2 \\ x - 3y + 4z = -13 \\ -2x + 5y + 6z = 29 \end{pmatrix}$

19. $\begin{pmatrix} 2x - y + 3z = 0 \\ 3x + 2y - 4z = 0 \\ 5x - 3y + 2z = 0 \end{pmatrix}$

20. $\begin{pmatrix} 3x - y + 4z = 9 \\ 3x + 2y - 8z = -12 \\ 9x + 5y - 12z = -23 \end{pmatrix}$

Solve each of the following problems by setting up and solving a system of three linear equations in three variables.

21. The sum of three numbers is 40. The third number is 10 less than the sum of the first two numbers. The second number is 1 larger than the first. Find the numbers.

22. The sum of three numbers is 20. The sum of the first and third numbers is 2 more than twice the second number. The third number minus the first yields three times the second number. Find the numbers.

23. A box contains $2 in nickels, dimes, and quarters. There are 19 coins in all, with twice as many nickels as dimes. How many coins of each kind are there?

24. The sum of the measures of the angles of a triangle is 180°. The largest angle is twice the smallest angle. The sum of the smallest and the largest angle is twice the other angle. Find the measure of each angle.

25. The perimeter of a triangle is 45 centimeters. The longest side is 4 centimeters less than twice the shortest side. The sum of the lengths of the shortest and longest sides is 7 centimeters less than three times the length of the remaining side. Find the lengths of all three sides of the triangle.

26. Part of $3000 is invested at 12%, another part at 13%, and the remainder at 14%. The total yearly income from the three investments is $400. The sum of the amounts invested at 12% and 13% equals the amount invested at 14%. Determine how much is invested at each rate.

27. The sum of the digits of a three-digit number is 13. The sum of the hundreds digit and the tens digit is 1 less than the units digit. The sum of three times the hundreds digit and four times the units digit is 26 more than twice the tens digit. Find the number.

28. The sum of the digits of a three-digit number is 14. The number is 14 larger than twenty times the tens digit. The sum of the tens digit and the units digit is 12 larger than the hundreds digit. Find the number.

29. Five pounds of potatoes, 1 pound of onions, and 2 pounds of apples cost $1.26. Two pounds of potatoes, 3 pounds of onions, and 4 pounds of apples cost $1.88. Three pounds of potatoes, 4 pounds of onions, and 1 pound of apples cost $1.24. Find the price per pound for each item.

30. Two bottles of catsup, 2 jars of peanut butter, and 1 jar of pickles cost $4.20. Three bottles of catsup, 4 jars of peanut butter, and 2 jars of pickles cost $7.70. Four bottles of catsup, 3 jars of peanut butter, and 5 jars of pickles cost $9.80. Find the cost per bottle of catsup and per jar for peanut butter and pickles.

11.4 Systems Involving Nonlinear Equations and Systems of Inequalities

Thus far in this chapter we have solved systems of linear equations. In this section, we will consider some systems in which at least one of the equations is nonlinear; we will also consider some systems of **linear inequalities**. Let's begin by considering a system of one linear equation and one quadratic equation.

Example 1

Solve the system $\begin{pmatrix} x^2 + y^2 = 17 \\ x \; + y \; = 5 \end{pmatrix}$.

Solution

First, we will graph the system so that we can predict approximate solutions. From our previous graphing experiences in Chapters 7 and 8, we should recognize $x^2 + y^2 = 17$ as a circle and $x + y = 5$ as a straight line (Figure 11.7). The graph indicates that the solutions for this system should be two ordered pairs with positive components (the points of intersection occur in the first quadrant). In fact, we can guess that these solutions are (1, 4) and (4, 1) and then verify our guess by checking these solutions in the given equations.

Figure 11.7

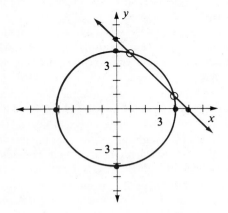

Let's also solve the system analytically using the substitution method as follows. First we change the form of $x + y = 5$ to $y = 5 - x$ and substitute $5 - x$ for y in the first equation.

$$x^2 + y^2 = 17$$
$$x^2 + (5 - x)^2 = 17$$
$$x^2 + 25 - 10x + x^2 = 17$$
$$2x^2 - 10x + 8 = 0$$
$$x^2 - 5x + 4 = 0$$
$$(x - 4)(x - 1) = 0$$
$$x - 4 = 0 \quad \text{or} \quad x - 1 = 0$$
$$x = 4 \quad \text{or} \quad x = 1$$

Substituting 4 for x and then 1 for x in the second equation of the system produces the following.

$$\begin{array}{ll} x + y = 5 & x + y = 5 \\ 4 + y = 5 & 1 + y = 5 \\ y = 1 & y = 4 \end{array}$$

Therefore, the solution set is $\{(1, 4), (4, 1)\}$.

Example 2   Solve the system $\begin{pmatrix} y = -x^2 + 1 \\ y = x^2 - 2 \end{pmatrix}$.

Solution   Again, let's get an idea of approximate solutions by graphing the system. Both equations produce parabolas, as indicated in Figure 11.8. From the graph we can predict two nonintegral ordered-pair solutions, one in the third quadrant and the other in the fourth quadrant.

Figure 11.8

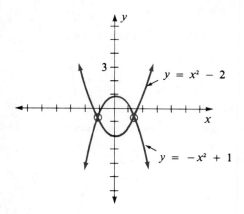

Substituting $-x^2 + 1$ for y in the second equation, we obtain

$$y = x^2 - 2$$
$$-x^2 + 1 = x^2 - 2$$
$$3 = 2x^2$$
$$\frac{3}{2} = x^2$$
$$\pm\sqrt{\frac{3}{2}} = x$$
$$\pm\frac{\sqrt{6}}{2} = x. \qquad \left(\sqrt{\frac{3}{2}} = \frac{\sqrt{3}}{\sqrt{2}} = \frac{\sqrt{3}}{\sqrt{2}}\frac{\sqrt{2}}{\sqrt{2}} = \frac{\sqrt{6}}{2} \right)$$

Substituting $\dfrac{\sqrt{6}}{2}$ for x in the second equation yields

$$y = x^2 - 2$$
$$y = \left(\frac{\sqrt{6}}{2} \right)^2 - 2$$
$$= \frac{6}{4} - 2 = -\frac{1}{2}.$$

Substituting $-\dfrac{\sqrt{6}}{2}$ for x in the second equation yields

$$y = x^2 - 2$$

$$y = \left(-\frac{\sqrt{6}}{2}\right)^2 - 2$$

$$= \frac{6}{4} - 2 = -\frac{1}{2}.$$

The solution set is $\left\{\left(-\dfrac{\sqrt{6}}{2}, -\dfrac{1}{2}\right), \left(\dfrac{\sqrt{6}}{2}, -\dfrac{1}{2}\right)\right\}$. (You should check both of these solutions in the original equations!) ●

Example 3

Solve the system $\begin{pmatrix} x^2 + y^2 = 16 \\ -x^2 + y^2 = 4 \end{pmatrix}$.

Solution

Graphing the system produces Figure 11.9, which indicates that there are four solutions to the system.

Figure 11.9

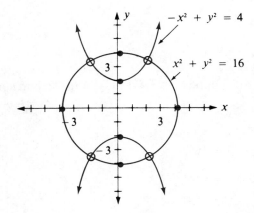

The elimination method works nicely. We can simply add the two equations, which eliminates the x's.

$$
\begin{aligned}
x^2 + y^2 &= 16 \\
-x^2 + y^2 &= 4 \\
\hline
2y^2 &= 20 \\
y^2 &= 10 \\
y &= \pm\sqrt{10}
\end{aligned}
$$

Substituting $\sqrt{10}$ for y in the first equation yields

$$x^2 + y^2 = 16$$

$$x^2 + (\sqrt{10})^2 = 16$$

$$x^2 + 10 = 16$$
$$x^2 = 6$$
$$x = \pm\sqrt{6}.$$

Thus, $(\sqrt{6}, \sqrt{10})$ and $(-\sqrt{6}, \sqrt{10})$ are solutions. Substituting $-\sqrt{10}$ for y in the first equation yields

$$x^2 + y^2 = 16$$
$$x^2 + (-\sqrt{10})^2 = 16$$
$$x^2 + 10 = 16$$
$$x^2 = 6$$
$$x = \pm\sqrt{6}.$$

Thus, $(\sqrt{6}, -\sqrt{10})$ and $(-\sqrt{6}, -\sqrt{10})$ are solutions. The solution set is $\{(-\sqrt{6}, \sqrt{10}), (-\sqrt{6}, -\sqrt{10}), (\sqrt{6}, \sqrt{10}), (\sqrt{6}, -\sqrt{10})\}$. ●

Remark The graphing of systems has been given additional emphasis in this section. Perhaps you will need to return to Chapters 7 and 8 to review the various graphing techniques covered there. Be sure to reread Section 7.2; the ideas from that section will be used in a moment.

Systems of Linear Inequalities

Finding solution sets for systems of **linear inequalities** relies heavily on the graphing approach. The solution set of a system of linear inequalities such as

$$\begin{pmatrix} x + y > 2 \\ x - y < 2 \end{pmatrix}$$

is the intersection of the solution sets of the individual inequalities. Figure 11.10(a) shows the solution set for $x + y > 2$, and Figure 11.10(b) shows the solution set for $x - y < 2$. Then the shaded region in Figure 11.10(c) represents the intersection of the

Figure 11.10

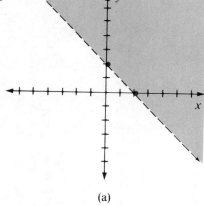

(a)

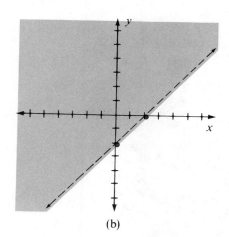

(b)

Figure 11.10
(continued)

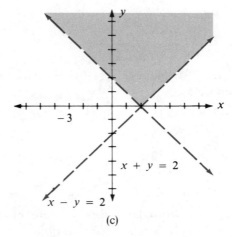

(c)

two solution sets and therefore is the graph of the system. Remember that dashed lines are used to indicate that points on the lines are not included in the solution set.

In the next examples only the final solution set for the systems is indicated.

Example 4   Solve the following system by graphing: $\begin{pmatrix} 2x - y \geq 4 \\ x + 2y < 2 \end{pmatrix}$.

Solution   The graph of $2x - y \geq 4$ consists of all points *on or below* the line $2x - y = 4$. The graph of $x + 2y < 2$ consists of all points *below* the line $x + 2y = 2$. The graph of the system is indicated by the shaded region in Figure 11.11. Notice that all points in the shaded region are on or below the line $2x - y = 4$ and below the line $x + 2y = 2$.

Figure 11.11

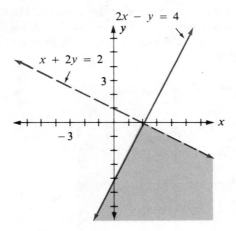

Example 5   Solve the following system by graphing: $\begin{pmatrix} x \leq 2 \\ y \geq -1 \end{pmatrix}$.

Solution

Remember that even though each inequality contains only one variable, we are working in a rectangular coordinate system involving ordered pairs. That is to say, the system could be written as

$$\begin{pmatrix} x + 0(y) \le 2 \\ 0(x) + \ \ y \ge -1 \end{pmatrix}.$$

The graph of this system is the shaded region in Figure 11.12. Notice that all points in the shaded region are on or to the left of the line $x = 2$ and on or above the line $y = -1$.

Figure 11.12

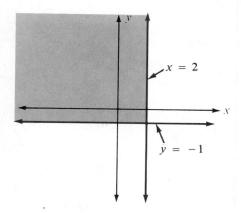

For the final example of this section, let's consider a system containing four inequalities.

Example 6

Solve the following system by graphing: $\begin{pmatrix} x \ge 0 \\ y \ge 0 \\ 2x + 3y \le 12 \\ 3x + \ \ y \le 6 \end{pmatrix}.$

Solution

The solution set for the system is the intersection of the solution sets of the four inequalities. The shaded region in Figure 11.13 indicates the solution set for the system. Notice that all points in the shaded region are on or to the right of the y-axis, on or above the x-axis, on or below the line $2x + 3y = 12$, and on or below the line $3x + y = 6$.

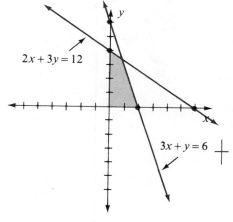

Figure 11.13

Problem Set 11.4

For each of the following systems, (a) graph the system so that approximate solutions can be predicted and (b) solve the system by the substitution or elimination method.

1. $\begin{pmatrix} x^2 + y^2 = 26 \\ x + y = 6 \end{pmatrix}$ 2. $\begin{pmatrix} x^2 + y^2 = 13 \\ 3x + 2y = 0 \end{pmatrix}$

3. $\begin{pmatrix} y = x^2 \\ y = x + 2 \end{pmatrix}$ 4. $\begin{pmatrix} y = (x + 2)^2 \\ y = -2x - 4 \end{pmatrix}$

5. $\begin{pmatrix} y = x^2 - 4x + 5 \\ -x + y = 1 \end{pmatrix}$ 6. $\begin{pmatrix} y = x^2 + 6x + 7 \\ 2x + y = -5 \end{pmatrix}$

7. $\begin{pmatrix} x - y = 2 \\ x^2 - y^2 = 16 \end{pmatrix}$ 8. $\begin{pmatrix} x + y = -8 \\ x^2 - y^2 = 16 \end{pmatrix}$

9. $\begin{pmatrix} y = -x^2 + 3 \\ y = x^2 + 1 \end{pmatrix}$ 10. $\begin{pmatrix} y = x^2 \\ y = x^2 - 4x + 4 \end{pmatrix}$

11. $\begin{pmatrix} x^2 - y^2 = 4 \\ x^2 + y^2 = 4 \end{pmatrix}$ 12. $\begin{pmatrix} x^2 + 2y^2 = 8 \\ x^2 - y^2 = 1 \end{pmatrix}$

13. $\begin{pmatrix} y = x^2 + 2 \\ y = 2x^2 + 1 \end{pmatrix}$ 14. $\begin{pmatrix} xy = 4 \\ y = x \end{pmatrix}$

15. $\begin{pmatrix} 2x^2 + y^2 = 8 \\ x^2 + y^2 = 4 \end{pmatrix}$ 16. $\begin{pmatrix} y = x^2 + 2x - 1 \\ y = x^2 + 4x + 5 \end{pmatrix}$

Indicate the solution set for each of the following systems of inequalities by graphing the system and shading the appropriate region.

17. $\begin{pmatrix} x + y > 1 \\ x - y < 1 \end{pmatrix}$ 18. $\begin{pmatrix} x + y < 4 \\ x - y > 2 \end{pmatrix}$ 19. $\begin{pmatrix} 2x - y \le 4 \\ 2x + y > 4 \end{pmatrix}$

20. $\begin{pmatrix} x - 3y < 6 \\ x + 2y \ge 4 \end{pmatrix}$ 21. $\begin{pmatrix} 3x + 2y \le 6 \\ 2x - 3y \ge 6 \end{pmatrix}$ 22. $\begin{pmatrix} 3x - 4y \ge 0 \\ 2x + 3y \le 0 \end{pmatrix}$

23. $\begin{pmatrix} y > x - 3 \\ y < x \end{pmatrix}$ 24. $\begin{pmatrix} y < x + 1 \\ y \ge x \end{pmatrix}$ 25. $\begin{pmatrix} 2x + 5y > 10 \\ 5x + 2y > 10 \end{pmatrix}$

26. $\begin{pmatrix} 3x - 2y < 6 \\ 2x - 3y < 6 \end{pmatrix}$ 27. $\begin{pmatrix} x + y > 4 \\ x + y > 6 \end{pmatrix}$ 28. $\begin{pmatrix} 2x - y < 4 \\ 2x - y > 0 \end{pmatrix}$

29. $\begin{pmatrix} x < 3 \\ y > 2 \end{pmatrix}$ 30. $\begin{pmatrix} x \ge -1 \\ y < 4 \end{pmatrix}$ 31. $\begin{pmatrix} 2x + y > 6 \\ 2x + y < 2 \end{pmatrix}$

32. $\begin{pmatrix} y > x \\ y > 2 \end{pmatrix}$ 33. $\begin{pmatrix} x \ge 0 \\ y \ge 0 \\ x + y \le 4 \\ 2x + y \le 6 \end{pmatrix}$ 34. $\begin{pmatrix} x \ge 0 \\ y \ge 0 \\ x - y \le 5 \\ 4x + 7y \le 28 \end{pmatrix}$

35. $\begin{pmatrix} x \geq 0 \\ y \geq 0 \\ 2x + y \leq 4 \\ 2x - 3y \leq 6 \end{pmatrix}$

36. $\begin{pmatrix} x \geq 0 \\ y \geq 0 \\ 3x + 5y \geq 15 \\ 5x + 3y \geq 15 \end{pmatrix}$

For each of the following problems, write a system of equations and solve.

37. The sum of the squares of two numbers is 13. One number is 1 larger than the other number. Find the numbers.

38. The sum of the squares of two numbers is 34. The difference of the squares of the same two numbers is 16. Find the numbers.

39. A number is 1 larger than the square of another number. The sum of the two numbers is 7. Find the numbers.

40. The area of a rectangular region is 54 square meters, and its perimeter is 30 meters. Find the length and the width of the rectangle.

Miscellaneous Problems

41. The following figure shows a graph of the system

$$\begin{pmatrix} y = x^2 \\ y = -1 \end{pmatrix}.$$

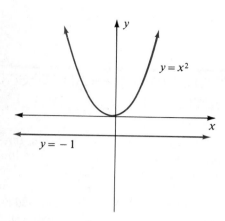

The graph indicates that the system has no solutions, but we must realize that we are graphing in the *real number plane*. Thus, the fact that the parabola and the line do not intersect simply indicates that the system has no *real number solutions*.

Using the substitution method, we can find some complex solutions as follows. Substitute x^2 for y in the second equation.

$$y = -1$$

becomes

$$x^2 = -1$$
$$x = \pm\sqrt{-1} = \pm i.$$

Substituting $\pm i$ for x in the first equation yields the following.

$$
\begin{array}{ll}
y = x^2 & y = x^2 \\
y = i^2 & y = (-i)^2 \\
y = -1 & y = i^2 \\
& y = -1
\end{array}
$$

The solution set is $\{(i, -1), (-i, -1)\}$.

For each of the following systems, (a) graph the system to show that there are no real number solutions and (b) solve the system by substitution or elimination to find the complex solutions.

(a) $\begin{pmatrix} y = x^2 + 1 \\ y = -3 \end{pmatrix}$

(b) $\begin{pmatrix} y = -x^2 + 1 \\ y = 3 \end{pmatrix}$

(c) $\begin{pmatrix} y = x^2 \\ x - y = 4 \end{pmatrix}$

(d) $\begin{pmatrix} y = x^2 + 1 \\ y = -x^2 \end{pmatrix}$

(e) $\begin{pmatrix} x^2 + y^2 = 1 \\ x + y = 2 \end{pmatrix}$

(f) $\begin{pmatrix} x^2 + y^2 = 2 \\ x^2 - y^2 = 6 \end{pmatrix}$

11.5 Linear Programming — Another Look at Problem Solving (Optional)

Problem solving is a unifying theme throughout this text. Therefore, it seems appropriate at this time to take a brief look at an area of mathematics that was developed in the 1940s specifically as a problem-solving tool. Many applied problems involve the idea of maximizing or minimizing a certain function that is subject to various constraints; these can be expressed as linear inequalities. **Linear programming** was developed as one method for solving such problems.

Remark The term "programming" refers to the distribution of limited resources in order to maximize or minimize a certain function such as cost, profit, or distance. Thus, it is not synonymous with its meaning in computer programming. The constraints under which the distribution of resources is to be made are the linear inequalities and equations; thus, the term "linear programming" is used.

Before introducing a linear-programming type of problem, we need to extend one mathematical concept a bit. A **linear function in two variables** x and y is a function of the form $f(x, y) = ax + by + c$, where a, b, and c are real numbers. In other words, with each ordered pair (x, y) we associate a third number by the rule $ax + by + c$. For example, suppose the function f is described by $f(x, y) = 4x + 3y + 5$. Then $f(2, 1) = 4(2) + 3(1) + 5 = 16$.

First, let's take a look at some mathematical ideas that form the basis for solving a linear-programming problem. Consider the shaded region in Figure 11.14 and the following linear functions in two variables:

$$f(x, y) = 4x + 3y + 5$$
$$f(x, y) = 2x + 7y - 1$$
$$f(x, y) = x - 2y.$$

Figure 11.14

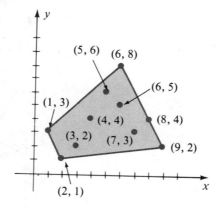

Suppose that we need to find the maximum and minimum values achieved by each of the functions in the indicated region. The following chart summarizes the values for the ordered pairs indicated in Figure 11.14.

Ordered pairs		Value of $f(x, y) = 4x + 3y + 5$		Value of $f(x, y) = 2x + 7y - 1$		Value of $f(x, y) = x - 2y$	
(2, 1)	(vertex)	16	(minimum)	10	(minimum)	0	
(3, 2)		23		19		−1	
(9, 2)	(vertex)	47		31		5	(maximum)
(1, 3)	(vertex)	18		22		−5	
(7, 3)		42		34		1	
(4, 4)		33		35		−4	
(8, 4)		49		43		0	
(6, 5)		44		46		−4	
(5, 6)		43		51		−7	
(6, 8)	(vertex)	53	(maximum)	67	(maximum)	−10	(minimum)

Notice that for each function the maximum and minimum values are obtained at a vertex of the region. To further substantiate the claim that maximum and minimum functional values are obtained at the vertices of the region, let's consider the family of lines $x - 2y = k$, where k is an arbitrary constant. [We are now working only with the function $f(x, y) = x - 2y$.] In slope-intercept form, $x - 2y = k$ becomes $y = \frac{1}{2}x - \frac{1}{2}k$; so we have a family of parallel lines, each having a slope of $\frac{1}{2}$. In Figure 11.15 some of these lines are sketched so that each line has at least one point in common with the given region. Note that $x - 2y$ reaches a minimum value of -10 at the vertex $(6, 8)$ and a maximum value of 5 at the vertex $(9, 2)$.

Figure 11.15

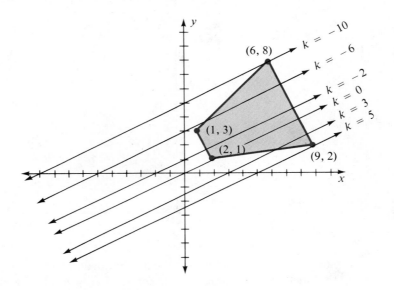

In general, suppose that f is a linear function in two variables x and y, and S is a region of the xy-plane. If f attains a maximum (minimum) value in S, then that maximum (minimum) value is obtained at a vertex of S.

Remark   A subset of the xy-plane is said to be **bounded** if there is a circle that contains all of its points; otherwise, the subset is said to be **unbounded**. A bounded set will contain a maximum and a minimum value for a function, but an unbounded set may not contain such values.

Now we will consider two examples that illustrate a general graphing approach for solving a linear-programming problem in two variables. The first example illustrates the general makeup of such a problem, without the "story" from which the function and inequalities evolve. The second example illustrates the type of setting from which such a problem is extracted.

Example 1

Find the maximum value and the minimum value of the function $f(x, y) = 9x + 13y$ in the region determined by the following system of inequalities:

$$\begin{pmatrix} x \geq 0 \\ y \geq 0 \\ 2x + 3y \leq 18 \\ 2x + y \leq 10 \end{pmatrix}.$$

Solution

First, let's graph the inequalities to determine the given region, as indicated in Figure 11.16. (Such a region is called the **set of feasible solutions**, and the inequalities are referred to as **constraints**.)

Figure 11.16

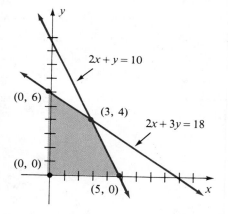

The point $(3, 4)$ is determined by solving the system $\begin{pmatrix} 2x + 3y = 18 \\ 2x + y = 10 \end{pmatrix}$. Second, we can determine the values of the given function at the vertices of the region. (Such a function to be maximized or minimized is called the **objective function**.)

Vertices	Value of $f(x, y) = 9x + 13y$	
$(0, 0)$	0	(minimum)
$(5, 0)$	45	
$(3, 4)$	79	(maximum)
$(0, 6)$	78	

Therefore, a minimum value of 0 is obtained at $(0, 0)$, and a maximum value of 79 is obtained at $(3, 4)$. ●

Example 2

A company that manufactures gidgets and gadgets has the following production information available:

1. Producing a gidget requires 3 hours of working time on machine A and 1 hour on machine B.

2. Producing a gadget requires 2 hours on machine A and 1 hour on machine B.

3. Machine A is available for no more than 120 hours per week; machine B is available for no more than 50 hours per week.

4. Gidgets can be sold at a profit of $3.75 each, whereas a profit of $3 each can be realized on a gadget.

How many gidgets and how many gadgets should be produced each week to maximize profit? What would the maximum profit be?

Solution

Let x be the number of gidgets and y the number of gadgets. Thus, the profit function is $P(x, y) = 3.75x + 3y$. The constraints for the problem can be represented by the following inequalities:

$3x + 2y \leq 120$ Machine A is available for no more than 120 hours.

$x + y \leq 50$ Machine B is available for no more than 50 hours.

$x \geq 0$ The number of gidgets and gadgets must be represented by a

$y \geq 0.$ nonnegative number.

Graphing these inequalities produces the set of feasible solutions indicated by the shaded region in Figure 11.17.

Figure 11.17

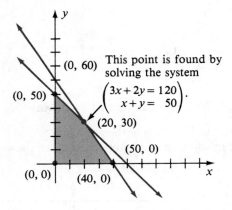

Finding the value of the profit function at the vertices of the region produces the following table.

Vertices	Value of $P(x, y) = 3.75x + 3y$
(0, 0)	0
(40, 0)	150
(20, 30)	165 (maximum)
(0, 50)	150

Thus, a maximum profit of $165 is realized by producing 20 gidgets and 30 gadgets. ●

Problem Set 11.5

In Problems 1–4, find the maximum and minimum values of the given function in the indicated region.

1. $f(x, y) = 3x + 5y$

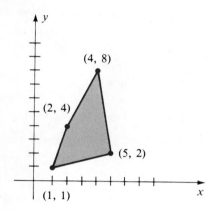

2. $f(x, y) = 8x + 3y$

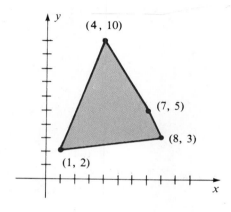

3. $f(x, y) = x + 4y$

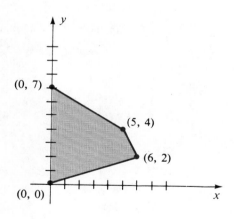

4. $f(x, y) = 2.5x + 3.5y$

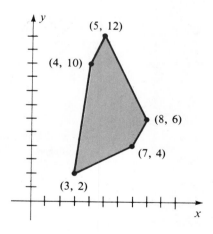

5. Maximize the function $f(x, y) = 3x + 7y$ in the region determined by the following constraints:

$$3x + 2y \leq 18$$
$$3x + 4y \geq 12$$
$$x \geq 0$$
$$y \geq 0.$$

6. Maximize the function $f(x, y) = 1.5x + 2y$ in the region determined by the following constraints:

$$3x + 2y \leq 36$$
$$3x + 10y \leq 60$$
$$x \geq 0$$
$$y \geq 0.$$

7. Maximize the function $f(x, y) = 40x + 55y$ in the region determined by the following constraints:

$$2x + y \leq 10$$
$$x + y \leq 7$$
$$2x + 3y \leq 18$$
$$x \geq 0$$
$$y \geq 0.$$

8. Maximize the function $f(x, y) = 0.08x + 0.09y$ in the region determined by the following constraints:

$$x + y \leq 8000$$
$$y \leq \frac{1}{3}x$$
$$y \geq 500$$
$$x \leq 7000$$
$$x \geq 0.$$

9. Minimize the function $f(x, y) = 0.2x + 0.5y$ in the region determined by the following constraints:

$$2x + y \geq 12$$
$$2x + 5y \geq 20$$
$$x \geq 0$$
$$y \geq 0.$$

10. Minimize the function $f(x, y) = 3x + 7y$ in the region determined by the following constraints:

$$x + y \geq 9$$
$$6x + 11y \geq 84$$
$$x \geq 0$$
$$y \geq 0.$$

11. Maximize the function $f(x, y) = 9x + 2y$ in the region determined by the following constraints:

$$5y - 4x \leq 20$$
$$4x + 5y \leq 60$$
$$x \geq 0$$
$$x \leq 10$$
$$y \geq 0.$$

12. Maximize the function $f(x, y) = 3x + 4y$ in the region determined by the following constraints:

$$2y - x \leq 6$$
$$x + y \leq 12$$
$$x \geq 2$$
$$x \leq 8$$
$$y \geq 0.$$

Solve each of the following linear-programming problems by using the graphical method. (See Example 2.)

13. Suppose that the company referred to in Example 2 also manufactures widgets and wadgets and has the following production information available:

Producing a widget requires 4 hours of working time on machine A and 2 hours on machine B.

Producing a wadget requires 5 hours of working time on machine A and 5 hours on machine B.

Machine A is available for no more than 200 hours in a month, machine B is available for no more than 150 hours per month.

Widgets can be sold at a profit of $7 each and wadgets at a profit of $8 each.

How many widgets and how many wadgets should be produced per month in order to maximize profit? What would the maximum profit be?

14. Products A and B are produced by a company that has the following production information available:

Producing one unit of product A requires 1 hour of working time on machine I, 2 hours on machine II, and 1 hour on machine III.

Producing one unit of product B requires 1 hour of working time on machine I, 1 hour on machine II, and 3 hours on machine III.

Machine I is available for no more than 40 hours per week, machine II for no more than 40 hours per week, and machine III for no more than 60 hours per week.

Product A can be sold at a profit of $2.75 per unit and product B at a profit of $3.50 per unit.

How many units of product A and how many of product B should be produced per week to maximize profit?

15. Suppose that an investor wants to invest up to $10,000. She plans to buy one speculative stock and one conservative stock. The speculative stock is paying a 12% return, and the conservative stock is paying a 9% return. The investor has decided to put at least $2000 in the conservative stock and no more than $6000 in the speculative stock. Furthermore, she does not want the speculative to exceed the conservative investment. How much should she invest at each rate to maximize her return?

16. A manufacturer of golf clubs makes a profit of $50 per set on a model A set and $45 per set on a model B set. Daily production of the model A clubs is between 30 and 50 sets, inclusive, and of the model B clubs is between 10 and 20 sets, inclusive. The total daily production is not to

exceed 50 sets. How many sets of each model should be manufactured per day to maximize the profit?

17. A company makes two types of calculators. Type A sells for \$12, and type B sells for \$10. It costs the company \$9 to produce one type A calculator and \$8 to produce one type B calculator. The company is equipped to produce between 200 and 300, inclusive, of the type A calculator per month and between 100 and 250, inclusive, of the type B calculator, but not more than 500 altogether. How many calculators of each type should be produced per month to maximize the difference between the selling prices and the costs of production?

18. A manufacturer of small copiers makes a profit of \$200 on a deluxe model and \$250 on a standard model. The company wants to produce at least 50 deluxe models per week and at least 75 standard models per week. However, the weekly production is not to exceed 150 copiers. How many copiers of each kind should be produced in order to maximize the profit?

Chapter 11 Summary

(11.1)   Graphing a system of two linear equations in two variables produces one of the following results:

1. The graphs of the two equations are two intersecting lines, which indicates that there is **one unique solution** of the system. Such a system is called a **consistent system**.

2. The graphs of the two equations are two parallel lines indicating that there is **no solution** for the system. Such a system is called an **inconsistent system**.

3. The graphs of the two equations are the same line, indicating **infinitely many solutions** for the system. The equations are called **dependent** equations.

The **elimination-by-addition method** of solving a system is based on the following property: If an ordered pair (p, q) satisfies each of the equations $A_1 x + B_1 y = C_1$ and $A_2 x + B_2 y = C_2$, then it will also satisfy the equation $k(A_1 x + B_1 y) + l(A_2 x + B_2 y) = kC_1 + lC_2$, where k and l are real numbers.

(11.2)   The **substitution method** of solving a system relies on substituting the value for x or y from one equation into the other equation, thereby producing an equation with one unknown.

The decision as to whether to use the elimination-by-addition or the substitution method may be influenced by the format of the original equations in the system.

(11.3)   Solving a system of three linear equations in three variables produces one of the following results:

1. There is **one ordered triple** that satisfies all three equations.

2. There are **infinitely many ordered triples** in the solution set, all of which are coordinates of points on a line common to the planes.

3. There are **infinitely many ordered triples** in the solution set, all of which are coordinates on a common plane.

4. The solution set is **empty; it is ∅**.

The **elimination method** can be used to solve a system of three linear equations in three variables.

(11.4)   The **substitution** and **elimination methods** can also be used to solve systems involving nonlinear equations.

The solution set of a system of **linear inequalities** is the intersection of the solution sets of the individual inequalities.

Chapter 11 Review Problem Set

For Problems 1–6, solve each of the systems using either the elimination or the substitution method.

1. $\begin{pmatrix} 3x - 2y = -6 \\ 2x + 5y = 34 \end{pmatrix}$

2. $\begin{pmatrix} x + 4y = 25 \\ \quad y = -3x - 2 \end{pmatrix}$

3. $\begin{pmatrix} x = \dfrac{y + 9}{2} \\ 2x + 3y = 5 \end{pmatrix}$

4. $\begin{pmatrix} 9x + 5y = 12 \\ 7x + 4y = 10 \end{pmatrix}$

5. $\begin{pmatrix} \dfrac{1}{2}x + \dfrac{1}{3}y = -4 \\ \dfrac{3}{4}x - \dfrac{2}{3}y = 22 \end{pmatrix}$

6. $\begin{pmatrix} -2x + 5y = 7 \\ \quad y = \dfrac{6x + 9}{15} \end{pmatrix}$

For Problems 7–11, solve each of the systems.

7. $\begin{pmatrix} x + y - z = -2 \\ 2x - 3y + 4z = 17 \\ -3x + 2y + 5z = -7 \end{pmatrix}$

8. $\begin{pmatrix} x - 2y + z = 7 \\ 2x - 5y + 2z = 17 \\ -3x + 7y + 5z = -32 \end{pmatrix}$

9. $\begin{pmatrix} x + 3y - 2z = -7 \\ 4x + 13y - 7z = -21 \\ 5x + 16y - 8z = -23 \end{pmatrix}$

10. $\begin{pmatrix} -x + 2y + 3z = 22 \\ 2x - 5y - 7z = -51 \\ -3x + 5y + 11z = 71 \end{pmatrix}$

11. $\begin{pmatrix} x - 3y - 4z = -1 \\ 2x + y - 2z = 3 \\ 5x - 8y - 14z = 0 \end{pmatrix}$

For Problems 12–15, graph the system and then solve it using either the elimination or the substitution method.

12. $\left(\begin{array}{c} y = 2x^2 - 1 \\ 2x + y = 3 \end{array}\right)$

13. $\left(\begin{array}{c} x^2 + y^2 = 7 \\ x^2 - y^2 = 1 \end{array}\right)$

14. $\left(\begin{array}{c} xy = -4 \\ x + y = 0 \end{array}\right)$

15. $\left(\begin{array}{c} y = x^2 + 4x + 7 \\ y = -x^2 - 4x + 1 \end{array}\right)$

For Problems 16–20, indicate the solution set of the system by graphing the system and shading the appropriate region.

16. $\left(\begin{array}{c} 3x + y > 6 \\ x - 2y \le 4 \end{array}\right)$

17. $\left(\begin{array}{c} 3x - 4y \ge 0 \\ 2x + 3y \le 0 \end{array}\right)$

18. $\left(\begin{array}{c} 3x - 2y < 6 \\ 2x - 3y < 6 \end{array}\right)$

19. $\left(\begin{array}{c} x - 4y < 4 \\ 2x + y \ge 2 \end{array}\right)$

20. $\left(\begin{array}{c} x \ge 0 \\ y \ge 0 \\ x + 2y \le 4 \\ 2x - y \le 4 \end{array}\right)$

12 Using Matrices and Determinants to Solve Linear Systems

In Chapter 11 we found that the techniques of elimination and substitution worked effectively with two equations and two unknowns, but they started to become a bit cumbersome with three equations and three unknowns. In this chapter we will continue to work primarily with small systems involving two or three equations, but we will consider some techniques for solving systems that can be easily extended to larger systems.

12.1 A Matrix Approach to Solving Systems

A matrix is simply an array of numbers arranged in horizontal rows and vertical columns. For example, the matrix

$$2 \text{ rows} \begin{array}{c} \longrightarrow \\ \longrightarrow \end{array} \begin{bmatrix} 2 & 1 & -4 \\ 5 & -7 & 6 \end{bmatrix}$$
$$\begin{array}{ccc} \uparrow & \uparrow & \uparrow \end{array}$$
$$3 \text{ columns}$$

has 2 rows and 3 columns and is referred to as a 2×3 (read "two by three") matrix. Each number in a matrix is called an **element** of the matrix. Some additional examples of matrices (matrices is the plural of matrix) are as follows.

$$3 \times 2 \qquad\qquad 2 \times 2 \qquad\qquad 1 \times 2 \qquad\qquad 4 \times 1$$

$$\begin{bmatrix} 2 & 1 \\ 1 & 0 \\ \dfrac{1}{2} & 3 \end{bmatrix} \qquad \begin{bmatrix} 17 & 16 \\ -14 & 12 \end{bmatrix} \qquad \begin{bmatrix} 2 & 13 \end{bmatrix} \qquad \begin{bmatrix} 3 \\ -2 \\ 1 \\ 8 \end{bmatrix}$$

In general, a matrix of m rows and n columns is called a matrix of **dimension** $m \times n$ or **order** $m \times n$.

With every system of linear equations we can associate a matrix that consists of the coefficients and constant terms. For example, with the system

$$\begin{pmatrix} a_1 x + b_1 y + c_1 z = d_1 \\ a_2 x + b_2 y + c_2 z = d_2 \\ a_3 x + b_3 y + c_3 z = d_3 \end{pmatrix}$$

we can associate the matrix

$$\begin{bmatrix} a_1 & b_1 & c_1 & \vdots & d_1 \\ a_2 & b_2 & c_2 & \vdots & d_2 \\ a_3 & b_3 & c_3 & \vdots & d_3 \end{bmatrix}$$

which is commonly called the **augmented matrix** of the system. The dashed line simply separates the coefficients from the constant terms and reminds us that we are working with an augmented matrix.

Our previous work with systems of linear equations was based on the following properties:

1. Any two equations of a system may be interchanged. *Example*: Interchanging the two equations makes the system $\begin{pmatrix} 2x - 5y = 9 \\ x + 3y = 4 \end{pmatrix}$ equivalent to the system $\begin{pmatrix} x + 3y = 4 \\ 2x - 5y = 9 \end{pmatrix}$.

2. Any equation of the system may be multiplied by a nonzero constant. *Example*: Multiplying the top equation by -2 makes the system $\begin{pmatrix} x + 3y = 4 \\ 2x - 5y = 9 \end{pmatrix}$ equivalent to the system $\begin{pmatrix} -2x - 6y = -8 \\ 2x - 5y = 9 \end{pmatrix}$.

3. Any equation of the system can be replaced by adding a nonzero multiple of another equation to that equation. *Example*: Adding -2 times the first equation to the second equation makes the system $\begin{pmatrix} x + 3y = 4 \\ 2x - 5y = 9 \end{pmatrix}$ equivalent to the system $\begin{pmatrix} x + 3y = 4 \\ -11y = 1 \end{pmatrix}$.

Each of the properties geared to solving a system of equations produces a corresponding property of the augmented matrix of the system. For example, exchanging two equations of a system corresponds to exchanging two rows of the augmented matrix that represents the system. The properties of an augmented matrix are usually referred to as elementary row operations and can be stated as follows.

For any augmented matrix of a system of linear equations, the following *elementary row operations* will produce a matrix of an equivalent system:

1. Any two rows of the matrix can be interchanged.

2. Any row of the matrix can be multiplied by a nonzero real number.

3. Any row of the matrix can be replaced by adding a nonzero multiple of another row to that row.

Using the elementary row operations on an augmented matrix provides a basis for solving systems of linear equations. Study the following examples very carefully; keep in mind that the general scheme, called **Gaussian elimination**, is one of using the elementary row operations on a matrix to continue replacing a system of equations with equivalent systems until a system is obtained for which the solutions can be easily determined.

Example 1   Solve the system $\begin{pmatrix} x - 3y = -17 \\ 2x + 7y = 31 \end{pmatrix}$.

Solution   The augmented matrix of the system is

$$\begin{bmatrix} 1 & -3 & | & -17 \\ 2 & 7 & | & 31 \end{bmatrix}.$$

Let's multiply row 1 by -2 and add this result to row 2 to produce a new row 2.

$$\begin{bmatrix} 1 & -3 & | & -17 \\ 0 & 13 & | & 65 \end{bmatrix}$$

The system represented by this matrix,

$$\begin{pmatrix} x - 3y = -17 \\ 13y = 65 \end{pmatrix},$$

is said to be in *triangular form*. We can easily solve the second equation for y and then substitute this value for y in the first equation in order to determine x.

$$13y = 65$$
$$y = 5$$

Substituting 5 for y in the first equation produces

$$x - 3(5) = -17$$
$$x - 15 = -17$$
$$x = -2.$$

The solution set is $\{(-2, 5)\}$.   ●

Example 2   Solve the system $\begin{pmatrix} 3x + 2y = 5 \\ 5x - y = -9 \end{pmatrix}$.

Solution   The augmented matrix of the system is

$$\begin{bmatrix} 3 & 2 & | & 5 \\ 5 & -1 & | & -9 \end{bmatrix}.$$

To get a 0 in the row 2–column 1 position, we multiply the first row by $-\dfrac{5}{3}$ and add this result to the second row to form a new second row.

$$\begin{bmatrix} 3 & 2 & | & 5 \\ 0 & -\dfrac{13}{3} & | & -\dfrac{52}{3} \end{bmatrix}$$

Now, from the second row we obtain

$$-\frac{13}{3}y = -\frac{52}{3}$$
$$y = 4.$$

Substituting 4 for y in the first equation produces

$$3x + 2y = 5$$
$$3x + 2(4) = 5$$
$$3x = -3$$
$$x = -1.$$

The solution set is $\{(-1, 4)\}$.   ●

It should be evident from Examples 1 and 2 that the matrix approach does not provide us with much extra power for solving systems of two linear equations in two unknowns. However, as the systems become larger, the compactness of the matrix approach becomes more convenient.

Suppose that we had the following system to solve.

$$\begin{pmatrix} 5x - 3y - 2z = -1 \\ 4y + 7z = 3 \\ 4z = -12 \end{pmatrix}$$

It would be easy to determine z from the last equation and then substitute that value in the second equation to determine y. The values for y and z could then be substituted in the first equation to determine x. In other words, the form of this system makes it convenient to solve. We say that the system is in **triangular form**; notice the location of zeros in its augmented matrix.

$$\begin{bmatrix} 5 & -3 & -2 & | & -1 \\ 0 & 4 & 7 & | & 3 \\ 0 & 0 & 4 & | & -12 \end{bmatrix}$$

Example 3   Solve the system $\begin{pmatrix} x - 2y + z = 16 \\ 2x - y - z = 14 \\ 3x + 5y - 4z = -10 \end{pmatrix}$.

Solution   The augmented matrix of the system is

$$\begin{bmatrix} 1 & -2 & 1 & | & 16 \\ 2 & -1 & -1 & | & 14 \\ 3 & 5 & -4 & | & -10 \end{bmatrix}.$$

Let's multiply row 1 by -2 and add this result to row 2 to form a new row 2. Also, let's multiply row 1 by -3 and add this result to row 3 to produce a new row 3.

$$\begin{bmatrix} 1 & -2 & 1 & | & 16 \\ 0 & 3 & -3 & | & -18 \\ 0 & 11 & -7 & | & -58 \end{bmatrix}$$

Now let's multiply row 2 by $\dfrac{1}{3}$.

$$\begin{bmatrix} 1 & -2 & 1 & | & 16 \\ 0 & 1 & -1 & | & -6 \\ 0 & 11 & -7 & | & -58 \end{bmatrix}$$

Now we can multiply row 2 by -11 and add to row 3 to produce a new row 3.

$$\begin{bmatrix} 1 & -2 & 1 & | & 16 \\ 0 & 1 & -1 & | & -6 \\ 0 & 0 & 4 & | & 8 \end{bmatrix}$$

The system represented by this matrix,

$$\begin{pmatrix} x - 2y + z = 16 \\ y - z = -6 \\ 4z = 8 \end{pmatrix},$$

is in triangular form. From the last equation we obtain

$$4z = 8$$
$$z = 2.$$

Substituting 2 for z in the second equation produces

$$y - z = -6$$
$$y - 2 = -6$$
$$y = -4.$$

Finally, substituting 2 for z and -4 for y in the first equation produces

$$x - 2y + z = 16$$
$$x - 2(-4) + 2 = 16$$
$$x + 10 = 16$$
$$x = 6.$$

The solution set is $\{(6, -4, 2)\}$.

●

Example 4   Solve the system $\begin{pmatrix} 3x - 2y + 4z = -16 \\ 2x - y + 3z = -10 \\ 5x + 3y + 9z = -10 \end{pmatrix}$.

Solution   The augmented matrix of the system is

$$\begin{bmatrix} 3 & -2 & 4 & | & -16 \\ 2 & -1 & 3 & | & -10 \\ 5 & 3 & 9 & | & -10 \end{bmatrix}.$$

Let's begin by multiplying row 2 by -1 and adding the result to row 1 to produce a new row 1.

$$\begin{bmatrix} 1 & -1 & 1 & | & -6 \\ 2 & -1 & 3 & | & -10 \\ 5 & 3 & 9 & | & -10 \end{bmatrix}$$

Notice that we now have a 1 at the top of the first column. This makes it easier to get zeros in the other two positions of the first column. Let's multiply row 1 by -2 and add the result to row 2 to produce a new row 2. Also, let's multiply row 1 by -5 and add the result to row 3 to produce a new row 3.

$$\begin{bmatrix} 1 & -1 & 1 & | & -6 \\ 0 & 1 & 1 & | & 2 \\ 0 & 8 & 4 & | & 20 \end{bmatrix}$$

Now we can multiply row 2 by -8 and add the result to row 3 to produce a new row 3.

$$\begin{bmatrix} 1 & -1 & 1 & | & -6 \\ 0 & 1 & 1 & | & 2 \\ 0 & 0 & -4 & | & 4 \end{bmatrix}$$

This matrix represents a system in triangular form. From the bottom row we obtain

$$-4z = 4$$
$$z = -1.$$

Substituting -1 for z in the equation represented by the second row produces

$y + z = 2$
$y - 1 = 2$
$\quad y = 3.$

Finally, substituting -1 for z and 3 for y in the equation represented by the top row produces

$x - y + z = -6$
$x - 3 - 1 = -6$
$\quad x - 4 = -6$
$\quad\quad x = -2.$

The solution set is $\{(-2, 3, -1)\}$.

Problem Set 12.1

Use a matrix approach to solve each of the following systems.

1. $\begin{pmatrix} x - 4y = 12 \\ 2x + 5y = -2 \end{pmatrix}$

2. $\begin{pmatrix} 3x + 7y = -45 \\ 5x - \ y = 1 \end{pmatrix}$

3. $\begin{pmatrix} 7x + \ y = -29 \\ 4x - 5y = -50 \end{pmatrix}$

4. $\begin{pmatrix} 3x + 2y = 17 \\ 11x - 4y = 85 \end{pmatrix}$

5. $\begin{pmatrix} 3x + 10y = -33 \\ 6x - \ 7y = 15 \end{pmatrix}$

6. $\begin{pmatrix} 2x + 9y = 98 \\ 6x - 5y = -26 \end{pmatrix}$

7. $\begin{pmatrix} 5x - 3y = -25 \\ 3x + 2y = 4 \end{pmatrix}$

8. $\begin{pmatrix} 9x - 2y = -6 \\ 4x + 5y = 15 \end{pmatrix}$

9. $\begin{pmatrix} x + 3y = 7 \\ 2x - 4y = 9 \end{pmatrix}$

10. $\begin{pmatrix} 7x - 3y = 4 \\ 2x + \ y = 5 \end{pmatrix}$

11. $\begin{pmatrix} x + 2y - 3z = -16 \\ 2x - \ y + 4z = 23 \\ 3x + 5y - 2z = -12 \end{pmatrix}$

12. $\begin{pmatrix} 2x - 9y + 4z = -27 \\ x - 5y - 2z = -11 \\ 5x + 2y - 3z = 19 \end{pmatrix}$

13. $\begin{pmatrix} 3x - \ y + 4z = 11 \\ -5x + 2y + \ z = 21 \\ 2x - 3y + 5z = 12 \end{pmatrix}$

14. $\begin{pmatrix} 2x + 3y - \ z = -3 \\ 5x - 2y - 4z = -1 \\ 3x - 8y + 3z = 17 \end{pmatrix}$

15. $\begin{pmatrix} 3x - 2y + 5z = -18 \\ 2x + 5y - 3z = 26 \\ 5x + \ y - 2z = 8 \end{pmatrix}$

16. $\begin{pmatrix} 2x + \ 5y - 2z = -8 \\ 3x - 11y + 4z = -5 \\ 4x + \ 3y - \ z = -13 \end{pmatrix}$

17. $\begin{pmatrix} 3x - \ y + \ z = 12 \\ 2x + 3y + 2z = 3 \\ 5x - 4y - \ z = 25 \end{pmatrix}$

18. $\begin{pmatrix} x + 5y - 2z = -5 \\ 3x - 2y - 3z = 24 \\ -2x - 9y + 5z = 3 \end{pmatrix}$

19. $\begin{pmatrix} x - 2y - 3z = 1 \\ -2x + 5y + 2z = 4 \\ 3x - 4y - 4z = 3 \end{pmatrix}$

20. $\begin{pmatrix} 2x - 3y - 4z = 1 \\ x - 2y - 5z = -1 \\ -3x + 5y + 2z = 2 \end{pmatrix}$

21. $\begin{pmatrix} -x + 2y - z = 7 \\ 4x - y + z = 1 \\ -3x + 5y + 3z = 28 \end{pmatrix}$

22. $\begin{pmatrix} -x + y - 4z = -15 \\ -2x + 3y + 5z = 25 \\ 4x - 6y - z = -14 \end{pmatrix}$

23. $\begin{pmatrix} 2x + 3z = -7 \\ -y - 4z = 0 \\ 3x + 2y - 5z = 7 \end{pmatrix}$

24. $\begin{pmatrix} 4x - 3y + 2z = 17 \\ 2y + 5z = 24 \\ 3x - z = -9 \end{pmatrix}$

25. $\begin{pmatrix} 2x - y + 3z = 8 \\ 3x + 2y + z = -2 \\ 3x - y - 5z = -26 \end{pmatrix}$

26. $\begin{pmatrix} 4x - y - 2z = -7 \\ 5x + y + 3z = -20 \\ 2x - 3y - z = 9 \end{pmatrix}$

Miscellaneous Problems

In some textbooks it is common to use subscripts to identify the different variables in an equation. For example, the equation $2x + 3y + 5z = 7$ would be written as $2x_1 + 3x_2 + 5x_3 = 7$. Solution sets are then expressed in terms of ordered triples of the form (x_1, x_2, x_3). Use a matrix approach to solve each of the following systems.

27. $\begin{pmatrix} x_1 + 2x_2 - x_3 = 5 \\ 2x_1 - x_2 + 4x_3 = -1 \\ 3x_1 + 3x_2 - 2x_3 = 11 \end{pmatrix}$

28. $\begin{pmatrix} 3x_1 + x_2 - x_3 = 5 \\ x_1 - 2x_2 + 3x_3 = -5 \\ -4x_1 + 3x_2 + 5x_3 = -14 \end{pmatrix}$

29. $\begin{pmatrix} 2x_1 + 3x_2 + x_3 = 6 \\ 3x_1 - 4x_2 - x_3 = 10 \\ 5x_1 + 2x_2 - 3x_3 = 4 \end{pmatrix}$

30. $\begin{pmatrix} 5x_1 - x_2 + x_3 = 5 \\ 4x_1 + 2x_2 + x_3 = 7 \\ 3x_1 - 4x_2 - 2x_3 = -3 \end{pmatrix}$

12.2   Reduced Echelon Form

In Example 3 of the previous section we took the augmented matrix of the system

$$\begin{pmatrix} x - 2y + z = 16 \\ 2x - y - z = 14 \\ 3x + 5y - 4z = -10 \end{pmatrix}$$

and changed it to the following triangular form.

$$\begin{bmatrix} 1 & -2 & 1 & | & 16 \\ 0 & 1 & -1 & | & -6 \\ 0 & 0 & 4 & | & 8 \end{bmatrix}$$

From the equation represented by the bottom row of this matrix ($4z = 8$), we found the value for z, and then we used back-substitution to find the values for y and x. It is possible to continue working with that matrix until the solution set can be read directly from the matrix. Let's try this process.

Multiply row 2 by 2 and add the result to row 1 to form a new row 1. (We want a zero in the row 1–column 2 position.)

$$\begin{bmatrix} 1 & 0 & -1 & | & 4 \\ 0 & 1 & -1 & | & -6 \\ 0 & 0 & 4 & | & 8 \end{bmatrix}$$

Multiply row 3 by $\dfrac{1}{4}$.

$$\begin{bmatrix} 1 & 0 & -1 & | & 4 \\ 0 & 1 & -1 & | & -6 \\ 0 & 0 & 1 & | & 2 \end{bmatrix}$$

Add row 3 to row 1 to form a new row 1. Also add row 3 to row 2 to form a new row 2. (We want zeros above the 1 in the third column.)

$$\begin{bmatrix} 1 & 0 & 0 & | & 6 \\ 0 & 1 & 0 & | & -4 \\ 0 & 0 & 1 & | & 2 \end{bmatrix}$$

Now, reading directly from the matrix, we have $x = 6$, $y = -4$, and $z = 2$. In other words, the solution set is $\{(6, -4, 2)\}$.

The matrix

$$\begin{bmatrix} 1 & 0 & 0 & | & 6 \\ 0 & 1 & 0 & | & -4 \\ 0 & 0 & 1 & | & 2 \end{bmatrix}$$

is said to be in **reduced echelon form**. In general, a matrix is in reduced echelon form if the following conditions are satisfied.

1. The first (reading from left to right) nonzero entry of each row is 1.
2. The remaining entries in the *column* containing the leftmost 1 of a row are all zeros.
3. The leftmost 1 of any row is to the right of the leftmost 1 of the preceding row.
4. Rows containing only zeros are below rows containing nonzero entries.

Each of the following matrices is in reduced echelon form.

$$\begin{bmatrix} 1 & 0 & | & 4 \\ 0 & 1 & | & -7 \end{bmatrix}, \qquad \begin{bmatrix} 1 & 0 & 0 & | & 9 \\ 0 & 1 & 0 & | & -10 \\ 0 & 0 & 1 & | & 14 \end{bmatrix},$$

$$\begin{bmatrix} 1 & 2 & | & -3 \\ 0 & 0 & | & 0 \end{bmatrix}, \qquad \begin{bmatrix} 1 & 0 & -2 & | & 5 \\ 0 & 1 & 4 & | & 7 \\ 0 & 0 & 0 & | & 0 \end{bmatrix}$$

In contrast, the following matrices are not in reduced echelon form for the reason indicated below each matrix.

$$\begin{bmatrix} 1 & 0 & 0 & \vdots & 11 \\ 0 & 3 & 0 & \vdots & -1 \\ 0 & 0 & 1 & \vdots & -2 \end{bmatrix},$$
violates 1

$$\begin{bmatrix} 1 & 2 & -3 & \vdots & 5 \\ 0 & 1 & 7 & \vdots & 9 \\ 0 & 0 & 1 & \vdots & -6 \end{bmatrix},$$
violates 2

$$\begin{bmatrix} 1 & 0 & 0 & \vdots & 7 \\ 0 & 0 & 1 & \vdots & -8 \\ 0 & 1 & 0 & \vdots & 14 \end{bmatrix},$$
violates 3

$$\begin{bmatrix} 1 & 0 & 0 & 0 & \vdots & -1 \\ 0 & 0 & 0 & 0 & \vdots & 0 \\ 0 & 0 & 1 & 0 & \vdots & 7 \\ 0 & 0 & 0 & 0 & \vdots & 0 \end{bmatrix}$$
violates 4

Once we have an augmented matrix in reduced echelon form, the solution set of the system is easily determined. Furthermore, the procedure for changing a given augmented matrix to reduced echelon form can be described in a very systematic way. For example, the augmented matrix of a system of three linear equations in three unknowns having a unique solution can be changed to reduced echelon form as follows.

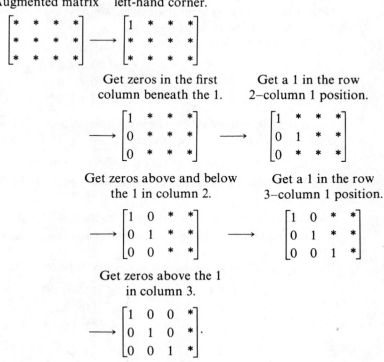

Inconsistent and dependent systems can be identified during the process of changing a matrix to reduced echelon form. We will look at some examples of such cases in a moment, but first let's consider some examples of systems that have a unique solution.

Example 1   Solve the system $\begin{pmatrix} 5x + 7y = -13 \\ x - 3y = -7 \end{pmatrix}$.

Solution   The augmented matrix

$$\begin{bmatrix} 5 & 7 & \vdots & -13 \\ 1 & -3 & \vdots & -7 \end{bmatrix}$$

does not have a 1 in the upper left-hand corner, but this problem can be remedied by exchanging the two rows.

$$\begin{bmatrix} 1 & -3 & \vdots & -7 \\ 5 & 7 & \vdots & -13 \end{bmatrix}$$

Now we can multiply row 1 by -5 and add the result to row 2 to form a new row 2.

$$\begin{bmatrix} 1 & -3 & \vdots & -7 \\ 0 & 22 & \vdots & 22 \end{bmatrix}$$

Next, let's multiply row 2 by $\frac{1}{22}$.

$$\begin{bmatrix} 1 & -3 & \vdots & -7 \\ 0 & 1 & \vdots & 1 \end{bmatrix}$$

Finally, we can multiply row 2 by 3 and add the result to row 1 to form a new row 1.

$$\begin{bmatrix} 1 & 0 & \vdots & -4 \\ 0 & 1 & \vdots & 1 \end{bmatrix}$$

The solution set of $\{(-4, 1)\}$ can be read directly from the last matrix.   ●

Example 2   Solve the system $\begin{pmatrix} x + 2y - 3z = 15 \\ -2x - 3y + z = -15 \\ 4x + 9y - 4z = 49 \end{pmatrix}$.

Solution   The augmented matrix of this system is

$$\begin{bmatrix} 1 & 2 & -3 & \vdots & 15 \\ -2 & -3 & 1 & \vdots & -15 \\ 4 & 9 & -4 & \vdots & 49 \end{bmatrix}.$$

Multiplying row 1 by 2 and adding this result to row 2 produces a new row 2. Likewise, multiplying row 1 by -4 and adding this result to row 3 produces a new row 3.

$$\begin{bmatrix} 1 & 2 & -3 & \vdots & 15 \\ 0 & 1 & -5 & \vdots & 15 \\ 0 & 1 & 8 & \vdots & -11 \end{bmatrix}$$

Now multiplying row 2 by -2 and adding to row 1 produces a new row 1. Also, multiplying row 2 by -1 and adding the result to row 3 produces a new row 3.

$$\begin{bmatrix} 1 & 0 & 7 & | & -15 \\ 0 & 1 & -5 & | & 15 \\ 0 & 0 & 13 & | & -26 \end{bmatrix}$$

Now let's multiply row 3 by $\dfrac{1}{13}$.

$$\begin{bmatrix} 1 & 0 & 7 & | & -15 \\ 0 & 1 & -5 & | & 15 \\ 0 & 0 & 1 & | & -2 \end{bmatrix}$$

Finally, we can multiply row 3 by -7 and add the result to row 1 to produce a new row 1, and multiply row 3 by 5 and add the result to row 2 for a new row 2.

$$\begin{bmatrix} 1 & 0 & 0 & | & -1 \\ 0 & 1 & 0 & | & 5 \\ 0 & 0 & 1 & | & -2 \end{bmatrix}$$

From this last matrix we can see that the solution set of the original system is $\{(-1, 5, -2)\}$.   ●

Example 3   Solve the system $\begin{pmatrix} 2x + 4y - 5z = 37 \\ x + 3y - 4z = 29 \\ 5x - y + 3z = -20 \end{pmatrix}$.

Solution   The augmented matrix

$$\begin{bmatrix} 2 & 4 & -5 & | & 37 \\ 1 & 3 & -4 & | & 29 \\ 5 & -1 & 3 & | & -20 \end{bmatrix}$$

does not have a 1 in the upper left-hand corner, but this problem can be remedied by exchanging rows 1 and 2.

$$\begin{bmatrix} 1 & 3 & -4 & | & 29 \\ 2 & 4 & -5 & | & 37 \\ 5 & -1 & 3 & | & -20 \end{bmatrix}$$

Now we can get zeros in the first column beneath the 1 by multiplying -2 times row 1 and adding the result to row 2 and multiplying -5 times row 1 and adding the result to row 3.

$$\begin{bmatrix} 1 & 3 & -4 & | & 29 \\ 0 & -2 & 3 & | & -21 \\ 0 & -16 & 23 & | & -165 \end{bmatrix}$$

Next, we can get a 1 as the first nonzero entry of the second row by multiplying the second row by $-\frac{1}{2}$.

$$\begin{bmatrix} 1 & 3 & -4 & \vline & 29 \\ 0 & 1 & -\frac{3}{2} & \vline & \frac{21}{2} \\ 0 & -16 & 23 & \vline & -165 \end{bmatrix}$$

Now we can get zeros above and below the 1 in the second column by multiplying -3 times row 2 and adding the result to row 1 and multiplying 16 times row 2 and adding the result to row 3.

$$\begin{bmatrix} 1 & 0 & \frac{1}{2} & \vline & -\frac{5}{2} \\ 0 & 1 & -\frac{3}{2} & \vline & \frac{21}{2} \\ 0 & 0 & -1 & \vline & 3 \end{bmatrix}$$

Next, we can get a 1 as the first nonzero entry of the third row by multiplying the third row by -1.

$$\begin{bmatrix} 1 & 0 & \frac{1}{2} & \vline & -\frac{5}{2} \\ 0 & 1 & -\frac{3}{2} & \vline & \frac{21}{2} \\ 0 & 0 & 1 & \vline & -3 \end{bmatrix}$$

Finally, we can get zeros above the 1 in the third column by multiplying $-\frac{1}{2}$ times row 3 and adding the result to row 1 and multiplying $\frac{3}{2}$ times row 3 and adding the result to row 2.

$$\begin{bmatrix} 1 & 0 & 0 & \vline & -1 \\ 0 & 1 & 0 & \vline & 6 \\ 0 & 0 & 1 & \vline & -3 \end{bmatrix}$$

From the last matrix, we see that the solution set of the original system is $\{(-1, 6, -3)\}$. ●

Example 3 illustrates that even though the process of changing to reduced echelon form can be systematically described, it can involve some rather messy calculations. However, with the aid of a computer, such calculations are not troublesome. For our purposes in this text, we have chosen as examples and problems those systems that minimize messy calculations. This should allow you to concentrate on developing an understanding of the procedures being discussed.

We want to call your attention to another issue in the solution of **Example 3.** Consider the matrix

$$\begin{bmatrix} 1 & 3 & -4 & \vdots & 29 \\ 0 & 1 & -\dfrac{3}{2} & \vdots & \dfrac{21}{2} \\ 0 & -16 & 23 & \vdots & -165 \end{bmatrix},$$

which is obtained about halfway through the solution. At this step it seems evident that the calculations are getting a little messy. Therefore, instead of continuing toward the reduced echelon form, let's multiply row 2 by 16 and add the result to row 3 to produce a new row 3.

$$\begin{bmatrix} 1 & 3 & -4 & \vdots & 29 \\ 0 & 1 & -\dfrac{3}{2} & \vdots & \dfrac{21}{2} \\ 0 & 0 & -1 & \vdots & 3 \end{bmatrix}$$

We now have a matrix that represents a system in triangular form. The last row determines the value of z, and then the values for y and x can be determined by back-substitution. For some problems it may be somewhat easier to stop at the triangular form rather than continue on to reduced echelon form.

Finally, let's consider two examples to illustrate what happens when we use the matrix approach on inconsistent and dependent systems.

Example 4

Solve the system $\begin{pmatrix} x - 2y + 3z = 3 \\ 5x - 9y + 4z = 2 \\ 2x - 4y + 6z = -1 \end{pmatrix}.$

Solution

The augmented matrix of the system is

$$\begin{bmatrix} 1 & -2 & 3 & \vdots & 3 \\ 5 & -9 & 4 & \vdots & 2 \\ 2 & -4 & 6 & \vdots & -1 \end{bmatrix}.$$

We can get zeros below the 1 in the first column by multiplying -5 times row 1 and adding the result to row 2 and multiplying -2 times row 1 and adding the result to row 3.

$$\begin{bmatrix} 1 & -2 & 3 & \vdots & 3 \\ 0 & 1 & -11 & \vdots & -13 \\ 0 & 0 & 0 & \vdots & -7 \end{bmatrix}$$

At this point we can stop, because the bottom row of the matrix represents the statement $0(x) + 0(y) + 0(z) = -7$, which is obviously a false statement for all values of x, y, and z. Thus, the original system is *inconsistent*; its solution set is Ø. ●

Example 5   Solve the system $\begin{pmatrix} x + 2y + 2z = 9 \\ x + 3y - 4z = 5 \\ 2x + 5y - 2z = 14 \end{pmatrix}$.

Solution   The augmented matrix of the system is

$$\begin{bmatrix} 1 & 2 & 2 & | & 9 \\ 1 & 3 & -4 & | & 5 \\ 2 & 5 & -2 & | & 14 \end{bmatrix}.$$

We can get zeros in the first column below the 1 in the upper left-hand corner by multiplying -1 times row 1 and adding the result to row 2 and multiplying -2 times row 1 and adding the result to row 3.

$$\begin{bmatrix} 1 & 2 & 2 & | & 9 \\ 0 & 1 & -6 & | & -4 \\ 0 & 1 & -6 & | & -4 \end{bmatrix}$$

Now we can get zeros in the second column above and below the 1 in the second row by multiplying -2 times row 2 and adding the result to row 1 and multiplying -1 times row 2 and adding the result to row 3.

$$\begin{bmatrix} 1 & 0 & 14 & | & 17 \\ 0 & 1 & -6 & | & -4 \\ 0 & 0 & 0 & | & 0 \end{bmatrix}$$

The bottom row of zeros represents the statement $0(x) + 0(y) + 0(z) = 0$, which is true for all values of x, y, and z. Thus, the original system is a *dependent system*, and its solution set consists of infinitely many ordered triples.

These ordered triples can be represented as follows. The second row of the previous matrix represents the statement $y - 6z = -4$, which can be written as $y = 6z - 4$. The top row represents the statement $x + 14z = 17$, which can be written as $x = -14z + 17$. Therefore, if we let $z = k$, where k is any real number, the solution set of infinitely many ordered triples can be represented by $\{(-14k + 17, 6k - 4, k)\}$. Specific solutions can be generated by letting k take on specific values. For example, if $k = 2$, then $6k - 4$ becomes $6(2) - 4 = 8$ and $-14k + 17$ becomes $-14(2) + 17 = -11$. Thus, the ordered triple $(-11, 8, 2)$ is a member of the solution set.   ●

Problem Set 12.2

For Problems 1–10, indicate whether each matrix is in reduced echelon form.

1. $\begin{bmatrix} 1 & 0 & | & -4 \\ 0 & 1 & | & 14 \end{bmatrix}$

2. $\begin{bmatrix} 1 & 2 & | & 8 \\ 0 & 0 & | & 0 \end{bmatrix}$

3. $\begin{bmatrix} 1 & 0 & 2 & | & 5 \\ 0 & 1 & 3 & | & 7 \\ 0 & 0 & 0 & | & 0 \end{bmatrix}$

4. $\begin{bmatrix} 1 & 0 & 0 & | & 5 \\ 0 & 3 & 0 & | & 8 \\ 0 & 0 & 1 & | & -11 \end{bmatrix}$

5. $\begin{bmatrix} 1 & 0 & 0 & | & 17 \\ 0 & 0 & 0 & | & 0 \\ 0 & 1 & 0 & | & -14 \end{bmatrix}$
6. $\begin{bmatrix} 1 & 0 & 0 & | & -7 \\ 0 & 1 & 0 & | & 0 \\ 0 & 0 & 1 & | & 9 \end{bmatrix}$

7. $\begin{bmatrix} 1 & 1 & 0 & | & -3 \\ 0 & 1 & 2 & | & 5 \\ 0 & 0 & 1 & | & 7 \end{bmatrix}$
8. $\begin{bmatrix} 1 & 0 & 3 & | & 8 \\ 0 & 1 & 2 & | & -6 \\ 0 & 0 & 0 & | & 0 \end{bmatrix}$

9. $\begin{bmatrix} 1 & 0 & 0 & 3 & | & 4 \\ 0 & 1 & 0 & 5 & | & -3 \\ 0 & 0 & 1 & -1 & | & 7 \\ 0 & 0 & 0 & 0 & | & 0 \end{bmatrix}$
10. $\begin{bmatrix} 1 & 0 & 0 & 0 & | & 2 \\ 0 & 0 & 1 & 0 & | & 4 \\ 0 & 1 & 0 & 0 & | & -3 \\ 0 & 0 & 0 & 1 & | & 9 \end{bmatrix}$

For Problems 11–30, use a matrix approach as presented in this section to solve each of the following systems.

11. $\left(\begin{matrix} x - 3y = 14 \\ 3x + 2y = -13 \end{matrix} \right)$
12. $\left(\begin{matrix} x + 5y = -18 \\ -2x + 3y = -16 \end{matrix} \right)$

13. $\left(\begin{matrix} 3x - 4y = 33 \\ x + 7y = -39 \end{matrix} \right)$
14. $\left(\begin{matrix} 2x + 7y = -55 \\ x - 4y = 25 \end{matrix} \right)$

15. $\left(\begin{matrix} x - 6y = -2 \\ 2x - 12y = 5 \end{matrix} \right)$
16. $\left(\begin{matrix} 2x - 3y = -12 \\ 3x + 2y = 8 \end{matrix} \right)$

17. $\left(\begin{matrix} 3x - 5y = 39 \\ 2x + 7y = -67 \end{matrix} \right)$
18. $\left(\begin{matrix} 3x + 9y = -1 \\ x + 3y = 10 \end{matrix} \right)$

19. $\left(\begin{matrix} x - 2y - 3z = -6 \\ 3x - 5y - z = 4 \\ 2x + y + 2z = 2 \end{matrix} \right)$
20. $\left(\begin{matrix} x + 3y - 4z = 13 \\ 2x + 7y - 3z = 11 \\ -2x - y + 2z = -8 \end{matrix} \right)$

21. $\left(\begin{matrix} -2x - 5y + 3z = 11 \\ x + 3y - 3z = -12 \\ 3x - 2y + 5z = 31 \end{matrix} \right)$
22. $\left(\begin{matrix} -3x + 2y + z = 17 \\ x - y + 5z = -2 \\ 4x - 5y - 3z = -36 \end{matrix} \right)$

23. $\left(\begin{matrix} x - 3y - z = 2 \\ 3x + y - 4z = -18 \\ -2x + 5y + 3z = 2 \end{matrix} \right)$
24. $\left(\begin{matrix} x - 4y + 3z = 16 \\ 2x + 3y - 4z = -22 \\ -3x + 11y - z = -36 \end{matrix} \right)$

25. $\left(\begin{matrix} x - y + 2z = 1 \\ -3x + 4y - z = 4 \\ -x + 2y + 3z = 6 \end{matrix} \right)$
26. $\left(\begin{matrix} x + 2y - 5z = -1 \\ 2x + 3y - 2z = 2 \\ 3x + 5y - 7z = 4 \end{matrix} \right)$

27. $\left(\begin{matrix} -2x + y + 5z = -5 \\ 3x + 8y - z = -34 \\ x + 2y + z = -12 \end{matrix} \right)$
28. $\left(\begin{matrix} 4x - 10y + 3z = -19 \\ 2x + 5y - z = -7 \\ x - 3y - 2z = -2 \end{matrix} \right)$

29. $\left(\begin{matrix} 2x + 3y - z = 7 \\ 3x + 4y + 5z = -2 \\ 5x + y + 3z = 13 \end{matrix} \right)$
30. $\left(\begin{matrix} 4x + 3y - z = 0 \\ 3x + 2y + 5z = 6 \\ 5x - y - 3z = 3 \end{matrix} \right)$

Subscript notation is frequently used in work with large systems of equations. Use a matrix approach to solve each of the systems in Problems 31–34. Express the solutions as 4-tuples of the form (x_1, x_2, x_3, x_4).

31. $\begin{pmatrix} x_1 - 3x_2 - 2x_3 + x_4 = -3 \\ -2x_1 + 7x_2 + x_3 - 2x_4 = -1 \\ 3x_1 - 7x_2 - 3x_3 + 3x_4 = -5 \\ 5x_1 + x_2 + 4x_3 - 2x_4 = 18 \end{pmatrix}$

32. $\begin{pmatrix} x_1 - 2x_2 + 2x_3 - x_4 = -2 \\ -3x_1 + 5x_2 - x_3 - 3x_4 = 2 \\ 2x_1 + 3x_2 + 3x_3 + 5x_4 = -9 \\ 4x_1 - x_2 - x_3 - 2x_4 = 8 \end{pmatrix}$

33. $\begin{pmatrix} x_1 + 3x_2 - x_3 + 2x_4 = -2 \\ 2x_1 + 7x_2 + 2x_3 - x_4 = 19 \\ -3x_1 - 8x_2 + 3x_3 + x_4 = -7 \\ 4x_1 + 11x_2 - 2x_3 - 3x_4 = 19 \end{pmatrix}$

34. $\begin{pmatrix} x_1 + 2x_2 - 3x_3 + x_4 = -2 \\ -2x_1 - 3x_2 + x_3 - x_4 = 5 \\ 4x_1 + 9x_2 - 2x_3 - 2x_4 = -28 \\ -5x_1 - 9x_2 + 2x_3 - 3x_4 = 14 \end{pmatrix}$

Each matrix in Problems 35–42 is the reduced echelon matrix for a system with variables x_1, x_2, x_3, and x_4. Find the solution set of each system.

35. $\begin{bmatrix} 1 & 0 & 0 & 0 & | & -2 \\ 0 & 1 & 0 & 0 & | & 4 \\ 0 & 0 & 1 & 0 & | & -3 \\ 0 & 0 & 0 & 1 & | & 0 \end{bmatrix}$

36. $\begin{bmatrix} 1 & 0 & 0 & 0 & | & 0 \\ 0 & 1 & 0 & 0 & | & -5 \\ 0 & 0 & 1 & 0 & | & 0 \\ 0 & 0 & 0 & 1 & | & 4 \end{bmatrix}$

37. $\begin{bmatrix} 1 & 0 & 0 & 0 & | & -8 \\ 0 & 1 & 0 & 0 & | & 5 \\ 0 & 0 & 1 & 0 & | & -2 \\ 0 & 0 & 0 & 0 & | & 1 \end{bmatrix}$

38. $\begin{bmatrix} 1 & 0 & 0 & 0 & | & 2 \\ 0 & 1 & 0 & 2 & | & -3 \\ 0 & 0 & 1 & 3 & | & 4 \\ 0 & 0 & 0 & 0 & | & 0 \end{bmatrix}$

39. $\begin{bmatrix} 1 & 0 & 0 & 3 & | & 5 \\ 0 & 1 & 0 & 0 & | & -1 \\ 0 & 0 & 1 & 4 & | & 2 \\ 0 & 0 & 0 & 0 & | & 0 \end{bmatrix}$

40. $\begin{bmatrix} 1 & 3 & 0 & 0 & | & 0 \\ 0 & 0 & 1 & 0 & | & 0 \\ 0 & 0 & 0 & 0 & | & 1 \\ 0 & 0 & 0 & 0 & | & 0 \end{bmatrix}$

41. $\begin{bmatrix} 1 & 3 & 0 & 0 & | & 9 \\ 0 & 0 & 1 & 0 & | & 2 \\ 0 & 0 & 0 & 1 & | & -3 \\ 0 & 0 & 0 & 0 & | & 0 \end{bmatrix}$

42. $\begin{bmatrix} 1 & 0 & 0 & 0 & | & 7 \\ 0 & 1 & 0 & 0 & | & -3 \\ 0 & 0 & 1 & -2 & | & 5 \\ 0 & 0 & 0 & 0 & | & 0 \end{bmatrix}$

Miscellaneous Problems

For Problems 43–48, change the augmented matrix of each system to reduced echelon form and then indicate the solutions of the system.

43. $\begin{pmatrix} x - 2y + 3z = 4 \\ 3x - 5y - z = 7 \end{pmatrix}$

44. $\begin{pmatrix} x + 3y - 2z = -1 \\ -2x - 5y + 7z = 4 \end{pmatrix}$

45. $\begin{pmatrix} 2x - 4y + 3z = 8 \\ 3x + 5y - z = 7 \end{pmatrix}$

46. $\begin{pmatrix} 3x + 6y - z = 9 \\ 2x - 3y + 4z = 1 \end{pmatrix}$

47. $\begin{pmatrix} x - 2y + 4z = 9 \\ 2x - 4y + 8z = 3 \end{pmatrix}$

48. $\begin{pmatrix} x + y - 2z = -1 \\ 3x + 3y - 6z = -3 \end{pmatrix}$

12.3 Determinants and Cramer's Rule

A **square matrix** is one that has the same number of rows as columns. Associated with each square matrix having real number entries is a real number called the **determinant** of the matrix. For a 2-by-2 matrix

$$\begin{bmatrix} a_1 & b_1 \\ a_2 & b_2 \end{bmatrix},$$

the determinant is written as

$$\begin{vmatrix} a_1 & b_1 \\ a_2 & b_2 \end{vmatrix}$$

and defined by

$$\begin{vmatrix} a_1 & b_1 \\ a_2 & b_2 \end{vmatrix} = a_1 b_2 - a_2 b_1. \tag{1}$$

Notice that a determinant is simply a number, the determinant notation used on the left side of equation (1) is a way of expressing the number on the right side. (Remember that a matrix has no such numerical value; it is simply a rectangular array of numbers.)

Example 1   Find the determinant of the matrix $\begin{bmatrix} 3 & -2 \\ 5 & 8 \end{bmatrix}$.

Solution   In this case, $a_1 = 3$, $b_1 = -2$, $a_2 = 5$, and $b_2 = 8$. Thus, we have

$$\begin{vmatrix} 3 & -2 \\ 5 & 8 \end{vmatrix} = 3(8) - 5(-2) = 24 + 10 = 34. \qquad \bullet$$

Finding the determinant of a square matrix is commonly called *evaluating the determinant*, and the matrix notation is sometimes omitted.

Example 2   Evaluate $\begin{vmatrix} -3 & 5 \\ 1 & 2 \end{vmatrix}$.

Solution $\begin{vmatrix} -3 & 5 \\ 1 & 2 \end{vmatrix} = -3(2) - 1(5) = -11. \qquad \bullet$

Cramer's Rule

Determinants provide the basis for another method of solving linear systems. Consider the system

$$\begin{pmatrix} a_1 x + b_1 y = c_1 \\ a_2 x + b_2 y = c_2 \end{pmatrix}.$$

(1)

(2)

We will solve this system by using the elimination method. To solve for x, we can multiply equation (1) by b_2 and equation (2) by $-b_1$ and then add.

$$a_1 b_2 x + b_1 b_2 y = c_1 b_2$$

$$\underline{-a_2 b_1 x - b_1 b_2 y = -c_2 b_1}$$

$$a_1 b_2 x - a_2 b_1 x = c_1 b_2 - c_2 b_1$$

$$(a_1 b_2 - a_2 b_1)x = c_1 b_2 - c_2 b_1$$

$$x = \frac{c_1 b_2 - c_2 b_1}{a_1 b_2 - a_2 b_1} \qquad \text{(if } a_1 b_2 - a_2 b_1 \neq 0\text{)}$$

To solve for y, we can multiply equation (1) by $-a_2$ and equation (2) by a_1 and add.

$$-a_1 a_2 x - a_2 b_1 y = -a_2 c_1$$

$$\underline{a_1 a_2 x + a_1 b_2 y = a_1 c_2}$$

$$a_1 b_2 y - a_2 b_1 y = a_1 c_2 - a_2 c_1$$

$$(a_1 b_2 - a_2 b_1)y = a_1 c_2 - a_2 c_1$$

$$y = \frac{a_1 c_2 - a_2 c_1}{a_1 b_2 - a_2 b_1} \qquad \text{(if } a_1 b_2 - a_2 b_1 \neq 0\text{)}$$

The solutions for x and y can be expressed in determinant form as follows.

$$x = \frac{c_1 b_2 - c_2 b_1}{a_1 b_2 - a_2 b_1} = \frac{\begin{vmatrix} c_1 & b_1 \\ c_2 & b_2 \end{vmatrix}}{\begin{vmatrix} a_1 & b_1 \\ a_2 & b_2 \end{vmatrix}}$$

$$y = \frac{a_1 c_2 - a_2 c_1}{a_1 b_2 - a_2 b_1} = \frac{\begin{vmatrix} a_1 & c_1 \\ a_2 & c_2 \end{vmatrix}}{\begin{vmatrix} a_1 & b_1 \\ a_2 & b_2 \end{vmatrix}}$$

For convenience, we will denote the three determinants in the solution as

$$\begin{vmatrix} a_1 & b_1 \\ a_2 & b_2 \end{vmatrix} = D,$$

$$\begin{vmatrix} c_1 & b_1 \\ c_2 & b_2 \end{vmatrix} = D_x,$$

$$\begin{vmatrix} a_1 & c_1 \\ a_2 & c_2 \end{vmatrix} = D_y.$$

Notice that the elements of D are the coefficients of the variables in the given system. In D_x, the elements are obtained by replacing the coefficients of x by the respective constants. In D_y, the coefficients of y are replaced by the respective constants.

This method of using determinants to solve a system of two linear equations in two variables, called **Cramer's rule**, can be stated as follows:

Cramer's Rule

Given the system

$$\begin{pmatrix} a_1 x + b_1 y = c_1 \\ a_2 x + b_2 y = c_2 \end{pmatrix} \quad \text{with } a_1 b_2 - a_2 b_1 \neq 0,$$

then

$$x = \frac{\begin{vmatrix} c_1 & b_1 \\ c_2 & b_2 \end{vmatrix}}{\begin{vmatrix} a_1 & b_1 \\ a_2 & b_2 \end{vmatrix}} = \frac{D_x}{D} \quad \text{and} \quad y = \frac{\begin{vmatrix} a_1 & c_1 \\ a_2 & c_2 \end{vmatrix}}{\begin{vmatrix} a_1 & b_1 \\ a_2 & b_2 \end{vmatrix}} = \frac{D_y}{D}.$$

Let's illustrate the use of Cramer's rule to solve some systems.

Example 3

Solve the system $\begin{pmatrix} x + 2y = 11 \\ 2x - y = 2 \end{pmatrix}$.

Solution

Let's find D, D_x, and D_y.

$$D = \begin{vmatrix} 1 & 2 \\ 2 & -1 \end{vmatrix} = -1 - 4 = -5$$

$$D_x = \begin{vmatrix} 11 & 2 \\ 2 & -1 \end{vmatrix} = -11 - 4 = -15$$

$$D_y = \begin{vmatrix} 1 & 11 \\ 2 & 2 \end{vmatrix} = 2 - 22 = -20$$

Thus, we have

$$x = \frac{D_x}{D} = \frac{-15}{-5} = 3,$$

$$y = \frac{D_y}{D} = \frac{-20}{-5} = 4.$$

The solution set is $\{(3, 4)\}$. This can be verified, as always, by substituting back into the original equations.   ●

Remark   Notice that Cramer's rule has a restriction in that $a_1 b_2 - a_2 b_1 \neq 0$; that is, $D \neq 0$. Thus, it is a good idea to find D first. If $D = 0$, Cramer's rule does not apply and you can revert to one of the other methods to determine whether the solution set is empty or has infinitely many solutions.

Example 4 Solve the system $\begin{pmatrix} 2x - 3y = -8 \\ 3x + 5y = 7 \end{pmatrix}$.

Solution
$$D = \begin{vmatrix} 2 & -3 \\ 3 & 5 \end{vmatrix} = 10 - (-9) = 19$$

$$D_x = \begin{vmatrix} -8 & -3 \\ 7 & 5 \end{vmatrix} = -40 - (-21) = -19$$

$$D_y = \begin{vmatrix} 2 & -8 \\ 3 & 7 \end{vmatrix} = 14 - (-24) = 38$$

Thus, we obtain

$$x = \frac{D_x}{D} = \frac{-19}{19} = -1,$$

$$y = \frac{D_y}{D} = \frac{38}{19} = 2.$$

The solution set is $\{(-1, 2)\}$.

Example 5 Solve the system $\begin{pmatrix} y = -2x - 2 \\ 4x - 5y = 17 \end{pmatrix}$.

Solution First, we must change the form of the first equation so that the system fits the form given in Cramer's rule. The equation $y = -2x - 2$ can be written as $2x + y = -2$. The system now becomes

$$\begin{pmatrix} 2x + y = -2 \\ 4x - 5y = 17 \end{pmatrix},$$

and we can proceed as before.

$$D = \begin{vmatrix} 2 & 1 \\ 4 & -5 \end{vmatrix} = -10 - 4 = -14$$

$$D_x = \begin{vmatrix} -2 & 1 \\ 17 & -5 \end{vmatrix} = 10 - 17 = -7$$

$$D_y = \begin{vmatrix} 2 & -2 \\ 4 & 17 \end{vmatrix} = 34 - (-8) = 42$$

Thus, the solutions are

$$x = \frac{D_x}{D} = \frac{-7}{-14} = \frac{1}{2},$$

$$y = \frac{D_y}{D} = \frac{42}{-14} = -3.$$

The solution set is $\left\{\left(\frac{1}{2}, -3\right)\right\}$.

Example 6   Solve the system $\begin{pmatrix} \dfrac{1}{2}x + \dfrac{2}{3}y = -4 \\ \dfrac{1}{4}x - \dfrac{3}{2}y = 20 \end{pmatrix}$.

Solution

$$D = \begin{vmatrix} \dfrac{1}{2} & \dfrac{2}{3} \\ \dfrac{1}{4} & -\dfrac{3}{2} \end{vmatrix} = \dfrac{1}{2}\left(-\dfrac{3}{2}\right) - \dfrac{1}{4}\left(\dfrac{2}{3}\right)$$

$$= -\dfrac{3}{4} - \dfrac{1}{6} = -\dfrac{11}{12}$$

$$D_x = \begin{vmatrix} -4 & \dfrac{2}{3} \\ 20 & -\dfrac{3}{2} \end{vmatrix} = -4\left(-\dfrac{3}{2}\right) - 20\left(\dfrac{2}{3}\right)$$

$$= 6 - \dfrac{40}{3} = -\dfrac{22}{3}$$

$$D_y = \begin{vmatrix} \dfrac{1}{2} & -4 \\ \dfrac{1}{4} & 20 \end{vmatrix} = \dfrac{1}{2}(20) - \dfrac{1}{4}(-4) = 11$$

Thus, the solutions are

$$x = \dfrac{D_x}{D} = \dfrac{-\dfrac{22}{3}}{-\dfrac{11}{12}} = \left(-\dfrac{22}{3}\right)\left(-\dfrac{12}{11}\right) = 8,$$

$$y = \dfrac{D_y}{D} = \dfrac{11}{-\dfrac{11}{12}} = (11)\left(-\dfrac{12}{11}\right) = -12.$$

The solution set is $\{(8, -12)\}$.

Problem Set 12.3

Evaluate each of the following determinants.

1. $\begin{vmatrix} 5 & 6 \\ 2 & 4 \end{vmatrix}$ **2.** $\begin{vmatrix} 4 & 8 \\ 3 & 4 \end{vmatrix}$ **3.** $\begin{vmatrix} -3 & 1 \\ 6 & 4 \end{vmatrix}$ **4.** $\begin{vmatrix} 6 & 12 \\ -1 & -3 \end{vmatrix}$

5. $\begin{vmatrix} -8 & 6 \\ 4 & -3 \end{vmatrix}$ **6.** $\begin{vmatrix} 1 & -2 \\ -9 & 7 \end{vmatrix}$ **7.** $\begin{vmatrix} -3 & -1 \\ -2 & -4 \end{vmatrix}$ **8.** $\begin{vmatrix} 1 & 1 \\ 2 & 3 \\ -3 & 4 \end{vmatrix}$

9. $\begin{vmatrix} \dfrac{1}{2} & \dfrac{2}{3} \\ \dfrac{3}{4} & -\dfrac{1}{3} \end{vmatrix}$ **10.** $\begin{vmatrix} \dfrac{2}{3} & \dfrac{1}{5} \\ -\dfrac{1}{4} & \dfrac{3}{2} \end{vmatrix}$

Use Cramer's rule to find the solution set for each of the following systems.

11. $\begin{pmatrix} x + y = 10 \\ x - y = 4 \end{pmatrix}$ **12.** $\begin{pmatrix} 2x - y = -2 \\ 3x + y = 22 \end{pmatrix}$ **13.** $\begin{pmatrix} -x + 2y = 10 \\ 3x - y = -10 \end{pmatrix}$

14. $\begin{pmatrix} 3x + 2y = -21 \\ 4x - 3y = 6 \end{pmatrix}$ **15.** $\begin{pmatrix} 2x + 5y = -17 \\ 5x - 4y = 40 \end{pmatrix}$ **16.** $\begin{pmatrix} 5x - 4y = 14 \\ -x + 2y = -4 \end{pmatrix}$

17. $\begin{pmatrix} 3x - 2y = 1 \\ 6x - 4y = -1 \end{pmatrix}$ **18.** $\begin{pmatrix} 7x - 5y = 0 \\ 3x + 4y = 0 \end{pmatrix}$ **19.** $\begin{pmatrix} 5x - y = 0 \\ 4x + y = 9 \end{pmatrix}$

20. $\begin{pmatrix} x - 2y = 4 \\ 2x - 4y = 8 \end{pmatrix}$ **21.** $\begin{pmatrix} x - 2y = -1 \\ x + 6y = 5 \end{pmatrix}$ **22.** $\begin{pmatrix} -4x + 3y = 3 \\ 4x - 6y = -5 \end{pmatrix}$

23. $\begin{pmatrix} 6x - 5y = 1 \\ 4x + 7y = 2 \end{pmatrix}$ **24.** $\begin{pmatrix} y = 3x + 5 \\ y = 6x + 6 \end{pmatrix}$ **25.** $\begin{pmatrix} 7x + 2y = -1 \\ y = -x + 2 \end{pmatrix}$

26. $\begin{pmatrix} 9x - y = -2 \\ 8x + y = 4 \end{pmatrix}$ **27.** $\begin{pmatrix} \dfrac{1}{2}x + \dfrac{2}{3}y = -6 \\ \dfrac{1}{4}x - \dfrac{1}{3}y = -1 \end{pmatrix}$ **28.** $\begin{pmatrix} -\dfrac{2}{3}x + \dfrac{1}{2}y = -7 \\ \dfrac{1}{3}x - \dfrac{3}{2}y = 6 \end{pmatrix}$

29. $\begin{pmatrix} 5x - 3y = 2 \\ y = 4 \end{pmatrix}$ **30.** $\begin{pmatrix} 2x + 7y = -1 \\ x = 2 \end{pmatrix}$

31. Verify each of the following. The variables represent real numbers.

(a) $\begin{vmatrix} a & b \\ a & b \end{vmatrix} = 0$ (b) $\begin{vmatrix} a & a \\ b & b \end{vmatrix} = 0$

(c) $\begin{vmatrix} a & b \\ c & d \end{vmatrix} = -\begin{vmatrix} b & a \\ d & c \end{vmatrix}$ (d) $\begin{vmatrix} a & b \\ c & d \end{vmatrix} = -\begin{vmatrix} c & d \\ a & b \end{vmatrix}$

(e) $k\begin{vmatrix} a & b \\ c & d \end{vmatrix} = \begin{vmatrix} ka & b \\ kc & d \end{vmatrix}$ (f) $k\begin{vmatrix} a & b \\ c & d \end{vmatrix} = \begin{vmatrix} ka & kb \\ c & d \end{vmatrix}$

Solve each of the following problems using a system of two equations and two unknowns. Use Cramer's rule to find the solution sets of the systems.

32. At a local confectionery, 7 pounds of cashews and 5 pounds of Spanish peanuts cost $88, and 3 pounds of cashews and 2 pounds of Spanish peanuts cost $37. Find the price per pound for cashews and for Spanish peanuts.

33. We bought 2 cartons of pop and 4 pounds of candy for $12. The next day we bought 3 cartons of pop and 2 pounds of candy for $9. Find the price of a carton of pop and the price of a pound of candy.

34. A mail-order company charges a fixed fee for shipping merchandise weighing 1 pound or less, plus an additional fee for each pound over 1 pound. If the shipping charge for 5 pounds is $2.40 and for 12 pounds is $3.10, find the fixed fee and the additional fee.

35. The sum of two numbers is 23, and the difference of the two numbers is 5. Find the numbers.

36. The sum of two numbers is 19. The larger number is 1 larger than twice the smaller number. Find the numbers.

37. A two-digit number is 4 times the sum of its digits. If the digits are interchanged, the new number is 36 larger than the original number. Find the original number.

38. A woman invested a sum of money at 8% and another amount at 9%. The total yearly interest from both investments is $101. If she interchanged her investments, the total interest would be $103. How much does she have invested at each rate?

12.4 3-by-3 Determinants and Cramer's Rule

In this section we will extend the concept of a determinant to include 3-by-3 determinants, and then we will also extend the use of determinants to solve systems of three linear equations in three variables.

For a 3-by-3 matrix

$$\begin{bmatrix} a_1 & b_1 & c_1 \\ a_2 & b_2 & c_2 \\ a_3 & b_3 & c_3 \end{bmatrix},$$

the determinant is written as

$$\begin{vmatrix} a_1 & b_1 & c_1 \\ a_2 & b_2 & c_2 \\ a_3 & b_3 & c_3 \end{vmatrix}$$

and is defined by

$$\begin{vmatrix} a_1 & b_1 & c_1 \\ a_2 & b_2 & c_2 \\ a_3 & b_3 & c_3 \end{vmatrix} = a_1 b_2 c_3 + b_1 c_2 a_3 + c_1 a_2 b_3 - a_3 b_2 c_1 - b_3 c_2 a_1 - c_3 a_2 b_1. \quad (1)$$

It is evident that the definition given by equation (1) is a bit too complicated to be very useful in practice. Fortunately, a method called **expansion of a determinant by minors** can be used to calculate such a determinant. The **minor** of an element in a determinant is the determinant that remains after the row and column in which the element appears are deleted. For example, consider the determinant of equation (1).

The minor of a_1 is $\begin{vmatrix} b_2 & c_2 \\ b_3 & c_3 \end{vmatrix}$,

the minor of a_2 is $\begin{vmatrix} b_1 & c_1 \\ b_3 & c_3 \end{vmatrix}$,

the minor of a_3 is $\begin{vmatrix} b_1 & c_1 \\ b_2 & c_2 \end{vmatrix}$, etc.

Now let's consider the terms of the right side of equation (1) in pairs, and show the tie-up with minors.

$$a_1 b_2 c_3 - b_3 c_2 a_1 = a_1 \underbrace{(b_2 c_3 - b_3 c_2)}$$

$$\downarrow$$

$$\begin{vmatrix} b_2 & c_2 \\ b_3 & c_3 \end{vmatrix}$$

$$c_1 a_2 b_3 - c_3 a_2 b_1 = -(c_3 a_2 b_1 - c_1 a_2 b_3)$$
$$= -a_2 \underbrace{(b_1 c_3 - b_3 c_1)}$$

$$\downarrow$$

$$\begin{vmatrix} b_1 & c_1 \\ b_3 & c_3 \end{vmatrix}$$

$$b_1 c_2 a_3 - a_3 b_2 c_1 = a_3 \underbrace{(b_1 c_2 - b_2 c_1)}$$

$$\downarrow$$

$$\begin{vmatrix} b_1 & c_1 \\ b_2 & c_2 \end{vmatrix}$$

Therefore, we have

$$\begin{vmatrix} a_1 & b_1 & c_1 \\ a_2 & b_2 & c_2 \\ a_3 & b_3 & c_3 \end{vmatrix} = a_1 \begin{vmatrix} b_2 & c_2 \\ b_3 & c_3 \end{vmatrix} - a_2 \begin{vmatrix} b_1 & c_1 \\ b_3 & c_3 \end{vmatrix} + a_3 \begin{vmatrix} b_1 & c_1 \\ b_2 & c_2 \end{vmatrix}.$$

This process is called the **expansion of the determinant by minors about the first column.**

Example 1   Evaluate $\begin{vmatrix} 1 & 2 & -1 \\ 3 & 1 & 2 \\ 2 & 4 & 3 \end{vmatrix}$ by expanding by minors about the first column.

Solution $\begin{vmatrix} 1 & 2 & -1 \\ 3 & 1 & 2 \\ 2 & 4 & 3 \end{vmatrix} = 1 \begin{vmatrix} 1 & 2 \\ 4 & 3 \end{vmatrix} - 3 \begin{vmatrix} 2 & -1 \\ 4 & 3 \end{vmatrix} + 2 \begin{vmatrix} 2 & -1 \\ 1 & 2 \end{vmatrix}$

$$= 1(3 - 8) - 3[6 - (-4)] + 2[4 - (-1)]$$
$$= 1(-5) - 3(10) + 2(5)$$
$$= -5 - 30 + 10$$
$$= -25$$

It is possible to expand a determinant by minors about *any row* or *any column*. The following sign array is very useful for determining the signs of the terms in the expansion.

$$
\begin{array}{ccc}
+ & - & + \\
- & + & - \\
+ & - & +
\end{array}
$$

For example, let's expand the determinant in Example 1 by minors about the *second row*. The second row in the sign array is $-\ +\ -$. Therefore,

$$
\begin{vmatrix} 1 & 2 & -1 \\ 3 & 1 & 2 \\ 2 & 4 & 3 \end{vmatrix} = -3\begin{vmatrix} 2 & -1 \\ 4 & 3 \end{vmatrix} + 1\begin{vmatrix} 1 & -1 \\ 2 & 3 \end{vmatrix} - 2\begin{vmatrix} 1 & 2 \\ 2 & 4 \end{vmatrix}
$$

$$
= -3[6 - (-4)] + 1[3 - (-2)] - 2(4 - 4)
$$

$$
= -3(10) + 1(5) - 2(0)
$$

$$
= -25.
$$

Your decision as to which row or column to use for expanding a particular determinant by minors may depend on the numbers involved in the determinant. A row or column with one or more zeros is frequently a good choice, as the next example illustrates.

Example 2   Evaluate $\begin{vmatrix} 3 & -1 & 4 \\ 5 & 2 & 0 \\ -2 & 6 & 0 \end{vmatrix}$.

Solution   Since the third column has two zeros, we will expand about it.

$$
\begin{vmatrix} 3 & -1 & 4 \\ 5 & 2 & 0 \\ -2 & 6 & 0 \end{vmatrix} = 4\begin{vmatrix} 5 & 2 \\ -2 & 6 \end{vmatrix} - 0\begin{vmatrix} 3 & -1 \\ -2 & 6 \end{vmatrix} + 0\begin{vmatrix} 3 & -1 \\ 5 & 2 \end{vmatrix}
$$

$$
= 4[30 - (-4)] - 0 + 0
$$

$$
= 136
$$

(Because of the zeros, there is no need to evaluate the last two minors.)   ●

Remark   The *expansion-by-minors method* can be extended to determinants of size 4-by-4, 5-by-5, and so on. However, it should be obvious that it becomes increasingly more tedious with bigger determinants. Fortunately, the computer handles the calculation of such determinants by using a different technique.

Cramer's Rule Expanded

Without showing all of the details, we will simply state that Cramer's rule also applies to solving systems of three linear equations in three variables. It can be stated as follows.

Cramer's Rule Expanded

Given the system

$$\begin{cases} a_1 x + b_1 y + c_1 z = d_1 \\ a_2 x + b_2 y + c_2 z = d_2 \\ a_3 x + b_3 y + c_3 z = d_3 \end{cases}$$

with

$$D = \begin{vmatrix} a_1 & b_1 & c_1 \\ a_2 & b_2 & c_2 \\ a_3 & b_3 & c_3 \end{vmatrix} \neq 0, \qquad D_x = \begin{vmatrix} d_1 & b_1 & c_1 \\ d_2 & b_2 & c_2 \\ d_3 & b_3 & c_3 \end{vmatrix},$$

$$D_y = \begin{vmatrix} a_1 & d_1 & c_1 \\ a_2 & d_2 & c_2 \\ a_3 & d_3 & c_3 \end{vmatrix}, \qquad D_z = \begin{vmatrix} a_1 & b_1 & d_1 \\ a_2 & b_2 & d_2 \\ a_3 & b_3 & d_3 \end{vmatrix},$$

then

$$x = \frac{D_x}{D}, \quad y = \frac{D_y}{D}, \quad \text{and} \quad z = \frac{D_z}{D}.$$

Notice that the elements of D are the coefficients of the variables in the given system. Then D_x, D_y, and D_z are formed by replacing the elements in the x, y, or z column, respectively, by the constants of the system d_1, d_2, and d_3. Again, note the restriction $D \neq 0$. As before, if $D = 0$, then Cramer's rule does not apply and you can use the elimination method to determine whether the system has *no solution* or *infinitely many* solutions, so calculate D first.

Example 3   Use Cramer's rule to solve the system $\begin{cases} x - 2y + z = -4 \\ 2x + y - z = 5 \\ 3x + 2y + 4z = 3 \end{cases}$.

Solution   To find D, we expand about row 1.

$$D = \begin{vmatrix} 1 & -2 & 1 \\ 2 & 1 & -1 \\ 3 & 2 & 4 \end{vmatrix} = 1 \begin{vmatrix} 1 & -1 \\ 2 & 4 \end{vmatrix} - (-2) \begin{vmatrix} 2 & -1 \\ 3 & 4 \end{vmatrix} + 1 \begin{vmatrix} 2 & 1 \\ 3 & 2 \end{vmatrix}$$

$$= 1[4 - (-2)] + 2[8 - (-3)] + 1(4 - 3)$$
$$= 1(6) + 2(11) + 1(1)$$
$$= 29$$

To find D_x, we expand about column 3.

$$D_x = \begin{vmatrix} -4 & -2 & 1 \\ 5 & 1 & -1 \\ 3 & 2 & 4 \end{vmatrix} = 1 \begin{vmatrix} 5 & 1 \\ 3 & 2 \end{vmatrix} - (-1) \begin{vmatrix} -4 & -2 \\ 3 & 2 \end{vmatrix} + 4 \begin{vmatrix} -4 & -2 \\ 5 & 1 \end{vmatrix}$$

$$= 1(10 - 3) + 1[-8 - (-6)] + 4[-4 - (-10)]$$
$$= 1(7) + 1(-2) + 4(6) = 29$$

To find D_y, we expand about row 1.

$$D_y = \begin{vmatrix} 1 & -4 & 1 \\ 2 & 5 & -1 \\ 3 & 3 & 4 \end{vmatrix} = 1 \begin{vmatrix} 5 & -1 \\ 3 & 4 \end{vmatrix} - (-4) \begin{vmatrix} 2 & -1 \\ 3 & 4 \end{vmatrix} + 1 \begin{vmatrix} 2 & 5 \\ 3 & 3 \end{vmatrix}$$

$$= 1[20 - (-3)] + 4[8 - (-3)] + 1(6 - 15)$$
$$= 1(23) + 4(11) + 1(-9)$$
$$= 58$$

To find D_z, we expand about column 1.

$$D_z = \begin{vmatrix} 1 & -2 & -4 \\ 2 & 1 & 5 \\ 3 & 2 & 3 \end{vmatrix} = 1 \begin{vmatrix} 1 & 5 \\ 2 & 3 \end{vmatrix} - 2 \begin{vmatrix} -2 & -4 \\ 2 & 3 \end{vmatrix} + 3 \begin{vmatrix} -2 & -4 \\ 1 & 5 \end{vmatrix}$$

$$= 1(3 - 10) - 2[-6 - (-8)] + 3[-10 - (-4)]$$
$$= 1(-7) - 2(2) + 3(-6)$$
$$= -29$$

Thus,

$$x = \frac{D_x}{D} = \frac{29}{29} = 1,$$

$$y = \frac{D_y}{D} = \frac{58}{29} = 2,$$

$$z = \frac{D_z}{D} = \frac{-29}{29} = -1.$$

The solution set is $\{(1, 2, -1)\}$. (Be sure to check it!)

Example 4

Use Cramer's rule to solve the system $\begin{pmatrix} 2x - y + 3z = -17 \\ 3y + z = 5 \\ x - 2y - z = -13 \end{pmatrix}$.

Solution

To find D, we expand about column 1.

$$D = \begin{vmatrix} 2 & -1 & 3 \\ 0 & 3 & 1 \\ 1 & -2 & -1 \end{vmatrix} = 2 \begin{vmatrix} 3 & 1 \\ -2 & -1 \end{vmatrix} - 0 \begin{vmatrix} -1 & 3 \\ -2 & -1 \end{vmatrix} + 1 \begin{vmatrix} -1 & 3 \\ 3 & 1 \end{vmatrix}$$

$$= 2[-3 - (-2)] - 0 + 1(-1 - 9)$$
$$= 2(-1) - 0 - 10 = -12$$

To find D_x, we expand about column 3.

$$D_x = \begin{vmatrix} -17 & -1 & 3 \\ 5 & 3 & 1 \\ -3 & -2 & -1 \end{vmatrix}$$

$$= 3 \begin{vmatrix} 5 & 3 \\ -3 & -2 \end{vmatrix} - 1 \begin{vmatrix} -17 & -1 \\ -3 & -2 \end{vmatrix} + (-1) \begin{vmatrix} -17 & -1 \\ 5 & 3 \end{vmatrix}$$

$$= 3[-10 - (-9)] - 1(34 - 3) - 1[-51 - (-5)]$$

$$= 3(-1) - 1(31) - 1(-46) = 12$$

To find D_y, we expand about column 1.

$$D_y = \begin{vmatrix} 2 & -17 & 3 \\ 0 & 5 & 1 \\ 1 & -3 & -1 \end{vmatrix}$$

$$= 2 \begin{vmatrix} 5 & 1 \\ -3 & -1 \end{vmatrix} - 0 \begin{vmatrix} -17 & 3 \\ -3 & -1 \end{vmatrix} + 1 \begin{vmatrix} -17 & 3 \\ 5 & 1 \end{vmatrix}$$

$$= 2[-5 - (-3)] - 0 + 1(-17 - 15)$$

$$= 2(-2) - 0 + 1(-32) = -36$$

To find D_z, we expand about column 1.

$$D_z = \begin{vmatrix} 2 & -1 & -17 \\ 0 & 3 & 5 \\ 1 & -2 & -3 \end{vmatrix} = 2 \begin{vmatrix} 3 & 5 \\ -2 & -3 \end{vmatrix} - 0 \begin{vmatrix} -1 & -17 \\ -2 & -3 \end{vmatrix} + 1 \begin{vmatrix} -1 & -17 \\ 3 & 5 \end{vmatrix}$$

$$= 2[-9 - (-10)] - 0 + 1[-5 - (-51)]$$

$$= 2(1) - 0 + 1(46) = 48$$

Thus

$$x = \frac{D_x}{D} = \frac{12}{-12} = -1,$$

$$y = \frac{D_y}{D} = \frac{-36}{-12} = 3,$$

$$z = \frac{D_z}{D} = \frac{48}{-12} = -4.$$

The solution set is $\{(-1, 3, -4)\}$.

●

Problem Set 12.4

Use *expansion by minors* to evaluate each of the following determinants.

1. $\begin{vmatrix} 1 & -1 & 2 \\ 2 & 1 & 3 \\ -1 & -2 & 1 \end{vmatrix}$ 2. $\begin{vmatrix} 3 & -2 & 1 \\ 2 & 1 & 4 \\ -1 & 3 & 5 \end{vmatrix}$

3. $\begin{vmatrix} 2 & 4 & 1 \\ -1 & 5 & 1 \\ -3 & 6 & 2 \end{vmatrix}$

4. $\begin{vmatrix} 2 & 7 & 5 \\ 1 & -1 & 1 \\ -4 & 3 & 2 \end{vmatrix}$

5. $\begin{vmatrix} -5 & 1 & -1 \\ 3 & 4 & 2 \\ 0 & 2 & -3 \end{vmatrix}$

6. $\begin{vmatrix} -3 & -2 & 1 \\ 5 & 0 & 6 \\ 2 & 1 & -4 \end{vmatrix}$

7. $\begin{vmatrix} -6 & 5 & 3 \\ 2 & 0 & -1 \\ 4 & 0 & 7 \end{vmatrix}$

8. $\begin{vmatrix} 3 & -4 & -2 \\ 5 & -2 & 1 \\ 1 & 0 & 0 \end{vmatrix}$

Use Cramer's rule to find the solutions for each of the following systems.

9. $\begin{pmatrix} x - 2y + z = 3 \\ 3x + 2y + z = -3 \\ 2x - 3y - 3z = -5 \end{pmatrix}$

10. $\begin{pmatrix} x - y + 2z = -8 \\ 2x + 3y - 4z = 18 \\ -x + 2y - z = 7 \end{pmatrix}$

11. $\begin{pmatrix} -x + y - z = 1 \\ 2x + 3y - 4z = 10 \\ -3x - y + z = -5 \end{pmatrix}$

12. $\begin{pmatrix} 2x - y + 3z = -10 \\ x + 2y - 3z = 2 \\ 3x - 2y + 5z = -16 \end{pmatrix}$

13. $\begin{pmatrix} 2x - 3y + 3z = -3 \\ -2x + 5y - 3z = 5 \\ 3x - y + 6z = -1 \end{pmatrix}$

14. $\begin{pmatrix} 3x - 2y - 3z = -5 \\ x + 2y + 3z = -3 \\ -x + 4y - 6z = 8 \end{pmatrix}$

15. $\begin{pmatrix} x - 2y + 3z = 1 \\ 2x + y + z = 4 \\ 4x - 3y + 7z = 6 \end{pmatrix}$

16. $\begin{pmatrix} -x + y + z = -1 \\ x - 2y + 5z = -4 \\ 3x + 4y - 6z = -1 \end{pmatrix}$

17. $\begin{pmatrix} 3x - 2y + z = 11 \\ 5x + 3y = 17 \\ x + y - 2z = 6 \end{pmatrix}$

18. $\begin{pmatrix} 2x - y + 3z = -5 \\ 3x + 4y - 2z = -25 \\ -x + z = 6 \end{pmatrix}$

19. $\begin{pmatrix} 6x - 5y + 2z = 7 \\ 2x + 3y - 4z = -21 \\ 2y + 3z = 10 \end{pmatrix}$

20. $\begin{pmatrix} 2y - z = 10 \\ 3x + 4y = 6 \\ x - y + z = -9 \end{pmatrix}$

21. $\begin{pmatrix} 2x - 3y + z = -7 \\ -3x + y - z = -7 \\ x - 2y - 5z = -45 \end{pmatrix}$

22. $\begin{pmatrix} 3x - y - z = 18 \\ 4x + 3y - 2z = 10 \\ -5x - 2y + 3z = -22 \end{pmatrix}$

23. $\begin{pmatrix} 4x + 5y - 2z = -14 \\ 7x - y + 2z = 42 \\ 3x + y + 4z = 28 \end{pmatrix}$

24. $\begin{pmatrix} -5x + 6y + 4z = -4 \\ -7x - 8y + 2z = -2 \\ 2x + 9y - z = 1 \end{pmatrix}$

25. $\begin{pmatrix} -4x - 5y + 3z = 4 \\ 3x + 2y - 2z = 6 \\ 3y + 7z = -8 \end{pmatrix}$

Miscellaneous Problems

26. Evaluate the following determinant by expanding about the *second column*.

$$\begin{vmatrix} a & e & a \\ b & f & b \\ c & g & c \end{vmatrix}$$

Make a conjecture about determinants that contain two identical columns.

27. Show that

$$\begin{vmatrix} 1 & -1 & 2 \\ 2 & 3 & -1 \\ -1 & 2 & 4 \end{vmatrix} = - \begin{vmatrix} -1 & 1 & 2 \\ 3 & 2 & -1 \\ 2 & -1 & 4 \end{vmatrix}.$$

Make a conjecture about the result of interchanging two columns of a determinant.

28. (a) Show that

$$\begin{vmatrix} 2 & 1 & 2 \\ 4 & -1 & -2 \\ 6 & 3 & 1 \end{vmatrix} = 2 \begin{vmatrix} 1 & 1 & 2 \\ 2 & -1 & -2 \\ 3 & 3 & 1 \end{vmatrix}.$$

Make a conjecture about the result of factoring a common factor from each element of a column in a determinant.

(b) Use your conjecture from part **(a)** to help you evaluate the following determinant.

$$\begin{vmatrix} 2 & 4 & -1 \\ -3 & -4 & -2 \\ 5 & 4 & 3 \end{vmatrix}$$

Chapter 12 Summary

(12.1)   A rectangular array of numbers such as

$$\begin{bmatrix} 1 & 3 & 7 \\ 2 & 8 & -4 \end{bmatrix}$$

is called a **matrix**. With every system of linear equations we can associate a matrix consisting of the coefficients and constant terms. For example, with the system

$$\left(\begin{matrix} a_1 x + b_1 y = c_1 \\ a_2 x + b_2 y = c_2 \end{matrix} \right)$$

we can associate the matrix

$$\begin{bmatrix} a_1 & b_1 & \vdots & c_1 \\ a_2 & b_2 & \vdots & c_2 \end{bmatrix},$$

which is called the **augmented matrix** of the system.

Systems of linear equations can be solved by applying the following **elementary row operations** to the augmented matrices:

1. Any two rows of an augmented matrix can be interchanged.
2. Any row can be multiplied by a nonzero constant.
3. Any row can be replaced by adding a nonzero multiple of another row to that row.

A system of linear equations can be solved by changing the augmented matrix to a **triangular form** and then using back-substitution. See Examples 1–4 of this section for a review of this process.

(12.2)   The form

$$\begin{bmatrix} 1 & 0 & 0 & * \\ 0 & 1 & 0 & * \\ 0 & 0 & 1 & * \end{bmatrix}$$

is called the **reduced echelon form** for a system of three linear equations in three variables. The solution set of a system is obvious once the augmented matrix has been put in reduced echelon form.

(12.3)   A rectangular array of numbers is called a **matrix**. A **square matrix** has the same number of rows as columns. For a 2-by-2 matrix

$$\begin{bmatrix} a_1 & b_1 \\ a_2 & b_2 \end{bmatrix},$$

the **determinant** of the matrix is written as

$$\begin{vmatrix} a_1 & b_1 \\ a_2 & b_2 \end{vmatrix}$$

and is defined by

$$\begin{vmatrix} a_1 & b_1 \\ a_2 & b_2 \end{vmatrix} = a_1 b_2 - a_2 b_1.$$

Cramer's rule for solving a system of two linear equations in two variables is stated as follows: Given the system

$$\begin{pmatrix} a_1 x + b_1 y = c_1 \\ a_2 x + b_2 y = c_2 \end{pmatrix},$$

with

$$D = \begin{vmatrix} a_1 & b_1 \\ a_2 & b_2 \end{vmatrix} \neq 0, \qquad D_x = \begin{vmatrix} c_1 & b_1 \\ c_2 & b_2 \end{vmatrix}, \qquad D_y = \begin{vmatrix} a_1 & c_1 \\ a_2 & c_2 \end{vmatrix},$$

then

$$x = \frac{D_x}{D} \quad \text{and} \quad y = \frac{D_y}{D}.$$

(12.4) A 3-by-3 determinant is defined by

$$\begin{vmatrix} a_1 & b_1 & c_1 \\ a_2 & b_2 & c_2 \\ a_3 & b_3 & c_3 \end{vmatrix} = a_1b_2c_3 + b_1c_2a_3 + c_1a_2b_3 - a_3b_2c_1 - b_3c_2a_1 - c_3a_2b_1.$$

The **minor** of an element in a determinant is the determinant that remains after the row and column in which the element appears are deleted. A determinant can be evaluated by **expansion by minors** of the elements of any row or any column.

Cramer's rule for solving a system of three linear equations in three variables is stated as follows: Given the system

$$\begin{pmatrix} a_1x + b_1y + c_1z = d_1 \\ a_2x + b_2y + c_2z = d_2 \\ a_3x + b_3y + c_3z = d_3 \end{pmatrix},$$

with

$$D = \begin{vmatrix} a_1 & b_1 & c_1 \\ a_2 & b_2 & c_2 \\ a_3 & b_3 & c_3 \end{vmatrix} \neq 0, \qquad D_x = \begin{vmatrix} d_1 & b_1 & c_1 \\ d_2 & b_2 & c_2 \\ d_3 & b_3 & c_3 \end{vmatrix},$$

$$D_y = \begin{vmatrix} a_1 & d_1 & c_1 \\ a_2 & d_2 & c_2 \\ a_3 & d_3 & c_3 \end{vmatrix}, \qquad D_z = \begin{vmatrix} a_1 & b_1 & d_1 \\ a_2 & b_2 & d_2 \\ a_3 & b_3 & d_3 \end{vmatrix},$$

then

$$x = \frac{D_x}{D}, \quad y = \frac{D_y}{D}, \quad \text{and} \quad z = \frac{D_z}{D}.$$

Chapter 12 Review Problem Set

For Problems 1–10, use a matrix approach to solve each of the systems.

1. $\begin{pmatrix} 2x + 5y = 2 \\ 4x - 3y = 30 \end{pmatrix}$

2. $\begin{pmatrix} 6x + 5y = -21 \\ x = 4y + 11 \end{pmatrix}$

3. $\begin{pmatrix} 4x - 3y = 34 \\ 3x + 2y = 0 \end{pmatrix}$

4. $\begin{pmatrix} \dfrac{1}{2}x - \dfrac{2}{3}y = 1 \\ \dfrac{3}{4}x + \dfrac{1}{6}y = -1 \end{pmatrix}$

5. $\begin{pmatrix} x + y - 2z = -7 \\ 3x + 4y + z = 6 \\ 5x - y - 3z = -1 \end{pmatrix}$

6. $\begin{pmatrix} x - 2y - z = 11 \\ 2x - 3y + 5z = 4 \\ -3x + 5y - 3z = -17 \end{pmatrix}$

7. $\begin{pmatrix} 3x - 4y - 3z = 10 \\ -2x + 9y - 2z = 14 \\ x - 5y + z = -8 \end{pmatrix}$
8. $\begin{pmatrix} 2x - y + 3z = -19 \\ 3x + 2y - 4z = 21 \\ 5x - 4y - z = -8 \end{pmatrix}$

9. $\begin{pmatrix} 3x - 2y - 5z = 2 \\ -4x + 3y + 11z = 3 \\ 2x - y + z = -1 \end{pmatrix}$
10. $\begin{pmatrix} 3x + 2y - 4z = 4 \\ 5x + 3y - z = 2 \\ 4x - 2y + 3z = 11 \end{pmatrix}$

For Problems 11–17, evaluate each of the determinants.

11. $\begin{vmatrix} -2 & 6 \\ 3 & 8 \end{vmatrix}$
12. $\begin{vmatrix} 5 & -4 \\ 7 & -3 \end{vmatrix}$
13. $\begin{vmatrix} 5 & -3 \\ -4 & -2 \end{vmatrix}$

14. $\begin{vmatrix} 4 & -1 & -3 \\ 2 & 1 & 4 \\ -3 & 2 & 2 \end{vmatrix}$
15. $\begin{vmatrix} 2 & 3 & -1 \\ 3 & 4 & -5 \\ 6 & 4 & 2 \end{vmatrix}$
16. $\begin{vmatrix} 3 & -2 & 4 \\ 1 & 0 & 6 \\ 3 & -3 & 5 \end{vmatrix}$

17. $\begin{vmatrix} 5 & 4 & 3 \\ 2 & -7 & 0 \\ 3 & -2 & 0 \end{vmatrix}$

For Problems 18–29, use Cramer's rule to solve each of the systems.

18. $\begin{pmatrix} 7x - 2y = -53 \\ x + 5y = 40 \end{pmatrix}$
19. $\begin{pmatrix} 2x - 3y = 12 \\ 3x + 5y = -20 \end{pmatrix}$

20. $\begin{pmatrix} y = 3x - 16 \\ 5x + 7y = -34 \end{pmatrix}$
21. $\begin{pmatrix} \dfrac{3}{4}x - \dfrac{1}{2}y = -15 \\ \dfrac{2}{3}x + \dfrac{1}{4}y = -5 \end{pmatrix}$

22. $\begin{pmatrix} 7x - 3y = -49 \\ y = \dfrac{3}{5}x - 1 \end{pmatrix}$
23. $\begin{pmatrix} 0.2x + 0.3y = 2.6 \\ 0.5x + 0.1y = 1.4 \end{pmatrix}$

24. $\begin{pmatrix} -x - y + z = -3 \\ 3x + 2y - 4z = 12 \\ 5x + y + 2z = 5 \end{pmatrix}$
25. $\begin{pmatrix} 3x + y - z = -6 \\ 3x + 2y + 3z = 9 \\ 6x - 3y + 2z = 9 \end{pmatrix}$

26. $\begin{pmatrix} x - 2y + z = -7 \\ 2x - 3y + 4z = -14 \\ -3x + y - 2z = 10 \end{pmatrix}$
27. $\begin{pmatrix} -2x - 7y + z = 9 \\ x + 3y - 4z = -11 \\ 4x + 5y - 3z = -11 \end{pmatrix}$

28. $\begin{pmatrix} 2x - 3y - 3z = 25 \\ 3x + y + 2z = -5 \\ 5x - 2y - 4z = 32 \end{pmatrix}$
29. $\begin{pmatrix} 3x - y + z = -10 \\ 6x - 2y + 5z = -35 \\ 7x + 3y - 4z = 19 \end{pmatrix}$

For Problems 30–33, solve each problem by setting up and solving a system of linear equations.

30. The sum of the digits of a two-digit number is 9. If the digits are reversed, the newly formed number is 45 less than the original number. Find the original number.

31. Sara invested $2500, part at 10% and the rest at 12% yearly interest. The income on the 12% investment is $102 more than the income on the 10% investment. How much money did Sara invest at each rate?

32. A box contains $17.70 in nickels, dimes, and quarters. The number of dimes is 8 less than twice the number of nickels. The number of quarters is 2 more than the sum of the number of nickels and dimes. How many coins of each kind are there in the box?

33. The measure of the largest angle of a triangle is 10° more than four times that of the smallest angle. The sum of the smallest and the largest angles is three times the measure of the other angle. Find the measure of each angle of the triangle.

13 Sequences and Series

Suppose that Math University has an enrollment of 7500 students in 1978 and that each year thereafter through 1982 the enrollment increased by 500 students. The numbers

$$7500, \quad 8000, \quad 8500, \quad 9000, \quad 9500$$

represent the enrollment figures for the years 1978 through 1982, respectively. This list of numbers, for which there is a constant difference of 500 between any two successive numbers in the list, is called an **arithmetic sequence**.

Suppose a woman's present yearly salary is $20,000 and she expects a 10% raise for each of the next four years. The numbers

$$20000, \quad 22000, \quad 24200, \quad 26620, \quad 29282$$

represent her present salary and her projected salary for each of the next four years. This list of numbers, for which each number after the first one is 1.1 times the previous number in the list, is called a **geometric sequence**. Arithmetic and geometric sequences are the focus of our attention in this chapter.

13.1 Arithmetic Sequences

An **infinite sequence** is a function whose domain is the set of positive integers. For example, consider the function defined by the equation

$$f(n) = 3n + 2,$$

where the domain is the set of positive integers. Let's substitute the numbers of the domain, in order, starting with 1. The resulting ordered pairs can be listed as

$$(1, 5), (2, 8), (3, 11), (4, 14), (5, 17),$$

and so on. Since we have agreed to use the domain of positive integers, in order, starting with 1, there is no need to use ordered pairs. We can simply express the infinite sequence as

$$5, 8, 11, 14, 17, \ldots .$$

Frequently, the letter a is used to represent sequential functions, and the functional value at n is written as a_n (read "a sub n") instead of $a(n)$. The sequence is then expressed as

$$a_1, a_2, a_3, a_4, \ldots ,$$

where a_1 is the *first term*, a_2 the *second term*, a_3 the *third term*, and so on. The expression a_n, which defines the sequence, is called the **general term** of the sequence. Knowing the general term of a sequence allows us to find as many terms of the sequence as needed and also to find any specific terms. Consider the following examples.

Example 1   Find the first five terms of each of the following sequences:

(a) $a_n = n^2 + 1$ (b) $a_n = 2^n$.

Solution   (a) The first five terms are found by replacing n with 1, 2, 3, 4, and 5.

$$a_1 = 1^2 + 1 = 2$$
$$a_2 = 2^2 + 1 = 5$$
$$a_3 = 3^2 + 1 = 10$$
$$a_4 = 4^2 + 1 = 17$$
$$a_5 = 5^2 + 1 = 26$$

Thus, the first five terms are 2, 5, 10, 17, and 26.

(b) $a_1 = 2^1 = 2$
$$a_2 = 2^2 = 4$$
$$a_3 = 2^3 = 8$$
$$a_4 = 2^4 = 16$$
$$a_5 = 2^5 = 32$$

The first five terms are 2, 4, 8, 16, and 32.

Example 2   Find the 12th and 25th terms of the sequence $a_n = 5n - 1$.

Solution $a_{12} = 5(12) - 1 = 59$
$$a_{25} = 5(25) - 1 = 124$$

The 12th term is 59, and the 25th term is 124.

An *arithmetic sequence* (also called an arithmetic progression) is a sequence for which there is a common difference between successive terms. The following are examples of arithmetic sequences.

1. 1, 4, 7, 10, 13, . . .
2. 6, 11, 16, 21, 26, . . .
3. 14, 25, 36, 47, 58, . . .
4. 4, 2, 0, -2, -4, . . .
5. -1, -7, -13, -19, -25, . . .

The common difference in sequence **1** is 3. That is to say, $4 - 1 = 3$, $7 - 4 = 3$, $10 - 7 = 3$, $13 - 10 = 3$, and so on. The common differences for sequences **2**, **3**, **4**, and **5** are 5, 11, -2, and -6, respectively. It is sometimes stated that arithmetic sequences exhibit *constant growth*. This is an accurate description if we keep in mind that the *growth* may be in a negative direction, as illustrated by sequences **4** and **5** above.

In a more general setting we say that the sequence

$$a_1, a_2, a_3, a_4, \ldots, a_n, \ldots$$

is an arithmetic sequence if and only if there is a real number d such that

$$a_{k+1} - a_k = d \tag{1}$$

for every positive integer k. The number d is called the **common difference**.

From equation (1) we see that $a_{k+1} = a_k + d$. In other words, we can generate an arithmetic sequence having a common difference of d by starting with a first term of a_1 and then simply adding d to each successive term as follows.

first term	a_1
second term	$a_1 + d$
third term	$a_1 + 2d$ $[(a_1 + d) + d = a_1 + 2d]$
fourth term	$a_1 + 3d$ $[(a_1 + 2d) + d = a_1 + 3d]$
$\vdots$	
nth term	$a_1 + (n - 1)d$

Thus, the general term of an arithmetic sequence is given by

$$\boxed{a_n = a_1 + (n - 1)d,}$$

where a_1 is the first term and d is the common difference. This general-term formula provides the basis for doing a variety of problems involving arithmetic sequences.

Example 3

Find the general term for each of the following arithmetic sequences:

(a) 1, 5, 9, 13, . . .   **(b)** 5, 2, -1, -4, ...

Solution (a) The common difference is 4, and the first term is 1. Substituting these values into $a_n = a_1 + (n - 1)d$ and simplifying, we obtain

$$a_n = a_1 + (n - 1)d$$
$$= 1 + (n - 1)4$$
$$= 1 + 4n - 4$$
$$= 4n - 3.$$

(Perhaps you should verify that the general term $a_n = 4n - 3$ does produce the sequence 1, 5, 9, 13, ...)

(b) Since the first term is 5 and the common difference is -3, we obtain

$$a_n = a_1 + (n - 1)d$$
$$= 5 + (n - 1)(-3)$$
$$= 5 - 3n + 3$$
$$= -3n + 8.$$

●

Example 4 Find the 50th term of the arithmetic sequence 2, 6, 10, 14, ...

Solution Certainly we could simply continue to write the terms of the given sequence until we reached the 50th term. However, let's use the general-term formula, $a_n = a_1 + (n - 1)d$, to find the 50th term without determining all of the other terms.

$$a_{50} = 2 + (50 - 1)4$$
$$= 2 + 49(4) = 2 + 196 = 198$$

●

Example 5 Find the first term of the arithmetic sequence for which the 3rd term is 13 and the 10th term is 62.

Solution Using $a_n = a_1 + (n - 1)d$ with $a_3 = 13$ (the 3rd term is 13) and $a_{10} = 62$ (the 10th term is 62), we have

$$13 = a_1 + (3 - 1)d = a_1 + 2d$$
$$62 = a_1 + (10 - 1)d = a_1 + 9d.$$

Solving the system of equations

$$\begin{pmatrix} a_1 + 2d = 13 \\ a_1 + 9d = 62 \end{pmatrix}$$

yields $a_1 = -1$. Thus, the first term is -1. ●

Remark Perhaps you can think of another way to solve the problem in Example 5 without using a system of equations. (Hint: How many "differences" are there between the 3rd and 10th terms of an arithmetic sequence?)

Phrases such as "the set of odd whole numbers," "the set of even whole numbers," and "the set of positive multiples of 5" are commonly used in mathematical literature to refer to various subsets of the whole numbers. Though no specific ordering of the numbers is implied by these phrases, most of us would probably react with a natural ordering. For example, if we were asked to list the set of *odd whole numbers*, our answer probably would be 1, 3, 5, 7, ... So ordered, the set of odd whole numbers can be thought of as an arithmetic sequence. Therefore, a general representation for the set of odd whole numbers can be formulated by using the general-term formula. Thus,

$$a_n = a_1 + (n - 1)d$$
$$= 1 + (n - 1)2$$
$$= 1 + 2n - 2$$
$$= 2n - 1.$$

The final example of this section illustrates the use of an arithmetic sequence to solve a problem dealing with the *constant growth* of a man's yearly salary.

Example 6   A man started to work in 1960 at an annual salary of $9500. He received a $700 raise each year. How much was his annual salary in 1981?

Solution   The following arithmetic sequence represents the annual salary beginning in 1960.

9500, 10200, 10900, 11600, ...

The general term of this sequence is

$$a_n = a_1 + (n - 1)d$$
$$= 9500 + (n - 1)700$$
$$= 9500 + 700n - 700$$
$$= 700n + 8800.$$

The man's 1981 salary is the 22nd term of this sequence. Thus,

$$a_{22} = 700(22) + 8800 = 24200.$$

His salary in 1981 was $24,200.   ●

Problem Set 13.1

For Problems 1–10, find the first five terms of the sequence having the indicated general term.

1. $a_n = 2n + 1$

2. $a_n = 3n - 4$

3. $a_n = -3n - 1$

4. $a_n = -5n - 1$

5. $a_n = n^2 - 1$

6. $a_n = 2n^2 - 3$

7. $a_n = -n^2 + 3$

8. $a_n = -n^2 - 1$

9. $a_n = 3^{n-1}$

10. $a_n = 2^{n+1}$

11. Find the 10th and 15th terms of the sequence for which $a_n = n^2 - 2n - 1$.

12. Find the 17th and 24th terms of the sequence for which $a_n = -7n - 3$.

For Problems 13–20, find the general term (nth term) for each of the arithmetic sequences.

13. 1, 4, 7, 10, 13, . . .   **14.** 1, 7, 13, 19, 25, . . .

15. 2, 7, 12, 17, 22, . . .   **16.** 3, 5, 7, 9, 11, . . .

17. 5, 3, 1, -1, -3, . . .   **18.** 4, -1, -6, -11, -16, . . .

19. -3, -6, -9, -12, -15, . . .   **20.** $1, \dfrac{3}{2}, 2, \dfrac{5}{2}, 3, \ldots$

For Problems 21–26, find the indicated term for each of the arithmetic sequences.

21. The 20th term of 2, 5, 8, 11, . . .   **22.** The 35th term of 1, 6, 11, 16, . . .

23. The 45th term of 3, 7, 11, 15, . . .   **24.** The 50th term of -1, -3, -5, -7, . . .

25. The 75th term of 2, -1, -4, -7, . . .   **26.** The 150th term of 1, 3, 5, 7, . . .

27. Find the number of terms of each of the following finite arithmetic sequences.

    (a) 2, 4, 6, 8, . . . , 96

    (b) 1, 3, 5, 7, . . . , 119

    (c) 3, 6, 9, 12, . . . , 171

    (d) 4, 8, 12, 16, . . . , 380

28. If the 4th term of an arithmetic sequence is 19 and the 15th term is 52, find the first term.

29. If the 8th term of an arithmetic sequence is 37 and the 20th term is 97, find the 50th term.

30. In the arithmetic sequence 0.97, 1.00, 1.03, 1.06, . . . , which term is 5.02?

31. Use $a_n = a_1 + (n - 1)d$ to find a general representation for each of the following subsets of the set of whole numbers.

    (a) The set of nonzero even numbers

    (b) The set of multiples of 3 ($\{0, 3, 6, 9, \ldots\}$)

    (c) The set of multiples of 5

    (d) The set of nonzero multiples of 6

    (e) The set of whole numbers that are all 1 larger than a multiple of 3 ($\{1, 4, 7, 10, \ldots\}$)

Set up an arithmetic sequence and use $a_n = a_1 + (n - 1)d$ to solve each of the following problems.

32. A woman started to work in 1965 at an annual salary of \$12,500. She received a \$900 raise each year. How much was her annual salary in 1982?

33. Math University had an enrollment of 8500 students in 1966. Each year the enrollment has increased by 350 students. What was the enrollment in 1980?

34. Suppose you are offered a job starting at \$900 a month, with a guaranteed increase of \$30 a month every 6 months for the next 5 years. What will your monthly salary be for the last 6 months of the 5th year of your employment?

35. Between 1966 and 1980 a person invested \$500 at 14% simple interest at the beginning of each year. By the end of 1980, how much interest had been earned by the \$500 that was invested at the beginning of 1972?

13.2 Arithmetic Series

Let's begin this section by solving a problem. Study the solution very carefully.

Problem 1   Find the sum of the first 100 positive integers.

Solution   We are being asked to find the sum of $1 + 2 + 3 + 4 + \cdots + 100$. Rather than using a calculator, let's find the sum in the following way.

$$
\begin{array}{l}
1 + 2 + 3 + 4 + \cdots + 100 \\
100 + 99 + 98 + 97 + \cdots + 1 \\
\hline
101 + 101 + 101 + 101 + \cdots + 101
\end{array}
$$

$$50$$
$$\frac{\cancel{100}(101)}{2} = 5050$$

Note that we simply wrote the indicated sum *forward and backward* and then added the results. So doing produces 100 sums of 101, but half of them are "repeats." For example, $(100 + 1)$ and $(1 + 100)$ are both counted in this process. Thus, we divided the product $(100)(101)$ by 2, yielding the final result of 5050.   ●

Now let's look at some terminology that applies to problems such as the one above. The indicated sum of a sequence is called a **series**. Associated with the finite sequence

$$a_1, a_2, a_3, \ldots, a_n$$

is a **finite series**

$$a_1 + a_2 + a_3 + \cdots + a_n.$$

Likewise, from the infinite sequence

$$a_1, a_2, a_3, \ldots$$

we can form the **infinite series**

$$a_1 + a_2 + a_3 + \cdots.$$

In this section we will direct our attention to working with **arithmetic series**; that is, the indicated sums of arithmetic sequences.

Problem 1 above could have been stated as "find the sum of the first 100 terms of the arithmetic series having a general term (nth term) of $a_n = n$." In fact, the *forward-backward approach* used to solve that problem can be applied to the general arithmetic series

$$a_1 + a_2 + a_3 + \cdots + a_n$$

to produce a formula for finding the sum of the first n terms of any arithmetic series. Using S_n to represent the sum of the first n terms, we can proceed as follows:

$$S_n = a_1 + (a_1 + d) + (a_1 + 2d) + \cdots + (a_n - 2d) + (a_n - d) + a_n.$$

Now we write this sum in reverse as

$$S_n = a_n + (a_n - d) + (a_n - 2d) + \cdots + (a_1 + 2d) + (a_1 + d) + a_1.$$

Adding the two equations produces

$$2S_n = (a_1 + a_n) + (a_1 + a_n) + (a_1 + a_n) + \cdots + (a_1 + a_n)$$
$$+ (a_1 + a_n) + (a_1 + a_n).$$

That is, we have n sums of $(a_1 + a_n)$, so

$$2S_n = n(a_1 + a_n),$$

from which we obtain

$$\boxed{S_n = \frac{n(a_1 + a_n)}{2}.}$$

Using the nth-term formula $a_n = a_1 + (n - 1)d$ along with the sum formula $S_n = \dfrac{n(a_1 + a_n)}{2}$, we can solve a variety of problems involving arithmetic series.

Example 1   Find the sum of the first 50 terms of the series

$2 + 5 + 8 + 11 + \cdots$.

Solution   Using $a_n = a_1 + (n - 1)d$, we find the 50th term to be

$a_{50} = 2 + 49(3) = 149.$

Then using $S_n = \dfrac{n(a_1 + a_n)}{2}$, we obtain

$$S_{50} = \frac{50(2 + 149)}{2} = 3775.$$

●

Example 2   Find the sum of all odd numbers between 7 and 433, inclusive.

Solution   We need to find the sum $7 + 9 + 11 + \cdots + 433$. To use $S_n = \dfrac{n(a_1 + a_n)}{2}$, we need

the number of terms, n. Perhaps we could figure this out without a formula (try it), but suppose we use the nth-term formula.

$a_n = a_1 + (n - 1)d$
$433 = 7 + (n - 1)2$
$433 = 7 + 2n - 2$
$433 = 2n + 5$
$428 = 2n$
$214 = n$

Then, using $n = 214$ in the sum formula yields

$$S_{214} = \frac{214(7 + 433)}{2} = 47080.$$ ●

Example 3   Find the sum of the first 75 terms of the series having a general term of $a_n = -5n + 9$.

Solution   Using $a_n = -5n + 9$, we can generate the series as follows:

$$a_1 = -5(1) + 9 = 4$$
$$a_2 = -5(2) + 9 = -1$$
$$a_3 = -5(3) + 9 = -6$$
$$\vdots$$
$$a_{75} = -5(75) + 9 = -366.$$

Thus, we have the series

$$4 + (-1) + (-6) + \cdots + (-366).$$

Using the sum formula, we obtain

$$S_{75} = \frac{75[4 + (-366)]}{2} = -13575.$$ ●

Example 4   Sue is saving quarters. She saves 1 quarter the first day, 2 quarters the second day, 3 quarters the third day, and so on. How much money will she have saved in 30 days?

Solution   The total number of quarters will be the sum of the series

$$1 + 2 + 3 \ldots + 30.$$

Using the sum formula yields

$$S_{30} = \frac{30(1 + 30)}{2} = 465.$$

So Sue will have saved $(465)(\$.25) = \$116.25$ at the end of 30 days.   ●

The sum formula, $S_n = \dfrac{n(a_1 + a_n)}{2}$, was developed using the *forward-backward technique* that we had previously used on a specific problem. Now that you know the sum formula, you have two choices. You can either memorize the formula and use it as it applies or disregard the formula and use the forward-backward technique. However, if you choose to use the sum formula and then some day your memory fails you and you forget it, you can still use the forward-backward approach. In other words, understanding the development of a formula may allow you to solve some problems even though you have forgotten the formula itself.

Problem Set 13.2

1. Find the sum of the first 25 terms of the series $1 + 3 + 5 + 7 + \cdots$.
2. Find the sum of the first 40 terms of the series $2 + 4 + 6 + 8 + \cdots$.
3. Find the sum of the first 50 terms of the series $2 + 6 + 10 + 14 + \cdots$.
4. Find the sum of the first 75 terms of the series $3 + 8 + 13 + 18 + \cdots$.
5. Find the sum of the first 80 terms of the series $5 + 2 + (-1) + (-4) + \cdots$.
6. Find the sum of the first 45 terms of the series $\dfrac{1}{2} + 1 + \dfrac{3}{2} + 2 + \cdots$.
7. Find the sum of the first 250 odd whole numbers.
8. Find the sum of the first 375 positive even whole numbers.
9. Find the sum of each of the following arithmetic series.

   (a) $7 + 9 + 11 + 13 + \cdots + 179$
   (b) $4 + 8 + 12 + 16 + \cdots + 212$
   (c) $5 + 10 + 15 + 20 + \cdots + 495$
   (d) $-4 + (-1) + 2 + 5 + \cdots + 173$
   (e) $1 + (-6) + (-13) + (-20) + \cdots + (-202)$
   (f) $2.5 + 3.0 + 3.5 + 4.0 + \cdots + 18.5$

10. Find the sum of all even numbers between 12 and 400, inclusive.
11. Find the sum of all odd numbers between 13 and 179, inclusive.
12. (a) Find the sum of the first 35 terms of the series having a general term of $a_n = 4n + 3$.
    (b) Find the sum of the first 25 terms of the series having a general term of $a_n = 5n - 7$.
    (c) Find the sum of the first 1000 terms of the series having a general term of $a_n = n$.
    (d) Find the sum of the first 30 terms of the series having a general term of $a_n = -2n - 1$.
13. A pile of wood has 15 logs in the bottom row, 14 logs in the next-to-the-bottom row, and so on, with 1 less log in each row until the top row, which consists of 1 log. How many logs are there in the pile?
14. An auditorium has 20 seats in the front row, 24 seats in the second row, 28 seats in the third row, and so on, for 15 rows. How many seats are there in the last row? How many seats are there in the auditorium?
15. A woman invests $700 at 13% simple interest at the beginning of each year for a period of 15 years. Find the total accumulated value of all of the investments at the end of the 15-year period.
16. A raffle is organized so that the amount paid for each ticket is determined by a number on the ticket. The tickets are numbered with the consecutive odd whole numbers $1, 3, 5, 7, \ldots$. Each participant draws a ticket, and the number on the ticket is the number of cents he or she pays. How much money will the raffle take in if 1000 tickets are sold?
17. A well-driller charges $9.00 per foot for the first 10 feet, $9.10 per foot for the next 10 feet, $9.20 per foot for the next 10 feet, and so on; he continues to increase the price by $0.10 per

foot for succeeding intervals of 10 feet. How much would it cost to drill a well with a depth of 150 feet?

18. A man started to work in 1970 at an annual salary of \$18,500. He received a \$1500 raise each year through 1982. What were his total earnings for the 13-year period?

19. Suppose that a person saves a dime the first day of August, \$0.20 the second day, \$0.30 the third day, and so on, always saving \$0.10 more per day than the previous day. How much could be saved in the 31 days of August?

20. A display in a grocery store has cans stacked, with 25 cans in the bottom row, 23 cans in the second row from the bottom, 21 cans in the third row from the bottom, and so on, until there is only 1 can in the top row. How many cans are there in the display?

21. A series for which the general term is known can be expressed in a convenient and compact form using the symbol Σ along with the general-term expression. For example, consider the finite arithmetic series

$$1 + 3 + 5 + 7 + 9 + 11,$$

for which the general term is $a_n = 2n - 1$. This series can be expressed in *summation notation* as

$$\sum_{i=1}^{6} (2i - 1),$$

where the letter i is used as the *index of summation*. The individual terms of the series can be generated by successively replacing i in the expression $(2i - 1)$ with the numbers 1, 2, 3, 4, 5, and 6. Thus, the first term is $2(1) - 1 = 1$, the second term is $2(2) - 1 = 3$, the third term is $2(3) - 1 = 5$, and so on. Write out the terms, and find the sum of each of the following series.

(a) $\sum_{i=1}^{3} (5i + 2)$ **(b)** $\sum_{i=1}^{4} (6i - 7)$

(c) $\sum_{i=1}^{6} (-2i - 1)$ **(d)** $\sum_{i=1}^{5} (-3i + 4)$

(e) $\sum_{i=1}^{5} 3i$ **(f)** $\sum_{i=1}^{6} -4i$

22. Write each of the following in summation notation. For example, since $3 + 8 + 13 + 18 + 23 + 28$ is an arithmetic series, the general-term formula $a_n = a_1 + (n - 1)d$ yields

$$a_n = 3 + (n - 1)5$$
$$= 3 + 5n - 5$$
$$= 5n - 2.$$

Now, using i as an index of summation, we can write

$$\sum_{i=1}^{6} (5i - 2).$$

(a) $2 + 5 + 8 + 11 + 14$

(b) $8 + 15 + 22 + 29 + 36 + 43$

(c) $1 + (-1) + (-3) + (-5) + (-7)$

(d) $(-5) + (-9) + (-13) + (-17) + (-21) + (-25) + (-29)$

13.3 Geometric Sequences and Series

A **geometric sequence** or **geometric progression** is a sequence in which each term after the first is obtained by multiplying the preceding term by a common multiplier. The common multiplier is called the **common ratio** of the sequence. The following geometric sequences have common ratios of $2, 3, \frac{1}{2}$, and -4, respectively.

$$1, 2, 4, 8, 16, \ldots; \qquad 3, 9, 27, 81, 243, \ldots;$$

$$8, 4, 2, 1, \frac{1}{2}, \ldots; \qquad 1, -4, 16, -64, 256, \ldots$$

The common ratio of a geometric sequence can be found by dividing a term (other than the first term) by the preceding term.

In a more general setting we say that the sequence

$$a_1, a_2, a_3, a_4, \ldots, a_n, \ldots$$

is a geometric sequence if and only if there is a nonzero real number r such that

$$a_{k+1} = ra_k \tag{1}$$

for every positive integer k. The nonzero real number r is called the **common ratio**.

Equation (1) can be used to generate a general geometric sequence that has a_1 as a first term and r as a common ratio. We can proceed as follows.

first term	a_1	
second term	$a_1 r$	
third term	$a_1 r^2$	$[(a_1 r)(r) = a_1 r^2]$
fourth term	$a_1 r^3$	$[(a_1 r^2)(r) = a_1 r^3]$
$\vdots$		
nth term	$a_1 r^{n-1}$	

Thus, the general term of a geometric sequence is given by

$$\boxed{a_n = a_1 r^{n-1},}$$

where a_1 is the first term and r is the common ratio.

Example 1 Find the general term for the geometric sequence $2, 4, 8, 16, \ldots$.

Solution Using $a_n = a_1 r^{n-1}$, we obtain

$$a_n = 2(2)^{n-1} \qquad \left(r = \frac{4}{2} = \frac{8}{4} = \frac{16}{8} = 2 \right)$$

$$= 2^n. \qquad [2^1 (2)^{n-1} = 2^{1+n-1} = 2^n]$$

●

Example 2

Find the 10th term of the geometric sequence 9, 3, 1, ...

Solution

Using $a_n = a_1 r^{n-1}$, we can find the 10th term as follows.

$$a_{10} = 9\left(\frac{1}{3}\right)^{10-1}$$

$$= 9\left(\frac{1}{3}\right)^9$$

$$= 9\left(\frac{1}{19683}\right)$$

$$= \frac{1}{2187}$$

A **geometric series** is the indicated sum of a geometric sequence. The following are examples of geometric series.

$$1 + 2 + 4 + 8 + \cdots$$
$$3 + 9 + 27 + 81 + \cdots$$
$$8 + 4 + 2 + 1 + \cdots$$
$$1 + (-4) + 16 + (-64) + \cdots$$

Before developing a general formula for finding the sum of a geometric series, let's consider a specific example.

Example 3

Find the sum of $1 + 2 + 4 + 8 + \cdots + 512$.

Solution

Letting S represent the sum, we can proceed as follows:

$$S = 1 + 2 + 4 + 8 + \cdots + 512 \tag{2}$$
$$2S = \quad 2 + 4 + 8 + \cdots + 512 + 1024 \tag{3}$$

Equation (3) is the result of multiplying both sides of equation (2) by 2. Subtracting equation (2) from equation (3) yields

$$S = 1024 - 1$$
$$= 1023.$$

Now let's consider the general geometric series

$$a_1 + a_1 r + a_1 r^2 + \cdots + a_1 r^{n-1}.$$

By applying a procedure similar to the one used in Example 3, we can develop a formula for finding the sum of the first n terms of any geometric series.

Let S_n represent the sum of the first n terms. Thus,

$$S_n = a_1 + a_1 r + a_1 r^2 + \cdots + a_1 r^{n-1}. \tag{4}$$

Multiplying both sides of equation (4) by the common ratio r produces

$$rS_n = a_1 r + a_1 r^2 + a_1 r^3 + \cdots + a_1 r^{n-1} + a_1 r^n. \tag{5}$$

Subtracting equation (4) from equation (5) yields

$$rS_n - S_n = a_1 r^n - a_1.$$

Applying the distributive property on the left side and then solving for S_n, we obtain

$$S_n(r - 1) = a_1 r^n - a_1$$

$$S_n = \frac{a_1 r^n - a_1}{r - 1}, \qquad r \neq 1.$$

Therefore, the sum of the first n terms of a geometric series having a first term of a_1 and a common ratio of r is given by

$$\boxed{S_n = \frac{a_1 r^n - a_1}{r - 1}, \qquad r \neq 1.}$$

Example 4   Find the sum of the first 7 terms of the geometric series $2 + 6 + 18 + \cdots$.

Solution   Using the sum formula, we obtain

$$S_7 = \frac{2(3)^7 - 2}{3 - 1}$$

$$= \frac{2(3^7 - 1)}{2}$$

$$= 3^7 - 1$$

$$= 2187 - 1$$

$$= 2186.$$

●

If the common ratio of a geometric series is less than 1, it may be more convenient to change the form of the sum formula. That is, the fraction $\dfrac{a_1 r^n - a_1}{r - 1}$ can be changed to $\dfrac{a_1 - a_1 r^n}{1 - r}$ by multiplying both the numerator and the denominator by -1. Thus, using $S_n = \dfrac{a_1 - a_1 r^n}{1 - r}$ when $r < 1$ can sometimes allow us to avoid unnecessary work with negative numbers, as the next example illustrates.

Example 5   Find the sum of the geometric series $1 + \dfrac{1}{2} + \dfrac{1}{4} + \cdots + \dfrac{1}{256}$.

Solution A   To use the sum formula, we need to know the number of terms, which can be found by simply counting them or by applying the nth-term formula, as follows.

$$a_n = a_1 r^{n-1}$$

$$\frac{1}{256} = 1\left(\frac{1}{2}\right)^{n-1}$$

$$\left(\frac{1}{2}\right)^8 = \left(\frac{1}{2}\right)^{n-1}$$

$$8 = n - 1 \qquad \text{Remember that "if } b^n = b^m \text{, then } n = m \text{."}$$

$$9 = n$$

Using $n = 9$, $a_1 = 1$, and $r = \dfrac{1}{2}$ in the form of the sum formula

$$S_n = \frac{a_1 - a_1 r^n}{1 - r},$$

we obtain

$$S_9 = \frac{1 - 1\left(\dfrac{1}{2}\right)^9}{1 - \dfrac{1}{2}}$$

$$= \frac{1 - \dfrac{1}{512}}{\dfrac{1}{2}}$$

$$= \frac{\dfrac{511}{512}}{\dfrac{1}{2}}$$

$$= \left(\frac{511}{512}\right)\left(\frac{2}{1}\right) = \frac{511}{256} = 1\frac{255}{256}.$$

You should realize that a problem such as Example 5 can be done *without* using the sum formula; we can simply apply the general technique used to develop the formula. Solution B illustrates this approach.

Solution B   Let S represent the desired sum. Thus,

$$S = 1 + \frac{1}{2} + \frac{1}{4} + \cdots + \frac{1}{256}.$$

Multiply both sides by $\dfrac{1}{2}$ (the common ratio):

$$\frac{1}{2}S = \frac{1}{2} + \frac{1}{4} + \cdots + \frac{1}{256} + \frac{1}{512}.$$

Subtract the second equation from the first equation, producing

$$\frac{1}{2}S = 1 - \frac{1}{512}$$

$$\frac{1}{2}S = \frac{511}{512}$$

$$S = \frac{511}{256} = 1\frac{255}{256}.$$

●

Example 6   Suppose your employer agrees to pay you a penny for your first day's wages and then double your pay on each succeeding day. How much will you earn on the 15th day? What will be your total earnings for the first 15 days?

Solution   The terms of the geometric series $1 + 2 + 4 + 8 + \cdots$ depict your daily wages, and the sum of the first 15 terms is your total earnings for the 15 days. The formula $a_n = a_1 r^{n-1}$ can be used to find the 15th day's wages.

$$a_{15} = (1)(2)^{14} = 16384$$

Since the terms of the series are expressed in cents, your wages for the 15th day would be $163.84. Now, using the sum formula, we can find your total earnings as follows.

$$S_n = \frac{a_1 r^n - a_1}{r - 1}$$

$$S_{15} = \frac{1(2)^{15} - 1}{1}$$

$$= 32768 - 1 = 32767$$

Thus, for the 15 days you would earn a total of $327.67.   ●

Problem Set 13.3

1. Find the general term of each of the following geometric sequences.

(a) 1, 2, 4, 8, . . .

(b) 1, 3, 9, 27, . . .

(c) $\frac{1}{2}, \frac{1}{4}, \frac{1}{8}, \frac{1}{16}, \ldots$

(d) 2, 8, 32, 128, . . .

(e) 1, −2, 4, −8, . . .

(f) $6, 2, \frac{2}{3}, \frac{2}{9}, \ldots$

(g) 1, 0.3, 0.09, 0.027, . . .

(h) $9, 6, 4, \frac{8}{3}, \ldots$

2. Find the indicated term for each of the following geometric sequences.

(a) The 9th term of 2, 4, 8, 16, . . .

(b) The 12th term of $\dfrac{1}{9}, \dfrac{1}{3}, 1, 3, \ldots$

(c) The 8th term of $\dfrac{1}{2}, \dfrac{1}{8}, \dfrac{1}{32}, \dfrac{1}{128}, \ldots$

(d) The 10th term of $1, -2, 4, -8, \ldots$

(e) The 11th term of $1, \dfrac{2}{3}, \dfrac{4}{9}, \dfrac{8}{27}, \ldots$

(f) The 9th term of $-1, -\dfrac{3}{2}, -\dfrac{9}{4}, -\dfrac{27}{8}, \ldots$

3. Find the first term of a geometric sequence if the 5th term is $\dfrac{32}{3}$ and the common ratio is 2.

4. Find the common ratio of a geometric sequence if the 2nd term is $\dfrac{1}{6}$ and the 5th term is $\dfrac{1}{48}$.

5. Find the sum of the first 9 terms of the geometric series $1 + 2 + 4 + 8 + \cdots$.

6. Find the sum of the first 10 terms of the geometric series $\dfrac{1}{2} + \dfrac{3}{2} + \dfrac{9}{2} + \dfrac{27}{2} + \cdots$.

7. Find the sum of the first 10 terms of the geometric series $-4 + 8 + (-16) + 32 + \cdots$.

8. Find the sum of the first 9 terms of the geometric series $-2 + 6 + (-18) + 54 + \cdots$.

9. Find the sum of the first 7 terms of the geometric series for which $a_n = \left(\dfrac{3}{4}\right)^n$.

10. Find the sum of the first 8 terms of the geometric series for which $a_n = \left(\dfrac{3}{2}\right)^n$.

11. Find the sum of the first 16 terms of the geometric series for which $a_n = (-1)^n$. Also find the sum of the first 19 terms.

12. Find the sum of each of the following geometric series.

(a) $1 + 3 + 9 + \cdots + 729$

(b) $2 + 8 + 32 + \cdots + 2048$

(c) $1 + \dfrac{1}{2} + \dfrac{1}{4} + \cdots + \dfrac{1}{1024}$

(d) $1 + (-2) + 4 + \cdots + (-128)$

(e) $8 + 4 + 2 + \cdots + \dfrac{1}{32}$

(f) $2 + 6 + 18 + \cdots + 4374$

13. Use your calculator to help you find the indicated term of each of the following geometric sequences.

(a) The 20th term of $2, 4, 8, 16, \ldots$

(b) The 15th term of $3, 9, 27, 81, \ldots$

(c) The 12th term of $\dfrac{2}{3}, \dfrac{4}{9}, \dfrac{8}{27}, \dfrac{16}{81}, \ldots$

(d) The 10th term of $-\dfrac{3}{4}, \dfrac{9}{16}, -\dfrac{27}{64}, \dfrac{81}{256}, \ldots$

(e) The 11th term of $-\dfrac{3}{2}, \dfrac{9}{4}, -\dfrac{27}{8}, \dfrac{81}{16}, \cdots$

(f) The 6th term of the sequence for which $a_n = (0.1)^n$

14. A tank contains 16,000 liters of water. Each day one-half of the water in the tank is removed and not replaced. How much water remains in the tank at the end of the 7th day?

15. A fungus culture growing under controlled conditions doubles in size each day. How many units will the culture contain after 7 days if it originally contained 5 units?

16. Suppose you save a nickel the first day of a month, a dime the second day, $0.20 the third day, and continue to double your savings each day. How much will you save on the 12th day of the month? What will be your total savings for the first 12 days?

17. Suppose that you save $0.25 the first day of a week, $0.50 the second day, $1 the third day, and continue to double your savings each day. How much will you save on the 7th day? What will be your total savings for the week?

18. A rubber ball is dropped from a height of 486 meters, and each time it rebounds one-third of the height from which it last fell. How far has the ball traveled by the time it strikes the ground for the 7th time?

19. Suppose an element has a half-life of 3 hours. This means that if n grams of it exist at a specific time, then only $\dfrac{1}{2}n$ grams remain 3 hours later. If at a particular moment we have 40 grams of the element, how much of it will remain 24 hours later?

20. A pump is attached to a container for the purpose of creating a vacuum. With each stroke of the pump, one-fourth of the air remaining in the container is removed.

    (a) Form a geometric sequence for which each term represents the fractional part of the air *that remains* in the container after each stroke. Then use this sequence to find out how much of the air remains after 6 strokes.

    (b) Form a geometric sequence for which each term represents the fractional part of the air *that is removed* from the container on each stroke of the pump. Then use this sequence (or the associated *series*) to find out how much of the air remains after 6 strokes.

21. Suppose an employer offered to pay you only a penny for the first day of employment, but to double your wages each succeeding day. How much would you be earning on the 31st day of your employment?

22. If you pay $9500 for a car and its value depreciates 10% per year, how much will it be worth in 5 years?

23. The summation notation introduced in Problem 21 of Problem Set 13.2 can also be used with geometric series. For example, the series

$$2 + 4 + 8 + 16 + 32$$

can be expressed as

$$\sum_{i=1}^{5} 2^i.$$

Write out the terms and find the sum of each of the following.

(a) $\displaystyle\sum_{i=1}^{6} 2^i$ **(b)** $\displaystyle\sum_{i=1}^{5} 3^i$ **(c)** $\displaystyle\sum_{i=1}^{5} 2^{i-1}$

(d) $\displaystyle\sum_{i=1}^{6} \left(\dfrac{1}{2}\right)^{i+1}$ **(e)** $\displaystyle\sum_{i=1}^{4} \left(\dfrac{2}{3}\right)^i$ **(f)** $\displaystyle\sum_{i=1}^{5} \left(-\dfrac{3}{4}\right)^i$

13.4 Infinite Geometric Series

In Section 13.3 we used the formula

$$S_n = \frac{a_1 - a_1 r^n}{1 - r}, \qquad r \neq 1 \tag{1}$$

to find the sum of the first n terms of a geometric series. By using the property $\frac{a - b}{c} = \frac{a}{c} - \frac{b}{c}$, we can express the right side of equation (1) in terms of two fractions as follows.

$$S_n = \frac{a_1 - a_1 r^n}{1 - r} = \frac{a_1}{1 - r} - \frac{a_1 r^n}{1 - r}, \qquad r \neq 1 \tag{2}$$

Now let's examine the behavior of r^n for $|r| < 1$, that is, for $-1 < r < 1$. For example, suppose that $r = \frac{1}{3}$; then

$$r^2 = \left(\frac{1}{3}\right)^2 = \frac{1}{9}, \qquad r^3 = \left(\frac{1}{3}\right)^3 = \frac{1}{27},$$

$$r^4 = \left(\frac{1}{3}\right)^4 = \frac{1}{81}, \qquad r^5 = \left(\frac{1}{3}\right)^5 = \frac{1}{243},$$

and so on. We can make $\left(\frac{1}{3}\right)^n$ as close to zero as we please by taking sufficiently large values for n. In general, for values of r such that $|r| < 1$, the expression r^n will approach zero as n increases. Therefore, in equation (2), the fraction $\frac{a_1 r^n}{1 - r}$ will approach zero as n increases, and we say that the *sum of an infinite geometric series* is given by

$$\boxed{S_\infty = \frac{a_1}{1 - r}, \qquad |r| < 1.}$$

Example 1   Find the sum of the infinite geometric series

$$1 + \frac{2}{3} + \frac{4}{9} + \frac{8}{27} + \cdots.$$

Solution   Since $a_1 = 1$ and $r = \frac{2}{3}$,

$$S_\infty = \frac{1}{1 - \dfrac{2}{3}} = \frac{1}{\dfrac{1}{3}} = 3.$$

In Example 1, $S_\infty = 3$ means that as we add more and more terms, the sum approaches 3, as follows.

first term: 1

sum of first 2 terms: $1 + \dfrac{2}{3} = 1\dfrac{2}{3}$

sum of first 3 terms: $1 + \dfrac{2}{3} + \dfrac{4}{9} = 2\dfrac{1}{9}$

sum of first 4 terms: $1 + \dfrac{2}{3} + \dfrac{4}{9} + \dfrac{8}{27} = 2\dfrac{11}{27}$

sum of first 5 terms: $1 + \dfrac{2}{3} + \dfrac{4}{9} + \dfrac{8}{27} + \dfrac{16}{81} = 2\dfrac{49}{81}$

etc.

Example 2 Find the sum of the infinite geometric series

$$\frac{1}{2} - \frac{1}{4} + \frac{1}{8} - \frac{1}{16} + \cdots.$$

Solution Since $a_1 = \dfrac{1}{2}$ and $r = -\dfrac{1}{2}$, we obtain

$$S_\infty = \frac{\dfrac{1}{2}}{1 - \left(-\dfrac{1}{2}\right)} = \frac{\dfrac{1}{2}}{\dfrac{3}{2}} = \frac{1}{3}.$$

If $|r| \geq 1$, the absolute value of r^n increases without bound as n increases. Notice the unbounded growth of the absolute value of r^n in the following two examples.

Let $r = 2$	Let $r = -3$	
$r^2 = 2^2 = 4$	$r^2 = (-3)^2 = 9$	
$r^3 = 2^3 = 8$	$r^3 = (-3)^3 = -27$	$|-27| = 27$
$r^4 = 2^4 = 16$	$r^4 = (-3)^4 = 81$	
$r^5 = 2^5 = 32$	$r^5 = (-3)^5 = -243$	$|-243| = 243$
etc.	etc.	

Therefore, any infinite geometric series for which $|r| \geq 1$ *has no sum.*

Repeating Decimals as Infinite Geometric Series

In Section 1.1 we stated that a rational number is a number that has either a terminating or a repeating decimal representation. For example,

$$0.23, 0.147, 0.3, 0.\overline{14}, \text{ and } 0.5\overline{81}$$

are examples of rational numbers. (Remember that $0.\overline{3}$ means $0.333\ldots$.) Our knowledge of place value provides the basis for changing terminating decimals such as 0.23 and 0.147 to $\dfrac{a}{b}$ form, where a and b are integers, $b \neq 0$.

$$0.23 = \frac{23}{100}$$

$$0.147 = \frac{147}{1000}$$

However, changing repeating decimals to $\dfrac{a}{b}$ form requires a different technique, and our work with infinite geometric series provides the basis for one such approach. Consider the following examples.

Example 3

Change $0.\overline{3}$ to $\dfrac{a}{b}$ form, where a and b are integers, $b \neq 0$.

Solution

The repeating decimal $0.\overline{3}$ can be written as the infinite geometric series

$0.3 + 0.03 + 0.003 + 0.0003 + \cdots,$

with $a_1 = 0.3$ and $r = 0.1$. Therefore, we can use the sum formula and obtain

$$S_\infty = \frac{a_1}{1 - r} = \frac{0.3}{1 - 0.1} = \frac{0.3}{0.9} = \frac{3}{9} = \frac{1}{3}.$$

So, $0.\overline{3} = \dfrac{1}{3}$.

Example 4

Change $0.\overline{14}$ to $\dfrac{a}{b}$ form, where a and b are integers, $b \neq 0$.

Solution

The repeating decimal $0.\overline{14}$ can be written as the infinite geometric series

$0.14 + 0.0014 + 0.000014 + \cdots,$

with $a_1 = 0.14$ and $r = 0.01$. The sum formula produces

$$S_\infty = \frac{0.14}{1 - 0.01} = \frac{0.14}{0.99} = \frac{14}{99}.$$

Thus, $0.\overline{14} = \dfrac{14}{99}$.

If the repeating block of digits does not begin immediately after the decimal point, we can make a slight adjustment, as the final example illustrates.

Example 5   Change $0.5\overline{81}$ to $\dfrac{a}{b}$ form, where a and b are integers, $b \neq 0$.

Solution   The repeating decimal $0.5\overline{81}$ can be written as

$$(0.5 + 0.081 + 0.00081 + 0.0000081 + \cdots),$$

where

$$0.081 + 0.00081 + 0.0000081 + \cdots$$

is an infinite geometric series with $a_1 = 0.081$ and $r = 0.01$. Thus,

$$S_{\infty} = \frac{0.081}{1 - 0.01} = \frac{0.081}{0.99} = \frac{81}{990} = \frac{9}{110}.$$

Therefore,

$$
\begin{aligned}
0.5\overline{81} &= 0.5 + \frac{9}{110} \\
&= \frac{5}{10} + \frac{9}{110} \\
&= \frac{55}{110} + \frac{9}{110} \\
&= \frac{64}{110} = \frac{32}{55}.
\end{aligned}
$$

Problem Set 13.4

Find the sum of each of the following geometric series. If the series has no sum, so state.

1. $1 + \dfrac{1}{2} + \dfrac{1}{4} + \dfrac{1}{8} + \cdots$

2. $\dfrac{1}{2} + \dfrac{1}{4} + \dfrac{1}{8} + \dfrac{1}{16} + \cdots$

3. $\dfrac{1}{3} + \dfrac{1}{9} + \dfrac{1}{27} + \dfrac{1}{81} + \cdots$

4. $\dfrac{2}{3} + \dfrac{4}{9} + \dfrac{8}{27} + \dfrac{16}{81} + \cdots$

5. $1 + 2 + 4 + 8 + \cdots$

6. $1 - \dfrac{1}{2} + \dfrac{1}{4} - \dfrac{1}{8} + \cdots$

7. $4 + 2 + 1 + \dfrac{1}{2} + \cdots$

8. $2 - 6 + 18 - 54 + \cdots$

9. $9 - 3 + 1 - \dfrac{1}{3} + \cdots$

10. $1 - \dfrac{3}{4} + \dfrac{9}{16} - \dfrac{27}{64} + \cdots$

11. $5 + 3 + \dfrac{9}{5} + \dfrac{27}{25} + \cdots$

12. $8 - 4 + 2 - 1 + \cdots$

Change each of the following repeating decimals to $\dfrac{a}{b}$ form, where a and b are integers, $b \neq 0$.

Express $\dfrac{a}{b}$ in reduced form.

13. $0.\overline{6}$ **14.** $0.\overline{5}$ **15.** $0.\overline{23}$ **16.** $0.\overline{47}$ **17.** $0.\overline{427}$ **18.** $0.\overline{129}$

19. $0.1\overline{6}$ **20.** $0.7\overline{3}$ **21.** $0.3\overline{27}$ **22.** $0.5\overline{14}$ **23.** $4.\overline{3}$ **24.** $2.\overline{15}$

13.5 Binomial Expansions

In Chapter 3, we used the pattern $(x + y)^2 = x^2 + 2xy + y^2$ to square binomials and the pattern $(x + y)^3 = x^3 + 3x^2y + 3xy^2 + y^3$ to cube binomials. At this time, we simply want to extend those ideas to arrive at a pattern that will allow us to write the expansion of $(x + y)^n$, where n is *any* positive integer. Let's begin by looking at some specific expansions, which can be verified by direct multiplication.

$$(x + y)^1 = x + y$$
$$(x + y)^2 = x^2 + 2xy + y^2$$
$$(x + y)^3 = x^3 + 3x^2y + 3xy^2 + y^3$$
$$(x + y)^4 = x^4 + 4x^3y + 6x^2y^2 + 4xy^3 + y^4$$
$$(x + y)^5 = x^5 + 5x^4y + 10x^3y^2 + 10x^2y^3 + 5xy^4 + y^5$$

First, note the patterns of the exponents for x and y on a term-by-term basis. The exponents of x begin with the exponent of the binomial, and term by term they decrease by 1, until the last term has x^0, which is 1. The exponents of y begin with 0 ($y^0 = 1$), and term by term they increase by 1, until the last term contains y to the power of the original binomial. In other words, the variables in the expansion of $(x + y)^n$ have the following pattern:

$$x^n, \ x^{n-1}y, \ x^{n-2}y^2, \ x^{n-3}y^3, \ldots, \ xy^{n-1}, \ y^n.$$

Notice that the sum of the exponents of x and y for each term is n.

Next, let's arrange the **coefficients** in the following triangular formation, which yields an easy-to-remember pattern.

$$
\begin{array}{ccccccccc}
 & & & & 1 & & 1 & & \\
 & & & 1 & & 2 & & 1 & \\
 & & 1 & & 3 & & 3 & & 1 \\
 & 1 & & 4 & & 6 & & 4 & & 1 \\
1 & & 5 & & 10 & & 10 & & 5 & & 1
\end{array}
$$

The number of the row of the formation contains the coefficients of the expansion of $(x + y)$ to that power. For example, the fifth row contains $1 \quad 5 \quad 10 \quad 10 \quad 5 \quad 1$, which are the coefficients of the terms of the expansion of $(x + y)^5$. Furthermore, each row can be formed from the previous row as follows:

1. Start and end each row with 1.

2. All other entries result from adding the two numbers in the row immediately above, one number to the left and one number to the right.

Thus, from row 5 we can form row 6 as follows:

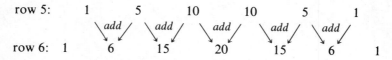

row 5:   1   5   10   10   5   1

row 6: 1   6   15   20   15   6   1

Using the row-6 coefficients and our previous observations relative to the exponents, we can write out the expansion for $(x + y)^6$.

$$(x + y)^6 = x^6 + 6x^5y + 15x^4y^2 + 20x^3y^3 + 15x^2y^4 + 6xy^5 + y^6$$

Remark   The triangular formation of numbers that we have been discussing is often referred to as *Pascal's triangle* in honor of Blaise Pascal, the seventeenth-century mathematician to whom the discovery of this pattern is attributed.

Although Pascal's triangle will work for any positive integral power of a binomial, it does become somewhat impractical for large powers. So we need another technique for determining the coefficients. For this purpose, it is convenient to make the following notational agreements.

$n!$ (read "n factorial") means $n(n - 1)(n - 2) \cdots 1$, where n is any positive integer. For example,

$$3! \text{ means } 3 \cdot 2 \cdot 1 = 6,$$
$$5! \text{ means } 5 \cdot 4 \cdot 3 \cdot 2 \cdot 1 = 120.$$

We also agree that $0! = 1$. (Note that both $0!$ and $1!$ equal 1.)

Let us now use the factorial notation to state the expansion of the general case $(x + y)^n$, where n is any positive integer.

$$(x + y)^n = x^n + nx^{n-1}y + \frac{n(n - 1)}{2!} x^{n-2}y^2 + \frac{n(n - 1)(n - 2)}{3!} x^{n-3}y^3 + \cdots + y^n$$

The binomial expansion for the general case may look a little confusing, but actually it is quite easy to apply once you try it a few times on some specific examples. Remember the decreasing pattern for the exponents of x and the increasing pattern for the exponents of y. Furthermore, note the following pattern of the coefficients.

$$1, \quad n, \quad \frac{n(n - 1)}{2!}, \frac{n(n - 1)(n - 2)}{3!}, \quad \text{etc.}$$

Keep these ideas in mind as you study the following examples.

Example 1   Expand $(x + y)^7$.

Solution
$$(x + y)^7 = x^7 + 7x^6y + \frac{7 \cdot 6}{2!} x^5y^2 + \frac{7 \cdot 6 \cdot 5}{3!} x^4y^3$$
$$+ \frac{7 \cdot 6 \cdot 5 \cdot 4}{4!} x^3y^4 + \frac{7 \cdot 6 \cdot 5 \cdot 4 \cdot 3}{5!} x^2y^5$$
$$+ \frac{7 \cdot 6 \cdot 5 \cdot 4 \cdot 3 \cdot 2}{6!} xy^6 + y^7$$
$$= x^7 + 7x^6y + 21x^5y^2 + 35x^4y^3 + 35x^3y^4 + 21x^2y^5 + 7xy^6 + y^7$$

Example 2   Expand $(x - y)^5$.

Solution   We shall treat $(x - y)^5$ as $[x + (-y)]^5$.
$$[x + (-y)]^5 = x^5 + 5x^4(-y) + \frac{5 \cdot 4}{2!} x^3(-y)^2 + \frac{5 \cdot 4 \cdot 3}{3!} x^2(-y)^3$$
$$+ \frac{5 \cdot 4 \cdot 3 \cdot 2}{4!} x(-y)^4 + (-y)^5$$
$$= x^5 - 5x^4y + 10x^3y^2 - 10x^2y^3 + 5xy^4 - y^5$$

Example 3   Expand and simplify $(2a + 3b)^4$.

Solution   Let $x = 2a$ and $y = 3b$.
$$(2a + 3b)^4 = (2a)^4 + 4(2a)^3(3b) + \frac{4 \cdot 3}{2!}(2a)^2(3b)^2 + \frac{4 \cdot 3 \cdot 2}{3!}(2a)(3b)^3 + (3b)^4$$
$$= 16a^4 + 96a^3b + 216a^2b^2 + 216ab^3 + 81b^4$$

Finding Specific Terms

Sometimes it is convenient to be able to find a specific term of a binomial expansion without writing out the entire expansion. For example, suppose that we need the 6th term of the expansion $(x + y)^{12}$. We could proceed as follows.

The 6th term will contain y^5. Note that in the general expansion the *exponent of y is always 1 less than the number of the term*. Since the sum of the exponents for x and y must be 12 (the exponent of the binomial), the 6th term will also contain x^7. Again looking back at the general binomial expansion, note that each *denominator of a coefficient* is of the form $r!$, where the value of r agrees with the exponent of y for each term. Thus, if we have y^5, the denominator of the coefficient is 5!. In the general expansion each *numerator of a coefficient* contains r factors, where the first factor is the exponent of the binomial and each succeeding factor is 1 less than the preceding one. Thus, the 6th term of $(x + y)^{12}$ is $\dfrac{12 \cdot 11 \cdot 10 \cdot 9 \cdot 8}{5!} x^7 y^5$, which simplifies to $792x^7 y^5$.

Example 4   Find the 4th term of $(3a + 2b)^7$.

Solution   The 4th term will contain $(2b)^3$ and therefore $(3a)^4$. The coefficient is $\dfrac{7 \cdot 6 \cdot 5}{3!}$.

Therefore, the 4th term is $\dfrac{7 \cdot 6 \cdot 5}{3!}(3a)^4 (2b)^3$, which simplifies to $22680a^4 b^3$.   ●

Problem Set 13.5

Use Pascal's triangle to help expand each of the following.

1. $(x + y)^7$ 2. $(x + y)^8$ 3. $(x + 2y)^4$
4. $(3x + y)^4$ 5. $(x - y)^4$ 6. $(x - y)^5$

Expand and simplify each of the following.

7. $(x + y)^9$ 8. $(x + y)^{10}$ 9. $(x + 3y)^5$ 10. $(2x + y)^6$
11. $(2x - y)^6$ 12. $(x - 3y)^5$ 13. $(2a - 3b)^4$ 14. $(3a - 2b)^5$
15. $(x^2 + y)^5$ 16. $(x + y^3)^6$ 17. $(x + 3)^6$ 18. $(x + 2)^7$
19. $(x - 1)^9$ 20. $(x - 3)^4$

Write the first four terms of each of the following expansions.

21. $(x + y)^{12}$ 22. $(x + y)^{15}$ 23. $(x - y)^{20}$ 24. $(a - 2b)^{13}$

Find the specified term for each of the following.

25. $(x + y)^8$, 4th term   26. $(x + y)^{11}$, 7th term   27. $(x - y)^9$, 5th term
28. $(x - 2y)^6$, 4th term   29. $(3a + b)^7$, 6th term   30. $(2x - 5y)^5$, 3rd term

Chapter 13 Summary

(13.1)   An **infinite sequence** is a function whose domain is the set of positive integers. Frequently, a general infinite sequence is expressed as

$$a_1, a_2, a_3, \ldots, a_n, \ldots,$$

where a_1 is the first term, a_2 the second term, and so on, with a_n representing the general, or nth, term.

An **arithmetic sequence** is a sequence for which there is a common difference between successive terms.

The general term of an arithmetic sequence is given by

$$a_n = a_1 + (n - 1)d,$$

where a_1 is the first term and d is the common difference.

(13.2) The indicated sum of a sequence is called a **series**. The sum of the first n terms of an arithmetic series is given by

$$S_n = \frac{n(a_1 + a_n)}{2}.$$

(13.3) A **geometric sequence** is a sequence in which each term after the first is obtained by multiplying the preceding term by a common multiplier. The common multiplier is called the **common ratio** of the sequence.

The general term of a geometric sequence is given by

$$a_n = a_1 r^{n-1},$$

where a_1 is the first term and r is the common ratio.

The sum of the first n terms of a geometric series is given by

$$S_n = \frac{a_1 r^n - a_1}{r - 1}, \qquad r \neq 1.$$

(13.4) The sum of an infinite geometric series is given by

$$S_\infty = \frac{a_1}{1 - r}, \qquad |r| < 1.$$

Any infinite geometric series for which $|r| \geq 1$ has no sum.

(13.5) The expansion of $(x + y)^n$, where n is a positive integer, is given by

$$(x + y)^n = x^n + nx^{n-1}y + \frac{n(n - 1)}{2!} x^{n-2}y^2$$

$$+ \frac{n(n - 1)(n - 2)}{3!} x^{n-3}y^3 + \cdots + y^n.$$

Chapter 13 Review Problem Set

For Problems 1–10, find the general term (nth term) for each of the following sequences. These problems contain a mixture of arithmetic and geometric sequences.

1. $3, 9, 15, 21, \ldots$ **2.** $\frac{1}{3}, 1, 3, 9, \ldots$ **3.** $10, 20, 40, 80, \ldots$

4. $5, 2, -1, -4, \ldots$ 5. $-5, -3, -1, 1, \ldots$ 6. $9, 3, 1, \dfrac{1}{3}, \ldots$

7. $-1, 2, -4, 8, \ldots$ 8. $12, 15, 18, 21, \ldots$ 9. $\dfrac{2}{3}, 1, \dfrac{4}{3}, \dfrac{5}{3}, \ldots$

10. $1, 4, 16, 64, \ldots$

For Problems 11–16, find the indicated term of each of the sequences.

11. The 19th term of $1, 5, 9, 13, \ldots$
12. The 28th term of $-2, 2, 6, 10, \ldots$
13. The 9th term of $8, 4, 2, 1, \ldots$
14. The 8th term of $\dfrac{243}{32}, \dfrac{81}{16}, \dfrac{27}{8}, \dfrac{9}{4}, \ldots$
15. The 34th term of $7, 4, 1, -2, \ldots$
16. The 10th term of $-32, 16, -8, 4, \ldots$
17. If the 5th term of an arithmetic sequence is -19 and the 8th term is -34, find the common difference of the sequence.
18. If the 8th term of an arithmetic sequence is 37 and the 13th term is 57, find the 20th term.
19. Find the first term of a geometric sequence if the 3rd term is 5 and the 6th term is 135.
20. Find the common ratio of a geometric sequence if the 2nd term is $\dfrac{1}{2}$ and the 6th term is 8.
21. Find the sum of the first 9 terms of the sequence $81, 27, 9, 3, \ldots$.
22. Find the sum of the first 70 terms of the sequence $-3, 0, 3, 6, \ldots$.
23. Find the sum of the first 75 terms of the sequence $5, 1, -3, -7, \ldots$.
24. Find the sum of the first 10 terms of the sequence for which $a_n = 2^{5-n}$.
25. Find the sum of the first 95 terms of the sequence for which $a_n = 7n + 1$.
26. Find the sum $5 + 7 + 9 + \cdots + 137$.
27. Find the sum $64 + 16 + 4 + \cdots + \dfrac{1}{64}$.
28. Find the sum of all even numbers between 8 and 384, inclusive.
29. Find the sum of all multiples of 3 between 27 and 276, inclusive.
30. Find the sum of the infinite geometric sequence $64, 16, 4, 1, \ldots$.
31. Change $0.\overline{36}$ to reduced $\dfrac{a}{b}$ form, where a and b are integers, $b \neq 0$.
32. Change $0.4\overline{5}$ to reduced $\dfrac{a}{b}$ form, where a and b are integers, $b \neq 0$.

Solve Problems 37–40 by using your knowledge of arithmetic and geometric sequences.

33. Suppose that at the beginning of the year your savings account contains $3750. If you withdraw $250 per month from the account, how much will it contain at the end of the year?
34. Sonya has decided to start saving dimes. She plans to save 1 dime the first day of April, 2 dimes the second day, 3 dimes the third day, 4 dimes the fourth day, and so on, for the 30 days of April. How much money will she save in April?

35. Nancy has decided to start saving dimes. She plans to save 1 dime the first day of April, 2 dimes the second day, 4 dimes the third day, 8 dimes the fourth day, and so on, for the first 15 days of April. How much will she save in 15 days?

36. A tank contains 61,440 gallons of water. Each day one-fourth of the water is to be drained out. How much will remain in the tank at the end of 6 days?

37. An object falling from rest in a vacuum falls 16 feet the first second, 48 feet the second second, 80 feet the third second, 112 feet the fourth second, and so on. How far will the object fall in 15 seconds?

14 Counting Techniques and Probability

When you are dealt a five-card hand from an ordinary deck of 52 playing cards, there is 1 chance out of 54,145 that you will be dealt four aces. The weatherman is predicting a 40% chance of severe thunderstorms by late afternoon. The odds in favor of the Cubs' winning the pennant are 2 to 3. Suppose that in a box containing 50 light bulbs, 45 are good and 5 are burned out; if 2 bulbs are chosen at random, the probability of getting at least 1 good bulb is $\frac{243}{245}$.

Historically, many basic probability concepts were developed as a result of studying various games of chance. However, in recent years probability applications have been surfacing at a phenomenal rate in a large variety of fields, such as physics, biology, psychology, economics, insurance, military science, manufacturing, and politics. It is the purpose in this chapter first to introduce some counting techniques and then to use those ideas to introduce some basic concepts of probability.

14.1 Fundamental Principle of Counting

One very useful counting strategy is referred to as the **Fundamental Principle of Counting**. After looking at some examples of the property, we will state the property and then use it to solve a variety of counting-type problems.

Example 1   A woman has 4 skirts and 5 blouses. Assuming that each blouse can be worn with each skirt, how many different skirt-blouse outfits does the woman have?

Solution   For each of the 4 skirts, the woman has a choice of 5 blouses. Therefore, she has 4(5) = 20 different skirt-blouse outfits from which to choose. ●

Example 2   Eric is shopping for a new bicycle. He has 2 different models (five-speed or ten-speed) and 4 different colors (red, white, blue, or silver) from which to choose. How many different choices does he have?

Solution   Eric's different choices can be "counted" through a **tree diagram** as follows.

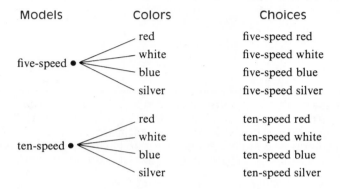

For each of the 2 model choices, there are 4 choices of color. So altogether Eric has $2 \cdot 4 = 8$ choices.   ●

The two previous examples motivate the following general principle.

Fundamental Principle of Counting

If one task can be accomplished in x different ways, and following this task, a second task can be accomplished in y different ways, then the first task followed by the second task can be accomplished in $x \cdot y$ different ways. (This counting principle can be extended to any finite number of tasks.)

In order to apply the Fundamental Principle of Counting, it is often helpful to systematically analyze a problem in terms of the tasks to be accomplished. Let's consider some examples illustrating this idea.

Example 3   How many numbers made up of 3 different digits each can be formed by choosing from the digits 1, 2, 3, 4, 5, and 6?

Solution   Let's analyze this problem in terms of three tasks as follows.

1. Choose the hundred's digit, for which there are 6 choices.
2. Now choose the ten's digit, for which there are only 5 choices, since 1 digit was used in the hundred's place.
3. Now choose the unit's digit, for which there are only 4 choices, since 2 digits have been used for the other places.

Therefore, Task 1 followed by Task 2 followed by Task 3 can be accomplished in $(6)(5)(4) = 120$ ways. In other words, there are 120 numbers made up of 3 different digits that can be formed by choosing from the 6 given digits. ●

To ensure that you understand the principle, answer the following questions, looking back over the solution for Example 3 as necessary.

1. Could the problem be solved by choosing the unit's digit first, then the ten's digit, and finally the hundred's digit?

2. How many 3-digit numbers could be formed from 1, 2, 3, 4, 5, and 6 if it were not required that each number have three *different* digits? (Your answer should be 216.)

3. Suppose that the available digits were 0, 1, 2, 3, 4, and 5. Now how many numbers made up of 3 different digits each could be formed, assuming that 0 cannot be in the hundred's place? (Your answer should be 100.)

4. How many *even* numbers made up of 3 different digits each could be formed by choosing from the digits 1, 2, 3, 4, 5, and 6? (Your answer should be 60.)

Example 4 Employee ID numbers at a certain factory consist of one capital letter followed by a 3-digit number containing no repeat digits. (For example, A-014 is an ID number.) How many such ID numbers can be formed? How many could be formed if repeat digits were allowed?

Solution Again, let's analyze in terms of tasks to be completed.

1. Choose the letter part of the ID number; there are 26 choices.

2. Choose the first digit of the three-digit number; there are 10 choices.

3. Choose the second digit; there are 9 choices.

4. Choose the third digit; there are 8 choices.

Therefore, applying the Fundamental Principle of Counting, we obtain $(26)(10)(9)(8) = 18{,}720$ possible ID numbers.

If repeat digits were allowed, then there would be $(26)(10)(10)(10) = 26{,}000$ possible ID numbers. ●

Example 5 How many ways can Al, Barb, Chad, Dan, and Edna be seated in a row of 5 seats so that Al and Barb are seated side by side?

Solution This problem can be analyzed in terms of the following tasks.

1. Choose 2 adjacent seats for Al and Barb. An illustration such as will help you to visualize that there are 4 choices for the 2 adjacent seats.

2. Determine the number of ways that Al and Barb can be seated. Since Al can be seated on the left and Barb on the right, or vice versa, there are 2 ways to seat Al and Barb for each of the 4 choices of the 2 adjacent seats.

3. The remaining 3 people must be seated in the remaining 3 seats. This can be done in $(3)(2)(1) = 6$ different ways.

Therefore, by the Fundamental Principle of Counting, Task 1 followed by Task 2 followed by Task 3 can be done in $(4)(2)(6) = 48$ ways.

Suppose that in Example 5 we wanted the number of ways in which the 5 people could be seated so that Al and Barb were *not* side by side. We could determine this number either by (1) analyzing and counting the number of "not side-by-side positions" for Al and Barb or (2) subtracting the side-by-side combinations determined in Example 5 from the total number of ways that 5 people can be seated in 5 seats. Do this problem both ways and see if you come up with the answer of 72 ways.

As you apply the Fundamental Principle of Counting, you may find that for certain problems simply thinking about an appropriate tree diagram is helpful, even though the size of the problem prohibits writing out the diagram in detail. Consider the following example.

Example 6   Suppose that the undergraduate students in three departments—geography, history, and psychology—are to be classified according to sex and year in school. How many categories are needed?

Solution   The various classifications can be symbolically represented as follows.

M: Male   1. Freshman   G: Geography
F: Female   2. Sophomore   H: History
            3. Junior   P: Psychology
            4. Senior

We can mentally picture a tree diagram in which each of the 2 sex classifications branches into 4 year-in-school classifications, which in turn branch into 3 department classifications. Thus, we have

$(2)(4)(3) = 24$ different categories.   ●

Another technique that works on certain problems involves what some people call the "back-door" approach. Suppose we know that our classroom contains 50 seats. To determine the number of students present on a given day, it may be easier to count the number of empty seats and subtract from 50, rather than to count the number of students in attendance. We suggested this back-door approach as one way to determine the "not side-by-side" seating arrangements back in the discussion following Example 5, and the next example further illustrates this approach.

Example 7   When a pair of dice is rolled, in how many ways can a sum greater than 4 be obtained?

Solution   For purposes of clarification, let's use a red die and a white die. (It is not necessary to use different colored dice, but it does help us analyze the different possible outcomes.)

It should be intuitively obvious that there are more ways of getting a sum greater than 4 than there are ways of getting a sum of 4 or less. Therefore, let's determine the possibilities for getting a sum of 4 or less, and then subtract from the total number of possible outcomes when a pair of dice is rolled.

First, we can simply list and count the ways of getting a sum of 4 or less.

Red die	White die
1	1
1	2
1	3
2	1
2	2
3	1

There are 6 ways of getting a sum of 4 or less.

Second, since there are 6 possible outcomes on the red die and 6 possible outcomes on the white die, there is a total of $(6)(6) = 36$ possible outcomes when a pair of dice is rolled.

Therefore, subtracting the number of ways of getting a sum of 4 or less from the total number of possible outcomes, we obtain $36 - 6 = 30$ ways of getting a sum greater than 4. ●

Problem Set 14.1

1. If a woman has 2 skirts and 10 blouses, how many different skirt-blouse combinations does she have?

2. If a man has 8 shirts, 5 pairs of slacks, and 3 pairs of shoes, how many different shirt-slack-shoe combinations does he have?

3. In how many ways can 4 people be seated in a row of 4 seats?

4. How many numbers of 2 different digits can be formed by choosing from the digits 1, 2, 3, 4, 5, 6, and 7?

5. How many even numbers of 3 different digits can be formed by choosing from the digits 2, 3, 4, 5, 6, 7, 8, and 9?

6. How many odd numbers of 4 different digits can be formed by choosing from the digits 1, 2, 3, 4, 5, 6, 7, and 8?

7. Suppose that the students at a certain university are to be classified according to college (College of Applied Science, College of Arts and Sciences, College of Business, College of Education, College of Fine Arts, College of Health and Physical Education), sex (female, male), and year in school (1, 2, 3, 4). How many categories are possible?

8. A medical researcher classifies subjects according to sex (female, male), smoking habits, (smoker, nonsmoker), and weight (below average, average, above average). How many different combined classifications are used?

9. A pollster classifies voters according to sex (female, male), party affiliation (Democrat, Republican, Independent) and family income (below $10,000, $10,000–$19,999, $20,000–$29,999, $30,000–$39,999, $40,000–$49,999, $50,000 and above). How many combined classifications does the pollster use?

10. A couple is planning to have 4 children. How many different arrangements of boys and girls are possible? (For example, *BBBG* would indicate that the first 3 children were boys and the last was a girl.)

11. In how many ways can 3 officers (president, secretary, and treasurer) be selected from a club that has 20 members?

12. In how many ways can 3 officers (president, secretary, and treasurer) be selected from a club with 15 female and 10 male members so that the president is female and the secretary and the treasurer are males?

13. A disc jockey has 6 tunes to be played once each in a half-hour program. How many different orderings of these tunes are possible?

14. State officials have decided to have their state's automobile license plates consist of 2 letters followed by 4 digits. They do not want to repeat any letters or digits in any license number. How many different license numbers do they have available?

15. In how many ways can 6 people be seated in a row of 6 seats?

16. In how many ways can Al, Bob, Carl, Don, Ed, and Fern be seated in a row of 6 seats if Al and Bob are to be seated side by side?

17. In how many ways can Amy, Bob, Cindy, Dan, and Elmer be seated in a row of 5 seats so that neither Amy nor Bob occupies an end seat?

18. In how many ways can Al, Bob, Carl, Don, Ed, and Fern be seated in a row of 6 seats if Al and Bob are not to be seated side by side? (Hint: Al and Bob will either be seated side by side or not be seated side by side.)

19. In how many ways can Al, Bob, Carol, Dawn, and Ed be seated in a row of 5 chairs if Al is to be seated in the middle chair?

20. In how many ways can 3 letters be dropped into 5 mail boxes?

21. In how many ways can 5 letters be dropped into 3 mail boxes?

22. In how many ways can 4 letters be dropped into 6 mail boxes so that no 2 letters are dropped into the same box?

23. In how many ways can 6 letters be dropped into 4 mail boxes so that no 2 letters are dropped into the same box?

24. If 5 coins are tossed, in how many ways can they fall?

25. If 3 dice are tossed, in how many ways can they fall?

26. In how many ways can a sum less than 10 be obtained when a pair of dice is tossed?

27. In how many ways can a sum greater than 5 be obtained when a pair of dice is tossed?

28. In how many ways can a sum greater than 4 be obtained when 3 dice are tossed?

29. How many numbers greater than 400, where no number contains repeated digits, can be formed from the digits 2, 3, 4, and 5? (Hint: Consider both 3-digit and 4-digit numbers.)

30. How many numbers greater than 5000, where no number contains repeated digits, can be formed by choosing from the digits 1, 2, 3, 4, 5, and 6?

31. In how many ways can 4 boys and 3 girls be seated in a row of 7 seats so that boys and girls occupy alternate seats?

32. In how many ways can 3 different mathematics books and 4 different history books be exhibited on a shelf so that all of the books in a subject area are side by side?

33. In how many ways can a true-false test of 10 questions be answered?

34. How many even numbers greater than 3000, where no number contains repeated digits, can be formed from the digits 1, 2, 3, and 4?

35. How many odd numbers greater than 40,000, where no number contains repeated digits, can be formed from the digits 1, 2, 3, 4, and 5?

36. In how many ways can Al, Bob, Carol, Don, Ed, Faye, and George be seated in a row of 7 seats so that Al, Bob, and Carol occupy consecutive seats in some order?

37. The license plates for a certain state consist of 2 letters followed by a 4-digit number such that the first digit of the number is not zero. An example would be PK-2446.

    (a) How many different license plates can be produced?

    (b) How many different plates would not have the same letter repeated?

    (c) How many plates would not have any digits repeated in the number part of the plate?

    (d) How many plates would not have the same letter repeated and would not have any digits repeated in the number part of the plate?

14.2 Permutations and Combinations

As we develop the material in this section, **factorial notation** will become very useful. The notation $n!$ (read "n factorial") is used with positive integers as follows.

$1! = 1$

$2! = 2 \cdot 1 = 2$

$3! = 3 \cdot 2 \cdot 1 = 6$ Notice that the factorial notation refers to an *indicated product.*

$4! = 4 \cdot 3 \cdot 2 \cdot 1 = 24$

In general, we write

$$n! = n(n-1)(n-2) \cdot \cdots \cdot 3 \cdot 2 \cdot 1.$$

Now, as a lead-in to the first concept of this section, let's consider a counting problem that closely resembles some from the previous section.

Example 1   In how many ways can the 3 letters A, B, and C be arranged in a row?

Solution A   Certainly one approach to the problem is to simply list and count the arrangements.

$ABC, ACB, BAC, BCA, CAB, CBA$

There are 6 arrangements of the 3 letters.

Solution B   Another approach, which can be generalized for more difficult problems, uses the Fundamental Principle of Counting. Since there are 3 choices for the first letter of an arrangement, 2 choices for the second letter, and 1 choice for the third letter, there are $(3)(2)(1) = 6$ different arrangements. ●

Ordered arrangements are called **permutations**. In general, a permutation of a set of n elements is an ordered arrangement of the n elements, and we will use the symbol $P(n, n)$ to note the number of these permutations. For example, from Example 1 we know that $P(3, 3) = 6$. Furthermore, by using the same basic approach as in Solution B of Example 1, we can obtain the following results.

$$
\begin{aligned}
P(1, 1) &= & 1 & = 1! \\
P(2, 2) &= & 2 \cdot 1 & = 2! \\
P(4, 4) &= & 4 \cdot 3 \cdot 2 \cdot 1 & = 4! \\
P(5, 5) &= 5 \cdot 4 \cdot 3 \cdot 2 \cdot 1 & = 5!
\end{aligned}
$$

The following formula becomes evident.

$$
P(n, n) = n!
$$

Now suppose that we are interested in the number of 2-letter permutations that can be formed by choosing from the 4 letters A, B, C, and D. (Some such permutations are AB, BA, AC, CA, BC, and CB.) In other words, we want to find the number of 2-element permutations that can be formed from a set of 4 elements. We denote this number by $P(4, 2)$. To find $P(4, 2)$, we can reason as follows: First, choose any one of the 4 letters to occupy the first position in the permutation, and then choose any one of the 3 remaining letters for the second position. Therefore, by the Fundamental Principle of Counting, we have $4(3) = 12$ different 2-letter permutations; that is, $P(4, 2) = 12$. The following can be determined using a similar line of reasoning. (Make sure that you agree with each of these.)

$$
\begin{aligned}
P(4, 3) &= 4 \cdot 3 \cdot 2 = 24 \\
P(5, 2) &= 5 \cdot 4 = 20 \\
P(6, 4) &= 6 \cdot 5 \cdot 4 \cdot 3 = 360 \\
P(7, 3) &= 7 \cdot 6 \cdot 5 = 210
\end{aligned}
$$

In general, we say that the *number of r-element permutations that can be formed from a set of n elements is given by*

$$
P(n, r) = \underbrace{n(n - 1)(n - 2) \cdots}_{r \text{ factors}}.
$$

Notice that the indicated product for $P(n, r)$ begins with n, thereafter each factor is 1 less than the previous one, and there is a total of r factors. Following are some examples.

$$
\begin{aligned}
P(6, 2) &= 6 \cdot 5 = 30 \\
P(8, 3) &= 8 \cdot 7 \cdot 6 = 336 \\
P(9, 4) &= 9 \cdot 8 \cdot 7 \cdot 6 = 3024
\end{aligned}
$$

Let's consider two examples illustrating the use of $P(n, n)$ and $P(n, r)$.

Example 2   In how many ways can 5 students be seated in a row of 5 seats?

Solution   The problem is asking for the number of 5-element permutations that can be formed from a set of 5 elements. Thus, we can apply $P(n, n) = n!$.

$$P(5, 5) = 5! = 5 \cdot 4 \cdot 3 \cdot 2 \cdot 1 = 120$$ ●

Example 3   Suppose that 7 people enter a swimming race. In how many ways can the first-, second-, and third-place ribbons be awarded?

Solution   This problem is asking for the number of 3-element permutations that can be formed from a set of 7 elements. Therefore, using the formula for $P(n, r)$, we obtain

$$P(7, 3) = 7 \cdot 6 \cdot 5 = 210.$$ ●

It should be evident that both Example 2 and Example 3 could have been solved by applying the Fundamental Principle of Counting. In fact, it should be noted that the formulas for $P(n, n)$ and $P(n, r)$ do not really give us much additional problem-solving power. However, as we will see in a moment, they do provide the basis for developing a formula that is very useful as a problem-solving tool.

Permutations Involving Nondistinguishable Objects

Suppose that we have 2 identical H's and 1 T in an arrangement such as HTH. Obviously, if we switched the 2 identical H's, the newly formed arrangement, HTH, would not be distinguishable from the original. In other words, if some of the n elements are identical, there are fewer distinguishable permutations than there would be of n distinctly different elements.

To see the effect of identical elements on the number of distinguishable permutations, let's look at some specific examples.

2 identical H's:	1 permutation—namely, HH
2 different letters:	2! permutations

Therefore, having 2 different letters affects the number of permutations by a *factor of 2!*.

3 identical H's:	1 permutation—namely, HHH
3 different letters:	3! permutations

Therefore, having 3 different letters affects the number of permutations by a *factor of 3!*.

4 identical H's:	1 permutation—namely, $HHHH$
4 different letters:	4! permutations

Therefore, having 4 different letters affects the number of permutations by a *factor of 4!*.

Now let's solve a specific problem.

Example 4   How many distinguishable permutations can be formed from 3 identical H's and 2 identical T's?

Solution   If we had 5 distinctly different letters, we could form 5! permutations. But the 3 identical H's affect the number of distinguishable permutations by a factor of 3!, and the 2 identical T's affect the number of permutations by a factor of 2!. Therefore, we must divide 5! by 3! and 2!. Thus, we obtain

$$\frac{5!}{(3!)(2!)} = \frac{5 \cdot \overset{2}{\cancel{4}} \cdot \cancel{3} \cdot \cancel{2} \cdot 1}{\cancel{3} \cdot \cancel{2} \cdot 1 \cdot \cancel{2} \cdot 1} = 10$$

distinguishable permutations of $3H$'s and $2T$'s.   ●

The type of reasoning used in Example 4 leads us to the following general counting technique: If there are n elements to be arranged, where there are r_1 of one kind, r_2 of another kind, r_3 of another kind, ... r_k of a kth kind, then the total number of distinguishable permutations is given by the expression

$$\frac{n!}{(r_1!)(r_2!)(r_3!)\cdots(r_k!)}.$$

Example 5   How many different 11-letter permutations can be formed from the 11 letters of the word $MISSISSIPPI$?

Solution   Since there are 4 I's, 4 S's, and 2 P's, we can form

$$\frac{11!}{(4!)(4!)(2!)} = \frac{11 \cdot 10 \cdot 9 \cdot 8 \cdot 7 \cdot 6 \cdot 5 \cdot 4 \cdot 3 \cdot 2 \cdot 1}{4 \cdot 3 \cdot 2 \cdot 1 \cdot 4 \cdot 3 \cdot 2 \cdot 1 \cdot 2 \cdot 1} = 34,650$$

distinguishable permutations.   ●

Combinations (Subsets)

Permutations are ordered arrangements; however, frequently *order* is not a consideration. For example, suppose that we want to determine the number of 3-person committees that can be formed from the 5 people Al, Barb, Carol, Dawn, and Eric. Certainly the committee consisting of Al, Barb, and Eric is the same as the committee consisting of Barb, Eric, and Al. In other words, the order in which we choose or list the members is not to be considered. Therefore, we are really dealing with subsets; that is, we are looking for the number of 3-element subsets that can be formed from a set of 5 elements. Traditionally, in this context, subsets are called **combinations**. So, stated another way, we are looking for the number of combinations of 5 things taken 3 at a time. In general, *r*-element subsets taken from a set of *n* elements are called *combinations of n things taken r at a time*. The symbol $C(n, r)$ will denote the number of these combinations.

Now let's restate the previous committee problem and look at a detailed solution that can be generalized to handle a variety of problems dealing with combinations.

Example 6   How many 3-person committees can be formed from the five people Al, Barb, Carol, Dawn, and Eric?

Solution   Let's use the set $\{A, B, C, D, E\}$ to represent the 5 people. Consider one possible 3-person committee (subset), such as $\{A, B, C\}$. There are 3! permutations of these 3 letters. Now take another committee, such as $\{A, B, D\}$. There are also 3! permutations of these 3 letters. If we were to continue this process with all of the 3-letter subsets that can be formed from the 5 letters, we would obtain all possible 3-letter permutations of the 5 letters. That is, we would obtain $P(5, 3)$. Therefore, if we let $C(5, 3)$ represent the number of 3-element subsets, then

$$[C(5, 3)](3!) = P(5, 3).$$

Solving this equation for $C(5, 3)$ yields

$$C(5, 3) = \frac{P(5, 3)}{3!} = \frac{5 \cdot 4 \cdot 3}{3 \cdot 2 \cdot 1} = 10.$$

So there are 10 3-person committees that can be formed from the 5 people.   ●

In general, $C(n, r)$ times $r!$ yields $P(n, r)$. Thus,

$$[C(n, r)](r!) = P(n, r),$$

and solving this equation for $C(n, r)$ produces

$$C(n, r) = \frac{P(n, r)}{r!}.$$

In other words, we can find the number of combinations of *n* things taken *r* at a time by dividing the number of permutations of *n* things taken *r* at a time by *r*!. The following examples illustrate this process.

$$C(7, 3) = \frac{P(7, 3)}{3!} = \frac{7 \cdot 6 \cdot 5}{3 \cdot 2 \cdot 1} = 35$$

$$C(9, 2) = \frac{P(9, 2)}{2!} = \frac{9 \cdot 8}{2 \cdot 1} = 36$$

$$C(10, 4) = \frac{P(10, 4)}{4!} = \frac{10 \cdot 9 \cdot 8 \cdot 7}{4 \cdot 3 \cdot 2 \cdot 1} = 210$$

Example 7   How many different 5-card hands can be dealt from a deck of 52 playing cards?

Solution   Since the order in which the cards are dealt is not an issue, we are working with a combination (subset) problem. Thus, using the formula for $C(n, r)$, we obtain

$$C(52, 5) = \frac{P(52, 5)}{5!} = \frac{52 \cdot 51 \cdot 50 \cdot 49 \cdot 48}{5 \cdot 4 \cdot 3 \cdot 2 \cdot 1} = 2{,}598{,}960.$$

There are 2,598,960 different 5-card hands that can be dealt from a deck of 52 playing cards.   ●

Some counting problems can be solved by using the Fundamental Principle of Counting along with the combination formula, as the next example illustrates.

Example 8   How many committees consisting of 3 women and 2 men can be formed from a group of 5 women and 4 men?

Solution   Let's think of this problem in terms of two tasks.

1. Choose a subset of 3 women from the 5 women. This can be done in

$$C(5, 3) = \frac{P(5, 3)}{3!} = \frac{5 \cdot 4 \cdot 3}{3 \cdot 2 \cdot 1} = 10 \text{ ways.}$$

2. Choose a subset of 2 men from the 4 men. This can be done in

$$C(4, 2) = \frac{P(4, 2)}{2!} = \frac{4 \cdot 3}{2 \cdot 1} = 6 \text{ ways.}$$

Task 1 followed by Task 2 can be done in $(10)(6) = 60$ ways. Therefore, there are 60 committees consisting of 3 women and 2 men that can be formed.   ●

Sometimes it takes a little thought to decide whether permutations or combinations should be used. Remember that if order is to be considered, permutations should be used, but if order does not matter, then combinations should be used. It may be helpful to think of combinations as subsets.

Example 9   A small accounting firm has 12 computer programmers. Three of these people are to be promoted to systems analysts. In how many ways can the 3 people to be promoted be selected?

Solution   Let's call the people $A, B, C, D, E, F, G, H, I, J, K$, and L. Suppose A, B, and C are chosen to be promoted. Is this any different than choosing B, C, and A? Obviously not. Therefore, order does not matter, and the problem is asking about combinations. Specifically, we need to find the number of combinations of 12 people taken 3 at a time.

Thus, we obtain

$$C(12, 3) = \frac{P(12, 3)}{3!} = \frac{12 \cdot 11 \cdot 10}{3 \cdot 2 \cdot 1} = 220$$

different ways that the 3 people to be promoted can be chosen.   ●

Example 10   A club is to elect 3 officers—president, secretary, and treasurer—from a group of 6 people who are each willing to serve in any office. How many different ways can the officers be chosen?

Solution   Let's call the people A, B, C, D, E, and F. Is electing A as president, B as secretary, and C as treasurer different from electing B as president, C as secretary, and A as treasurer? Obviously it is, and therefore we are working with permutations. Thus, we obtain

$$P(6, 3) = 6 \cdot 5 \cdot 4 = 120$$

different ways of filling the offices.   ●

Problem Set 14.2

In Problems 1–12, evaluate the expression.

1. $P(5, 3)$ **2.** $P(8, 2)$ **3.** $P(6, 4)$ **4.** $P(9, 3)$

5. $C(7, 2)$ **6.** $C(8, 5)$ **7.** $C(10, 5)$ **8.** $C(12, 4)$

9. $C(15, 2)$ **10.** $P(5, 5)$ **11.** $C(5, 5)$ **12.** $C(11, 1)$

13. How many permutations of the 4 letters A, B, C, and D can be formed by using all letters in each permutation?

14. In how many ways can 6 students be seated in a row of 6 seats?

15. How many 3-person committees can be formed from a group of 9 people?

16. How many 2-card hands can be dealt from a deck of 52 playing cards?

17. How many 3-letter permutations can be formed from the first 8 letters of the alphabet if

    (a) repetitions are not allowed?

    (b) repetitions are allowed?

18. In a 7-team baseball league, in how many ways can the top 3 positions in the final standings be filled?

19. In how many ways can the manager of a baseball team arrange his batting order if he wants his 4 best hitters in the top 4 positions?

20. In a baseball league of 9 teams, how many games are needed to complete the schedule if each team must play 12 games with every other team?

21. How many committees consisting of 4 women and 4 men can be chosen from a group of 7 women and 8 men?

22. How many 3-element subsets containing 1 vowel and 2 consonants can be formed from the set $\{a, b, c, d, e, f, g, h, i\}$?

23. There are 5 associate professors being considered for promotion to the rank of full professor, but only 3 will be promoted. How many different ways are there of selecting the 3 to be promoted?

24. From the digits 1, 2, 3, 4, 5, 6, 7, 8, and 9, how many numbers of 4 different digits each can be formed if each number consists of 2 odd and 2 even digits?

25. How many 3-element subsets containing the letter A can be formed from the set $\{A, B, C, D, E, F\}$?

26. How many 4-person committees can be chosen from a group of 5 women and 3 men so that each committee contains at least 1 man?

27. How many different 7-letter permutations can be formed from 4 identical H's and 3 identical T's?

28. How many different 8-letter permutations can be formed from 6 identical H's and 2 identical T's?

29. How many different 9-letter permutations can be formed from 3 identical A's, 4 identical B's, and 2 identical C's?

30. How many different 10-letter permutations can be formed from 5 identical A's, 4 identical B's, and a C?

31. How many different 7-letter permutations can be formed from the 7 letters of the word *ALGEBRA*?

32. How many different 11-letter permutations can be formed from the 11 letters of the word *MATHEMATICS*?

33. In how many ways can x^4y^2 be written without using exponents? (Hint: One way is *xxxxyy*.)

34. In how many ways can $x^3y^4z^3$ be written without using exponents?

35. In how many ways can 10 basketball players be divided into 2 teams of 5 players each?

36. In how many ways can 10 basketball players be divided into 2 teams of 5 players so that the 2 best players are on opposite teams?

37. A box contains 9 good light bulbs and 4 defective bulbs. How many samples of 3 bulbs would contain 1 defective bulb? How many samples of 3 bulbs would contain *at least* 1 defective bulb?

38. How many 5-person committees consisting of 2 juniors and 3 seniors can be formed from a group of 6 juniors and 8 seniors?

39. In how many ways can 6 people be divided into 2 groups so that there are 4 in one group and 2 in the other group? How many ways could 6 people be divided into groups of 3 each?

40. How many 5-element subsets containing both A and B can be formed from the set $\{A, B, C, D, E, F, G, H\}$?

41. How many 4-element subsets containing A or B, but not both A and B, can be formed from the set $\{A, B, C, D, E, F, G\}$?

42. How many different 5-person committees can be selected from a group of 9 people if 2 of those people refuse to serve together on a committee?

43. How many different line segments are determined by 5 points? By 6 points? By 7 points? By n points?

44. (a) How many 5-card hands consisting of 2 kings and 3 aces can be dealt from a deck of 52 playing cards?

(b) How many 5-card hands consisting of 3 kings and 2 aces can be dealt from a deck of 52 playing cards?

(c) How many 5-card hands consisting of 3 cards of one face value and 2 cards of another face value can be dealt from a deck of 52 playing cards?

Miscellaneous Problems

45. In how many ways can 6 people be seated at a circular table? [Hint: Moving each person one place to the right (or left) does not create a new seating arrangement.]

46. The meaning of $P(8, 3)$ can be expressed completely in factorial notation as follows.

$$P(8, 3) = \frac{P(8, 3) \cdot 5!}{5!} = \frac{(8 \cdot 7 \cdot 6)(5 \cdot 4 \cdot 3 \cdot 2 \cdot 1)}{5!} = \frac{8!}{5!}$$

Express each of the following in terms of factorial notation.

(a) $P(7, 3)$ (b) $P(9, 2)$ (c) $P(10, 7)$

(d) $P(n, r)$, where $r \leq n$ and $0!$ is defined to be 1

47. Sometimes the formula $C(n, r) = \dfrac{n!}{r!(n - r)!}$ is used to find the number of combinations of n things taken r at a time. Using the result from Problem 46(d), develop this formula.

48. Compute $C(7, 3)$ and $C(7, 4)$. Compute $C(8, 2)$ and $C(8, 6)$. Compute $C(9, 8)$ and $C(9, 1)$. Now argue that $C(n, r) = C(n, n - r)$ for $r \leq n$.

14.3 Probability

For the purpose of introducing some terminology and symbolism, let's consider a simple experiment of tossing a regular 6-sided die. There are 6 possible outcomes to this experiment; either 1, 2, 3, 4, 5, or 6 will be on the side of the die facing up. The set of all possible outcomes of a given experiment is called a **sample space**, and the individual elements of the sample space are called **sample points**. (In this text we will be working only with finite sample spaces.) We will use S (sometimes with subscripts for identification purposes) to refer to a particular sample space of an experiment; the number of sample points will be denoted by $n(S)$. Thus, $S = \{1, 2, 3, 4, 5, 6\}$, where $n(S) = 6$, can be used for the experiment of tossing a die.

Now suppose that we are interested in some of the various possible outcomes of the die-tossing experiment. For example, we might be interested in the event "an even number comes up." In this case we are satisfied if 2, 4, or 6 appears on the up-face of the die; therefore, the event "an even number comes up" is the subset $E = \{2, 4, 6\}$, where $n(E) = 3$. Or we might be interested in the event "a multiple of 3 comes up." This would be the subset $F = \{3, 6\}$, where $n(F) = 2$.

In general, any subset of a sample space is called an **event** or **event space**. If the event consists of exactly one element of the sample space, then it is called a **simple event**.

Any nonempty event that is not simple is called a **compound event**. A compound event can be represented as the union of simple events.

It is now possible to give a very simple definition for **probability** as we will use it in this text.

Definition 14.1	The *probability* of an event E in an experiment in which all of the possible outcomes in the sample space S are equally likely to occur is defined by $$P(E) = \frac{n(E)}{n(S)},$$ where $n(E)$ denotes the number of elements in the event E and $n(S)$ denotes the number of elements in the sample space S.

Many probability problems can be solved by applying Definition 14.1. We merely need to be able to determine the number of elements in the sample space and the number of elements in the event space. For example, returning to the die-tossing experiment, the probability of getting an even number with 1 toss of a die is given by

$$P(E) = \frac{n(E)}{n(S)} = \frac{3}{6} = \frac{1}{2}.$$

Let's consider two examples for which the number of elements in both the sample space and the event space is quite easy to determine.

Example 1   A coin is tossed. Find the probability that a head (H) turns up.

Solution   Let the sample space be $S = \{H, T\}$, where $n(S) = 2$. The event of turning up a head is the subset $E = \{H\}$, where $n(E) = 1$. Therefore, we say that the probability of getting a head with 1 flip of a coin is given by

$$P(E) = \frac{n(E)}{n(S)} = \frac{1}{2}.$$

Example 2   What is the probability that at least 1 head will turn up if 2 coins are tossed?

Solution   For purposes of clarification, let the coins be a penny and a nickel. The possible outcomes of this experiment are (1) a head on both coins, (2) a head on the penny and a tail on the nickel, (3) a tail on the penny and a head on the nickel, and (4) a tail on both coins. Using ordered-pair symbolism in which the first entry of a pair records the result of the penny and the second entry records the result of the nickel, we have a sample space of

$$S = \{(H, H), (H, T), (T, H), (T, T)\},$$

where $n(S) = 4$.

Let E be the event of getting at least 1 head. Thus,

$E = \{(H, H), (H, T), (T, H)\}$

and $n(E) = 3$. Therefore, the probability of getting at least 1 head with one toss of 2 coins is

$$P(E) = \frac{n(E)}{n(S)} = \frac{3}{4}.$$ ●

As you might expect, the counting techniques discussed in the first two sections of this chapter can frequently be used to help solve probability problems.

Example 3   Find the probability of getting 3 heads and 1 tail if 4 coins are tossed.

Solution   The sample space consists of the possible outcomes for tossing 4 coins. Since there are 2 things that can happen on each coin, by the Fundamental Principle of Counting there are $2 \cdot 2 \cdot 2 \cdot 2 = 16$ possible outcomes for tossing 4 coins. So we know that $n(S) = 16$ without taking the time to list all of the elements. The event of getting 3 heads and 1 tail is the subset

$E = \{(H, H, H, T), (H, H, T, H), (H, T, H, H), (T, H, H, H)\},$

where $n(E) = 4$. Therefore, the requested probability is

$$P(E) = \frac{n(E)}{n(S)} = \frac{4}{16} = \frac{1}{4}.$$ ●

Example 4   Al, Bob, Chad, Dawn, Eve, and Francis are randomly seated in a row of 6 chairs. What is the probability that Al and Bob are seated in the end seats?

Solution   The sample space consists of all possible ways of seating 6 people in 6 chairs—in other words, the permutations of 6 things taken 6 at a time. Thus,

$n(S) = P(6, 6) = 6! = 6 \cdot 5 \cdot 4 \cdot 3 \cdot 2 \cdot 1 = 720.$

The event space consists of all possible ways of seating the 6 people so that Al and Bob occupy end seats. The number of these possibilities can be counted as follows.

1. Put Al and Bob in the end seats. This can be done in 2 ways, since Al can be on the left end and Bob on the right end, or vice versa.

2. Put the other 4 people in the remaining 4 seats. This can be done in $4! = 4 \cdot 3 \cdot 2 \cdot 1 = 24$ different ways.

Therefore, Task 1 followed by Task 2 can be done in $(2)(24) = 48$ different ways; so $n(E) = 48$.

Thus, the requested probability is

$$P(E) = \frac{n(E)}{n(S)} = \frac{48}{720} = \frac{1}{15}.$$ ●

Notice that in Example 3 we used the Fundamental Principle of Counting to determine the number of elements in the sample space without actually listing all of the elements. For the event space, we listed the elements and counted them in the usual way. In Example 4 we used the permutation formula $P(n, n) = n!$ to determine the number of elements in the sample space, and then we used the Fundamental Principle of Counting to help determine the number of elements in the event space. There are no definite rules as to when to list the elements and when to apply some sort of counting technique. In general, if you do not see a counting pattern for a particular problem, you should begin the listing process. If a counting pattern then emerges as you are listing the elements, use the pattern at that time.

The combination (subset) formula, $C(n, r) = \dfrac{P(n, r)}{r!}$, developed in Section 14.2, is also a very useful tool for solving certain kinds of probability problems. The next three examples illustrate some problems involving combinations.

Example 5   A committee of 3 people is randomly selected from a group composed of Alice, Barb, Chad, Dee, and Eric. What is the probability that Alice is on the committee?

Solution   The sample space S consists of all possible 3-person committees that can be formed from the 5 people. Therefore,

$$n(S) = C(5, 3) = \frac{P(5, 3)}{3!} = \frac{5 \cdot 4 \cdot 3}{3 \cdot 2 \cdot 1} = 10.$$

The event space E consists of all of the 3-person committees that have Alice as a member. Each of these committees contains Alice and 2 other people chosen from the 4 remaining people. Thus, the number of these committees is given by $C(4, 2)$. So we obtain

$$n(E) = C(4, 2) = \frac{P(4, 2)}{2!} = \frac{4 \cdot 3}{2 \cdot 1} = 6.$$

The requested probability is

$$P(E) = \frac{n(E)}{n(S)} = \frac{6}{10} = \frac{3}{5}.$$

Example 6   From a group of 5 seniors and 4 juniors, a committee of 4 is chosen at random. Find the probability that the committee will contain 2 seniors and 2 juniors.

Solution   The sample space S consists of all possible 4-person committees that can be formed from the 9 people. Thus,

$$n(S) = C(9, 4) = \frac{P(9, 4)}{4!} = \frac{9 \cdot 8 \cdot 7 \cdot 6}{4 \cdot 3 \cdot 2 \cdot 1} = 126.$$

The event space E consists of all of those 4-person committees that contain 2 seniors and 2 juniors. They can be counted as follows.

1. Choose 2 seniors from the 5 available seniors in $C(5, 2) = 10$ ways.

2. Choose 2 juniors from the 4 available juniors in $C(4, 2) = 6$ ways.

Therefore, there are $10 \cdot 6 = 60$ committees consisting of 2 seniors and 2 juniors. The requested probability is

$$P(E) = \frac{n(E)}{n(S)} = \frac{60}{126} = \frac{10}{21}.$$

   ●

Example 7   Find the probability of getting 2 heads and 6 tails if 8 coins are tossed.

Solution   Since 2 things can happen to each coin, the total number of possible outcomes, $n(S)$, is $2^8 = 256$.

We can select the 2 coins that must fall heads up in $C(8, 2) = 28$ ways. For each of these ways, there is only 1 way to select the other 6 coins that must fall tails up. Therefore, there are $28 \cdot 1 = 28$ ways of getting 2 heads and 6 tails; so $n(E) = 28$. The requested probability is

$$P(E) = \frac{n(E)}{n(S)} = \frac{28}{256} = \frac{7}{64}.$$

   ●

Problem Set 14.3

For Problems 1–4, find the probability of each of the following events if 2 coins are tossed.

1. Getting 1 head and 1 tail

2. Getting 2 tails

3. Getting at least 1 tail

4. Getting no tails

For Problems 5–8, find the probability of each of the following events if 3 coins are tossed.

5. Getting 3 heads

6. Getting 2 heads and 1 tail

7. Getting at least 1 head

8. Getting exactly 1 tail

For Problems 9–12, find the probability of each of the following events if 4 coins are tossed.

9. Getting 4 heads

10. Getting 3 heads and 1 tail

11. Getting 2 heads and 2 tails

12. Getting at least 1 head

For Problems 13–16, find the probability of each of the following events if 1 die is tossed.

13. Getting a multiple of 3

14. Getting a prime number

15. Getting an even number

16. Getting a multiple of 7

For Problems 17–22, find the probability of each of the following events if 2 dice are tossed.

17. Getting a sum of 6

18. Getting a sum of 11

19. Getting a sum less than 5

20. Getting a 5 on exactly 1 die

21. Getting a 4 on at least 1 die

22. Getting a sum greater than 4

For Problems 23–26, find the probability of each of the following events if 1 card is drawn from a standard deck of 52 playing cards.

23. A heart is drawn.

24. A king is drawn.

25. A spade or a diamond is drawn.

26. A red jack is drawn.

For Problems 27–30, find the probability of each of the following events if 25 slips of paper numbered 1 to 25, inclusive, are put in a hat and then 1 is drawn out at random.

27. The slip with the 5 on it is drawn.

28. A slip with an even number on it is drawn.

29. A slip with a prime number on it is drawn.

30. A slip with a multiple of 6 on it is drawn.

For Problems 31–34, find the probability of each of the following events if a committee of 2 boys is chosen at random from a group composed of the 5 boys Al, Bill, Carl, Dan, and Elmer.

31. Dan is on the committee.

32. Dan and Elmer are both on the committee.

33. Bill and Carl are not both on the committee.

34. Either Dan or Elmer, but not both, is on the committee.

For Problems 35–38, find the probability of each of the following events if a 5-person committee is selected at random from a group composed of the 8 people Al, Barb, Chad, Dawn, Eric, Fern, George, and Harriet.

35. Al and Barb are both on the committee.

36. George is not on the committee.

37. Either Chad or Dawn, but not both, is on the committee.

38. Neither Al nor Barb is on the committee.

For Problems 39–41, suppose that a box of 10 items from a manufacturing process is known to contain 2 defective and 8 nondefective items. Find the probability of each of the following events if a sample of 3 is selected at random.

39. The sample contains all nondefective items.

40. The sample contains 1 defective and 2 nondefective items.

41. The sample contains 2 defective and 1 nondefective items.

42. A building has 5 doors. Find the probability that 2 persons entering the building at random will choose the same door.

43. Bill, Carol, and Alice are seated at random in a row of 3 seats. Find the probability that Bill and Carol are seated side by side.

44. April, Bill, Carl, and Denise are to be seated at random in a row of 4 chairs. What is the probability that April and Bill will occupy the end seats?

45. A committee of 4 girls is to be chosen at random from a group composed of the 5 girls Alice, Becky, Candy, Dee, and Elaine. Find the probability that Elaine will not be on the committee.

46. Three boys and 2 girls are randomly seated in a row of 5 seats. What is the probability that the boys and girls are in alternate seats?

47. Four different mathematics books and 5 different history books are randomly placed on a shelf. What is the probability that all of the books on a subject are side by side?

48. Each of 3 letters is to be mailed in any of 5 different mail boxes. What is the probability that all will be mailed in the same mail box?

49. Randomly form a 4-digit number by using the digits 2, 3, 4, and 6 once each. What is the probability that the number formed will be greater than 4000?

50. Randomly select 1 of the 120 permutations of the letters a, b, c, d, and e. Find the probability that in the permutation chosen the letter a precedes the letter b (that is, a is to the left of b).

51. From a group of 6 women and 5 men, a committee of 4 is chosen at random. Find the probability that the committee contains 2 women and 2 men.

52. From a group of 4 women and 5 men, a committee of 3 is chosen at random. Find the probability that the committee contains at least 1 man.

53. Al, Bob, Carl, Dan, Ed, Frank, Gino, Harry, Jerry, and Mike are randomly divided into 2 5-man teams for a basketball game. What is the probability that Al, Bob, and Carl are on the same team?

54. Find the probability of getting 4 heads and 3 tails if 7 coins are tossed.

55. Find the probability of getting 3 heads and 6 tails if 9 coins are tossed.

56. Find the probability of getting at least 4 heads if 6 coins are tossed.

57. Find the probability of not getting more than 3 heads if 5 coins are tossed.

58. Each arrangement of the 11 letters of the word *MISSISSIPPI* is put on a slip of paper and placed in a hat. One slip is drawn at random from the hat. Find the probability that the slip contains an arrangement of the letters with the 4 *S*'s at the very beginning of the arrangement.

59. Each arrangement of the 7 letters of the word *OSMOSIS* is put on a slip of paper and placed in a hat. One slip is drawn at random from the hat. Find the probability that the slip contains an arrangement of the letters with an *O* at the beginning and an *O* at the end.

60. Consider all possible arrangements of 3 identical *H*'s and 3 identical *T*'s. Suppose that 1 of these arrangements is selected at random. What is the probability that the arrangement selected has the 3 *H*'s in consecutive positions?

Miscellaneous Problems

In Example 7 of Section 14.2 we found that there are 2,598,960 different 5-card hands that can be dealt from a deck of 52 playing cards. Therefore, probabilities for certain kinds of 5-card poker hands can be calculated using 2,598,960 as the number of elements in the sample space.

For Problems 61–69, determine the number of different 5-card poker hands of the indicated type that can be obtained.

61. Straight flush (5 cards in sequence and of the same suit; aces are used as both low and high cards—A2345 and 10JQKA are both acceptable)

62. Four of a kind (4 cards of the same face value, such as 4 kings)

63. Full house (3 cards of the same face value and 2 cards of another face value)

64. Flush (5 cards of the same suit but not in sequence)

65. Straight (5 cards in sequence but not all of the same suit)

66. Three of a kind (exactly 3 cards of the same face value and 2 other cards of different face values)

67. Two pairs

68. One pair

69. No pairs

14.4 Some Properties of Probability and Tree Diagrams.

In this section we will investigate some basic properties that are helpful in the computation of probabilities. The first property may state the obvious, but it still needs to be emphasized.

Property 14.1	For all events E, $0 \leq P(E) \leq 1$.

Property 14.1 merely states that probabilities must fall in the range from 0 to 1, inclusive. This should seem reasonable, since $P(E) = \dfrac{n(E)}{n(S)}$ and E is a subset of S. The next two examples illustrate circumstances for which $P(E) = 0$ and $P(E) = 1$.

Example 1   What is the probability of getting a 7 when a regular 6-sided die is tossed?

Solution   The sample space is $S = \{1, 2, 3, 4, 5, 6\}$, and the event space is $E = \emptyset$. Since $n(E) = 0$, we have $P(E) = \dfrac{n(E)}{n(S)} = \dfrac{0}{6} = 0$.   ●

Example 2   What is the probability of getting a head or a tail when a coin is tossed?

Solution   The sample space is $S = \{H, T\}$, and the event space is $E = \{H, T\}$. Therefore, $n(S) = n(E) = 2$ and $P(E) = \dfrac{n(E)}{n(S)} = \dfrac{2}{2} = 1$.   ●

An event that has a probability of 1 is sometimes called *certain success*, and an event with a probability of 0 is called *certain failure*.

It should be noted that Property 14.1 also serves as a check for the *reasonableness* of answers. In other words, when computing probabilities we know that our answer must fall in the range from 0 to 1, inclusive. Any other probability answer is simply not reasonable.

Complementary events are complementary sets for which S, the sample space, serves as the universal set. The following examples illustrate this idea.

Sample space	Event space	Complementary event space
$S = \{1, 2, 3, 4, 5, 6\}$	$E = \{1, 2\}$	$E' = \{3, 4, 5, 6\}$
$S = \{H, T\}$	$E = \{T\}$	$E' = \{H\}$
$S = \{2, 3, 4, \ldots, 12\}$	$E = \{2, 3, 4\}$	$E' = \{5, 6, 7, \ldots, 12\}$
$S = \{1, 2, 3, \ldots, 25\}$	$E = \{3, 4, 5, \ldots, 25\}$	$E' = \{1, 2\}$

In each row, note that E' (the complement of E) consists of all elements of S that *are not* in E. Thus, E and E' are called *complementary events*. Also note that for each row, the statement $P(E) + P(E') = 1$ can be made. For example, in the first row $P(E) = \frac{2}{6} = \frac{1}{3}$ and $P(E') = \frac{4}{6} = \frac{2}{3}$. Likewise, in the last row $P(E) = \frac{23}{25}$ and $P(E') = \frac{2}{25}$. The following general property can be stated.

Property 14.2

If E is any event of a sample space S and if E' is the complementary event, then
$$P(E) + P(E') = 1.$$

From a computational viewpoint, Property 14.2 provides us with a double-barreled approach to some probability problems. That is to say, if we are able to compute either $P(E)$ or $P(E')$, then the other one can be determined by subtracting from 1. For example, suppose that for a particular problem we can find that $P(E) = \frac{3}{13}$. Then we immediately know that $P(E') = 1 - P(E) = 1 - \frac{3}{13} = \frac{10}{13}$. The following examples further illustrate the use of Property 14.2.

Example 3   Find the probability of getting a sum greater than 3 if two dice are tossed.

Solution   Let S be the familiar sample space of ordered pairs for this problem, where $n(S) = 36$. Let E be the event of obtaining a sum greater than 3. Then E', the event of obtaining a sum less than or equal to 3, is $\{(1, 1), (1, 2), (2, 1)\}$. Thus, $P(E') = \dfrac{n(E')}{n(S)} = \dfrac{3}{36} = \dfrac{1}{12}$, and thus

$$P(E) = 1 - P(E') = 1 - \frac{1}{12} = \frac{11}{12}.$$

Example 4　　Find the probability of getting at least 1 head if three coins are tossed.

Solution　　The sample space S consists of all possible outcomes for tossing 3 coins. Using the Fundamental Principle of Counting, we know that there are $(2)(2)(2) = 8$ outcomes, so $n(S) = 8$. Let E be the event of getting at least 1 head; then E' is the complementary event of not getting at least 1 head. The set E' is easy to list—namely, $E' = \{(T, T, T)\}$. Thus, $n(E') = 1$ and $P(E') = \dfrac{1}{8}$. From this $P(E)$ can be determined to be

$$P(E) = 1 - P(E') = 1 - \frac{1}{8} = \frac{7}{8}.$$

Example 5　　From a group of 5 women and 4 men a 3-person committee is chosen at random. Find the probability that the committee contains at least 1 woman.

Solution　　Let the sample space S be the set of all possible 3-person committees that can be formed from 9 people. There are $C(9, 3) = 84$ such committees; thus, $n(S) = 84$.

Let E be the event that the committee contains at least 1 woman. Then E' is the complementary event that the committee contains all men. Thus, E' consists of all 3-men committees that can be chosen from 4 men. There are $C(4, 3) = 4$ such committees; thus, $n(E') = 4$. Therefore,

$$P(E') = \frac{n(E')}{n(S)} = \frac{4}{84} = \frac{1}{21},$$

which determines $P(E)$ to be

$$P(E) = 1 - P(E') = 1 - \frac{1}{21} = \frac{20}{21}.$$

Tree Diagrams

Tree diagrams provide a convenient way of analyzing certain kinds of probability problems involving *drawing with replacement* and *drawing without replacement*. Suppose that a marble is randomly drawn from a bag containing 2 red and 3 white marbles. Since there are 5 marbles, of which 2 are red, the probability of drawing a red marble is $\dfrac{2}{5}$. Likewise, there are 3 white marbles; so the probability of drawing a white

Figure 14.1

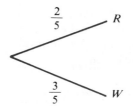

marble is $\frac{3}{5}$. These probabilities can be recorded on a tree diagram as in Figure 14.1.

Now suppose that after we draw that first marble at random and record its color, we replace the marble in the bag before drawing a second marble and recording its color. (This is called **drawing with replacement**.) The tree diagram in Figure 14.2 shows this two-step experiment. Each path along the tree corresponds to a possible outcome of this experiment. For example, the RW path means that we obtain a red marble on the first draw and a white marble on the second draw.

Figure 14.2

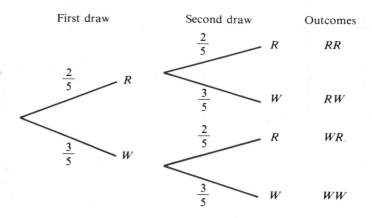

Probabilities for each of the outcomes of the two-step experiment can be calculated from the tree diagram. Let's analyze the RW outcome. [We will use $P(RW)$ to symbolize the probability of the RW outcome.]

The probability of drawing a red marble on the first draw is $\frac{2}{5}$, and the probability of drawing a white marble on the second draw is $\frac{3}{5}$. This means that we expect to draw a red marble on the first draw $\frac{2}{5}$ of the time and to draw a white marble on the second draw $\frac{3}{5}$ *of the times that we obtained a red marble on the first draw.* Thus,

$$P(RW) = \frac{2}{5} \text{ of } \frac{3}{5} = \frac{2}{5} \cdot \frac{3}{5} = \frac{6}{25}.$$

In other words, the probability for each outcome of the two-step experiment can be found from the tree diagram by following the paths to each of the outcomes and then *multiplying the probabilities along the paths*. Thus, for the other three outcomes in Figure 14.2, we obtain

$$P(RR) = \frac{2}{5} \cdot \frac{2}{5} = \frac{4}{25}, \qquad P(WR) = \frac{3}{5} \cdot \frac{2}{5} = \frac{6}{25},$$

$$P(WW) = \frac{3}{5} \cdot \frac{3}{5} = \frac{9}{25}.$$

Notice that $P(RR) + P(RW) + P(WR) + P(WW) = \frac{4}{25} + \frac{6}{25} + \frac{6}{25} + \frac{9}{25} = 1.$

From the tree diagram in Figure 14.2 we can also answer some other probability questions relative to this experiment. For example, suppose that we want to know the probability of drawing a red marble and a white marble. The two outcomes RW and WR both contain a red and a white marble. Since $P(RW) = \frac{6}{25}$ and $P(WR) = \frac{6}{25}$, the probability that RW or WR will occur is the *sum* of the probabilities. Thus, $P(RW$ or $WR) = \frac{6}{25} + \frac{6}{25} = \frac{12}{25}.$

Tree diagrams also provide a convenient way of analyzing **drawing without replacement**. For example, suppose that we randomly draw a marble from a bag containing 5 red and 2 white marbles and record its color. Then, *without replacing* the marble, we draw another marble and record its color. The tree diagram in Figure 14.3 shows this two-step experiment.

Figure 14.3

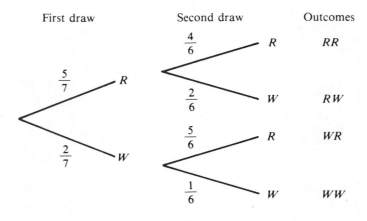

Notice that the denominators of the probabilities on the second draw are all 6. Since after the first draw the marble is *not* replaced, there are only 6 marbles remaining for the second draw. We can calculate probabilities for the various outcomes from the tree diagram as we did before.

$$P(RR) = \frac{5}{7} \cdot \frac{4}{6} = \frac{20}{42} = \frac{10}{21}, \qquad P(RW) = \frac{5}{7} \cdot \frac{2}{6} = \frac{10}{42} = \frac{5}{21},$$

$$P(WR) = \frac{2}{7} \cdot \frac{5}{6} = \frac{10}{42} = \frac{5}{21}, \qquad P(WW) = \frac{2}{7} \cdot \frac{1}{6} = \frac{2}{42} = \frac{1}{21},$$

Notice again that $P(RR) + P(RW) + P(WR) + P(WW) = \frac{10}{21} + \frac{5}{21} + \frac{5}{21} + \frac{1}{21} = \frac{21}{21} = 1.$

Problem Set 14.4

For Problems 1–11, compute the probability by using $P(E) = \dfrac{n(E)}{n(S)}$. For some of these problems, the property $P(E) + P(E') = 1$ will be helpful.

1. Find the probability of each of the following events if 2 dice are tossed.

    (a) Getting a sum of 6
    (b) Getting a sum greater than 2
    (c) Getting a sum less than 8
    (d) Getting a sum greater than 1

2. Find the probability of each of the following events if 3 dice are tossed.

    (a) Getting a sum of 3
    (b) Getting a sum greater than 4
    (c) Getting a sum less than 17
    (d) Getting a sum greater than 18

3. What is the probability of not getting a double when a pair of dice is tossed?

4. Find the probability of each of the following events if 4 coins are tossed.

    (a) Getting 4 heads
    (b) Getting 3 heads and 1 tail
    (c) Getting at least 1 tail

5. Find the probability of each of the following events if 5 coins are tossed.

    (a) Getting 5 tails
    (b) Getting 4 heads and 1 tail
    (c) Getting at least 1 tail

6. The probability that a certain horse will win the Kentucky Derby is $\dfrac{1}{20}$. What is the probability that it will lose the race?

7. One card is randomly drawn from a deck of 52 playing cards. What is the probability that it is not an ace?

8. Find the probability of getting at least 2 heads if 6 coins are tossed.

9. A subset of 2 letters is chosen at random from the set $\{a, b, c, d, e, f, g, h, i\}$. Find the probability that the subset will contain at least 1 vowel.

10. From a group of 4 men and 3 women, a 2-person committee is chosen at random. Find the probability that the committee contains at least 1 man.

11. From a group of 7 women and 5 men, a 3-person committee is chosen at random. Find the probability that the committee contains at least 1 man.

For Problems 12–18, use tree diagrams to help analyze the probability situations.

12. A bag contains 3 red and 2 white marbles. One marble is drawn and its color recorded. The marble is then replaced in the bag, and a second marble is drawn and its color recorded. Find the probability of each of the following.

    (a) The first marble drawn is red, and the second marble drawn is white.

    (b) The first marble drawn is white, and the second marble drawn is red.

    (c) Both marbles drawn are white.

    (d) At least 1 marble drawn is red.

    (e) A red and a white marble are drawn.

13. A bag contains 5 blue and 6 gold marbles. Two marbles are drawn in succession, without replacement. Find the probability of each of the following.

    (a) Both marbles drawn are blue.

    (b) One marble drawn is blue, and the other is gold.

    (c) At least 1 marble drawn is gold.

    (d) Both marbles drawn are gold.

14. A bag contains 1 red and 2 white marbles. Two marbles are drawn in succession without replacement. Find the probability of each of the following.

    (a) One marble drawn is red, and the other is white.

    (b) Both marbles drawn are white.

    (c) Both marbles drawn are red.

15. A bag contains 5 red and 12 white marbles. Two marbles are drawn in succession without replacement. Find the probability of each of the following.

    (a) Both marbles drawn are red.

    (b) Both marbles drawn are white.

    (c) A red and a white marble are drawn.

    (d) At least 1 marble drawn is red.

16. A bag contains 2 red, 3 white, and 4 blue marbles. Two marbles are drawn in succession without replacement. Find the probability of each of the following.

    (a) Both marbles drawn are white.

    (b) One marble drawn is white, and the other is blue.

    (c) Both marbles drawn are blue.

    (d) At least 1 red marble is drawn.

    (e) No white marbles are drawn.

17. Two boxes with red and white marbles are shown below. A marble is drawn at random from Box 1, then a marble is drawn from Box 2, and the colors are recorded in order. Find the probability of each of the following.

3 red 4 white	2 red 1 white
Box 1	Box 2

(a) Both marbles drawn are white.

(b) Both marbles drawn are red.

(c) A red and a white marble are drawn.

18. Three boxes containing red and white marbles are shown below. Randomly draw a marble from Box 1 and put it in Box 2. Then draw a marble from Box 2 and put it in Box 3. Then draw a marble from Box 3. What is the probability that the last marble drawn, from Box 3, is red? What is the probability that it is white?

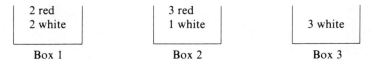

2 red	3 red	
2 white	1 white	3 white
Box 1	Box 2	Box 3

Miscellaneous Problem

19. The term *odds* is sometimes used to express a probability statement. For example, we might say that "the odds in favor of the Cubs' winning the pennant are 5 to 1" or "the odds against the White Sox's winning the pennant are 50 to 1." "Odds in favor" and "odds against" for equally likely outcomes can be defined as follows:

$$\text{odds in favor} = \frac{\text{number of favorable outcomes}}{\text{number of unfavorable outcomes}}$$

$$\text{odds against} = \frac{\text{number of unfavorable outcomes}}{\text{number of favorable outcomes}}$$

We have used the fraction form to state the definitions for odds; however, in practice the "to" vocabulary is commonly used. Thus, the odds in favor of rolling a 4 with 1 roll of a die are usually stated as 1 to 5 instead of $\frac{1}{5}$. The odds against rolling a 4 are stated as 5 to 1.

The "odds in favor of" statement about the Cubs means that there are 5 favorable outcomes compared to 1 unfavorable, or a total of 6 possible outcomes. So the "5 to 1 in favor of" statement also means that the probability of the Cubs' winning the pennant is $\frac{5}{6}$. Likewise, the "50 to 1 against" statement about the White Sox means that the probability that the White Sox will not win the pennant is $\frac{50}{51}$.

(a) What are the odds in favor of getting 2 heads with a toss of 2 coins?

(b) What are the odds against getting 2 heads and 1 tail with a toss of 3 coins?

(c) What are the odds against rolling a double with a pair of dice?

(d) Suppose that $P(E) = \frac{4}{7}$ for some event E. What are the odds against E's happening?

(e) Suppose that $P(E) = \frac{5}{9}$ for some event E. What are the odds in favor of E's happening?

(f) If the odds against an event's occurring are 5 to 2, find the probability that the event will occur.

Chapter 14 Summary

(14.1)   The Fundamental Principle of Counting can be stated as "if one task can be accomplished in x different ways, and following this task, a second task can be accomplished in y different ways, then the first task followed by the second task can be accomplished in $x \cdot y$ different ways." (This counting principle can be extended to any finite number of tasks.)

In order to apply the Fundamental Principle of Counting, it is often helpful to analyze the problem in terms of the tasks to be accomplished.

Simply thinking about an appropriate tree diagram, without writing out the diagram in full, can also be helpful.

(14.2)   The number of permutations of **n** things taken **n** at a time is given by

$$P(n, n) = n!.$$

The number of **permutations** of **n** things taken **r** at a time is given by

$$P(n, r) = \underbrace{n(n - 1)(n - 2) \ldots .}_{r \text{ factors}}$$

The number of **combinations** (subsets) of **n** things taken **r** at a time is given by

$$C(n, r) = \frac{P(n, r)}{r!}.$$

Remember: If order matters, use permutations, but if order does not matter, use combinations.

It is often very helpful to think of combinations as subsets.

(14.3)   The set of all possible outcomes of a given experiment is called the **sample space**, and the individual elements of the sample space are called **sample points**. Any subset of a sample space is called an **event** or **event space**.

The probability of an event E, in an experiment for which all of the possible outcomes in the sample space S are equally likely to occur, is defined by

$$P(E) = \frac{n(E)}{n(S)}.$$

(14.4)   For all events E, $0 \le P(E) \le 1$.

If E is any event of a sample space S and E' is the complement of E, then

$$P(E) + P(E') = 1.$$

Tree diagrams provide a convenient way of analyzing probability problems dealing with **drawing with replacement** and **drawing without replacement**.

Chapter 14 Review Problem Set

Problems 1–14 are counting-type problems.

1. How many different arrangements of the letters A, B, C, D, E, and F can be made?

2. How many different 9-letter arrangements can be formed from the 9 letters of the word *APPARATUS*?

3. How many odd numbers of 3 different digits each can be formed by choosing from the digits 1, 2, 3, 5, 7, 8, and 9?

4. In how many ways can Arlene, Brent, Cindy, Dave, Ernie, Frank, and Gladys be seated in a row of 7 seats so that Arlene and Cindy are side by side?

5. In how many ways can a committee of 3 people be chosen from a group of 6 people?

6. How many committees consisting of 3 men and 2 women can be formed from a group of 7 men and 6 women?

7. How many different 5-card hands consisting of all hearts can be formed from a deck of 52 playing cards?

8. How many numbers greater than 500, where no number contains repeated digits, can be formed by choosing from the digits 2, 3, 4, 5, and 6?

9. How many 3-person committees can be formed from a group of 4 men and 5 women so that each committee contains at least 1 man?

10. How many different 4-person committees can be formed from a group of 8 people if a certain pair of people refuse to serve together on a committee?

11. How many 4-element subsets containing A or B but not both A and B can be formed from the set $\{A, B, C, D, E, F, G, H\}$?

12. How many different 6-letter permutations can be formed from 4 identical H's and 2 identical T's?

13. How many 4-person committees consisting of 2 seniors, 1 sophomore, and 1 junior can be formed from a group of 3 seniors, 4 juniors, and 5 sophomores?

14. In a baseball league of 6 teams, how many games are needed to complete a schedule if each team must play 8 games with each other team?

Problems 15–32 pose some probability questions.

15. Find the probability of getting 2 heads and 1 tail if 3 coins are tossed.

16. Find the probability of getting 3 heads and 2 tails if 5 coins are tossed.

17. What is the probability of getting a sum of 8 with 1 roll of a pair of dice?

18. What is the probability of getting a sum more than 5 with 1 roll of a pair of dice?

19. Amy, Brenda, Chuck, Dave, and Elmer are randomly seated in a row of 5 seats. Find the probability that Amy and Chuck are not seated side by side.

20. Four girls and 3 boys are randomly seated in a row of 7 seats. Find the probability that the girls and boys are seated in alternate seats.

21. Find the probability of getting at least 2 heads if 6 coins are tossed.

22. Two cards are randomly chosen from a deck of 52 playing cards. What is the probability that 2 jacks are drawn?

23. Each arrangement of the 6 letters of the word *CYCLIC* is put on a slip of paper and placed in a hat. One slip is drawn at random. Find the probability that the slip contains an arrangement with the *Y* at the beginning.

24. A committee of 3 is randomly chosen from a group of 1 man and 6 women. What is the probability that the man is not on the committee chosen?

25. A 4-person committee is selected at random from among the 8 people Alice, Bob, Carl, Dee, Edna, Fred, Gina, and Hilda. Find the probability that Alice or Bob, but not both, is on the committee.

26. From a group of 5 men and 4 women, a committee of 3 is chosen at random. Find the probability that the committee contains 2 men and 1 woman.

27. From a group of 6 men and 7 women, a committee of 4 is chosen at random. Find the probability that the committee contains at least 1 woman.

28. A bag contains 5 red and 8 white marbles. Two marbles are drawn in succession with replacement. Find the probability that at least 1 red marble is drawn.

29. A bag contains 4 red, 5 white, and 3 blue marbles. Two marbles are drawn in succession with replacement. Find the probability that 1 red and 1 blue marble are drawn.

30. A bag contains 4 red and 7 blue marbles. Two marbles are drawn in succession without replacement. Find the probability of drawing 1 red and 1 blue marble.

31. A bag contains 3 red, 2 white, and 2 blue marbles. Two marbles are drawn in succession without replacement. Find the probability of drawing at least 1 red marble.

32. Each of 3 letters is to be mailed in any of 4 different mail boxes. What is the probability that all 3 letters will be mailed in the same mail box?

Answers to Odd-Numbered Exercises

Chapter 1

Problem Set 1.1

1. True   **3.** False   **5.** False   **7.** True   **9.** False
11. 46   **13.** $0, -14, 46$ **15.** $\sqrt{5}, -\sqrt{2}, -\pi$ **17.** $0, -14$ **19.** $\subseteq$
21. $\subseteq$ **23.** $\not\subseteq$ **25.** $\subseteq$ **27.** $\not\subseteq$ **29.** $\subseteq$
31. $\not\subseteq$ **33.** $\{1\}$ **35.** $\{0, 1, 2, 3\}$ **37.** $\{\ldots -2, -1, 0, 1\}$ **39.** $\varnothing$
41. $\{0, 1, 2\}$ **43.** 5   **45.** 4   **47.** $z + 2$ **49.** $5x + 7$
51. 14   **53.** 32   **55.** 62   **57.** 8   **59.** 80
61. 220   **63.** 728   **65.** 223   **67.** 168   **69.** 54
71. 36   **73.** 3

Problem Set 1.2

1. -15 **3.** -6 **5.** 3   **7.** 3   **9.** -11 **11.** -8 **13.** 23
15. -5 **17.** -9 **19.** -55 **21.** -18 **23.** -65 **25.** 27   **27.** -18
29. 19   **31.** 60   **33.** -12 **35.** 28   **37.** 2   **39.** -4 **41.** 5
43. -1 **45.** 17   **47.** -12 **49.** -38 **51.** 7   **53.** -10 **55.** -37
57. -7 **59.** -44 **61. (a)** 2   **(c)** -3 **(e)** -6 **(g)** -8 **(i)** 9

Problem Set 1.3

1. 28   **3.** -32 **5.** -84 **7.** 3   **9.** -12 **11.** -2
13. 30   **15.** -45 **17.** -80 **19.** 0   **21.** undefined   **23.** 12
25. -1 **27.** -3 **29.** 2   **31.** -8 **33.** 19   **35.** -3
37. -1 **39.** -61 **41.** -2 **43.** 30   **45.** 55   **47.** -11
49. 52   **51.** 4   **53.** -748 **55.** -14 **57.** 23

Problem Set 1.4

1. Commutative property of multiplication   **3.** Identity property of multiplication
5. Multiplication property of negative one   **7.** Distributive property

9. Associative property of multiplication   **11.** Associative property of addition
13. Multiplicative inverse property
15. 42   **17.** 73   **19.** -1700 **21.** 1300   **23.** 8   **25.** -400 **27.** 1
29. 9   **31.** -54 **33.** -50 **35.** 13   **37.** 57   **39.** 49   **41.** -5
43. -64 **45.** 100   **47.** 5   **49.** -23

Problem Set 1.5

1. $3x$ **3.** $-a^2$ **5.** $-9n$ **7.** $-x + 2y$ **9.** $-7a^2 - b^2$
11. $13x - 13$ **13.** $8a^2b - 2ab^2$ **15.** $5x + 8$ **17.** $-6a + 10$ **19.** $-2n^2 + 29$
21. $-4x^2 + 7$ **23.** $22x - 21$ **25.** $7x + 7$ **27.** $-8n^2 - 13$ **29.** $-3x + 38y$
31. -22 **33.** -6 **35.** 9   **37.** 2   **39.** -10
41. 36   **43.** 19   **45.** 130   **47.** 11   **49.** -18
51. 13.8   **53.** 90.0   **55.** 526.6   **57.** $n + 2$ **59.** $n - 15$
61. $50 - n$ **63.** $\frac{1}{2}n$ **65.** $2n + 5$ **67.** $\frac{100}{n}$ **69.** $\frac{1}{2}n + 8$
71. $10(n + 5)$ **73.** $y - 15$ **75.** $65 - x$ **77.** $\frac{98}{n}$ **79.** $\frac{c}{4}$
81. $5n + 10d + 25q$ **83.** $n + 1$ **85.** $n + 2$ **87.** $\frac{d}{12}$ **89.** $\frac{i}{12}$
91. $0.1d$ or $\frac{d}{10}$

Chapter 1 Review Problem Set

1. **(a)** 67   **(b)** $0, -8$, and 67   **(c)** 0 and 67   **(d)** $0, \frac{3}{4}, -\frac{5}{6}, 8\frac{1}{3}, 0.34, 0.2\overline{3}, 67,$ and $\frac{9}{7}$
(e) $\sqrt{2}$ and $-\sqrt{3}$
2. Associative property for addition   **3.** Substitution property of equality
4. Multiplication property of negative one   **5.** Distributive property
6. Associative property for multiplication   **7.** Commutative property for addition
8. Distributive property   **9.** Multiplicative inverse property
10. Symmetric property of equality
11. -6 **12.** -6 **13.** -10 **14.** -15 **15.** 20
16. 49   **17.** -56 **18.** -24 **19.** 6   **20.** 4
21. -1000 **22.** 8   **23.** $-4a^2 - 5b^2$ **24.** $3x - 2$ **25.** $5ab^2 + 3a^2b$
26. $6x + 31$ **27.** $10n^2 - 17$ **28.** $-13a + 4$ **29.** $-2n + 2$ **30.** $-7x - 29y$
31. $-7a - 9$ **32.** $-9x^2 + 7$ **33.** -31 **34.** -5 **35.** -55
36. 144   **37.** -16 **38.** -44 **39.** 53   **40.** 53
41. 6   **42.** -221 **43.** $4 + 2n$ **44.** $3n - 50$ **45.** $\frac{2}{3}n - 6$
46. $10(n - 14)$ **47.** $5n + 8$ **48.** $\frac{n}{n - 3}$ **49.** $5(n + 2) - 3$ **50.** $\frac{3}{4}(n + 12)$
51. $37 - n$ **52.** $\frac{w}{60}$ **53.** $2y - 7$ **54.** $n + 3$ **55.** $p + 5n + 25q$
56. $\frac{i}{48}$ **57.** $24f + 72y$ **58.** $10d$ **59.** $12f + i$ **60.** $25 - c$

Chapter 2

Problem Set 2.1

1. $\{2\}$ **3.** $\{-4\}$ **5.** $\{-14\}$ **7.** $\{6\}$ **9.** $\left\{\dfrac{21}{2}\right\}$ **11.** $\{-4\}$

13. $\left\{-\dfrac{5}{2}\right\}$ **15.** $\{5\}$ **17.** $\left\{-\dfrac{1}{4}\right\}$ **19.** $\{2\}$ **21.** $\{11\}$ **23.** $\{2\}$

25. $\{-1\}$ **27.** $\{-5\}$ **29.** $\left\{-\dfrac{10}{3}\right\}$ **31.** $\{-5\}$ **33.** $\left\{\dfrac{1}{2}\right\}$ **35.** $\left\{\dfrac{5}{3}\right\}$

37. $\{11\}$ **39.** $\left\{\dfrac{1}{14}\right\}$ **41.** $\{-17\}$ **43.** $\left\{-\dfrac{1}{6}\right\}$ **45.** 6   **47.** 15, 16, and 17

49. 14, 16, and 18   **51.** 22 and 87   **53.** $26 per hour
55. 7 nickels, 15 dimes, and 19 quarters   **57.** 41 females and 21 males   **59.** 30 bicycles

Problem Set 2.2

1. $\{9\}$ **3.** $\{-2\}$ **5.** $\left\{\dfrac{13}{6}\right\}$ **7.** $\{9\}$ **9.** $\{-10\}$ **11.** $\left\{\dfrac{12}{7}\right\}$ **13.** $\left\{\dfrac{12}{7}\right\}$

15. $\{1\}$ **17.** $\{-28\}$ **19.** $\left\{\dfrac{33}{2}\right\}$ **21.** $\{3\}$ **23.** $\{27\}$ **25.** $\{19\}$ **27.** $\left\{\dfrac{74}{39}\right\}$

29. $\left\{\dfrac{5}{6}\right\}$ **31.** $\{0\}$ **33.** $\{7\}$ **35.** $\{3\}$ **37.** $\left\{\dfrac{38}{17}\right\}$ **39.** $\left\{-\dfrac{22}{13}\right\}$ **41.** 24

43. The length is 8 centimeters and the width is 6 centimeters.
45. $8 per hour
47. 48 dimes, 31 quarters, and 37 silver dollars
49. $15,000 for Tim, $25,000 for Ann, and $32,000 for Eric
51. Jesse 18 and Donna 12
53. $56°$ **55.** A is $64°$, B is $24°$, and C is $92°$

Problem Set 2.3

1. $\{20\}$ **3.** $\{400\}$ **5.** $\{5\}$ **7.** $\left\{\dfrac{3}{2}\right\}$ **9.** $\{3\}$ **11.** $\{200\}$

13. $\{200\}$ **15.** $\left\{-\dfrac{66}{37}\right\}$ **17.** $\{10\}$ **19.** $\{500\}$ **21.** $\{2000\}$ **23.** $\{5000\}$

25. $\{13.5\}$ **27.** $\{8\}$ **29.** $40   **31.** $39.20   **33.** $3.90   **35.** $0.75
37. 30%   **39.** $12,900
41. $1000 at 8% and $3000 at 9%   **43.** $1200 at 8% and $800 at 7%
45. 15 nickels, 25 dimes, and 50 quarters   **47.** 12 dimes and 25 quarters
49. $\{1.5\}$ **51.** $\{225\}$ **53.** $\{3.6\}$ **55.** $\{14.4\}$ **57.** $\{12.9\}$

Problem Set 2.4

1. $72   **3.** 3 years   **5.** 8%
7. $1425   **9.** 8 years   **11.** 8 cm; 6 cm; 4 cm; 10 cm; 20 cm

13. $l = \dfrac{A}{w}$ **15.** $B = \dfrac{3V}{h}$ **17.** $h = \dfrac{A - 2\pi r^2}{2\pi r}$

19. $h = \dfrac{2A}{b_1 + b_2}$ **21.** $F = \dfrac{9C + 160}{5}$ or $F = \dfrac{9}{5}C + 32$ **23.** $x = \dfrac{c + b}{a}$

25. $x = \dfrac{c}{a + b}$ **27.** $x = ab - a$ **29.** $x = \dfrac{c - ab}{a}$

31. $x = \dfrac{mc}{m - a + b}$ **33.** $x = \dfrac{9 - 5y}{3}$ **35.** $y = \dfrac{1 + 7x}{2}$

37. $x = \dfrac{y - b}{m}$ **39.** $x = \dfrac{by - ab + ac}{c}$ **41.** $y = \dfrac{x + 2a + 2}{a + 1}$

43. The length is 17 meters and the width is 5 meters.

45. $11\dfrac{1}{9}$ years   **47.** 20 years   **49.** $3\dfrac{1}{2}$ hours   **51.** 4 hours   **53.** 40 miles   **55.** $2\dfrac{2}{9}$ cups

57. 10 gallons   **59.** \$437.50   **61.** 3.5 years   **63.** 11.5%   **65.** \$1300

Problem Set 2.5

1. $\{x \mid x > -1\}$

3. $\{x \mid x \le -2\}$

5. $\{x \mid x \le -1\}$

7. $\{x \mid x < 2\}$

9. $\{x \mid x > 0\}$

11. $\{x \mid x \ge -1\}$

13. $\{x \mid x > -1\}$

15. $\{x \mid x > -2\}$

17. $\{x \mid x > 3\}$

19. $\{x \mid x \ge 1\}$

21. $\left\{x \mid x > \dfrac{7}{3}\right\}$ **23.** $\left\{x \mid x > \dfrac{16}{7}\right\}$ **25.** $\left\{x \mid x \ge \dfrac{4}{3}\right\}$ **27.** $\{x \mid x \ge 1\}$ **29.** $\{a \mid a > -2\}$

31. $\{n \mid n > 3\}$ **33.** $\left\{n \mid n \ge -\dfrac{1}{4}\right\}$ **35.** $\left\{n \mid n > \dfrac{13}{7}\right\}$ **37.** $\left\{x \mid x < \dfrac{5}{4}\right\}$ **39.** $\{x \mid x \le -16\}$

41. $\{y \mid y < 1\}$ **43.** $\{x \mid x > -42\}$ **45.** $\left\{x \mid x > -\dfrac{7}{4}\right\}$ **47.** $\left\{t \mid t \le -\dfrac{2}{11}\right\}$ **49.** $\{x \mid x > 4\}$

Problem Set 2.6

1. $\{x \mid x > 3\}$ **3.** $\left\{x \mid x < \dfrac{15}{7}\right\}$ **5.** $\left\{n \mid n \ge \dfrac{26}{11}\right\}$ **7.** $\{a \mid a \le 13\}$ **9.** $\left\{x \mid x < \dfrac{8}{3}\right\}$

11. $\{s \mid s \ge 3\}$ **13.** $\{x \mid x > 200\}$ **15.** $\{x \mid x \ge 500\}$

17.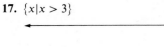

19.

21.

23.

25.

27. $\varnothing$

29.

31.

33. $\{x | 1 < x < 5\}$

35. $\{x | x > 3 \text{ or } x < -5\}$

37. $\{x | x \geq 2\}$

39. $\left\{ x \,\middle|\, \dfrac{1}{2} < x < \dfrac{2}{3} \right\}$

41. $\left\{ x \,\middle|\, x < -\dfrac{3}{5} \text{ or } x > -\dfrac{1}{5} \right\}$

43. $\{x | -1 < x < 4\}$

45. $\left\{ t \,\middle|\, -\dfrac{11}{2} \leq t \leq -3 \right\}$ **47.** $\left\{ x \,\middle|\, -\dfrac{5}{3} < x < 1 \right\}$ **49.** $\{x | -22 < x < 26\}$

51. $\{x | 1 < x < 7\}$ **53.** A score of 75 or better   **55.** More than $200

57. $-20° \leq C \leq -5°$ **59.** $8.8 \leq M \leq 15.4$ **61.** 96 or better

63. 77 or less

Problem Set 2.7

1. $\{x | -3 < x < 3\}$

3. $\{x | -1 \leq x \leq 1\}$

5. $\{x | x < -3 \text{ or } x > 3\}$

7. $\{x | x \leq -2 \text{ or } x \geq 2\}$

9. $\{x | -3 < x < 1\}$

11. $\{x | -2 < x < 6\}$

13. $\{x | x \leq -5 \text{ or } x \geq 1\}$

15. $\{x | x < 3 \text{ or } x > 5\}$

17. $\{-4, 4\}$ **19.** $\{-2, 4\}$ **21.** $\{-11, 5\}$

23. $\{-5, 6\}$ **25.** $\left\{ -\dfrac{5}{2}, 1 \right\}$ **27.** $\{x | -4 < x < 5\}$

29. $\{x | x < -7 \text{ or } x > 2\}$ **31.** $\left\{ x \,\middle|\, -1 \leq x \leq \dfrac{1}{2} \right\}$ **33.** $\left\{ x \,\middle|\, x \leq -1 \text{ or } x \geq \dfrac{7}{3} \right\}$

35. $\{x | 1 < x < 5\}$ **37.** $\left\{ -\dfrac{1}{12}, \dfrac{17}{12} \right\}$ **39.** $\left\{ x \,\middle|\, x < -\dfrac{7}{2} \text{ or } x > \dfrac{7}{6} \right\}$

41. $\varnothing$ **43.** $\varnothing$ **45.** All reals

47. $\{0\}$ **49.** $\varnothing$ **51.** $\{-2\}$

53. $\{0\}$ **55.** $\left\{ -\dfrac{4}{5}, 2 \right\}$

Chapter 2 Review Problem Set

1. $\{18\}$ **2.** $\{-14\}$ **3.** $\{0\}$ **4.** $\left\{\dfrac{1}{2}\right\}$ **5.** $\{10\}$

6. $\left\{\dfrac{7}{3}\right\}$ **7.** $\left\{\dfrac{28}{17}\right\}$ **8.** $\left\{-\dfrac{1}{38}\right\}$ **9.** $\left\{\dfrac{27}{17}\right\}$ **10.** $\left\{-\dfrac{10}{3}, 4\right\}$

11. $\{50\}$ **12.** $\left\{-\dfrac{39}{2}\right\}$ **13.** $\{200\}$ **14.** $\{-8\}$ **15.** $\left\{-\dfrac{7}{2}, \dfrac{1}{2}\right\}$

16. $x = \dfrac{2b + 2}{a}$ **17.** $x = \dfrac{c}{a - b}$ **18.** $x = \dfrac{pb - ma}{m - p}$ **19.** $x = \dfrac{11 + 7y}{5}$ **20.** $x = \dfrac{by + b + ac}{c}$

21. $s = \dfrac{A - \pi r^2}{\pi r}$ **22.** $b_2 = \dfrac{2A - hb_1}{h}$ **23.** $R_1 = \dfrac{RR_2}{R_2 - R}$ **24.** $R = \dfrac{R_1 R_2}{R_1 + R_2}$

25. $\{x \mid x \geq -5\}$ **26.** $\{x \mid x > 4\}$ **27.** $\left\{x \mid x > -\dfrac{7}{3}\right\}$ **28.** $\left\{x \mid x \geq \dfrac{17}{2}\right\}$

29. $\left\{n \mid n < \dfrac{1}{3}\right\}$ **30.** $\left\{n \mid n > \dfrac{5}{11}\right\}$ **31.** $\{s \mid s \geq 6\}$ **32.** $\{x \mid x \leq 100\}$

33. $\{x \mid -5 < x < 6\}$ **34.** $\left\{x \mid x < -\dfrac{11}{3} \text{ or } x > 3\right\}$ **35.** $\{t \mid t < -17\}$ **36.** $\left\{x \mid x < -\dfrac{15}{4}\right\}$

37. *number line from −6 to 6, open circle at 0 to open circle at 5*

38. *number line from −6 to 6, closed dot at −2 and open circle at 2*

39. *number line from −6 to 6, open circle at 3*

40. *number line from −6 to 6, all reals*
all reals

41. *number line from −6 to 6, open circle at −1 and open circle at 2*

42. *number line from −6 to 6, closed dot at −3 to closed dot at 1*

43. *number line from −6 to 6, open circle at $\frac{1}{2}$ to closed dot at 3*
$\frac{1}{2}$

44. The length is 15 and the width is 7 meters.
45. $200 at 7% and $300 at 8%
46. 88 or better
47. 4, 5, and 6
48. $10.50 per hour
49. 20 nickels, 50 dimes, and 75 quarters
50. 80°
51. $45.60
52. 30 or more
53. 55 miles per hour
54. Sonya for $3\dfrac{1}{4}$ hours and Rita for $4\dfrac{1}{2}$ hours
55. $6\dfrac{1}{4}$ cups

Chapter 3

Problem Set 3.1

1. (a) 3   **(c)** 4   **(e)** 3   **(g)** 7   **(i)** 0   **3.** $4x + 4$ **5.** $9x^2 - x + 8$
7. $-6x^2y - 6xy$ **9.** $6x^2 - 11x - 2$ **11.** $-5x^2 + 13x + 1$ **13.** $-2x - 6$
15. $-7a - 3$ **17.** $-2x^2 - 4x + 1$ **19.** $2a^2 + 2a + 3$ **21.** $-3x^3 + 3x^2 - 5x + 1$
23. $19x - 11$ **25.** $-2a^2 + 13a + 5$ **27.** $4x^4 - 3x^3 + x^2 - 2$ **29.** $6x - 11$
31. $6x^2 - 5x - 7$ **33.** $x^3 - 7x^2 - 3x + 1$ **35.** $-2t^3 + 3t^2 + 2t - 3$ **37.** $x^2 + 3x - 8$

39. $-6n^2 - 2n - 15$ **41.** $-x - 2$ **43.** $4x + 12$ **45.** $-3x^2 + x + 4$

47. $-4n^2 + 3n - 1$ **49.** 6

Problem Set 3.2

1. $10x^5$ **3.** $-24x^4$ **5.** $42a^3b^5$ **7.** $-2x^3y^3z^6$ **9.** $-3x^5y^4$

11. $-12a^3b^5$ **13.** $-m^3n^4$ **15.** $\dfrac{8}{15}x^6y^3$ **17.** $-\dfrac{1}{2}a^3b^3$ **19.** $-\dfrac{1}{6}x^3y^2$

21. $-30x^6$ **23.** $80n^9$ **25.** $-3x^6y^7$ **27.** $-6y^7$ **29.** $-60a^4b^2$

31. $-6a^5b^7$ **33.** $-3x^5y^5$ **35.** $-32x^5y^5$ **37.** $32a^5b^{10}$ **39.** $-8x^6y^3$

41. $a^{10}b^{10}c^5$ **43.** $64x^4y^6$ **45.** $16x^4y^8$ **47.** $-a^6b^9$ **49.** $25m^4n^6$

51. $-x^4y^6$ **53.** $27x^9y^3z^6$ **55.** $2x^2y$ **57.** $-3xy^2$ **59.** $7abc$

61. $-15xz^2$ **63.** $-r^2st^3$ **65.** 7 **67.** -1 **69.** $24x^4$

71. x^{5n} **73.** a^{7n+2} **75.** x^{5n-1} **77.** a^{6n+1} **79.** $-6x^{2n}$

81. $6a^{3n+3}$ **83.** $24x^{4n-1}$ **85.** $2x^{8n-3}$

Problem Set 3.3

1. $12x^3y^2 + 15x^2y^3$ **3.** $-6a^3b^3 + 8a^2b^5$ **5.** $30a^4b^3 - 24a^5b^3 + 18a^4b^4$

7. $-5x^3y^5 + 4x^2y^6 - 3x^3y^6$ **9.** $ac + ad + bc + bd$ **11.** $s^2 - t^2$

13. $x^2 + 20x + 96$ **15.** $y^2 + 5y - 36$ **17.** $n^2 - 5n - 24$

19. $x^2 - 25$ **21.** $x^2 - 18x + 81$ **23.** $x^2 - 11x + 30$

25. $x^3 - 2x^2 - 5x + 6$ **27.** $x^3 - 6x^2 - 9x + 54$ **29.** $t^2 + 12t + 36$

31. $y^2 - 10y + 25$ **33.** $12x^2 + 31x + 20$ **35.** $4y^2 - 1$

37. $5x^2 + 14x - 3$ **39.** $2t^2 - 3t - 5$ **41.** $9t^2 + 30t + 25$

43. $16x^2 - 9$ **45.** $36x^2 - 60x + 25$ **47.** $21x^2 - 47x + 20$

49. $6t^2 - 5ts - 6s^2$ **51.** $9x^2 - a^2$ **53.** $t^3 - t^2 - 6t - 4$

55. $x^3 + 4x^2 - 15x + 6$ **57.** $2x^3 + 7x^2 + 2x - 3$ **59.** $6x^3 + 5x^2 + 6x - 8$

61. $x^4 + 5x^3 + 11x^2 + 11x + 4$ **63.** $2x^4 + 9x^3 - 8x^2 - 13x + 10$ **65.** $x^3 + 3x^2 + 3x + 1$

67. $x^3 - 6x^2 + 12x - 8$ **69.** $8x^3 + 12x^2 + 6x + 1$ **71.** $27x^3 - 27x^2 + 9x - 1$

73. $64x^3 + 144x^2 + 108x + 27$

75. (a) $a^6 + 6a^5b + 15a^4b^2 + 20a^3b^3 + 15a^2b^4 + 6ab^5 + b^6$

(c) $a^8 - 8a^7b + 28a^6b^2 - 56a^5b^3 + 70a^4b^4 - 50a^3b^5 + 28a^2b^6 - 8ab^7 + b^8$

(e) $243x^5 + 810x^4y + 1080x^3y^2 + 720x^2y^3 + 240xy^4 + 32y^5$

(g) $128x^7 - 448x^6 + 672x^5 - 560x^4 + 280x^3 - 84x^2 + 14x - 1$

77. $x^{4a} - 9$ **79.** $x^{2a} + 5x^a - 14$ **81.** $12x^{2a} + 23x^a + 5$

83. $x^{3m} + 3x^{2m} - x^m - 3$ **85.** $9x^{2m} - 12x^m + 4$

Problem Set 3.4

1. (a) Prime   **(c)** Composite   **(e)** Composite   **(g)** Prime   **(i)** Composite

3. $2(2x + 3y)$ **5.** $3x(2x + 3)$ **7.** $5y(y - 5)$ **9.** $7x(2y - 3)$

11. $2x^2(3x + 4)$ **13.** $9ab(2a + 3b)$ **15.** $3x^3y^3(4y - 13x)$ **17.** $4x^2(2x^2 + 3x - 6)$

19. $x(5 + 7x + 9x^3)$ **21.** $5xy^2(3xy + 4 + 7x^2y^2)$ **23.** $(z + 3)(x + y)$ **25.** $(x - y)(2a + 3b)$

27. $(2a + b)(3x - 2y)$ **29.** $(x + y)(3 + a)$ **31.** $(2x - y)(a + b)$ **33.** $(x + 1)(2y + 3a)$

35. $(x + y)(1 + a)$ **37.** $(x - y)(a - b)$ **39.** $(a - 2)(x^2 + 3)$ **41.** $(2b + c)(x + y)$

43. $(a - b)(2a + 3c)$ **45.** $(x + 4)(x + 2)$ **47.** $(x - 3)(x - 2)$ **49.** $\{-3, 0\}$

51. $\{0, 20\}$ **53.** $\{0, 2\}$ **55.** $\{0, 3\}$ **57.** $\{0, 2\}$ **59.** $\{-7, 0\}$ **61.** $\left\{0, \dfrac{7}{3}\right\}$

63. $\left\{-\dfrac{2}{7}, 0\right\}$ **65.** $\left\{-\dfrac{b}{a}, 0\right\}$ **67.** $\left\{0, \dfrac{b}{3a}\right\}$ **69.** $\{-a, -b\}$ **71.** 16 **73.** 0 or $\dfrac{4}{3}$

75. 0 or -4 **77.** 2 units

79. The square is 100 feet by 100 feet and the rectangle is 50 feet by 100 feet.

81. $x^a(2x^a - 3)$ **83.** $y^{2m}(y^m + 5)$ **85.** $x^{4a}(2x^{2a} - 3x^a + 7)$

Problem Set 3.5

1. $(x + 2)(x - 2)$

3. $(3x + 5)(3x - 5)$

5. $(2x + 5y)(2x - 5y)$

7. $(3xy + 8)(3xy - 8)$

9. $(x + y^2)(x - y^2)$

11. $(1 + 9n)(1 - 9n)$

13. $(x + 4 + y)(x + 4 - y)$

15. $(x + y - 1)(x - y + 1)$

17. $(2a + 3b + 1)(2a - 3b - 1)$

19. $-3(2x + 9)$

21. $5(x^2 - 5)$

23. $2(x^2 + 9)$

25. $6(2y + 3)(2y - 3)$

27. $ab(a + 4)(a - 4)$

29. not factorable

31. $(x^2 + 9)(x + 3)(x - 3)$

33. $2x^2(3x + 4)$

35. $3x(2x + 3y)(2x - 3y)$

37. $(1 + 4x^2)(1 + 2x)(1 - 2x)$

39. $5x(2 + x)(2 - x)$

41. $9(x + 2y)(x - 2y)$

43. $(x + 2)(x^2 - 2x + 4)$

45. $(a - 3)(a^2 + 3a + 9)$

47. $(2x + 3y)(4x^2 - 6xy + 9y^2)$

49. $(1 - 2x)(1 + 2x + 4x^2)$

51. $(5x + 3y)(25x^2 - 15xy + 9y^2)$

53. $(x^2 + y^2)(x^4 - x^2y^2 + y^4)$

55. $\{-7, 7\}$

57. $\left\{-\dfrac{5}{2}, \dfrac{5}{2}\right\}$

59. $\{-2, 2\}$

61. $\{0, 7\}$

63. $\{-2, 0, 2\}$

65. $\{-1, 1\}$

67. $\{-5, 0, 5\}$

69. 0 or 1

71. A 1-by-1 square and a 5-by-5 square

73. The length is 8 centimeters and the width is 6 centimeters.

75. 6 feet   **77.** 2 centimeters

Problem Set 3.6

1. $(x + 4)(x + 6)$

3. $(x - 5)(x - 1)$

5. $(a + 7)(a - 2)$

7. $(y + 10)(y + 9)$

9. $(x - 7)(x + 2)$

11. not factorable

13. $(5 + x)(3 - x)$

15. $(x + 2y)(x + 3y)$

17. $(a - 7b)(a + 6b)$

19. $(2x + 3)(x + 2)$

21. $(5x - 2)(x - 1)$

23. $(2a + 3)(3a - 1)$

25. not factorable

27. $(2x + 5)(x - 6)$

29. $(5n + 6)(4n - 3)$

31. $(5x + 1)(2x - 7)$

33. $(6x + 1)(5x - 1)$

35. $(4y - 3)(2y + 7)$

37. $(4n - 7)(3n + 5)$

39. $(7n + 3)(n + 4)$

41. $(x + 9)(x + 12)$

43. $(n - 12)(n - 14)$

45. $(t + 11)(t - 13)$

47. $(t^2 + 4)(t^2 + 6)$

49. $(3x^2 - 2)(x^2 + 3)$

51. $(x + 2)(x - 2)(x^2 + 3)$

53. $(2n + 3)(2n - 3)(n^2 + 3)$

55. $(x + 2)(x - 2)(x + 3)(x - 3)$

57. $(x + 9)(x - 6)$

59. $4(x^2 + 4)$

61. $x(x + 3)(x - 3)$

63. $(3a - 7)^2$

65. $2n(n^2 + 3n + 5)$

67. $(5x - 3)(2x + 9)$

69. $x(x^2 + 1)(x + 1)(x - 1)$

71. not factorable

73. $(x + 5 + y)(x + 5 - y)$

75. $(n + 11)(n + 7)$

77. $(4x - y)(2x + y)$

79. $(7w + 3)(2w - 5)$

81. $25(t + 2)(t - 2)$

83. $(4x - 5)(x - 8)$

85. $5x(2x^2 + 3x + 4)$

87. $(2x^2 + 3)(3x^2 - 7)$

89. $2(3x - 1)^2$

91. $(x + 5)(x^2 - 5x + 25)$

93. $(x^a - 4)(x^a + 6)$

95. $(2x^a - 1)(3x^a - 2)$

97. $(3x^n + 4)(4x^n - 3)$

99. $(x + 1)(x + 3)$

101. $2(x + 4)(2x - 3)$

103. $(2x - 5)(3x - 13)$

Problem Set 3.7

1. $\{-2, -1\}$

3. $\{-7, -4\}$

5. $\{2, 8\}$

7. $\{-7, 3\}$

9. $\{-5, 10\}$

11. $\{-12, -8\}$

13. $\left\{-7, -\dfrac{5}{2}\right\}$

15. $\left\{-\dfrac{7}{2}, \dfrac{2}{3}\right\}$

17. $\left\{-\dfrac{1}{2}, \dfrac{5}{4}\right\}$

19. $\left\{\dfrac{3}{4}, \dfrac{5}{6}\right\}$

21. $\left\{-9, \dfrac{1}{7}\right\}$

23. $\left\{-\dfrac{5}{4}, -\dfrac{4}{5}\right\}$

25. $\{-2, 1\}$

27. $\{-5, 0, 5\}$

29. $\left\{\dfrac{9}{4}\right\}$

31. $\{0, 2, 6\}$

33. $\{-1, 2\}$

35. $\{-3, 1\}$

37. $\{-1, 1\}$

39. $\left\{-\dfrac{7}{6}, \dfrac{1}{3}\right\}$

41. $\{-6, 4\}$ **43.** $\left\{0, \dfrac{2}{3}\right\}$ **45.** $\{-2, 2\}$ **47.** $\{-3, 0, 3\}$ **49.** $\left\{-\dfrac{3}{2}, -1, 1, \dfrac{3}{2}\right\}$

51. $\{-14, 13\}$ **53.** $\left\{\dfrac{3}{5}, 8\right\}$

55. -8 and -7 or 7 and 8 **57.** 7 and 13

59. The length is 13 centimeters and the width is 3 centimeters.

61. 6 rows and 15 trees per row   **63.** 14 feet

65. 6, 8, and 10 units   **67.** 2 inches   **69.** 4 feet and 7 feet

Chapter 3 Review Problem Set

1. $5x - 3$ **2.** $3x^2 + 12x - 2$ **3.** $12x^2 - x + 5$

4. $-20x^5y^7$ **5.** $-6a^5b^5$ **6.** $15a^4 - 10a^3 - 5a^2$

7. $24x^2 + 2xy - 15y^2$ **8.** $3x^3 + 7x^2 - 21x - 4$ **9.** $256x^8y^{12}$

10. $9x^2 - 12xy + 4y^2$ **11.** $-8x^6y^9z^3$ **12.** $-13x^2y$

13. $2x + y - 2$ **14.** $x^4 + x^3 - 18x^2 - x + 35$ **15.** $21 + 26x - 15x^2$

16. $-12a^5b^7$ **17.** $-8a^7b^3$ **18.** $7x^2 + 19x - 36$

19. $6x^3 - 11x^2 - 7x + 2$ **20.** $6x^{4n}$ **21.** $(x + 7)(x - 4)$

22. $2(t + 3)(t - 3)$ **23.** Not factorable   **24.** $(4n - 1)(3n - 1)$

25. $x^2(x^2 + 1)(x + 1)(x - 1)$ **26.** $x(x - 12)(x + 6)$ **27.** $2a^2b(3a + 2b - c)$

28. $(x - y + 1)(x + y - 1)$ **29.** $4(2x^2 + 3)$ **30.** $(4x + 7)(3x - 5)$

31. $(4n - 5)^2$ **32.** $4n(n - 2)$ **33.** $3w(w^2 + 6w - 8)$

34. $(5x + 2y)(4x - y)$ **35.** $16a(a - 4)$ **36.** $3x(x + 1)(x - 6)$

37. $(n + 8)(n - 16)$ **38.** $(t + 5)(t - 5)(t^2 + 3)$ **39.** $(5x - 3)(7x + 2)$

40. $(3 - x)(5 - 3x)$ **41.** $(4n - 3)(16n^2 + 12n + 9)$ **42.** $2(2x + 5)(4x^2 - 10x + 25)$

43. $\{-3, 3\}$ **44.** $\{-6, 1\}$ **45.** $\left\{\dfrac{2}{7}\right\}$ **46.** $\left\{-\dfrac{2}{5}, \dfrac{1}{3}\right\}$ **47.** $\left\{-\dfrac{1}{3}, 3\right\}$

48. $\{-3, 0, 3\}$ **49.** $\{-1, 0, 1\}$ **50.** $\{-7, 9\}$ **51.** $\left\{-\dfrac{4}{7}, \dfrac{2}{7}\right\}$ **52.** $\left\{-\dfrac{4}{5}, \dfrac{5}{6}\right\}$

53. $\{-2, 2\}$ **54.** $\left\{\dfrac{5}{3}\right\}$ **55.** $\{-8, 6\}$ **56.** $\left\{-5, \dfrac{2}{7}\right\}$ **57.** $\{-8, 5\}$

58. $\{-12, 1\}$ **59.** $\varnothing$ **60.** $\left\{-5, \dfrac{6}{5}\right\}$ **61.** $\{0, 1, 8\}$ **62.** $\left\{-10, \dfrac{1}{4}\right\}$

63. 8, 9, and 10 or -1, 0, and 1   **64.** -6 and 8   **65.** 13 and 15

66. 12 miles and 16 miles   **67.** 4 meters by 12 meters   **68.** 9 rows and 16 chairs per row

69. The side is 13 feet long and the altitude is 6 feet.   **70.** 3 feet

71. 5 centimeters by 5 centimeters and 8 centimeters by 8 centimeters   **72.** 6 inches

Chapter 4

Problem Set 4.1

1. $\dfrac{2}{3}$ **3.** $\dfrac{2}{3}$ **5.** $-\dfrac{8}{21}$ **7.** $\dfrac{4}{11}$ **9.** $\dfrac{4}{9y}$ **11.** $\dfrac{3a}{13b}$

13. $-\dfrac{y}{3x}$ **15.** $-\dfrac{11c}{12b}$ **17.** $\dfrac{2x}{3y^3}$ **19.** $\dfrac{x}{x + 2}$ **21.** $\dfrac{2x + 3}{2(3x + 1)}$ **23.** $\dfrac{a + 4}{a - 9}$

25. $\dfrac{n-5}{3n-1}$ **27.** $\dfrac{x^2}{x^2+4}$ **29.** $\dfrac{3x+5}{5x-2}$ **31.** $\dfrac{n-4}{n+4}$ **33.** $\dfrac{x(x-4)}{3(x+1)}$ **35.** $\dfrac{3y+1}{y+4}$

37. $\dfrac{x^2+4}{3(x+2)}$ **39.** $\dfrac{2x(x+2y)}{3(x+y)}$ **41.** $\dfrac{5n-2}{3n+1}$ **43.** $\dfrac{3-2x}{2-x}$ **45.** $\dfrac{x^2-5}{2x^2+3}$ **47.** $\dfrac{n-10}{n+16}$

49. $\dfrac{2x-1}{2(3x-5)}$ **51.** $\dfrac{y+3}{y+4}$ **53.** $\dfrac{x+a}{x+3a}$ **55.** $\dfrac{x+4}{2x-1}$ **57.** $\dfrac{r-3}{r+5}$ **59.** -1

61. $-n-5$ **63.** $-\dfrac{x}{x+3}$ **65.** $-\dfrac{x+6}{x+5}$ **67.** -1

Problem Set 4.2

1. $\dfrac{4}{25}$ **3.** $-\dfrac{4}{15}$ **5.** $\dfrac{8}{27}$ **7.** $\dfrac{1}{3}$ **9.** $-\dfrac{2}{3}$

11. $\dfrac{3}{14}$ **13.** $\dfrac{x}{2y^3}$ **15.** $\dfrac{15b}{a^2}$ **17.** $-\dfrac{8x^3y^3}{15}$ **19.** $\dfrac{3b^4c}{7}$

21. $\dfrac{14}{27x}$ **23.** $\dfrac{a^2}{4}$ **25.** $5y$ **27.** $\dfrac{a(a^2+3)}{2(2a-1)}$ **29.** 2

31. $\dfrac{x(x+2)}{2(x-2)}$ **33.** $\dfrac{(x+6y)^2(2x+3y)}{y^3(x+4y)}$ **35.** $\dfrac{(t^2+9)(6t+7)}{5t-7}$ **37.** $\dfrac{7t-2}{t-9}$ **39.** $\dfrac{2x^2+3}{2x^2+1}$

41. $\dfrac{6-5n}{1+n}$ **43.** $\dfrac{(y+c)(x-2)}{2(y-2c)(3x-1)}$ **45.** $\dfrac{3n-2}{n(n-8)}$ **47.** $\dfrac{6x^6y}{7}$ **49.** $\dfrac{x-9}{42x^2}$

Problem Set 4.3

1. $\dfrac{11}{12}$ **3.** $\dfrac{19}{40}$ **5.** $\dfrac{7}{12}$ **7.** $\dfrac{62}{75}$

9. $\dfrac{1}{24}$ **11.** $-\dfrac{11}{45}$ **13.** $\dfrac{5x+3}{x-2}$ **15.** 3

17. $-\dfrac{3}{y}$ **19.** $\dfrac{7x-18}{12}$ **21.** $\dfrac{2(4a+1)}{15}$ **23.** $\dfrac{n+17}{18}$

25. $\dfrac{-3x+19}{30}$ **27.** $\dfrac{7x}{24}$ **29.** $\dfrac{23}{24x}$ **31.** $\dfrac{32y-45x}{36xy}$

33. $\dfrac{9y+8x-12xy}{12xy}$ **35.** $\dfrac{18+25x}{30x^2}$ **37.** $\dfrac{15-12n}{20n^2}$ **39.** $\dfrac{12+9n-10n^2}{12n^2}$

41. $\dfrac{2x-15}{10x^2}$ **43.** $\dfrac{22t+21}{28t^2}$ **45.** $\dfrac{9y-8x^3}{14x^2y^2}$ **47.** $\dfrac{35b+12a^3}{80a^2b^2}$

49. $\dfrac{3x^2+2x+2}{x(x+1)}$ **51.** $\dfrac{a^2+1}{a(a+1)}$ **53.** $\dfrac{13x+14}{(2x+1)(3x+4)}$ **55.** $\dfrac{7x-12}{(x-1)(2x-3)}$

57. $\dfrac{14x+30}{(2x-5)(4x+3)}$ **59.** $\dfrac{-10n+30}{(n-6)(2n+3)}$ **61. (a)** -1 **(c)** 0

Problem Set 4.4

1. $\dfrac{5x-12}{x(x-6)}$ **3.** $\dfrac{-3x-18}{x(x+8)}$ **5.** $\dfrac{5x+12}{(x+3)(x-3)}$

7. $\dfrac{1}{a-2}$ **9.** $\dfrac{n+8}{3(n+4)(n-4)}$ **11.** $\dfrac{1}{x+1}$

13. $\dfrac{9x + 73}{(x + 3)(x + 7)(x + 9)}$

15. $\dfrac{-7a - 2}{(a - 5)(a + 2)(a + 8)}$

17. $\dfrac{5a^2 - 10a}{(2a + 5)(3a - 2)(a - 4)}$

19. $\dfrac{3x^2 + 30x - 78}{(x + 1)(x - 1)(x + 8)(x - 2)}$

21. $\dfrac{-y + 30}{(y + 7)(y - 5)}$

23. $\dfrac{-x^2 - x + 1}{(x + 1)(x - 1)}$

25. $\dfrac{2x^2 + 9x + 1}{(x + 2)(x + 4)}$

27. $\dfrac{2n + 3}{n + 6}$

29. $\dfrac{t + 7}{2t + 1}$

31. $\dfrac{-4x^2}{(4x + 3)(3x - 2)}$

33. $\dfrac{-8}{(n^2 + 4)(n + 2)(n - 2)}$

35. $\dfrac{4x^2 + 26x - 103}{(x - 3)(x + 6)(x - 2)}$

37. $\dfrac{2}{5}$

39. $-\dfrac{13}{3}$

41. $\dfrac{3y}{10}$

43. $\dfrac{2y + 7x}{3y - 10x}$

45. $\dfrac{4b^2 - 5a^2b}{9ab^2 + 8a^2}$

47. $\dfrac{y + 3xy}{2x + 4xy}$

49. $\dfrac{2n + 7}{4n + 7}$

51. $\dfrac{n - 1}{n + 1}$

53. $\dfrac{5x - y}{5x - 5y - 1}$

55. $\dfrac{-6x - 4}{3x + 9}$

57. $\dfrac{x^2 + x + 1}{x + 1}$

59. $\dfrac{a^2 + 4a + 1}{4a + 1}$

Problem Set 4.5

1. $2x^2 + 3x$

3. $-6x^4 + 9x^2$

5. $7a^2 - 6a - 4$

7. $-15x^4 + 17x^2 + 19$

9. $-5x^2y^2 + 9xy^4$

11. $-9ab^2 + 7a^2b + 5b$

13. $x - 9$

15. $2x + 5$

17. $4x + 5$

19. $y^2 - 4y - 1$

21. $t^2 + 2t - 4$

23. $x - 5 + \dfrac{2}{x + 8}$

25. $3x + 1 + \dfrac{6}{x - 1}$

27. $2x + 5 + \dfrac{1}{3x + 2}$

29. $t^2 - 2t + 4$

31. $x^2 + 4x + 16 + \dfrac{56}{x - 4}$

33. $3x - 4 + \dfrac{3x - 1}{x^2 + 2x}$

35. $5y - 1 + \dfrac{-8y - 2}{y^2 - y}$

37. $4a + 6 + \dfrac{7a - 19}{a^2 - 2a + 3}$

39. $3x + 4y$

Problem Set 4.6

1. $\{1\}$

3. $\left\{-\dfrac{80}{13}\right\}$

5. $\{9\}$

7. $\{6\}$

9. $\left\{-\dfrac{1}{15}\right\}$

11. $\{6\}$

13. $\{39\}$

15. $\left\{\dfrac{1}{2}, 2\right\}$

17. $\left\{\dfrac{2}{3}, \dfrac{3}{2}\right\}$

19. $\{11\}$

21. $\left\{\dfrac{1}{4}\right\}$

23. $\varnothing$

25. $\{3\}$

27. $\{-5, 0\}$

29. $\left\{-3, -\dfrac{2}{3}\right\}$

31. $\left\{-1, \dfrac{13}{15}\right\}$

33. $\{-4, 3\}$

35. $\{10\}$

37. $\{17\}$

39. $\left\{-\dfrac{4}{27}\right\}$

41. $\{-5, 10\}$

43. $\left\{\dfrac{3}{4}, \dfrac{4}{3}\right\}$

45. $\dfrac{2}{5}$ or $\dfrac{5}{2}$

47. $\dfrac{27}{72}$

49. 6 and 63

51. $17\dfrac{1}{2}$ feet by $28\dfrac{3}{4}$ feet

53. 6750 males and 9450 females

55. $225,000

57. 21 centimeters by 36 centimeters

59. $59,500

61. $60°$

Problem Set 4.7

1. $\{-8\}$

3. $\left\{-\dfrac{7}{3}\right\}$

5. $\left\{-\dfrac{2}{3}\right\}$

7. $\{-2\}$

9. $\{-1\}$

11. $\left\{\dfrac{6}{29}\right\}$

13. $\{-3\}$

15. $\{-8, 1\}$

17. $\{-11, 5\}$

19. $\left\{-\dfrac{5}{3}, 1\right\}$

21. $\{-2\}$ **23.** $\left\{-\dfrac{14}{29}\right\}$ **25.** $\left\{-\dfrac{1}{4}\right\}$ **27.** $\left\{-\dfrac{1}{7}\right\}$ **29.** $\left\{-\dfrac{10}{3}\right\}$

31. $T = \dfrac{N(C - V)}{C}$ **33.** $R = \dfrac{ST}{T + S}$ **35.** $x = \dfrac{12y + 8}{9}$ **37.** $y = \dfrac{7x + 16}{3}$ **39.** $y = \dfrac{3x - 41}{7}$

41. $y = mx + b$ **43.** $x = \dfrac{c - by}{a}$

45. 10 miles per hour for Kim and 15 miles per hour for Wendy

47. Ann rides for 5 hours at 10 miles per hour and Sue rides for 3 hours at 20 miles per hour.

49. Walks at $2\dfrac{1}{2}$ miles per hour and jogs at 5 miles per hour

51. $1\dfrac{7}{8}$ hours   **53.** 3 hours   **55.** $10\dfrac{10}{11}$ minutes

57. 10 golf balls at $2 each and 15 golf balls at $1.50 each

Chapter 4 Review Problem Set

1. $\dfrac{2y}{3x^2}$ **2.** $\dfrac{a - 3}{a}$ **3.** $\dfrac{n - 5}{n - 1}$ **4.** $\dfrac{x^2 + 1}{x}$

5. $\dfrac{2x + 1}{3}$ **6.** $\dfrac{x^2 - 10}{2x^2 + 1}$ **7.** $\dfrac{3}{22}$ **8.** $\dfrac{18y + 20x}{48y - 9x}$

9. $\dfrac{3x + 2}{3x - 2}$ **10.** $\dfrac{x - 1}{2x - 1}$ **11.** $\dfrac{2x}{7y^2}$ **12.** $3b$

13. $\dfrac{n(n + 5)}{n - 1}$ **14.** $\dfrac{x(x - 3y)}{x^2 + 9y^2}$ **15.** $\dfrac{23x - 6}{20}$ **16.** $\dfrac{57 - 2n}{18n}$

17. $\dfrac{3x^2 - 2x - 14}{x(x + 7)}$ **18.** $\dfrac{2}{x - 5}$ **19.** $\dfrac{5n - 21}{(n - 9)(n + 4)(n - 1)}$ **20.** $\dfrac{6y - 23}{(2y + 3)(y - 6)}$

21. $6x - 1$ **22.** $3x^2 - 7x + 22 - \dfrac{90}{x + 4}$ **23.** $\left\{\dfrac{4}{13}\right\}$ **24.** $\left\{\dfrac{3}{16}\right\}$

25. $\varnothing$ **26.** $\{-17\}$ **27.** $\left\{\dfrac{2}{7}, \dfrac{7}{2}\right\}$ **28.** $\{22\}$

29. $\left\{-\dfrac{6}{7}, 3\right\}$ **30.** $\left\{\dfrac{3}{4}, \dfrac{5}{2}\right\}$ **31.** $\left\{\dfrac{9}{7}\right\}$ **32.** $\left\{-\dfrac{5}{4}\right\}$

33. $y = \dfrac{3x + 27}{4}$ **34.** $y = \dfrac{bx - ab}{a}$ **35.** $525 and $875

36. 20 minutes for Don and 30 minutes for Dan

37. 50 miles per hour and 55 miles per hour

38. 9 hours   **39.** 80 hours   **40.** 13 miles per hour

Cumulative Review Problem Set for Chapters 1, 2, 3 and 4

1. -28 **2.** -10 **3.** $\dfrac{64}{15}$ **4.** $\dfrac{11}{3}$

5. $\dfrac{1}{6}$ **6.** $-\dfrac{44}{5}$ **7.** $2x - 11$ **8.** $-24a^4b^5$

9. $2x^3 + 5x^2 + x + 12$

10. $\dfrac{3x^2y^2}{8}$

11. $\dfrac{a(a+1)}{2a-1}$

12. $\dfrac{-x+14}{18}$

13. $\dfrac{5x+19}{x(x+3)}$

14. $28x^2 + 23x - 15$

15. $\dfrac{2}{n+8}$

16. $\dfrac{x-14}{(x+1)(5x-2)(x-4)}$

17. $\left\{-\dfrac{12}{7}\right\}$

18. $\{150\}$

19. $\{-2, 2\}$

20. $\{-7\}$

21. $\left\{-6, \dfrac{4}{3}\right\}$

22. $\left\{\dfrac{5}{4}\right\}$

23. $\left\{\dfrac{1}{5}\right\}$

24. $\left\{\dfrac{5}{7}\right\}$

25. $\{-2, 2\}$

26. $\{0\}$

27. $\{-6, 19\}$

28. $\{-9, 4\}$

29. $\{x \mid x \le -2\}$

30. $\left\{x \mid x < \dfrac{19}{5}\right\}$

31. $\left\{n \mid n > \dfrac{1}{4}\right\}$

32. $\{x \mid -2 < x < 3\}$

33. $\left\{x \mid x < -\dfrac{13}{3} \text{ or } x > 3\right\}$

34. $\{x \mid x \le 29\}$

Chapter 5

Problem Set 5.1

1. $\dfrac{1}{8}$

3. $-\dfrac{1}{1000}$

5. 27

7. 4

9. $-\dfrac{27}{8}$

11. 1

13. $\dfrac{16}{25}$

15. 4

17. $\dfrac{1}{100}$ or 0.01

19. $\dfrac{1}{100000}$ or 0.00001

21. 81

23. $\dfrac{1}{16}$

25. $\dfrac{3}{4}$

27. $\dfrac{256}{25}$

29. $\dfrac{16}{25}$

31. $\dfrac{64}{81}$

33. 64

35. $\dfrac{1}{100000}$ or 0.00001

37. $\dfrac{17}{72}$

39. $\dfrac{1}{6}$

41. $\dfrac{48}{19}$

43. $\dfrac{1}{x^4}$

45. $\dfrac{1}{a^2}$

47. $\dfrac{1}{a^6}$

49. $\dfrac{y^4}{x^3}$

51. $\dfrac{c^3}{a^3b^6}$

53. $\dfrac{y^2}{4x^4}$

55. $\dfrac{x^4}{y^6}$

57. $\dfrac{9a^2}{4b^4}$

59. $\dfrac{1}{x^3}$

61. $\dfrac{a^3}{b}$

63. $\dfrac{6}{x^3y}$

65. $\dfrac{6}{a^2y^3}$

67. $\dfrac{4x^3}{y^5}$

69. $-\dfrac{5}{a^2b}$

71. $\dfrac{1}{4x^2y^4}$

73. $\dfrac{x+1}{x^2}$

75. $\dfrac{y-x^2}{x^2y}$

77. $\dfrac{3b^3+2a^2}{a^2b^3}$

79. $\dfrac{y^2-x^2}{xy}$

Problem Set 5.2

1. 9

3. -7

5. 4

7. -3

9. 2

11. $\dfrac{2}{5}$

13. 2

15. $\dfrac{1}{3}$

17. $\dfrac{2}{3}$

19. $-\dfrac{1}{4}$

21. $2\sqrt{3}$

23. $2\sqrt{6}$

25. $2\sqrt{13}$

27. $4\sqrt{7}$

29. $10\sqrt{2}$

31. $-6\sqrt{11}$

33. $\sqrt{2}$

35. $\dfrac{10\sqrt{3}}{3}$

37. $\dfrac{\sqrt{17}}{3}$ **39.** $\dfrac{\sqrt{3}}{4}$ **41.** $\dfrac{2\sqrt{6}}{7}$ **43.** $\dfrac{\sqrt{15}}{5}$ **45.** $\dfrac{\sqrt{6}}{2}$ **47.** $\dfrac{\sqrt{21}}{6}$

49. $\dfrac{\sqrt{6}}{2}$ **51.** $\sqrt{7}$ **53.** $\dfrac{3\sqrt{14}}{7}$ **55.** $\dfrac{6\sqrt{15}}{5}$ **57.** $\dfrac{\sqrt{15}}{4}$ **59.** $-\dfrac{9}{20}$

61. $2\sqrt[3]{4}$ **63.** $3\sqrt[3]{5}$ **65.** $\dfrac{5\sqrt[3]{9}}{3}$ **67.** $\dfrac{3\sqrt[3]{2}}{2}$ **69.** $\dfrac{\sqrt[3]{20}}{2}$ **71. (a)** 1.414

(c) 12.490 **(e)** 57.000 **(g)** 0.374 **(i)** 0.930

Problem Set 5.3

1. $12\sqrt{3}$ **3.** $9\sqrt{2}$ **5.** $-12\sqrt{5}$ **7.** $12\sqrt{3}$ **9.** $3\sqrt{7}$

11. $\dfrac{13\sqrt{2}}{15}$ **13.** $\dfrac{11\sqrt{3}}{6}$ **15.** $\dfrac{-7\sqrt{5}}{12}$ **17.** $5\sqrt[3]{2}$ **19.** $-17\sqrt[3]{3}$

21. 0 **23.** $-5\sqrt{3n}$ **25.** $-7x\sqrt{2x}$ **27.** $2\sqrt{2x}$ **29.** $2y\sqrt{3}$

31. $2x\sqrt{5y}$ **33.** $7xy^2\sqrt{xy}$ **35.** $5a^2b^3\sqrt{2a}$ **37.** $5x^3y^4\sqrt{3}$ **39.** $12a\sqrt{2a}$

41. $\dfrac{9y^3\sqrt{5x}}{7}$ **43.** $\dfrac{\sqrt{6xy}}{2y}$ **45.** $\dfrac{\sqrt{14}}{4x}$ **47.** $\dfrac{\sqrt{3x}}{2x}$ **49.** $\dfrac{\sqrt{10xy}}{6x^2}$

51. $\dfrac{x\sqrt{2xy}}{3y}$ **53.** $\dfrac{2\sqrt{15a}}{5ab}$ **55.** $2\sqrt[3]{2x^2}$ **57.** $3x\sqrt[3]{2}$ **59.** $3xy^2\sqrt[3]{3x^2}$

61. $\dfrac{\sqrt[3]{20x^2}}{2x}$ **63.** $\dfrac{\sqrt[3]{18x^2y}}{3x}$ **65.** $\dfrac{5\sqrt[3]{3x^2y}}{3xy}$ **67.** $2\sqrt{x+y}$ **69.** $3\sqrt{3x+2y}$

71. (a) 0.707 **(c)** 0.707

Problem Set 5.4

1. $4\sqrt{3}$ **3.** $6\sqrt{10}$ **5.** 140

7. $48\sqrt{6}$ **9.** 30 **11.** $\sqrt{15}+\sqrt{21}$

13. $6\sqrt{35}-9\sqrt{55}$ **15.** $20\sqrt{6}+30\sqrt{3}$ **17.** $6\sqrt[3]{12}-8\sqrt[3]{10}$

19. $36+15\sqrt[3]{21}$ **21.** $2\sqrt{2xy}-3\sqrt{10x}$ **23.** $15x\sqrt{y}+12x$

25. $3x\sqrt{2y}-2\sqrt{6xy}$ **27.** $13+7\sqrt{3}$ **29.** $\sqrt{10}-\sqrt{14}+\sqrt{15}-\sqrt{21}$

31. $30+11\sqrt{6}$ **33.** $24\sqrt{2}-4\sqrt{30}-24\sqrt{6}+12\sqrt{10}$ **35.** -2

37. 16 **39.** -4 **41.** 6 **43.** $a-b$

45. $4x-9y$ **47.** $3\sqrt{5}-6$ **49.** $\sqrt{6}+1$ **51.** $\sqrt{5}-\sqrt{3}$

53. $\dfrac{\sqrt{42}+3\sqrt{2}}{4}$ **55.** $\dfrac{2\sqrt{15}+\sqrt{5}}{11}$ **57.** $\dfrac{6\sqrt{3}+3\sqrt{5}}{14}$ **59.** $\dfrac{-2\sqrt{10}+3\sqrt{14}}{43}$

61. $\dfrac{3\sqrt{x}+6}{x-4}$ **63.** $\dfrac{x+\sqrt{x}}{x-1}$ **65.** $\dfrac{x+\sqrt{x}-6}{x-4}$ **67.** $\dfrac{x-\sqrt{xy}}{x-y}$

69. $\dfrac{6x+7\sqrt{xy}+2y}{9x-4y}$

Problem Set 5.5

1. $\{8\}$ **3.** $\varnothing$ **5.** $\left\{\dfrac{25}{4}\right\}$ **7.** $\{5\}$ **9.** $\varnothing$ **11.** $\left\{\dfrac{39}{5}\right\}$

13. $\varnothing$ **15.** $\{3\}$ **17.** $\left\{\dfrac{10}{3}\right\}$ **19.** $\left\{\dfrac{7}{9}\right\}$ **21.** $\{-1,1\}$ **23.** $\{-4,6\}$

25. $\varnothing$ **27.** $\{-2\}$ **29.** $\{5\}$ **31.** $\{-1,-2\}$ **33.** $\{9\}$ **35.** $\{-4\}$

37. $\{13\}$ **39.** $\{-21\}$ **41.** $\{0\}$ **43.** $\left\{\dfrac{3}{8}\right\}$ **45.** $\{5\}$ **47.** $\{13\}$
49. $\{4\}$

Problem Set 5.6

1. 8   **3.** 4   **5.** -2 **7.** -4 **9.** $\dfrac{1}{7}$

11. 2   **13.** 32   **15.** 16   **17.** 16   **19.** -9

21. $\dfrac{16}{81}$ **23.** 9   **25.** $\dfrac{1}{32}$ **27.** -1024 **29.** 27

31. $\sqrt[3]{x^2}$ **33.** $2\sqrt{x}$ **35.** $\sqrt{3y}$ **37.** $\sqrt[3]{x-2y}$ **39.** $\sqrt[5]{(2a+b)^3}$
41. $\sqrt[3]{xy^2}$ **43.** $-3\sqrt[4]{xy^3}$ **45.** $2^{1/2}x^{1/2}$ **47.** $2x^{1/2}$ **49.** $x^{2/3}y^{1/3}$
51. $a^{1/4}b^{3/4}$ **53.** $(2x-y)^{2/3}$ **55.** $2xy^{1/2}$ **57.** $-(a+b)^{1/3}$ **59.** $15x^{7/12}$

61. $y^{1/4}$ **63.** $\dfrac{2}{x^{1/6}}$ **65.** $64x^{3/4}y^{3/2}$ **67.** $3xy^2$ **69.** $2x^{1/6}$

71. $\dfrac{7}{a^{1/12}}$ **73.** $\dfrac{16x^{4/3}}{81y}$ **75.** $\dfrac{y^{3/2}}{x}$ **77.** $\sqrt[4]{8}$ **79.** $\sqrt[6]{3125}$

81. $\sqrt[6]{2}$ **83.** $\sqrt[3]{9}$ **85.** 2   **89. (a)** 13.391   **(c)** 2.702
(e) 4.304

Problem Set 5.7

1. $(8.9)(10)^1$ **3.** $(4.29)(10)^3$ **5.** $(6.12)(10)^6$ **7.** $(4)(10)^7$
9. $(3.764)(10)^2$ **11.** $(3.47)(10)^{-1}$ **13.** $(2.14)(10)^{-2}$ **15.** $(5)(10)^{-5}$
17. $(1.94)(10)^{-9}$ **19.** 23   **21.** 4190   **23.** 500,000,000
25. 31,400,000,000   **27.** 0.43   **29.** 0.000914   **31.** 0.00000005123
33. 0.000000074   **35.** 0.77   **37.** 300,000,000,000   **39.** 0.000000004
41. 1000   **43.** 1000   **45.** 3000   **47.** 20
49. 27,000,000   **51. (a)** 7000   **(c)** 120   **(e)** 30
53. (a) $(4.385)(10)^{14}$ **(c)** $(2.322)(10)^{17}$ **(e)** $(3.052)(10)^{12}$

Chapter 5 Review Problem Set

1. $\dfrac{1}{64}$ **2.** $\dfrac{9}{4}$ **3.** 3   **4.** -2 **5.** $\dfrac{2}{3}$

6. 32   **7.** 1   **8.** $\dfrac{4}{9}$ **9.** -64 **10.** 32

11. 1   **12.** 27   **13.** $3\sqrt{6}$ **14.** $4x\sqrt{3xy}$ **15.** $2\sqrt{2}$

16. $\dfrac{\sqrt{15x}}{6x^2}$ **17.** $2\sqrt[3]{7}$ **18.** $\dfrac{\sqrt[3]{6}}{3}$ **19.** $\dfrac{9\sqrt{5}}{5}$ **20.** $\dfrac{x\sqrt{21x}}{7}$

21. $3xy^2\sqrt[3]{4xy^2}$ **22.** $\dfrac{15\sqrt{6}}{4}$ **23.** $2y\sqrt{5xy}$ **24.** $2\sqrt{x}$ **25.** $24\sqrt{10}$

26. 60   **27.** $24\sqrt{3}-6\sqrt{14}$ **28.** $x-2\sqrt{x}-15$ **29.** 17

30. $12-8\sqrt{3}$ **31.** $6a-5\sqrt{ab}-4b$ **32.** 70   **33.** $\dfrac{2(\sqrt{7}+1)}{3}$

34. $\dfrac{2\sqrt{6}-\sqrt{15}}{3}$ **35.** $\dfrac{3\sqrt{5}-2\sqrt{3}}{11}$ **36.** $\dfrac{6\sqrt{3}+3\sqrt{5}}{7}$ **37.** $\dfrac{x^6}{y^8}$

38. $\dfrac{27a^3b^{12}}{8}$ **39.** $20x^{\frac{7}{10}}$ **40.** $7a^{\frac{5}{12}}$ **41.** $\dfrac{y^{\frac{4}{3}}}{x}$

42. $\dfrac{x^{12}}{9}$ **43.** $\sqrt{5}$ **44.** $5\sqrt[3]{3}$ **45.** $\dfrac{29\sqrt{6}}{5}$

46. $-15\sqrt{3x}$ **47.** $\dfrac{y+x^2}{x^2y}$ **48.** $\dfrac{b-2a}{a^2b}$ **49.** $\left\{\dfrac{19}{7}\right\}$

50. $\{4\}$ **51.** $\{8\}$ **52.** $\varnothing$ **53.** $\{14\}$
54. $\{-10,1\}$ **55.** $\{2\}$ **56.** $\{8\}$

Chapter 6

Problem Set 6.1

1. $\{0,7\}$ **3.** $\{-9,0\}$ **5.** $\{-3,0\}$ **7.** $\left\{0,\dfrac{7}{3}\right\}$ **9.** $\{-4,5\}$

11. $\{-2,9\}$ **13.** $\left\{-6,-\dfrac{1}{3}\right\}$ **15.** $\left\{-\dfrac{5}{4},\dfrac{2}{3}\right\}$ **17.** $\left\{-3,\dfrac{5}{6}\right\}$ **19.** $\left\{\dfrac{1}{3}\right\}$

21. $\{2,8\}$ **23.** $\{4\}$ **25.** $\{20\}$ **27.** $\{-3k,0\}$ **29.** $\{0,16k\}$
31. $\{k,6k\}$ **33.** $\{3k\}$ **35.** $\{-5,5\}$ **37.** $\{-1,1\}$ **39.** $\{-\sqrt{5},\sqrt{5}\}$

41. $\{-2\sqrt{5},2\sqrt{5}\}$ **43.** $\{-2\sqrt{3},2\sqrt{3}\}$ **45.** $\left\{\dfrac{-2\sqrt{6}}{3},\dfrac{2\sqrt{6}}{3}\right\}$

47. $\left\{\dfrac{-2\sqrt{70}}{7},\dfrac{2\sqrt{70}}{7}\right\}$ **49.** $\left\{\dfrac{-3\sqrt{10}}{4},\dfrac{3\sqrt{10}}{4}\right\}$ **51.** $\left\{\dfrac{-5\sqrt{6}}{12},\dfrac{5\sqrt{6}}{12}\right\}$

53. $\{-3,1\}$ **55.** $\varnothing$ **57.** $\left\{\dfrac{1}{3},1\right\}$

59. $\{-2-\sqrt{7},-2+\sqrt{7}\}$ **61.** $\{-4-2\sqrt{2},-4+2\sqrt{2}\}$ **63.** $\left\{\dfrac{1-3\sqrt{2}}{2},\dfrac{1+3\sqrt{2}}{2}\right\}$

65. $\left\{\dfrac{3-4\sqrt{2}}{4},\dfrac{3+4\sqrt{2}}{4}\right\}$ **67.** $\{-12,0\}$ **69.** $\left\{\dfrac{1-\sqrt{5}}{4},\dfrac{1+\sqrt{5}}{4}\right\}$

71. $2\sqrt{5}$ centimeters **73.** $2\sqrt{21}$ feet
75. 7 meters **77.** $5\sqrt{2}$ inches

79. $a=b=\dfrac{7\sqrt{2}}{2}$ feet **81.** $b=2\sqrt{3}$ inches and $c=4$ inches

83. $a=8$ centimeters and $b=8\sqrt{3}$ centimeters **85.** $a=\dfrac{8\sqrt{3}}{3}$ feet and $c=\dfrac{16\sqrt{3}}{3}$ feet

Problem Set 6.2

1. $\{-6,2\}$ **3.** $\{2,8\}$ **5.** $\{-5,2\}$

7. $\left\{-\dfrac{1}{3},2\right\}$ **9.** $\left\{-4,\dfrac{1}{2}\right\}$ **11.** $\{-3-\sqrt{10},-3+\sqrt{10}\}$

13. $\{-4-3\sqrt{2},-4+3\sqrt{2}\}$ **15.** $\{-3-2\sqrt{6},-3+2\sqrt{6}\}$ **17.** $\varnothing$

19. $\{-1 - 3\sqrt{2}, -1 + 3\sqrt{2}\}$

21. $\left\{\dfrac{3 + \sqrt{13}}{2}, \dfrac{3 - \sqrt{13}}{2}\right\}$

23. $\left\{\dfrac{-5 - \sqrt{21}}{2}, \dfrac{-5 + \sqrt{21}}{2}\right\}$

25. $\left\{\dfrac{-1 - \sqrt{33}}{2}, \dfrac{-1 + \sqrt{33}}{2}\right\}$

27. $\left\{\dfrac{-6 - \sqrt{42}}{3}, \dfrac{-6 + \sqrt{42}}{3}\right\}$

29. $\left\{\dfrac{2 - \sqrt{2}}{2}, \dfrac{2 + \sqrt{2}}{2}\right\}$

31. $\{0, 16\}$

33. $\{-7, 2\}$

35. $\{-3 - 2\sqrt{3}, -3 + 2\sqrt{3}\}$

37. $\left\{-\dfrac{2}{3}, \dfrac{1}{2}\right\}$

39. $\left\{\dfrac{1 - \sqrt{3}}{2}, \dfrac{1 + \sqrt{3}}{2}\right\}$

41. $\left\{-\dfrac{4}{3}, \dfrac{7}{2}\right\}$

43. $\{10, 16\}$

45. $\{-2 + \sqrt{3}, -2 - \sqrt{3}\}$

47. $\left\{\dfrac{-3 + 2\sqrt{3}}{3}, \dfrac{-3 - 2\sqrt{3}}{3}\right\}$

49. $\left\{\dfrac{1}{4}, \dfrac{1}{3}\right\}$

51. $\left\{\dfrac{-b + \sqrt{b^2 - 4ac}}{2a}, \dfrac{-b - \sqrt{b^2 - 4ac}}{2a}\right\}$

53. $t = \dfrac{\sqrt{2gs}}{g}$

55. $y = \dfrac{b\sqrt{x^2 - a^2}}{a}$

57. $\{-3a, -5a\}$

59. $\left\{-\dfrac{2a}{5}, \dfrac{7a}{2}\right\}$

61. $\left\{-\dfrac{b}{2}\right\}$

Problem Set 6.3

1. (a) two real solutions   **(c)** no real solutions   **(e)** two real solutions   **(g)** two real solutions

3. (a) sum of -9 and product of 8   **(c)** sum of $\dfrac{9}{2}$ and product of -9 **(e)** sum of 2 and product of -2

(g) sum of $\dfrac{4}{3}$ and product of $-\dfrac{1}{3}$

5. $\left\{\dfrac{-3 - \sqrt{17}}{2}, \dfrac{-3 + \sqrt{17}}{2}\right\}$

7. $\{-2 - \sqrt{5}, -2 + \sqrt{5}\}$

9. $\{3 - \sqrt{11}, 3 + \sqrt{11}\}$

11. $\varnothing$

13. $\{-14, -5\}$

15. $\left\{\dfrac{7 - \sqrt{33}}{2}, \dfrac{7 + \sqrt{33}}{2}\right\}$

17. $\left\{\dfrac{-5 - \sqrt{41}}{4}, \dfrac{-5 + \sqrt{41}}{4}\right\}$

19. $\left\{\dfrac{-5 - \sqrt{13}}{6}, \dfrac{-5 + \sqrt{13}}{6}\right\}$

21. $\left\{\dfrac{3 - \sqrt{7}}{2}, \dfrac{3 + \sqrt{7}}{2}\right\}$

23. $\varnothing$

25. $\left\{\dfrac{5}{2}, 6\right\}$

27. $\left\{\dfrac{2 - \sqrt{14}}{5}, \dfrac{2 + \sqrt{14}}{5}\right\}$

29. $\left\{-\dfrac{1}{2}, \dfrac{2}{3}\right\}$

31. $\left\{-\dfrac{14}{3}, 0\right\}$

33. $\left\{\dfrac{-\sqrt{3}}{2}, \dfrac{\sqrt{3}}{2}\right\}$

35. $\left\{\dfrac{-3 + \sqrt{15}}{2}, \dfrac{-3 - \sqrt{15}}{2}\right\}$

37. $\{-12, 16\}$

39. $\left\{-\dfrac{15}{2}, \dfrac{17}{3}\right\}$

41. $\left\{\dfrac{2 + \sqrt{2}}{3}, \dfrac{2 - \sqrt{2}}{3}\right\}$

43. $\left\{\dfrac{2 + \sqrt{6}}{2}, \dfrac{2 - \sqrt{6}}{2}\right\}$

45. (a) $\{-1.359, 7.359\}$

(c) $\{-10.230, 4.280\}$

(e) $\{-0.257, 2.591\}$

(g) $\{0.191, 1.309\}$

(i) $\{-0.422, 5.922\}$

Problem Set 6.4

1. $\left\{\dfrac{3 - \sqrt{37}}{2}, \dfrac{3 + \sqrt{37}}{2}\right\}$

3. $\left\{-4, \dfrac{7}{2}\right\}$

5. $\{-15, -7\}$

7. $\left\{\dfrac{1 - \sqrt{13}}{3}, \dfrac{1 + \sqrt{13}}{3}\right\}$

9. $\left\{-\dfrac{9}{4}, \dfrac{1}{3}\right\}$

11. $\{8 - 5\sqrt{2}, 8 + 5\sqrt{2}\}$

13. $\varnothing$

15. $\left\{\dfrac{-5-3\sqrt{5}}{10},\dfrac{-5+3\sqrt{5}}{10}\right\}$

17. $\{-10-5\sqrt{5},-10+5\sqrt{5}\}$

19. $\left\{\dfrac{3}{4},\dfrac{4}{3}\right\}$

21. $\left\{\dfrac{5-\sqrt{41}}{2},\dfrac{5+\sqrt{41}}{2}\right\}$

23. $\left\{-\dfrac{24}{11},2\right\}$

25. $\{-15,12\}$

27. $\left\{\dfrac{7-\sqrt{129}}{4},\dfrac{7+\sqrt{129}}{4}\right\}$

29. $\{-\sqrt{6},-1,1,\sqrt{6}\}$

31. $\{-\sqrt{6},-\sqrt{2},\sqrt{2},\sqrt{6}\}$

33. $\left\{-\dfrac{2\sqrt{3}}{3},-\dfrac{\sqrt{2}}{2},\dfrac{\sqrt{2}}{2},\dfrac{2\sqrt{3}}{3}\right\}$

35. $\left\{-\dfrac{\sqrt{3}}{2},\dfrac{\sqrt{3}}{2}\right\}$

37. 9 and 14   **39.** 5 and 7

41. A length of 12 meters and a width of 9 meters

43. $1\dfrac{1}{2}$ inches   **45.** 9 units by 11 units   **47.** $3-\sqrt{2}$ and $3+\sqrt{2}$

49. 54 miles per hour for Mike and 52 miles per hour for Larry

51. 20 miles per hour

53. 40 minutes with the power mower and 80 minutes with the push mower

55. 25 people at $4 each

57. 12 sides   **59.** 30 yards   **61.** 12%   **63.** $\{1,9\}$ **65.** $\{-27,8\}$ **67.** $\left\{-\dfrac{1}{6},\dfrac{1}{2}\right\}$

Problem Set 6.5

1.

3.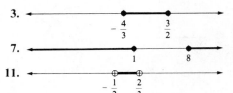

5.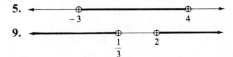

7.

9.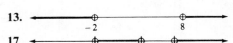

11.

all reals

13.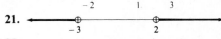

15.

17.

19.

21.

23.

25.

27.

29.

31.

33.

35.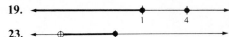

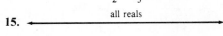

Problem Set 6.6

1. True   **3.** False   **5.** True   **7.** False   **9.** $13+8i$

11. $-5+8i$ **13.** $3+4i$ **15.** $-10-2i$ **17.** $-11+i$ **19.** $-1-i$

21. $-1-2i$ **23.** $\dfrac{5}{6}+\dfrac{13}{20}i$ **25.** $-\dfrac{1}{4}+\dfrac{3}{10}i$ **27.** $3i$ **29.** $i\sqrt{19}$

31. $\dfrac{2}{3}i$ **33.** $2i\sqrt{2}$ **35.** $3i\sqrt{3}$ **37.** $3i\sqrt{6}$ **39.** $18i$

41. $12i\sqrt{2}$ **43.** -8 **45.** $-\sqrt{6}$ **47.** $-2\sqrt{5}$ **49.** $-2\sqrt{15}$

51. $-2\sqrt{14}$ **53.** 3 **55.** $\sqrt{6}$ **57.** -21 **59.** $8 + 12i$

61. $0 + 26i$ **63.** $53 - 26i$ **65.** $10 - 24i$ **67.** $-14 - 8i$ **69.** $-7 + 24i$

71. $-3 + 4i$ **73.** $113 + 0i$ **75.** $13 + 0i$ **77.** $-\dfrac{8}{13} + \dfrac{12}{13}i$ **79.** $1 - \dfrac{2}{3}i$

81. $0 - \dfrac{3}{2}i$ **83.** $\dfrac{22}{41} - \dfrac{7}{41}i$ **85.** $-1 + 2i$ **87.** $-\dfrac{17}{10} + \dfrac{1}{10}i$ **89.** $\dfrac{5}{13} - \dfrac{1}{13}i$

Problem Set 6.7

1. $\{-6i, 6i\}$

3. $\{-4, -3i, -4 + 3i\}$

5. $\{2 - i\sqrt{5}, 2 + i\sqrt{5}\}$

7. $\{1 - 2i\sqrt{3}, 1 + 2i\sqrt{3}\}$

9. $\left\{0, \dfrac{4}{3}\right\}$

11. $\{-1 - 2i, -1 + 2i\}$

13. $\{3 - i, 3 + i\}$

15. $\{-7, 4\}$

17. $\{1 - \sqrt{2}, 1 + \sqrt{2}\}$

19. $\left\{\dfrac{3}{4} - \dfrac{i\sqrt{23}}{4}, \dfrac{3}{4} + \dfrac{i\sqrt{23}}{4}\right\}$

21. $\left\{\dfrac{1}{3} - \dfrac{i\sqrt{2}}{3}, \dfrac{1}{3} + \dfrac{i\sqrt{2}}{3}\right\}$

23. $\left\{-\dfrac{5}{2}, \dfrac{1}{3}\right\}$

25. $\left\{\dfrac{1}{8} - \dfrac{i\sqrt{15}}{8}, \dfrac{1}{8} + \dfrac{i\sqrt{15}}{8}\right\}$

27. $\{3 - 4i, 3 + 4i\}$

29. $\left\{\dfrac{7}{2}\right\}$

31. (a) two unequal real solutions   **(c)** two equal real solutions   **(e)** two complex but nonreal solutions
(g) two unequal real solutions   **(i)** two equal real solutions   **33.** $k = 4$ or $k = -4$

Chapter 6 Review Problem Set

1. $\{0, 17\}$

2. $\{-4, 8\}$

3. $\{-3, 7\}$

4. $\{-1 - \sqrt{10}, -1 + \sqrt{10}\}$

5. $\{25\}$

6. $\left\{-4, \dfrac{2}{3}\right\}$

7. $\{-10, 20\}$

8. $\left\{\dfrac{-1 - \sqrt{61}}{6}, \dfrac{-1 + \sqrt{61}}{6}\right\}$

9. $\left\{\dfrac{-2 - \sqrt{14}}{2}, \dfrac{-2 + \sqrt{14}}{2}\right\}$

10. $\{-9, 4\}$

11. $\{1 - \sqrt{10}, 1 + \sqrt{10}\}$

12. $\{-6, 12\}$

13. $\left\{-2\sqrt{2}, -\dfrac{\sqrt{14}}{2}, \dfrac{\sqrt{14}}{2}, 2\sqrt{2}\right\}$

14. $\left\{\dfrac{-3 - \sqrt{97}}{2}, \dfrac{-3 + \sqrt{97}}{2}\right\}$

15. Two equal real solutions

16. No real solutions

17. Two unequal real solutions

18. Two unequal real solutions

19.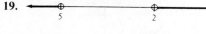

20.

21.

22.

23. $3 + \sqrt{7}$ and $3 - \sqrt{7}$ **24.** 20 shares at \$15 per share
25. 45 miles per hour and 52 miles per hour
26. 8 units   **27.** 8 and 10
28. 7 inches by 12 inches
29. 4 hours for Janet and 6 hours for Billy
30. 10 meters   **31.** $2 - 2i$ **32.** $-3 - i$

33. $30 + 15i$ **34.** $86 - 2i$ **35.** $-32 + 4i$

36. $\dfrac{9}{20} + \dfrac{13}{20}i$ **37.** $\left\{\dfrac{1}{2} - 4i, \dfrac{1}{2} + 4i\right\}$ **38.** $\{3 - 5i, 3 + 5i\}$

39. $\left\{\dfrac{1}{2} - \dfrac{\sqrt{11}}{2}i, \dfrac{1}{2} + \dfrac{\sqrt{11}}{2}i\right\}$ **40.** $\left\{\dfrac{5}{4} - \dfrac{\sqrt{23}}{4}i, \dfrac{5}{4} + \dfrac{\sqrt{23}}{4}i\right\}$ **41.** $\{-2 - i\sqrt{5}, -2 + i\sqrt{5}\}$

Chapter 7

Problem Set 7.1

1. 10   **3.** $\sqrt{34}$ **5.** $4\sqrt{2}$ **7.** $2\sqrt{13}$ **9.** 10   **11.** Two sides are of length $2\sqrt{10}$.

13. The distance between $(3,1)$ and $(-2,6)$ equals the distance between $(3,1)$ and $(8,-4)$, which is $5\sqrt{2}$ units.

15. $\dfrac{3}{4}$ **17.** $\dfrac{2}{3}$ **19.** $-\dfrac{1}{5}$ **21.** 0   **23.** $-\dfrac{b}{a}$ **25.** Not defined   **27.** -3 **29.** 9

Problems 31, 33, 35, and 37: answers will vary.

39. $\dfrac{1}{2}$ **41.** $\dfrac{5}{2}$ **43.** $-\dfrac{3}{5}$ **45.** 2   **47. (a)** 32 centimeters

Problem Set 7.2

1.
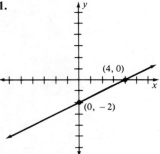

3.
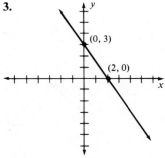

5.
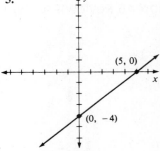

7.
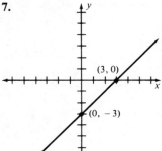

9.
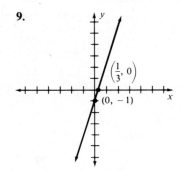

11.
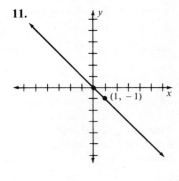

13.

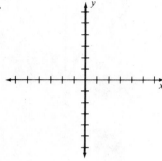

The graph is the *y*-axis.

15.

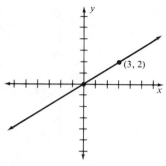

(3, 2)

17.

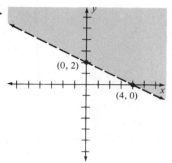

(0, 2)

(4, 0)

19.

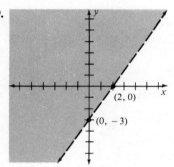

(2, 0)

(0, −3)

21.

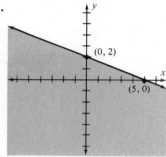

(0, 2)

(5, 0)

23.

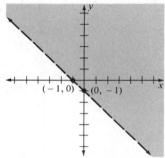

(−1, 0)   (0, −1)

25.

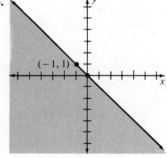

(−1, 1)

27.

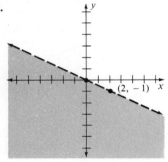

(2, −1)

29.

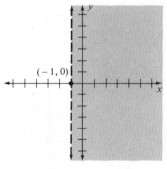

(−1, 0)

31.

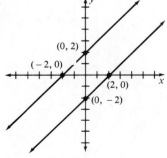

(0, 2)

(−2, 0)

(2, 0)

(0, −2)

33.

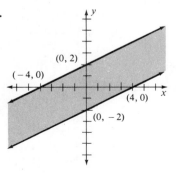

(0, 2)

(−4, 0)

(4, 0)

(0, −2)

35.

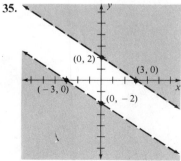

(0, 2)

(3, 0)

(−3, 0)

(0, −2)

37.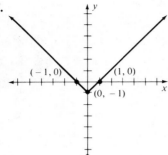

39.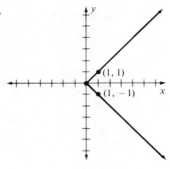

41.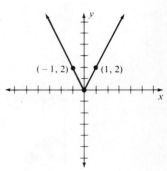

Problem Set 7.3

1. $3x - 4y = -14$ **3.** $2x - y = -6$ **5.** $3x + 5y = -21$ **7.** $y = -4$

9. $2x - 3y = -4$ **11.** $2x - y = 1$ **13.** $x + 5y = 14$ **15.** $2x + y = -5$

17. $x = 2$ **19.** $x - 5y = -10$ **21.** $4x - y = 1$ **23.** $2x + 7y = 28$

25. $4x + y = -2$ **27.** $2x - 3y = 6$ **29.** $y = 4$ **31.** $x = 2$

33. parallel   **35.** perpendicular

37. intersecting lines that are not perpendicular

39. parallel   **41.** perpendicular   **43.** $x + 3y = 23$ **45.** $3x - 4y = 10$

47. $3x - 2y = -5$ **49.** $7x - 5y = 0$ **51.** $y = \dfrac{3}{4}x - 2$ **53.** $y = \dfrac{2}{5}x - \dfrac{26}{5}$

55. (a)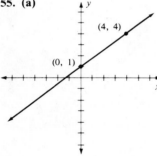

(c)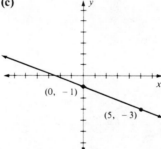

(e)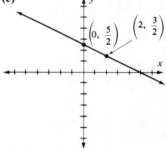

(g)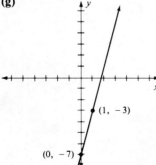

(i)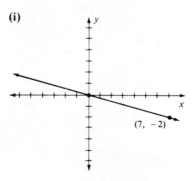

(k)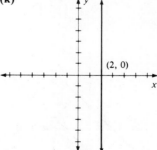

57. (a) $x - 4y = -3$ **(c)** $2x + 3y = 9$

59. (a) $2x + 3y = 26$ **(c)** $5x + 3y = -11$

Problem Set 7.4

1.

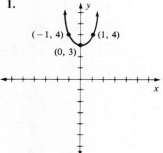

3.

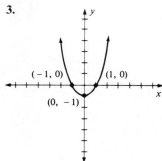

5.

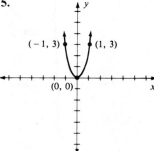

7.

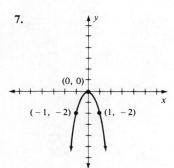

9.

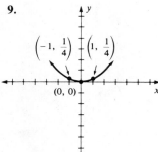

11.

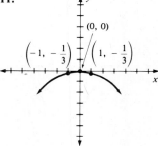

13.

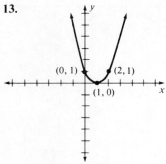

15.

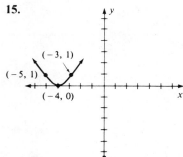

17.

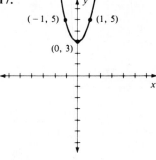

19.

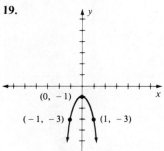

21.

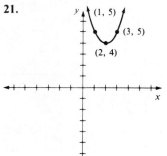

23.

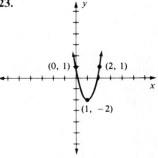

25.

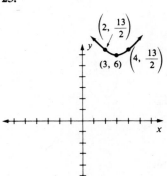

27.

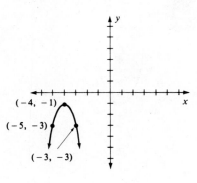

29.

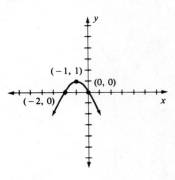

Problem Set 7.5

1.

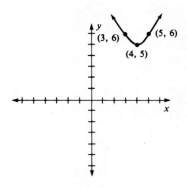

3.

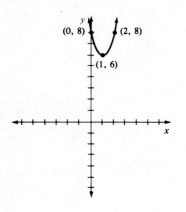

5.

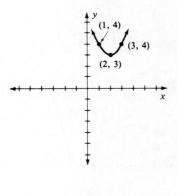

7.

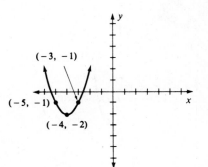

9.

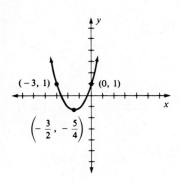

11.

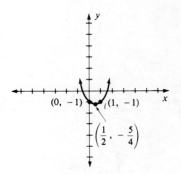

13.

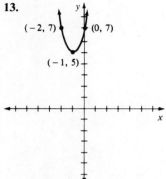

$(-2, 7)$ $(0, 7)$

$(-1, 5)$

15.

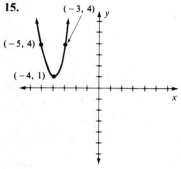

$(-3, 4)$

$(-5, 4)$

$(-4, 1)$

17.

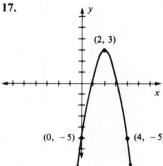

$(2, 3)$

$(0, -5)$ $(4, -5)$

19.

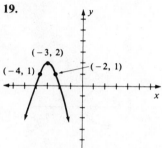

$(-3, 2)$

$(-4, 1)$ $(-2, 1)$

21.

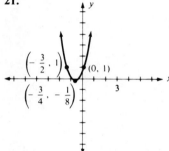

$\left(-\frac{3}{2}, 1\right)$ $(0, 1)$

$\left(-\frac{3}{4}, -\frac{1}{8}\right)$ 3

23.

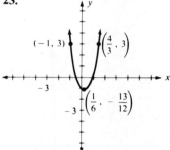

$(-1, 3)$ $\left(\frac{4}{3}, 3\right)$

-3

$\left(\frac{1}{6}, -\frac{13}{12}\right)$

-3

25.

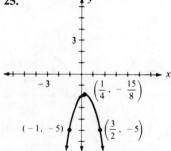

3

-3

$\left(\frac{1}{4}, -\frac{15}{8}\right)$

$(-1, -5)$ $\left(\frac{3}{2}, -5\right)$

27. $(0,0), r = 1$

29. $(0,0)\ r = \sqrt{10}$

31. $x^2 + y^2 - 4x - 8y + 11 = 0$

33. $x^2 + y^2 + 6x - 8y - 24 = 0$

35. $x^2 + y^2 - 4y = 0$

37. $x^2 + y^2 = 64$

39. $(1, -2), r = 3$

41. $(-3,4), r = 5$

43. $(3,5), r = \sqrt{10}$

45. $x^2 + y^2 + 6x = 0$

47.

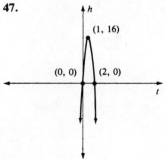

$(1, 16)$

$(0, 0)$ $(2, 0)$

49. (a)

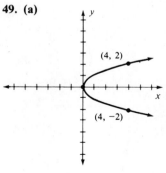

$(4, 2)$

$(4, -2)$

(c)

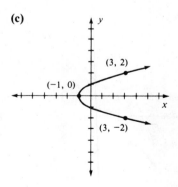

$(3, 2)$

$(-1, 0)$

$(3, -2)$

(e)

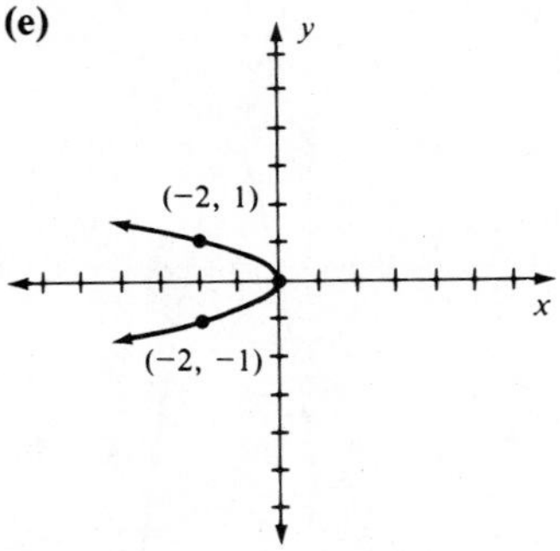

(g)

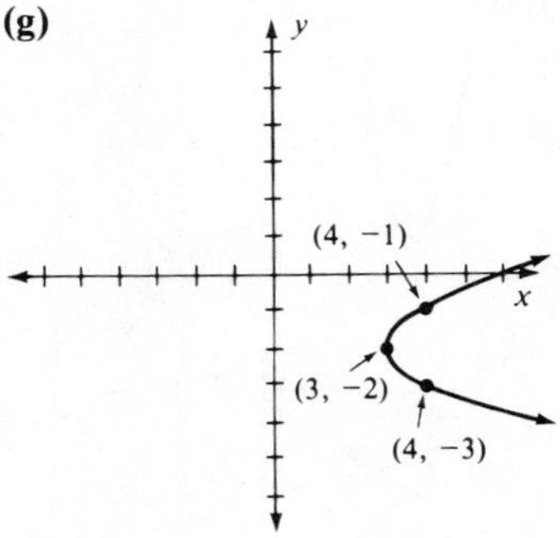

Problem Set 7.6

1.

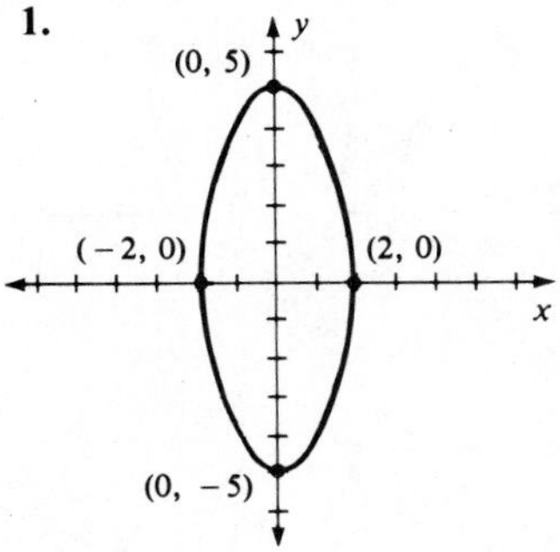

3.

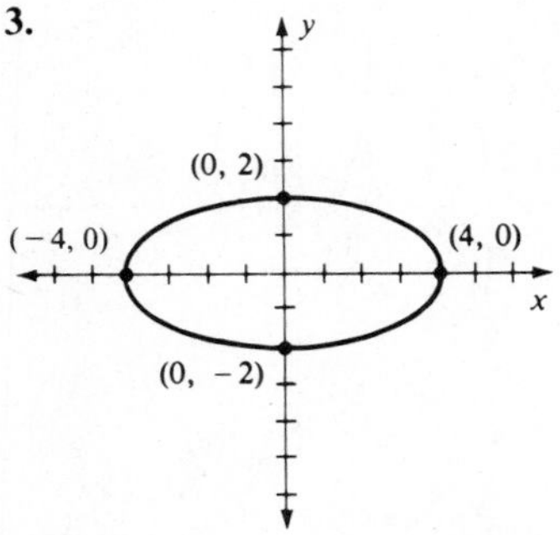

5.

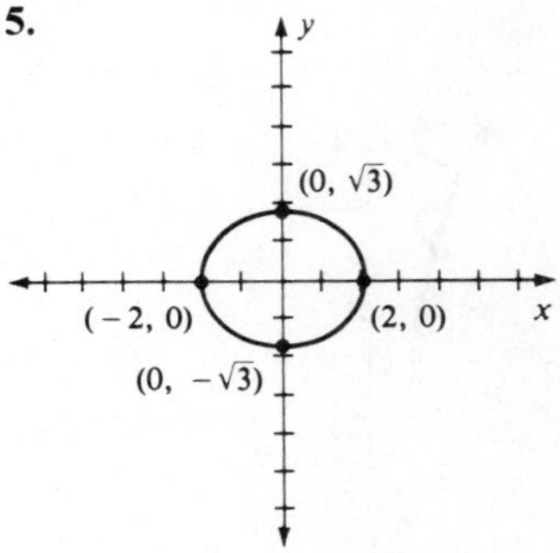

7.

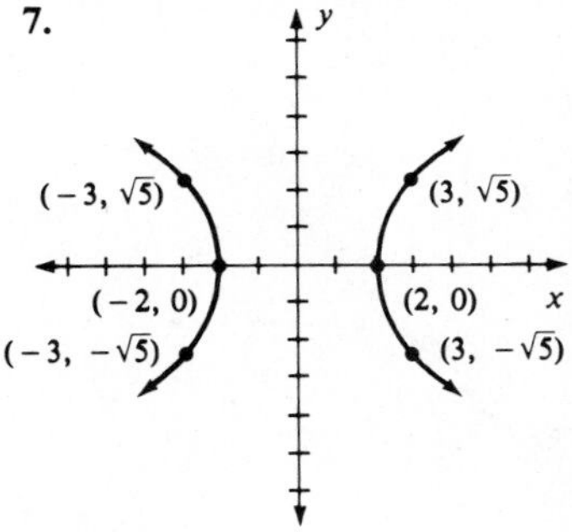

9.

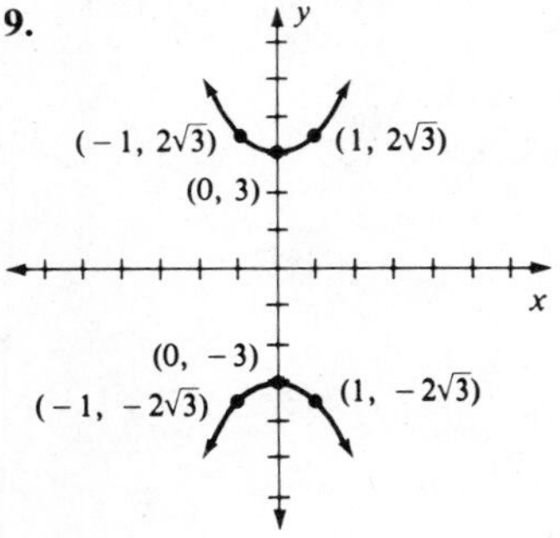

11.

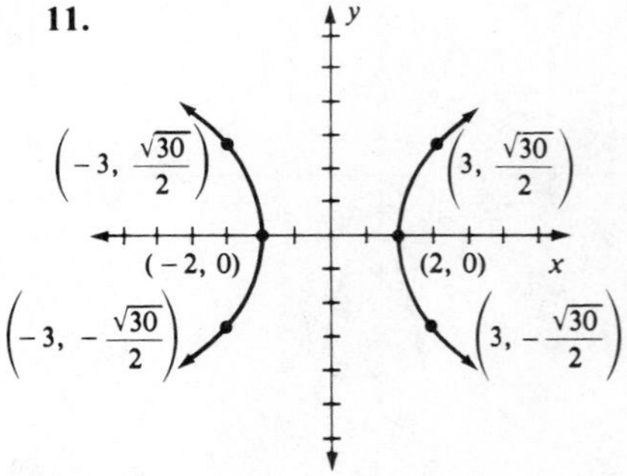

13.

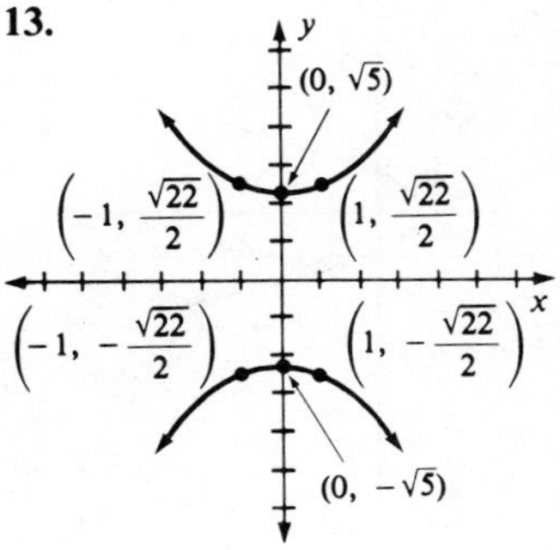

15. (a)

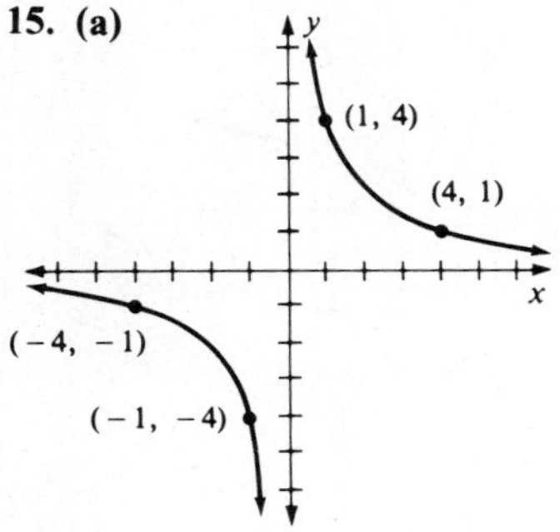

(c)

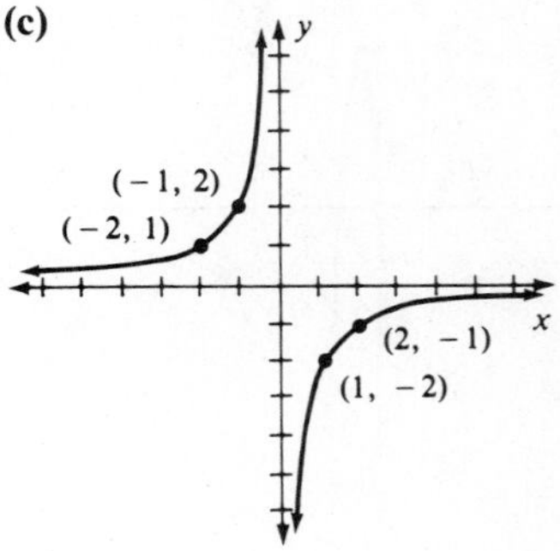

17. **(a)** Origin

(c)
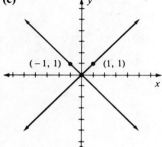

19.
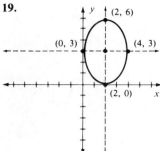

21.
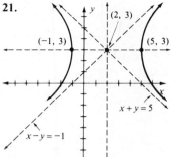

23.
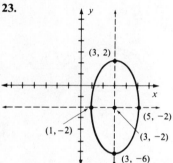

25.
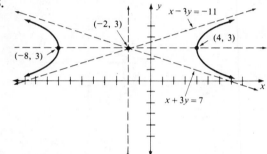

Problem Set 7.7

1. $(4, -3)$; $(-4, 3)$; $(-4, -3)$
3. $(-6, 1)$; $(6, -1)$; $(6, 1)$
5. $(0, -4)$; $(0, 4)$; $(0, -4)$
7. y-axis
9. x-axis
11. x-axis, y-axis, and origin
13. None   **15.** Origin   **17.** None   **19.** y-axis

21.
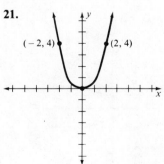

23.
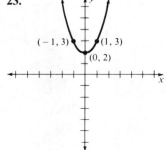

25.
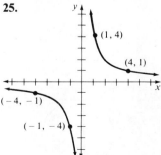

27.

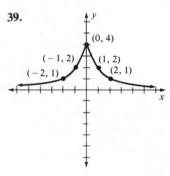

$(-2, 8)$

$(-1, 1)$

$(1, -1)$

$(2, -8)$

29.

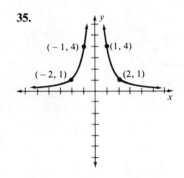

$(2, 2\sqrt{2})$

$(1, 1)$

$(1, -1)$

$(2, -2\sqrt{2})$

31.

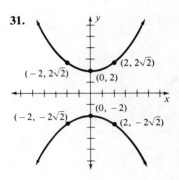

$(-2, 2\sqrt{2})$ $(2, 2\sqrt{2})$

$(0, 2)$

$(0, -2)$

$(-2, -2\sqrt{2})$ $(2, -2\sqrt{2})$

33.

$(1, -1)$

$(4, -2)$

35.

$(-1, 4)$ $(1, 4)$

$(-2, 1)$ $(2, 1)$

37.

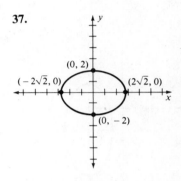

$(0, 2)$

$(-2\sqrt{2}, 0)$ $(2\sqrt{2}, 0)$

$(0, -2)$

39.

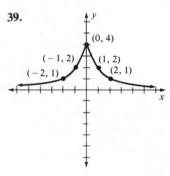

$(0, 4)$

$(-1, 2)$ $(1, 2)$

$(-2, 1)$ $(2, 1)$

41.

$(-1, 0)$

$(0, -2)$

43.

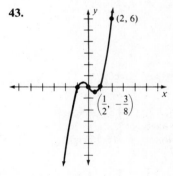

$(2, 6)$

$\left(\frac{1}{2}, -\frac{3}{8}\right)$

45.

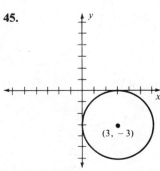

$(3, -3)$

47.

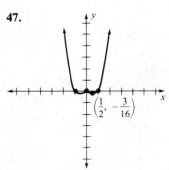

$\left(\frac{1}{2}, -\frac{3}{16}\right)$

49.

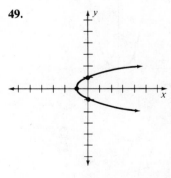

Chapter 7 Review Problem Set

1. (a) $\dfrac{6}{5}$ **(b)** $-\dfrac{2}{3}$ **2. (a)** -4 **(b)** $\dfrac{2}{7}$ **3.** 5, 10, and $\sqrt{97}$ **4.** $7x + 4y = 1$

5. $3x + 7y = 28$ **6.** $2x - 3y = 16$ **7.** $x - 2y = -8$ **8.** $2x - 3y = 14$

9.

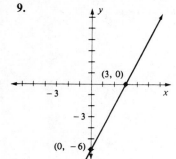

10.

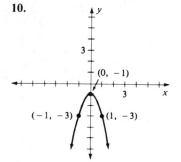

11.

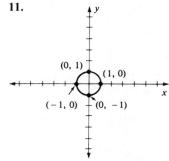

12.

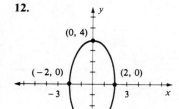

13.

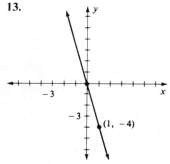

14.

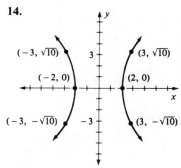

15.

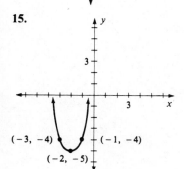

16.

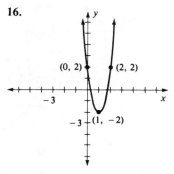

17.

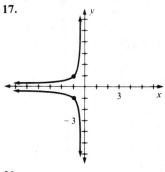

18.

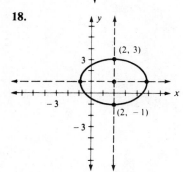

19.

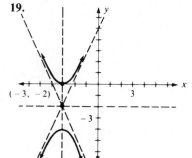

20.

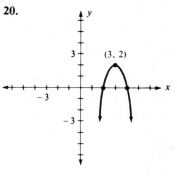

21.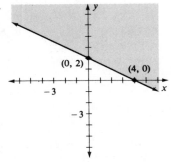

22.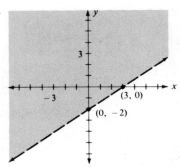

23. $(-3, 4), r = 3$

24. $(-4, -6)$

25. $y = \dfrac{3}{2}x$ and $y = -\dfrac{3}{2}x$

26. 6

Chapter 8

Problem Set 8.1

1. (a) $D = \{1, 2, 3, 4\}$; $R = \{4, 6, 12, 17\}$; It is a function.
   (c) $D = \{0, 1, -1\}$; $R = \{0, 1, -1\}$; It is not a function.
   (e) $D = \{1\}$; $R = \{3, 4, -1, -2\}$; It is not a function.
   (g) $D = \{$all reals$\}$; $R = \{$all reals$\}$; It is a function.
   (i) $D = \{x | -1 \le x \le 1\}$; $R = \{y | -1 \le y \le 1\}$; It is not a function.

3. All reals

5. $D = \{x | x \ne -6\}$

7. $D = \left\{x | x \ne \dfrac{2}{5}\right\}$

9. $D = \{x | x \ne 1 \text{ and } x \ne -5\}$

11. $D = \{x | x \ne 5 \text{ and } x \ne -3\}$

13. $D = \{x | x \ne 0 \text{ and } x \ne 6\}$

15. All reals

17. $D = \{x | x \ne -5 \text{ and } x \ne 5\}$

19. $D = \left\{x | x \ge \dfrac{3}{2}\right\}$

21. $D = \{x | x \ge -4\}$

23. $D = \{x | x \ge 3 \text{ or } x \le -3\}$

25. $f(-1) = 13$; $f(-2) = 18$; $f(3) = -7$; $f(5) = -17$

27. $g(1) = 3$; $g(-1) = -3$; $g(3) = 17$; $g(-4) = 3$

29. $h(1) = -4$; $h(-1) = -4$; $h(-3) = -12$; $h(5) = -28$

31. $f(1) = 0$; $f(5) = 2$; $f(13) = 2\sqrt{3}$; $f(26) = 5$

33. $f(3) = 3$; $f(0) = -\dfrac{3}{2}$; $f(-1) = -1$; $f(-5) = -\dfrac{3}{7}$

35. $f(-2) = 1$; $f(3) = 11$; $g(-4) = 11$; $g(5) = 29$

37. $f(1) = 1$; $f(-1) = 5$; $g(2) = 4$; $g(-3) = 5$.

39. 3   **41.** $2a + h$ **43.** $4a + 2h + 1$

45. $h(1) = 48$; $h(2) = 64$; $h(3) = 48$; $h(4) = 0$

47. $C(75) = \$74$; $C(150) = \$98$; $C(225) = \$122$; $C(650) = \$258$

49. $I(0.11) = 55$; $I(0.12) = 60$; $I(0.135) = 67.5$; $I(0.15) = 75$

Problem Set 8.2

1.
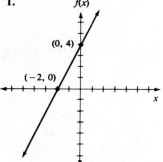

3.
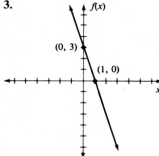

5.
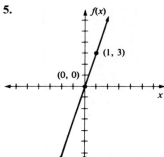

7.
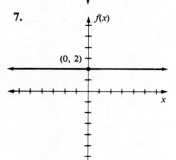

9.
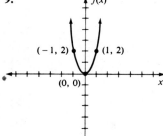

11.
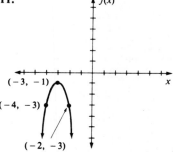

13.
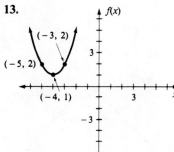

15.
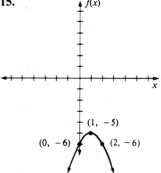

17.
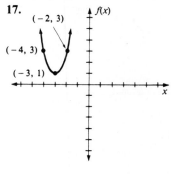

19.
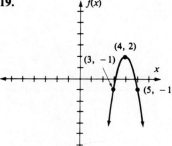

21.
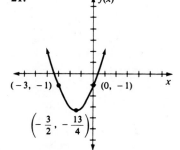

23.
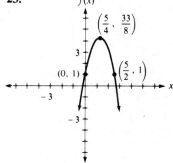

25.

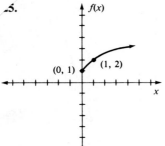

(0, 1) (1, 2)

27.

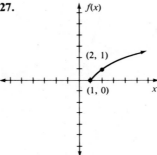

(2, 1) (1, 0)

29.

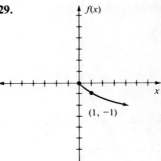

(1, −1)

31.

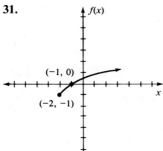

(−1, 0) (−2, −1)

33.

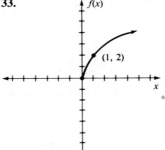

(1, 2)

35.

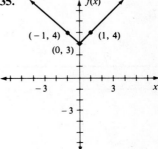

(−1, 4) (1, 4) (0, 3) −3 3 −3

37.

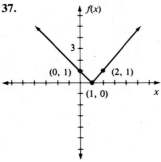

3 (0, 1) (2, 1) (1, 0)

39.

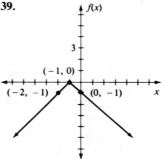

3 (−1, 0) (−2, −1) (0, −1)

41.

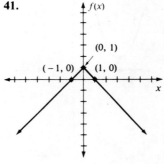

(0, 1) (−1, 0) (1, 0)

43.

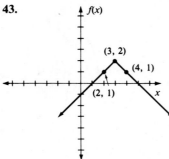

(3, 2) (4, 1) (2, 1)

45.

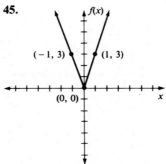

(−1, 3) (1, 3) (0, 0)

47.

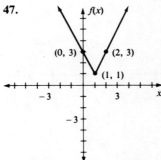

(0, 3) (2, 3) (1, 1) −3 3 −3

49. (a)

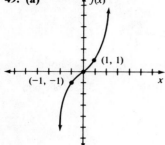

Problem Set 8.3

1. 70   **3.** 144   **5.** 25 and 25   **7.** 65 couples

9. $(f \circ g)(x) = 6x - 2, D = \{\text{all reals}\}$
$(g \circ f)(x) = 6x - 1, D = \{\text{all reals}\}$

11. $(f \circ g)(x) = 10x + 2, D = \{\text{all reals}\}$
$(g \circ f)(x) = 10x - 5, D = \{\text{all reals}\}$

13. $(f \circ g)(x) = 3x^2 + 7, D = \{\text{all reals}\}$
$(g \circ f)(x) = 9x^2 + 24x + 17, D = \{\text{all reals}\}$

15. $(f \circ g)(x) = 3x^2 + 9x - 16, D = \{\text{all reals}\}$
$(g \circ f)(x) = 9x^2 - 15x, D = \{\text{all reals}\}$

17. $(f \circ g)(x) = \dfrac{1}{2x + 7}, D = \left\{x \mid x \neq -\dfrac{7}{2}\right\}$
$(g \circ f)(x) = \dfrac{7x + 2}{x}, D = \{x \mid x \neq 0\}$

19. $(f \circ g)(x) = \sqrt{3x - 3}, D = \{x \mid x \geq 1\}$
$(g \circ f)(x) = 3\sqrt{x - 2} - 1, D = \{x \mid x \geq 2\}$

21. $(f \circ g)(x) = \dfrac{x}{2 - x}, D = \{x \mid x \neq 0 \text{ and } x \neq 2\}$
$(g \circ f)(x) = 2x - 2, D = \{x \mid x \neq 1\}$

23. 4; 50   **25.** 9; 0   **27.** $\sqrt{11}$; 5

Problem Set 8.4

1. Function   **3.** Not a function   **5.** Function   **7.** One-to-one function

9. Not a one-to-one function   **11.** One-to-one function

13. Domain of f: $\{1, 2, 5\}$,   Range of f: $\{5, 9, 21\}$, $f^{-1} = \{(5, 1), (9, 2), (21, 5)\}$,
Domain of f^{-1}: $\{5, 9, 12\}$,   Range of f^{-1}: $\{1, 2, 5\}$

15. Domain of f: $\{0, 2, -1, -2\}$,   Range of f: $\{0, 8, -1, -8\}$,
$f^{-1} = \{(0, 0), (8, 2), (-1, -1), (-8, -2)\}$,
Domain of f^{-1}: $\{0, 8, -1, -8\}$,   Range of f^{-1}: $\{0, 2, -1, -2\}$

17. $f^{-1}(x) = \dfrac{x - 3}{2}$ **19.** $f^{-1}(x) = 2x$ **21.** $f^{-1}(x) = \dfrac{x - 9}{5}$

23. $f^{-1}(x) = 3x + 12$ **25.** $f^{-1}(x) = \dfrac{3x - 15}{2}$ **27.** $f^{-1}(x) = \dfrac{x}{4}$

29. $f^{-1}(x) = \dfrac{x - 9}{2}$ **31.** $f^{-1}(x) = -\dfrac{3}{2}x$ **33.** $f^{-1}(x) = \dfrac{-x - 4}{3}$

35. $f^{-1}(x) = \dfrac{12x + 10}{9}$ **37.** $f^{-1}(x) = \dfrac{1}{6}x$ **39.** $f^{-1}(x) = -4x$

41. $f^{-1}(x) = \dfrac{x + 1}{2}$ **43.** $f^{-1}(x) = \dfrac{-x + 3}{4}$ **45.** $f^{-1}(x) = \sqrt{x}$ where $x \geq 0$

47. Every nonconstant linear function is a one-to-one function.

Problem Set 8.5

1. $y = kx^2$ **3.** $V = klw$ **5.** $V = \dfrac{k}{p}$ **7.** $I = \dfrac{k}{d^2}$ **9.** $V = \dfrac{kT}{P}$ **11.** 8

13. $\dfrac{22}{7}$ **15.** $\dfrac{1}{2}$ **17.** 18 **19.** 4 **21.** 16 **23.** 5 hours

25. 400 feet **27.** $210 **29.** 3080 cm^3 **31. (a)** $210 **(c)** $1050 **33.** 3560.76 m^3

35. 0.048

Chapter 8 Review Problem Set

1. $D = \{1,2,4\}$ **2.** $D = \{x \mid x \neq 5\}$
3. $D = \{x \mid x \neq 0 \text{ and } x \neq -4\}$ **4.** $D = \{x \mid x \geq 5 \text{ or } x \leq -5\}$
5. $f(2) = -1; f(-3) = 14; f(a) = a^2 - 2a - 1$ **6.** $4a + 2h + 1$

7.
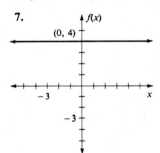

8.
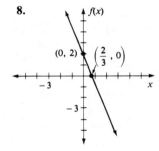

9.
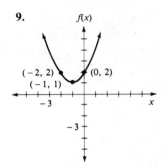

10.
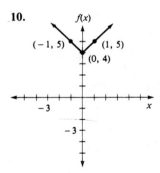

11.
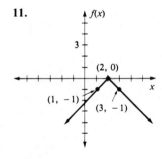

12.
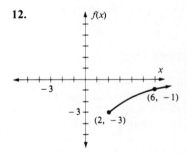

13.
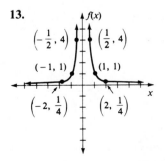

14.
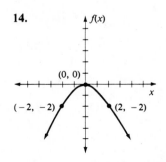

15.
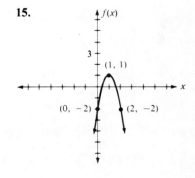

16.

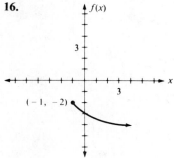

$(-1, -2)$

17. (a) $(-5, -28); x = -5$

(b) $\left(-\dfrac{7}{2}, \dfrac{67}{2}\right); x = -\dfrac{7}{2}$

18. $(f \circ g)(x) = 6x - 11$ and $(g \circ f)(x) = 6x - 13$
19. $(f \circ g)(x) = x^2 - 2x - 1$ and $(g \circ f)(x) = x^2 - 10x + 27$
20. $(f \circ g)(x) = 4x^2 - 20x + 20$ and $(g \circ f)(x) = -2x^2 + 15$

21. $f^{-1}(x) = \dfrac{x + 1}{6}$ **22.** $f^{-1}(x) = \dfrac{3x - 21}{2}$ **23.** $f^{-1}(x) = \dfrac{-35x - 10}{21}$ **24.** $k = 9$

25. $y = 120$ **26.** 128 pounds   **27.** 20 and 20   **28.** 3 and 47
29. 25 students   **30.** 600 square inches

Chapter 9

Problem Set 9.1

1. $Q: 3x + 4$ **3.** $Q: x + 6$ **5.** $Q: 4x - 3$ **7.** $Q: x^2 - 1$
$R: 0$ $R: 14$ $R: 2$ $R: 0$
9. $Q: 3x^3 - 4x^2 + 6x - 13$ **11.** $Q: x^2 - 2x - 3$ **13.** $Q: x^3 + 7x^2 + 21x + 56$
$R: 12$ $R: 0$ $R: 167$
15. $Q: x^2 + 3x + 2$ **17.** $Q: x^4 + x^3 + x^2 + x + 1$ **19.** $Q: x^4 + x^3 + x^2 + x + 1$
$R: 0$ $R: 0$ $R: 2$
21. $Q: 2x^4 - x^3 - 2x^2 + 3x - 1$ **23.** $Q: 3x^4 - 2x^3 + x^2 + 4x - 1$
$R: 0$ $R: 2$
25. $Q: 2x^2 + 2x - 3$ **27.** $Q: 4x^3 + 2x^2 - 4x - 2$
$R: \dfrac{9}{2}$ $R: 0$

Problem Set 9.2

1. $f(2) = -2$ **3.** $f(-4) = -105$ **5.** $f(-2) = 9$ **7.** $f(6) = 74$
9. $f(3) = 200$ **11.** $f(-1) = 5$ **13.** $f(7) = -5$ **15.** $f(-2) = -27$
17. $f\left(\dfrac{1}{2}\right) = -2$ **19.** Yes   **21.** Yes   **23.** No
25. Yes   **27.** Yes   **29.** $(x + 2)(x + 6)(x - 1)$ **31.** $(x - 3)(2x - 1)(3x + 2)$
33. $(x + 1)^2(x - 4)$ **35.** $k = 6$ **37.** $k = -30$
39. Let $f(x) = x^{12} - 4096$; then $f(-2) = 0$. Therefore, $x + 2$ is a factor of $f(x)$.
41. Let $f(x) = x^n - 1$. Since $1^n = 1$ for all positive integral values of n, $f(1) = 0$ and $x - 1$ is a factor.
43. (a) Let $f(x) = x^n - y^n$. Therefore, $f(y) = y^n - y^n = 0$ and $x - y$ is a factor of $f(x)$.
  (c) Let $f(x) = x^n + y^n$. Therefore, $f(-y) = (-y)^n + y^n = -y^n + y^n = 0$ when n is odd, and $x - (-y) = x + y$ is a factor of $f(x)$.

45. $f(1 + i) = 2 + 6i$

49. (a) $f(4) = 137; f(-5) = 11; f(7) = 575$
    (c) $f(4) = -79; f(5) = -162; f(-3) = 110$

Problem Set 9.3

1. $\{-2, -1, 2\}$ **3.** $\left\{-\dfrac{3}{2}, \dfrac{1}{3}, 1\right\}$ **5.** $\left\{-7, \dfrac{2}{3}, 2\right\}$ **7.** $\{-1, 4\}$

9. $\{-3, 1, 2, 4\}$ **11.** $\{-2, 1 \pm \sqrt{7}\}$ **13.** $\left\{-\dfrac{2}{3}, 1, \pm\sqrt{2}\right\}$ **15.** $\left\{-\dfrac{4}{3}, 0, \dfrac{1}{2}, 3\right\}$

17. $\{-1, 2, 1 \pm i\}$ **19.** $\left\{-1, \dfrac{3}{2}, 2, \pm i\right\}$ **27. (a)** $\{-4, -2, 1\}$ **(c)** $\left\{-4, -2, \dfrac{3}{2}\right\}$

29. 2 positives or 2 nonreal complex solutions

31. 1 negative and 2 nonreal complex solutions

33. 1 positive solution and 2 negative solutions
    or
    1 positive solution and 2 nonreal complex solutions

35. 1 negative solution and 2 positive solutions and 2 nonreal complex solutions
    or
    1 negative solution and 4 nonreal complex solutions

37. 1 positive solution and 1 negative solution and 4 nonreal complex solutions

41. (a) upper bound of 3 and lower bound of -1
    (c) upper bound of 3 and lower bound of -6
    (e) upper bound of 5 and lower bound of -3

Problem Set 9.4

1.

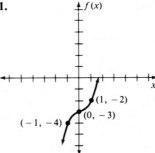

3.

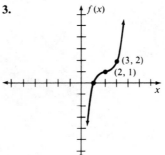

5.

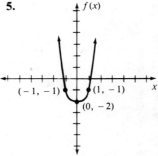

7.

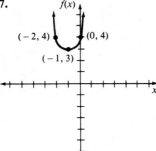

9.

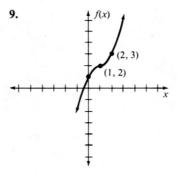

11.

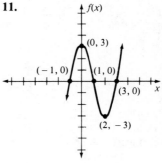

13.

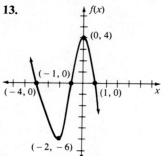

15.

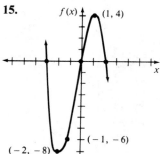

17.

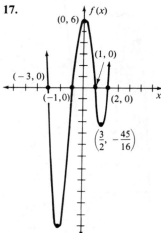

19.

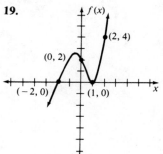

21.

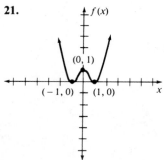

23.

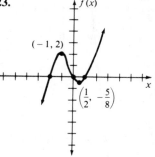

25.

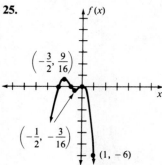

27.

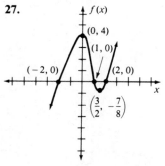

29.

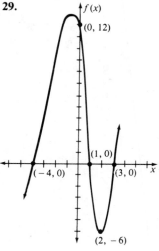

31.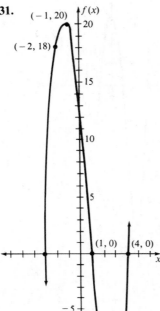

33.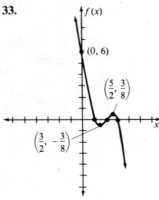

35. (a) 60   **(c)** $f(x) > 0$ for $\{x \mid -4 < x < 3 \text{ or } x > 5\}$
                      $f(x) < 0$ for $\{x \mid x < -4 \text{ or } 3 < x < 5\}$

37. (a) 432   **(c)** $f(x) > 0$ for $\{x \mid -3 < x < 4 \text{ or } x > 4\}$
                      $f(x) < 0$ for $\{x \mid x < -3\}$

39. (a) 8   **(c)** $f(x) > 0$ for $\{x \mid x < -2 \text{ or } -2 < x < 1 \text{ or } x > 2\}$
                      $f(x) < 0$ for $\{x \mid 1 < x < 2\}$

41. (a) 512   **(c)** $f(x) > 0$ for $\{x \mid -2 < x < 4 \text{ or } x > 4\}$
                      $f(x) < 0$ for $\{x \mid x < -2\}$

Problem Set 9.5

1.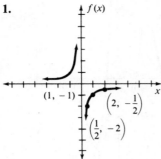

3.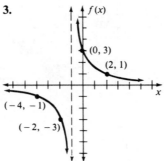

5.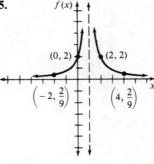

7.

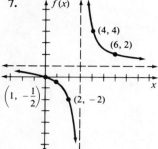

$(4, 4)$
$(6, 2)$
$\left(1, -\dfrac{1}{2}\right)$
$(2, -2)$

9.

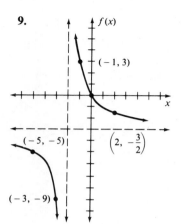

$(-1, 3)$
$\left(2, -\dfrac{3}{2}\right)$
$(-5, -5)$
$(-3, -9)$

11.

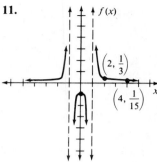

$\left(2, \dfrac{1}{3}\right)$
$\left(4, \dfrac{1}{15}\right)$

13.

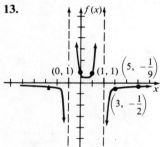

$(0, 1)$
$(1, 1)$
$\left(5, -\dfrac{1}{9}\right)$
$\left(3, -\dfrac{1}{2}\right)$

15.

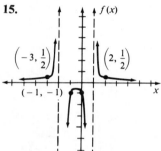

$\left(-3, \dfrac{1}{2}\right)$
$\left(2, \dfrac{1}{2}\right)$
$(-1, -1)$

17.

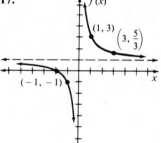

$(1, 3)$
$\left(3, \dfrac{5}{3}\right)$
$(-1, -1)$

19.

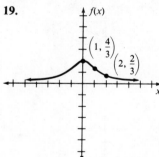

$\left(1, \dfrac{4}{3}\right)$
$\left(2, \dfrac{2}{3}\right)$

21.

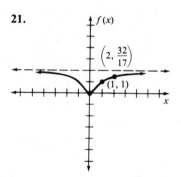

$\left(2, \dfrac{32}{17}\right)$
$(1, 1)$

23. (a)

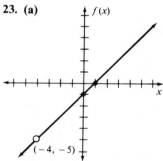

$(-4, -5)$

23. (c)

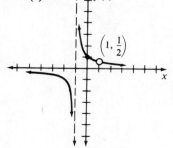

$\left(1, \dfrac{1}{2}\right)$

Chapter 9 Review Problem Set

1. Q: $3x^2 - x + 5$
   R: 3

2. Q: $5x^2 - 3x - 3$
   R: 16

3. Q: $-2x^3 + 9x^2 - 38x + 151$
   R: -605

4. Q: $-3x^3 - 9x^2 - 32x - 96$
   R: -279

5. $f(1) = 1$

6. $f(-3) = -197$

7. $f(-2) = 20$

8. $f(8) = 0$

9. Yes

10. No

11. Yes

12. Yes

13. $\{-3, 1, 5\}$

14. $\left\{-\dfrac{7}{2}, -1, \dfrac{5}{4}\right\}$

15. $\{1, 2, 1 \pm 5i\}$

16. $\{-2, 3 \pm \sqrt{7}\}$

17. 2 positive solutions and 2 negative solutions
   or
   2 positive solutions and 2 nonreal complex solutions
   or
   2 negative solutions and 2 nonreal complex solutions
   or
   4 nonreal complex solutions

18. 1 negative solution and 4 nonreal complex solutions

19.
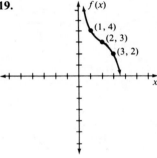

20.
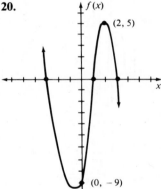

21.
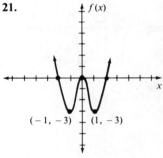

22.
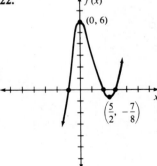

23.
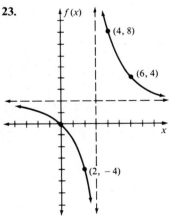

24.
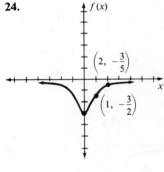

Chapter 10

Problem Set 10.1

1. $\{3\}$ **3.** $\{3\}$ **5.** $\{4\}$ **7.** $\{2\}$ **9.** $\{-2\}$ **11.** $\left\{\dfrac{5}{3}\right\}$ **13.** $\left\{\dfrac{3}{2}\right\}$ **15.** $\left\{\dfrac{4}{9}\right\}$

17.

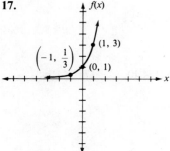

19.

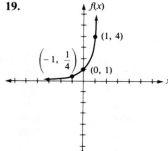

21.

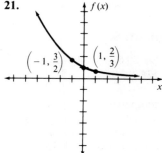

23.

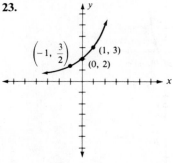

25.

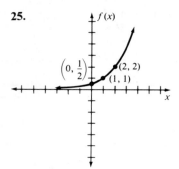

27.

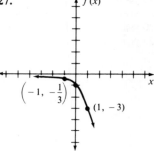

29.

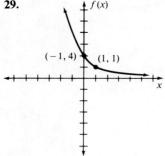

31.

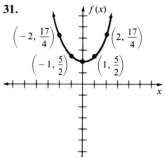

33.

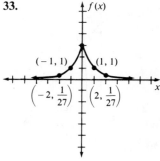

35.

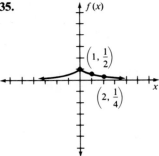

Problem Set 10.2

1. (a) $0.67   **(c)** $2.31   **(e)** $12,623   **(g)** $803   **3.** $384.66   **5.** $480.31

7. $2479.35   **9.** $1816.70   **11.** $1356.59   **13.** $745.88   **15.** $2174.40   **17.** $4416.52

19.

	8%	10%	12%	14%
compounded annually	$2159	2594	3106	3707
compounded semiannually	2191	2653	3207	3870
compounded quarterly	2208	2685	3262	3959
compounded monthly	2220	2707	3300	4022
compounded continuously	2225	2718	3320	4055

21.

	8%	10%	12%	14%
5 years	$1492	1649	1822	2014
10 years	2225	2718	3320	4055
15 years	3320	4481	6049	8164
20 years	4952	7388	11020	16440
25 years	7388	12179	20079	33103

23.

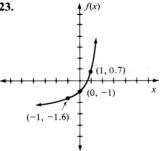

25.

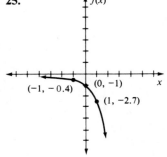

27.

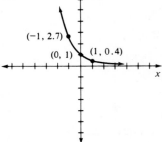

29. 8243; 22405; 100396

33. (a) 82,887   **(c)** 96,299

35. 500; 210

31. 203; 27; 0.5

37. $-\dfrac{1}{25}$

39.

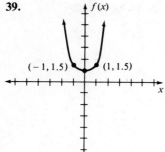

41.

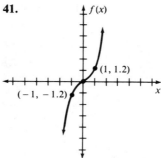

Problem Set 10.3

1. $\log_3 9 = 2$

3. $\log_5 125 = 3$

5. $\log_2\left(\dfrac{1}{16}\right) = -4$

7. $\log_{10} 0.01 = -2$

9. $2^6 = 64$

11. $10^{-1} = 0.1$

13. $2^{-4} = \dfrac{1}{16}$

15. 2

17. -1

19. 1

21. $\dfrac{1}{2}$

23. $\dfrac{1}{2}$

25. $-\dfrac{1}{8}$

27. 7

29. 0

31. $\{25\}$

33. $\{32\}$

35. $\{9\}$

37. $\{1\}$

39. 1.1461

41. 0.6020

43. 2.5353

45. 0.1505

47. 1.1268

49. 1.4471

51. 1.9912

53. 2.3010

55. 3.1461

57. $\log_b x + \log_b y + \log_b z$

59. $2\log_b x + 3\log_b y$

61. $\dfrac{1}{2}\log_b x + \dfrac{1}{2}\log_b y$

63. $\dfrac{1}{2}\log_b x - \dfrac{1}{2}\log_b y$

65. $\log_b\left(\dfrac{xy}{z}\right)$

67. $\log_b\left(\dfrac{x}{yz}\right)$

69. $\log_b(x\sqrt{y})$

71. $\log_b\left(\dfrac{x^2\sqrt{x-1}}{(2x+5)^4}\right)$

73. $\{2\}$

75. $\{2\}$

77. $\left\{\dfrac{2}{9}\right\}$

79. $\{6\}$

Problem Set 10.4

1.

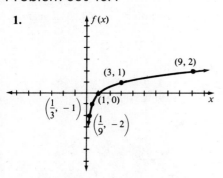

3.

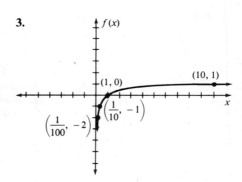

5.

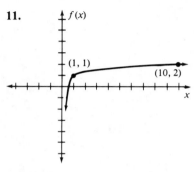

7. Same graph as Problem 1   **9.** Same graph as Problem 5

11.

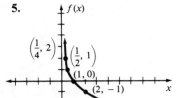

13.

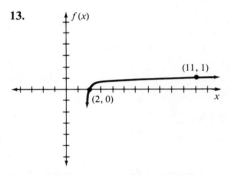

15. 0.9754   **17.** 1.5393   **19.** 3.6741   **21.** -0.2132 **23.** -2.3279
25. 0.9405   **27.** 1.7551   **29.** 2.8500   **31.** $.6920 + (-1)$ **33.** $.7226 + (-3)$
35. 5.8825   **37.** 33.597   **39.** 8580.3   **41.** 3.5620   **43.** 0.71581
45. 0.0022172   **47.** 3.55   **49.** 30.4   **51.** 983   **53.** 640,000
55. 0.542   **57.** 0.00731   **59.** 0.179   **61.** 0.6931   **63.** 3.0634
65. 6.0210   **67.** -1.1394 **69.** -2.6381 **71.** -7.1309 **73.** 1.3083
75. 4.3567   **77.** 6.4297   **79.** -0.8675 **81.** -4.7677

Problem Set 10.5

1. $\{3.17\}$ **3.** $\{1.95\}$ **5.** $\{1.81\}$ **7.** $\{1.41\}$ **9.** $\{1.41\}$
11. $\{3.10\}$ **13.** $\{1.82\}$ **15.** $\{7.84\}$ **17.** $\{10.32\}$ **19.** $\{2\}$
21. $\left\{\dfrac{29}{8}\right\}$ **23.** $\left\{\dfrac{-1 + \sqrt{65}}{2}\right\}$ **25.** $\{\sqrt{2}\}$ **27.** $\{6\}$ **29.** $\{1, 100\}$
31. 2.402   **33.** 0.461   **35.** 2.657   **37.** 1.211   **39.** 7.9 years
41. 12.2 years   **43.** 11.8%   **45.** 6.6 years   **47.** 1.5 hours   **49.** 34.7 years
51. 0.17   **55.** $\{1.13\}$ **57.** $x = \ln\left(y + \sqrt{y^2 + 1}\right)$

Problem Set 10.6

1. 0.6362   **3.** 2.1103   **5.** $.9426 + (-1)$ **7.** $.7148 + (-3)$ **9.** 2.945
11. 155.8   **13.** 0.3181   **15.** 20,930   **17.** 5.749   **19.** 1,549,000
21. 614.1   **23.** 0.9614   **25.** 13.79   **27.** 15.91

Chapter 10 Review Problem Set

1. 32   **2.** -125 **3.** 81   **4.** 3   **5.** -2 **6.** $\dfrac{1}{3}$

7. $\dfrac{1}{4}$ **8.** -5 **9.** 1   **10.** 12   **11.** $\{5\}$ **12.** $\left\{\dfrac{1}{9}\right\}$

13. $\left\{\dfrac{7}{2}\right\}$ **14.** $\{3.40\}$ **15.** $\{8\}$ **16.** $\left\{\dfrac{1}{11}\right\}$ **17.** $\{1.95\}$ **18.** $\{1.41\}$

19. $\{1.56\}$ **20.** $\{20\}$ **21.** $\{10^{100}\}$ **22.** $\{2\}$ **23.** $\left\{\dfrac{11}{2}\right\}$ **24.** $\{0\}$

25. 0.3680   **26.** 1.3222   **27.** 1.4313   **28.** 0.5634

29. (a) $\log_b x - 2\log_b y$ **(b)** $\dfrac{1}{4}\log_b x + \dfrac{1}{2}\log_b y$ **(c)** $\dfrac{1}{2}\log_b x - 3\log_b y$

30. (a) $\log_b x^3 y^2$ **(b)** $\log_b\left(\dfrac{\sqrt{y}}{x^4}\right)$ **(c)** $\log_b\left(\dfrac{\sqrt{xy}}{z^2}\right)$

31. 1.58   **32.** 0.63   **33.** 3.79   **34.** -2.12

35.

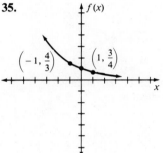

36.

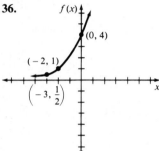

37.

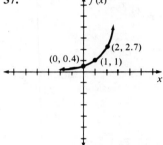

38.

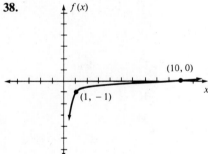

39.

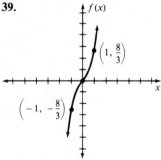

40.

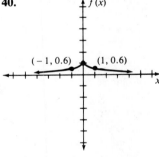

41.

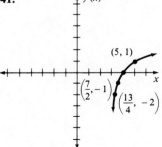

42.

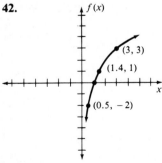

43. $2219.91   **44.** $4797.55   **45.** $15,999.31
46. Approximately 5.3 years   **47.** Approximately 12.1 years   **48.** Approximately 8.7%
49. 61,070; 67,493; 74,591   **50.** Approximately 4.8 hours

Chapter 11

Problem Set 11.1

1. $\{(8, 3)\}$ **3.** Inconsistent   **5.** $\{(-1, -5)\}$ **7.** Dependent

9. $\{(4, 1)\}$ **11.** $\{(4, 5)\}$ **13.** $\{(-4, -3)\}$ **15.** $\{(2, -5)\}$

17. $\left\{\left(\dfrac{1}{2}, \dfrac{2}{3}\right)\right\}$ **19.** $a = \dfrac{9}{43}$ and $b = \dfrac{10}{43}$ **21.** $a = 8$ and $b = 12$ **23.** $\left\{\left(\dfrac{37}{15}, -\dfrac{61}{15}\right)\right\}$

25. $\{(12, 9)\}$ **27.** $\left\{\left(\dfrac{13}{11}, \dfrac{45}{44}\right)\right\}$ **29.** $\left\{\left(-\dfrac{6}{37}, \dfrac{98}{37}\right)\right\}$ **31.** 4 and -6

33. $300 at 12% and $200 at 13%   **35.** 30 20-cent stamps and 25 17-cent stamps
37. 16 dimes and 14 quarters   **39.** 5 large boxes and 20 small boxes
41. 14 books at $12 each and 21 books at $14 each   **43.** $.89

Problem Set 11.2

1. $\{(9, 1)\}$ **3.** $\{(2, -7)\}$ **5.** $\{(-2, 10)\}$ **7.** $\left\{\left(\dfrac{21}{11}, \dfrac{8}{11}\right)\right\}$

9. $t = 9$ and $u = 2$ **11.** $\left\{\left(-\dfrac{11}{2}, -\dfrac{73}{8}\right)\right\}$ **13.** $\{(5, 16)\}$ **15.** $\{(1, 4)\}$

17. $\{(2, -5)\}$ **19.** $\{(8, -7)\}$ **21.** $a = \dfrac{32}{13}$ and $b = -\dfrac{21}{13}$ **23.** Dependent

25. $\left\{\left(\dfrac{11}{71}, \dfrac{14}{71}\right)\right\}$ **27.** $\left\{\left(\dfrac{53}{16}, \dfrac{35}{24}\right)\right\}$ **29.** $\left\{\left(\dfrac{66}{17}, \dfrac{24}{17}\right)\right\}$ **31.** $\left\{\left(-1, -\dfrac{14}{3}\right)\right\}$

33. $\{(4, -10)\}$ **35.** $\{(-3, -6)\}$ **37.** $\{(6, 2)\}$ **39.** $\{(1, 2)\}$

41. $250 at 12% and $750 at 14%   **43.** 37
45. 35 double rooms and 15 single rooms   **47.** 94
49. 3.5 liters of the 50% solution and 7 liters of the 80% solution

51. $\dfrac{3}{4}$ **53.** 9 feet   **55.** 18 centimeters by 24 centimeters

57. (a) $\{(4, 6)\}$ **(c)** $\{(2, -3)\}$ **(e)** $\left\{\left(\dfrac{1}{4}, -\dfrac{2}{3}\right)\right\}$

Problem Set 11.3

1. $\{(3, 1, 2)\}$ **3.** $\{(-2, -1, 3)\}$ **5.** $\varnothing$ **7.** $\{(0, 2, 4)\}$

9. $\left\{\left(\dfrac{1}{2}, 3, -4\right)\right\}$ **11.** Infinitely many   **13.** $\{(-1, 0, 6)\}$ **15.** $\{(1, 2, -6)\}$

17. $\left\{\left(\dfrac{1}{2}, \dfrac{1}{3}, \dfrac{1}{6}\right)\right\}$ **19.** $\{(0, 0, 0)\}$ **21.** 12, 13, and 15

23. 10 nickels, 5 dimes, and 4 quarters

25. 12 centimeters, 13 centimeters, and 20 centimeters

27. 247

29. $0.10 per pound for potatoes, $0.16 per pound for onions, and $0.30 per pound for apples

Problem Set 11.4

1. $\{(1, 5), (5, 1)\}$ **3.** $\{(2, 4), (-1, 1)\}$ **5.** $\{(1, 2), (4, 5)\}$ **7.** $\{(5, 3)\}$

9. $\{(1, 2), (-1, 2)\}$ **11.** $\{(2, 0), (-2, 0)\}$ **13.** $\{(1, 3), (-1, 3)\}$ **15.** $\{(2, 0), (-2, 0)\}$

17.

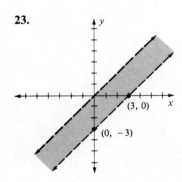

19.

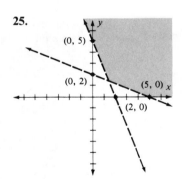

21.

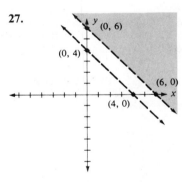

23.

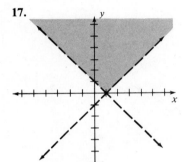

25.

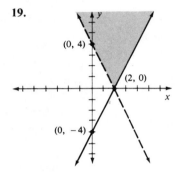

27.

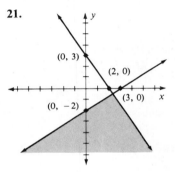

29.

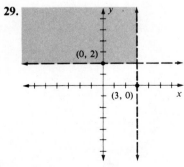

31. $\varnothing$

33.

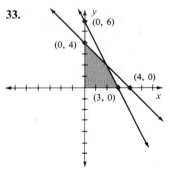

35.

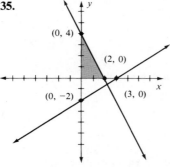

37. 2 and 3, or -3 and -2 **39.** 2 and 5, or -3 and 10

41. (a) $\{(2i, -3), (-2i, -3)\}$

    (c) $\left\{ \left(\dfrac{1 + i\sqrt{15}}{2}, \dfrac{-7 + i\sqrt{15}}{2} \right), \left(\dfrac{1 - i\sqrt{15}}{2}, \dfrac{-7 - i\sqrt{15}}{2} \right) \right\}$

    (e) $\left\{ \left(\dfrac{2 + i\sqrt{2}}{2}, \dfrac{2 - i\sqrt{2}}{2} \right), \left(\dfrac{2 - i\sqrt{2}}{2}, \dfrac{2 + i\sqrt{2}}{2} \right) \right\}$,

Problem Set 11.5

1. Minimum of 8 and maximum of 52   **3.** Minimum of 0 and maximum of 28
5. 63   **7.** 340   **9.** 2   **11.** 98
13. $350 profit can be made by producing 50 widgets and no wadgets.
15. Investing $5000 at 9% and $5000 at 12% will produce a maximum of $1050.
17. 300 of type A and 200 of type B

Chapter 11 Review Problem Set

1. $\{(2, 6)\}$

2. $\{(-3, 7)\}$

3. $\{(4, -1)\}$

4. $\{(-2, 6)\}$

5. $\{(8, -24)\}$

6. $\varnothing$

7. $\{(2, -3, 1)\}$

8. $\{(2, -3, -1)\}$

9. $\{(-3, 2, 5)\}$

10. $\{(-4, 3, 4)\}$

11. Infinitely many

12. $\{(1, 1), (-2, 7)\}$

13. $\{(2, \sqrt{3}), (2, -\sqrt{3}), (-2, \sqrt{3}), (-2, -\sqrt{3})\}$

14. $\{(2, -2), (-2, 2)\}$

15. $\{(-1, 4), (-3, 4)\}$

16.

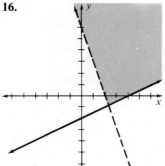

17.

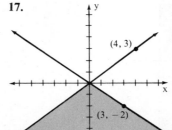

18. 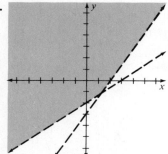 **19.** 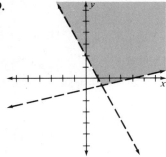 **20.**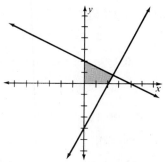

Chapter 12

Problem Set 12.1

1. $\{(4, -2)\}$ **3.** $\{(-5, 6)\}$ **5.** $\{(-1, -3)\}$ **7.** $\{(-2, 5)\}$

9. $\left\{\left(\dfrac{11}{2}, \dfrac{1}{2}\right)\right\}$ **11.** $\{(1, -1, 5)\}$ **13.** $\{(-2, 3, 5)\}$ **15.** $\{(0, 4, -2)\}$

17. $\{(4, -1, -1)\}$ **19.** $\left\{\left(\dfrac{37}{13}, \dfrac{30}{13}, -\dfrac{12}{13}\right)\right\}$ **21.** $\{(1, 5, 2)\}$ **23.** $\{(-2, 4, -1)\}$

25. $\{(-2, 0, 4)\}$ **27.** $\{(2, 1, -1)\}$ **29.** $\{(3, -1, 3)\}$

Problem Set 12.2

1. Yes **3.** Yes **5.** No **7.** No

9. Yes **11.** $\{(-1, -5)\}$ **13.** $\{(3, -6)\}$ **15.** $\varnothing$

17. $\{(-2, -9)\}$ **19.** $\{(-1, -2, 3)\}$ **21.** $\{(3, -1, 4)\}$ **23.** $\{(0, -2, 4)\}$

25. $\{(-7k + 8, -5k + 7, k)\}$ **27.** $\{(-4, -3, -2)\}$ **29.** $\{(4, -1, -2)\}$

31. $\{(1, -1, 2, -3)\}$ **33.** $\{(2, 1, 3, -2)\}$ **35.** $\{(-2, 4, -3, 0)\}$

37. $\varnothing$ **39.** $\{(-3k + 5, -1, -4k + 2, k)\}$ **41.** $\{(-3k + 9, k, 2, -3)\}$

43. $\{(17k - 6, 10k - 5, k)\}$ **45.** $\left\{\left(-\dfrac{1}{2}k + \dfrac{34}{11}, \dfrac{1}{2}k - \dfrac{5}{11}, k\right)\right\}$ **47.** $\varnothing$

Problem Set 12.3

1. 8 **3.** -18 **5.** 0 **7.** 10 **9.** $-\dfrac{2}{3}$ **11.** $\{(7, 3)\}$ **13.** $\{(-2, 4)\}$

15. $\{(4, -5)\}$ **17.** $\varnothing$ **19.** $\{(1, 5)\}$ **21.** $\left\{\left(\dfrac{1}{2}, \dfrac{3}{4}\right)\right\}$ **23.** $\left\{\left(\dfrac{17}{62}, \dfrac{4}{31}\right)\right\}$

25. $\{(-1, 3)\}$ **27.** $\{(-8, -3)\}$ **29.** $\left\{\left(\dfrac{14}{5}, 4\right)\right\}$

33. \$1.50 for a carton of pop and \$2.25 for a pound of candy **35.** 9 and 14 **37.** 48

Problem Set 12.4

1. 6   **3.** 13   **5.** 83   **7.** -90 **9.** $\{(-1, -1, 2)\}$ **11.** $\{(1, 0, -2)\}$

13. $\{(0, 1, 0)\}$ **15.** Infinitely many solutions   **17.** $\left\{\left(\dfrac{15}{4}, -\dfrac{7}{12}, -\dfrac{17}{12}\right)\right\}$

19. $\{(-1, -1, 4)\}$ **21.** $\{(2, 6, 7)\}$ **23.** $\{(4, -4, 5)\}$ **25.** $\{(6, -5, 1)\}$

Chapter 12 Review Problem Set

1. $\{(6, -2)\}$ **2.** $\{(-1, -3)\}$ **3.** $\{(4, -6)\}$ **4.** $\left\{\left(-\dfrac{6}{7}, -\dfrac{15}{7}\right)\right\}$

5. $\{(2, -1, 4)\}$ **6.** $\{(1, -4, -2)\}$ **7.** $\{(4, 2, -2)\}$ **8.** $\{(-1, 2, -5)\}$

9. $\varnothing$ **10.** $\{(2, -3, -1)\}$ **11.** -34 **12.** 13

13. -22 **14.** -29 **15.** -40 **16.** 16

17. 51   **18.** $\{(-5, 9)\}$ **19.** $\{(0, -4)\}$ **20.** $\{(3, -7)\}$

21. $\{(-12, 12)\}$ **22.** $\{(-10, -7)\}$ **23.** $\{(4, 6)\}$ **24.** $\{(2, -1, -2)\}$

25. $\left\{\left(-\dfrac{1}{3}, -1, 4\right)\right\}$ **26.** $\{(-2, 2, -1)\}$ **27.** $\{(0, -1, 2)\}$ **28.** $\{(2, -3, -4)\}$

29. $\{(-1, 2, -5)\}$ **30.** 72   **31.** \$900 at 10% and \$1600 at 12%

32. 20 nickels, 32 dimes, and 54 quarters   **33.** $25°$, $45°$, and $110°$

Chapter 13

Problem Set 13.1

1. 3, 5, 7, 9, 11   **3.** $-4, -7, -10, -13, -16$ **5.** 0, 3, 8, 15, 24

7. $2, -1, -6, -13, -22$ **9.** 1, 3, 9, 27, 81   **11.** $a_{10} = 79$ and $a_{15} = 194$

13. $a_n = 3n - 2$ **15.** $a_n = 5n - 3$ **17.** $a_n = -2n + 7$ **19.** $a_n = -3n$

21. 59   **23.** 179   **25.** -220 **27. (a)** 48   **(c)** 57

29. 247   **31. (a)** $2n$ **(c)** $5n - 5$ **(e)** $3n - 2$ **33.** 13,400   **35.** \$630

Problem Set 13.2

1. 625   **3.** 5000   **5.** -9080 **7.** 62500

9. (a) 8091   **(c)** 24750   **(e)** -3015

11. 8064   **13.** 120 logs   **15.** \$21,420   **17.** \$1455   **19.** \$49.60

21. (a) $7 + 12 + 17$; 36

     (c) $-3 + (-5) + (-7) + (-9) + (-11) + (-13)$; -48

     (e) $3 + 6 + 9 + 12 + 15$; 45

Problem Set 13.3

1. (a) $a_n = 2^{n-1}$ **(c)** $a_n = \left(\dfrac{1}{2}\right)^n$ **(e)** $a_n = (-2)^{n-1}$ **(g)** $a_n = (0.3)^{n-1}$

3. $\dfrac{2}{3}$ **5.** 511   **7.** 1364   **9.** $\dfrac{42591}{16384}$, or $2\dfrac{9823}{16384}$ **11.** $0; -1$

13. (a) 1,048,576   **(c)** $\dfrac{4096}{531441}$ **(e)** $-\dfrac{177147}{2048}$

15. 640 units   **17.** $16; $31.75   **19.** $\dfrac{5}{32}$ gram   **21.** $10,737,418

23. (a) $2 + 4 + 8 + 16 + 32 + 64; 126$
(c) $1 + 2 + 4 + 8 + 16; 31$
(e) $\dfrac{2}{3} + \dfrac{4}{9} + \dfrac{8}{27} + \dfrac{16}{81}; 1\dfrac{49}{81}$

Problem Set 13.4

1. 2   **3.** $\dfrac{1}{2}$ **5.** No sum   **7.** 8   **9.** $\dfrac{27}{4}$ **11.** $\dfrac{25}{2}$

13. $\dfrac{2}{3}$ **15.** $\dfrac{23}{99}$ **17.** $\dfrac{427}{999}$ **19.** $\dfrac{1}{6}$ **21.** $\dfrac{18}{55}$ **23.** $\dfrac{13}{3}$

Problem Set 13.5

1. $x^7 + 7x^6y + 21x^5y^2 + 35x^4y^3 + 35x^3y^4 + 21x^2y^5 + 7xy^6 + y^7$
3. $x^4 + 8x^3y + 24x^2y^2 + 32xy^3 + 16y^4$
5. $x^4 - 4x^3y + 6x^2y^2 - 4xy^3 + y^4$
7. $x^9 + 9x^8y + 36x^7y^2 + 84x^6y^3 + 126x^5y^4 + 126x^4y^5 + 84x^3y^6 + 36x^2y^7 + 9xy^8 + y^9$
9. $x^5 + 15x^4y + 90x^3y^2 + 270x^2y^3 + 405xy^4 + 243y^5$
11. $64x^6 - 192x^5y + 240x^4y^2 - 160x^3y^3 + 60x^2y^4 - 12xy^5 + y^6$
13. $16a^4 - 96a^3b + 216a^2b^2 - 216ab^3 + 81b^4$
15. $x^{10} + 5x^8y + 10x^6y^2 + 10x^4y^3 + 5x^2y^4 + y^5$
17. $x^6 + 18x^5 + 135x^4 + 540x^3 + 1215x^2 + 1458x + 729$
19. $x^9 - 9x^8 + 36x^7 - 84x^6 + 126x^5 - 126x^4 + 84x^3 - 36x^2 + 9x - 1$
21. $x^{12} + 12x^{11} + 66x^{10}y^2 + 220x^9y^3$
23. $x^{20} - 20x^{19}y + 190x^{18}y^2 - 1140x^{17}y^3$
25. $56x^5y^3$ **27.** $126x^5y^4$ **29.** $189a^2b^5$

Chapter 13 Review Problem Set

1. $a_n = 6n - 3$ **2.** $a_n = 3^{n-2}$ **3.** $a_n = 5 \cdot 2^n$ **4.** $a_n = -3n + 8$
5. $a_n = 2n - 7$ **6.** $a_n = 3^{3-n}$ **7.** $a_n = -(-2)^{n-1}$ **8.** $a_n = 3n + 9$

9. $a_n = \dfrac{n+1}{3}$ **10.** $a_n = 4^{n-1}$ **11.** 73   **12.** 106

13. $\dfrac{1}{32}$ **14.** $\dfrac{4}{9}$ **15.** -92 **16.** $\dfrac{1}{16}$

17. -5 **18.** 85   **19.** $\dfrac{5}{9}$ **20.** 2 or -2

21. $121\dfrac{40}{81}$ **22.** 7035   **23.** $-10,725$ **24.** $31\dfrac{31}{32}$

25. 32,015   **26.** 4757   **27.** $85\dfrac{21}{64}$ **28.** 37,044

29. 12,726   **30.** $85\dfrac{1}{3}$ **31.** $\dfrac{4}{11}$ **32.** $\dfrac{41}{90}$

33. $750   **34.** $46.50   **35.** $3276.70   **36.** 10,935 gallons
37. 3600 feet

Chapter 14

Problem Set 14.1

1. 20	**3.** 24	**5.** 168	**7.** 48	**9.** 36
11. 6840	**13.** 720	**15.** 720	**17.** 36	**19.** 24
21. 243	**23.** Impossible	**25.** 216	**27.** 26	**29.** 36
31. 144	**33.** 1024	**35.** 30		

37. (a) 6,084,000   **(c)** 3,066,336

Problem Set 14.2

1. 60	**3.** 360	**5.** 21	**7.** 252	**9.** 105	**11.** 1
13. 24	**15.** 84	**17. (a)** 336	**19.** 2880	**21.** 2450	**23.** 10
25. 10	**27.** 35	**29.** 1260	**31.** 2520	**33.** 15	**35.** 126

37. 144; 202   **39.** 15; 10   **41.** 20   **43.** 10; 15; 21; $\dfrac{n(n-1)}{2}$ **45.** 120

Problem Set 14.3

1. $\dfrac{1}{2}$	**3.** $\dfrac{3}{4}$	**5.** $\dfrac{1}{8}$	**7.** $\dfrac{7}{8}$	**9.** $\dfrac{1}{16}$	**11.** $\dfrac{3}{8}$	**13.** $\dfrac{1}{3}$
15. $\dfrac{1}{2}$	**17.** $\dfrac{5}{36}$	**19.** $\dfrac{1}{6}$	**21.** $\dfrac{11}{36}$	**23.** $\dfrac{1}{4}$	**25.** $\dfrac{1}{2}$	**27.** $\dfrac{1}{25}$
29. $\dfrac{9}{25}$	**31.** $\dfrac{2}{5}$	**33.** $\dfrac{9}{10}$	**35.** $\dfrac{5}{14}$	**37.** $\dfrac{15}{28}$	**39.** $\dfrac{7}{15}$	**41.** $\dfrac{1}{15}$
43. $\dfrac{2}{3}$	**45.** $\dfrac{1}{5}$	**47.** $\dfrac{1}{63}$	**49.** $\dfrac{1}{2}$	**51.** $\dfrac{5}{11}$	**53.** $\dfrac{1}{6}$	**55.** $\dfrac{21}{128}$

57. $\dfrac{13}{16}$ **59.** $\dfrac{1}{21}$ **61.** 40   **63.** 3744   **65.** 10,200   **67.** 123,552   **69.** 1,302,540

Problem Set 14.4

1. (a) $\dfrac{5}{36}$ **(c)** $\dfrac{7}{12}$ **3.** $\dfrac{5}{6}$ **5. (a)** $\dfrac{1}{32}$ **(c)** $\dfrac{31}{32}$ **7.** $\dfrac{12}{13}$ **9.** $\dfrac{7}{12}$ **11.** $\dfrac{37}{44}$

13. (a) $\dfrac{2}{11}$ **(c)** $\dfrac{9}{11}$ **15. (a)** $\dfrac{5}{68}$ **(c)** $\dfrac{15}{34}$ **17. (a)** $\dfrac{4}{21}$ **(c)** $\dfrac{11}{21}$

19. (a) 1 to 3   **(c)** 5 to 1   **(e)** 5 to 4

Chapter 14 Review Problem Set

1. 720	**2.** 30,240	**3.** 150	**4.** 1440	**5.** 20	**6.** 525	**7.** 1287	**8.** 264
9. 74	**10.** 55	**11.** 40	**12.** 15	**13.** 60	**14.** 120	**15.** $\dfrac{3}{8}$	**16.** $\dfrac{5}{16}$

17. $\dfrac{5}{36}$ **18.** $\dfrac{13}{18}$ **19.** $\dfrac{3}{5}$ **20.** $\dfrac{1}{35}$ **21.** $\dfrac{57}{64}$ **22.** $\dfrac{1}{221}$ **23.** $\dfrac{1}{6}$ **24.** $\dfrac{4}{7}$

25. $\dfrac{4}{7}$ **26.** $\dfrac{10}{21}$ **27.** $\dfrac{140}{143}$ **28.** $\dfrac{105}{169}$ **29.** $\dfrac{1}{6}$ **30.** $\dfrac{28}{55}$ **31.** $\dfrac{5}{7}$ **32.** $\dfrac{1}{16}$

Index

Table of Common Logarithms

N	0	1	2	3	4	5	6	7	8	9
1.0	.0000	.0043	.0086	0.128	.0170	.0212	.0253	.0294	.0334	.0374
1.1	.0414	.0453	.0492	.0531	.0569	.0607	.0645	.0682	.0719	.0755
1.2	.0792	.0828	.0864	.0899	.0934	.0969	.1004	.1038	.1072	.1106
1.3	.1139	.1173	.1206	.1239	.1271	.1303	.1335	.1367	.1399	.1430
1.4	.1461	.1492	.1523	.1553	.1584	.1614	.1644	.1673	.1703	.1732
1.5	.1761	.1790	.1818	.1847	.1875	.1903	.1931	.1959	.1987	.2014
1.6	.2041	.2068	.2095	.2122	.2148	.2175	.2201	.2227	.2253	.2279
1.7	.2304	.2330	.2355	.2380	.2405	.2430	.2455	.2480	.2504	.2529
1.8	.2553	.2577	.2601	.2625	.2648	.2672	.2695	.2718	.2742	.2765
1.9	.2788	.2810	.2833	.2856	.2878	.2900	.2923	.2945	.2967	.2989
2.0	.3010	.3032	.3054	.3075	.3096	.3118	.3139	.3160	.3181	.3201
2.1	.3222	.3243	.3263	.3284	.3304	.3324	.3345	.3365	.3385	.3404
2.2	.3424	.3444	.3464	.3483	.3502	.3522	.3541	.3560	.3579	.3598
2.3	.3617	.3636	.3655	.3674	.3692	.3711	.3729	.3747	.3766	.3784
2.4	.3802	.3820	.3838	.3856	.3874	.3892	.3909	.3927	.3945	.3962
2.5	.3979	.3997	.4014	.4031	.4048	.4065	.4082	.4099	.4116	.4133
2.6	.4150	.4166	.4183	.4200	.4216	.4232	.4249	.4265	.4281	.4298
2.7	.4314	.4330	.4346	.4362	.4378	.4393	.4409	.4425	.4440	.4456
2.8	.4472	.4487	.4502	.4518	.4533	.4548	.4564	.4579	.4594	.4609
2.9	.4624	.4639	.4654	.4669	.4683	.4698	.4713	.4728	.4742	.4757
3.0	.4771	.4786	.4800	.4814	.4829	.4843	.4857	.4871	.4886	.4900
3.1	.4914	.4928	.4942	.4955	.4969	.4983	.4997	.5011	.5024	.5038
3.2	.5051	.5065	.5079	.5092	.5105	.5119	.5132	.5145	.5159	.5172
3.3	.5185	.5198	.5211	.5224	.5237	.5250	.5263	.5276	.5289	.5302
3.4	.5315	.5328	.5340	.5353	.5366	.5378	.5391	.5403	.5416	.5428
3.5	.5441	.5453	.5465	.5478	.5490	.5502	.5514	.5527	.5539	.5551
3.6	.5563	.5575	.5587	.5599	.5611	.5623	.5635	.5647	.5658	.5670
3.7	.5682	.5694	.5705	.5717	.5729	.5740	.5752	.5763	.5775	.5786
3.8	.5798	.5809	.5821	.5832	.5843	.5855	.5866	.5877	.5888	.5899
3.9	.5911	.5922	.5933	.5944	.5955	.5966	.5977	.5988	.5999	.6010
4.0	.6021	.6031	.6042	.6053	.6064	.6075	.6085	.6096	.6107	.6117
4.1	.6128	.6138	.6149	.6160	.6170	.6180	.6191	.6201	.6212	.6222
4.2	.6232	.6243	.6253	.6263	.6274	.6284	.6294	.6304	.6314	.6325
4.3	.6335	.6345	.6355	.6365	.6375	.6385	.6395	.6405	.6415	.6425
4.4	.6435	.6444	.6454	.6464	.6474	.6484	.6493	.6503	.6513	.6522
4.5	.6532	.6542	.6551	.6561	.6571	.6580	.6590	.6599	.6609	.6618
4.6	.6628	.6637	.6646	.6656	.6665	.6675	.6684	.6693	.6702	.6712
4.7	.6721	.6730	.6739	.6749	.6758	.6767	.6776	.6785	.6794	.6803
4.8	.6812	.6821	.6830	.6839	.6848	.6857	.6866	.6875	.6884	.6893
4.9	.6902	.6911	.6920	.6928	.6937	.6946	.6955	.6964	.6972	.6981
5.0	.6990	.6998	.7007	.7016	.7024	.7033	.7042	.7050	.7059	.7067
5.1	.7076	.7084	.7093	.7101	.7110	.7118	.7126	.7135	.7143	.7152
5.2	.7160	.7168	.7177	.7185	.7193	.7202	.7210	.7218	.7226	.7235
5.3	.7243	.7251	.7259	.7267	.7275	.7284	.7292	.7300	.7308	.7316
5.4	.7324	.7332	.7340	.7348	.7356	.7364	.7372	.7380	.7388	.7396